HISTORICAL GEOLOGY

EVOLUTION OF THE EARTH AND LIFE THROUGH TIME

SECOND EDITION

SECOND EDITION

HISTORICAL GEOLOGY

EVOLUTION OF THE EARTH AND LIFE THROUGH TIME

Reed Wicander
James S. Monroe

CENTRAL MICHIGAN UNIVERSITY

WEST PUBLISHING COMPANY

Minneapolis/Saint Paul   New York   Los Angeles   San Francisco

Production Credits

COPYEDITOR	Patricia Lewis
COVER DESIGN	Diane Beasley
INTERIOR DESIGN	K. M. Weber
COMPOSITION	Carlisle Communications, Ltd.
ARTWORK	Carlyn Iverson, Precision Graphics, Publication Services, and Alice B. Thiede and William A. Thiede, Carto-Graphics. Individual credits follow index.
COVER IMAGE	Cover illustration of *Sauropelta* by Eleanor M. Kish. Reproduced with permission of the Canadian Museum of Nature, Ottawa, Canada.

West's Commitment to the Environment

In 1906, West Publishing Company began recycling materials left over from the production of books. This began a tradition of efficient and responsible use of resources. Today, up to 95 percent of our legal books and 70 percent of our college and school texts are printed on recycled, acid-free stock. West also recycles nearly 22 million pounds of scrap paper annually—the equivalent of 181,717 trees. Since the 1960s, West has devised ways to capture and recycle waste inks, solvents, oils, and vapors created in the printing process. We also recycle plastics of all kinds, wood, glass, corrugated cardboard, and batteries, and have eliminated the use of styrofoam book packaging. We at West are proud of the longevity and the scope of our commitment to our environment.

Production, prepress, printing, and binding by West Publishing Company.

COPYRIGHT © 1993   By WEST PUBLISHING COMPANY
610 Opperman Drive
P.O. Box 64526
St. Paul, MN 55164-0526

Printed in the United States of America

00 99 98 97 96 95 94 93   8 7 6 5 4 3 2 1

Library of Congress Cataloging-in-Publication Data

Wicander Reed, 1946–
    Historical geology : evolution of the earth and life through time / Reed Wicander, James S. Monroe — 2nd ed.
        p. cm.
    Includes index.
    ISBN 0-314-01240-0
    1. Historical geology.   I. Monroe, James S. (James Stewart), 1938- . II. Title.
QE28.3.W53 1993
551.7—dc20   92-39257
                                                                    CIP

Brief Table of Contents

CONTENTS

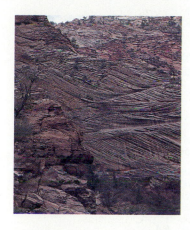

CHAPTER 3

GEOLOGIC TIME: CONCEPTS AND PRINCIPLES 52

CHAPTER 4

ROCKS, FOSSILS, AND TIME 72

CHAPTER 5

ORIGIN AND INTERPRETATION OF SEDIMENTARY ROCKS 108

CHAPTER 6

EVOLUTION 140

CHAPTER 7

PLATE TECTONICS: A UNIFYING THEORY 172

CHAPTER 8

THE HISTORY OF THE UNIVERSE, SOLAR SYSTEM, AND PLANETS 202

CHAPTER 11

GEOLOGY OF THE EARLY PALEOZOIC ERA
290

CHAPTER 12

GEOLOGY OF THE LATE PALEOZOIC ERA
318

CHAPTER 13

LIFE OF THE PALEOZOIC ERA: INVERTEBRATES 348

CHAPTER 14

LIFE OF THE PALEOZOIC ERA: VERTEBRATES AND PLANTS 372

CHAPTER 15

GEOLOGY OF THE MESOZOIC ERA 400

CHAPTER 16

LIFE OF THE MESOZOIC ERA 432

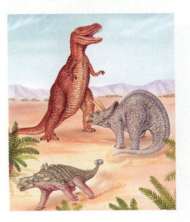

CHAPTER 17

Cenozoic Geologic History: Tertiary Period 472

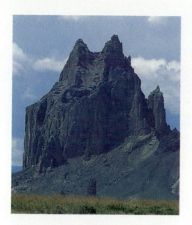

CHAPTER 18

Cenozoic Geologic History: Quaternary Period 516

PREFACE

The Earth is a dynamic planet that has changed continuously during its 4.6 billion years of existence. The size, shape, and geographic distribution of the continents and oceans have changed through time, as have the atmosphere and biota. Historical geologists are concerned with all aspects of the Earth and its life history. They seek to determine what events occurred during the past, place those events into an orderly chronological sequence, and provide conceptual frameworks for explaining such events. All of this makes the study of the history of the Earth and its life a fascinating endeavor and one that has relevance today.

Historical Geology: Evolution of the Earth and Life Through Time, second edition, is designed for a one-semester geology course and is written with the student in mind. One of the goals of this book is to provide students with an understanding of the principles of historical geology and how these principles can be applied to unraveling the history of the Earth. It is also our intention to provide students with an overview of the geologic and biologic history of the Earth, not as a set of encyclopedic facts to memorize, but rather as a continuum of interrelated events that reflect the underlying geologic and biologic principles and processes that have shaped our planet.

Because of the nature of the subject, all historical geology textbooks share some broad similarities. Most begin with several chapters on concepts and principles, followed by a chronological discussion of Earth history. In this respect we have not departed from convention. We have, however, attempted to place greater emphasis on basic concepts and principles, their historical development, and their importance in deciphering the history of the Earth. For instance, stratigraphic terminology is used sparingly with the idea that understanding the causes and consequences of events is more important than knowing the names of numerous formations. Additionally, we have strived for a balanced coverage of both physical and biological events because we feel that such balance is lacking in many historical geology textbooks.

❖ NEW FEATURES IN THE SECOND EDITION

This second edition has undergone considerable rewriting and updating to produce a book that is easier to read and completely current. Drawing on the comments and suggestions of reviewers, we have incorporated many new features into this edition. The first and most visible change is the use of full color in most of the illustrations and photographs. We feel this adds dramatically to the visual appeal of the book and makes the illustrations easier to comprehend.

In response to comments that some students have not had any geology before taking a historical geology course, we have added a new chapter on Minerals and Rocks. This should be of great benefit to students who have never had a geology or Earth science course and will provide a review for students who have.

Other important changes include a new Perspective on cladistics and cladograms in Chapter 6 for instructors who would like to cover this topic; a new Prologue on planetary plate tectonics and igneous activity, a new Perspective on the supercontinent cycle, and expanded coverage of plate tectonics in Chapter 7; a new section covering the planets in Chapter 8; a revised discussion of the Precambrian in Chapters 9 and 10 to include the latest information about ideas on the origin of life and the origin of eukaryotes; substantial revisions of the treatment of Paleozoic geology in Chapters 11 and 12; the division of the discussion of Paleozoic life into two chapters, one dealing with invertebrates (Chapter 13) and the other with the vertebrates and plants (Chapter 14); revision of Chapter 16 on Mesozoic life, including new artwork and a new Perspective about mosasaurs and paleopathology; revision of Chapter 17 on Cenozoic geology; a new chapter on Pleistocene geology (Chapter 18); revisions and new artwork for Chapter 19 (Cenozoic life); and a new chapter on Primate and Human Evolution that includes the latest information and controversies concerning human origins.

We feel these changes are a marked improvement over the first edition and make the book easier to read and comprehend, as well as a more effective teaching tool.

❖ THEMATIC APPROACH

As in the first edition, we develop three major themes in this textbook that are essential to the interpretation and appreciation of historical geology, introduce these themes early, and reinforce them throughout the book. These themes are *time,* the dimension that sets historical geology apart from most of the other sciences; *evolutionary theory,* the explanation for inferred relationships among living and fossil organisms; and *plate tectonics,* a unifying theory for interpreting much of the Earth's physical history and, to a large extent, its biological history. In addition, we have emphasized the intimate interrelationship that exists between physical and biological events.

The concept of geologic time and its historical development is presented in Chapter 3. The concept is more fully developed in the following chapter on Rocks, Fossils, and Time. In Chapter 6, we discuss the development of evolutionary theory in the context of the historical period during which it was formulated. Included in this discussion are sections that emphasize the evidence for evolution and the importance of the fossil record in evolutionary studies. Evolutionary theory serves as the unifying theme for subsequent discussions of life history and is constantly reinforced throughout the book.

Particular emphasis is given to the evidence that substantiates plate tectonic theory, why this theory is one of the cornerstones of geology, and why plate tectonic theory serves as a unifying paradigm in explaining many apparently unrelated geologic phenomena. Several examples of how plate tectonics is responsible for seemingly disparate aspects of geology are given in this context; events are viewed as parts of an ever-changing continuum, rather than as discrete events separated from one another in time and space. While we emphasize the geologic evolution of North America, each chapter on geologic history contains an overview of global tectonic events so that the history of North America is discussed in a global context.

❖ TEXT ORGANIZATION

This book was written for a one-semester course in historical geology to serve both majors and nonmajors in geology and in the Earth sciences. It may also be used in a one-quarter course by selecting the appropriate material the instructor wishes to cover or emphasize. Many students taking a historical geology course will have had an introductory physical geology course. In this case, Chapter 1 can be used as a review of the principles and concepts of geology and as an introduction to the science of historical geology and the three themes (time, organic evolution, and plate tectonics) that the book emphasizes. The text is written at an appropriate level for those students taking historical geology with no prerequisite, but the instructor may have to spend more time expanding some of the concepts and terminology discussed in Chapter 1. Furthermore, Chapter 2 on Minerals and Rocks has been added to this edition to provide students with a basic understanding of the classification and formation of minerals and rocks. This will be of benefit to both groups of students—a review for students who have taken a physical geology course and an introduction for students who have not had a geology or Earth science course.

Chapter 3 explores the first major theme of the book, the concepts and principles of geologic time. Chapter 4 expands on that theme by integrating geologic time with rocks and fossils. Depositional environments are sometimes covered rather superficially (perhaps little more than a summary table) in some historical geology textbooks. Chapter 5, "The Origin and Interpretation of Sedimentary Rocks," is completely devoted to this topic; it contains sufficient detail to be meaningful, but avoids an overly detailed discussion more appropriate for advanced courses.

The second major theme of the book, evolution, is examined in Chapter 6. Chapter 7 discusses the third major theme, plate tectonics. For those students who have had a physical geology course, much of the chapter will be review. It is, however, an important review because we will often refer to this theory throughout the rest of the book as we discuss how plate tectonics has played a major role in the evolution of our planet.

Chapter 8, "The History of the Universe, Solar System, and Planets," provides a comprehensive, yet easily readable treatment of this topic. The chapter includes discussions of the origin of the universe from the time when physics as we know it began, the theories for the origin of the solar system, the significance of meteorites, and the competing theories for the formation and early history of the Earth.

Precambrian time, fully 87% of all geologic time, is sometimes considered in a single chapter. Chapters 9 and 10 are devoted to the geologic and biologic histories of the Archean and Proterozoic eons, respectively.

Chapters 11 through 20 constitute our chronological treatment of the Phanerozoic geologic and biologic history of the Earth. These chapters are arranged so that the geologic history of an era is followed by a discussion of the biologic history of that era. We believe that this format allows easier integration of life history with geologic history.

Instructors who have used our first edition will notice some significant changes in the second half of the text. The discussion of Paleozoic life is now divided into two chapters with invertebrates in Chapter 13 followed by vertebrates and plants in Chapter 14. This allows for more thorough coverage without overwhelming students. The new Chapter 18, "Cenozoic Geologic History: Quaternary Period," is an updated, expanded discussion of the material covered in Chapter 17 of the first edition. Chapter 20, "Evolution of Primates and Humans," now provides more extensive coverage of human evolution.

❖ CHAPTER ORGANIZATION

All chapters have the same organizational format. Each chapter opens with a photograph relating to the chapter content, a detailed outline, and a Prologue, which is intended to stimulate interest in the chapter by discussing some aspect of the material.

The text is written in a clear informal style, making it easy for students to comprehend the subject. Numerous color diagrams, many of them block diagrams, and photographs complement the text, providing a visual representation of the concepts and information presented. Each chapter, except Chapter 2, contains at least one Perspective that briefly examines an interesting aspect of historical geology or geological research. Each of the chapters on geologic history in the second half of the text contains a final section on mineral resources characteristic of that time period. These sections provide applied economic material of interest to students.

The end-of-chapter materials begin with a concise review of important concepts and ideas in the Chapter Summary. The Important Terms, which are printed in boldface type in the chapter text, are listed at the end of each chapter for easy review, and a full glossary of important terms appears at the end of the text. The Review Questions are another important feature of this book; they contain between 20 and 30 questions per chapter including multiple-choice questions with answers as well as short answer and essay questions. Each chapter concludes with an up-to-date list of Additional Readings, most of which are written at a level appropriate for beginning students interested in pursuing a particular topic.

❖ SPECIAL FEATURES

Figures

Many of the figures are original artwork especially designed for this book. Our color paleogeographic maps are designed to illustrate clearly and accurately the geography during the various geologic periods. Many of the illustrations depicting geologic processes or events are block diagrams rather than cross sections so that students can more easily visualize the salient features of these processes and events. Full color scenes showing associations of plants and animals are based on the most current interpretations.

Instructors who have used the previous edition will notice that the entire illustration program has been revamped. New full color photographs, block diagrams, maps and charts, and paleobiological art have been prepared for this edition. Great care was taken to ensure that the art and captions provide an attractive, informative, and accurate illustration program for the second edition.

Figure Reference System

A color cue (◆) will be found in the text next to the first reference for each figure. This system is designed to help students quickly return to their place in the text when they interrupt their reading to examine an illustration.

Prologues

The introductory prologues are designed to provide an introduction to the chapter material by focusing on a particularly interesting aspect of the chapter. Prologue topics include planetary plate tectonics and igneous activity (Chapter 7), the Mazon Creek fauna and flora (Chapter 12), nesting behavior in dinosaurs (Chapter

16), and the controversial Eve hypothesis concerning the origin of modern humans (Chapter 20).

Summary Tables

A number of summary tables appear throughout the book. Of particular assistance to students are the end-of-chapter summary tables. These tables are designed to give an overall perspective of the geologic and biologic events that occurred during a particular time interval and to show how these events are interrelated. The emphasis in these tables is on the geologic evolution of North America. Global tectonic events and sea-level changes are also incorporated into these tables to provide global insights.

Perspectives

The chapter Perspectives focus on aspects of current geologic research, interesting features related to the material in the chapter, or national parks or monuments. They were chosen to provide students with an overview of the many fascinating aspects of geology. Examples include cladistics and cladograms (Perspective 6.1), the supercontinent cycle (Perspective 7.1), Dinosaur National Monument (Perspective 15.2), and mosasaurs and paleopathology (Perspective 16.2). The Perspectives can be assigned as part of the chapter reading, used as the basis for lecture or discussion topics, or even as the starting point for student papers.

Review Questions

Most historical geology texts have a set of review questions at the end of each chapter. This book, however, includes not only the usual essay and thought-provoking questions, but also a set of multiple-choice questions, a feature not found in other historical geology textbooks. The answers to the multiple-choice questions are at the end of the book so that students can check their answers and increase their confidence before taking an examination.

Appendixes

With few exceptions, units of length, mass, and volume are given in metric units in the text. Appendix A is an *English-Metric Conversion Chart*. The terminology of biological classification is complex, and though we have minimized it in the text, some is necessary. A *Classification of Organisms* appears in Appendix B.

Glossary

It is important for students to comprehend geologic terminology if they are to understand geologic principles and concepts. Accordingly, an alphabetical and comprehensive glossary of all important terms appears at the back of the book.

❖ INSTRUCTOR ANCILLARY MATERIALS

To assist you in teaching this course and supplying your students with the best in teaching aids, West Publishing Company has prepared a complete supplemental package available to all adopters.

The Comprehensive Instructor's Manual includes teaching ideas, lecture outlines (including notes on figures and photographs available as slides), teaching tips, and an expanded list of references.

A slide set will be provided that includes the most important and attractive figures and photographs from the book. Transparency acetates of the important charts and figures are available to provide clear and effective illustrations of important artwork from the text.

A Newsletter will be provided to adopters each year to update the book with recent and relevant research disclosures. This will ensure that your students have the most current information available.

Videotapes on relevant topics in the book are available to qualified adopters. Contact your West sales representative for details.

❖ ACKNOWLEDGMENTS

As the authors, we are of course responsible for the organization, style, and accuracy of the text, and any mistakes, omissions, or errors are our responsibility. We wish to express our sincere appreciation to the following geologists who reviewed the first edition. The second edition greatly benefited from their comments, advice, and suggestions.

Thomas W. Broadhead
    University of Tennessee at Knoxville
Richard Fluegeman, Jr.
    Ball State University
Susan Goldstein
    University of Georgia
Bryan Gregor
    Wright State University

Jonathan D. Karr
    North Carolina State University
R. L. Langenheim, Jr.
    University of Illinois at Urbana-Champaign
Michael McKinney
    University of Tennessee at Knoxville
Arthur Mirsky
    Indiana University, Purdue University at Indianapolis
William C. Parker
    Florida State University
Anne Raymond
    Texas A & M University
G. J. Retallack
    University of Oregon
Mark Rich
    University of Georgia
Barbara L. Ruff
    University of Georgia
W. Bruce Saunders
    Bryn Mawr College
Ronald D. Stieglitz
    University of Wisconsin, Green Bay
Michael J. Tevesz
    Cleveland State University
Art Troell
    San Antonio College

We also wish to thank Professor Emeritus Richard V. Dietrich of Central Michigan University for providing us with several photographs, and discussing various aspects of the text with us on numerous occasions. In addition, we are grateful to the other members of the Geology Department of Central Michigan University for also providing us with photographs and answering our questions concerning various topics. They are Eric Johnson, David J. Matty, Jane M. Matty, Wayne E. Moore, and Stephen D. Stahl. We also thank Mrs. Martha Brian of the Geology Department for her cheerfulness and general efficiency, which were indispensable during the preparation of the second edition, and Bruce M. C. Pape of the Geography Department for providing photographs. We are also grateful for the generosity of the various agencies and individuals from many countries who provided photographs.

Special thanks must go to Jerry Westby, college editorial manager for West Publishing Company, a friend and taskmaster who made many valuable suggestions while patiently guiding us through the entire project. His continued encouragement and perspective helped us produce the best possible book. We are also indebted to our team of production editors whose attention to detail and consistency made our job so much easier. Team members included Mark Jacobsen; Laura K. Evans; Kara Zum-Bahlen, our photo researcher; Chris Hurney; and Tom Modl. We would also like to thank Patricia Lewis for her excellent copyediting and indexing skills. We sincerely appreciate her help in improving the text of our second edition and particularly for catching small errors that we should have caught in the first place. Because historical geology is largely a visual science, we extend special thanks to Carlyn Iverson who rendered the reflective art in this edition. In addition, we thank the artists at Precision Graphics and Publication Services who are responsible for much of the rest of the art program and Alice B. Thiede who rendered many of the excellent paleogeographic maps throughout the text. They all did an excellent job, and we enjoyed working with them. We also need to thank Charles Marshall of U.C.L.A. who reviewed the new paleobiological art for accuracy and was a great help in improving the art for this edition. We would also like to acknowledge our promotion manager, Ann Hillstrom, for her help in the development of the promotional material associated with this book.

As always, our families were patient and encouraging when most of our spare time and energy were devoted to this book. We thank them again for their support and understanding.

HISTORICAL GEOLOGY

EVOLUTION OF THE EARTH AND LIFE THROUGH TIME

SECOND EDITION

CHAPTER 1

Apollo 17 view of the Earth. Almost the entire coastline of Africa is clearly shown in this view with Madagascar visible off its eastern coast. The Arabian Peninsula can be seen at the northeastern edge of Africa, while the Asian mainland is on the horizon toward the northeast. The present location of continents and ocean basins is the result of plate movement. The interaction of plates through time has affected the physical and biological history of the Earth. (Photo courtesy of NASA.)

THE DYNAMIC EARTH

Prologue

If we could film the history of the Earth from its beginning 4.6 billion years ago to the present, and speed up that film so it could be shown as a feature movie, we would see a planet undergoing remarkable change. Its geography would change as continents moved about its surface, and as a result of these movements, ocean basins would open and close. Mountain ranges would form along continental margins or where continents collided with each other. Oceanic and atmospheric circulation patterns would shift in response to the movement of continents. Massive ice sheets would form, grow, and then melt away. At other times, our film would show extensive swamps or vast interior deserts.

If we focused our imaginary camera closer to the Earth's surface, we would capture a breathtaking panorama of different organisms. We would witness the first living cells evolving from a primordial organic soup sometime between 4.6 and 3.5 billion years ago. About 2 billion years later, cells with a nucleus evolved. Next (approximately 700 million years ago), the first multicelled soft-bodied organisms evolved in the oceans, then animals with hard parts, and then animals with backbones. As we scanned the various continents, we would see an essentially bare landscape until about 450 million years ago. At that time, the landscape would seem to come to life as plants and animals moved out of the water and onto the land. From then on, the landscape would be dominated by insects, amphibians, reptiles, birds, and mammals. In our film, humans and their history would occupy only the last second or so of the movie.

Three interrelated themes run through the preceding discussion. The first is that the Earth is composed of a series of moving plates whose interactions have affected its physical and biological history. The second is that the Earth's biota has evolved. The third is that the physical and biological changes that occurred took place over long periods of time.

The first of these themes, *plate tectonics,* provides geologists with a unifying model that explains the Earth's internal workings and accounts for a variety of apparently unrelated geologic features and events. The second theme, the theory of *organic evolution,* explains how life has changed through time, based on the idea that all living organisms are the evolutionary descendants of life-forms that existed in the past. Finally, the third theme, the concept of geologic time, allows geologists to show how small, almost imperceptible changes over vast lengths of time have resulted in significant changes. For example, the formation of the Grand Canyon is the result of erosion by the Colorado River over the past 2 to 3 million years.

These three interrelated themes, plate tectonics, organic evolution, and geologic time, are central to our understanding and appreciation of the history of our planet. Each will be discussed in depth in later chapters.

❖ INTRODUCTION

Geology, from the Greek *geo* and *logos,* is defined as the study of the Earth. It is generally divided into two broad areas—physical geology and historical geology. *Physical geology* is the study of Earth materials, such as minerals and rocks, as well as the processes that operate beneath and upon its surface. *Historical geology* examines the

THE TRAGIC LYSENKO AFFAIR

An unquestioning obedience to the dictates of some pseudoscience, or to a discredited scientific theory, can have devastating results. One of the most tragic examples of adherence to a disproved scientific theory involved Trofim Denisovich Lysenko (1898–1976) who became president of the Soviet Academy of Agricultural Sciences in 1938. Lysenko endorsed the theory of inheritance of acquired characteristics according to which plants and animals could be changed in desirable ways simply by exposing them to a new environment. For example, according to Lysenko, seeds exposed to dry conditions would acquire a resistance to drought, and this trait would be inherited by future generations. Lysenko accepted inheritance of acquired characteristics because of its apparent compatibility with Marxist-Leninist philosophy. As president of the academy, Lysenko did not allow Soviet scientists to conduct any other research concerning inheritance mechanisms, and those who publicly disagreed with him lost their jobs or were even sent to labor camps.

Unfortunately for the Soviet people, inheritance of acquired characteristics had been discredited as a scientific theory more than 50 years before. The results of Lysenko's belief in the political correctness of this theory were widespread crop failures, starvation, and misery for millions of people. In fact, Lysenko's dismantling of genetic research was so complete that the former Soviet Union is still striving to catch up to the rest of the world in this field.

An interesting sidenote to this tragedy is that the only reputable Western biologist to support Lysenko was England's J. B. S. Haldane. Haldane, whose numerous contributions to genetics and evolutionary theory won him international recognition and respect, discovered Marxism during the 1930s and joined the Communist party in 1942. Unfortunately, his belief in the political correctness of the Marxist-Leninist doctrine appears to have blinded Haldane to the scientific absurdity of Lysenko's claims.

The lesson to be drawn from this example is that scientific research must be based on scientific realities. Science proceeds on the basis of the scientific method, not philosophical or political beliefs.

origin and evolution of the Earth, its continents, atmosphere, oceans, and life.

Historical geology is, however, more than just a recitation of past events. It is the study of a dynamic planet that has changed continuously over the past 4.6 billion years—although not always at the same rate. Geologists seek not only to place events in an orderly chronologic arrangement but, more importantly, to explain how and why past events happened. It is one thing to observe in the fossil record that dinosaurs went extinct but quite another to ask how they became extinct. Geologists have long noted evidence in rocks of numerous mountain-building episodes. Only in the last few decades, however, has a theory emerged to provide geologists with a comprehensive model explaining how mountain building occurs and how individual episodes of mountain building may be related to each other.

In addition to aiding in the interpretation of the Earth's history, the basic principles of historical geology have practical applications. William Smith, an English surveyor and engineer, recognized that by studying the sequences of rocks and the fossils they contained, he could predict the kinds and thicknesses of rocks that would have to be excavated in the construction of canals. The same principles Smith used in the late eighteenth and early nineteenth century are still used today in mineral and oil exploration.

❖ HISTORICAL GEOLOGY AND THE FORMULATION OF THEORIES

The term **theory** has various meanings. In colloquial usage, it means a speculative or conjectural view of something—hence the widespread belief that scientific theories are little more than unsubstantiated wild guesses. In scientific usage, however, a theory is a coherent explanation for one or several related natural phenomena that is supported by a large body of objective evidence. From

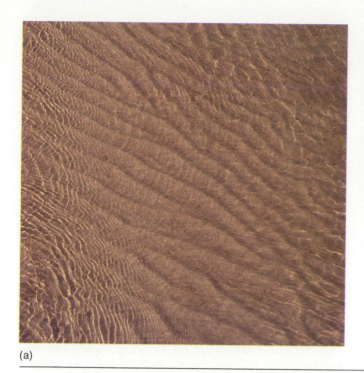

(a)

(b)

◆ **FIGURE 1.1** Sedimentary features of a nearshore environment and their counterparts in the rock record. (a) Modern wave-formed ripples at Ludington, Michigan. (b) Ripples preserved in Jurassic rocks at Gull Hill Park, South Dakota.

a theory are derived predictive statements that can be tested by observation and/or experiment so that their validity can be assessed. The law of universal gravitation is an example of a theory describing the attraction between masses (an apple and the Earth in the popularized account of Sir Isaac Newton and his discovery).

Theories are formulated through the process known as the **scientific method.** This method is an orderly, logical approach that involves gathering and analyzing the facts or data about the problem under consideration. Tentative explanations or **hypotheses** are then formulated to explain the observed phenomena. Next, the hypotheses are tested to see if what they predicted actually occurs in a given situation. Finally, if one of the hypotheses is found, after repeated tests, to explain the phenomena, then that hypothesis is proposed as a theory. One should remember, however, that in science, even a theory is still subject to further testing and refinement as new data become available.

The fact that a scientific theory can be tested and is subject to such testing separates science from other forms of human inquiry. Because scientific theories can be tested, they have the potential of being supported or even proved wrong (see Perspective 1.1). Accordingly, science

must proceed without any appeal to beliefs or supernatural explanations, not because such beliefs or explanations are necessarily untrue, but because we have no way to investigate them. For this reason, science makes no claim about the existence or nonexistence of a supernatural or spiritual realm.

Since scientific theories must be testable, how is it possible to formulate theories to explain events that occurred before there were human observers to record such events? To deal with such difficulties, geologists must make two assumptions. One assumption is that the evidence preserved in rocks, the geologic record, is an accurate record of prehistoric events. That is, rocks contain evidence that can be used to determine the processes by which they formed. The second assumption is that the physical, chemical, and biological processes operating today have operated the same way in the past; that is, the laws of nature have not changed.

The following example demonstrates that these two assumptions are justified. Modern nearshore sediments and ancient sedimentary rocks have several features in common (◆ Fig. 1.1). Having examined a present-day feature, geologists have determined that it forms when waves pass over sand. In short, these wave-formed ripple marks are the products of a physical process. The ancient ripples in Figure 1.1b were also produced by a physical process, and it seems reasonable to infer that that process was wave action.

As a matter of fact, people commonly interpret events that they did not witness. Although they may not propose a formal theory or go through a testing process, most people would have little difficulty interpreting a shattered, charred tree in the forest as the product of a lightning strike. Geologists use similar reasoning in their investigations of Earth history. However, they do propose formal theories, some of which are discussed at greater length in other parts of this book. For example, the theory of evolution (Chapter 6) serves as the theoretical framework to explain changes in life-forms through time, as well as the differences and similarities among living organisms. Plate tectonic theory is a well-supported explanation for a number of ancient and modern aspects of the Earth (Chapter 7).

❖ EARTH MATERIALS

The only evidence available for interpreting Earth history is preserved in the rock record. Consequently, the origin, distribution, and interrelationships among rocks are fundamental to an understanding of Earth history. In this chapter, we will review the three major rock groups emphasizing the sedimentary rocks because they constitute the majority (over 75%) of rocks exposed on the Earth's surface. A more detailed discussion on minerals, rocks, and their classification is provided in Chapter 2.

A **rock** is an aggregate of minerals. **Minerals** are nat-

urally occurring, inorganic, crystalline solids that have definite physical and chemical properties. Minerals are composed of elements such as oxygen or aluminum, and elements are made up of atoms, the smallest particles of matter that still retain the characteristics of an element. More than 3,500 minerals have been identified and described, but only about a dozen make up the bulk of the rocks in the Earth's crust (Table 1.1).

The Rock Cycle

The **rock cycle** is a way of viewing the interrelationships between the Earth's internal and external processes (◆ Fig. 1.2). It relates the three rock groups—igneous, sedimentary, and metamorphic—to each other; to surficial processes such as weathering, transportation, and deposition; and to internal processes such as magma generation and metamorphism.

Igneous rocks result from the crystallization of molten material called *magma*, or by the accumulation and consolidation of volcanic ejecta such as ash. As magma cools, minerals crystallize, and the resulting rock is characterized by interlocking mineral grains. Magma that cools slowly beneath the Earth's surface produces intrusive igneous rocks with a phaneritic texture (◆ Fig 1.3a). A phaneritic texture is one in which the individual minerals can be seen without magnification. Extrusive igneous rocks, such as those forming from lava, have an aphanitic texture that results from magma cooling rapidly at the Earth's surface. Aphanitic textures must be magnified for the individual grains to be seen (◆ Fig. 1.3b). When a magma cools so rapidly that crystallization cannot occur, the resulting rock is a natural glass. Igneous rocks are classified on the basis of their texture and mineral composition (see Fig. 2.17).

Rocks exposed at the Earth's surface are broken and dissolved by various weathering processes. The resulting particles and dissolved material may be transported by wind, water, or ice and eventually deposited. The process of converting this material, called *sediment*, into a sedimentary rock is termed *lithification*.

Two major categories of **sedimentary rocks** are recognized. *Detrital sedimentary rocks* are composed of solid particles that were once part of other rocks and are classified according to particle size (see Table 2.5). For example, conglomerate is composed of rounded particles larger than 2 millimeters (mm) (◆ Fig. 1.3c), while sand is made up of particles 1/16 to 2mm in diameter.

TABLE 1.1 Common Rock-Forming Minerals

RELATIVE ABUNDANCES OF MINERALS IN THE EARTH'S CRUST	
Plagioclase	39%
Quartz	12%
Orthoclase	12%
Pyroxenes	11%
Micas	5%
Amphiboles	5%
Clay minerals	5%
Olivine	3%
Others	8%

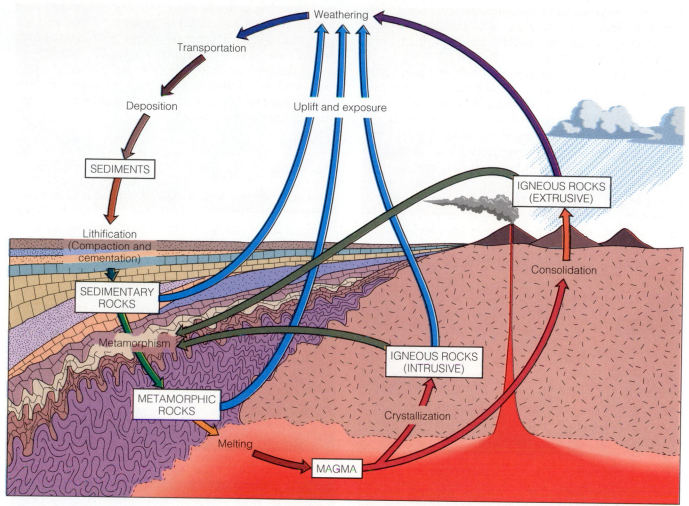

◆ **FIGURE 1.2** The rock cycle showing the interrelationships between the Earth's internal and external processes and how each of the three rock groups is related to the others.

Chemical sedimentary rocks are derived from dissolved material carried in solution (◆ Fig. 1.3d) and are classified on the basis of their chemical composition (see Table 2.5). Some chemical rocks are the result of precipitation of minerals by organisms or are accumulations of organic material such as shells and plants.

Sedimentary rocks are very useful for interpreting Earth history. These rocks form at the Earth's surface, and as layers of sediment accumulate, they record something about the environment of deposition, the type of transporting agent, and perhaps even something about the source from which the particles were derived (◆ Fig. 1.4). If the sedimentary layers contain fossils, geologists can use these fossils to reconstruct past environmental conditions and identify the different types of organisms that were living at that time; they may also be able to determine the age of the rocks and correlate them with rocks of the same age in different places (◆ Fig. 1.5).

Some rock layers contain structures formed at the time of deposition (sedimentary structures). For example, cross-bedding allows geologists to reconstruct the environment at the time the cross-beds formed (◆ Fig. 1.6).

(a)

(b)

(c)

(d)

(e)

(f)

◆ **FIGURE 1.3** Hand specimens of common igneous (a, b), sedimentary (c, d), and metamorphic (e, f) rocks. (a) Granite, a phaneritic igneous rock. (b) Basalt, an aphanitic igneous rock. (c) Conglomerate, a detrital sedimentary rock. (d) Limestone, a chemical sedimentary rock. (e) Gneiss, a foliated metamorphic rock. (f) Marble, a nonfoliated metamorphic rock. (Photos courtesy of Sue Monroe.)

◆ **FIGURE 1.4** Sedimentary rocks preserve evidence of the processes responsible for deposition of the sediment. This rock specimen was most likely deposited in a stream environment as indicated by its reddish color and the small ripple marks showing that the current flowed from left to right. (Photo courtesy of Sue Monroe.)

◆ **FIGURE 1.5** This fossiliferous limestone was deposited in a marine environment as indicated by the numerous fossil brachiopods. (Photo courtesy of Sue Monroe.)

Furthermore, some rocks by their very nature are good indicators of past environmental conditions. Tillites, for example, indicate glacial conditions (◆ Fig. 1.7), while coals signify swampy environments. In Chapters 4 and 5, we will discuss in more detail how sedimentary rocks and sedimentary structures can be used to interpret Earth history.

Metamorphic rocks, the third rock group, result from the alteration of other rocks, usually beneath the Earth's surface, by heat, pressure, and the chemical activity of fluids (see Table 2.6). Metamorphic rocks are either foliated or nonfoliated. Foliation is the parallel alignment of minerals due to pressure. It gives the rock a layered or banded appearance. If the metamorphic rock is foliated,

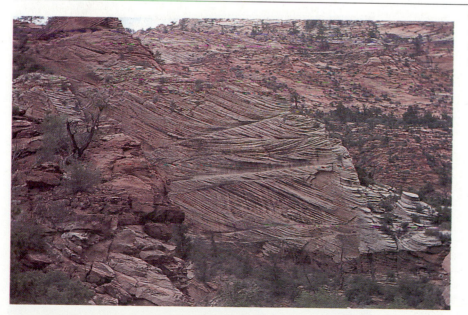

◆ **FIGURE 1.6** Cross-bedding in this sandstone in Zion National Park, Utah, indicates that it was deposited as sand dunes.

the type of foliation determines its name, while nonfoliated metamorphic rocks are named on the basis of their composition (Figs. 1.3e and f).

The sequence of processes shown in the rock cycle (Fig. 1.2) illustrates the interrelationship of the three rock groups and typifies an idealized cycle. However, modifications of this idealized cycle are common. For example, an igneous rock may never be exposed at the surface. After forming, it may be subjected to heat and pressure and converted directly to metamorphic rock without ever going through the intermediate sedimentary phase. Sedimentary rocks may never be buried deeply enough to be converted to metamorphic rock or melted to form magma, but instead may undergo several cycles of weathering, erosion, and deposition.

The concept of the rock cycle was first outlined in the late eighteenth century by James Hutton as part of his overall theory of the Earth (see Fig. 3.5). Hutton believed that the Earth's internal heat provided the mechanism for magma generation, uplift, and mountain building. Today, plate tectonic theory offers an explanation for how and where magma is generated, and how and why mountain building occurs. Following a brief discussion of the Earth's interior and plate tectonics, we will return to the relationship between the rock cycle and plate tectonics.

◆ **FIGURE 1.7** Glacial till is an unsorted mixture of many different sediment sizes deposited at the front of a melting glacier. The rock tillite (which is lithified till) provides evidence that glaciation once occurred at this location. (Photo courtesy of Wayne E. Moore.)

❖ THE EARTH'S SURFACE AND INTERIOR

◆ Figure 1.8 is a cross section showing the internal structure of the Earth. The **core** of the Earth occupies about

◆ **FIGURE 1.8** A cross section of the Earth illustrating the core, mantle, and crust. The enlarged portion shows the relationship between the lithosphere, composed of the continental crust, oceanic crust, and upper mantle, and the underlying asthenosphere and lower mantle.

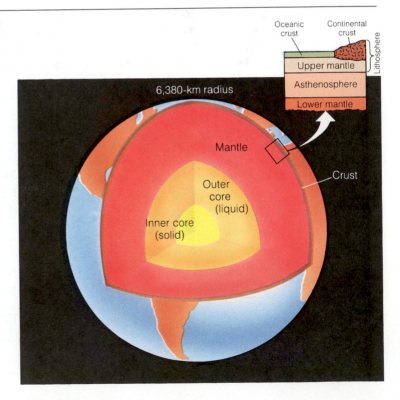

16% of its total volume and is divided into an outer and an inner core. The inner core is believed to be solid while the outer core is apparently liquid. Both are inferred to consist largely of iron with a small amount of nickel.

Surrounding the core is the **mantle.** It comprises about 83% of the Earth's volume and is thought to be composed largely of *peridotite,* a dark, dense rock containing abundant iron and magnesium. The mantle can be divided into three distinct zones based on physical characteristics. The lower mantle is solid and forms most of the volume of the Earth's interior. Above the lower mantle, between approximately 100 to 250 kilometers (km) in depth, is the **asthenosphere.** It has the same composition as the lower mantle but behaves plastically and slowly flows. It is in this zone that partial melting takes place, generating magma that migrates toward the Earth's surface because it is less dense than the rock from which it was derived. Above the asthenosphere is the solid upper mantle. The upper mantle and the overlying crust form the **lithosphere.** The lithosphere is broken into several individual pieces called *plates.* These plates are believed to move over the asthenosphere as a result of underlying mantle convection cells (◆ Fig. 1.9).

The **crust** is the outermost layer of the Earth and consists of two types. The *continental crust* is thick (20 to 90 km), has an average density of 2.7 grams per cubic centimeter (g/cm^3), and is commonly referred to as *sialic,* meaning that it contains considerable *si*licon and *al*uminum. The oceanic crust is thin (5 to 10 km), denser than continental crust (3.0 g/cm^3), and is commonly referred to as *sima,* a term derived from the words *si*lica and *ma*gnesium.

❖ PLATE TECTONICS

Plate tectonic theory is generally accepted by most geologists because it provides a comprehensive model that accounts for the Earth's internal workings and explains the relationships among many seemingly unrelated geological phenomena. Although the concept of continental movement goes back to the early 1900s when Alfred Wegener developed the continental drift hypothesis, the theory of plate tectonics was not widely accepted until the 1960s. Here we will present a brief overview of plate tectonics and will provide a more thorough treatment of the theory and its significance in reconstructing Earth history in Chapter 7.

According to plate tectonic theory, the lithosphere (crust and upper mantle) is divided into plates that move over the asthenosphere (◆ Fig. 1.10). Zones of volcanic activity, earthquake activity, or both, mark most plate

---

◆ **FIGURE 1.9** The Earth's plates are believed to move as a result of underlying mantle convection cells in which warm material from deep within the Earth rises toward the surface, cools, and then, upon losing heat, descends back downward into the interior. The movement of these convection cells is believed to be the mechanism responsible for the movement of the Earth's plates as shown in this diagrammatic cross section.

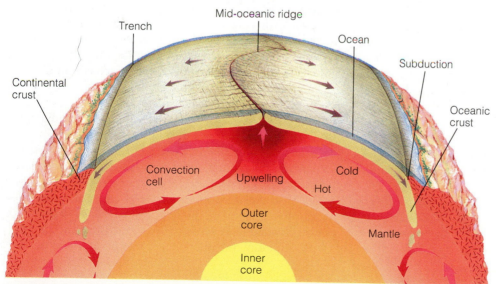

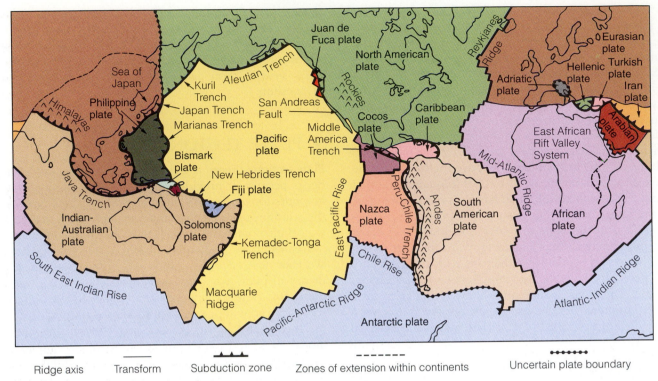

▬▬	▬▬▬	▲▲▲	– – – –	•◆•◆•◆•◆
Ridge axis	Transform	Subduction zone	Zones of extension within continents	Uncertain plate boundary

◆ **FIGURE 1.10** Plates of the world. The Earth's lithosphere is divided into rigid plates of various sizes that move over the asthenosphere.

boundaries. Along these boundaries, plates diverge, converge, or slide sideways past each other.

At **divergent plate boundaries,** plates move apart as magma rises to the surface from the asthenosphere. Divergent boundaries in oceanic crust are marked by mid-oceanic ridges (◆ Fig. 1.11). Beneath the continental crust, divergent boundaries are recognized by linear rift valleys.

Plates move toward one another along **convergent plate boundaries.** When an oceanic plate meets a continental plate, for example, the oceanic plate sinks along what is known as a **subduction zone** because it is denser than the continental plate (◆ Fig. 1.12). As it descends into the Earth, the subducting oceanic plate becomes hotter and finally melts, generating magma. As this magma rises, it may erupt at the Earth's surface, forming a chain of volcanoes.

Crustal production and consumption occur at divergent and convergent plate boundaries, respectively. In contrast, **transform plate boundaries** are sites where plates slide sideways past each other, and crust is neither

produced nor destroyed. The San Andreas fault in California is an excellent example of a transform plate boundary. It separates the North American from the Pacific plate (◆ Fig. 1.13).

Although geologists can describe and explain what is happening at plate boundaries, there is still debate about a possible driving mechanism that causes plate movement. Most geologists, however, accept a mechanism involving some type of convection system in which mantle rock is heated and flows like a very viscous liquid.

The importance of plate tectonic theory is that it allows geologists to explain many seemingly unrelated events. For example, the Paleozoic mountain-building episodes in North America and Europe were actually part of a larger event involving the closing of the Iapetus Ocean and the formation of the supercontinent Pangaea (see Chapter 12). Plate tectonic theory also allows geologists to correlate periods of rapid seafloor spreading along divergent plate boundaries to times when the oceans flooded large parts of the continents (see Chapter 4). In addition, the theory helps explain some of the

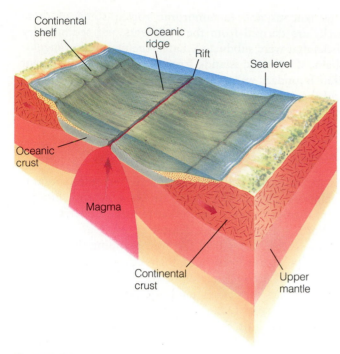

♦ **FIGURE 1.11** A divergent plate boundary forms when magma wells up beneath oceanic crust and forms a mid-oceanic ridge.

geographic distribution patterns of fossil and living organisms (see Chapter 6).

A revolutionary concept when it was proposed in the 1960s, plate tectonic theory has had significant and far-reaching consequences in all fields of geology because it provides the basis for relating many seemingly unrelated phenomena. Its impact has been particularly notable on the reconstruction and interpretation of Earth history.

The Rock Cycle and Plate Tectonics

The preceding discussion of the rock cycle was presented without reference to a mechanism responsible for recycling rock materials. Plate tectonic theory provides such a recycling mechanism.

Interactions among plates determine, to a certain extent, which of the three rock groups will form (♦ Fig. 1.14). For example, weathering and erosion produce sediments that are transported by various agents such as running water from the continents to the oceans where they are deposited and accumulate. These sediments, some of which may be lithified and become sedimentary rock, along with the underlying oceanic crust become part of a moving plate. When plates converge, heat and pressure generated at the plate boundary may lead to igneous activity and metamorphism, thus producing var-

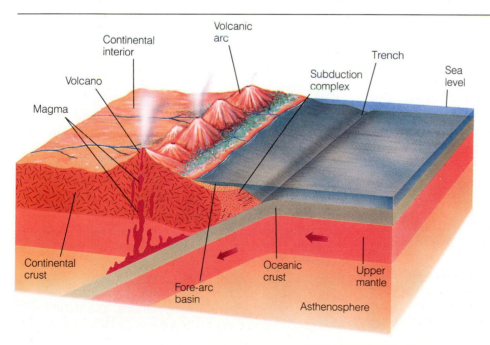

♦ **FIGURE 1.12** Convergent plate boundary between oceanic crust and continental crust. Because the oceanic crust is denser, it is subducted under the continental crust. A chain of volcanic mountains forms on the continental plate as a result of rising magma.

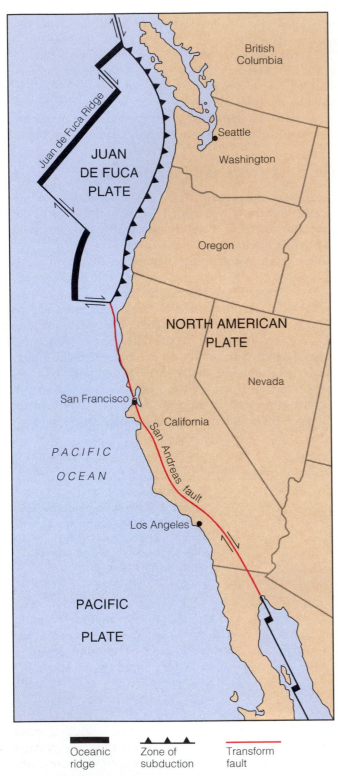

Oceanic ridge   Zone of subduction   Transform fault

◆ **FIGURE 1.13** The San Andreas fault is a transform fault separating the Pacific plate from the North American plate.

ious igneous and metamorphic rocks; some of these rocks are derived from the sediments and sedimentary rocks that were subducted. Later, the mountain range or chain of volcanic islands formed along the convergent plate boundary will once again be weathered and eroded, and the new sediments will be transported to the ocean to begin yet another cycle.

❖ ORGANIC EVOLUTION

As we stated in the Prologue, plate tectonics, organic evolution, and geologic time are three central, interrelated themes that are fundamental to an understanding of Earth history. As just demonstrated, plate tectonics provides us with a model for understanding the internal workings of the Earth and their effect on the Earth's surface. The second theme, the theory of **organic evolution,** provides the conceptual framework for understanding the history of life.

The concept of evolution is not new. In fact, some naturalists had seriously considered it long before Charles Darwin published *On The Origin of Species by Means of Natural Selection* in 1859. The appearance of Darwin's book, however, revolutionized biology and marked the beginning of modern evolutionary biology. With its publication, most naturalists recognized that evolution provided a unifying theory that explained an otherwise encyclopedic collection of biologic facts.

The central thesis of organic evolution is that all living organisms are related, and that they have descended with modifications from organisms that lived during the past. When Darwin proposed his theory of organic evolution, he cited a wealth of supporting evidence such as the way organisms are classified, embryology, comparative anatomy, the geographic distribution of organisms, and, to a limited extent, the fossil record. Furthermore, Darwin proposed that *natural selection,* which results in the survival to reproductive age of those organisms best adapted to their environment, is the mechanism that accounts for evolution.

In spite of overwhelming evidence to support evolution, Darwin could not explain how variation within a population arose, nor how it was maintained. We now know that mutations occur within genes (the basic units of inheritance) causing changes within an organism and that these changes are passed on to the next generation during reproduction. Thus, mutations provide one source of variation within populations, and natural selection acts upon them in determining which variants are favorable for survival. Another important source of vari-

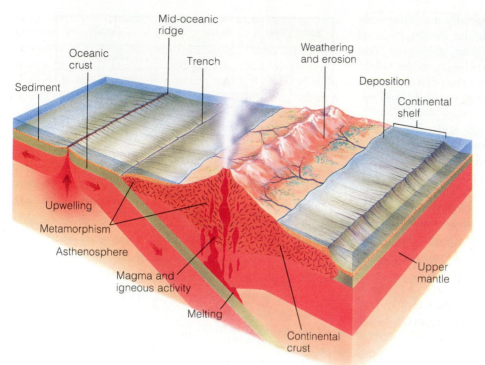

Sediment

Oceanic crust

Mid-oceanic ridge

Trench

Weathering and erosion

Deposition

Continental shelf

Upwelling

Metamorphism

Asthenosphere

Magma and igneous activity

Melting

Continental crust

Upper mantle

◆ **FIGURE 1.14** Plate tectonics and the rock cycle. The cross section shows how the three rock groups, igneous, sedimentary, and metamorphic, are recycled through both the continental and oceanic regions.

ation occurs during sexual reproduction when the offspring receive one-half of their genes from each parent. This allows new mutations to be introduced into the population in many different combinations because potentially the parent with the mutation may mate with a variety of partners.

Perhaps the most compelling evidence in favor of evolution can be found in the fossil record. Just as the rock record allows geologists to interpret physical events and conditions in the geologic past, fossils provide us with a record of life in the past and evidence that organisms have evolved through time.

Fossils, which are the remains or traces of once-living organisms, not only provide evidence that evolution has occurred, but also demonstrate that the Earth has a history extending beyond that recorded by humans. The succession of fossils in the rock record provides geologists with a means for dating rocks and allowed a relative geologic time scale to be constructed in the 1800s.

❖ GEOLOGIC TIME AND UNIFORMITARIANISM

The concept of vast amounts of geologic time is central to understanding the evolution of the Earth and its biota.

The discovery of radioactivity in 1895 provided geologists with the means to determine absolute ages of rock units in years. Prior to this, a geologic calendar had been developed using the principles of relative-age dating. These principles allow geologists to place events in a sequential chronology without knowing how many years ago a particular event occurred. These principles are still used today for interpreting the geologic history of an area and will be discussed in Chapter 3.

The development of the **geologic time scale** (◆ Fig. 1.15) in the nineteenth century was based primarily on the sequence of fossils found in the rock record. After many years of collecting fossils from numerous locations, geologists discovered that fossil species succeeded one another in a definite order, and that any time period in the geologic past could be recognized by the fossils left behind. Once these were established, geologists could identify rocks of the same age in widely separated areas on the basis of their fossil content. The construction of the geologic time scale resulted from the work of many geologists who pieced together information from numerous rock exposures and reconstructed a chronology based on changes in the Earth's biota through time.

The division of the geologic time scale into eras, periods, and epochs was primarily based on changes in the fossil record; therefore, it was originally a relative time scale. Subsequently, however, with the development of various radiometric dating techniques, geologists have been able to assign absolute dates in years to the subdivisions of the geologic time scale and thus can determine how many years ago the various eras, periods, and epochs began.

One of the cornerstones of geology is the **principle of uniformitarianism.** It is based on the premise that present-day processes have operated throughout geologic time. What this means is that even though the rates and intensities of geological processes have varied during the past, the physical and chemical laws of nature have remained the same and cannot be violated. Uniformitarianism is thus a powerful principle that allows us to use present-day processes as the basis for interpreting the past and predicting potential future events.

Although the Earth is in a dynamic state of change and has been ever since it formed, the processes that shaped it during the past are the same ones that are in operation today. Therefore, in order to understand and interpret the rock record, we must first understand present-day processes and their results.

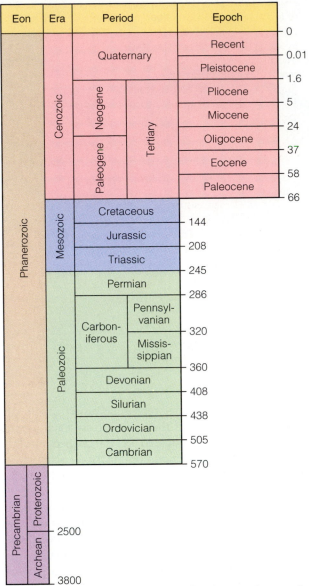

◆ **FIGURE 1.15** The geologic time scale. Numbers to the right of the columns are ages in millions of years before the present.

Chapter Summary

1. Historical geology deals with the origin and history of the Earth's continents, atmosphere, oceans, and life.
2. The scientific method is an orderly, logical approach that involves gathering and analyzing facts about a particular phenomenon, formulating hypotheses to explain the phenomenon, testing the hypotheses, and finally proposing a theory. A theory is an explanation for some natural phenomenon that has a large body of supporting evidence and can be tested.
3. The rock and fossil record provide the only evidence for interpreting Earth history.
4. Igneous, sedimentary, and metamorphic rocks comprise the three major rock groups. Igneous rocks result from the crystallization of magma or consolidation of volcanic

ejecta. Sedimentary rocks are formed from the lithification of any type of sediment. Metamorphic rocks are produced from other rocks by heat, pressure, and chemically active fluids beneath the Earth's surface.

5. The rock cycle illustrates the interaction between the internal and external forces of the Earth and shows how the three rock groups are interrelated.

6. The Earth is differentiated into layers. The outermost layer is the crust, which is divided into continental and oceanic crust. Below the crust is the upper mantle. The crust and upper mantle overlie the asthenosphere, a zone that behaves plastically. The asthenosphere is underlain by the lower mantle. The Earth's core is divided into an outer liquid portion and an inner solid portion.

7. The Earth's crust and upper mantle form the lithosphere, which is broken into a series of plates that diverge, converge, and slide sideways past one another.

8. Plate tectonic theory provides a unifying explanation for many geological features and events. It also provides a mechanism for recycling rock material.

9. The central thesis of the theory of organic evolution is that all living organisms evolved (descended with modification) from organisms that existed in the past. Mutations and sexual reproduction provide the source of variation within a population that natural selection acts upon.

10. The concept of geologic time is central to understanding the evolution of the Earth and its biota. Geologic time is subdivided into eras, periods, and epochs. The subdivisions of the geologic time scale combine relative ages based on fossils with more recently determined absolute ages in years.

11. The principle of uniformitarianism is basic to the interpretation of Earth history. This principle holds that the laws of nature have been constant through time and that the same processes at work today have operated in the past, albeit at different rates.

Important Terms

asthenosphere
convergent plate boundary
core
crust
divergent plate boundary
fossil
geologic time scale
geology

hypothesis
igneous rock
lithosphere
mantle
metamorphic rock
mineral
organic evolution
plate tectonic theory

principle of uniformitarianism
rock
rock cycle
scientific method
sedimentary rock
subduction zone
theory
transform plate boundary

Review Questions

1. The study of Earth materials is:
   a. _____ paleontology; b. _____ stratigraphy; c. _____ physical geology; d. _____ historical geology; e. _____ environmental geology.

2. Which of the following statements about a scientific theory is not true?
   a. _____ it is an explanation for some natural phenomenon; b. _____ it is testable; c. _____ it involves a large body of supporting evidence; d. _____ it is a conjecture or a guess; e. _____ predictive statements can be derived from it.

3. Which of the following statements about a mineral is not true?
   a. _____ it is organic; b. _____ it has definite physical and chemical properties; c. _____ it is naturally occurring; d. _____ it is a crystalline solid; e. _____ none of these.

4. The rock cycle implies that:
   a. _____ only sedimentary rocks can be derived from igneous rocks; b. _____ igneous rocks only form beneath the Earth's surface; c. _____ only sedimentary rocks can be weathered; d. _____ any rock group can be derived from any other rock group; e. _____ metamorphic rocks form from the cooling of magma.

5. A phaneritic texture results from:
   a. _____ rapid cooling of a magma; b. _____ slow cooling of a magma; c. _____ weathering of a detrital sedimentary rock; d. _____ pressure applied to a fine-grained sedimentary rock; e. _____ the two-stage cooling of a magma.

6. What type of rock is produced by the precipitation of the dissolved material carried in solution by surface waters?
   a. _____ intrusive igneous; b. _____ extrusive igneous;

c. _____ detrital sedimentary; d. _____ chemical sedimentary; e. _____ foliated metamorphic.

7. Which rock group results from the application of pressure?
   a. _____ intrusive igneous; b. _____ extrusive igneous; c. _____ detrital sedimentary; d. _____ nonfoliated metamorphic; e. _____ foliated metamorphic.

8. The layer immediately below the crust is the:
   a. _____ upper mantle; b. _____ asthenosphere; c. _____ lower mantle; d. _____ outer core; e. _____ inner core.

9. Which layer has the same composition as the mantle, but behaves plastically?
   a. _____ continental crust; b. _____ oceanic crust; c. _____ outer core; d. _____ inner core; e. _____ asthenosphere.

10. Which layer is composed predominantly of iron with a small amount of nickel?
    a. _____ crust; b. _____ mantle; c. _____ asthenosphere; d. _____ core; e. _____ lithosphere.

11. A plate is composed of which of the following pairs?
    a. _____ outer core–lower mantle; b. _____ continental crust–oceanic crust; c. _____ crust–upper mantle; d. _____ upper mantle–asthenosphere; e. _____ crust-asthenosphere.

12. A plate boundary at which plates slide horizontally past each other is a:
    a. _____ divergent boundary; b. _____ convergent boundary; c. _____ transform boundary; d. _____ subduction boundary; e. _____ none of these.

13. Plate movements are believed to be driven by:
    a. _____ magnetism; b. _____ convection cells; c. _____ gravity; d. _____ the rock cycle; e. _____ geothermal activity.

14. The development of the geologic time scale was based primarily on:
    a. _____ absolute dating techniques; b. _____ sedimentary structures; c. _____ plate movement; d. _____ the succession of fossil assemblages; e. _____ all of these.

15. That all living organisms are the descendants of different life-forms that existed in the past is the central claim of:
    a. _____ organic evolution; b. _____ the principle of fossil succession; c. _____ the principle of uniformitarianism; d. _____ plate tectonics; e. _____ none of these.

16. Explain how historical and physical geology differ and how they are related.

17. Name and define the three major rock groups.

18. Explain the scientific method and how it relates to a theory.

19. Name the major layers of the Earth and describe their composition.

20. Why are the concepts of plate tectonics, organic evolution, and geologic time so important to historical geology?

21. Briefly describe plate tectonic theory. Why is it such a unifying theory in geology?

22. Name and describe the three types of plate boundaries.

23. Describe the rock cycle and explain how it is related to plate tectonics.

24. Briefly discuss the theory of organic evolution.

25. How was the geologic time scale originally developed?

26. Reproduce the geologic time scale and give the dates for the boundaries of each era.

27. Define uniformitarianism and briefly discuss why it is important and how it allows geologists to interpret the past.

Additional Readings

Berry, W. B. N. 1987. *Growth of a prehistoric time scale*, 2d ed. Palo Alto, Calif.: Blackwell Scientific Publications.

Dietrich, R. V., and B. J. Skinner. 1990. *Gems, granites, and gravels: Knowing and using rocks and minerals.* New York: Cambridge University Press.

Ernst, W. G. 1990. *The dynamic planet.* Irvington, N.Y.: Columbia University Press.

Lane, N. G. 1992. *Life of the past*, 3d ed. New York: Macmillan.

Monroe, J. S., and R. Wicander. 1992. *Physical geology: Exploring the Earth.* St. Paul, Minn.: West Publishing Co.

CHAPTER 2

Chapter Outline

Intrusive igneous rocks in Yosemite National Park, California. The near vertical rock face, called El Capitan, is composed of granite and rises more than 900 m above the valley floor. (Photo courtesy of Richard L. Chambers.)

MINERALS AND ROCKS

Prologue

Many minerals and rocks (which are composed of minerals) are valuable. Indeed our industrialized society depends directly upon finding and using various mineral resources such as iron, copper, gold, and many others.

Geologists of the U.S. Geological Survey and U.S. Bureau of Mines define a *resource* as follows:

A concentration of naturally occurring solid, liquid, or gaseous material in or on the Earth's crust in such form and amount that economic extraction of a commodity from the concentration is currently or potentially feasible.

Accordingly, resources include such substances as metals (*metallic resources*); sand, gravel, crushed stone, and sulfur (*nonmetallic resources*); and uranium, coal, oil, and natural gas (*energy resources*). However, an important distinction must be made between a resource, which is the total amount of a commodity whether discovered or undiscovered, and a *reserve,* which is that part of the resource base that can be extracted economically.

The amount of mineral resources used has steadily increased since Europeans settled this continent; during 1988, for example, each resident of North America used about 14 metric tons, a large part of which was bulk items such as sand, gravel, and crushed stone. Thus, it is no exaggeration to say that our industrialized society is totally dependent on mineral resources. Unfortunately, they are being used at rates far faster than they form. In other words, mineral resources are nonrenewable, meaning that once the resources from a deposit have been exhausted, new supplies or suitable substitutes must be found.

For some mineral resources, adequate supplies are available for indefinite periods (sand and gravel, for example), whereas for others, supplies are limited or must be imported from other parts of the world (◆ Fig. 2.1). For example, the United States is almost totally dependent on imports for manganese, an essential component in the manufacture of steel. Even though the United States is a leading producer of gold, it still depends on imports for more than half of its gold needs. More than half of our crude oil is imported, much of it from the Middle East where more than 50% of the Earth's proven reserves exist. The United States' response to the takeover of Kuwait by Iraq during August 1990 was a poignant reminder of our dependence on the availability of resources.

Whether a given deposit constitutes a resource or a reserve depends on several factors. For example, iron-bearing minerals occur in many rocks, but in quantities or ways that make their recovery uneconomical. Geographic location is also an important consideration. A mineral resource in a remote region may not be mined because transportation costs are too high, and what may be considered a resource in the United States or Canada may be mined in a less industrialized country where labor costs are low. The market price of a commodity is, of course, important in evaluating a potential resource. From 1935 to 1968, the U.S. government maintained the price of gold at $35 per troy ounce (= 31.1 grams). When this restriction was removed and the price of gold became subject to supply and demand, the price rose (it reached an all-time high of $843 per troy ounce during January 1980). As a consequence, many marginal deposits became reserves and many abandoned mines were reopened.

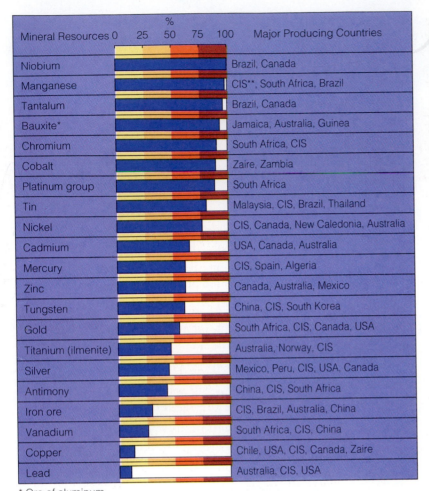

Mineral Resources	% 0 25 50 75 100	Major Producing Countries
Niobium		Brazil, Canada
Manganese		CIS**, South Africa, Brazil
Tantalum		Brazil, Canada
Bauxite*		Jamaica, Australia, Guinea
Chromium		South Africa, CIS
Cobalt		Zaire, Zambia
Platinum group		South Africa
Tin		Malaysia, CIS, Brazil, Thailand
Nickel		CIS, Canada, New Caledonia, Australia
Cadmium		USA, Canada, Australia
Mercury		CIS, Spain, Algeria
Zinc		Canada, Australia, Mexico
Tungsten		China, CIS, South Korea
Gold		South Africa, CIS, Canada, USA
Titanium (ilmenite)		Australia, Norway, CIS
Silver		Mexico, Peru, CIS, USA, Canada
Antimony		China, CIS, South Africa
Iron ore		CIS, Brazil, Australia, China
Vanadium		South Africa, CIS, China
Copper		Chile, USA, CIS, Canada, Zaire
Lead		Australia, CIS, USA

* Ore of aluminum.
** Commonwealth of Independent States (includes much of the former Soviet Union).

(a)

(b)

(c)

◆ **FIGURE 2.1** (a) The percentages of some mineral resources imported by the United States. The lengths of the blue bars correspond to the amounts of resources imported. (b) Bauxite, the ore of aluminum. (Photo courtesy of Sue Monroe.) (c) A copper mine at Butte, Montana.

Technological developments can also change the status of a resource. For example, the rich iron ores of the Great Lakes region of the United States and Canada had been depleted by World War II. Then, the development of a method of separating the iron from previously unusable rocks and shaping it into pellets that are ideal for use in blast furnaces made it feasible to mine poorer-grade ores.

Most of the largest and richest mineral deposits have probably already been discovered and, in some cases, depleted. To ensure continued supplies of essential minerals, geologists are using increasingly sophisticated geophysical and geochemical mineral exploration techniques. The U.S. Geological Survey and the U.S. Bureau of Mines continually assess the status of resources in view of changing economic and political conditions and developments in science and technology.

❖ INTRODUCTION

The term "mineral" commonly brings to mind dietary substances such as calcium, iron, potassium, and magnesium. These are actually chemical elements, not minerals in the geologic sense. Mineral is also sometimes used to refer to any substance that is neither animal nor vegetable. Such usage implies that minerals are inorganic substances, which is correct, but not all inorganic substances are minerals. In fact, geologists have a very specific definition of the term **mineral;** a naturally occurring, inorganic crystalline solid. Crystalline means it has a regular internal structure. Furthermore, a mineral has a narrowly defined chemical composition and characteristic physical properties such as density, color, and hardness. Most rocks are solid aggregates of one or more minerals, and thus minerals are the building blocks of rocks.

Obviously, minerals are important to geologists as the constituents of rocks, which provide our only record of Earth history, but they are important for other reasons as well. Many gemstones such as diamond and topaz are actually minerals, the sand used to make glass is composed of the mineral quartz, and ore deposits are concentrations of economically valuable minerals.

❖ MATTER AND ITS COMPOSITION

Anything that has mass and occupies space is *matter.* Matter occurs in one of three states or phases, all of which are important in geology: *solids, liquids,* and *gases.* Here, we are concerned chiefly with solids because all minerals are solids.

Elements and Atoms

All matter is made up of chemical **elements,** each of which is composed of incredibly small particles called **atoms.** Atoms are the smallest units of matter that retain the characteristics of an element. Ninety-one naturally occurring elements have been discovered, each of which has a name and a chemical symbol (Table 2.1).

Atoms consist of a compact **nucleus** composed of one or more *protons,* which are particles with a positive electrical charge, and electrically neutral *neutrons* (◆ Fig. 2.2). Encircling the nucleus are negatively charged particles called *electrons.* Electrons orbit rapidly around the nucleus at specific distances in one or more *electron shells* (Fig. 2.2). The number of protons in the nucleus of an atom determines what the element is and determines the **atomic number** for that element. For example, each atom of the element hydrogen (H) has one proton in its nucleus and thus has an atomic number of 1, whereas carbon (C), with 6 protons, has an atomic number of 6 (Table 2.1).

Atoms are also characterized by their **atomic mass number,** which is determined by adding together the number of protons and neutrons in the nucleus (electrons contribute negligible mass to an atom). However, not all atoms of the same element have the same number of

neutrons. For example, different carbon (C) atoms have atomic mass numbers of 12, 13, and 14. All of these atoms possess 6 protons—otherwise they would not be carbon—but the number of neutrons varies (◆ Fig. 2.3). Forms of the same element with different atomic mass numbers are *isotopes.*

A number of elements have a single isotope but many, such as uranium and carbon, have several; all isotopes of an element behave the same chemically. Some isotopes are unstable and spontaneously change to a stable form. This process, called *radioactive decay,* occurs because the forces binding the nucleus together are not strong enough. Such decay occurs at known rates and is the basis for several techniques for determining age that will be discussed in Chapter 3.

Bonding and Compounds

The process whereby atoms are joined to other atoms is called **bonding.** When atoms of two or more different elements are bonded, the resulting substance is a **compound.** Thus, a chemical substance such as gaseous oxygen, which consists entirely of oxygen atoms, is an element, whereas ice, which consists of hydrogen and oxygen, is a compound. Most minerals are compounds although there are several important exceptions.

To understand bonding, it is necessary to delve deeper into the structure of atoms. With the exception of hydrogen, which has only one proton and one electron, the innermost electron shell of an atom contains no more than two electrons, and the outermost shell never contains more than eight (Table 2.1). The electrons in the outermost shell are those that are usually involved in chemical bonding. Two types of chemical bonds are particularly important in minerals, *ionic* and *covalent,* and many minerals contain both types of bonds.

Notice in Table 2.1 that the atoms of a few elements, such as neon and argon, have complete outer shells containing eight electrons; they are known as the *noble gases.* The noble gases do not react readily with other elements to form compounds because of this electron configuration. Interactions among atoms tend to produce electron configurations similar to those of the noble gases, unless the first shell (with two electrons) is also the outermost electron shell as in helium.

One way in which the noble gas configuration can be attained is by the transfer of one or more electrons from one atom to another. Common salt is composed of sodium (Na) and chlorine (Cl); sodium has only one electron in its outermost shell whereas chlorine has seven

TABLE 2.1 Symbols, Atomic Numbers, and Electron Configurations for Some of the Naturally Occurring Elements

ELEMENT	SYMBOL	ATOMIC NUMBER	NUMBER OF ELECTRONS IN EACH SHELL			
			1	2	3	4
Hydrogen	H	1	1			
Helium	He	2	2			
Lithium	Li	3	2	1		
Beryllium	Be	4	2	2		
Boron	B	5	2	3		
Carbon	C	6	2	4		
Nitrogen	N	7	2	5		
Oxygen	O	8	2	6		
Fluorine	F	9	2	7		
Neon	Ne	10	2	8		
Sodium	Na	11	2	8	1	
Magnesium	Mg	12	2	8	2	
Aluminum	Al	13	2	8	3	
Silicon	Si	14	2	8	4	
Phosphorus	P	15	2	8	5	
Sulfur	S	16	2	8	6	
Chlorine	Cl	17	2	8	7	
Argon	Ar	18	2	8	8	
Potassium	K	19	2	8	8	1
Calcium	Ca	20	2	8	8	2

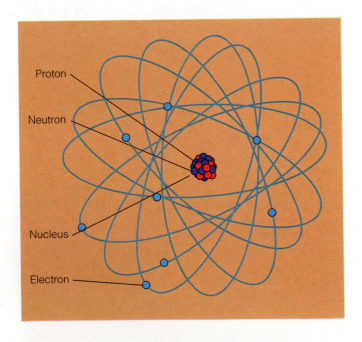

◆ **FIGURE 2.2** The structure of an atom. The dense nucleus consisting of protons and neutrons is surrounded by a cloud of orbiting electrons.

(◆ Fig. 2.4). In order to attain a stable electron configuration, sodium loses the electron in its outermost shell, leaving its next shell with eight electrons as the outermost one (Fig. 2.4a). However, sodium now has one fewer electron (negative charge) than it has protons (positive charge) so it is an electrically charged particle. Such a particle is an *ion* and, in the case of sodium, is symbolized as Na^{+1}.

The electron lost by sodium is transferred to the outermost electron shell of chlorine, which had seven electrons to begin with. Thus, the addition of one more electron gives chlorine an outermost electron shell of eight electrons, the configuration of a noble gas. Its total number of electrons, however, is now 18, which exceeds by one the number of protons. Accordingly, chlorine also becomes an ion, but it is negatively charged (Cl^{-1}). Because an attractive force exists between the positively charged sodium ion and the negatively charged chlorine ion, an **ionic bond** forms between sodium and chlorine yielding the compound sodium chloride (the mineral halite) (Fig. 2.4).

Covalent bonds form between atoms when their electron shells overlap and electrons are shared. For example, a carbon atom in diamond shares all four of its

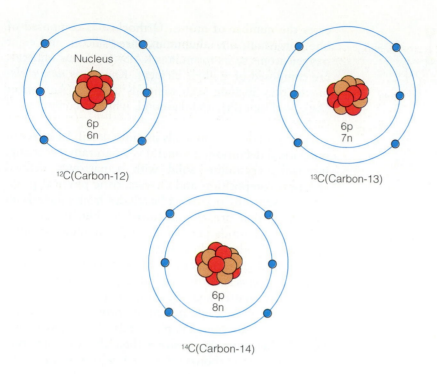

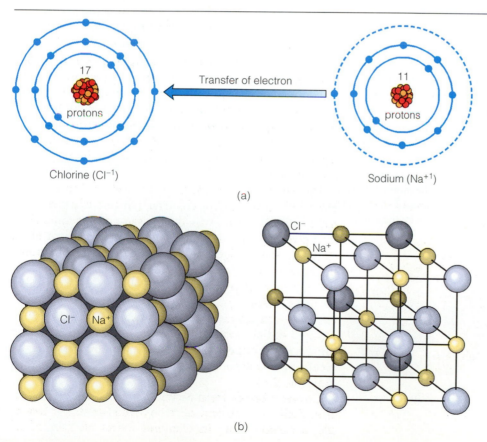

◆ **FIGURE 2.3** Schematic representation of isotopes of carbon. A carbon atom has an atomic number of 6 and an atomic mass number of 12, 13, or 14 depending on the number of neutrons in its nucleus.

◆ **FIGURE 2.4** (a) Ionic bonding. The electron in the outermost shell of sodium is transferred to the outermost electron shell of chlorine. Once the transfer has occurred, sodium and chlorine are positively and negatively charged ions, respectively. (b) The crystal structure of sodium chloride, the mineral halite. The diagram on the left shows the relative sizes of sodium and chlorine ions, and the diagram on the right shows the locations of the ions in the crystal structure.

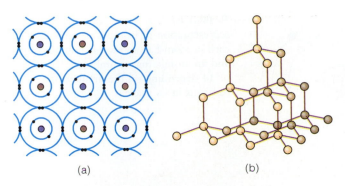

(a)   (b)

◆ **FIGURE 2.5** (a) Covalent bonds formed by adjacent atoms sharing electrons in diamond. (b) The three-dimensional framework of carbon atoms in diamond.

outermost electrons with a neighbor to produce a stable noble gas configuration (◆ Fig. 2.5). Among the most common minerals, the silicates (discussed later in this chapter), the element silicon forms partly covalent and partly ionic bonds with oxygen.

❖ MINERALS

Most minerals are compounds of two or more elements, and their composition is generally shown by a chemical formula, which is a shorthand way of indicating the number of atoms of different elements composing a mineral. The mineral quartz, for example, consists of one silicon (Si) atom for every two oxygen (O) atoms, and thus has the formula SiO_2; the subscript number indi-

cates the number of atoms. Orthoclase is composed of one potassium, one aluminum, three silicon, and eight oxygen atoms so its formula is $KAlSi_3O_8$. Some minerals are composed of a single element. Known as *native elements,* they include such minerals as gold (Au), silver (Ag), platinum (Pt), and diamond, which is composed of carbon (C).

Before we discuss minerals in more detail, let us recall our formal definition: a mineral is a naturally occurring, inorganic, crystalline solid with a narrowly defined chemical composition and characteristic physical properties. "Naturally occurring" excludes from minerals all substances that are manufactured by humans such as synthetic diamonds and rubies and a number of other artificially synthesized substances.

Some geologists think the term "inorganic" in the mineral definition is superfluous. It does, however, remind us that animal matter and vegetable matter are not minerals. Nevertheless, some organisms produce compounds that are minerals. For example, corals, clams, and a number of other animals construct their shells of the compound calcium carbonate ($CaCO_3$), which is either aragonite or calcite, both of which are minerals.

By definition minerals are **crystalline solids** in which the constituent atoms are arranged in a regular, three-dimensional framework (◆ Fig. 2.6). Under ideal conditions, mineral crystals can grow and form perfect crystals that possess planar surfaces (crystal faces), sharp corners, and straight edges (Fig. 2.6). In other words, the regular geometric shape of a well-formed mineral crystal is the exterior manifestation of an ordered internal atomic arrangement. In many cases numerous minerals

◆ **FIGURE 2.6** Mineral crystals occur in a variety of shapes, several of which are shown here. (a) Cubic crystals typically develop in the minerals halite, galena, and pyrite. (b) Dodecahedron crystals such as those of garnet have 12 sides. (c) Diamond has octahedral or 8-sided crystals. (d) A prism terminated by pyramids is found in quartz.

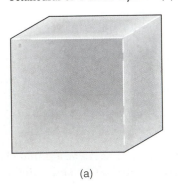

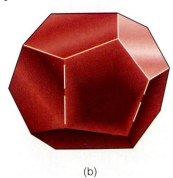

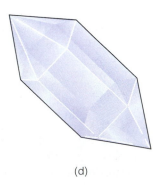

(a)   (b)   (c)   (d)

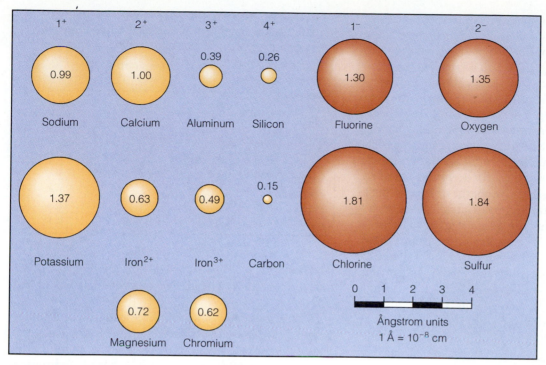

FIGURE 2.7 Electrical charges and relative sizes of ions common in minerals. The numbers within the ions are the radii shown in Ångstrom units.

grow in proximity, as in a cooling lava flow, and thus do not have an opportunity to develop well-formed crystals.

The definition of a mineral contains the phrase "a narrowly defined chemical composition," because some minerals have a limited range of compositions. For many minerals the chemical composition is constant: quartz is always composed of silicon and oxygen (SiO_2), and halite contains only sodium and chlorine ($NaCl$). Other minerals have a range of compositions because one element can substitute for another if the atoms of two or more elements are nearly the same size and have the same electrical charge. Iron and magnesium atoms are about the same size and can therefore substitute for one another (◆ Fig. 2.7). The mineral olivine $(Mg,Fe)_2SiO_4$ may contain, in addition to silicon and oxygen, only magnesium, only iron, or a combination of both.

The last criterion in our definition of a mineral, "characteristic physical properties," refers to such properties as hardness, color, and crystal form. Such properties are controlled by composition and structure.

❖ MINERAL GROUPS

More than 3,500 minerals have been identified and described, but only a very few—perhaps two dozen—are particularly common. The common minerals are composed mostly of the eight chemical elements making up the bulk of the Earth's crust (Table 2.2). Geologists recognize mineral classes or groups, each of which is composed of members that share the same negatively charged ion or ion group (Table 2.3).

Silicate Minerals

Since silicon and oxygen are the two most abundant elements in the Earth's crust (Table 2.2), it is not surprising that many minerals contain these elements. A combination of silicon and oxygen is called *silica*, and the minerals containing silica are **silicates**. Quartz (SiO_2) is composed entirely of silicon and oxygen so it is pure silica. Most silicates, however, have one or more additional elements, as in orthoclase ($KAlSi_3O_8$) and olivine

[(Mg,Fe)$_2$SiO$_4$]. Silicate minerals include about one-third of all known minerals, but their abundance is even more impressive when one considers that they make up perhaps as much as 95% of the Earth's crust.

The basic building block of all silicate minerals is the **silica tetrahedron,** which consists of one silicon atom and four oxygen atoms (◆ Fig. 2.8). These atoms are arranged such that the four oxygen atoms surround a silicon atom, which occupies the space between the oxygen atoms. The silica tetrahedron is symbolized as (SiO$_4$)$^{-4}$, because the silicon atom has a positive charge of 4, while each of the four oxygen atoms has a negative charge of 2, resulting in an ion group with a total negative charge of 4.

Since the silica tetrahedron has a negative charge, it does not exist in nature as an isolated ion group; rather it combines with positively charged ions or shares its oxygen atoms with other silica tetrahedra. In the simplest silicate minerals, the silica tetrahedra exist as single units bonded to positively charged ions. In minerals containing isolated tetrahedra, the silica ion is balanced by positive ions (◆ Fig. 2.9a).

Silica tetrahedra may also be arranged so that they join together to form chains of indefinite length (◆ Fig. 2.9b). Single chains, as in the pyroxene minerals, form when each tetrahedron shares two of its oxygens with an adjacent tetrahedron. The amphibole group of minerals is characterized by a double-chain structure in which alternate tetrahedra in two parallel rows are cross-linked (Fig. 2.9b).

In sheet structure silicates, three oxygens of each tetrahedron are shared by adjacent tetrahedra (◆ Fig. 2.9c). This structure is what accounts for the characteristic sheet structure of the *micas,* such as biotite and muscovite, and the *clay minerals.* Three-dimensional frameworks of silica tetrahedra form when all four oxygens of the silica tetrahedron are shared by adjacent tetrahedra (◆ Fig. 2.9d). Quartz is a common framework silicate.

TABLE 2.2 Common Elements in the Earth's Crust

ELEMENT	SYMBOL	PERCENTAGE OF CRUST (BY WEIGHT)	PERCENTAGE OF CRUST (BY ATOMS)
Oxygen	O	46.6%	62.6%
Silicon	Si	27.7	21.2
Aluminum	Al	8.1	6.5
Iron	Fe	5.0	1.9
Calcium	Ca	3.6	1.9
Sodium	Na	2.8	2.6
Potassium	K	2.6	1.4
Magnesium	Mg	2.1	1.8
All others		1.5	0.1

TABLE 2.3 Some of the Mineral Groups Recognized by Geologists

MINERAL GROUP	NEGATIVELY CHARGED ION OR ION GROUP	EXAMPLES	COMPOSITION
Carbonate	(CO$_3$)$^{-2}$	Calcite	CaCO$_3$
		Dolomite	CaMg(CO$_3$)$_2$
Halide	Cl^{-1}, F^{-1}	Halite	NaCl
		Fluorite	CaF$_2$
Native element	—	Gold	Au
		Silver	Ag
		Diamond	C
		Graphite	C
Oxide	O^{-2}	Hematite	Fe$_2$O$_3$
Silicate	(SiO$_4$)$^{-4}$	Quartz	SiO$_2$
		Potassium feldspar	KAlSi$_3$O$_8$
		Olivine	(Mg,Fe)$_2$SiO$_4$
Sulfate	(SO$_4$)$^{-2}$	Anhydrite	CaSO$_4$
		Gypsum	CaSO$_4 \cdot$ 2H$_2$O
Sulfide	S^{-2}	Galena	PbS
		Pyrite	FeS$_2$

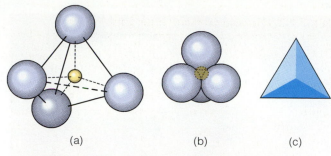

◆ FIGURE 2.8 The silica tetrahedron. (a) Expanded view showing oxygen atoms at the corners of a tetrahedron and a small silicon atom at the center. (b) View of the silica tetrahedron as it really exists with the oxygen atoms touching. (c) The silica tetrahedron represented diagramatically; the oxygen atoms are at the four points of the tetrahedron.

The *ferromagnesian silicates* are silicate minerals containing iron (Fe), magnesium (Mg), or both. Such minerals are commonly dark colored and more dense than nonferromagnesian silicates. Some of the common ferromagnesian silicate minerals are olivine, the pyroxenes, the amphiboles, and biotite mica (◆ Fig. 2.10).

The *nonferromagnesian silicates,* as their name implies, lack iron and magnesium, are generally light colored, and are less dense than ferromagnesian silicates (◆ Fig. 2.11). Quartz (SiO_2), an abundant nonferromagnesian silicate, is common in all three major rock groups. The most common minerals in the Earth's crust are nonferromagnesian silicates known as *feldspars.* Two distinct groups are recognized, each of which includes several species. The *potassium feldspars* are represented by microcline and orthoclase ($KAlSi_3O_8$). The second group of feldspars, the *plagioclase feldspars,* range from

◆ FIGURE 2.9 Structures of some of the common silicate minerals shown by various arrangements of silica tetrahedra: (a) isolated tetrahedra; (b) continuous chains; (c) continuous sheets; and (d) networks. The arrows adjacent to single-chain, double-chain, and sheet silicates indicate that these structures continue indefinitely in the directions shown.

			Formula of negatively charged ion group	Silicon to oxygen ratio	Example
(a)	Isolated tetrahedra		$(SiO_4)^{-4}$	1:4	Olivine
(b)	Continuous chains of tetrahedra	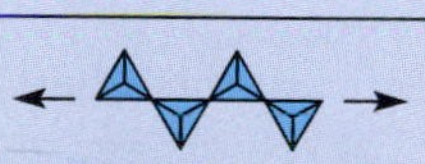 Single chain	$(SiO_3)^{-2}$	1:3	Pyroxene group
		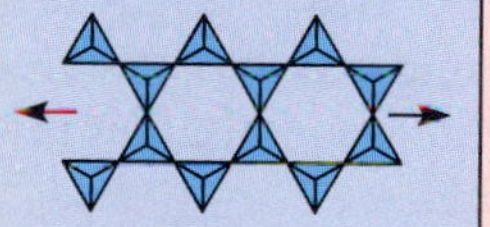 Double chain	$(Si_4O_{11})^{-6}$	4:11	Amphibole group
(c)	Continuous sheets	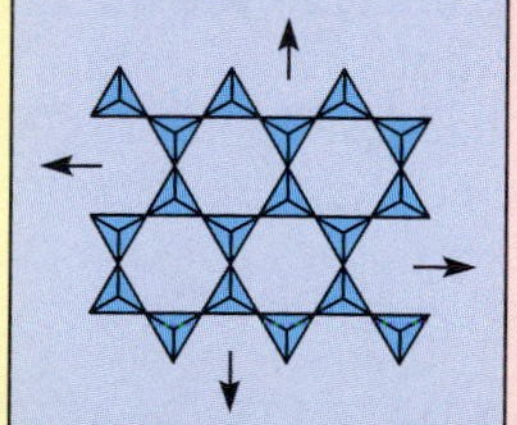	$(Si_4O_{10})^{-4}$	2:5	Micas
(d)	Three-dimensional networks	Too complex to be shown by a simple two-dimensional drawing	$(SiO_2)^0$	1:2	Quartz

◆ **FIGURE 2.10** Common ferromagnesian silicates: (a) olivine; (b) augite, a pyroxene group mineral; (c) hornblende, an amphibole group mineral; and (d) biotite mica. (Photo courtesy of Sue Monroe.)

◆ **FIGURE 2.11** Common nonferromagnesian silicates: (a) quartz; (b) the potassium feldspar orthoclase; (c) plagioclase feldspar; and (d) muscovite mica. (Photo courtesy of Sue Monroe.)

calcium-rich ($CaAl_2Si_2O_8$) to sodium-rich ($NaAlSi_3O_8$) varieties. Another common nonferromagnesian silicate is muscovite mica.

Other Mineral Groups

Carbonate minerals are those that contain the negatively charged carbonate ion $(CO_3)^{-2}$. An example is calcium carbonate ($CaCO_3$), which is the mineral *calcite* (Table 2.3). Calcite is the main constituent of the sedimentary rock *limestone*. A number of other carbonate minerals are known, but only one of these need concern us: *dolomite* [$CaMg(CO_3)_2$] is formed by the chemical alteration of calcite by the addition of magnesium. Sedimentary rock composed of the mineral dolomite is *dolostone*.

The *oxides* consist of an element combined with oxygen as in *hematite* (Fe_2O_3) (Table 2.3). The *sulfides* have a positively charged ion combined with sulfur (S^{-2}) such as in the mineral galena (PbS), which contains lead (Pb) and sulfur. *Sulfates* contain an element combined with the complex sulfate ion $(SO_4)^{-2}$; gypsum ($CaSO_4 \cdot 2H_2O$) is a good example. The *halides* contain halogen elements such as chlorine (Cl^{-1}) and fluorine (F^{-1}); examples include the minerals halite (NaCl) and fluorite (CaF_2).

ROCK-FORMING MINERALS

Rocks are generally defined as solid aggregates of grains of one or more minerals. Although it is true that many minerals occur in various kinds of rocks, only a few varieties are common enough to be designated as **rock-forming minerals.** Most of the others occur in such small amounts that they can be disregarded in the identification and classification of rocks.

Since the Earth's crust is composed largely of silicate minerals, one would suspect that most rocks are also composed of silicate minerals, and this is correct. Only a few of the hundreds of known silicates, however, are common in rocks (Table 2.4). The most common nonsilicate rock-forming minerals are the two carbonates, calcite ($CaCO_3$) and dolomite [$CaMg(CO_3)_2$]. Among the sulfates and the halides, gypsum ($CaSO_4 \cdot 2H_2O$) and halite (NaCl), respectively, are the only ones common in rocks.

ROCKS AND THE ROCK CYCLE

The only evidence available for interpreting Earth history is preserved in rocks. Consequently, the origin, distribution, and interrelationships of rocks are fundamental to an understanding of Earth history. We will review the three major rock groups in this chapter and will consider sedimentary rocks in more detail in Chapters 4 and 5 because they are particularly important in historical geology.

The **rock cycle** is one way of viewing the interaction between the Earth's internal and external processes (◆ Fig. 2.12). It relates the three major rock types to one another, to surface processes such as weathering, transport, and deposition, and to internal processes such as magma generation and mountain building.

❖ IGNEOUS ROCKS

Magma is molten rock material below the Earth's surface, and **lava** is magma at the surface. Some magma is erupted onto the surface as *lava flows,* and some is forcefully ejected into the atmosphere as particles known as **pyroclastic materials** (◆ Fig. 2.13). Geologists recognize two major categories of **igneous rocks:** (1) **volcanic** or **extrusive igneous rocks,** which form when magma extruded onto the Earth's surface cools and crystallizes (Fig. 2.13) or when pyroclastic materials become consolidated, and (2) **plutonic** or **intrusive igneous rocks,** which crystallize from magma intruded into or formed in place within the Earth's crust.

Intrusive igneous bodies known as **plutons** form when magma cools and crystallizes within the Earth's crust (◆ Fig. 2.14, p. 34). Several types of plutons are recognized, all of which are defined by their geometry (three-dimensional shape) and their relationship to the *country rock,* which is the previously existing rock with which they are associated (Fig. 2.14). Plutons are described as concordant or discordant. A *concordant* pluton has boundaries that are parallel to the layering in the country rock, whereas a *discordant* pluton has boundaries that cut across the layering of the country rock (Fig. 2.14).

Both *dikes* and *sills* are tabular or sheetlike plutons, but dikes are discordant whereas sills are concordant (◆ Fig. 2.15, p. 34). *Laccoliths* are similar to sills in that they are concordant, but instead of being tabular, they

TABLE 2.4 Rock-Forming Minerals

MINERAL	COMPOSITION	PRIMARY OCCURRENCE
FERROMAGNESIAN SILICATES		
Olivine	$(Mg,Fe)_2SiO_4$	Igneous, metamorphic rocks
Pyroxene group		
Augite most common	Ca, Mg, Fe, Al silicate	Igneous, metamorphic rocks
Amphibole group		
Hornblende most common	Na, Ca, Mg, Fe, Al silicate	Igneous, metamorphic rocks
Biotite	Hydrous K, Mg, Fe silicate	All rock types
NONFERROMAGNESIAN SILICATES		
Quartz	SiO_2	All rock types
Potassium feldspar group		
Orthoclase, microcline	$KAlSi_3O_8$	All rock types
Plagioclase feldspar group	Varies from $CaAl_2Si_2O_8$ to $NaAlSi_3O_8$	All rock types
Muscovite	Hydrous K, Al silicate	All rock types
Clay mineral group	Varies	Soils and sedimentary rocks
CARBONATES		
Calcite	$CaCO_3$	Sedimentary rocks
Dolomite	$CaMg(CO_3)_2$	Sedimentary rocks
SULFATES		
Anhydrite	$CaSO_4$	Sedimentary rocks
Gypsum	$CaSO_4 \cdot 2H_2O$	Sedimentary rocks
HALIDES		
Halite	$NaCl$	Sedimentary rocks

Igneous Rocks **31**

have a mushroomlike geometry (Fig. 2.14). The conduit connecting the crater of a volcano with an underlying magma chamber is a *volcanic pipe* (Fig. 2.14). When a volcano ceases to erupt, it eventually erodes away, but the magma that solidified in the pipe is more resistant to weathering and erosion and is often left as an erosional remnant, a *volcanic neck* (Fig. 2.14).

Batholiths are the largest intrusive bodies. By definition they must have at least 100 km² of surface area, and most are much larger than this (Fig. 2.14). *Stocks* have the same general features as batholiths but are smaller.

Batholiths are generally discordant, and most consist of multiple intrusions of granitic magma.

Igneous Textures

Several textures of igneous rocks are related to the cooling history of a magma or lava. For example, rapid cooling, as occurs in lava flows, results in a fine-grained texture termed *aphanitic* in which individual mineral grains are too small to be observed without magnification (◆ Fig. 2.16a). In contrast, igneous rocks with a coarse-

◆ **FIGURE 2.12** The rock cycle showing the interrelationships between the Earth's internal and external processes and how each of the three major rock groups is related to the others.

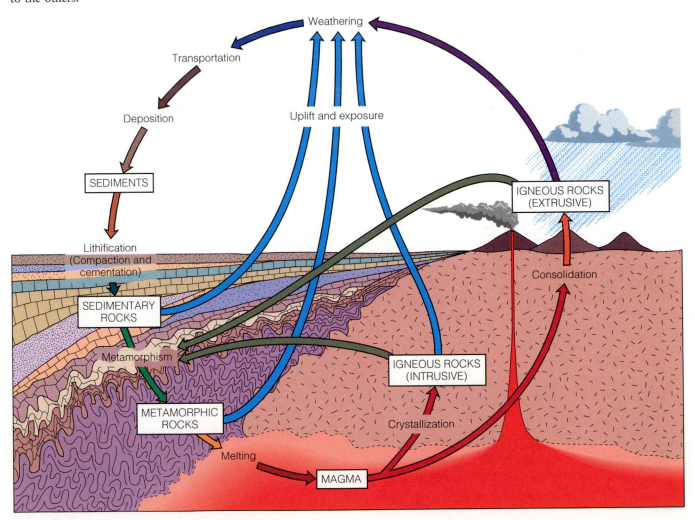

(a)

(b)

◆ **FIGURE 2.13** (a) Pyroclastic materials and volcanic ash, being erupted from Mount Ngaurauhoe, New Zealand, during January 1974. (Photo courtesy of the University of Colorado.) (b) Lava flow in the east rift zone of Kilauea volcano in 1972. (Photo courtesy of D.W. Peterson, USGS.) Cooling of lava flows is one of the ways igneous rocks form.

grained or *phaneritic* texture have mineral grains that are easily visible without magnification (◆ Fig. 2.16b, p. 35). Such large mineral grains indicate slow cooling and generally an intrusive origin.

Rocks with a combination of mineral grains of markedly different sizes have a *porphyritic* texture. The larger grains are *phenocrysts,* and the smaller ones are referred to as *groundmass* (◆ Fig. 2.16c, p. 35). Igneous rocks with a porphyritic texture have a two-stage cooling history that might involve partial crystallization as a pluton, followed by extrusion of the remaining magma and mineral crystals and rapid cooling. The resulting igneous rock would have large mineral grains (phenocrysts) suspended in a finely crystalline groundmass, and the rock would be characterized as a *porphyry.*

A lava may cool so rapidly that its constituent atoms do not have time to become arranged in the ordered, three-dimensional frameworks typical of minerals. As a consequence of such rapid cooling, a *natural glass* such as *obsidian* forms. Even though obsidian is not composed of minerals, it is still considered to be an igneous rock.

Some magmas contain large amounts of water vapor and other gases. These gases may be trapped in cooling lava where they form numerous small holes or cavities known as *vesicles;* rocks possessing numerous vesicles are termed *vesicular,* as in vesicular basalt (◆ Fig. 2.16d, p. 35). A *pyroclastic* or fragmental texture characterizes igneous rocks formed by explosive volcanic activity (Fig. 2.13).

Composition of Igneous Rocks

Although silica is the primary constituent of nearly all magmas, silica content varies and serves to distinguish felsic (>65% silica), *intermediate* (53–65% silica), and *mafic* (45–52% silica) *magmas.* In addition to being at

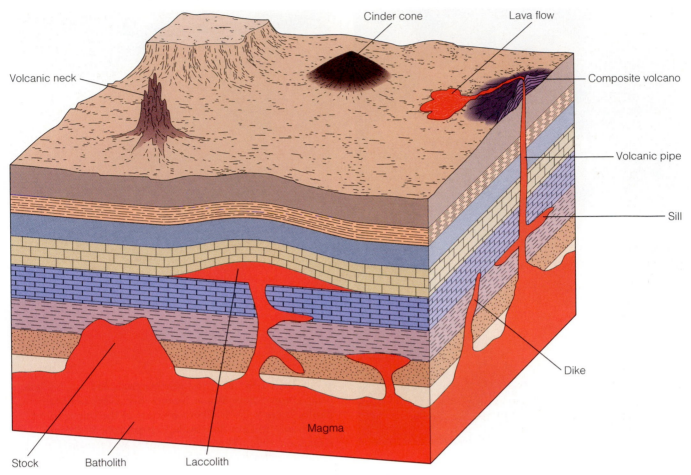

Cinder cone

Lava flow

Volcanic neck

Composite volcano

Volcanic pipe

Sill

Dike

Magma

Stock Batholith Laccolith

◆ **FIGURE 2.14** Block diagram showing the various types of plutons. Notice that some of these plutons cut across the layering in the country rock and are thus discordant, whereas others parallel the layering and are concordant.

◆ **FIGURE 2.15** The dark layer cutting diagonally across the rock layers is a dike. The other dark layer is a sill because it parallels the layering. (Photo © Martin G. Miller/Visuals Unlimited.)

least 65% silica, **felsic magmas** contain considerable sodium, potassium, and aluminum, but little calcium, iron, and magnesium. Cooling of felsic magma yields igneous rocks, such as rhyolite and granite, which are composed largely of nonferromagnesian silicates.

In contrast, **mafic magmas** are silica poor and contain proportionately more calcium, iron, and magnesium. When such magmas cool and crystallize, they yield igneous rocks such as basalt and gabbro, which contain high percentages of ferromagnesian silicates. As one would expect, igneous rocks that crystallize from **intermediate magmas** have mineral compositions intermediate between those of mafic and felsic rocks.

Classification of Igneous Rocks

Most igneous rocks are classified on the basis of textural features and composition (◆ Fig. 2.17). Notice in Figure

◆ **FIGURE 2.16** Textures of igneous rocks. (a) Aphanitic or fine-grained texture in which individual minerals are too small to be seen without magnification. (b) Phaneritic or coarse-grained texture in which minerals are easily discerned without magnification. (c) Porphyritic texture consisting of minerals of markedly different sizes. (d) Vesicular texture. (Photos courtesy of Sue Monroe.)

2.17 that all of the rocks, except periodotite, constitute pairs; the members of a pair have the same composition but different textures. Thus, basalt and gabbro, andesite and diorite, and rhyolite and granite are compositional (mineralogical) equivalents, but basalt, andesite, and rhyolite are aphanitic and most commonly extrusive, whereas gabbro, diorite, and granite have phaneritic textures that generally indicate an intrusive origin. The igneous rocks shown in Figure 2.17 are also differentiated by composition. Reading across the chart from rhyolite to andesite to basalt, for example, the proportions of nonferromagnesian and ferromagnesian minerals differ.

Rocks referred to as ultramafic are composed largely of ferromagnesian silicate minerals (◆ Fig. 2.18). For ex-

ample, *peridotite* contains mostly olivine, lesser amounts of pyroxene, and generally a little plagioclase feldspar (Fig. 2.17). *Basalt* and *gabbro* are the fine-grained and coarse-grained rocks, respectively, that crystallize from mafic magmas (◆ Fig. 2.19a and b). Thus, both have the same composition—mostly calcium-rich plagioclase and pyroxene, with smaller amounts of olivine and amphibole (Fig. 2.17). Because they contain a large proportion of ferromagnesian minerals, they are dark colored.

Magmas of intermediate composition crystallize to form *andesite* and *diorite,* which are compositionally equivalent fine- and coarse-grained igneous rocks (◆ Fig. 2.19c and d). Andesite and diorite are composed predominantly of plagioclase feldspar, with the typical

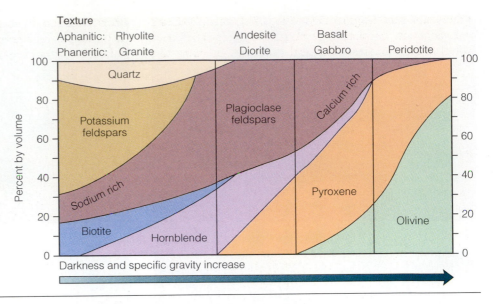

◆ **FIGURE 2.17** Classification of igneous rocks. The diagram illustrates the relative proportions of the chief mineral components of common igneous rocks.

ferromagnesian component being amphibole or biotite (Fig. 2.17). Andesite is generally medium to dark gray, but diorite has a salt and pepper appearance because of its white to light gray plagioclase and dark ferromagnesian minerals (Fig. 2.19c and d).

Rhyolite and *granite* crystallize from felsic magmas and are therefore silica-rich rocks (◆ Fig. 2.19e and f). These rocks consist largely of potassium feldspar, sodium-rich plagioclase, and quartz, with perhaps some biotite and rarely amphibole (Fig. 2.17). Because nonferromagnesian minerals predominate, rhyolite and granite

are generally light colored. Rhyolite is fine grained, although most often it contains phenocrysts of potassium feldspar or quartz, and granite is coarse grained.

Some igneous rocks, including tuff, volcanic breccia, obsidian, and pumice, are identified solely by their textures. Much of the fragmental material erupted by volcanoes is *ash,* a designation for pyroclastic materials less than 2.0 mm in diameter, much of which is broken pieces or shards of volcanic glass. The consolidation of ash forms the pyroclastic rock *tuff* (◆ Fig. 2.20a). Some ash flows are so hot that as they come to rest, the ash particles fuse together and form a *welded tuff.* Consolidated deposits of larger pyroclastic materials, such as cinders, blocks, and bombs, are *volcanic breccia.*

Both *obsidian* and *pumice* are varieties of volcanic glass (◆ Fig. 2.20b and c). Obsidian is generally black, red, or brown and has the appearance of glass. Pumice is a variety of volcanic glass containing numerous bubble-shaped vesicles that develop when gas escapes through lava and forms a froth.

◆ **FIGURE 2.18** The ultramafic rock peridotite. (Photo courtesy of Sue Monroe.)

❖ SEDIMENTARY ROCKS

Any rocks exposed at the Earth's surface are subjected to mechanical and chemical weathering processes that disintegrate and decompose them, thereby yielding the raw materials for **sedimentary rocks** (Fig. 2.12). All *sediment* is derived from preexisting rocks and can be characterized as (1) **detrital sediment,** which consists of mineral grains and rock fragments (◆ Fig. 2.21), and (2) **chemical sediment,** consisting of minerals precipitated from solution by inorganic chemical processes or extracted from solution by organisms. In any case, sediment is deposited as an aggregate of loose solids such as sand on a beach or gravel in a stream channel.

◆ **FIGURE 2.19** Mafic igneous rocks: (a) basalt and (b) gabbro. Intermediate igneous rocks: (c) andesite and (d) diorite. Felsic igneous rocks: (e) rhyolite and (f) granite. (Photos courtesy of Sue Monroe.)

(a)

(b)

(c)

(d)

(e)

(f)

(a)

(b)

(c)

◆ **FIGURE 2.20** (a) Exposure of tuff. (Photo courtesy of David J. Matty); (b) Obsidian and (c) pumice. (Photos courtesy of Sue Monroe.)

Sediment Transport and Deposition

Sediment can be transported by any geologic agent possessing enough energy to move particles of a given size. Glaciers are very effective agents of transport and can move particles of any size. Wind, on the other hand, can transport only sand-sized and smaller sediment. Waves and marine currents also transport sediment, but by far the most effective way to erode sediment from the weathering site and transport it elsewhere is by water in streams.

Sediment may be transported a considerable distance from its source area, but eventually it is deposited. Some of the sand and mud being deposited at the mouth of the Mississippi River at the present time came from such distant places as Ohio, Minnesota, and Wyoming. Any geographic area in which sediment is deposited is a depositional environment (see Chapter 5).

◆ **FIGURE 2.21** Mechanically weathered granite. The sandy material consists of small pieces of granite (rock fragments) and minerals such as quartz and feldspars liberated from the parent material.

Lithification: Sediment to Sedimentary Rock

The process by which sediment is transformed into sedimentary rock is **lithification**. When either detrital or chemical sediment is deposited, it consists of solid particles and pore spaces, which are the voids between particles. When sediment is buried, *compaction*, resulting from the pressure exerted by the weight of the overlying sediments, reduces the amount of pore space and thus the volume of the deposit (◆ Fig. 2.22).

Compaction alone is generally sufficient for lithification of mud, but for sand (particles measuring 1/16–2 mm) and gravel (particles >2 mm), *cementation* is necessary to convert the sediment into sedimentary rock. In cementation, dissolved compounds are precipitated in the pore spaces to form a chemical cement that effectively binds particles together.

Calcium carbonate ($CaCO_3$) and silica (SiO_2) are the most common cements, but iron oxides and hydroxides, such as hematite (Fe_2O_3) and limonite [$FeO(OH)$] also form cement in some rocks. The yellow, brown, and red sedimentary rocks exposed in the walls of the vast canyons of Utah and Arizona are colored by small amounts of iron oxide or hydroxide cement (◆ Fig. 2.23).

Types of Sedimentary Rocks

Even though about 95% of the Earth's crust is composed of igneous and metamorphic rocks, sedimentary rocks are most common at or near the surface. About 75% of the surface exposures on continents consist of sediments or sedimentary rocks, and they cover most of the seafloor. Sedimentary rocks are generally classified as detrital or chemical (Table 2.5).

Detrital Sedimentary Rocks

Detrital sedimentary rocks consist of *detritus*, the solid particles of preexisting rocks. Such rocks have a *clastic texture*, meaning that they are composed of fragments or particles also known as *clasts*. Several varieties of detrital sedimentary rocks are recognized, each of which is characterized by the size of its constituent particles (Table 2.5).

Both *conglomerate* and *sedimentary breccia* consist of gravel-sized particles (Table 2.5; ◆ Fig. 2.24a and b, p. 42). The only difference between them is the shape of the gravel particles; conglomerate consists of particles whose corners and edges have been smoothed or rounded, whereas sedimentary breccia is composed of angular gravel called *rubble*.

The term *sand* is simply a size designation, so *sandstone* may be composed of grains of any type of mineral or rock fragment. However, most sandstones consist primarily of the mineral quartz (◆ Fig. 2.24c, p. 42). Geologists recognize several types of sandstone, each characterized by its composition. *Quartz sandstone* is the most common and, as its name suggests, is composed mostly of quartz. *Arkose*, which contains more than 25% feldspars, is also a fairly common variety of sandstone (Table 2.5).

The *mudrocks* include all detrital sedimentary rocks composed of silt- and clay-sized particles (◆ Fig. 2.24d, p. 42). Among the mudrocks we can differentiate between *siltstone, mudstone*, and *claystone*. Siltstone, as the name implies, is composed of silt-sized particles; mudstone contains a mixture of silt- and clay-sized particles; and claystone is composed mostly of clay (Table 2.5). Some mudstones and claystones are designated as *shale* if they are fissile, which means they break along closely spaced parallel planes.

Chemical Sedimentary Rocks

Chemical sedimentary rocks originate from the ions and salts taken into solution during chemical weathering (Table 2.5). Such dissolved materials are transported to lakes and the oceans where they become concentrated. These materials may be removed from solution by inorganic chemical processes or by the chemical processes of organisms and then accumulate as minerals. Many chemical sedimentary rocks are composed of an interlocking mosaic of mineral crystals and are said to have a *crystalline texture*. Some, however, have a clastic texture; many limestones, for example, consist of broken pieces of shells.

◆ **FIGURE 2.22** Lithification of sand. (a) When initially deposited, sand has considerable pore space between grains. (b) Compaction resulting from the weight of overlying sediments reduces the amount of pore space. (c) Sand is converted to sandstone as cement is precipitated in pore spaces from groundwater.

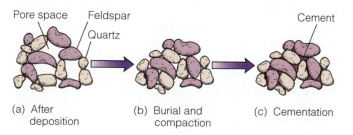

(a) After deposition

(b) Burial and compaction

(c) Cementation

◆ **FIGURE 2.23** These sedimentary rocks in the Valley of the Gods, Utah, are red because they contain iron oxide cement. (Photo courtesy of Sue Monroe.)

Calcite (the main component of limestone) and dolomite (the mineral composing dolostone) are both carbonate minerals; calcite is a calcium carbonate ($CaCO_3$), whereas dolomite [$CaMg(CO_3)_2$] is a calcium magnesium carbonate. Thus, limestone and dolostone are **carbonate rocks.** Many limestones are conveniently classified as biochemical sedimentary rocks, a subcategory of chemical sedimentary rocks, because organisms play a significant role in their origin (Table 2.5; ◆ Fig. 2.25). For example, the limestone known as *coquina* consists entirely of broken shells cemented by calcium carbonate, and *chalk* is a soft variety of biochemical limestone composed largely of microscopic shells of organisms. One distinctive type of limestone contains small spherical grains called *ooids*. Ooids have a small nucleus—a sand grain or shell fragment perhaps—around which concentric layers of calcite precipitate; lithified deposits of ooids form *oolitic limestones* (◆ Fig. 2.25b).

The near-absence of recent dolostone and evidence from chemistry and studies of rocks indicate that most dolostone was originally limestone that has been altered to dolostone. Many geologists think most dolostones originated through the replacement of some of the calcium in calcite by magnesium.

If the volume of a solution is reduced by evaporation, the amount of dissolved mineral matter increases in proportion to the volume of the solution and eventually reaches the saturation limit, the point at which precipitation must occur. **Evaporites** include such rocks as *rock salt* and *rock gypsum,* both of which form by inorganic chemical precipitation of minerals from solutions that become saturated as a result of evaporation (Table 2.5). Rock salt, composed of the mineral halite (NaCl), is simply sodium chloride that was precipitated from seawater or, more rarely, lake water (◆ Fig. 2.25c). Rock gypsum, the most common evaporite rock, is composed of the mineral gypsum ($CaSO_4 \cdot H_2O$), which also precipitates from evaporating solutions (◆ Fig. 2.25d). A number of other evaporite rocks and minerals are known, but most of these are rare.

Chert is a hard rock composed of microscopic crystals of quartz (SiO_2) (Table 2.5; ◆ Fig. 2.25e). Several color varieties of chert occur including *flint,* which is black because of inclusions of organic matter, and *jasper,* which is red or brown because of iron oxide inclusions. Some chert occurs as oval to spherical masses within other rocks, especially limestones, and is clearly secondary; that is, it has replaced part of the host rock. Bedded chert, which occurs in distinct layers, results from precipitation from solution or from accumulations of the shells of silica-secreting organisms.

Coal is a biochemical sedimentary rock composed of the compressed, altered remains of organisms, especially land plants (Table 2.5; ◆ Fig. 2.25f). It forms in swamps and bogs where the water is oxygen deficient or where organic matter accumulates faster than it decomposes.

TABLE 2.5 Classification of Detrital Chemical and Biochemical Sedimentary Rocks

DETRITAL ROCKS		
Sediment Name and Size	Description	Rock Name
Gravel (>2 mm)	Rounded gravel	Conglomerate
	Angular gravel	Sedimentary breccia
Sand (1/16–2 mm)	Mostly quartz	Quartz sandstone
	Quartz with >25% feldspar	Arkose
Mud (<1/16 mm)	Mostly silt	Siltstone ⎤
	Silt and clay	Mudstone* ⎬ Mudrocks
	Mostly clay	Claystone* ⎦

CHEMICAL AND BIOCHEMICAL ROCKS		
Texture	Composition	Rock Name
Clastic or crystalline	Calcite ($CaCO_3$)	Limestone (includes coquina, ⎤
		chalk, and oolitic limestone) ⎬ Carbonates
	Dolomite [$CaMg(CO_3)_2$]	Dolostone ⎦
Crystalline	Gypsum ($CaSO_4 \cdot 2H_2O$)	Rock gypsum ⎤
	Halite (NaCl)	Rock salt ⎬ Evaporites
Usually crystalline	Microscopic SiO_2 shells	Chert
	Altered plant remains	Coal

*Mudrocks possessing the property of fissility, meaning they break along closely spaced, parallel planes, are commonly called *shale*.

When buried, this organic matter is converted to *peat,* which looks rather like coarse pipe tobacco. Peat that is buried more deeply and compressed, especially if it is also heated, is altered to coal.

❖ METAMORPHISM AND METAMORPHIC ROCKS

Metamorphic rocks result from the transformation of other rocks, including previously formed metamorphic rocks (Fig. 2.12). *Metamorphism* usually takes place beneath the Earth's surface where rocks are subjected to sufficient heat, pressure, and fluid activity to change their mineral composition and/or texture, thus forming new rocks. These transformations take place in the solid state, and the type of metamorphic rock formed depends on the composition and texture of the parent rock, the agents of metamorphism, and the amount of time that the parent rock was subjected to the effects of metamorphism.

A large portion of the Earth's continental crust is composed of metamorphic and igneous rocks. Together, they form the crystalline basement rocks that underlie the sedimentary rocks of a continent's surface. This basement rock is exposed widely in regions of the continents known as *shields* (see Chapter 9), areas that have been very stable during the past 600 million years. Metamorphic rocks also constitute a sizable portion of the crystalline core of large mountain ranges.

The study of metamorphic rocks provides information about geological processes operating within the Earth and about the way these processes have varied through time. From the presence of certain minerals in metamorphic rocks, geologists can determine the approximate temperatures and pressures that parent rocks were subjected to during metamorphism and can thus gain insights into the physical and chemical changes that occur at different depths within the Earth's crust.

The Agents of Metamorphism

During metamorphism, rocks undergo change so as to come into equilibrium with a new environment. The changes may result in the formation of new minerals and/or a change in texture by the reorientation of the

(a)

(b)

(c)

(d)

◆ **FIGURE 2.24** Detrital sedimentary rocks: (a) conglomerate; (b) sedimentary breccia; (c) sandstone; and (d) the mudrock shale. (Photos courtesy of Sue Monroe.)

original minerals. In some instances the change is minor, and features of the parent rock can still be recognized. In other cases the rock changes so much that the identity of the parent rock can be determined only with great difficulty, if at all. Depending on the degree of alteration, metamorphic rocks can be characterized as resulting from low-, medium-, or high-grade metamorphism.

Heat is an important agent of metamorphism because it increases the rate of chemical reactions that may produce mineral assemblages different from those in the parent rock. The heat may come from intrusive magmas or result from deep burial such as occurs during subduction along a convergent plate boundary.

When rocks are buried, they are subjected to increasingly greater *lithostatic pressure;* this pressure, which re-

sults from the weight of the overlying rocks, is applied equally in all directions. When subjected to increasing lithostatic pressure with depth, mineral grains within a rock may become more closely packed. Under such conditions, the minerals may *recrystallize;* that is, form

◆ **FIGURE 2.25** (on the opposite page) (a) Chalk cliffs in Denmark. Chalk is made up of microscopic shells. (Photo courtesy of R.V. Dietrich.) (b) Photomicrograph of ooids in an oolitic limestone. These ooids measure about 1 mm in diameter. (c) Core of rock salt from a well in Michigan. (d) Rock gypsum. (e) Chert. (f) Coal is a biochemical sedimentary rock composed of the altered remains of land plants. (Photos c, d, e, and f courtesy of Sue Monroe.)

(a)

(b)

(c)

(d)

(e)

(f)

smaller and denser minerals that may or may not retain the chemical composition of the original minerals.

In addition to lithostatic pressure, rocks may also experience *differential pressures*. In this case, the pressure is not equal on all sides, and the rock is consequently distorted. Differential pressures typically occur during deformation associated with mountain building.

In almost every region where metamorphism occurs, water is present in varying amounts along mineral grain boundaries or in the pore spaces of rocks. This water, which may contain ions in solution, enhances metamorphism by increasing the rate of chemical reactions. Even small amounts of water are enough to speed up reaction rates, mainly because ions can move readily through a fluid and thus enhance chemical reactions and the formation of new minerals. Accordingly, *fluid activity* is also an important agent of metamorphism.

Types of Metamorphism

Two major types of metamorphism are recognized: *contact metamorphism* and *regional metamorphism*. **Contact metamorphism** takes place when a body of magma alters the surrounding country rock (◆ Fig. 2.26). At shallow depths an intruding magma raises the temperature of the surrounding rock, causing thermal alteration. Temperatures can reach nearly 900°C adjacent to an intrusion, but they gradually decrease with distance. Furthermore, the release of hot fluids into the country rock by the cooling intrusion can also aid in the formation of new minerals.

The size of an intrusion is also important. In the case of small intrusions, such as dikes and sills, usually only those rocks in immediate contact with them are affected. Because large intrusions, such as batholiths, take a long time to cool, the increased temperature in the surrounding rock may last long enough for a larger area to be affected.

Fluids also play an important role in contact metamorphism when they react with the surrounding rock and aid in the formation of new minerals. In addition, the country rock may contain pore fluids that, when heated by the magma, also increase reaction rates.

Most metamorphic rocks are the result of **regional metamorphism,** which occurs over large areas and is usually the result of tremendous temperatures, pressures, and deformation within the deeper portions of the crust. Regional metamorphism is most obvious along convergent plate margins where rocks are intensely deformed during convergence and subduction. Regional metamor-

phism also occurs in areas where plates diverge, though usually at much shallower depths because of the high temperatures associated with spreading ridges.

Classification of Metamorphic Rocks

Metamorphic rocks are commonly divided into two groups: those exhibiting a *foliated texture* and those with a *nonfoliated texture* (Table 2.6). Rocks subjected to heat and differential pressure during metamorphism typically have minerals arranged in a parallel fashion that gives them a **foliated texture** (◆ Fig. 2.27a and b). Some metamorphic rocks do not show a discernible preferred orientation of their mineral grains. Instead, they consist of a mosaic of roughly equidimensional minerals and are characterized as **nonfolidated** (◆ Fig. 2.27c).

Foliated Metamorphic Rocks

Slate is a very fine-grained metamorphic rock resulting from low-grade regional metamorphism of shale or, more rarely, volcanic ash (Table 2.6; ◆ Fig. 2.28a). *Phyllite* is similar in composition to slate, but is coarser grained. However, the minerals are still too small to be identified without magnification. Phyllite represents an intermediate grain size between slate and schist.

Schist is most commonly produced by regional metamorphism (Table 2.6; ◆ Fig. 2.28b). All schists contain more than 50% platy and elongated minerals, all of which are large enough to be clearly visible. Their mineral composition imparts a *schistosity* or *schistose foliation* to the rock that commonly produces a wavy type of parting when split. Metamorphism of many rock types can yield schist, but most schist appears to have formed from clay-rich sedimentary rocks.

Gneiss is a metamorphic rock that is streaked or has segregated bands of light and dark minerals (Table 2.6; ◆ Fig. 2.28c). Gneisses are composed mostly of granular minerals such as quartz and feldspars with lesser amounts of platy and elongated minerals such as biotite and amphiboles. Quartz and feldspar are the principal light-colored minerals, while biotite and hornblende are the typical dark-colored minerals. Most gneiss probably results from regional metamorphism of clay-rich sedimentary rocks. Gneiss can also form from crystalline igneous rocks such as granite and older metamorphic rocks.

In some areas of regional metamorphism, exposures of "mixed rocks" having both igneous and high-grade metamorphic characteristics are present. These rocks, called *migmatites,* usually consist of steaks or lenses of

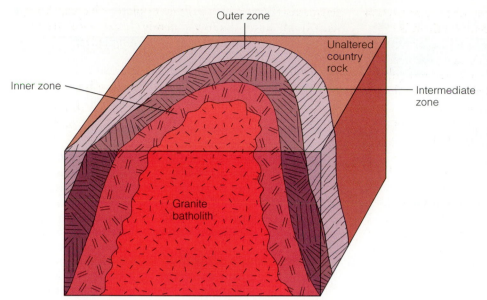

Outer zone

Unaltered country rock

Inner zone

Intermediate zone

Granite batholith

◆ **FIGURE 2.26** The metamorphic effects of an igneous intrusion usually occur in concentric zones. The metamorphic zones associated with this idealized granite batholith contain three distinct mineral assemblages reflecting the decrease in temperature with distance from the intrusion.

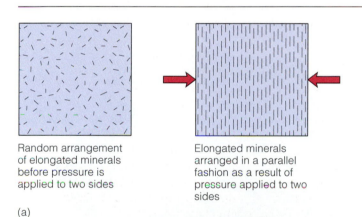

Random arrangement of elongated minerals before pressure is applied to two sides

Elongated minerals arranged in a parallel fashion as a result of pressure applied to two sides

(a)

◆ **FIGURE 2.27** (a) When rocks are subjected to differential pressure, the mineral grains are typically arranged in a parallel fashion, producing a foliated texture. (b) Photomicrograph of a metamorphic rock with a foliated texture showing the parallel arrangement of mineral grains. (c) Photomicrograph of marble showing the mosaic of roughly equidimensional minerals that is characteristic of nonfoliated textures.

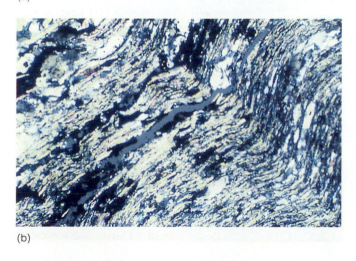

(b)

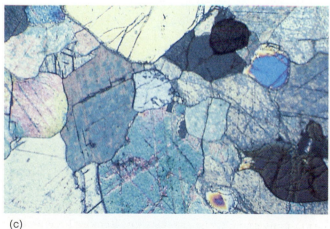

(c)

TABLE 2.6 Classification of Common Metamorphic Rocks

TEXTURE	METAMORPHIC ROCK	TYPICAL MINERALS	METAMORPHIC GRADE	CHARACTERISTICS OF ROCKS	PARENT ROCK
Foliated	Slate	Clays, micas, chlorite	Low	Fine grained, splits easily into flat pieces	Shale, claystones, volcanic ash
	Phyllite	Fine-grained quartz, micas, chlorite	Low to medium	Fine grained, glossy or lustrous sheen	Mudrocks
	Schist	Micas, chlorite, quartz, talc, hornblende, garnet, staurolite, graphite	Low to high	Distinct foliation, minerals visible	Variable mudrocks, impure carbonates, mafic igneous rocks
	Gneiss	Quartz, feldspars, hornblende, micas	High	Segregated light and dark bands visible	Mudrocks, sandstones, felsic igneous rocks
	Migmatite	Quartz, feldspars, hornblende, micas	High	Streaks or lenses of granite intermixed with gneiss	Felsic igneous rocks mixed with sedimentary rocks

table continues on next page

◆ **FIGURE 2.28** Hand specimens of (a) slate, (b) schist, (c) gneiss, and (d) migmatite. (Photos courtesy of Sue Monroe.)

(a)

(b)

(c)

(d)

TABLE 2.6 Classification of Common Metamorphic Rocks (continued)

TEXTURE	METAMORPHIC ROCK	TYPICAL MINERALS	METAMORPHIC GRADE	CHARACTERISTICS OF ROCKS	PARENT ROCK
Nonfoliated	Marble	Calcite, dolomite	Low to high	Interlocking grains of calcite or dolomite, reacts with HCl	Limestone or dolostone
	Quartzite	Quartz	Medium to high	Interlocking quartz grains, hard, dense	Quartz sandstone
	Greenstone	Chlorite, epidote, hornblende	Low to high	Fine grained, green color	Mafic igneous rocks
	Hornfels	Micas, garnets, andalusite, cordierite, quartz	Low to medium	Fine grained, equidimensional grains, hard, dense	Shale
	Anthracite	Carbon	High	Black, lustrous, subconcoidal fracture	Low-grade coal

granite intermixed with high-grade ferromagnesian-rich metamorphic rocks (◆ Fig. 2.28d).

Nonfoliated Metamorphic Rocks

Most nonfoliated metamorphic rocks result from contact or regional metamorphism of rocks in which no platy or prismatic minerals are present. Nonfoliated metamorphic rocks are generally of two types: those composed largely of one mineral, for example, marble and quartzite; and those in which the mineral grains are too small to be seen without magnification, such as in greenstone and hornfels.

Marble is a well-known metamorphic rock composed mostly of calcite or dolomite; it results from either contact or regional metamorphism of limestone or dolostone (Table 2.6; ◆ Fig. 2.29a). *Quartzite* is a hard, compact rock typically formed from quartz sandstone under medium-grade metamorphic conditions during contact or regional metamorphism (Table 2.6; ◆ Fig. 2.29b).

The name *greenstone* is applied to any compact, dark green, altered, mafic igneous rock that formed under

◆ FIGURE 2.29 (a) Marble results from the metamorphism of the sedimentary rock limestone or dolostone. (b) Quartzite results from the metamorphism of quartz sandstone. (Photos courtesy of Sue Monroe.)

(a)

(b)

low- to high-grade metamorphic conditions. The green color results from the presence of such green minerals as chlorite and epidote.

Hornfels is a fine-grained, nonfoliated metamorphic rock resulting from contact metamorphism; it is composed of various equidimensional mineral grains. Many compositional varieties of hornfels are known, but the majority of them are apparently formed by contact meta-

morphism of clay-rich sedimentary rocks or impure dolostones.

Anthracite is a black, lustrous, hard coal that contains a high percentage of fixed carbon and a low percentage of volatile matter. It usually forms from the metamorphism of lower-grade coals by heat and pressure and is thus considered by many geologists to be a metamorphic rock.

Chapter Summary

1. All matter is composed of chemical elements, each of which consists of atoms. Individual atoms consist of a nucleus, containing protons and neutrons, and electrons that circle the nucleus in electron shells.

2. Atoms are characterized by their atomic number (the number of protons in the nucleus) and their atomic mass number (the number of protons plus the number of neutrons in the nucleus).

3. Bonding is the process whereby atoms are joined to other atoms. If atoms of different elements are bonded, they form compounds. Ionic and covalent bonds are most common in minerals.

4. All minerals are crystalline solids, meaning that they possess an orderly internal arrangement of atoms.

5. Of the more than 3,500 known minerals, most are silicates. Ferromagnesian silicates contain iron (Fe) and magnesium (Mg), and nonferromagnesian silicates lack these elements.

6. In addition to silicates, several other mineral groups are recognized, including carbonates, oxides, sulfides, sulfates, and halides.

7. The composition of igneous rocks is determined largely by the composition of the parent magma. Magmas are characterized as mafic (45–52% silica), intermediate (52–65% silica), and felsic (>65% silica).

8. Most igneous rocks are classified on the basis of their texture and composition. Two groups of igneous rocks are recognized: volcanic or extrusive igneous rocks include rhyolite, andesite, and basalt, all of which are aphanitic, and tuff; and plutonic or intrusive igneous rocks such as granite, diorite, and gabbro, which are phaneritic.

9. Plutons are igneous bodies that formed from magma that cooled and crystallized within the Earth's crust. Various types of plutons are identified by their geometry and whether they are concordant or discordant.

10. Detrital sediment consists of mechanically weathered solid particles, whereas chemical sediment consists of minerals extracted from solution by inorganic chemical processes or by the activities of organisms.

11. Compaction and cementation are the processes of sediment lithification in which sediment is converted into sedimentary rock. Silica (SiO_2) and calcium carbonate ($CaCO_3$) are the most common cements, but iron oxide and iron hydroxide cements are important in some rocks.

12. Sedimentary rocks are generally classified as detrital or chemical:

    a. Detrital sedimentary rocks consist of solid particles derived from preexisting rocks. They include conglomerate/sedimentary breccia, sandstone, and various mudrocks.

    b. Chemical sedimentary rocks are derived from ions in solution by organisms or inorganic chemical processes. They include various types of limestone, dolostone, evaporites, chert, and coal.

13. Metamorphic rocks result from the transformation of other rocks, usually beneath the surface, as a consequence of heat, pressure, or fluid activity or a combination of these.

14. The two major types of metamorphism are contact and regional.

15. Metamorphic rocks are classified primarily according to their texture. In a foliated texture, platy minerals have a preferred orientation. A nonfoliated texture does not exhibit any discernible preferred orientation of the mineral grains.

16. Foliated metamorphic rocks include slate, phyllite, schist, and gneiss. Nonfoliated metamorphic rocks are marble, quartzite, greenstone, and hornfels.

Important Terms

atom

atomic mass number

atomic number

bonding

carbonate mineral

carbonate rock

chemical sediment/sedimentary rock
compound
contact metamorphism
covalent bond
crystalline solid
detrital sediment/sedimentary rock
element
evaporite
felsic magma
foliated texture
igneous rock

intermediate magma
ionic bond
lava
lithification
mafic magma
magma
metamorphic rock
mineral
nonfoliated texture
nucleus

pluton
plutonic (intrusive igneous) rock
pyroclastic material
regional metamorphism
rock cycle
rock-forming mineral
sedimentary rock
silicate
silica tetrahedron
volcanic (extrusive igneous) rock

Review Questions

1. The atomic number of an element is determined by the:
   a. _____ number of electrons in its outermost shell;
   b. _____ number of protons in its nucleus; c. _____
   diameter of its most common isotope; d. _____ number
   of neutrons plus electrons in its nucleus; e. _____ total
   number of neutrons orbiting its nucleus.
2. In which type of bonding do adjacent atoms share
   electrons?
   a. _____ silicate; b. _____ ionic; c. _____ tetrahedral;
   d. _____ carbon; e. _____ covalent.
3. A chemical element is a substance made up of atoms, all
   of which have the same:
   a. _____ atomic mass number; b. _____ number of
   neutrons; c. _____ number of protons; d. _____ size;
   e. _____ weight.
4. The basic building block of all silicate minerals is the:
   a. _____ silicon sheet; b. _____ oxygen-silicon cube;
   c. _____ silica tetrahedron; d. _____ silicate double
   chain; e. _____ silica framework.
5. An example of a common nonferromagnesian silicate is:
   a. _____ calcite; b. _____ quartz; c. _____ biotite;
   d. _____ hematite; e. _____ halite.
6. Minerals are solids possessing an orderly internal
   arrangement of atoms, meaning that they are:
   a. _____ natural glasses; b. _____ crystalline; c. _____
   composed of at least three different elements; d. _____
   composed of a single element; e. _____ ionic
   compounds.
7. Volcanic rocks can usually be distinguished from
   plutonic rocks by:
   a. _____ color; b. _____ composition; c. _____
   iron-magnesium content; d. _____ the size of their
   mineral grains; e. _____ weight.
8. An example of a concordant pluton having a tabular or
   sheetlike geometry is a:
   a. _____ sill; b. _____ batholith; c. _____ volcanic
   neck; d. _____ lava flow; e. _____ dike.
9. Which of the following pairs of igneous rocks have the
   same mineral composition?

   a. _____ granite-peridotite; b. _____ andesite-rhyolite;
   c. _____ pumice-diorite; d. _____ basalt-gabbro;
   e. _____ tuff-andesite.
10. An igneous rock possessing mineral grains large enough
    to be seen without magnification is said to have a _____
    texture.
    a. _____ porphyritic; b. _____ aphanitic; c. _____
    fragmental; d. _____ phaneritic; e. _____ vesicular.
11. Which of the following is detrital sediment?
    a. _____ broken sea shells; b. _____ ions in solution;
    c. _____ quartz sand; d. _____ conglomerate;
    e. _____ limestone.
12. The process whereby dissolved mineral matter precipitates
    in the pore spaces of sediment and binds it together is:
    a. _____ compaction; b. _____ weathering; c. _____
    arkose; d. _____ cementation; e. _____ ionic bonding.
13. Most limestones have a large component of calcite that
    was extracted from seawater by:
    a. _____ inorganic chemical reactions; b. _____
    organisms; c. _____ evaporation; d. _____ chemical
    weathering; e. _____ lithification.
14. Dolostone is formed by the addition of _____ to calcite
    in limestone.
    a. _____ calcium; b. _____ carbonate; c. _____
    magnesium; d. _____ iron; e. _____ sodium.
15. The metamorphic rock formed from limestone is:
    a. _____ quartzite; b. _____ hornfels; c. _____
    marble; d. _____ slate; e. _____ greenstone.
16. Pressure exerted equally in all directions on an object is:
    a. _____ differential; b. _____ directional; c. _____
    lithostatic; d. _____ shear; e. _____ none of these.
17. In which type of metamorphism are magmatic heat and
    fluids most important in causing change?
    a. _____ contact; b. _____ local; c. _____ regional;
    d. _____ volcanic; e. _____ lithostatic.
18. Which of the following metamorphic rocks displays a
    foliated texture?
    a. _____ marble; b. _____ quartzite; c. _____
    greenstone; d. _____ hornfels; e. _____ schist.

19. Mixed rocks with the characteristics of both igneous and high-grade metamorphic rocks are: a. _____ amphibolites; b. _____ gneisses; c. _____ greenstones; d. _____ migmatites; e. _____ granites.
20. What is a crystalline solid?
21. Compare ionic bonding with covalent bonding.
22. What is a silicate mineral? How do the two subgroups of silicate minerals differ from one another?
23. What are the two major kinds of igneous rocks, and how do they differ?
24. Explain how natural glass forms.
25. How are granite and diorite similar and dissimilar in composition?
26. What are the common chemical cements in sedimentary rocks?
27. Distinguish between clastic and crystalline textures. Give an example of a sedimentary rock with each texture.
28. What are the common evaporites, and how do they originate?
29. Name the agents of metamorphism, and explain how each contributes to metamorphism.
30. Where does contact metamorphism occur, and what type of change does it produce?
31. What is regional metamorphism, and under what conditions does it occur?
32. Describe the two types of metamorphic texture, and explain how they are produced.

Additional Readings

Baker, D. S. 1983. *Igneous rocks*. Englewood Cliffs, N.J.: Prentice-Hall.

Best, M. G. 1982. *Igneous and metamorphic petrology*. San Francisco, Calif.: W. H. Freeman.

Blackburn, W. H., and W. H. Dennen. 1988. *Principles of mineralogy*. Dubuque, Iowa: William C. Brown.

Blatt, H., G. Middleton, and R. Murray. 1980. *Origin of sedimentary rocks*. New York: W. H. Freeman.

Bowes, D. R., ed. 1989. *The encyclopedia of igneous and metamorphic petrology*. New York: Van Nostrand Reinhold.

Dietrich, R. V., and B. J. Skinner. 1990. *Gems, granites, and gravels: Knowing and using rocks and minerals*. New York: Cambridge University Press.

Gillen, C. 1982. *Metamorphic geology*. London: Chapman and Hall.

Hess, P. C. 1989. *Origins of igneous rocks*. Cambridge, Mass.: Harvard University Press.

Hyndman, D. W. 1985. *Petrology of igneous and metamorphic rocks*, 2d ed. New York: McGraw-Hill.

MacKenzie, W. S., C. H. Donaldson, and C. Guilford. 1982. *Atlas of igneous rocks and their textures*. New York: Halsted Press.

Middlemost, E. A. K. 1985. *Magma and magmatic rocks*. London: Longman Group.

Pough, F. H. 1987. *A field guide to rocks and minerals*, 4th ed. Boston, Mass.: Houghton Mifflin.

Selley, R. C. 1982. *An introduction to sedimentology*, 2d ed. New York: Academic Press.

CHAPTER 3

Horizontal sedimentary rocks exposed at Inspiration Point in Bryce Canyon National Park, Utah. (Photo © Peter Kresan.)

GEOLOGIC TIME: CONCEPTS AND PRINCIPLES

Prologue

In some respects, time is defined by the methods used to measure it. In geology, we are accustomed to thinking in terms of deep time, or incredibly long intervals of time. Geologists talk of events occurring thousands, millions, and even billions of years ago. At the other extreme, computer engineers measure time in nanoseconds, that is, in billionths of a second. Between the two extremes, most humans think of time in the more familiar terms of minutes, hours, days, and years, which relate to marked periodicities such as the change from day to night and the change of seasons.

Over the centuries, humans have used various devices and mechanisms to measure time more precisely. Many prehistoric monuments are oriented to detect the summer solstice. Sundials were used to divide the day into measurable units. As civilizations advanced, mechanical devices were invented to measure time, the earliest being the waterclock, or clepsydra, first used by the ancient Egyptians and further developed by the Greeks and Romans. The pendulum clock was invented in the seventeenth century and provided the most accurate timekeeping for the next two and a half centuries.

Major advances in accurately measuring time came with the discovery that the oscillatory motions of a vibrating solid can be used to measure time. The property of piezoelectricity (which literally means "pressure" electricity) in crystals like quartz is what enables them to be such accurate timekeepers. When pressure is applied to a quartz crystal, an electric current is generated. If an electric current is applied to a quartz crystal, the crystal expands and compresses extremely rapidly and regularly (about 100,000 times per second). It

is these vibrations from the electricity supplied by a watch's battery that make a quartz watch so accurate.

The first clock driven by a quartz crystal was developed in 1928. Today quartz clocks and watches are commonplace. Even inexpensive quartz timepieces are extremely accurate, and precision-manufactured quartz clocks used in observatories do not gain or lose more than one second every 10 years.

Greater accuracy than can be obtained with a quartz timepiece can be achieved with atomic clocks. This accuracy arises from the magnetic interaction of the nucleus of an atom with its electrons. Cesium atomic clocks were used to prove Einstein's prediction that a clock will slow down as its speed increases. Even more accurate clocks than atomic clocks are being developed based on lasers and related quantum devices.

At first glance, such precise subdivisions of time might seem unnecessary, yet many applications require extraordinary accuracy. Precision clocks are needed to measure the periods of pulsars, which are stars that emit radiation in short pulses. One pulsar is so stable that it may be suitable as a standard of time over long periods.

Accurate measurements of time also permit measurements of things that were once thought to be constant. For example, the rotation of the Earth is now known to vary from winter to summer and from year to year. Some of the variation is regular and some is as yet unpredictable. The reason for the variation is still not known.

Time is a fascinating topic that has been the subject of numerous essays and books. And while we can comprehend concepts like milliseconds and understand how a quartz watch works, deep time, or geologic time, is still very difficult for most people to comprehend. As we will see in this chapter, geology could not have advanced as a science until the foundation of geologic time was firmly established and accepted.

❖ INTRODUCTION

Time is what sets geology apart from most of the other sciences, and an appreciation of the immensity of geologic time is fundamental to an understanding of both the physical and biologic history of our planet. "So vast is the span of time recorded in the history of the Earth that it is generally distinguished from the more modest kinds of time by being called 'geologic time.' "[1]

Most people have difficulty comprehending geologic time because they tend to view time from the perspective of their own existence. Ancient history is what occurred hundreds or perhaps thousands of years ago, and yet when geologists talk in terms of ancient geologic history, they mean events that happened millions or even billions of years ago.

Geologists use two different frames of reference when speaking of geologic time. **Relative dating** involves placing geologic events in a sequential order as determined from their position in the rock record. Relative dating will not tell us how long ago a particular event occurred, only that one event preceded another. The various principles used to determine relative dating, such as superposition and cross-cutting relationships, were discovered hundreds of years ago and since then have been used to develop the relative geologic time scale. These principles are still widely used and will be discussed later in this chapter.

Absolute dating results in specific dates for rock units expressed in years before the present. Radiometric dating is the most common method of obtaining absolute age dates. It is based on the natural decay of various radioactive elements that occur in trace amounts in some rocks. It was not until the discovery of radioactivity near the end of the last century that absolute ages could be accurately applied to the relative geologic time scale. Today the geologic time scale is really a dual scale: a relative scale based on rock sequences fitted to an absolute scale based on radiometric dates expressed as years before present (❖ Fig. 3.1).

This chapter has two objectives. The first is to establish the concept of geologic time as one of the three corner-

stones of historical geology; the other two, organic evolution and plate tectonics, will be discussed in Chapters 6 and 7. The second objective is to introduce the major underlying principles of historical geology that allow geologists to interpret Earth history. Without an appreciation of the vastness of time and a firm understanding of basic geologic principles, the history of the Earth becomes nothing more than a recitation of seemingly unrelated facts.

❖ EARLY DEVELOPMENT OF THE CONCEPT OF GEOLOGIC TIME

The concept of geologic time and its measurement have changed through human history. For example, early Christian theologians were largely responsible for formulating the idea that time is linear rather than circular. When St. Augustine of Hippo (354–430 A.D.) stated that the Crucifixion was a unique event from which all other events could be measured, he helped establish the idea of the B.C. and A.D. time scale. This prompted many religious scholars and clerics to try to establish the date of creation by analyzing historical records and the genealogies found in Scripture.

A famous and influential Christian scholar, James Ussher (1581–1665), archbishop of Armagh, in Ireland, is popularly credited as being the first to calculate the age of the Earth based on recorded history and the genealogies described in Genesis. In 1650 Ussher announced that the Earth had been created on October 22, 4004 B.C. This date was later reproduced in many editions of the Bible and incorporated into the dogma of the Christian church. For nearly a century thereafter, it was considered heresy to assume that the Earth and all its features were more than about 6,000 years old. Thus, the idea of a very young Earth provided the basis for most chronologies of Earth history prior to the eighteenth century.

During the eighteenth and nineteenth centuries, several attempts were made to determine the age of the Earth based on scientific evidence rather than on revelation. For example, the French zoologist Georges Louis de Buffon (1707–1788), assumed the Earth formed as a molten ball and gradually cooled to its present condition. He heated iron balls of various diameters to their melting point, measured how long the balls took to cool to the surrounding temperature, and extrapolated his results to account for the cooling rate of the Earth. From these experiments he determined that the Earth had required at least 96,000 years to cool to its present temperature.

---

[1] A. Knopf, *Time and Its Mysteries* (New York: New York University Press, Ser. 3, 1949), p. 33.

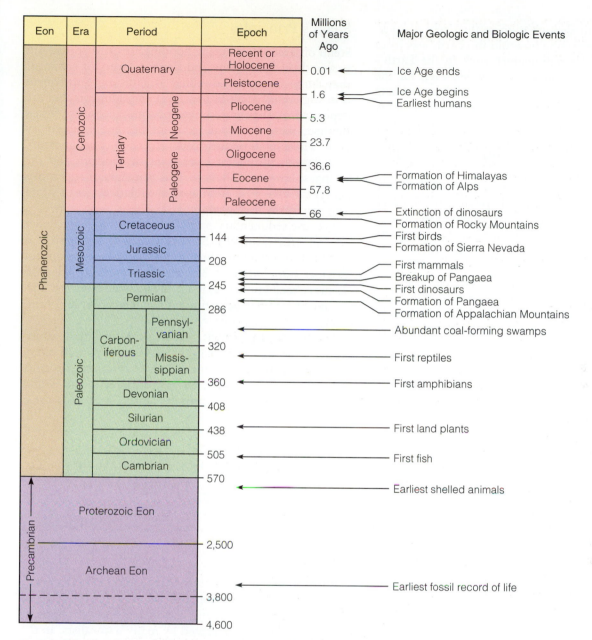

Eon	Era	Period		Epoch	Millions of Years Ago	Major Geologic and Biologic Events
Phanerozoic	Cenozoic	Quaternary		Recent or Holocene	0.01	Ice Age ends
				Pleistocene	1.6	Ice Age begins / Earliest humans
		Tertiary	Neogene	Pliocene	5.3	
				Miocene	23.7	
			Paleogene	Oligocene	36.6	
				Eocene	57.8	Formation of Himalayas / Formation of Alps
				Paleocene	66	
	Mesozoic	Cretaceous			144	Extinction of dinosaurs / Formation of Rocky Mountains / First birds / Formation of Sierra Nevada
		Jurassic			208	
		Triassic			245	First mammals / Breakup of Pangaea / First dinosaurs / Formation of Pangaea / Formation of Appalachian Mountains
	Paleozoic	Permian			286	
		Carboniferous	Pennsylvanian		320	Abundant coal-forming swamps
			Mississippian		360	First reptiles
		Devonian			408	First amphibians
		Silurian			438	First land plants
		Ordovician			505	First fish
		Cambrian			570	Earliest shelled animals
Precambrian		Proterozoic Eon			2,500	
		Archean Eon			3,800	Earliest fossil record of life
					4,600	

◆ **FIGURE 3.1** The geologic time scale. Some of the important events in Earth and life history are indicated at the right.

Later he revised this estimate downward to 75,000 years based on experiments using mixtures of metallic and nonmetallic substances.

This age contrasted sharply with the much younger age based on Scripture. During Buffon's time, however, it was very risky to publish views that seemed to contradict Church doctrine. To escape censure, Buffon acknowl-edged that his theory was pure philosophical speculation. His close connection with the French court also helped protect him from the Church.

Others were equally ingenious in attempting to calculate the age of the Earth. For example, geologists reasoned that if they could determine rates of deposition for various sedimentary rocks, they might be able to calculate the time required to deposit a given thickness of rock. They could then determine how old the Earth was from the total thickness of sedimentary rock in the Earth's crust. Rates of deposition vary, however, even for the same rock type. Furthermore, it is impossible to estimate how much rock has been removed by erosion, or how much a rock sequence has been reduced by compaction. As a result of these uncertainties, estimates ranged widely—from less than a million years to over a billion years.

Another attempt at determining the age of the Earth involved calculating the age of the oceans. If the ocean basins were filled very soon after the origin of the planet, then they would be only slightly younger than the Earth itself. The best-known calculations for the oceans' age were made by the Irish geologist John Joly in 1899. He reasoned that the Earth's ocean waters were originally fresh and their present salinity was the result of dissolved salt being carried into the ocean basins by rivers. By measuring the present amount of salt in the world's rivers, and knowing the volume of ocean water and its salinity, Joly calculated that it would have taken at least 90 million years for the oceans to reach their present salinity level. This was still much younger than the now-accepted age of 4.6 billion years for the Earth, mainly because Joly had no way to calculate how much salt had been recycled or the amount of salt stored in continental salt deposits or in seafloor clay deposits.

Although each of these methods yielded ages considerably younger than we now know the Earth to be, such attempts did change the way naturalists perceived the age of the Earth and represented a significant milestone in our understanding of Earth history.

❖ FUNDAMENTAL GEOLOGIC PRINCIPLES

The seventeenth century was an important time in the development of geology as a science because of the widely circulated writings of the Danish anatomist Nicolas Steno (1638–1686). Steno observed the present-day processes of sediment transport and deposition during stream flooding near Florence, Italy. These observations not only allowed him to determine the manner in which sedimentary rock layers formed, but also served as the basis for three fundamental principles of geology. While these principles may now seem self-evident, their discovery was an important scientific achievement and absolutely essential for interpreting geologic history.

Steno's Principles

Steno observed that during flooding, streams spread out across their floodplains and deposit layers of sediment that bury floodplain-dwelling organisms. Subsequent flooding events produce new layers of sediments that are deposited or superposed over previous deposits. When lithified, these layers of sediment become sedimentary rock. Thus, in an undisturbed succession of sedimentary rock layers, the oldest layer is at the bottom and the youngest layer is at the top. This **principle of superposition** is the basis for determining the relative ages of strata and the fossils they contain.

Because sedimentary particles settle from water under the influence of gravity, Steno reasoned that sediment is deposited in essentially horizontal layers (the **principle of original horizontality**). Therefore, a sequence of sedimentary rock layers that is steeply inclined from the horizontal must have been tilted after deposition and lithification.

Steno's third principle, the **principle of lateral continuity,** states that sediment extends laterally in all directions until it thins and pinches out or terminates against the edge of the depositional basin. Rocks in the Grand Canyon in Arizona beautifully illustrate Steno's three principles (◆ Fig. 3.2).

❖ ESTABLISHMENT OF GEOLOGY AS A SCIENCE—THE TRIUMPH OF UNIFORMITARIANISM OVER NEPTUNISM AND CATASTROPHISM

Steno's principles were significant contributions to early geologic thought, but the prevailing concepts of Earth history continued to be those that could be easily reconciled with a literal interpretation of Scripture. Two of these ideas, neptunism and catastrophism, were particularly appealing and accepted by many naturalists. In the final analysis, however, another concept, uniformitarianism, became the underlying philosophy of geology because it provided a better explanation for observed geologic phenomena than either neptunism or catastrophism.

Neptunism and Catastrophism

The concept of **neptunism** was proposed in 1787 by a German professor of mineralogy, Abraham Gottlob Werner (1749–1817). Although Werner was an excellent mineralogist, he is best remembered for his incorrect in-

◆ **FIGURE 3.2** The rocks in the Grand Canyon of Arizona illustrate Steno's principles. The sedimentary rocks of the Grand Canyon were originally deposited horizontally in a variety of marine and terrestrial environments (principle of original horizontality). The oldest rocks are therefore at the bottom of the canyon and the youngest rocks are at the top, forming the rim (principle of superposition). The exposed rock layers extend laterally for some distance (principle of lateral continuity).

terpretation of Earth history. He believed that all rocks, including granite and basalt, were precipitated in an orderly sequence from a primeval, worldwide ocean (Table 3.1). The oldest, or Primitive rocks, were all unfossiliferous igneous and metamorphic rocks that supposedly formed entirely by precipitation from seawater. Since they are found in the cores of mountain ranges, Werner reasoned they must have been the earliest rocks precipitated from the sea. As the ocean waters subsided, the Transition rocks were deposited. These rocks contain fossils and include the first detrital and chemical rocks. Werner believed that the fossils in the Transition rocks marked the time the Earth became suitable for habitation. The next rocks in Werner's sequence, the Secondary rocks, included a variety of fossiliferous detrital and chemical rocks, as well as basalt layers. The youngest, or Alluvial rocks, consisted of unconsolidated sediments. A fifth category of rocks included volcanics; these rocks, such as pumice and lava, were still being produced by volcanoes, but Werner did not consider them important.

Werner's subdivision of the Earth's crust by supposed relative age attracted a large following in the late 1700s and became almost universally accepted as the standard geologic column. Two factors account for this. First, Werner's charismatic personality, enthusiasm for geology, and captivating lectures popularized the concept. Equally important was the fact that neptunism included a worldwide ocean that could easily be reconciled with the biblical deluge.

In spite of Werner's personality and arguments, his neptunian theory failed to explain what happened to the tremendous amount of water that once covered the Earth. An even greater problem was Werner's insistence that all igneous rocks were precipitated from seawater. To Werner, all volcanoes were recent and had no importance in Earth history. He believed that volcanic eruptions resulted from the combustion of buried coal seams; therefore volcanoes could not have occurred until after the deposition of coals, which were categorized among his Secondary rocks. It was this failure to recognize the igneous origin of basalt that led to the downfall of neptunism.

From the late eighteenth century to the mid-nineteenth century, the concept of **catastrophism,** proposed by the French zoologist Baron Georges Cuvier (1769–1832), dominated European geologic thinking. Cuvier explained the physical and biologic history of the Earth as resulting from a series of sudden widespread catastrophes. Each catastrophe accounted for significant and rapid changes in the Earth, exterminating existing life in the affected area. Following a catastrophe, new organisms were either created or migrated in from elsewhere.

According to Cuvier, six major catastrophes had occurred in the past. These conveniently corresponded to the six days of biblical creation. Furthermore, the last catastrophe was taken to be the biblical deluge, so catastrophism had wide appeal, especially among theologians. While not all catastrophists accepted a 6,000-year

TABLE 3.1 Werner's Subdivision of the Rocks of the Earth's Crust (Standard Geologic Column, c. 1800)

Alluvial rocks	Unconsolidated sediments
Secondary rocks	Various fossiliferous, detrital, and chemical rocks such as sandstones, limestones, and coal as well as basalts
Transition rocks	Chemical and detrital rocks including fossiliferous rocks
Primitive rocks	Oldest rocks of the Earth (igneous and metamorphic)

age for the Earth, catastrophism as expounded by Cuvier was consistent with such a young age.

Eventually, both neptunism and catastrophism were abandoned as untenable hypotheses because their basic assumptions could not be supported by field evidence. The simplistic sequence of rocks predicted by neptunism (Table 3.1) was simply contradicted by field observations from many different areas. Moreover, basalt was shown to be of igneous origin, and subsequent discoveries of volcanic rocks interbedded with secondary and primitive deposits proved that volcanic activity had occurred throughout Earth history. As more and more field observations from widely separated areas were made, naturalists began to realize that far more than six catastrophes were needed to account for Earth history. With the demise of neptunism and catastrophism, the principle of uniformitarianism, advocated by James Hutton and Charles Lyell, became the guiding philosophy of geology.

◆ **FIGURE 3.3** James Hutton originated the principle of uniformitarianism and through his writings profoundly influenced the course of geologic thinking.

Uniformitarianism

James Hutton (1726–1797) is considered by many to be the father of historical geology (◆ Fig. 3.3). His detailed studies and observations of rock exposures and present geologic processes served as the basis for two important geologic principles. The **principle of cross-cutting relationships** holds that an igneous intrusion or a fault must be younger than the rocks it intrudes or displaces (◆ Fig. 3.4). This principle is very important in relative dating of geologic events and in interpreting Earth history.

Hutton observed the processes of wave action, erosion by running water, and sediment transport and concluded that given enough time these processes could account for the geologic features in his native Scotland. He believed that "the past history of our globe must be explained by what can be seen to be happening now." This assumption that present-day processes have operated throughout geologic time was the basis for the **principle of uniformitarianism.**

Although Hutton developed a comprehensive theory of uniformitarian geology, it was Charles Lyell (1797–1875) who became the principal advocate and interpreter of uniformitarianism. The term itself was coined by William Whewell in 1832.

Hutton viewed Earth history as cyclical: the continents are worn down by erosional processes, the eroded sediment is deposited in the sea, and uplift of the seafloor creates new continents, thus completing a cycle (◆ Fig. 3.5). He believed the mechanism for uplift was thermal expansion from the Earth's hot interior. Hutton's field observations, and experiments performed by his contemporaries involving the melting of basalt samples, convinced him that igneous rocks were the result of cooling magmas. This interpretation of the origin of igneous rocks, called *plutonism*, eventually displaced the neptunian view that igneous rocks precipitated from seawater.

Hutton also recognized the importance of unconformities in his cyclical view of Earth history. At Siccar Point, Scotland, he observed steeply inclined rocks that had been eroded and covered by flat-lying younger rocks (◆ Fig. 3.6). It was clear to him that severe upheavals had tilted the lower rocks and formed mountains. These were then worn away and covered by younger, flat-lying rocks. The erosional surface meant there was a gap in the rock record, and the rocks above and below this surface provided evidence that both mountain building and erosion had occurred. Although Hutton did not use the word "unconformity," he was the first to understand and explain the significance of such gaps in the rock record.

 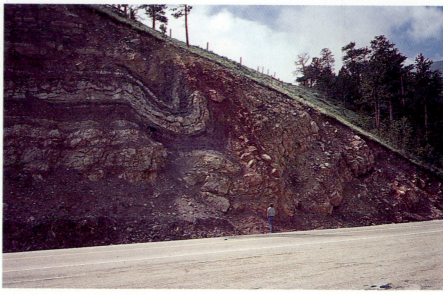

(a)   (b)

◆ **FIGURE 3.4** The principle of cross-cutting relationships. (a) A dark-colored dike has been intruded into older light-colored granite, north shore of Lake Superior, Ontario, Canada. (b) A fault, shown by the zone of rubble directly above the geologist, cutting through strata in the Bighorn Mountains, Wyoming.

Hutton was also instrumental in establishing the concept that geologic processes had vast amounts of time in which to operate. Because Hutton relied on known processes to account for Earth history, he concluded that the Earth must be very old. However, he estimated neither how old the Earth was nor how long it took to complete a cycle of erosion, deposition, and uplift. He merely allowed that "we find no vestige of a beginning, and no prospect of an end," which was in keeping with a cyclical view of Earth history.

Unfortunately, Hutton was not a particularly good writer, so his ideas were not widely disseminated or accepted. In fact, neptunism and catastrophism continued to be the dominant geologic concepts well into the 1800s. In 1830, however, Charles Lyell (◆ Fig. 3.7) published a landmark book, *Principles of Geology*, in which he championed Hutton's concept of uniformitarianism.

Lyell clearly recognized that small, imperceptible changes brought about by present-day processes could, over long periods of time, have tremendous cumulative effects. Not only did Lyell effectively reintroduce and establish the concept of unlimited geologic time, but he also discredited catastrophism as a viable explanation of geologic phenomena and firmly established uniformitarianism as the guiding philosophy of geology. The recog-

nition of virtually limitless time was also instrumental in the acceptance of Darwin's theory of evolution (see Chapter 6).

Perhaps because uniformitarianism is such a general concept, scientists have interpreted it in a number of different ways. Lyell's concept of uniformitarianism embodied the idea of a steady-state Earth in which present-day processes have operated at the same rate in the past as they do today. For example, according to Lyell, the frequency of earthquakes and volcanic eruptions for any given period of time in the past was the same as it is today. If the climate in one part of the world became warmer, another area would have to become cooler so that overall the climate could remain the same. By such reasoning, Lyell claimed that conditions for the Earth as a whole had remained essentially constant and unchanging through time.

Modern View of Uniformitarianism

Geologists today assume that the principles, or laws, of nature are constant but the rates and intensities of change have varied through time. For example, volcanic activity was more intense in North America during the Miocene Epoch than it is today, while glaciation has been

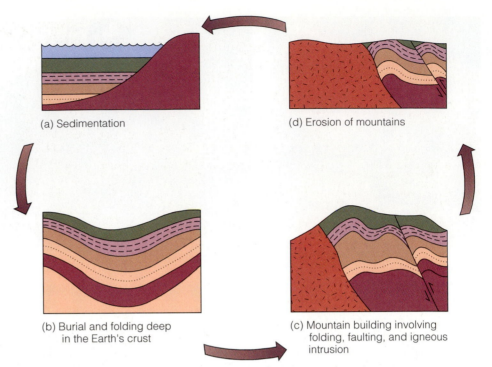

(a) Sedimentation

(d) Erosion of mountains

(b) Burial and folding deep in the Earth's crust

(c) Mountain building involving folding, faulting, and igneous intrusion

◆ **FIGURE 3.5** The geologic cycle as formulated by James Hutton. (a) Sediments are deposited in the sea, (b) deeply buried in the Earth's crust, (c) pushed up by thermal expansion from the Earth's hot interior, and then (d) eroded by such processes as wind, wave, and stream action. The eroded sediment is then deposited in the sea, and the cycle begins again.

more prevalent in the last 1.6 million years than in the previous 200 million years. Because the rates and intensities of geological processes have varied through time, some geologists prefer to use the term *actualism* rather than uniformitarianism to remove the idea of "uniformity" from the concept. Most geologists, though, still use the term uniformitarianism, because it indicates that even though the rates and intensities of change have varied in the past, the laws of nature have remained the same.

Uniformitarianism is a powerful concept that allows us through analogy and inductive reasoning to use present-day processes as the basis for interpreting the past and predicting potential future events. It does not eliminate the occurrence of occasional, sudden, short-term events such as volcanic eruptions, earthquakes, floods, or even meteorite impacts as forces that shape our modern world. In fact, some geologists view Earth history as a series of such short-term, or punctuated, events; and this view is certainly in keeping with the modern principle of uniformitarianism. The Earth is in a state of dynamic change and has been since it formed. While the rates of change may have varied in the past, the natural laws governing the processes have not.

❖ LORD KELVIN AND A CRISIS IN GEOLOGY

Lord Kelvin (1824–1907), an English physicist, claimed in a paper written in 1866 to have destroyed the uniformitarian foundation on which Huttonian-Lyellian geology was based. Kelvin did not accept Lyell's strict uniformitarianism, in which chemical reactions in the Earth's interior were supposed to continually produce heat, allowing for a steady-state Earth. Kelvin rejected this idea as perpetual motion—impossible according to the known laws of physics—and accepted instead the assumption that the Earth was originally molten.

Kelvin knew from deep mines in Europe that the Earth's temperature increases with depth, and he reasoned that the Earth is losing heat from its interior. By knowing the melting temperature of the Earth's rocks, the size of the Earth, and the rate of heat loss, Kelvin was able to calculate back to the time when the Earth was entirely molten. From these calculations, he concluded that the Earth could not be older than 400 million years or younger than 20 million years. This wide discrepancy in age reflected uncertainties in average temperature increases with depth and the various melting points of the Earth's constituent materials.

◆ **FIGURE 3.6** Angular unconformity at Siccar Point, Scotland. James Hutton first realized the significance of unconformities at this site in 1788. (Photo courtesy of Dorthy L. Stout.)

◆ **FIGURE 3.7** Portrait of Sir Charles Lyell, author of *Principles of Geology,* in which uniformitarianism was established as the guiding philosophy of geology.

After finally establishing that the Earth was very old and showing how present-day processes can be extrapolated over long periods of time to explain geologic features, geologists were in a quandary. Either they had to accept Kelvin's dates and squeeze events into a shorter time frame, or they had to abandon the concept of seemingly limitless time that was the underpinning of Huttonian-Lyellian geology and one of the foundations of Darwinian evolution. Some geologists objected to such a young age for the Earth, but their objections seem to have been based more on faith than on hard facts. Kelvin's quantitative measurements and arguments seemed flawless and unassailable to many scientists.

Kelvin's reasoning and calculations were sound, but his basic premises were false, thereby invalidating his conclusions. Kelvin was unaware that the Earth has an internal heat source, radioactivity, that has allowed it to maintain a fairly constant temperature through time.[2] Kelvin's 40-year campaign for a young Earth ended with the discovery of radioactivity near the end of the nineteenth century. His "unassailable calculations" were no longer valid and his proof for a geologically young Earth collapsed. Kelvin's theory, like neptunism, catastrophism, and a worldwide flood, became an interesting footnote in the history of geology.

---

[2] Actually, the Earth's temperature has decreased through time, but at a considerably slower rate than would lend any credence to Kelvin's calculations (see p. 245, Chapter 9).

While the discovery of radioactivity destroyed Kelvin's arguments, it provided geologists with a clock that could measure the Earth's age and validate what geologists had been saying—that the Earth was indeed very old! Less than 10 years after the discovery that radium generated heat, radiometric calculations were providing ages of billions of years for some of Earth's oldest rocks.

❖ ABSOLUTE DATING METHODS

Our preceding discussions have largely concerned the concept of vast amounts of geologic time and the formulation of principles used to determine relative ages. Ironically, the very process that invalidated Lord Kelvin's calculations of absolute dates serves as the basis for the radiometric calculations used to determine absolute ages.

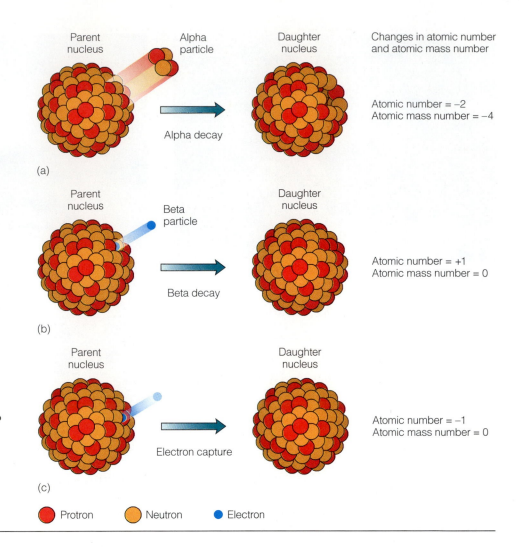

Parent nucleus

Alpha particle

Daughter nucleus

Changes in atomic number and atomic mass number

Alpha decay

Atomic number = –2
Atomic mass number = –4

(a)

Parent nucleus

Beta particle

Daughter nucleus

Beta decay

Atomic number = +1
Atomic mass number = 0

(b)

Parent nucleus

Daughter nucleus

Electron capture

Atomic number = –1
Atomic mass number = 0

(c)

◆ Protron   ◆ Neutron   ◆ Electron

◆ **FIGURE 3.8** Three types of radioactive decay.
(a) Alpha decay, in which an unstable parent nucleus emits two protons and two neutrons.
(b) Beta decay, in which an electron is emitted from the nucleus. (c) Electron capture, in which a proton captures an electron and is thereby converted to a neutron.

Some of the 91 naturally occurring elements are radioactive and spontaneously decay to other elements, releasing energy in the process. The discovery, in 1903 by Pierre and Marie Curie, that radioactive decay produces heat as a by-product meant that geologists no longer needed to assume the Earth had cooled from a molten state. The geologic importance of radioactive decay was soon recognized, and in 1907 Bertram B. Boltwood, a chemist-physicist at Yale University, proposed that lead is an end product of the radioactive decay of uranium and calculated the Earth's age as somewhere between 400 million and 2.2 billion years.

Geologists now had a mechanism for explaining the internal heat of the Earth that did not rely on residual cooling from a molten origin. With these discoveries, geologists and paleontologists had the long time periods they needed for a Huttonian-Lyellian-Darwinian view of the Earth and could ignore Lord Kelvin's discredited calculations. Furthermore, geologists now had a powerful tool to date geologic events accurately.

Atoms and Isotopes

As we discussed in Chapter 2, all matter is made up of chemical elements, each of which is composed of extremely small particles called *atoms*. The nucleus of an atom is composed of *protons* (positively charged particles) and *neutrons* (neutral particles) (see Fig. 2.2). The number of protons determines the element's *atomic number* and its properties and characteristics. Therefore, each change in the number of protons (and hence the atomic number) forms a new element with a different

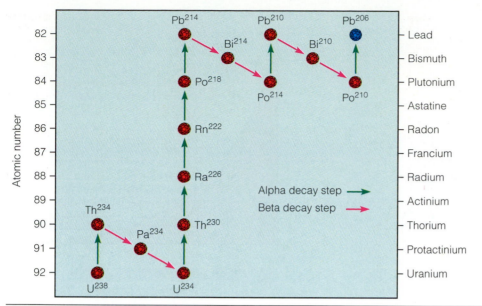

◆ **FIGURE 3.9** Radioactive decay series for uranium 238 to lead 206. Radioactive uranium 238 decays to its stable end product, lead 206, by eight alpha and six beta decay emissions. A number of different isotopes are produced as intermediate steps in the decay series.

atomic structure and thus different physical and chemical properties.

The combined number of protons and neutrons in an atom is its *atomic mass number*. Carbon, for example, has an atomic number of 6 and an atomic mass number of 12, 13, or 14, depending on the number of neutrons present. Forms of the same element with different atomic mass numbers are *isotopes* (see Fig. 2.3). Most isotopes, such as carbon 12 and 13, for example, are stable, but others, such as carbon 14, are unstable. It is the unstable isotopes that are radioactive, and their decay rate is what geologists measure to determine absolute ages.

Radioactive Decay and Half-Lives

Radioactive decay is the process whereby an unstable atomic nucleus spontaneously transforms into another atomic nucleus. Three types of radioactive decay, all of which result in a change of atomic structure, are recognized: alpha decay, beta decay, and electron capture decay (◆ Fig. 3.8). In **alpha decay,** two protons and two neutrons are emitted from the nucleus, resulting in a loss of two atomic numbers and four atomic mass numbers. In **beta decay,** a fast-moving electron is emitted from a neutron in the nucleus, changing that neutron to a proton and consequently increasing the atomic number by one, with no resultant atomic mass number change. **Electron capture decay** results when a proton captures an

electron from an electron shell and thereby converts to a neutron, resulting in a loss of one atomic number and no change in the atomic mass number.

Some elements undergo only one decay step in converting from an unstable form to a stable form. For example, rubidium 87 decays to strontium 87 by a single beta emission, while potassium 40 decays to argon 40 by a single electron capture. Other radioactive elements undergo several decay steps. Uranium 235 decays to lead 207 by seven alpha steps and six beta steps, while uranium 238 decays to lead 206 by eight alpha and six beta steps (◆ Fig. 3.9).

When discussing decay rates, it is convenient to refer to them in terms of half-lives. The **half-life** of a radioactive element is the time it takes for one-half of the atoms of the original unstable **parent element** to decay to atoms of a new, more stable **daughter element.** The half-life of a given radioactive element is constant, regardless of external conditions, and can be precisely measured in the laboratory. Half-lives of various radioactive elements range from less than a billionth of a second to 49 billion years.

Radioactive decay occurs at a geometric rate rather than a linear rate. Therefore, a graph of the decay rate produces a curve rather than a straight line (◆ Fig. 3.10). For example, after one half-life, an element with *1,000,000* parent atoms will have *500,000* parent atoms and *500,000* daughter atoms; after two half-lives, it will

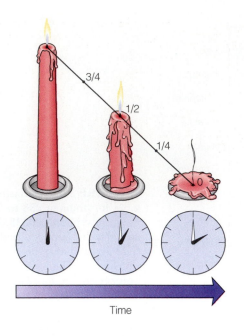

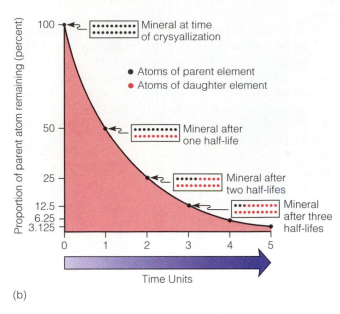

◆ **FIGURE 3.10** Uniform, straight-line depletion (a) is characteristic of many familiar processes. (b) Geometric radioactive decay curve, in which each time unit represents one half-life, and each half-life is the time it takes for one-half of the parent element to decay to the daughter element.

have *250,000* parent atoms (one-half of the previous parent atoms, which is equivalent to one-fourth of the original parent atoms) and 750,000 daughter atoms; after three half-lives, it will have *125,000* parent atoms (one-half of the previous parent atoms, or one-eighth of the original parent atoms) and 875,000 daughter atoms, and so on, until the number of parent atoms remaining is so few that they cannot be accurately measured by present-day instruments.

By measuring the parent-daughter ratio and knowing the half-life of the parent (which has been determined in the laboratory), geologists can calculate the age of a sample containing the radioactive element. The parent-daughter ratio is usually determined by a *mass spectrometer,* an instrument that measures the proportions of elements of different masses.

Sources of Uncertainty

The most accurate radiometric dates are obtained from igneous rocks. As a magma cools and begins to crystal-lize, radioactive parent atoms are separated from previously formed daughter atoms. Because they are the right size, some radioactive parent atoms are incorporated into the crystal structure of certain minerals. The stable daughter atoms, however, are a different size than the radioactive parent atoms and consequently cannot fit into the crystal structure of the same mineral as the parent atoms. Therefore, when the magma begins to crystallize, the mineral will contain radioactive parent atoms but no stable daughter atoms. Thus, the time that is being measured is the time the mineral containing the radioactive atoms crystallized, not the time when the radioactive atoms formed.

To obtain accurate radiometric dates, geologists must be sure that they are dealing with a closed system, meaning that neither parent nor daughter atoms have been added or removed from the system since crystallization and that the ratio between them results only from radioactive decay. Otherwise, an inaccurate date will result. If daughter atoms have leaked out of the mineral being analyzed, the calculated age will be too young; if parent atoms have been removed, the calculated age will be too great.

Leakage may occur if the rock is heated or subjected to intense pressure such as occurs during metamorphism. If this happens, some of the parent or daughter atoms may be driven from the mineral, and analysis will result in an inaccurate age determination. If the daughter product was selectively removed, then one would be measuring the time since metamorphism (a useful measurement itself) and not the time since crystallization of the mineral (♦ Fig. 3.11). For example, when dating rocks using the potassium-argon technique, one must be very careful because argon, the measured stable daughter element, is an inert gas that is easily driven out of a mineral when the rock is metamorphosed.

There is also a source of error inherent in measuring the minute amounts of the different elements and isotopes used for age dating. Mass spectrometers are ± 0.2 to 2.0% accurate in their measurements. Thus, for a rock 10 million years old, the possible error is 20,000 to 200,000 years. For a rock one billion years old, however, this difference amounts to 2 million to 20 million years. Therefore, when a radiometric age is given, a plus or minus factor in number of years is usually appended to the age to indicate the limits of error in the dating technique.

Long-Lived Radioactive Isotope Pairs

The most reliable dates are obtained by analyzing at least two different radioactive decay series in the same rock. Naturally occurring uranium consists of both uranium 235 and uranium 238 isotopes in the ratio of 138 to 1. Through various decay steps, uranium 235 decays to lead 207, whereas uranium 238 decays to lead 206. The age of a rock can be calculated using both of these decay series. If the minerals containing them have remained closed systems, then the ages obtained from each parent-daughter ratio should be in close agreement. If they are, they are said to be *concordant,* reflecting the time of crystallization of the magma. If the ages do not closely agree, then they are said to be *discordant,* and other samples must be taken and the ratios measured to see which, if either, date is correct.

Table 3.2 shows five different long-lived parent-daughter isotope pairs used in radiometric dating. Long-lived pairs have half-lives of millions or billions of years. All of these were present when the Earth formed and are still present in measurable quantities. Other shorter-lived

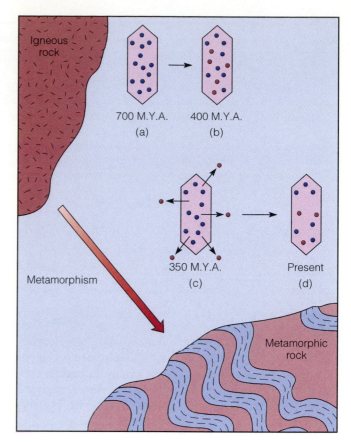

♦ **FIGURE 3.11** The effect of metamorphism in driving daughter atoms out of a mineral that crystallized 700 million years ago (M.Y.A.). (a) The mineral is shown immediately after crystallization, then (b) at 400 M.Y.A. when some of the parent atoms have decayed to daughter atoms. (c) Metamorphism at 350 M.Y.A. drives the daughter atoms out of the mineral into the surrounding rock. (d) Assuming the rock has remained a closed chemical system throughout its history, dating the mineral today yields the time of metamorphism, while dating the rock provides the time of the mineral's crystallization, and hence, the rock's formation, 700 M.Y.A.

radioactive isotope pairs have decayed to the point that only small quantities near the limit of detection remain.

The most commonly used isotope pairs are the uranium-lead and thorium-lead series, which are used principally to date ancient igneous intrusives, lunar samples, and some meteorites. The rubidium-strontium pair is also used for very old samples and has been effective in

TABLE 3.2 Five of the Principal Long-Lived Radioactive Isotope Pairs used in Radiometric Dating

ISOTOPES		Half-Life of Parent (Years)	Effective Dating Range (Years)	Minerals and Rocks That Can Be Dated
Parent	Daughter			
Uranium 238	Lead 206	4.5 billion	10 million to 4.6 billion	Zircon Uraninite
Uranium 235	Lead 207	704 million		
Thorium 232	Lead 208	14 billion		
Rubidium 87	Strontium 87	48.8 billion	10 million to 4.6 billion	Muscovite Biotite Potassium feldspar Whole metamorphic or igneous rock
Potassium 40	Argon 40	1.3 billion	100,000 to 4.6 billion	Glauconite Hornblende Muscovite Whole volcanic rock Biotite

dating the oldest rocks on Earth as well as meteorites. The potassium-argon method is typically used for dating fine-grained volcanic rocks from which individual crystals cannot be separated; hence the whole rock is analyzed. Other long-lived radioactive isotope pairs exist, but they are rather rare and used only in special situations.

Dating by Fission Tracks

When a uranium isotope in a mineral emits an alpha decay particle, the heavy, rapidly moving alpha particle damages the crystal structure. The damage appears as small linear tracks that are visible only after etching the mineral with hydrofluoric acid. The age of the sample is determined on the basis of the number of fission tracks present and the amount of uranium the sample contains. The older the sample, the greater the number of tracks (◆ Fig. 3.12).

Fission track dating is of particular interest to geologists because the technique can be used to date samples ranging from only a few hundred to hundreds of millions of years in age. It is most useful for dating samples between about 40,000 and one million years ago, a period for which other dating techniques are not particularly suitable. One of the problems in fission track dating occurs when the rocks have been subjected to high temperatures. If this happens, the damaged crystal structures are repaired by annealing, and consequently the tracks disappear. In such instances, the calculated age will be younger than the actual age.

Radiocarbon and Tree-Ring Dating Methods

Carbon is an important element in nature and is one of the basic elements found in all forms of life. It has three isotopes; two of these, carbon 12 and 13, are stable, whereas carbon 14 is radioactive. Carbon 14 has a half-life of 5,730

◆ **FIGURE 3.12** Each fission track (about 16 μm in length) in this apatite crystal is the result of the radioactive decay of a uranium atom. The crystal, which has been etched with hydrofluoric acid to make the fission tracks visible, comes from one of the dikes at Shiprock, New Mexico, and has a calculated age of 27 million years. (Photo courtesy of Charles W. Naeser, U.S. Geological Survey.)

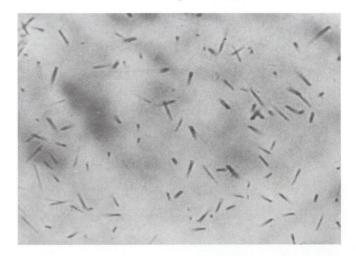

FOSSILS AND THE HISTORY OF THE EARTH'S ROTATION

Astronomers generally agree that the period of the Earth's revolution around the Sun has been constant through time, but its rotation around its axis has been gradually slowing down. Based on modern and ancient astronomical data, it appears that during the past 3,000 years, the length of the day has been increasing at a constant rate of approximately 2.5 seconds per 100,000 years. If this is true, then it follows that the number of days in the year has been decreasing through time (◆ Fig. 1). How can such a hypothesis be tested?

Many living organisms—for example, corals and clams—show daily growth increments, or rings, much like the growth-rings of a tree. These daily growth-rings can be grouped into larger yearly growth-bands. Modern organisms exhibit approximately 365 growth-rings per year. If the Earth's rotation has changed in the past, that change should be reflected in the number of daily growth-rings per yearly cycle.

In 1963 Professor John Wells of Cornell University published "Coral Growth and Geochronometry." In this pioneering study, he counted the number of daily growth-rings deposited per year by living and fossil corals. He found that modern corals add approximately 365 daily growth-rings per year, while Pennsylvanian corals added about 390 growth-rings per year and Devonian corals added around 400 growth-rings per year (◆ Fig. 2). These results indicate that the number of days per year have been decreasing since at least the Devonian and seemingly confirm the astronomical data. Geologists studying the growth increments of other fossil organisms have generally confirmed Wells's conclusions.

Wells's study is important because it revealed the only known way to directly measure the rate of the Earth's rotation in the geologic past. Paleontological data can thus be used to verify theoretical calculations based on astronomical and geophysical assumptions. Furthermore, a time scale based on growth-rings found in fossils provides a means for absolute dating of sedimentary rocks.

◆ **FIGURE 1** Length of the year for various geologic periods based on calculations indicating that the Earth's rotation has been slowing at an average rate of 2.5 seconds per 100,000 years. Counts of the bands on corals and other invertebrates have agreed with these estimates.

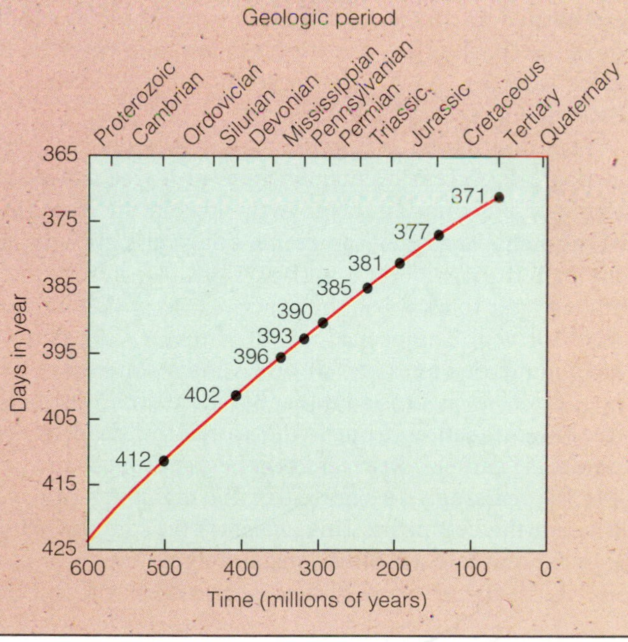

Geologic period

Proterozoic Cambrian Ordovician Silurian Devonian Mississippian Pennsylvanian Permian Triassic Jurassic Cretaceous Tertiary Quaternary

Days in year

365 · 375 · 385 · 395 · 405 · 415 · 425

371 · 377 · 381 · 385 · 390 · 393 · 396 · 402 · 412

Time (millions of years): 600 · 500 · 400 · 300 · 200 · 100 · 0

◆ **FIGURE 2** Devonian solitary coral showing growth-rings. The fine lines are presumed to be daily growth-rings. Counts of these growth-rings indicate there were approximately 400 days in the year during the Devonian Period. (Photo courtesy of Sue Monroe.)

years plus or minus 30 years. **Carbon 14 dating** is based on the ratio of carbon 14 to carbon 12 and is generally used to date once-living material.

The short half-life of carbon 14 makes this dating technique practical only for specimens younger than about 70,000 years. Consequently, the carbon 14 dating method is especially useful in archaeology and has greatly aided in unraveling the events of the Late Pleistocene Epoch.

Carbon 14 is constantly formed in the upper atmosphere through the activity of cosmic rays, which are high-energy particles (mostly protons). These high-energy particles strike the atoms of upper-atmospheric gases, splitting their nuclei into protons and neutrons. When a neutron strikes the nucleus of a nitrogen atom (atomic number 7, atomic mass number 14), it may be absorbed into the nucleus and a proton emitted. Thus, the atomic number of the atom decreases by one, while the atomic mass number stays the same. Because the atomic number has changed, a new element, carbon 14 (atomic number 6, atomic mass number 14), is formed. The newly formed carbon 14 is quickly assimilated into the carbon cycle and, along with carbon 12 and 13, is absorbed in a nearly constant ratio in all living organisms (◆ Fig. 3.13).

Because carbon 14 is continually created in the upper atmosphere and distributed throughout the carbon cycle, the ratio of carbon 14 to carbon 12 remains constant in all living things. When an organism dies, however, carbon 14 is not replenished, and hence the ratio of carbon 14 to carbon 12 decreases as the carbon 14 decays.

Currently, the ratio of carbon 14 to carbon 12 is remarkably constant in both the atmosphere and living organisms. There is good evidence, however, that the production of carbon 14, and thus the ratio of carbon 14 to carbon 12, has varied somewhat over the past several thousand years. This was determined by comparing ages established by carbon 14 dating of wood samples against those established by counting annual tree-rings in the same samples. As a result, carbon 14 ages have been corrected to reflect such variations in the past.

Tree-ring dating is another useful method for dating geologically recent events. The age of a tree can be determined by counting the growth-rings in the lower part of the stem. Each ring represents one year's growth, and the pattern of wide and narrow rings can be compared among trees to establish the exact year in which the rings were formed. The procedure of matching ring patterns from numerous trees and wood fragments in a given area is referred to as cross-dating. By correlating distinctive tree-ring sequences from living to nearby dead trees, a time scale can be constructed that extends back to about 14,000 years ago (◆ Fig. 3.14). By matching ring patterns to the composite ring scale, wood samples whose ages are not known can be accurately dated.

◆ **FIGURE 3.13** The carbon cycle showing the formation, dispersal, and decay of carbon 14.

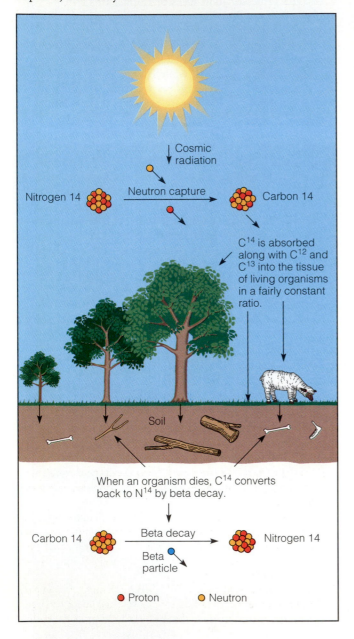

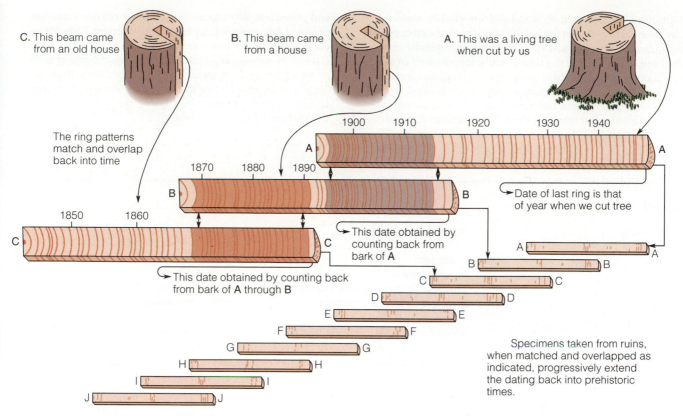

FIGURE 3.14 In the cross-dating method, tree-ring patterns from different woods are matched against each other to establish a ring-width chronology backward in time.

Chapter Summary

1. Early Christian theologians were responsible for formulating the idea that time is linear and that the Earth was very young. Archbishop Ussher calculated an age of about 6,000 years based on his interpretation of Scripture.

2. During the eighteenth and nineteenth centuries, attempts were made to determine the age of the Earth based on scientific evidence rather than revelation. While some attempts were quite ingenious, they all yielded ages that we now know are much too young.

3. Considering the religious, political, and social climate of the seventeenth, eighteenth, and early nineteenth centuries, it is easy to see why concepts such as neptunism, catastrophism, and a very young Earth were eagerly embraced. As geologic data accumulated, it soon became apparent that these concepts were not supported by evidence, and that the Earth must be older than 6,000 years.

4. James Hutton believed that present-day processes operating over long periods of time could explain all the geologic features of the Earth. He also viewed Earth history ascyclical and believed the Earth to be very old. Hutton was instrumental in establishing the basis for the principle of uniformitarianism.

5. Uniformitarianism as articulated by Charles Lyell soon displaced neptunism and catastrophism as the guiding doctrine of geology. The principle of uniformitarianism holds that the laws of nature have been constant through time and that the same processes operating today also operated in the past.

6. In addition to uniformitarianism, the principles of superposition, original horizontality, lateral continuity, and cross-cutting relationships are basic for determining relative geologic ages and for interpreting the geologic history of the Earth.

7. Radioactivity was discovered in the late nineteenth century, and soon thereafter radiometric dating techniques allowed geologists to determine absolute ages for geologic events.

8. Absolute age dates for rock samples are usually obtained by determining how many half-lives of the radioactive parent element have elapsed since the sample originally crystallized. A half-life is the time it takes for one-half of the radioactive parent element to decay to a daughter element.

9. The most accurate radiometric dates are obtained from igneous rocks. The five common long-lived radioactive isotope pairs are uranium 238–lead 206, uranium 235–lead 207, thorium 232–lead 208, rubidium 87–strontium 87, and potassium 40–argon 40. The most reliable dates are obtained by analyzing at least two different radioactive decay series in the same rock.

10. Carbon 14 dating can be used only for organic matter such as wood, bones, and shells and is effective back to about 70,000 years ago. Unlike the long-lived isotopic pairs, the carbon 14 method determines age by the ratio of radioactive carbon 14 to stable carbon 12.

Important Terms

absolute dating
alpha decay
beta decay
carbon 14 dating
catastrophism
daughter element
electron capture decay

fission track dating
half-life
neptunism
parent element
principle of cross-cutting relationships
principle of lateral continuity

principle of original horizontality
principle of superposition
radioactive decay
relative dating
tree-ring dating
uniformitarianism

Review Questions

1. In what type of dating method are events placed in sequential order as determined from their position in the rock record?
   a. _____ uniformitarianism; b. _____ absolute;
   c. _____ historical; d. _____ relative; e. _____ specific.

2. Which of the following dating methods will not yield an absolute date?
   a. _____ tree-ring; b. _____ superposition; c. _____ fission track; d. _____ carbon 14; e. _____ uranium-lead.

3. Which fundamental geologic principle states that sediment extends in all directions until it thins and pinches out or terminates against the edge of the depositional basin?
   a. _____ superposition; b. _____ original horizontality;
   c. _____ lateral continuity; d. _____ cross-cutting relationships; e. _____ uniformitarianism.

4. Who proposed the concept of catastrophism?
   a. _____ Cuvier; b. _____ Hutton; c. _____ Lyell;
   d. _____ Kelvin; e. _____ Werner.

5. Who is generally considered to be the father of historical geology?
   a. _____ Cuvier; b. _____ Hutton; c. _____ Lyell;
   d. _____ Whewell; e. _____ Werner.

6. The atomic number of an element is determined by the number of:
   a. _____ protons; b. _____ neutrons; c. _____ electrons; d. _____ protons and neutrons; e. _____ protons and electrons.

7. The type of decay in which a fast-moving electron is emitted from a neutron is:
   a. _____ alpha; b. _____ beta; c. _____ gamma;
   d. _____ electron capture; e. _____ none of these.

8. If a radioactive element has a half-life of 2 million years, what fraction of the original amount of parent material will remain after 6 million years of decay?
   a. _____ $1/32$ _____ $1/16$; c. _____ $1/8$; d. _____ $1/4$;
   e. _____ $1/2$.

9. How many half-lives are required to yield a mineral with 1,250 atoms of uranium 235 and 18,750 atoms of lead 206?
   a. _____ 10; b. _____ 8; c. _____ 6; d. _____ 5;
   e. _____ 4.

10. If daughter atoms have leaked out of a mineral being analyzed, the calculated age will be:
    a. _____ too old; b. _____ too young; c. _____ the same as if the daughter atoms hadn't leaked out;
    d. _____ impossible to determine; e. _____ none of these.

11. Which of the following is not a long-lived radioactive isotope pair?
    a. _____ uranium-lead; b. _____ rubidium-strontium;
    c. _____ thorium-lead; d. _____ potassium-argon;
    e. _____ carbon 14–carbon 12.

12. The author of *Principles of Geology* and the principal advocate and interpreter of uniformitarianism was:

a. _____ Hutton; b. _____ Steno; c. _____ Kelvin; d. _____ Lyell; e. _____ Werner.

13. In carbon 14 dating, which ratio is being measured?
a. _____ C^{14}/N^{14}; b. _____ C^{12}/C^{13}; c. _____ C^{12}/C^{14}; d. _____ C^{12}/N^{14}; e. _____ parent-daughter ratio.

14. If the radiometric ages obtained from two different radioactive isotope pairs in the same mineral are in close agreement, they are said to be:
a. _____ concordant; b. _____ discordant; c. _____ harmonic; d. _____ recumbent; e. _____ conformable.

15. Briefly discuss the concept of neptunism and how its basic tenets seem to defy the principle of superposition.

16. Briefly discuss Hutton's view of Earth history.

17. Discuss Lord Kelvin's method of determining the age of the Earth and explain what led to its rejection.

18. Why is it difficult to date sedimentary and metamorphic rocks radiometrically?

19. Why is the principle of uniformitarianism important to geologists?

20. Who was Archbishop Ussher? What contribution did he make to determining the age of the Earth?

21. What is the difference between relative and absolute dating of geologic events?

22. What are the fundamental principles used in relative age dating? Why are they so important in historical geology?

23. Describe the contributions each of the following made to the development of geology: Georges Louis de Buffon, James Hutton, Lord Kelvin, Charles Lyell, Nicolas Steno.

24. Compare the modern view of uniformitarianism to Lyell's view.

25. If you wanted to calculate the absolute age of an intrusive body, what information would you need?

26. What are some of the uncertainties that must be considered in radiometric dating?

27. How can geologists be sure the absolute age dates they obtain from igneous rocks are accurate?

28. How does the carbon 14 dating technique differ from uranium-lead dating methods?

Additional Readings

Albritton, C. C., Jr. 1980. *The abyss of time.* San Francisco, Calif.: Freeman, Cooper & Co.

_____. 1984. Geologic time. *Journal of Geological Education* 32, no. 1: 29–37.

Berry, W. B. N. 1987. *Growth of a prehistoric time scale,* 2d ed. Palo Alto, Calif.: Blackwell Scientific Publications.

Boslough, J. 1990. The enigma of time. *National Geographic* 177, no. 3: 109–32.

Faul, H. 1978. A history of geologic time. *American Scientist* 66, no. 2: 159–65.

Geyh, M. A., and H. Schleicher. 1990. *Absolute age determination.* New York: Springer-Verlag.

Gould, S. J. 1987. *Time's arrow, time's cycle.* Cambridge, Mass.: Harvard University Press.

Harland, W. B., R. L. Armstrong, A. V. Cox, L. E. Craig, A. G. Smith, and D. G. Smith. 1990. *A geologic time scale 1989.* New York: Cambridge University Press.

Laudan, R. 1987. *From mineralogy to geology: The foundations of a science, 1650–1830.* Chicago, Ill.: University of Chicago Press.

Ramsey, N. F. 1988. Precise measurement of time. *American Scientist* 76, no. 1: 42–49.

Shea, J. H. 1982. Twelve fallacies of uniformitarianism. *Geology* 10, no. 9: 455–60.

Wetherill, G. W. 1982. Dating very old objects. *Natural History* 91, no. 9: 14–20.

CHAPTER 4

Chapter Outline

Fossil fish from the Green River Formation of Wyoming. (Photo courtesy of Sue Monroe.)

ROCKS, FOSSILS, AND TIME

Prologue

Most people today know that sand deposited on a beach may become sandstone, and that shells and bones in rocks are the remains of once-living animals. These concepts are well entrenched in modern thinking. During most of human existence, however, people had little or no idea of how rocks formed or what fossils were.

Humans have used rocks and fossils since prehistoric times for tools, weapons, and ornaments. Fossils, for example, were a trade item in the ancient world, and even during the Middle Ages some fossils were believed to have curative powers (◆ Fig. 4.1). The Greek philosopher Xenophanes, who lived during the sixth century B.C., was apparently the first to recognize that fossil seashells and bones represent the remains of once-living organisms, while the Chinese came to the same conclusion by the first century B.C. Through most of historic time in the West, however, fossils were variously believed to be inorganic objects formed inside rocks by spontaneous generation or growing seeds; images produced in rocks by some kind of molding force; or even objects placed in the rocks by the Creator to test our faith or by Satan to sow seeds of doubt.

One of the reasons fossils were misunderstood was that no one knew how rocks formed. Some of the ancient Greeks and Egyptians had some idea of the rock-forming processes, and Leonardo da Vinci (1452–1519) realized that sand transported in rivers could become sandstone. Nevertheless, it was not until Nicolas Steno (see Chapter 3) elucidated his principles in 1669 that a framework for interpreting rocks existed. In short, before Steno's work, little was known about Earth history.

By the late 1700s and early 1800s, scientists had gained considerable insight into Earth history. Old ideas are difficult to displace, however, and the idea of a young Earth shaped by catastrophes (perhaps a single catastrophe) persisted. One of the most extreme views of Earth history was proposed in 1857 by Philip Henry Gosse, a British naturalist. According to Gosse, the Earth was created a few thousands of years ago with the appearance of a lengthy history. In other words, rocks that appeared to have been deposited by rivers or that appeared to have formed from lava flows were created with that appearance. The events denoted by their appearance had not really happened. Fossils were not the remains of organisms that once lived on Earth but were simply another created aspect of rocks.

◆ **FIGURE 4.1** Fossil teeth of rays, a type of fish, were called toadstones during the Middle Ages. The stones were believed to come from the heads of living toads *(left)* and were considered to be a powerful cure for several ailments *(right).*

Gosse's proposal was immediately rejected by most people, who reasoned that if he were correct, the historical record preserved in rocks was the product of a deceptive Creator. In addition to the philosophical implications of Gosse's proposal, it is important to understand why it cannot be considered scientific. There is no conceivable observation or test that could demonstrate whether it is right or wrong. In short, it is completely untestable and thus not scientific.

Even before Gosse proposed his idea, scientists had concluded that the Earth had a long history and that that history could be read from the only available record—the geologic record.

❖ INTRODUCTION

The Earth is a dynamic planet. Volcanic eruptions, earthquakes, floods, and shoreline erosion are manifestations of some of the internal and surface processes that continuously modify it. The Earth has changed, or evolved, throughout its history, but our only record of that change is the evidence preserved in rocks, the **geologic record**. All rock types are important in analysis of the geologic record: igneous rocks record volcanic activity related to rifting and plate convergence, and studies of metamorphic rocks reveal something of the dynamic processes that occur within the Earth's crust.

Sedimentary rocks have a special place in historical geology. Sediments or sedimentary rocks cover most of the seafloor and about two-thirds of the continents, and they preserve evidence of the surficial processes responsible for their origin. Furthermore, many contain the remains or traces of prehistoric life that we call *fossils*.

Our main objective in this chapter is to explore the principles and concepts geologists use to make sense of the geologic record. Just as the written historical record

(a)

(b)

(c)

❖ **FIGURE 4.2** Stratification. (a) These igneous rocks in the San Luis Valley, Colorado, are stratified; the lowest exposed unit is a volcanic mudflow overlain by an ash bed (white) and a lava flow. (b) Stratification in metamorphic rocks, the Siamo Slate of Michigan. (c) Stratified sedimentary rocks in Capitol Reef National Park, Utah. (Photo courtesy of Wayne Moore.)

must be analyzed and interpreted, the geologic record requires analyses if we are to make inferences about geologic history—that is, if we are to read Earth history from rocks.

❖ STRATIGRAPHY

Stratigraphy is the branch of geology that is concerned with the composition, origin, age relationships, and areal extent of layered or stratified rocks. Stratification may occur in any of the major rock groups; a succession of lava flows or ash falls will be layered, metamorphosed layered rocks also show this property, and nearly all sedimentary rocks are stratified (◆ Fig. 4.2).

Our main concern here is with stratification in sedimentary rocks, but some of the concepts of stratigraphy apply to any sequence of layered rocks. Where sedimentary rocks are observed in surface exposures (Fig. 4.2), the vertical relationships among the strata are easily determined. Lateral relationships, equally important in stratigraphic studies, usually must be determined from a number of separate exposures.

Vertical Stratigraphic Relationships

The bounding surface separating one layer of strata from another is a **bedding plane** and may be gradational or sharp. At gradational bedding planes, one rock type gradually changes upward into another rock type. They represent gradually changing conditions of sedimentation. Sharp bedding planes are surfaces above and below which the rocks are different (◆ Fig. 4.3). Abrupt changes in composition indicate rapid changes in sedimentation or perhaps a period of nondeposition or erosion followed by renewed deposition. Regardless of the nature of bedding planes, it is essential that the correct vertical sequence of strata at the time of deposition be known.

Superposition

Nicolas Steno (see Chapter 3) realized that correct relative ages of horizontal (undeformed) strata could be determined by their position in a sequence of rocks (Fig. 3.2). If strata have been deformed, however, relative age determinations are more difficult (◆ Fig. 4.4). For example, the strata in Figure 4.4 have been tilted from their original horizontal orientation and now stand nearly vertical. Can you tell which bed is youngest?

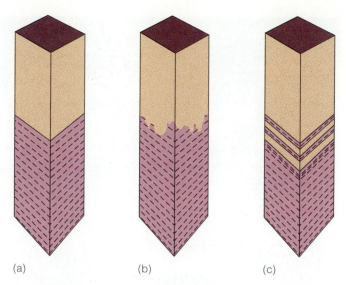

(a)   (b)   (c)

◆ **FIGURE 4.3** Vertical stratigraphic relationships. (a) Shale and sandstone separated by a sharp bedding plane. Such abrupt changes in composition indicate rapid changes in sedimentation or a time during which no deposition occurred. (b) and (c) Two examples of gradual upward changes from one rock type into another.

Geologists commonly solve relative age problems such as the one in Figure 4.4 by looking for sedimentary structures, structures that formed in sediments at the time of deposition or shortly thereafter. In Figure 4.4 the strata contain wave-formed ripple marks, the crests of which point in the direction of the youngest bed, or the top of the rock sequence. Several other sedimentary structures and fossil organisms in their original living positions may also be used to determine which bed is youngest in a sequence (see Chapter 5).

In addition to deformation, the association of sedimentary and igneous rocks may cause problems in relative dating. Buried lava flows and sills both look much the same in a sequence of strata (◆ Fig. 4.5). However, a buried lava flow is older than the rocks above it, while a sill is younger than the bed immediately above it. Similar problems arise where sedimentary rocks are associated with large, irregularly shaped igneous bodies such as batholiths (◆ Fig. 4.6).

To resolve relative age problems such as these, geologists look closely at the sedimentary rocks in contact with the igneous rocks to see if they show signs of baking by heat. Any sedimentary rock showing such effects is older than the igneous rock with which it is in contact. In Figure 4.5, for example, a sill produces a zone of baking

(a)

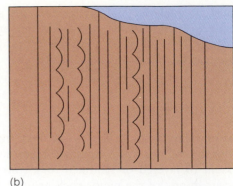
(b)

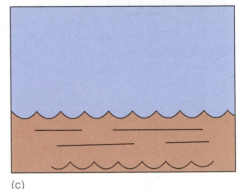
(c)

◆ **FIGURE 4.4** (a) Precambrian Mesnard Quartzite, Michigan. These beds were deposited horizontally, but after they were lithified, they were deformed, and now the layers are oriented vertically. (b) Diagramatic view of the internal structure of the Mesnard Quartzite beds. Note that the wave-formed ripple marks' crests point to the left, toward the youngest bed. (c) Orientation of the ripple-marked beds when they were deposited.

in the rocks at its upper and lower margins, while a lava flow bakes only those rocks below.

Another way to determine relative ages is by use of the **principle of inclusions.** This principle holds that inclusions, or fragments in a rock unit, are older than the rock unit itself. For example, the batholith shown in Figure 4.6a contains sandstone inclusions, and the sandstone unit shows the effects of baking. Accordingly, we conclude that the sandstone is older than the batholith. In Figure 4.6b, on the other hand, the sandstone contains

◆ **FIGURE 4.5** Relative ages of lava flows, sills, and associated sedimentary rocks may be difficult to determine. (a) A buried lava flow (4) baked the underlying bed, and bed 5 contains inclusions of the lava flow. The lava flow is younger than bed 3 and older than beds 5 and 6. (b) The rock units above and below the sill (3) have been baked, indicating that the sill is younger than beds 2 and 4, but its age relative to bed 5 cannot be determined.

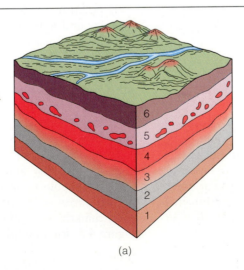
(a)

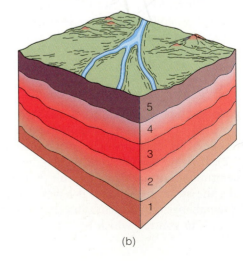
(b)

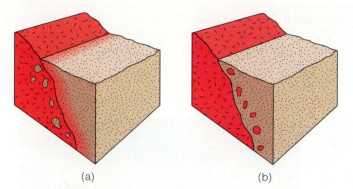

(a)   (b)

◆ **FIGURE 4.6** (a) The batholith is younger than the sandstone because the sandstone has been baked at its contact with the granite and the granite contains sandstone inclusions. (b) Granite inclusions in the sandstone indicate that the batholith was the source of the sandstone and therefore is older.

granite rock fragments, indicating that the batholith was a source rock for the sandstone and must be older than the sandstone.

Unconformities

Our discussion thus far has been concerned with **conformable** sequences of strata, sequences in which no depositional breaks of any consequence occur. A sharp bedding plane (Fig. 4.3) separating strata may represent a depositional break of minutes, hours, years, or even tens of years but is inconsequential when considered in the context of geologic time. Surfaces of discontinuity representing significant amounts of geologic time are **unconformities,** and any interval of geologic time not represented by strata in a particular area is a **hiatus** (◆ Fig. 4.7). Thus, an unconformity is a surface of nondeposition or erosion that separates younger strata from older rocks. As such, it represents a break in our record of geologic time.

Three major types of unconformities are recognized (◆ Fig. 4.8). A **disconformity** is a surface of erosion or nondeposition between younger and older beds that are parallel with one another. An **angular unconformity** is an erosional surface on tilted or folded strata over which younger strata have been deposited. Both younger and older strata may dip, but if their dip angles are different

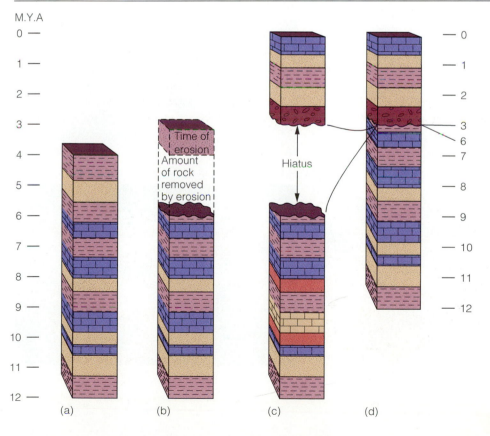

(a)   (b)   (c)   (d)

◆ **FIGURE 4.7** Simplified diagram showing the development of an unconformity and a hiatus. (a) Deposition began 12 million years ago (M.Y.A) and continued more or less uninterrupted until 4 M.Y.A. (b) A 1-million-year episode of erosion occurred, and during that time strata representing 2 million years of geologic time were eroded. (c) A hiatus of 3 million years exists between the older strata and the strata that formed during a renewed episode of deposition that began 3 M.Y.A. (d) The actual stratigraphic record. The unconformity is the surface separating the strata and represents a major break in our record of geologic time.

◆ **FIGURE 4.8** Unconformities. (a) Block diagram showing a disconformity. (b) Disconformity between Mississippian and Jurassic strata in Montana. The geologist at the upper left is sitting on Jurassic strata, and his right foot is resting upon Mississippian rocks. (c) Angular unconformity. (d) Angular unconformity in New Mexico. (Photo courtesy of R.V. Dietrich.) (e) Nonconformity. (f) Nonconformity between metamorphic rocks and the overlying Cambrian age sandstone in South Dakota.

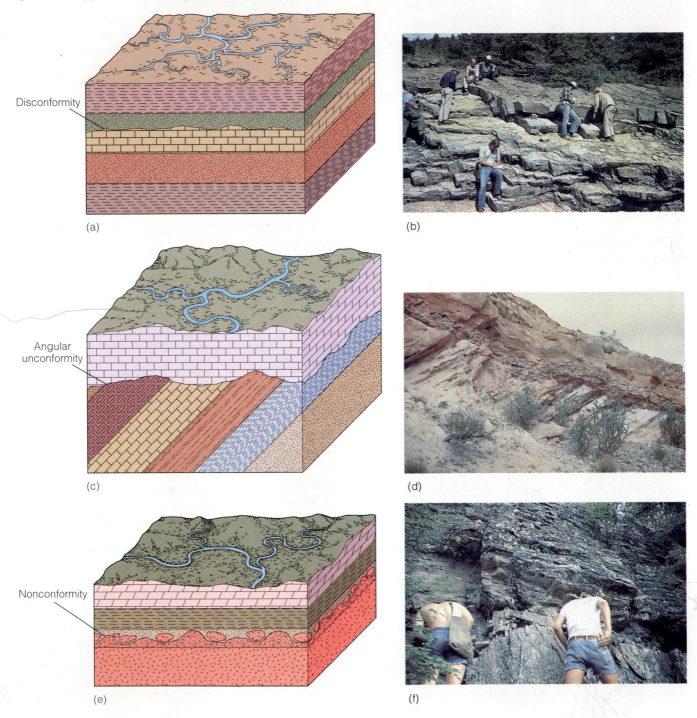

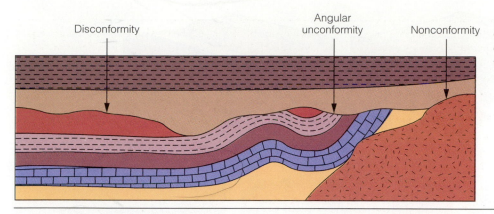

(generally the older strata dip more steeply), an angular unconformity is present. A **nonconformity** is a type of unconformity in which an erosion surface cut into metamorphic or intrusive rocks is covered by sedimentary rocks. Unconformities of regional extent may change laterally from one type to another and do not encompass equivalent amounts of time everywhere (◆ Fig. 4.9).

Lateral Relationships — Facies

In 1669 Nicolas Steno formulated the principle of lateral continuity (see Chapter 3). He recognized that sediment layers extend outward in all directions until they terminate. They may terminate abruptly where they abut the edge of a depositional basin, where eroded, or where truncated by faults (◆ Fig. 4.10). Lateral termination of sedimentary rocks may also occur when a rock unit becomes progressively thinner until it pinches out. *Intertonguing* occurs when a unit splits laterally into thinner units, each of which pinches out. And finally, a unit may change in composition or texture, or both, by lateral gradation until it is no longer recognizable (◆ Fig. 4.10e).

Intertonguing and lateral gradation result from different depositional processes operating in adjacent environments. For example, sand may be deposited in the high-energy nearshore part of the continental shelf, while mud and carbonate sediments accumulate simultaneously in the laterally adjacent low-energy, offshore area (◆ Fig. 4.11). Deposition in each of these environments produces a body of sediment, each of which is characterized by a distinctive set of physical, chemical, and biological attributes. Such distinctive bodies of sediment or sedimentary rock are **sedimentary facies.**

Any aspect of a sedimentary rock unit that makes it recognizably different from laterally adjacent rocks of the same age, or approximately the same age, can be used to establish a sedimentary facies. In Figure 4.11 we recognize three sedimentary facies, a sandstone facies, a shale facies, and a limestone facies. Armanz Gressly in 1838 was the first to use the term *facies*. Gressly carefully traced sedimentary rock units in the Jura Mountains of Switzerland and noticed lateral changes, from shale to limestone, for example. He concluded that such lateral changes resulted from deposition in different depositional environments that had existed simultaneously.

Transgressions and Regressions

Many rock units in the interiors of continents show clear evidence of having been deposited in marine depositional environments. Such deposits consist of vertical sequences of facies that resulted from deposition during times when sea level rose or fell with respect to continents. When sea level rises with respect to a continent, the shoreline moves inland, giving rise to a **marine transgression** (◆ Fig. 4.12). As the shoreline moves inland, the depositional environments located parallel to the shoreline do likewise. Remember that each laterally adjacent environment in Figure 4.12 is the depositional site of a different sedimentary facies. As a result of a marine transgression, the facies that formed in the offshore environments are superposed over the facies that was deposited in the nearshore environment.

Another important aspect of marine transgressions is that individual rock units become younger in a landward direction. The sandstone, shale, and limestone facies in Figure 4.12 were deposited continuously as the shoreline

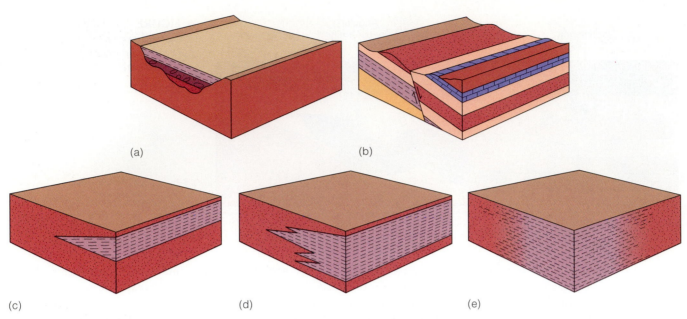

(a)

(b)

(c)

(d)

(e)

◆ **FIGURE 4.10** (a) Lateral termination of rock units at the edge of a depositional basin. (b) Faulting and erosion causing lateral termination. Notice that the beds above the sandstone do not appear on the left side of the fault. (c) Lateral termination by pinchout. (d) Intertonguing. (e) Lateral gradation.

moved landward, but none of the facies was deposited simultaneously over the entire area it now covers. In other words, the facies are **time transgressive,** meaning that their ages vary from place to place.

The opposite of a marine transgression is a **marine regression.** If sea level falls with respect to a continent, the shoreline and environments that parallel the shore-line move in a seaward direction (Fig. 4.12). The vertical sequence produced by a marine regression has facies of the nearshore environment superposed over facies of offshore environments. In addition, individual rock units become younger in a seaward direction.

The significance of lateral and vertical facies relationships was first recognized by Johannes Walther (1860–

◆ **FIGURE 4.11** Simultaneous deposition in different environments adjacent to one another. Each distinct depositional unit is recognized as a sedimentary facies.

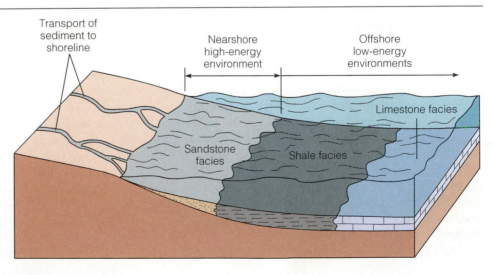

Transport of sediment to shoreline

Nearshore high-energy environment

Offshore low-energy environments

Limestone facies

Sandstone facies

Shale facies

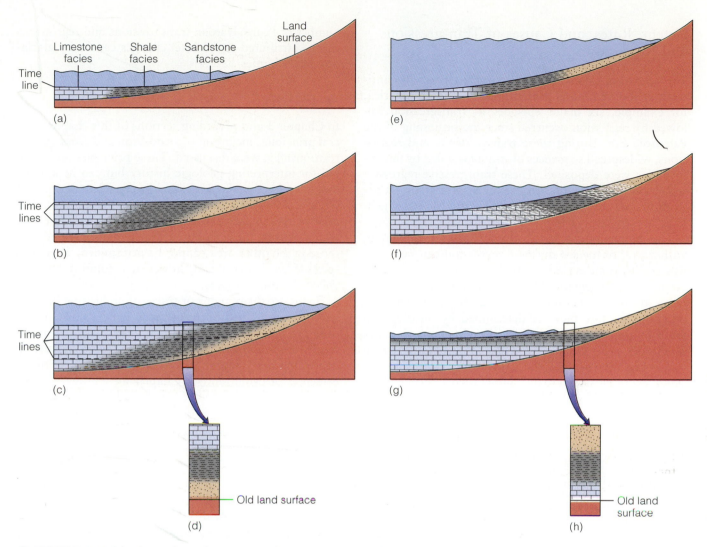

◆ **FIGURE 4.12** (a), (b), and (c) Three stages of a marine transgression. (d) Diagrammatic representation of the vertical sequence of facies resulting from the transgression. (e), (f), and (g) Three stages of a marine regression. (h) Diagrammatic representation of the vertical sequence of facies resulting from the regression.

1937). When Walther traced rock units laterally, he reasoned, as Gressly had, that each sedimentary facies he encountered had been deposited in laterally adjacent but different environments. Furthermore, he realized that the same facies that occurred laterally were also superposed upon one another in a vertical sequence. Walther's observations have since been formulated into **Walther's law,** which holds that the same facies following one another in a conformable vertical sequence will also replace one another laterally.

The value of Walther's law is well illustrated by its application to vertical facies relationships resulting from marine transgression or regression. In practice, one commonly cannot trace individual rock units far enough laterally to demonstrate facies changes. It is much easier to observe vertical facies relationships and then to work out the lateral relationships from them using Walther's law. Note, however, that Walther's law applies only to vertical successions of facies that contain no unconformities. Should an unconformity be present, the facies immediately above and below the surface of unconformity are unrelated, and Walther's law is not applicable.

Extent, Rates, and Causes of Transgressions and Regressions

The seas have clearly occupied large parts of all continents at several times during the past. In North America, for example, six major episodes of transgression followed by regression occurred since the beginning of the Paleozoic Era. During these transgressions and regressions, widespread sequences of strata bounded by unconformities were deposited. These transgressive-regressive sequences of strata will serve as the stratigraphic framework for our discussions of Paleozoic and Mesozoic geologic history (see Chapters, 11, 12, and 15).

The maximum extent of a transgression or maximum withdrawal of the sea during a regression can be established with some certainty, but rates and causes of such events are problematic. Shoreline movements probably occur at rates of centimeters per year, but these rates are deceptive because they are determined by dividing distance by time. For example, if a shoreline moved inland 1,000 km in 20 million years, the rate of transgression would average 5 cm per year. However, major transgressions and regressions are not simply events during which the shoreline moves steadily inland or seaward. As a matter of fact, both are characterized by numerous reversals in the overall transgressive or regressive trend (◆ Fig. 4.13).

Most geologists agree that uplift or subsidence of the continents, rates of seafloor spreading, and the amount of seawater contained in glaciers are sufficient to cause a transgression or regression. If a continent is uplifted, it rises with respect to sea level, the shoreline moves seaward, and a regression ensues. The opposite type of movement, subsidence, results in a transgression (◆ Fig. 4.14).

Seafloor spreading can cause transgressions and regressions by changing the volumes of the ocean basins. During relatively rapid episodes of extensive seafloor spreading, such as during the Cretaceous, the mid-oceanic ridges expand and displace seawater onto the continents. At other times, when seafloor spreading is comparatively slow, the ridges subside, the volume of the ocean basins increases, and the seas retreat from the continents (◆ Fig. 4.15). Worldwide changes in sea level related to additions or removal of water from the oceans may also cause marine transgressions and regressions. During a widespread glacial episode, large quantities of seawater are tied up on land as glacial ice, and, consequently, sea level falls. When glaciers melt, the water returns to the seas, and sea level rises (Fig. 4.14d). A

number of Pennsylvanian transgressions and regressions can be attributed to sea level changes caused by glaciation (see Chapter 12).

❖ FOSSILS

In Chapter 3 and preceding sections of this chapter, several principles, including superposition and cross-cutting relationships, were discussed. These principles are essential for interpreting geologic history but can be applied only on a local scale. For example, consider two vertical sequences of strata in different areas (◆ Fig. 4.16). The relative ages in each area are easily determined, but with the data given, the age of the rocks in one area relative to those in the other area cannot be determined. There is a solution to this problem, however, a solution that involves the use of fossils.

Fossils are the remains or traces of prehistoric life that have been preserved in rocks of the Earth's crust. In addition to their use in determining relative ages of strata, fossils are important in determining environments of deposition (see Chapter 5), and they provide some of the evidence for evolution (see Chapter 6).

◆ **FIGURE 4.13** Relative position of the shoreline during a transgressive-regressive event. Notice that while the overall trend may be one of transgression or regression, periodic reversals in the trends occur.

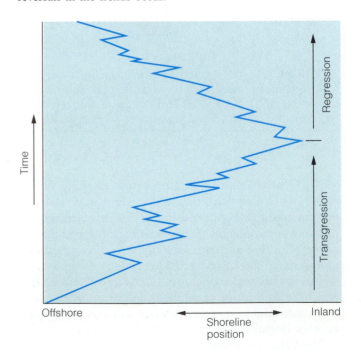

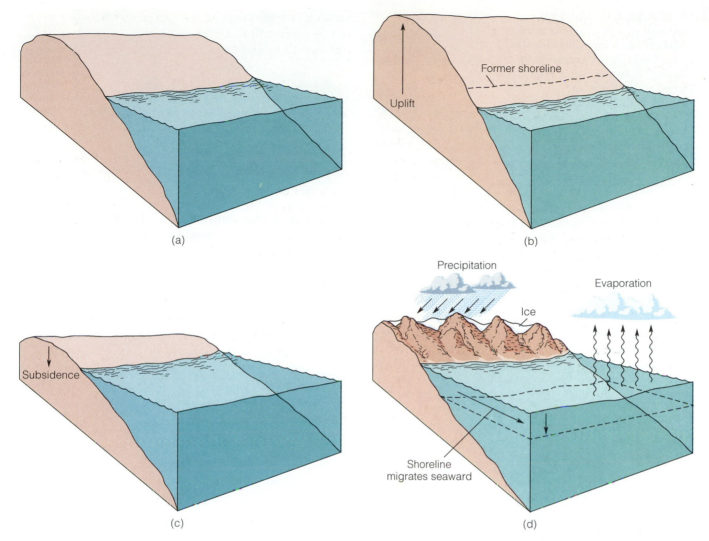

◆ FIGURE 4.14 Causes of transgressions and regressions. (a) Reference diagram showing position of shoreline. (b) Uplift of a continent causes the shoreline to migrate seaward resulting in a marine regression. (c) A marine transgression occurs when a continent subsides. (d) When a large volume of seawater is tied up on land as glacial ice, sea level is lowered, and a marine regression occurs. When glaciers melt, however, the water returns to the sea, sea level rises, and a transgression occurs.

During most of historic time, most people did not recognize that fossils were the remains of organisms (see the Prologue). Such perceptive observers as Leonardo da Vinci in 1508, Robert Hooke in 1665, and Nicolas Steno in 1667 recognized the true nature of fossils (◆ Fig. 4.17), but their views were largely ignored. By the late eighteenth century, however, the evidence had become overwhelming that fossils had indeed once been living organisms. By the

nineteenth century, it was also apparent that many fossils represented types of organisms now extinct.

Fossilization

Our definition of fossils includes the phrase "remains or traces." Remains, commonly called **body fossils** (◆ Fig. 4.18), are usually the hard skeletal elements such as

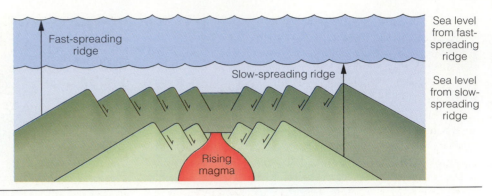

◆ **FIGURE 4.15** The cross-sectional area of a mid-ocean ridge depends on its spreading rate. The ridge forms due to expansion by heating. A fast-spreading ridge has a larger cross section and consequently displaces more seawater, resulting in a rise in sea level and marine transgressions onto the continents.

bones, teeth, and shells. In exceptional cases, preservation by freezing and mummification provides us with partial or nearly complete body fossils with flesh, hair, and internal organs. **Trace fossils** include tracks, trails, burrows, borings, nests, or any other indication of the

activities of organisms (◆ Fig. 4.19). Particularly interesting trace fossils are nests built by duckbilled dinosaurs and the so-called Devil's corkscrew (◆ Fig. 4.19c), which is the burrow of an extinct beaver. Fossilized feces, called **coprolites,** provide evidence of activity of organisms

◆ **FIGURE 4.16** The relative ages of strata in either section A or B can be determined by superposition; but with the data given, the ages of rocks in section A cannot be determined relative to those in section B. The rocks in section A may be older than any in B, the same age as some in B, or younger than the rocks in B.

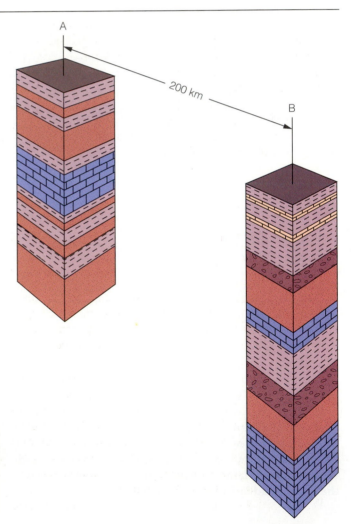

◆ **FIGURE 4.17** *Glossopterae* or tonguestones, as they were called during the Middle Ages, were thought to fall from the sky and kill people. Nicolas Steno determined that tonguestones were actually fossil shark teeth.

(◆ Fig. 4.20) and may provide important information on the diet and size of the animal that produced them.

Fossils in general are quite common. The remains of various microorganisms are by far the most abundant and in many ways the most useful. Shells of marine animals are also common and are easily collected in many areas (Figure 4.18a). Despite their abundance, however, fossils represent very few of the organisms that lived at any one time, since any potential fossil must escape the ravages of destructive processes such as waves or running water, scavengers, exposure to the atmosphere, and bacterial action. Hard skeletal elements are more resistant to destructive processes than soft parts, but even skeletal elements must be buried in some protective medium such as mud or sand if they are to be preserved; and even if buried, skeletal elements may be dissolved by groundwater or destroyed by alteration of the host rock.

Considering all the ways in which potential fossils can be destroyed, it comes as no surprise that the fossil record is biased toward those organisms with hard parts and those organisms that lived in areas of active sedimentation. Accordingly, most fossils are preserved in sediments or sedimentary rocks. Volcanic ash falls may also contain fossils (see Perspective 4.1), but fossils are rare to nonexistent in other types of igneous rocks and in metamorphic rocks. Despite its biases, the fossil record is our only record of prehistoric life. Indeed, it is only through fossils that we have any knowledge of such extinct life-forms as trilobites and dinosaurs.

Types of Preservation

Fossils may be preserved as **unaltered remains,** in which case they retain their original structure and composition. The pollen and spores of plants have a tough outer cuticle that is chemically resistant to change (◆ Fig. 4.21a). Partial or complete insects may be trapped in the resin secreted by some plants, especially by coniferous trees. When this resin is buried, it hardens to become amber and preserves in exquisite detail the remains contained therein (◆ Fig. 4.21b).

Unusual types of preservation of unaltered remains include preservation in tar, mummification, and freezing. The La Brea Tar Pits of southern California contain numerous tar-impregnated bones of Pleistocene animals that otherwise retain their original composition and structure (◆ Fig. 4.22). Mummification involves air drying and shriveling of soft parts such as muscles and tendons before burial (◆ Fig. 4.23a, p. 88).

Perhaps the most interesting example of preservation

(a)

(b)

◆ **FIGURE 4.18** Body fossils are the actual remains of organisms. (a) Corals in Devonian limestone from Kelly's Island in Lake Erie. (b) Miocene horse teeth from Montana. (Photos courtesy of Sue Monroe.)

of unaltered remains is by freezing. Only a few types of animals have been found frozen, including mammoths, woolly rhinoceroses, musk oxen, and a few others, and all are from Late Pleistocene age deposits. The frozen mammoths of Siberia are no doubt the best known example of this mode of preservation (◆ Fig. 4.23b, p. 88).

Altered remains are fossils that have been changed structurally, chemically, or both. Quite commonly, mineral matter is added to the pores and cavities of bones, teeth, and shells after burial. This process, called *permineralization,* increases the preservation potential of fossils by increasing their durability.

The shells of many marine invertebrates are composed of an unstable form of calcium carbonate known as ara-

(a)

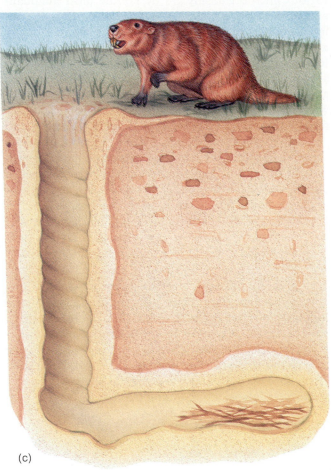

(c)

(b)

◆ **FIGURE 4.19** Trace fossils. (a) Crawling trace in the Lower Pennsylvanian Fentress Formation, Tennessee. (Photo courtesy of Molly Fritz Miller, Vanderbilt University.) (b) Dinosaur tracks in Cretaceous age rocks in the bed of the Paluxy River, Texas. (c) In Nebraska these spiral burrows, which may reach 2.5 m underground, are called "Devil's corkscrews." Restoration of *Palaeocastor,* the small Miocene beaver thought to have made the burrows.

gonite. When buried, these shells commonly recrystallize in the more stable form of calcium carbonate called calcite. A few plants and animals have shells or skeletal elements composed of opal, a variety of silicon dioxide

(SiO_2). These also commonly recrystallize. Although recrystallization does not change the chemical composition or outward appearance, the microscopic crystal structure of the shell is altered.

◆ **FIGURE 4.20** Fossilized feces (coprolite) of a carnivorous mammal. Specimen measures 5.5 cm long. (Photo courtesy of Sue Monroe.)

◆ **FIGURE 4.21** (a) Late Eocene pollen from the Eurla Basin, Australia. (b) Insects in amber. (Photo © J. Koivula/ Science Source, Photo Researchers, Inc.)

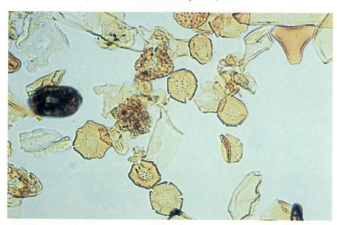

(a)

(b)

We know that clams, snails, and many other invertebrates have skeletons of calcium carbonate, yet in some strata we find these shells composed of silicon dioxide (SiO_2) or pyrite (FeS_2). These shells have been altered by *replacement,* during which the original skeletal material is replaced by a compound of a different composition. Insect skeletons replaced by silicon dioxide are a particularly interesting example of this type of preservation (◆ Fig. 4.24a, p. 90).

Wood may be preserved by the complete replacement of woody tissue by silicon dioxide in the form of chert or opal and, more rarely, by calcite. Wood preserved in this manner is often called *petrified,* a term that means to become stone. Particularly when silicon dioxide is involved, the replacement may be so exact that even details of cell structure are finely preserved (◆ Fig. 4.24b and c, p. 90). In many cases the form of the woody tissue actually remains unaltered, and the pore spaces have simply been filled in by silicon dioxide.

The altered remains of plants may be preserved by *carbonization.* During carbonization, the volatile elements of organic material, such as a leaf, vaporize, leav

◆ **FIGURE 4.22** Excavation of mammoth bones from tar pit 9 at Rancho La Brea in Los Angeles, California, in 1913. (Photo courtesy of Los Angeles County Museum of Natural History.)

A Miocene Catastrophe in Nebraska

Ten million years ago, in what is now northeastern Nebraska, a vast grassland was inhabited by short-legged, aquatic rhinoceroses, camels, horses, saber-toothed deer, land turtles, and many other animals. This was a temperate savanna habitat with life as varied and abundant as it is now on the savannas of East Africa. But many of these animals perished when a vast cloud of volcanic ash rolled in, probably from the southwest.

Michael Voorhies of the University of Nebraska State Museum and his crews have recovered the remains of hundreds of victims of this catastrophe (◆ Fig. 1). The magnitude of this event is difficult to imagine, since the source of the ash cloud may have been in New Mexico, more than 1,000 km away. We know from historic eruptions that an ash cloud can travel this far. For example, the April 11, 1815, eruption of Tambora in Indonesia was the deadliest volcanic eruption in recorded history. About 92,000 people perished, an ash layer 22 cm thick accumulated 400 km to the west, and some ash fell more than 1,500 km from the volcano.

The Nebraska ash fall was probably 10 to 20 cm thick, although it was redistributed by the wind and collected to 1 m deep in a depression where the fossils were recovered. This depression was the site of a water hole where animals congregated. Many animals, including three-toed horses, camels, deer, turtles, and birds, perished here when the ash fell. Their skeletons show signs of partial decomposition, scavenging, and trampling before they were completely buried.

Very soon after the initial ash fall, the depression was visited by herds of rhinoceroses and a few horses and camels. These, too, were suffocated by ash, but in this case the ash may simply have been blown into the depression by the wind. In any case, these animals were quickly buried, as indicated by the large number of complete skeletons.

One of the most remarkable things about these fossils is the preserved detail. According to Voorhies,

> Rarely found parts such as tongue bones, cartilages, tendons, and tiny bones in the middle ear all survive in exquisite detail and in their correct positions.[1]

---

[1]"Ancient Ashfall Creates a Pompeii of Prehistoric Animals," *National Geographic* 159, no. 1: (1981): 69.

◆ **FIGURE 4.23** (a) Close-up of the mummified remains of the dinosaur *Anatosaurus* showing detail of the surface pattern of the skin. (Photo courtesy of AMNH.) (b) Frozen baby mammoth found in Siberia in 1977. The baby was six or seven months old, 1.15 m long, 1.0 m tall, and when alive had a hairy coat. Most of the hair has fallen out except around the feet. The carcass is about 40,000 years old. (Photo © Sovfoto/V. Khristoforov.)

(a)

(b)

Additionally, the association of babies, juveniles, young adults, and mature adults gives us a better understanding of herd structure. The rhinoceros herds, for example, were probably made up of a single bull and several cows with their calves.

Only rarely are paleontologists fortunate enough to find and recover so many well-preserved, associated vertebrate animals. A Late Miocene catastrophe turns out to be our good fortune since it provides us with a unique glimpse of what life was like in Nebraska 10 million years ago.

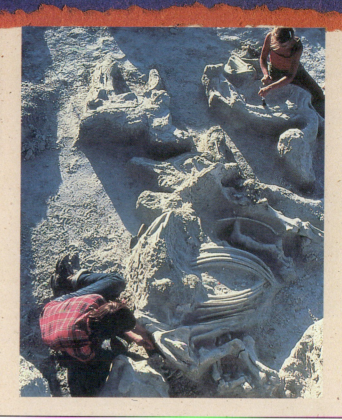

◆ **FIGURE 1** Paleontologists excavating rhinoceros (foreground) and horse (background) skeletons from volcanic ash near Orchard, Nebraska. (Photo courtesy of University of Nebraska State Museum.)

ing a thin carbon film (◆ Fig. 4.25). Fish and soft-bodied animals such as worms may also be carbonized.

In addition to unaltered and altered remains, fossils may be preserved as **molds** and **casts.** Molds and casts of shelled marine organisms such as clams and brachiopods are quite common (◆ Fig. 4.26). Molds are formed when the remains of a buried organism are dissolved, thereby leaving a cavity with the external shape of the organism (an external mold). If the mold is later filled with sediment or mineral matter, an external cast is formed. Internal molds are cavities showing the inner features of shells, and internal casts are formed when such cavities are filled with sediment or mineral matter.

Fossils and Uniformitarianism

Uniformitarianism in its simplest form can be expressed as "the present is the key to the past." While this is an oversimplification, it is true that observations of the present-day processes leading to the burial of organic remains give us a better understanding of how fossils were preserved.

Fossilization begins with the death of an organism and its burial in sediment or volcanic ash. Only rarely is the cause of death of a fossil organism apparent, but the conditions of burial commonly can be inferred by comparison with present-day processes. For example, Charles Darwin visited Uruguay during the 1830s and learned that millions of horses and cattle had died during a prolonged drought. When the drought broke, the rivers flooded and buried thousands of these animals in floodplain deposits. Similarly, many of the fossil mammals in the White River Badlands of South Dakota were buried in floodplain deposits.

Numerous additional examples exist of the burial of organic remains. Small animals are trapped and die in the sticky residue of southern California oil seeps, a single storm on the continental shelf may cause the burial of

(a)

(c)

(b)

 FIGURE 4.24 (a) Insects replaced by silicon dioxide in the Miocene Button Bed of the Barstow Formation, southern California. (b) Petrified wood of Miocene age in Ginkgo Petrified Forest, Washington. Replacement is by silicon dioxide. (c) Section of petrified wood showing preserved cell structure.

(a)

(b)

◆ **FIGURE 4.25** Altered remains preserved by carbonization. (a) Insect from the Oligocene Renova Formation, Montana. (b) Leaf from the Pennsylvanian Mazon Creek flora, Illinois. (Photo courtesy of Sue Monroe.)

(a)

(b)

◆ **FIGURE 4.26** (a) Mold. (b) Cast. (Photos courtesy of Sue Monroe.)

shallow marine invertebrates, and lake sediments contain leaves, insects, and occasionally fish. Following the 1980 eruption of Mount St. Helens in Washington, mudflows transported and buried logs and stumps and partially buried trees in growth position. The same or very similar processes occurred numerous times during the Eocene Epoch in what is now Yellowstone National Park, Wyoming (◆ Fig. 4.27). Events such as the burial of Pompeii by volcanic ash in A.D. 79 remind us that humans too are potential fossils.

It is important to note from these examples that the burial of organisms by natural processes is a common occurrence. The conditions of burial range from slow sedimentation in lakes to rapid burial by ash falls and mudflows, but all are consistent with the concept of uniformitarianism. We have every reason to believe that organisms in the fossil record were buried under comparable circumstances.

❖ FOSSILS AND TIME

The utility of fossils in relative dating and geologic mapping was clearly demonstrated in England and France in the early nineteenth century. William Smith (1769–1839), an English civil engineer involved in surveying and building canals in southern England, independently recognized the principle of superposition, reasoning that the lowest fossils in a sequence of strata are oldest and those higher in the sequence are younger. He made observations in many natural exposures, mines, and quarries and discovered that the relative sequence of fossils, and particularly groups or assemblages of fossils, is consistent from area to area. In short, he discovered a method whereby the relative ages of strata in widely separated areas regardless of their composition could be determined by their fossil content (◆ Fig. 4.28).

By recognizing the relationship between strata and fossils, Smith was able to predict the rock units that would be encountered in digging a canal. This knowledge also helped him determine the best route for a canal and the best areas for bridge foundations. His observations proved to be of economic significance, and his services were in great demand.

The discoveries of Smith, and those of other geologists such as Alexandre Brongniart in France, served as the basis for what is now called the **principle of fossil succession.** According to this principle, fossil assemblages succeed one another through time in a regular and determinable order. The validity and successful use of this principle depend on three factors: (1) life has varied

◆ **FIGURE 4.27** Trees buried in growth position. (a) This stump was partly buried by a 1980 mudflow in Smith Creek, Washington, following the eruption of Mount St. Helens. (b) Fossilized stump in the Eocene Lamar River Formation, Yellowstone National Park, Wyoming. (Photos courtesy of Amy Karowe, University of Oslo, Norway.)

(a)   (b)

through time, (2) fossil assemblages are recognizably different from one another, and (3) the relative ages of the fossil assemblages can be determined. Observations of fossils in older versus younger strata clearly demonstrate that life-forms have changed. And since this is true, fossil assemblages (point 2) are recognizably different. And finally, superposition can be used to demonstrate the relative ages of the fossil assemblages.

◆ THE RELATIVE TIME SCALE

Recall from Chapter 3 that many eighteenth-century geologists, such as the Neptunists, believed that the relative ages of rocks could be determined by their composition. This same thinking prevailed into the nineteenth century, but eventually it became apparent that composition is not a sufficient criterion for determining relative ages, and geologists in Great Britain and continental Europe began using stratigraphic position (superposition) and fossil content (fossil succession) for their subdivisions. Their efforts, largely between 1830 and 1842, resulted in the recognition of rock bodies called **systems** and the construction of a geologic column that is the basis for the relative time scale (◆ Fig. 4.29). A short discussion of how some of the systems were recognized and defined

will be helpful in understanding how the geologic column and relative time scale became established.

In the 1830s, Adam Sedgwick described and named the Cambrian System for rocks exposed in northern Wales, while Sir Roderick Impey Murchison, working in southern Wales, named the Silurian System. Unfortunately, Sedgwick's strata contain few fossils, and he paid little attention to the few present. His Cambrian System based on rock type could not be recognized beyond the area where it was described. Murchison, on the other hand, carefully described fossils characteristic of his Silurian System, so strata of Silurian age could be recognized elsewhere.

In 1835 Sedgwick and Murchison jointly published the results of their studies. Stratigraphic position clearly demonstrated that Silurian strata were younger than Cambrian strata, but it was also apparent that Sedgwick's and Murchison's systems partially overlapped (◆ Fig. 4.30). Each man claimed that the strata in the overlapping interval belonged to his system. This boundary dispute resulted in a feud that ended their long-standing friendship and was not resolved until 1879, when Charles Lapworth suggested that the disputed strata be assigned to a new system, the Ordovician. In the final analysis, three systems were named, their stratigraphic positions were known, and distinctive fossils of each system were recognized.

Before Sedgwick's and Murchison's feud, they named the Devonian System for rocks exposed near Devonshire, England (Fig. 4.29). Rocks defined as Devonian contained fossils that differed from those in underlying Silurian strata and were overlain by rocks of the Carboniferous System,[1] which had been described earlier. In 1840–41, Murchison visited western Russia, where he identified strata as Silurian, Devonian, and Carboniferous by their fossils. Overlying the Carboniferous strata were fossil-bearing rocks that he assigned to a Permian System (Fig. 4.29).

---

[1]In North America two systems, the Mississippian and the Pennsylvanian, are recognized and correspond to the Lower and Upper Carboniferous, respectively.

We need not discuss the specifics of where and when the other systems were defined, except to note that superposition and fossil content were the criteria used in their recognition. The important point is that geologists were piecing together a composite geologic column (Fig. 4.29). This geologic column is, in effect, a relative time scale because the systems are arranged in their correct chronological order. Thus, we can refer to the Devonian System when speaking of the stratigraphic position of rocks, and we can also use Devonian as a term to designate a particular interval of geologic time, the Devonian Period. We will have more to say about systems and periods in the following section.

One additional aspect of the geologic column should be mentioned. Notice in Figure 4.29 that all rocks beneath Cambrian strata are called Precambrian. Long ago geologists realized that these strata contain few fossils

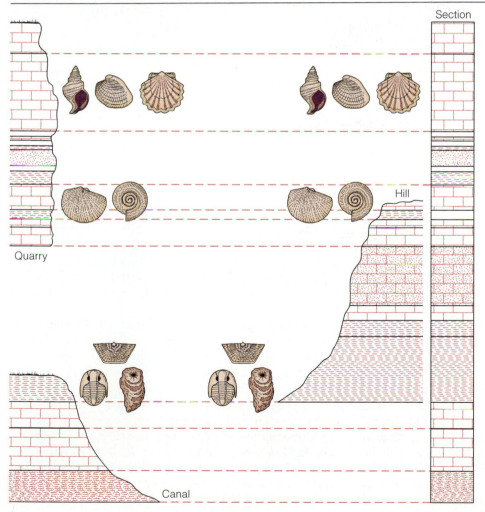

Section

Hill

Quarry

Canal

◆ FIGURE 4.28 This generalized diagram shows how William Smith used fossils to identify strata of the same age in different areas (principle of fossil succession). The composite section on the right shows the relative ages of all strata in this area.

(none were positively identified until the 1950s) and that they could not be effectively subdivided into systems. Terminology for Precambrian rocks and time is discussed in Chapters 9 and 10.

❖ STRATIGRAPHIC TERMINOLOGY

The recognition of systems and a relative time scale brought some order to stratigraphy. Problems remained, however, because many rock units are time transgressive, consequently, a rock unit may belong to one system in a particular area but also be included in another system elsewhere, or it may simply straddle the boundary be-

tween systems (◆ Fig. 4.31). In order to deal with both rocks and time, modern stratigraphic terminology includes two fundamentally different kinds of units: those defined by their content and those expressing or related to geologic time (Table 4.1).

Units defined by their content include **lithostratigraphic** and **biostratigraphic units.** Lithostratigraphic[2] units are defined by physical attributes of the rocks, such as rock type, with no consideration of time of origin. The basic lithostratigraphic unit is the **formation,** which is a

---

[2]*Lith-* and *litho-* are prefixes meaning stone or stonelike.

◆ **FIGURE 4.29** Development of the geologic column and relative time scale. (a) Geologists in Great Britain and continental Europe defined the geologic systems at the locations shown. (b) A geologic column was constructed by arranging the systems in their correct relative sequence, so the geologic column is in effect a relative time scale.

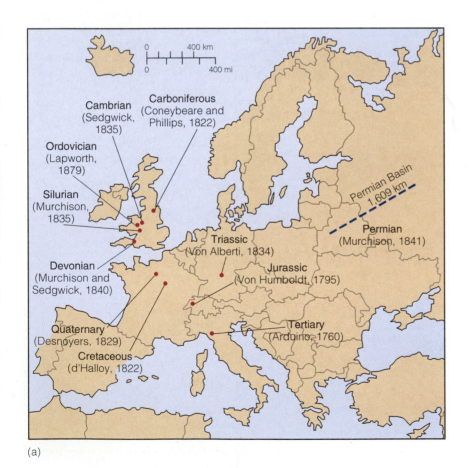

(a)

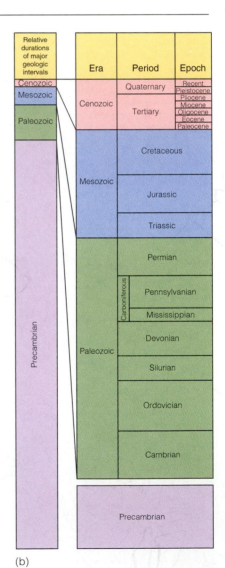

(b)

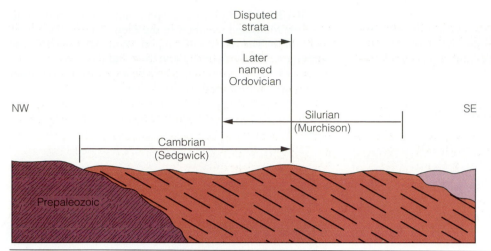

◆ **FIGURE 4.30** Simplified cross section showing the disputed interval of strata that Sedgwick and Murchison each claimed belonged to their respective systems. The dispute was resolved in 1879 when Charles Lapworth assigned the disputed strata to a new system, the Ordovician.

mappable rock unit with distinctive upper and lower boundaries. Formations may be lumped together into larger units called *groups* or *supergroups* or divided into smaller units called *members* and *beds* (◆ Fig. 4.32 and Table 4.1).

Fossil content is the only criterion used to define biostratigraphic units, which are bodies of strata containing recognizably distinct fossils. Biostratigraphic unit boundaries do not necessarily correspond to lithostratigraphic boundaries (◆ Fig. 4.33). The fundamental biostratigraphic unit is the **biozone.** Several types of biozones are recognized, three of which are discussed in the section on correlation.

The category of units expressing or related to geologic time includes **time-stratigraphic units** (also known as chronostratigraphic units) and **time units** (Table 4.1). Time-stratigraphic units are units of rock that were deposited during a specific interval of time. The **system** is

◆ **FIGURE 4.31** The Chattanooga Shale, shown here in Tennessee, straddles the boundary between the Devonian and Mississippian systems. (Photo courtesy of Molly Fritz Miller, Vanderbilt University.)

TABLE 4.1 Classification of Stratigraphic Units

UNITS DEFINED BY CONTENT		UNITS EXPRESSING OR RELATED TO GEOLOGIC TIME	
Lithostratigraphic Units	*Biostratigraphic Units*	*Time-Stratigraphic Units*	*Time Units*
Supergroup	Biozones	Eonothem . . . . Eon	
Group		Erathem . . . . Era	
Formation		System . . . . Period	
Member		Series . . . . Epoch	
Bed		Stage . . . . Age	

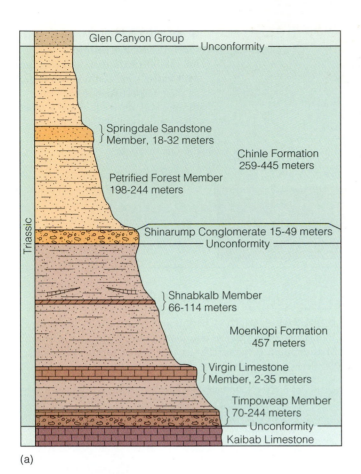

(a)

(b)

◆ **FIGURE 4.32** Lithostratigraphic units. (a) Illustration of the formations and members in the Zion National Park area of Utah. (b) The Madison Group in Montana consists of two formations, the Lodgepole Formation and the overlying Mission Canyon Formation. The Mission Canyon Formation is the rock unit exposed on the skyline. The underlying Lodgepole Formation is the rock covered on the slopes below. (c) Mesozoic formations exposed in Goblin Valley State Park, Utah.

(c)

Sommerville Formation
Chocolate-colored silty sandstone. Even bedding is distinctive.

Curtis Formation
Greenish-gray sandstone and siltstone.

Entrada Sandstone
Red siltstone and fine sandstone in alternating beds.

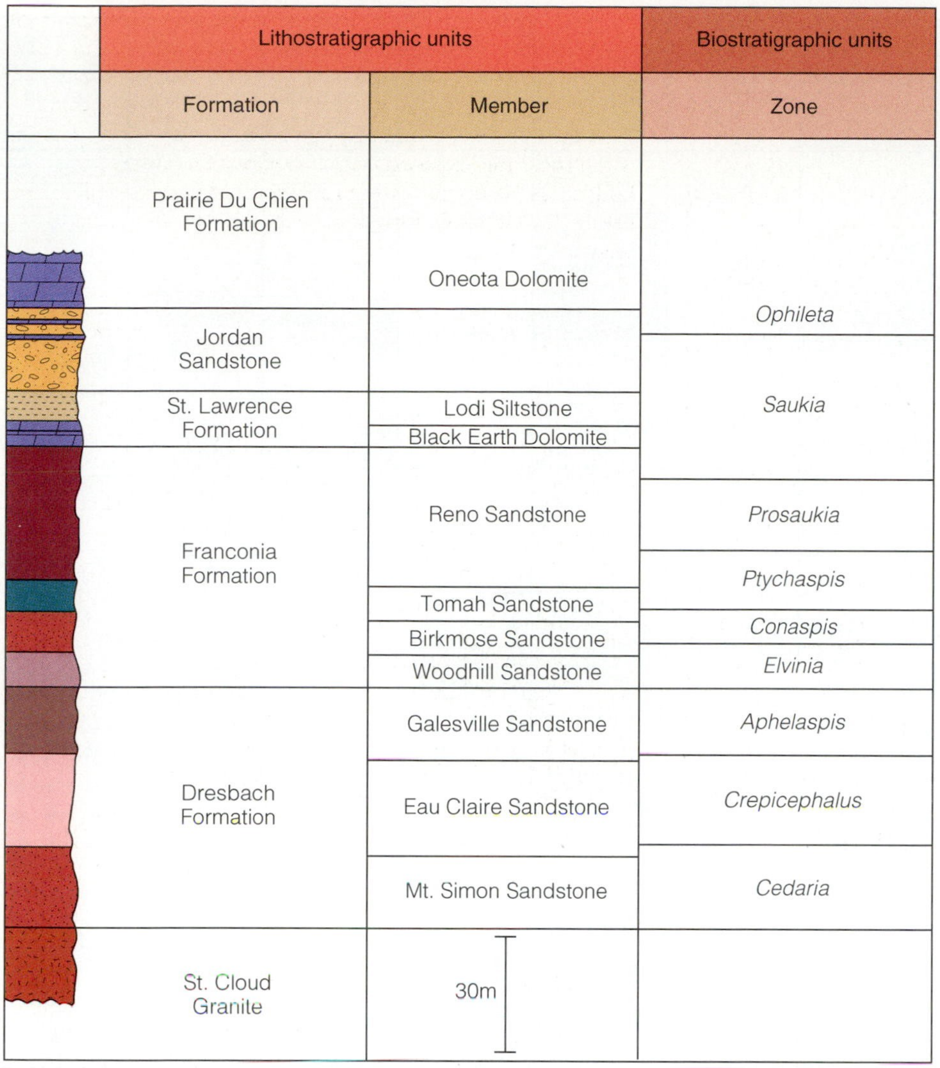

◆ FIGURE 4.33 Relationships of biostratigraphic units to lithostratigraphic units in southeastern Minnesota. Notice that the biozone boundaries do not necessarily correspond to lithostratigraphic boundaries.

the fundamental time-stratigraphic unit. It is based on rocks in a particular area, the stratotype, and is recognized beyond the stratotype area primarily on the basis of fossil content.

Time units are simply units designating specific intervals of geologic time. For example, the Cambrian Period is defined as the time during which strata of the Cambrian System were deposited. The basic time unit is the **period,** but smaller units including epoch and age are also recognized. The time units period, epoch, and age correspond to the time-stratigraphic units system, series, and stage, respectively (Table 4.1). Time units of higher rank than period also exist. Eras include several periods, while eons include two or more eras. The time–stratigraphic terms corresponding to era and eon are rarely used (Table 4.1).

Time-stratigraphic units and time units and their relationships are particularly confusing to students. Time-stratigraphic units are defined by their position in a sequence, so terms such as lower, medial, and upper may be applied (◆ Fig. 4.34). Thus, we may refer to Upper Devonian strata, meaning that these strata occur in the upper part of the Devonian System. In contrast, time units are simply subdivisions of geologic time and may be designated as early, middle, and late; the Late Devonian Period was the time during which rocks of the Upper Devonian System was deposited (Fig. 4.34).

❖ CORRELATION

If we could not demonstrate the equivalency of stratigraphic units in different areas, stratigraphic studies

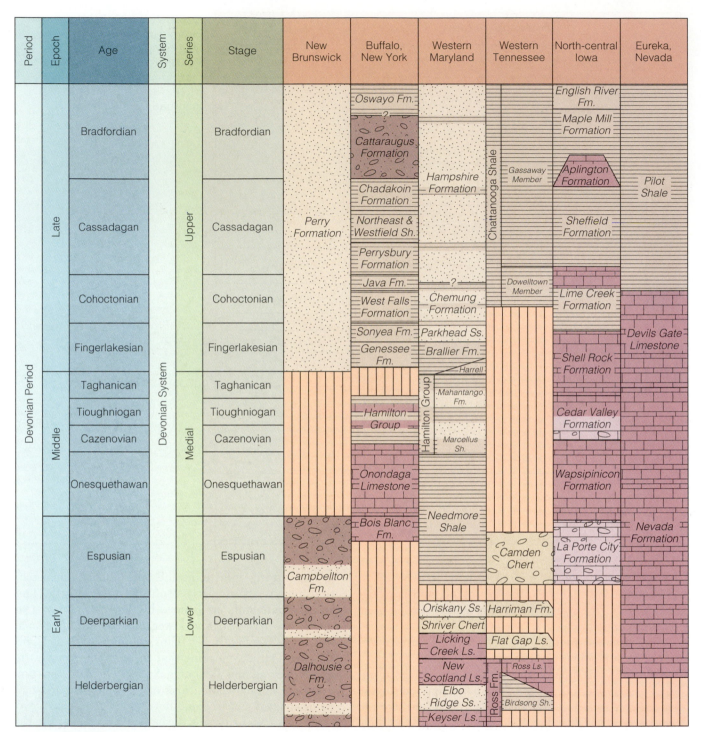

◆ **FIGURE 4.34** Correlation of Devonian strata of North America. Notice that the same-aged strata in different areas are not necessarily of the same rock types. Also notice the use of the time-stratigraphic terms lower, medial, and upper, and their corresponding time terms, early, middle, and late. The Sheffield Formation of Iowa, for example, occurs in the Upper Devonian System, and it was deposited during the Late Devonian Period.

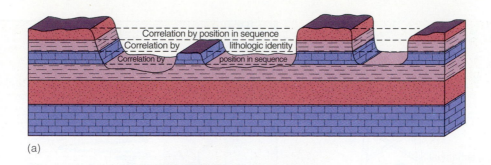

(a)

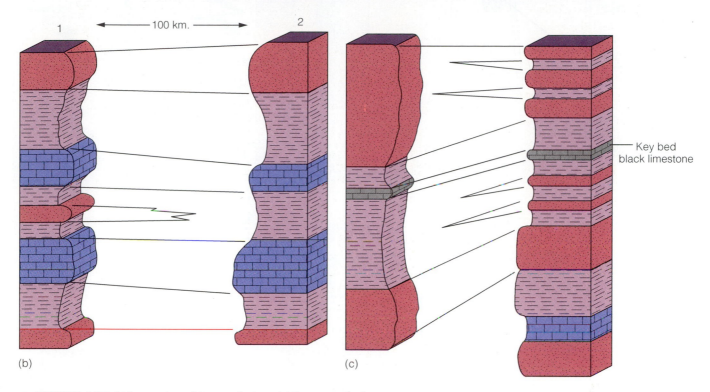

1   ←— 100 km. —→   2

Key bed
black limestone

(b)

(c)

◆ **FIGURE 4.35** Lithostratigraphic correlation. (a) In areas of adequate exposures, rock units can be traced laterally even if occasional gaps exist. (b) Correlation by similarities in rock type and position in a sequence. The sandstone in section 1 is assumed to intertongue or grade laterally into the shale at section 2. (c) Correlation using a key bed, a distinctive black limestone.

would be limited in their scope. **Correlation** is the process of demonstrating equivalency. Correlation of lithostratigraphic units is the recognition of units of similar lithology (composition and texture) and stratigraphic position over broad geographic areas. If exposures are adequate, units may simply be traced laterally, even if occasional gaps exist (◆ Fig. 4.35). Other criteria used to correlate are lithologic similarity, position in a sequence, and *key beds,* which are beds sufficiently distinctive to be identified in different areas (Fig. 4.35).

Drilling operations have provided a wealth of subsurface data. When drilling for oil or natural gas, cores or rock chips called *well cuttings,* are commonly recovered from the drill hole. In addition, geophysical instruments may be lowered down the drill hole to record such rock properties as electrical resistivity, density, and radioactivity, thus yielding a lithologic record, or *well log,* of the rocks penetrated. Cores, well cuttings, and well logs are all extremely useful in making subsurface lithostratigraphic correlations (◆ Fig. 4.36).

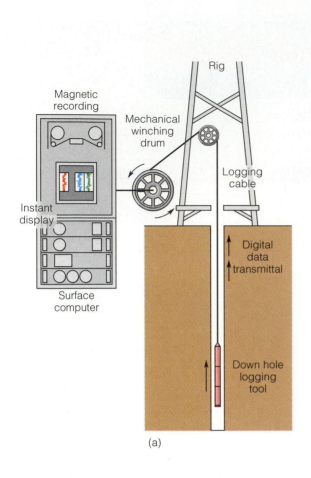

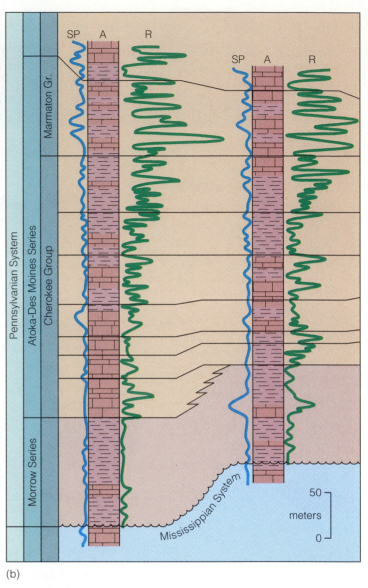

(a)

(b)

◆ **FIGURE 4.36** (a) Schematic diagram showing how well logs are made. A logging tool is lowered down the drill hole. As the tool is withdrawn from the drill hole, data are transmitted to the surface where they are recorded and printed out as a well log. (b) Electric logs and lithologic correlations of rocks in two wells in Colorado. The curves labeled SP are plots of self-potential (electrical potential caused by different conductors in a solution that conducts electricity) with depth, and the curves labeled R are plots of resistivity with depth.

Subsurface rock units may also be detected and traced by the study of seismic profiles. Energy pulses, such as those from explosions, travel through rocks at a velocity determined by rock density, and some of this energy is reflected from various horizons back to the surface, where it is recorded (◆ Fig. 4.37). Seismic stratigraphy is

particularly useful in areas such as the continental shelves where other stratigraphic techniques have limited use.

Lithostratigraphic correlation demonstrates the geographic extent of rock units, but it does not imply time equivalence. To demonstrate that strata in different areas

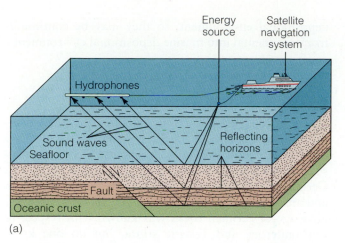

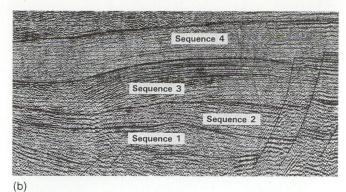

(a)

(b)

◆ **FIGURE 4.37** (a) A diagram showing the use of seismic reflections to detect buried rock units at sea. Sound waves are generated at the energy source. Some of the energy of these waves is reflected from various horizons back to the surface where it is detected by hydrophones. Buried rock units can also be detected on land, but here explosive charges are detonated as an energy source. (b) Seismic record and depositional sequences defined in the Beaufort Sea. (Photo courtesy of the American Association of Petroleum Geologists.) Boundaries of seismic sequences are shown by solid black lines. The scale on the right shows seismic wave travel time. Notice the sloping lines indicating faults on the right side of the seismic record.

◆ **FIGURE 4.38** Correlation between two sections based on the range zone of fossil A.

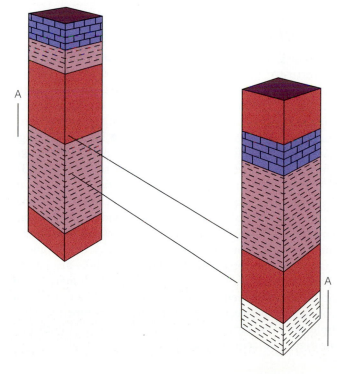

are the same age involves time-stratigraphic correlation. Since most rock units are time transgressive, we usually cannot use rock type in this kind of correlation. In most cases, biozones are used, although several other methods are useful as well.

One type of biozone, the **range zone,** is defined by the total geologic range of a particular fossil group (a species or a group of related species, called a *genus*, for example) (◆ Fig. 4.38). Fossils that are easily identified, are geographically widespread, and have rather short geologic ranges are particularly useful. The brachiopod genus *Lingula* is easily identified and widespread, but its geologic range of Ordovician to Recent makes it of little use in biostratigraphy. In contrast, the trilobite *Paradoxides* and the brachiopod *Atrypa* meet all of these criteria and are therefore good **guide fossils** (◆ Fig. 4.39). **Concurrent range zones** are established by plotting the overlapping ranges of fossils that have different geologic ranges. The first and last occurrences of fossils are used to establish zone boundaries (◆ Fig. 4.40).

Correlation of range zones or concurrent range zones generally yields correlation lines that are considered time equivalent. In other words, the strata encompassed by the correlation lines so established are thought to be the same age. Geologists are aware, however, that zones are

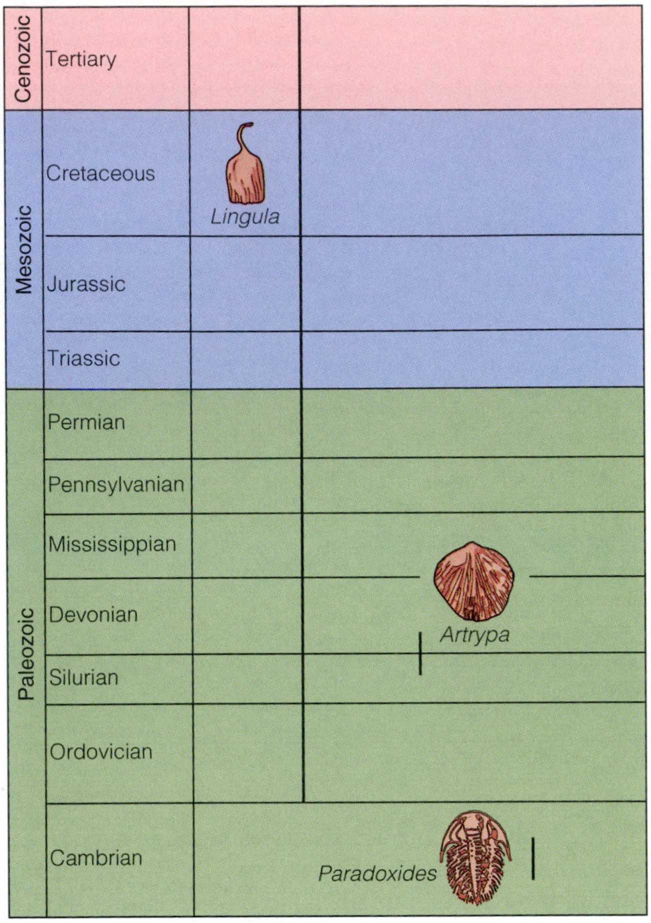

◆ **FIGURE 4.39** Comparison of the geologic ranges of three marine invertebrates. *Lingula* is of little use in biostratigraphy because of its long geologic range. *Atrypa* and *Paradoxides* are good guide fossils because they are widespread, are easily identified, and have short geologic ranges.

depositional environment, so they may be continuous from continental into marine depositional environments.

❖ QUANTIFYING THE RELATIVE TIME SCALE

Several of the correlation techniques discussed here demonstrate time equivalence between sections, but the ages determined are relative only. During the 1840s, Murchison used fossils to demonstrate that strata in Russia, Germany, and England belong to the Permian System and are therefore of the same age. Furthermore, the relative age of Permian rocks was known with respect to Carboniferous and Triassic strata, but no one knew when the Permian began or how long it lasted.

A number of minerals in sedimentary rocks can be dated radiometrically, but the ages tell only the age of the source rock that supplied the minerals to the deposit. All this tells us is that the sedimentary rock and its fossils are younger than the age determined. Glauconite, a mineral that forms in some sediments shortly after deposition, can be dated by the potassium-argon method (see Chapter 3). Unfortunately, glauconite easily loses argon, so absolute ages determined by this method must be considered minimum ages. In most cases, absolute ages of sedimentary rocks must be determined indirectly by dating associated igneous and metamorphic rocks.

Dating Associated Igneous and Metamorphic Rocks

According to the principle of cross-cutting relationships, an intrusive igneous body is younger than the rock it intrudes. Therefore, an accurately dated intrusive igneous body provides a minimum age for the sedimentary rock it intrudes and a maximum age for the sedimentary rocks unconformably overlying it (◆ Fig. 4.43). Age dates of regionally metamorphosed rocks give a maximum age for any overlying sedimentary rocks (Fig. 4.43).

One of the best ways to get good radiometric dates in a sedimentary sequence is by using interbedded volcanic rocks. An ash fall or lava flow provides an excellent marker bed that is a time-equivalent surface, providing a minimum age for the sedimentary rocks below and a maximum age for the rocks above. Ash falls are particularly useful because they may fall over both marine and nonmarine sedimentary environments and can provide a connection between these different environments. Multiple ash falls, lava flows, or a combination of both in a stratigraphic sequence are particularly useful in deter-

not of precisely the same age everywhere, because no fossil organism appeared and disappeared simultaneously over its entire geographic range. Even so, correlation of range zones can yield an error no greater than the range of the fossil used (◆ Fig. 4.41), and the use of concurrent range zones yields even more precise correlations.

Time equivalence can also be based on some physical events of short duration. Ash falls (◆ Fig. 4.42), for example, may cover large areas, and they occur instantaneously when considered in the context of geologic time. Furthermore, ash falls are not restricted to a particular

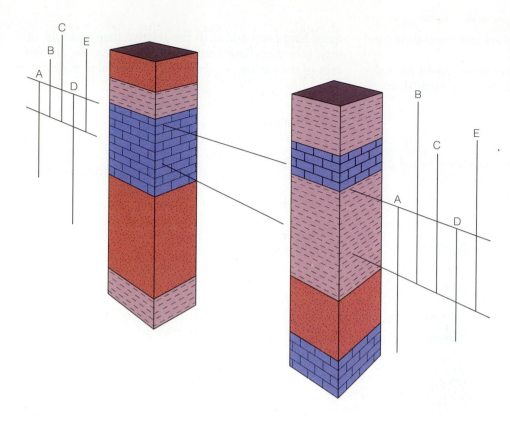

◆ **FIGURE 4.40** Correlation of two sections by using concurrent range zones, which are established by the overlapping ranges of fossils A through E.

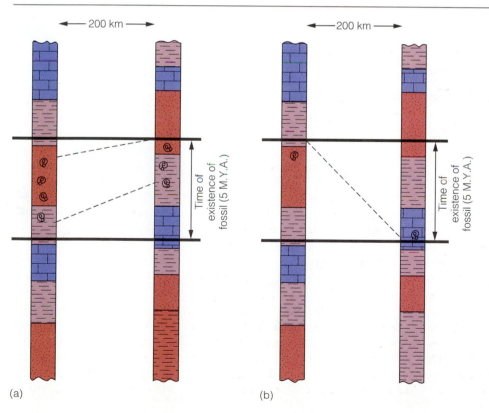

(a)

(b)

◆ **FIGURE 4.41** Simplified diagrams showing possible errors in time-stratigraphic correlations based on the range zone of a fossil that existed for five million years. (a) In these two stratigraphic sections, the fossil is found at the positions shown. The biostratigraphic correlation lines (dashed) obviously encompass strata only partly of the same age in two sections. (b) The maximum possible error would occur if the last and first occurrences of the fossil were assumed to represent time-equivalent points.

mining absolute ages of sedimentary rocks and their contained fossils.

Thousands of absolute ages are now known for sedimentary rocks of known relative ages, and these absolute dates have been added to the relative time scale. In this way, geologists have been able to determine the absolute ages of the various systems and to determine their durations.

◆ **FIGURE 4.42** Time-stratigraphic correlation by using an ash bed, the Upper Pliocene Bailey Ash in the Ventura Basin, California.

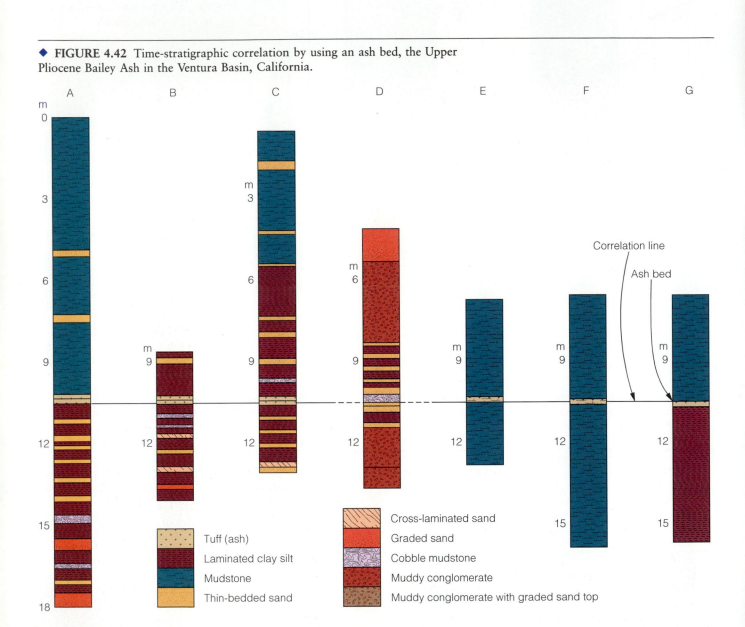

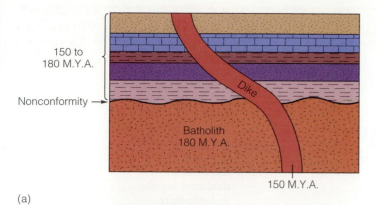

150 to
180 M.Y.A.

Nonconformity

Dike

Batholith
180 M.Y.A.

150 M.Y.A.

(a)

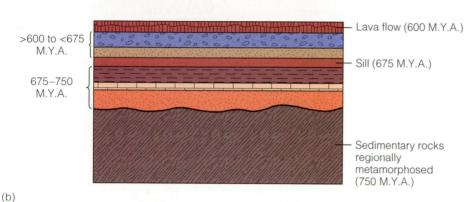

>600 to <675
M.Y.A.

675–750
M.Y.A.

Lava flow (600 M.Y.A.)

Sill (675 M.Y.A.)

Sedimentary rocks
regionally
metamorphosed
(750 M.Y.A.)

(b)

◆ **FIGURE 4.43** Absolute ages of sedimentary rocks determined by dating associated igneous and metamorphic rocks. In both (a) and (b), sedimentary rocks are bracketed by rock bodies for which absolute ages have been determined.

Chapter Summary

1. The relative ages of rock units in any sequence of strata can be determined by superposition even if the strata are complexly deformed.

2. Surfaces of discontinuity that encompassed significant amounts of geologic time are common in the geologic record. Such surfaces are unconformities and result from times of nondeposition or erosion.

3. Sedimentary facies are distinctive sedimentary rock units resulting from deposition in adjacent but different environments. The same facies that replace one another laterally may be superposed in a vertical sequence as a result of marine transgression or regression.

4. In a conformable sequence of facies, Walther's law can be applied to determine what the lateral facies relationships were at the time of deposition.

5. The causes of marine transgressions and regressions include uplift and subsidence of the continents, rates of seafloor spreading, and the amount of seawater contained in glaciers.

6. Most fossils are preserved in sedimentary rocks as unaltered remains, altered remains, molds and casts, or traces of organic activity.

7. Although fossils of some organisms are quite common,

very few of the organisms that lived at any one time were actually preserved as fossils. Furthermore, the fossil record is biased toward those organisms with hard skeletal elements and those organisms that lived in areas of active sedimentation.

8. Fossil assemblages succeed one another through time in a predictable sequence. William Smith's work with fossils in Great Britain is the basis for what is now called the principle of fossil succession.

9. The geologic column is a composite sequence of rock bodies called systems arranged in chronological order. Superposition and fossil succession were used to determine the relative ages of the systems.

10. Stratigraphic terminology includes two fundamentally different kinds of units: those based on content and those related to geologic time.

11. Correlation is the stratigraphic practice of demonstrating equivalency of units in different areas. Demonstrating time equivalence is most commonly done by correlating strata with similar fossils.

12. Most absolute ages of sedimentary rocks and their contained fossils are obtained indirectly by dating associated igneous or metamorphic rocks.

Important Terms

altered remains
angular unconformity
bedding plane
biostratigrahic unit
biozone
body fossil
cast
concurrent range zone
conformable
coprolite
correlation
disconformity

formation
fossil
geologic record
guide fossil
hiatus
lithostratigraphic unit
marine regression
marine transgression
mold
nonconformity
period
principle of fossil succession

principle of inclusions
range zone
sedimentary facies
stratigraphy
system
time-stratigraphic unit
time transgressive
time unit
trace fossil
unaltered remains
unconformity
Walther's law

Review Questions

1. The bounding surface separating one layer of sedimentary rock from another is a:
   a. _____ range zone; b. _____ bedding plane; c. _____ guide fossil; d. _____ mold; e. _____ lava flow.

2. Which of the following is younger than both the bed upon which it rests and the bed immediately overlying it?
   a. _____ batholith; b. _____ lava flow; c. _____ ash fall; d. _____ sill; e. _____ unconformity.

3. A hiatus is a(n):
   a. _____ type of biozone useful for correlation; b. _____ interval of geologic time not represented by strata; c. _____ type of sandstone containing granite fragments; d. _____ kind of fossil preservation; e. _____ geologic system.

4. The superposition of offshore facies over nearshore facies can be accounted for by a(n):
   a. _____ angular unconformity; b. _____ concurrent range zone; c. _____ marine transgression; d. _____ time-stratigraphic correlation; e. _____ none of these.

5. A sequence of strata containing no depositional breaks of any consequence is said to be:
   a. _____ conformable; b. _____ deformed; c. _____ time transgressive; d. _____ unaltered; e. _____ regressive.

6. Which of the following is a time unit?
   a. _____ Cambrian System; b. _____ Lower Cretaceous; c. _____ Eocene Series; d. _____ Pennsylvanian Period; e. _____ Upper Silurian.

7. A time-transgressive rock unit:
   a. _____ is unconformably overlain by limestone; b. _____ is not the same age throughout its extent; c. _____ is composed of alternating layers of shale and sandstone; d. _____ contains fossils useful for correlation; e. _____ is Paleozoic in age.

8. The basic lithostratigraphic unit is a:
   a. _____ formation; b. _____ biozone; c. _____ disconformity; d. _____ system; e. _____ period.

9. Which of the following is not a fossil?
   a. _____ dinosaur footprint; b. _____ Paleozoic clam shell; c. _____ Roman coin; d. _____ Pleistocene elephant tusk; e. _____ Mesozoic worm burrow.

10. An erosion surface cut into metamorphic rocks overlain by sedimentary rocks is a:
    a. _____ coprolite; b. _____ cast; c. _____ bedding plane; d. _____ time unit; e. _____ nonconformity.

11. If sedimentary rocks are cut by a dike 380 million years old, they must be _____ million years old.
    a. _____ about 400; b. _____ more than 380; c. _____ between 200 and 400; d. _____ younger than 10; e. _____ younger than 380.

12. Suppose that a clam shell is buried in sand, then dissolved leaving a cavity with the shape of a clam shell. This cavity is a:
    a. _____ formation; b. _____ coprolite; c. _____ cast; d. _____ system; e. _____ mold.

13. Which of the following holds that the same facies following one another in a conformable vertical sequence will also replace one another laterally?
    a. _____ Walther's law; b. _____ the principle of inclusions; c. _____ Steno's third law; d. _____ Smith's principle of fossil succession; e. _____ the principle of superposition.

14. Demonstrating that strata in different regions are the same age is:
    a. _____ time-stratigraphic correlation; b. _____ lithologic similarity; c. _____ seismic stratigraphy; d. _____ correlation based on rock type; e. _____ the principle of uniformitarianism.

15. Explain how a geologist can determine the relative ages of a granite batholith and an overlying sandstone formation.
16. Describe a sequence of events that could account for an angular unconformity.
17. A conformable vertical sequence of facies consists of sandstone followed upward by shale and limestone. All facies contain marine fossils. Diagram the lateral facies relationships, and show which facies was deposited nearest the shoreline.
18. Explain how geologists use the concept of uniformitarianism to understand the conditions leading to the preservation of fossils.
19. Compare the processes of permineralization, replacement, and recrystallization of fossils.
20. The principles of superposition and fossil succession were used to establish the geologic column. Define both principles, and explain how they were used to determine the relative ages of the systems.
21. What is a concurrent range zone? How can such zones be used to demonstrate time equivalency of strata in widely separated areas?
22. How do lithostratigraphic units differ from time-stratigraphic units?
23. In a vertical sequence of strata, the following are observed: lava flow (absolute age 255 million years) followed upward by a Permian sandstone, a Triassic shale, and an ash fall (absolute age 240 million years). What is the approximate absolute age of the Permian-Triassic boundary?
24. How does a disconformity differ from a nonconformity?
25. How are biostratigraphic units used to establish time-stratigraphic correlations?

Additional Readings

Boggs, S., Jr. 1987. *Principles of sedimentology and stratigraphy.* Columbus, Ohio: Merrill Publishing Co.

Brenner, R. L., and T. R. McHarque. 1988. *Integrative stratigraphy: Concepts and applications.* Englewood Cliffs, N.J.: Prentice-Hall.

Donovan, S. K., ed. 1991. *The processes of fossilization.* New York: Columbia University Press.

Friedman, G. M., J. E. Sanders, and D. C. Kopaska-Merkel. 1992. *Principles of sedimentary deposits.* New York: Macmillan.

Fritz, W. T., and N. J. Moore. 1988. *Basics of physical stratigraphy and sedimentology.* New York: John Wiley & Sons.

Harbaugh, J. W. 1974. *Stratigraphy and the geologic time scale.* Dubuque, Iowa: William C. Brown.

Moody, R. 1986. *Fossils.* New York: Macmillan.

Pinna, G. 1990. *Illustrated encyclopedia of fossils.* New York: Facts on File.

Simpson, G. G. 1983. *Fossils and the history of life.* New York: Scientific American Books.

CHAPTER 5

Sedimentary rocks in the Sheep Rock area of John Day Fossil Beds National Monument, Oregon. This small hill is capped by the remnants of a lava flow.

ORIGIN AND INTERPRETATION OF SEDIMENTARY ROCKS

Prologue

Today the Mediterranean Sea is in an arid region where the rate of evaporation exceeds the rate at which water is added to the sea by rainfall runoff. If it were not for the connection between the Mediterranean and the Atlantic Ocean at the Strait of Gibraltar (◆ Fig. 5.1), the Mediterranean would eventually dry up and become a vast desert basin lying far below sea level. Some geologists, particularly Kenneth Hsü of the Swiss Federal Institute of Technology, think this is precisely what happened to the Mediterranean during the Late Miocene.

Evaporite deposits up to 2 km thick have been discovered beneath the floor of the modern Mediterranean Sea. Sedimentological studies of these Late Miocene evaporites indicate that some were deposited in shallow-water environments rather than in a deep-ocean basin, as the Mediterranean is now. Observations such as this have led to the hypothesis that the Mediterranean Sea periodically lost its connection with the Atlantic Ocean during the Late Miocene. With its supply of water cut off, the Mediterranean Sea evaporated to near dryness in as little as 1,000 years and became a vast desert basin lying 3,000 m below sea level (◆ Fig. 5.2).

Apparently, periods of isolation of the Mediterranean Basin, with evaporation, alternated with periods when an oceanic connection was reestablished at the Strait of Gibraltar. During those times when an oceanic connection existed, gravels and silts were deposited around the basin margin, and deep-sea sedimentation occurred near the basin center. However, when the oceanic connection was lost, deposition occurred in a body of water that became progressively shallower.

Vast amounts of carbonates were deposited, followed by evaporites, especially gypsum and halite, as the waters became more and more saline.

A desert basin lying 3,000 m below sea level may seem preposterous, but additional evidence supports this view of Mediterranean history. The modern Mediterranean is the principal base level for rivers flowing from Africa and southern Europe, yet beneath these

---

◆ **FIGURE 5.1** View showing the western end of the Mediterranean Sea (upper right). The narrowly constricted area leading from the Mediterranean Sea to the Atlantic Ocean (lower right) is the Strait of Gibraltar. About 6 million years ago, the Mediterranean was probably a vast desert lying 3,000 m below sea level, where evaporite minerals were deposited. At the end of the Miocene, an oceanic connection was reestablished at the Strait of Gibraltar, and the basin rapidly refilled. (Photo courtesy of NASA.)

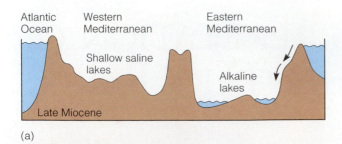

(a)

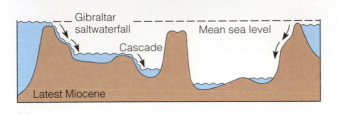

(b)

◆ **FIGURE 5.2** Proposed model to explain the deposition of evaporites in the Mediterranean basin. (a) Isolation of the Mediterranean and evaporation to near dryness. (b) Refilling stage. To account for the thick evaporite deposits present beneath the floor of the modern Mediterranean, this cycle of isolation, evaporation, and refilling must have occurred many times. The vertical scale is greatly exaggerated here.

rivers are buried river valleys deeply incised into bedrock far below present-day sea level. For example, a buried channel has been discovered by drilling more than 900 m into the Rhone delta in southern Europe. Several river-cut submarine canyons have been traced out into the Mediterranean Basin itself where they now lie more than 2,000 m below sea level.

At the Aswan High Dam, 465 km upstream on the Nile, Russian geologists discovered a buried river valley incised more than 200 m below present sea level. Apparently, this valley was cut when the water level in the Mediterranean was much lower; later it was filled with Pliocene marine sediments when sea level rose again and buried the canyon.

Although the model just described is widely accepted among geologists, it does not have unanimous support. For example, A. Debenedetti of the University of Torino in Torino, Italy, thinks that these thick evaporite deposits formed in a deep, water-filled basin (◆ Fig. 5.3). According to this model, the Mediterranean and Atlantic were connected continuously, so that the level of the former did not vary as in the other model. Inflow from the Atlantic replenished losses to

evaporation, but the Mediterranean eventually reached the saturation point and precipitation of evaporite minerals began.

❖ INTRODUCTION

Recall from Chapter 2 that sedimentary rocks are composed of solid particles called *detritus* and dissolved mineral matter derived from preexisting rocks. Such detritus and dissolved material may be transported and deposited as sediment; sand on a beach is detrital sediment, whereas chemically precipitated mineral matter is chemical sediment. Any sediment can be acted upon by the processes of lithification (compaction and cementation) and thus be converted into sedimentary rock.

In this chapter our main objective is to investigate the processes leading to the accumulation of sedimentary deposits. The sedimentary rocks we observe in the geologic record acquired their various properties, in part, as a result of the physical, chemical, and biological processes that operated in the original depositional environment. Thus, our task is to determine what the original environment was by examining the properties of the sedimentary rocks.

❖ FEATURES OF SEDIMENTARY ROCKS

The first step in any investigation of sedimentary rocks is observation and data gathering (Table 5.1). Accordingly,

◆ **FIGURE 5.3** Model for deposition of Mediterranean evaporites in a deep water-filled basin. Unlike the model depicted in Figure 5.2, this model holds that the Mediterranean was never isolated from the Atlantic Ocean.

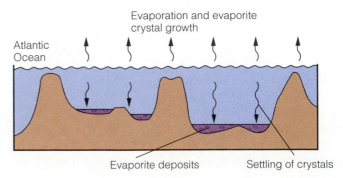

TABLE 5.1 Parameters of Environmental Significance

1. *Geometry:* Three-dimensional shape of rock bodies.
2. *Lithology:*
   a. Detrital—grain size as an index of energy level, texture is process controlled rather than environment controlled.
   b. Carbonates—sedimentary facies with characteristic grain types.
3. *Fossils:*
   a. Macrofossils as rock builders (reef, for example).
   b. Macro- and microfossils as environmental indices.
   c. Trace fossils—diagnostic assemblages in specific environments.
4. *Sedimentary structures:* Associations of sedimentary structures such as ripple marks and cross-beds useful in environmental interpretations.
5. *Paleocurrents:* Ancient current directions determined from sedimentary structures.

geologists visit the areas where sedimentary rocks are exposed, carefully examine the strata, describe and measure the thickness of the strata, trace units laterally, and collect samples for further analysis in the lab. While in the field, geologists commonly make some preliminary interpretations. For example, the color of sedimentary rocks is a useful environmental parameter: red-colored rocks may indicate deposition in a continental environment, whereas greenish rocks are more typical of marine environments. Exceptions are numerous, so color must be interpreted with caution. Geologists may also recognize types of sedimentary particles or fossils that indicate deposition in a particular environment.

After field studies have been completed, the data can be more fully analyzed in the lab. Such analyses may include microscopic and chemical examination of rock samples, identification of fossils, statistical determination of ancient current directions, and construction of diagrammatic representations of vertical and lateral facies relationships. And finally, when all available data have been analyzed, an environmental interpretation is made.

Lithology

The term **lithology** refers to the physical characteristics of rocks such as composition and texture. Sedimentary rocks may have a **clastic texture** meaning they are composed of clasts (the broken particles of preexisting rocks) (◆ Fig. 5.4a). All detrital sedimentary rocks and some chemical rocks are composed of clasts (see Table 2.5). Conglomerates and sandstones, for example, contain gravel- and sand-sized clasts, respectively, and clastic limestones may be composed of broken shell fragments. Many chemical sedimentary rocks are composed of an interlocking mosaic of mineral crystals and are said to have a **crystalline texture** (◆ Fig. 5.4b).

◆ **FIGURE 5.4** (a) Photomicrograph of a sandstone showing a clastic texture consisting of fragments of minerals, mostly quartz in this case. (b) Photomicrograph of the crystalline texture of a limestone showing a mosaic of calcite crystals.

(a)

(b)

◆ **FIGURE 5.5** Rounding and sorting of sedimentary particles. (a) A deposit consisting of well-sorted and well-rounded gravel. (b) Poorly sorted, angular gravel. (Photos courtesy of R. V. Dietrich.)

Other important textural features include grain size and sorting. **Grain size** is simply a measure of a sedimentary particle's size; gravel, sand, silt, and clay are size designations for sedimentary particles (Table 2.5). High-energy transport agents such as running water and waves are needed to transport large particles; thus, detrital gravel and sand tend to be deposited in stream channels or on beaches. Silt and clay, on the other hand, can be transported by weak currents and accumulate under low-energy conditions such as in lakes and lagoons. Clastic limestones are typical of beaches and offshore bars where wave energy is intense and may be composed of sand- and gravel-sized shell fragments.

During transport, the sharp corners and edges of sedimentary particles are abraded and smoothed. This process is particularly effective on sand- and gravel-sized particles, which become **rounded** as they collide with one another (◆ Fig. 5.5).

Sorting refers to the variation in grain sizes. Sedimentary rocks can be characterized as well sorted if the range of grain sizes is not great and as poorly sorted if the range is great (Fig. 5.5). Sorting results from processes that selectively transport and deposit particles by size. Wind-blown dunes are composed of well-sorted sand, because wind cannot effectively transport gravel, and it blows silt and clay beyond the areas of sand accumulation. Glaciers, however, are unselective, because their transport power allows them to move many sizes of sediment, and their deposits are typically poorly sorted.

◆ **FIGURE 5.6** This limestone is composed mostly of shells of brachiopods. (Photo courtesy of Sue Monroe.)

(a)

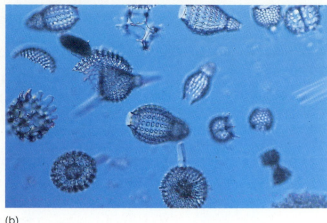
(b)

◆ **FIGURE 5.7** Skeletons of microorganisms such as (a) diatoms and (b) radiolarians are important constituents of some sedimentary rocks.

More than 100 different minerals have been identified in detrital sedimentary rocks, but only quartz, feldspars, and clays are very common. Unfortunately, the composition of detrital rocks reveals little about their transport or depositional history. Composition is, however, very important in determining the source area or areas that yielded the sedimentary particles.

Environmental inferences can be made from the composition of chemical sedimentary rocks. Limestones, for example, are composed of the mineral calcite ($CaCO_3$), a mineral that most commonly indicates a warm, shallow marine environment; some limestones, however, also form in lakes. Evaporites such as gypsum and rock salt indicate depositional environments in which evaporation rates were high—conditions that prevail in some lake and marine environments such as the Great Salt Lake, Utah, and the Persian Gulf area.

Fossils

Fossils or fossil fragments are among the most common sedimentary grains in limestone (◆ Fig. 5.6). Much of the sediment on the deep-sea floor is composed of microfossils, and the structural framework of reefs is made up of the skeletons of corals and other organisms. Diatomite, a rock composed of the shells of microscopic plants called *diatoms,* is mined and used for abrasives and as a filtering agent in gas purification (◆ Fig. 5.7).

Two factors must be considered when using fossils in environmental analyses. The first is whether the fossil organisms lived where they were buried or whether they were transported there. The second is what kind of habitat the organisms originally occupied. Studies of a fossil's structure and its living descendants, if there are any, are helpful in this endeavor. For example, clams with heavily constructed shells typically live in shallow, turbulent water. In contrast, organisms dwelling in low-energy environments commonly have thin, fragile shells (◆ Fig. 5.8). Also, organisms that carry on photosynthesis are restricted to the zone of sunlight penetration in the seas, which is generally less than 200 m. The amount of

◆ **FIGURE 5.8** The association of fossils shown here includes brachiopods, corals, and crinoids and indicates a shallow marine environment.

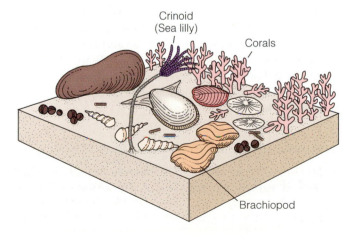

Crinoid (Sea lily)

Corals

Brachiopod

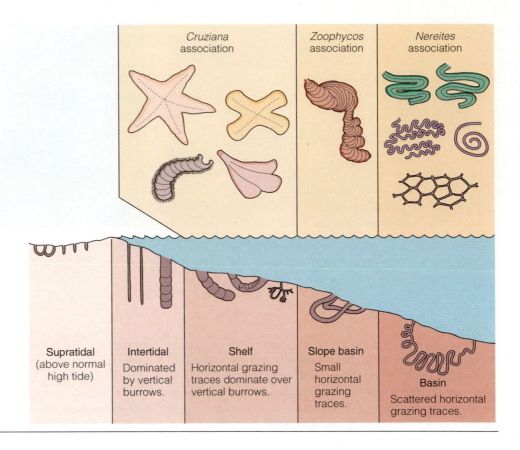

◆ **FIGURE 5.9** Common associations of trace fossils and their environmental significance, especially with regard to water depth.

suspended sediment is also a limiting factor on the distribution of organisms. Many corals, for example, live in shallow, clear seawater, because suspended sediment clogs their respiratory and food-gathering organs, and some have photosynthesizing algae living in their tissues.

Microfossils are particularly useful for environmental studies because the remains of numerous individual organisms can be recovered from small rock samples. In oil-drilling operations, for example, small rock chips called *well cuttings* are brought to the surface. Such samples rarely contain entire macrofossils but may contain numerous microfossils that aid in age and environmental interpretations. Trace fossils (see Fig. 4.19) are not transported from the place where they formed, and certain traces are known to be characteristic of particular environments (◆ Fig. 5.9).

Sedimentary Structures and Paleocurrents

Sedimentary rocks commonly contain structures that formed in sediment at the time of deposition or shortly thereafter. Such **sedimentary structures** are the manifestations of physical and biological processes that operated in the depositional environments and thus provide information on what the depositional processes were. The processes whereby many sedimentary structures form are well known because they can be observed forming in present-day depositional environments, and many can be formed experimentally.

Sedimentary rocks have a layered appearance called **bedding** or **stratification.** Individual layers are referred to as *laminae* if they are less than 1 cm thick and *beds* if they are thicker. Some beds show a decrease in grain size from bottom to top, a feature known as **graded bedding** (◆ Fig. 5.10). Graded bedding is common in turbidity current deposits. *Turbidity currents* are underwater flows of sediment-water mixtures that have greater densities than sediment-free water. They move downslope to the bottom of the sea or lake, where their velocity diminishes and the heaviest particles settle out first, followed by progressively smaller particles (Fig. 5.10). Turbidity currents also commonly produce scour marks in the muddy sediments over which they flow. When the turbidity current loses velocity, the scour marks fill with sand, producing *flute casts* in the sand layer at the base of the turbidity current deposit (◆ Fig. 5.11).

Cross-bedding is formed when layers are deposited at an angle to the surface upon which they are accumulating. Cross-beds invariably result from transport and deposition by wind or water currents and dip in the direction of flow (◆ Fig. 5.12). Since their orientation depends on the direction of flow, cross-beds are good indicators of ancient current directions, or **paleocurrents.**

Bedding planes may be marked by such sedimentary structures as **ripple marks** and **desiccation cracks** (◆ Fig. 5.13). Current ripples form in response to water or wind currents that move in one direction. The asymmetrical profiles of current ripples (Fig. 5.13a) allow geologists to determine paleocurrent directions. Wave-formed ripples result from the to-and-fro motion of waves and tend to be symmetrical in profile (Fig. 5.13b). Crests

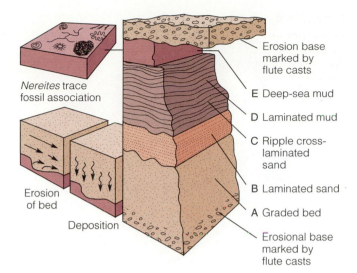

◆ **FIGURE 5.11** Associations of sedimentary structures are useful for environmental interpretations, especially when considered along with other rock properties such as textures and fossils. This example shows an idealized vertical sequence deposited by a turbidity current. The flute casts are produced by erosion, but as a turbidity current slows down, units A through D are deposited in succession. The deep-sea mud (unit E) forms as particles settle from seawater. It commonly contains the *Nereites* trace fossil association.

◆ **FIGURE 5.10** Deposition of a graded bed by a turbidity current. (a) A turbidity current generated on the continental slope moves downslope to the continental rise and seafloor. (b) Origin of a graded bed.

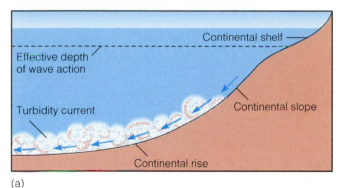

(a)

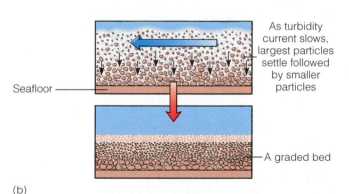

(b)

of wave-formed ripples are oriented more or less parallel to shorelines. When clay-rich sediments dry, they shrink and crack into polygonal forms bounded by desiccation cracks (Fig. 5.13c). These desiccation cracks indicate alternating periods of wetting and drying such as occur along lake margins or on river floodplains.

Biogenic sedimentary structures are produced by organisms and include tracks, trails, and burrows; such organically produced marks are also called *trace fossils* (see Chapter 4). As we pointed out in Figure 5.9, trace fossils are important in environmental analyses. Extensive burrowing by organisms, a process called **bioturbation,** may alter the physical and chemical properties of sediments and modify or destroy other sedimentary structures (Fig. 5.13d).

It is important to realize that no single sedimentary structure is unique to a particular depositional environment. Current ripples, for example, are common in stream channels but may also be found in tidal channels, on the seafloor, or near lake margins. Associations of sedimentary structures are particularly useful in environmental analyses, especially when considered with other

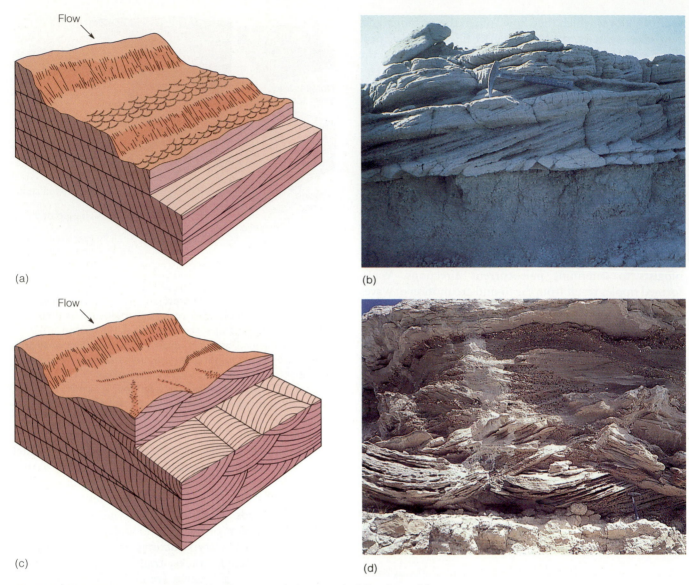

(a)

(b)

(c)

(d)

◆ **FIGURE 5.12** Two types of cross-bedding. (a) Tabular cross-bedding formed by migration of sand waves. (b) Tabular cross-bedding in the Upper Cretaceous Two Medicine Formation, Montana. (c) Trough cross-bedding formed by migrating dunes. (d) Trough cross-bedding in the Pliocene Six-Mile Creek Formation, Montana. This view shows the cross-beds illustrated on the right side of diagram (c).

aspects of a rock unit such as composition, texture, and fossil content (Fig. 5.11).

Geometry

The three-dimensional shape, or the geometry, of a rock body may be helpful in environmental analyses, but must be interpreted with caution. The same geometry may be produced in different sedimentary environments. Moreover, rock-body geometry may be modified by sediment compaction, erosion, and deformation. Nevertheless, geometry can be useful when considered in conjunction with other properties of a rock body.

(a)

(b)

(c)

(d)

◆ **FIGURE 5.13** Sedimentary structures from present-day depositional environments. (a) Current ripples in a stream channel. (b) Wave-formed ripple marks. (c) Desiccation cracks. (d) Bioturbation in siltstone caused by burrowing organisms. (Photo courtesy of Sue Monroe.)

Some of the most extensive sedimentary rock units in the stratigraphic record are those deposited during marine transgressions and regressions (Fig. 4.12). Such rock units may cover tens or hundreds of thousands of square kilometers, but they are not very thick compared to their other dimensions of length and width. Rock units with these dimensions are said to have a *blanket,* or *sheet, geometry* (◆ Fig. 5.14).

Some sand bodies have an *elongate,* or *shoestring, geometry* (Fig. 5.14) especially those deposited in stream channels or barrier islands. Delta deposits tend to be lens-shaped in cross profile or long profile but lobate when observed from above. Buried reefs are irregular, but many are elongated.

❖ DEPOSITIONAL ENVIRONMENTS

Any area in which sediment accumulates is called a **depositional environment.** No completely satisfactory classification of depositional environments exists; however, geologists generally recognize three major depositional settings: continental, transitional, and marine, each of which contains several specific environments (◆ Fig. 5.15). As a starting point in environmental interpretations, geologists rely upon the concept of uniformitarianism. We assume, for example, that current ripple marks and cross-beds formed in the past just as they do in present-day depositional environments. Thus, the study of depositional processes and environments is fundamental to our understanding of ancient processes and environments.

Continental Environments

Major depositional environments on the continents include stream (fluvial) systems, deserts, and glacial environments (Table 5.2). **Fluvial** is a term referring to stream activity and to the deposits that accumulate in stream systems. Two types of stream systems are recognized: braided and meandering (♦ Fig. 5.16). Both do most of their geologic work of erosion, sediment transport, and deposition when they flood. Consequently, their deposits do not record continuous day-to-day activities, but rather the periodic, large-scale events of sedimentation associated with flooding. The deposits of braided and meandering streams differ considerably, however, as shown in the idealized facies models in Figure 5.16b and d. Deposits of braided streams consist mostly of horizontally bedded gravel and cross-bedded sand, while mud is conspicuous by its near-absence (♦ Fig. 5.17a). In contrast, meandering stream deposits are dominated by mud with subordinate sand bodies with elongate, or shoestring, geometries (Fig. 5.16c and d).

One of the most distinctive aspects of meandering stream sand bodies is the **point bar** sequence. When meandering streams flood, they erode the steep, outer banks of meanders and simultaneously deposit sediment on the

TABLE 5.2 Depositional Environments Discussed in the Text

CONTINENTAL ENVIRONMENTS	Fluvial	Braided stream
		Meandering stream
	Desert	Sand dunes
		Alluvial fans
		Playa lakes
	Glacial	Ice deposition (moraines)
		Fluvial deposition (outwash)
		Glacial lakes
TRANSITIONAL ENVIRONMENTS	Delta	Stream-dominated
		Wave-dominated
		Tide-dominated
	Beach	
	Barrier island	Beach
		Sand dune
		Lagoon
MARINE ENVIRONMENTS	Continental shelf	Detrital deposition
		Carbonate deposition
	Carbonate platform	
	Continental slope and rise	Submarine fans (turbidities)
	Deep-ocean basin	Oozes
		Pelagic clay
	Evaporite environments*	

*Evaporites may be deposited in a variety of environments including playa lakes, saline lakes, lagoons, and tidal flats in arid regions, and in marine environments.

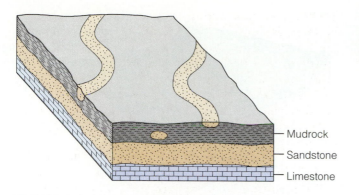

◆ **FIGURE 5.14** Two different geometries of sedimentary rock bodies. The limestone, sandstone, and mudrock all have blanket geometries. Within the mudrock, however, elongate sandstones have a shoestring geometry.

Mudrock

Sandstone

Limestone

◆ **FIGURE 5.15** Major depositional environments. The environments located along the marine shoreline are transitional from continental to marine. The shallow marine environment corresponds to the continental shelf and can be the site of either detrital or carbonate deposition.

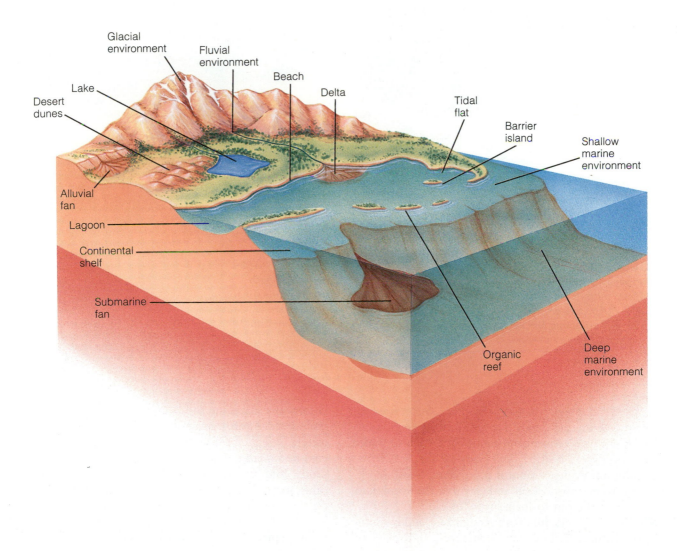

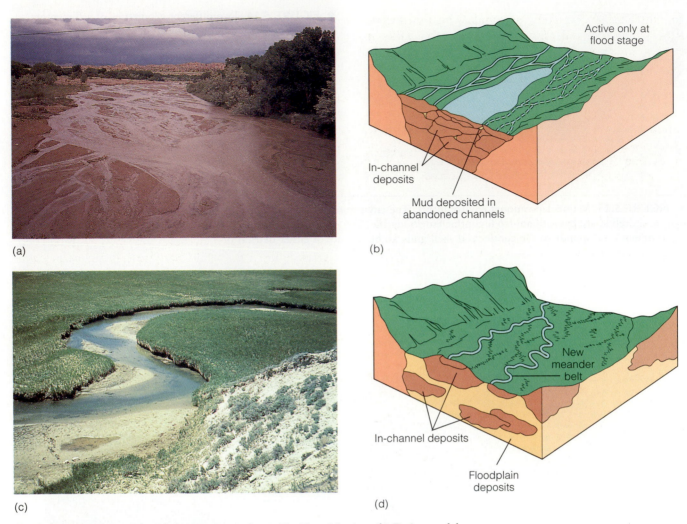

(a)

(b)

(c)

(d)

◆ **FIGURE 5.16** (a) A braided stream near Santa Fe, New Mexico. (b) Facies model for a braided stream system. Braided streams deposit mostly sand and gravel with subordinate mud. (c) A meandering stream, Otter Creek in Yellowstone National Park, Wyoming. Point bars are well developed on the inside banks of each meander. (d) Facies model for a meandering stream system. Meandering stream deposits are shoestring sands surrounded by mudstones.

inner, gently sloping banks as point bars. Successive episodes of erosion and deposition result in an eroded surface overlain by a sequence of point bars, each of which consists of a cross-bedded sand body (◆ Fig. 5.17b and c). In addition to deposition of point bars, flooding meandering rivers transport silt and clay into the floodplain, where they settle out to form layers of mud.

Windblown desert dunes are recognized largely on the basis of textures, sedimentary structures, and facies re-

lationships. All dune sands are typically well sorted and cross-bedded, on a scale measured in meters or tens of meters (◆ Fig. 5.18a). Large-scale cross-bedding also occurs in dunelike sands of the continental shelf, but such sands contain marine fossils and are associated with other marine rocks. Desert dunes, in contrast may contain the remains or traces of land animals.

The association of windblown dunes with other deposits typical of deserts is especially helpful. For example, **alluvial fans** commonly accumulate on the margins of desert basins, and *playa lakes,* temporary lakes in which mud and evaporites accumulate, may form on the desert floor (◆ Fig. 5.18b).

Many small glaciers currently exist in high mountains, but the only extensive ice sheets of continental proportions are in Greenland and Antarctica. Widespread glaciation occurred at several times during the past, however—during the Late Precambrian (Chapter 10), the Ordovician (Chapter 11) and Pennsylvanian (Chapter 12), and most recently during the Pleistocene (Chapter 18).

Glaciers are very effective agents of erosion, transport, and deposition. Many of the surficial deposits in the northern tier of states and in parts of Canada were deposited by Pleistocene glaciers or in associated subenvironments (◆ Fig. 5.19). All sediments deposited in gla-

cial environments are collectively called **drift,** but we must distinguish between two types of drift. **Till** is unsorted, unstratified drift deposited directly by glacial ice, mostly as ridgelike deposits called *moraines* (Fig. 5.19). A second type of drift, called **outwash,** is deposited by fluvial processes, mostly in braided streams, that issue from melting glaciers (Fig. 5.19). Such streams are heavily loaded with sediment, much of which is deposited as sheets of sand and gravel. Since the terminus of a glacier may advance or retreat, it is not uncommon for till and outwash to be interbedded.

Deposits of glacial lakes deserve special mention. Such lakes typically accumulate finely laminated muds consisting of alternating light and dark laminae; each light-dark couplet is a **varve** (Fig. 5.19d). Each varve represents an annual deposit: the light layers form in the spring and summer and consist of silt and clay; dark layers form in

(a)

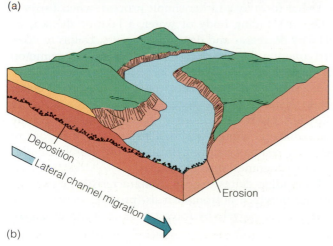

(b)

◆ **FIGURE 5.17** (a) Gravel bar in the braided Smith River of northern California. (b) Origin of a point bar in a meandering river. (c) Idealized point bar sequence overlain by floodplain deposits.

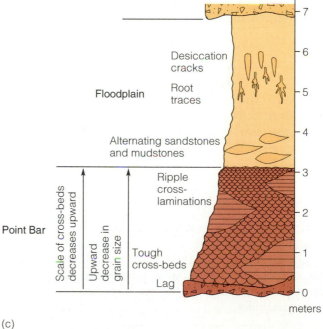

(c)

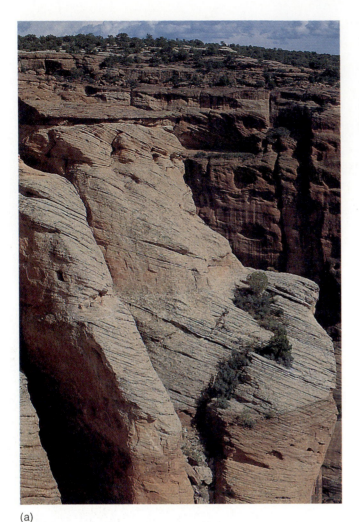

(a)

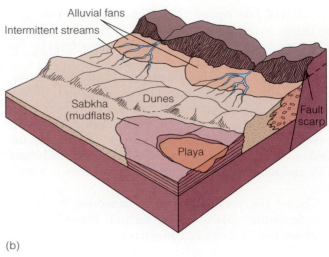

Alluvial fans

Intermittent streams

Dunes

Sabkha (mudflats)

Fault scarp

Playa

(b)

◆ **FIGURE 5.18** (a) Large-scale cross-bedding in Permian dune sands of the De Chelly Sandstone, Canyon de Chelly National Monument, Arizona. (Photo courtesy of Sue Monroe.) (b) Desert basin showing the association of alluvial fans, windblown dunes, and playa lakes.

the winter when fine-grained clay and organic matter settle from suspension as the lake freezes over. Another distinctive feature of varved deposits is *dropstones,* which are pieces of gravel dropped from floating ice (Fig. 5.19d).

Transitional Environments

Environments such as deltas, barrier islands, and tidal flats are transitional from continental to marine (Fig. 5.15; Table 5.2). The deposits of deltas, for example, accumulate where a fluvial system enters the sea, but the deposits are modified by such marine processes as waves and tides.

Deltas form by a rather simple process: when a stream enters a standing body of water, a lake or the sea, its velocity decreases and deposition occurs. Deltas in lakes are common, but marine deltas are even more common in the geologic record, much larger, and far more important economically (see Perspective 5.1). Marine deltas form where a fluvial system supplies sediments to a shoreline faster than marine processes can redistribute the sediments along the shoreline or carry them out to sea. As a result of deltaic deposition, the shoreline builds out, or **progrades,** into the sea.

Prograding deltas deposit a characteristic vertical sequence in which bottomset beds (prodelta deposits) are overlain successively by foreset beds (delta front depos-

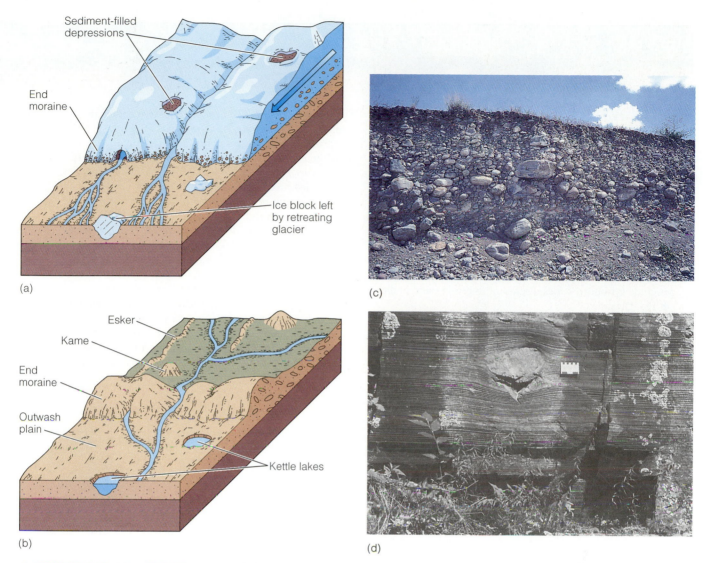

◆ FIGURE 5.19 (a) and (b) The origin of glacial drift. Deposits of drift at the terminus of a glacier are known as end moraines. (c) Glacial till in an end moraine. (d) Ice-rafted dropstone in Late Proterozoic varved sediments of the Gowganda Formation, Canada. (Photo courtesy of the Geological Survey of Canada, neg. 179852.)

Labels in figure (a): Sediment-filled depressions; End moraine; Ice block left by retreating glacier. (a)

Labels in figure (b): Esker; Kame; End moraine; Outwash plain; Kettle lakes. (b)

(c)

(d)

its) and topset beds (delta plain deposits) (◆ Fig. 5.20, p. 126). Such vertical sequences result from the lateral migration of environments, illustrating Walther's law (see Chapter 4). The flow velocity of a stream entering the sea diminishes rapidly, but the finest sediments are carried some distance beyond the stream's mouth, where they settle from suspension and form bottomset beds. Nearer the stream's mouth, sand and silt are deposited as

gently inclined layers on the foreset beds. The topset beds are traversed by a network of distributary channels in which fluvial deposits of silt and sand accumulate (◆ Fig. 5.21, p. 127). Between the distributary channels are low marshy areas where fine-grained sediment and organic matter are deposited.

It should be clear from the preceding discussion that deltas are composites of marine and continental deposits.

STREAM-, WAVE-, AND TIDE-DOMINATED DELTAS

As we have noted, deltas form in response to fluvial processes, but they may be modified by waves and tides, so geologists recognize stream-dominated, wave-dominated, and tide-dominated deltas. The Mississippi River delta, for example, is dominated by stream processes; the sediment supply rate is high, and it lies on a shallow shelf where it is protected from waves and tides. Distributary channels prograde far out to sea as a series of fingerlike sand bodies (◆ Fig. 1).

Distributary channels of stream-dominated deltas eventually prograde so far seaward that they are no longer effective avenues of sediment transport. When this occurs, the river abandons that channel and establishes a new one

elsewhere (Fig. 1). The new distributary channel then progrades until it, too, is abandoned. In short, stream-dominated deltas build a major deltaic lobe seaward by the progradation and abandonment of distributary channels. The active lobe of the present-day Mississippi River delta records a series of such events (Fig. 1).

In addition to abandoning individual distributary channels, stream-dominated deltas periodically abandon entire deltaic lobes. The Mississippi River is currently depositing sediment on a single lobe, but the delta is composed of several lobes that formed during the last few thousand years (◆ Fig. 2). As a matter of fact, if it were not for the efforts of the Army Corps of Engineers, the currently active lobe of the Mississippi River delta would probably have been abandoned by now.

Deltas that are strongly modified by marine processes are common; the Nile River delta of Egypt is wave-dominated, and the Ganges-Brahmaputra delta of

◆ **FIGURE 1** (a) Block diagram illustrating progradation of distributary channel sand bodies (bar finger sands). (b) The present lobe of the Mississippi River delta consists of several subdeltas that formed when the main distributary channel was abandoned. The area shown in this view corresponds to the Balize lobe in Figure 2.

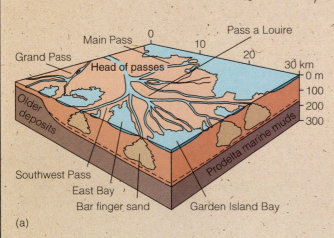

(a)

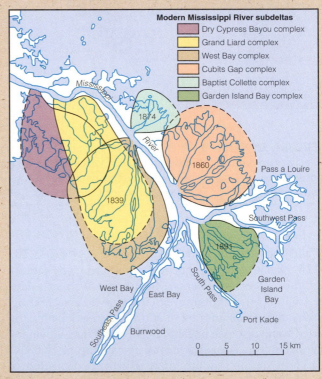

(b)

Bangladesh is tide-dominated (◆ Fig. 3). Waves continually modify the seaward margin of the Nile delta; therefore, progradation of fingerlike lobes does not occur. Rather, the river-deposited sediment is reworked and redistributed along the delta margin as a series of barrier sand bodies that grade seaward into silt and clay. The Nile delta's triangular shape resembles the Greek capital letter *delta* (Δ), which is why, in about 490 B.C., the Greek philosopher Herodotus coined the term.

Deltas on shorelines affected by strong tidal currents are shaped into elongate tidal sand bodies that parallel the direction of tidal flow (Fig. 3). When such deltas prograde, they produce a vertical sequence of bottomset beds overlain by sand bodies that look much like those deposited in braided stream channels.

Coal, oil, and gas are important resources found in deltaic deposits. Coal can form in several depositional environments, one of which is the marshes between

distributary channels (Fig. 1). Such marshes are dominated by nonwoody plants whose remains accumulate to form peat, the first stage in the formation of coal. If peat is buried, the volatile components of the plants are lost, and mostly carbon remains, thus forming coal.

Deltaic progradation is one way in which potential reservoirs for oil and gas form in marine basins. Distributary sand bodies, because of their porosity and permeability, together with their association with organic-rich marine sediments, commonly contain oil and gas. Much of the oil and gas production of the Gulf Coast of Texas comes from subsurface deltaic deposits. The Niger River delta of Africa and the Mississippi River delta are also known to contain reserves of oil and gas.

continued on next page

◆ **FIGURE 2** Major deltaic lobes deposited by the Mississippi River during the last 7,000 years. The oldest lobe, the Sale Cypremort lobe, was deposited for about 1,200 years more than 150 km west of the modern lobe. Deposition of the present-day Balize lobe began 600 to 800 years ago.

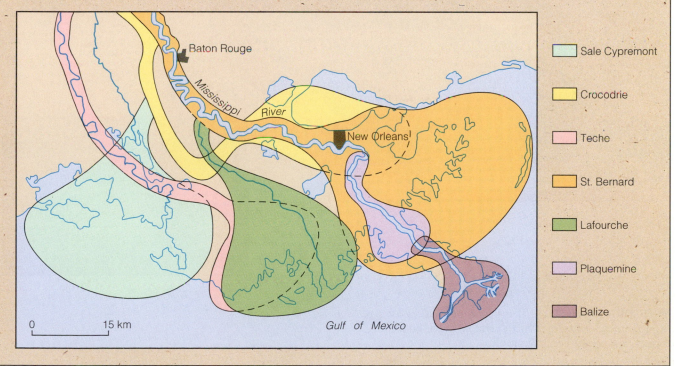

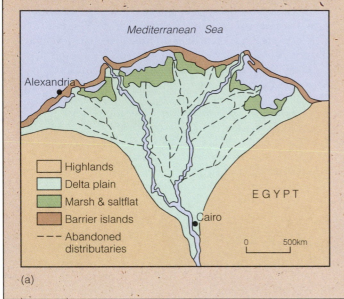

◆ **FIGURE 3** Wave- and tide-dominated deltas. (a) The Nile River delta of Egypt is wave-dominated. Sand supplied by distributary channels is redistributed along the front of the delta as a series of barrier sand bodies. (b) The Ganges-Brahmaputra delta is tide-dominated.

(a)

Mediterranean Sea

Alexandria

EGYPT

Cairo

Highlands
Delta plain
Marsh & saltflat
Barrier islands
Abandoned distributaries

0   500km

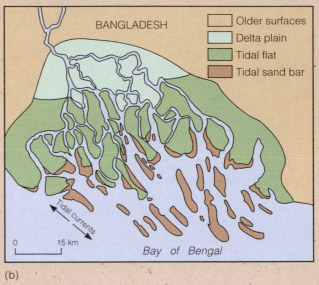

BANGLADESH

Older surfaces
Delta plain
Tidal flat
Tidal sand bar

Tidal currents

0   15 km

Bay of Bengal

(b)

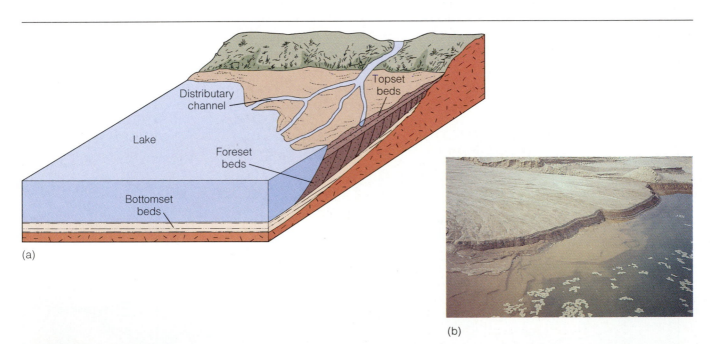

Topset beds

Distributary channel

Lake

Foreset beds

Bottomset beds

(a)

(b)

◆ **FIGURE 5.20** (a) Internal structure of the simplest type of prograding delta. (b) A small delta in which bottomset, foreset, and topset beds are visible.

The facies of the bottomset and foreset beds are marine and commonly contain marine fossils. However, the foreset beds may also contain pieces of waterlogged wood and other remains of land plants. The distributary channel deposits of the topset beds show clear evidence of deposition by fluvial processes, and any fossils they contain will be of land plants and animals.

Shorelines where marine processes effectively redistribute sediments as fast as they are supplied are characterized by linear beaches and barrier islands paralleling the shoreline (Fig. 5.15). Beaches are simply narrow sand bodies, constructed and modified by waves and nearshore currents, grading seaward into interbedded silt and sand. On their landward sides, beaches are commonly bordered by windblown sand dunes.

Barrier islands are elongate sand bodies separated from the mainland by a lagoon (◆ Fig. 5.22). These are high-energy environments where windblown dune and beach deposits accumulate; and during storms, waves may overtop barrier islands and deposit washover lobes in the lagoon (Fig. 5.22). Sand dunes consist of well-sorted, cross-bedded sand. Quartz is the dominant mineral in dune sands, but sand-sized fragments of marine shells are commonly present, too. In fact, the presence of marine fossils and facies relationships are the criteria used to distinguish these dunes from desert dunes. Beach sands are typically coarser grained than dune sands, contain fragmented marine fossils, and grade seaward into sediments of the shoreface environment (Fig. 5.22).

Because of their proximity to potential source rocks, barrier islands are good reservoirs for hydrocarbons. Accordingly, geologists have thoroughly studied such modern barrier islands as those along the Gulf Coast of Texas and the east coast of the United States, in their search for ancient hydrocarbon-bearing deposits.

Marine Environments

Marine environments lie seaward of the transitional environments (Fig. 5.15; Table 5.2). They include such

◆ **FIGURE 5.21** Schematic view of a delta showing facies deposited during progradation. Notice the branching network of distributary channels and the delta plain marsh. Progradation of distributary channels produces a vertical succession of facies like that in Figure 5.20.

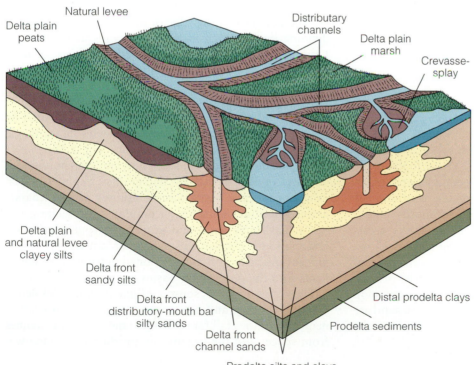

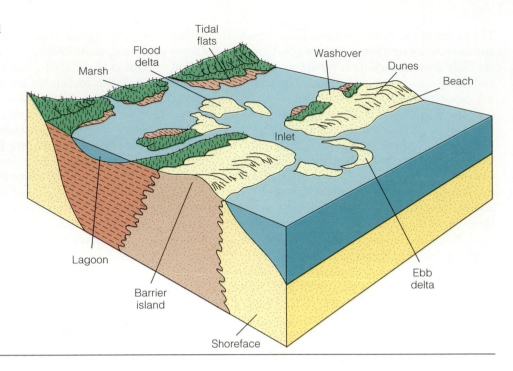

◆ **FIGURE 5.22** A barrier island and associated environments.

Labels in figure: Tidal flats, Flood delta, Marsh, Washover, Dunes, Beach, Inlet, Lagoon, Barrier island, Shoreface, Ebb delta

depositional settings as the continental shelf, slope, and rise and the deep-sea floor. Much of the detritus eroded from continents is deposited in marine environments, but other types of sediment may accumulate as well. For example, limestones are common in warm, shallow seas, and evaporites may be deposited in arid regions.

Detrital Marine Environments

The continental shelf is a gently seaward-sloping surface bounded at its outer edge by a shelf-slope break. Beyond this break lie the continental slope and continental rise and finally the deep-ocean basins (◆ Fig. 5.23). Continental shelves are subdivided into a shallow, high-energy inner shelf and a deeper, low-energy outer shelf (Fig. 5.23). Waves and tides more or less continuously stir up and sort inner-shelf sediments until mostly sand is left. This sand is commonly shaped into large, cross-bedded dunes and sand waves. The outer shelf is a low-energy area largely unaffected by tides and waves, except during major storms when surface waves may stir up sediments of the entire shelf. Mud is the most common sediment on the outer shelf, but layers of silt and sand are present too, especially near the boundary between the inner and outer shelf.

Much of the sediment derived from the continents crosses the continental shelf in submarine canyons and eventually comes to rest on the continental slope and rise as a series of overlapping **submarine fans** (◆ Fig. 5.24). Sedimentologically, the edge of the continental shelf is a dividing line between two major realms in the oceans. Landward from the shelf edge, sediments are at least periodically affected by waves and tides, but in a seaward direction these processes are no longer of any importance in sediment transport and deposition. Once sediment has passed the shelf edge, it is transported and deposited by gravity processes such as slumps, submarine debris flows, and particularly by turbidity currents (Fig. 5.10).

Beyond the continental slope and rise system, the sea-floor is covered mostly by fine-grained deposits. Local exceptions occur near oceanic islands where coarse-grained detrital sediments accumulate, and melting ice-bergs may carry sand and gravel far from continents. Otherwise no mechanisms exist that can transport significant quantities of coarse-grained detritus from continents into the deep-ocean basins.

There are, however, sources of fine-grained sediments. Windblown dust and ash from oceanic islands or continental volcanoes fall far out to sea; meteorite dust comes from space; and the organically productive surface wa-

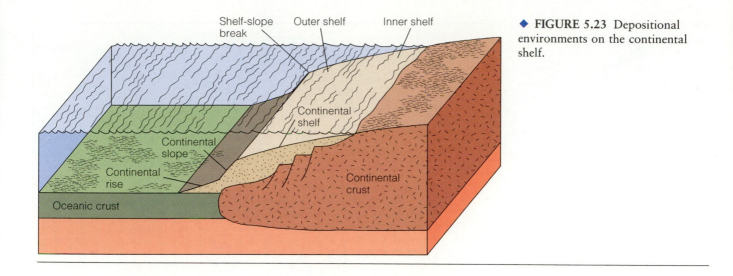

ters of the oceans contribute the shells of floating microorganisms. ◆ Figure 5.25 shows that the seafloor is covered mostly by three types of sediment—pelagic clay and two types of oozes composed of the shells of single-celled plants and animals.

Carbonate Depositional Environments

Limestone and dolostone are the only common carbonate rocks. Most dolostone, however, was originally limestone. When limestone is dolomitized, the mineral calcite, $CaCO_3$, changes to dolomite, $CaMg(CO_3)_2$. Accordingly, our discussion of carbonate depositional environments is mostly a consideration of the origin of limestones.

In some respects limestones are similar to detrital sedimentary rocks. For example, many limestones are composed of particles, or grains, and microcrystalline carbonate mud called *micrite*. Such grains and micrite are the carbonate equivalents of detrital gravel, sand, and mud. Limestones composed of shell fragments or spherical grains called *oolites* (◆ Fig. 5.26) are deposited where currents or waves are intense. In contrast, micrite and small grains such as fecal pellets accumulate in low-energy environments such as lagoons.

Despite their textural similarities, limestones and detrital rocks also have some fundamental differences. Composition is one obvious difference; but more importantly, carbonate grains and micrite originate within the basin of deposition. Such carbonate constituents may be locally transported and shaped into dunes and ripples,

but they do not indicate a remote source area. Furthermore, some limestones such as reef rock have no detrital equivalents.

Limestones may form in lakes, but by far the most extensive limestones have been deposited in warm, shallow seas. Such warm, shallow seas were widespread at several times during the past, occupying large parts of the continents. However, the continents stand high now, so limestone deposition is more restricted.

◆ FIGURE 5.24 Submarine canyons are one of the main avenues by which sediment is transported across the shelf. Turbidity currents carry sediments beyond the shelf into the deeper waters of the slope and rise where they accumulate as submarine fans.

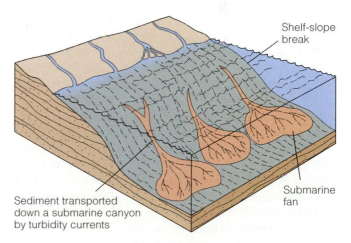

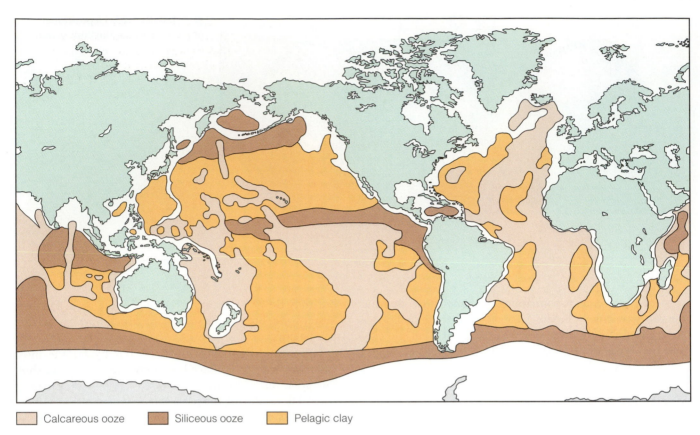

| | Calcareous ooze | | Siliceous ooze | | Pelagic clay |

◆ **FIGURE 5.25** The distribution of sediments on the deep-sea floor.

Modern carbonate environments and studies of ancient carbonate rocks indicate that most limestones form on carbonate shelves. Such shelves may be attached to a continent or may lie across the tops of offshore banks such as the Great Bahama Bank (◆ Fig. 5.27). In either case deposition occurs in shallow-water areas where there is little or no influx of detritus, especially mud.

Carbonate barriers (Fig. 5.27) form in high-energy environments and may be reefs or banks of skeletal carbonate sand or oolites. **Reefs** have a structural framework of coral skeletons, mollusks, sponges, and many other organisms and form wave-resistant structures. Within lagoons small, isolated reef masses called *patch reefs* may occur (Fig. 5.27). Reef rock is massive, structureless limestone, large pieces of which may be torn loose by storm waves and deposited in the forereef and backreef areas as reef breccia.

Except during major storms, the lagoon is a low-energy environment, where fossiliferous micrite and

◆ **FIGURE 5.26** Small, spherical or ovate calcium carbonate grains, such as those in this oolitic limestone, are common constituents of some limestones. Most oolites measure 0.5 to 1.0 mm in diameter.

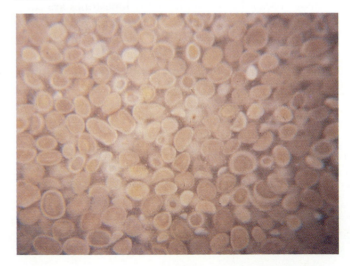

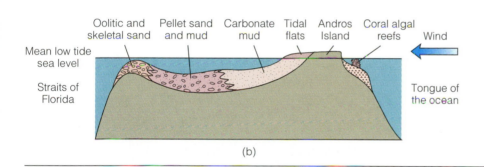

◆ **FIGURE 5.27** Carbonate depositional environments. (a) A shelf attached to a continent. Modern carbonate deposition is occurring on such attached shelves in southern Florida and the Persian Gulf. (b) Generalized cross section of the Great Bahama Bank. The carbonate shelf lies on top of a bank that rises from the seafloor.

fecal-pellet sands accumulate (Fig. 5.27). Fossils are typically preserved as unbroken shells. Much of the mud in lagoons is derived from organisms whose skeletal elements consist of mud-sized crystals. For example, some types of algae have calcium carbonate skeletal elements that help support the soft tissues during life. When these algae die, the skeletal elements are released and settle on the lagoon floor as carbonate mud.

The tidal-flat subenvironment of carbonate shelves is periodically flooded and exposed during high and low tides, respectively. Tidal flats are traversed by tidal channels in which the sediments can vary from carbonate sand to mud, and stromatolites (structures produced by photosynthesizing bacteria) may be present on the upper parts of some tidal flats (Fig. 5.27). On the landward side of tidal flats is a supratidal marsh, an area flooded only during exceptionally high tides (Fig. 5.27). Algal mats, desiccation cracks, and laminated micrite are common features of supratidal marshes. In arid regions supratidal areas are called *sabkhas*. High evaporation rates on sabkhas cause seawater contained within the sediments to evaporate. When this occurs, evaporites, especially gypsum ($CaSO_4 \cdot 2H_2O$), form within the sediments as irregular masses called *nodules*.

Carbonate rock units covering tens of thousands of square kilometers are common in the geologic record. Many of these carbonate units were deposited during major transgressions or regressions and therefore have overall sheetlike geometries. However, individual depositional units within carbonates may be circular (patch reefs) or shoestring-shaped (oolite and skeletal sand barriers).

Evaporite Environments

Evaporites, mostly rock gypsum[1] and rock salt, are not particularly common when compared with the abundance of sandstone, shale, and limestone. Nevertheless, some evaporite deposits are of impressive dimensions, and some are locally important as resources. Evaporite rocks can form in saline lakes such as the Dead Sea, in playa lakes, and in marine environments including sabkhas (see the Prologue).

---

[1]Gypsum ($CaSO_4 \cdot 2H_2O$) is the common sulfate mineral precipitated from seawater; but when deeply buried, gypsum loses its water and is converted to anhydrite ($CaSO_4$).

Marine evaporites are the most common type in the geologic record, but little agreement exists on the specific environment in which they formed. Geologists agree, however, on some of the conditions necessary for evaporite deposition. Obviously, evaporation rates would have to be high so that seawater would become increasingly saline until it reached a point at which minerals would begin to precipitate. Such conditions are best met in nearshore environments such as lagoons or embayments in arid regions. In an embayment, for example, the inflow of normal seawater is restricted, salinity increases because of evaporation, and evaporite minerals form.

Some marine evaporites appear to have been deposited in long, narrow seas that formed in response to continental rifting. When a continent is rifted, the crust is stretched and thinned and eventually subsides below sea level. Chapter 7 addresses this subject further. If this newly formed sea is in an arid region, evaporite deposition occurs. The Louann Salt (see Chapters 15 and 17) of the Gulf Coast of the United States and several other evaporite deposits are thought to have formed in such a plate tectonic setting.

◆ FIGURE 5.28 Simplified cross section showing the probable lateral relationships and environmental interpretation for Lower Silurian strata in the eastern United States.

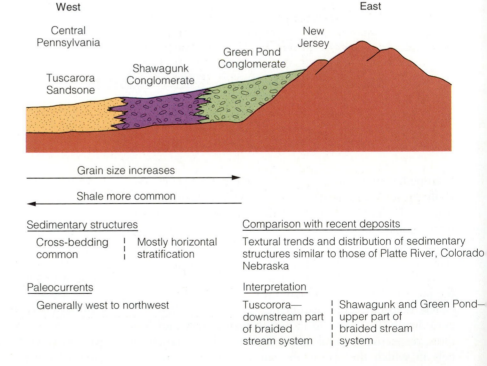

❖ ENVIRONMENTAL INTERPRETATIONS AND HISTORICAL GEOLOGY

The ultimate goal in historical geology is to determine (1) what events occurred during the past and (2) the order in which they occurred. Such determinations are based, in part, on analyses of sedimentary rocks and inferences that can be drawn from such analyses. In Chapter 4 we discussed the use of sedimentary rocks and their contained fossils in establishing a meaningful stratigraphic succession, and this chapter has been devoted to environmental analyses. In short, we now have the information necessary to determine what specific events occurred and when. For example, Lower Silurian strata in New Jersey and Pennsylvania possess characteristics of grain size, rock types, and sedimentary structures that indicate deposition in a braided stream environment ◆ Fig. 5.28). Vertical facies relationships, rock types, fossils, and sedimentary structures of Ordovician carbonate rocks in Arkansas are interpreted as transgressive carbonate shelf sediments ◆ Fig. 5.29). Depositional environments will figure importantly in the chapters on geologic history. Such interpretations are based on the

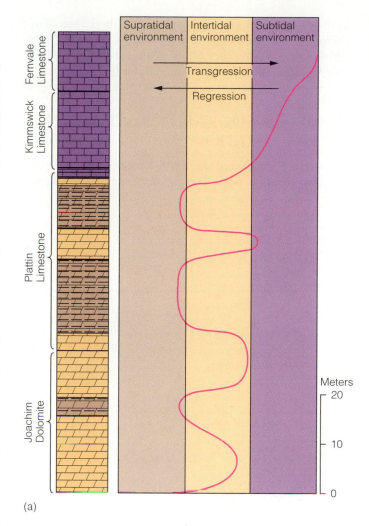

◆ **FIGURE 5.29** Middle and Upper Ordovician strata in northern Arkansas. (a) Vertical stratigraphic relationships and inferred environments of deposition. Notice that the trend was transgressive even though several minor transgressions and regressions occurred. (b) Simplified cross section showing lateral facies relationships.

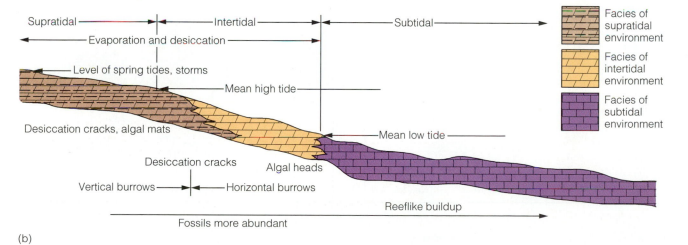

kinds of data and analyses covered in this chapter. Furthermore, environmental interpretations are not of academic interest only; they allow geologists to more effectively predict the locations and geometries of rock units that contain important mineral resources or oil and gas.

❖ PALEOGEOGRAPHY AND PALEOCLIMATES

The surface patterns of the Earth have not remained unchanged through time. During the Late Paleozoic, for example, a single supercontinent, Pangaea, existed, but it began breaking up during the Triassic and has continued to do so ever since. Thus, the present geographic distribution of continents evolved through time. **Paleogeography** involves a study of the Earth's or a region's surface patterns for a particular time in the past. A series of paleogeographic maps will show the relative positions of the continents through time.

The paleogeography of several Rocky Mountain states has been reconstructed in some detail in ◆ Figure 5.30. Such reconstructions must be made for a particular time,

◆ **FIGURE 5.30** Eocene paleogeography of some of the Rocky Mountain states. (a) Paleogeography in late Early to Middle Eocene time. (b) Paleogeography in Late Eocene time.

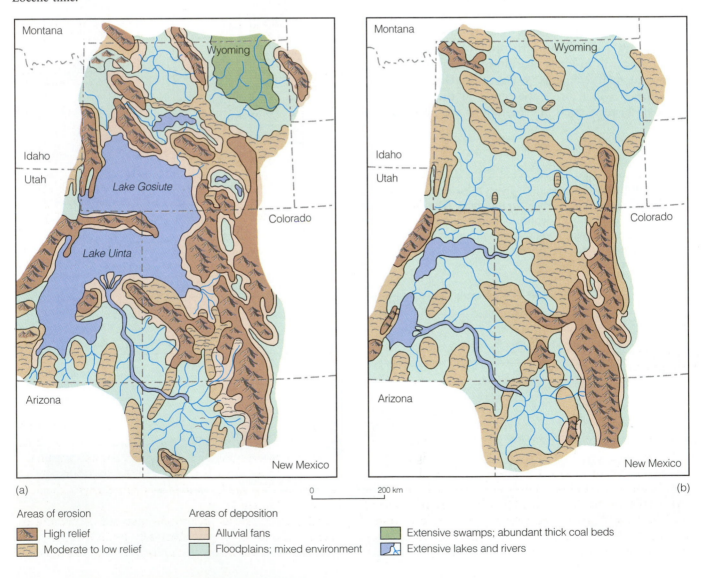

Areas of erosion

- High relief
- Moderate to low relief

Areas of deposition

- Alluvial fans
- Floodplains; mixed environment
- Extensive swamps; abundant thick coal beds
- Extensive lakes and rivers

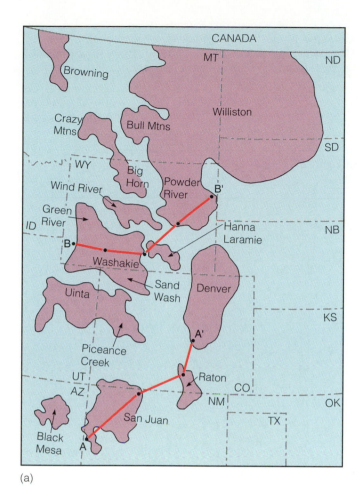

(a)

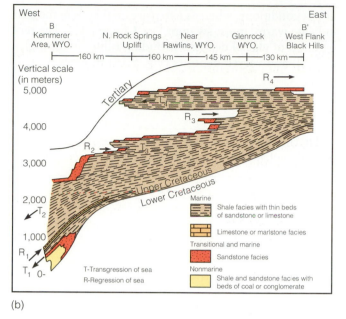

(b)

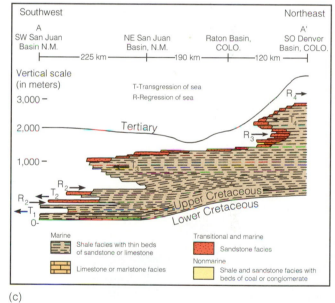

(c)

◆ **FIGURE 5.31** (a) Map showing the major basins in which Upper Cretaceous strata are present in the western United States. (b) and (c) Cross sections along the lines indicated. Three major facies are recognized, and the stratigraphic sequence of facies records transgressions and regressions.

so geologists must determine depositional environments, work out the physical continuity of rock units, and determine time-stratigraphic relationships. ◆ Figure 5.31 shows the major basins in the western United States in which Upper Cretaceous strata are preserved. Notice in the cross sections that three major facies are recognized, and that the stratigraphic relationships indicate transgressions and regressions. Several paleogeographic maps have been constructed for this region, two of which are shown in ◆ Figure 5.32. According to these reconstructions, a broad coastal plain sloped gently eastward from a mountainous region to the sea. This coastal plain habitat was occupied by lush vegetation, dinosaurs, flying reptiles, lizards, turtles, and early mammals (see the Prologue to Chapter 16).

Sedimentary rocks and fossils also tell something of ancient climates, or **paleoclimates.** Pennsylvanian and Permian strata of the southwestern United States indicate desert conditions, whereas extensive coal beds in the east indicate warm, moist swamps (see Chapter 12). The recognition of glacial deposits including tillite (lithified glacial till) and varved deposits (Fig. 5.19) in Precambrian, Ordovician, Pennsylvanian, and Pleistocene strata has allowed geologists to outline several areas of glacial climates. Deposits of evaporites indicate hot, arid climates.

◆ **FIGURE 5.32** Paleogeographic maps prepared from the data in Figure 5.31. Both maps show Late Cretaceous paleogeography but (a) is older than (b).

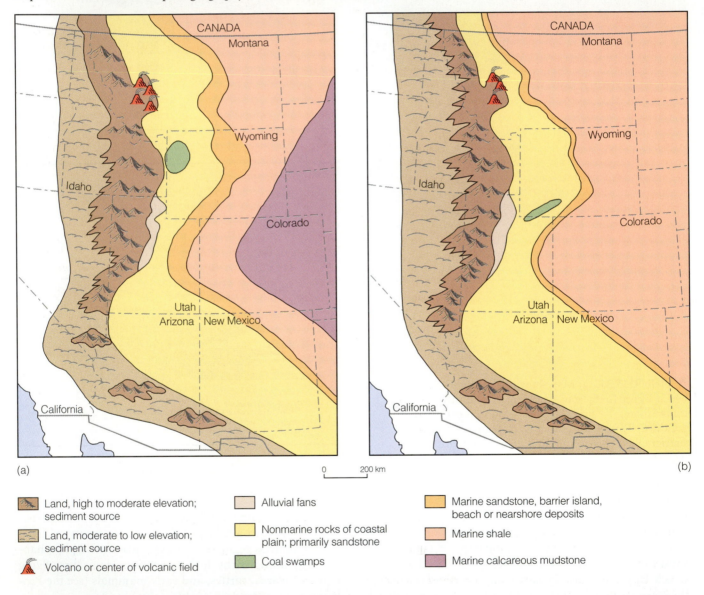

(a)　　　　　　　　　　　　　　　　　　　　　　　　　　　　　　　　　　(b)

0　　200 km

	Land, high to moderate elevation; sediment source		Alluvial fans		Marine sandstone, barrier island, beach or nearshore deposits
	Land, moderate to low elevation; sediment source		Nonmarine rocks of coastal plain; primarily sandstone		Marine shale
	Volcano or center of volcanic field		Coal swamps		Marine calcareous mudstone

Chapter Summary

1. In order to determine what environment sedimentary rocks were deposited in, geologists must observe and interpret the physical and biological features of those rocks.

2. Sedimentary structures and fossils are the most useful sedimentary rock features for environmental analyses, but composition, texture, and rock-body geometry are helpful as well.

3. Three major depositional settings are recognized: continental, transitional, and marine. Each setting includes several specific depositional environments.

4. Fluvial systems may be braided, where deposits are mostly gravel and sand with subordinate mud, or meandering where deposits are mostly mud with elongate sand bodies.

5. The association of alluvial fans, playa lake deposits, and windblown dunes is typical of desert depositional environments. Deposits in areas affected by glaciation are mostly till and outwash.

6. Marine deltas form where a stream supplies sediments to a shoreline. Deltas prograde seaward and produce vertical sequences of bottomset beds (prodelta deposits) overlain by foreset beds and topset beds (delta-front and delta-plain deposits).

7. A barrier island complex includes lagoon, beach, and dune subenvironments, each characterized by a unique association of lithologies, fossils, and sedimentary structures.

8. Waves and tides are important agents on continental shelves, but beyond the shelf-slope break, gravity processes—especially turbidity currents—predominate. Detrital shelf deposits consist of inner-shelf sands and outer-shelf muds. Deposits on the continental slope-rise systems are mostly submarine fans.

9. Sediments on the deep-sea floor are mostly fine grained because no mechanism exists to transport large amounts of coarse detritus from continents into deep seas.

10. Most carbonates are deposited in warm, shallow seas. Many limestone textural features are indicators of depositional conditions. Carbonate shelves may be attached to a landmass or lie on top of banks rising from the seafloor.

11. The best conditions for the origin of evaporites occur along marine shorelines in arid regions. Marine embayments, lagoons, and sabkhas are areas in which many present-day evaporites are deposited.

Important Terms

alluvial fan
barrier island
bedding (stratification)
biogenic sedimentary structure
bioturbation
clastic texture
cross-bedding
crystalline texture
delta
depositional environment

desiccation crack
drift
fluvial
graded bedding
grain size
lithology
outwash
paleoclimate
paleocurrent
paleogeography

point bar
progradation
reef
ripple mark
rounding
sedimentary structure
sorting
submarine fan
till
varve

Review Questions

1. A type of sedimentary deposit characteristic of the margins of desert basins is a(n):
   a. _____ playa lake; b. _____ till; c. _____ alluvial fan; d. _____ ooze; e. _____ point bar.

2. Which of the following sedimentary structures is a good paleocurrent indicator?
   a. _____ wave-formed ripple marks; b. _____ graded bedding; c. _____ laminations; d. _____ trace fossils; e. _____ cross-bedding.

3. A detrital sedimentary rock in which all particles are approximately the same size is said to be:
   a. _____ well sorted; b. _____ moderately rounded; c. _____ graded; d. _____ an evaporite; e. _____ poorly consolidated.

4. The most common evaporite rocks are:
   a. _____granite-basalt; b. _____sandstone-conglomerate; c. _____ limestone-dolostone; d. _____ rock gypsum–rock salt; e. _____ chert-micrite.

5. Pelagic clay and oozes are:
   a. _____ deposited in high-energy environments such as stream channels; b. _____ the types of sediments covering most of the deep-sea floor; c. _____ annual deposits that accumulate in glacial lakes; d. _____ deposited by turbidity currents; e. _____ largely restricted to the continental shelves.

6. The process whereby burrowing organisms alter the physical and chemical properties of sediments is:
   a. _____ bioturbation; b. _____ meandering; c. _____ desiccation; d. _____ turbidities; e. _____ lamination.

7. Sedimentary rocks composed of broken particles of other rocks or broken shells have a _____ texture.
   a. _____ poorly sorted; b. _____ crystalline; c. _____ paleocurrent; d. _____ stratified; e. _____ clastic.

8. Small, spherical to ovate calcium carbonate grains deposited where currents or waves are intense are:
   a. _____ quartz; b. _____ micrite; c. _____ oolites; d. _____ varves; e. _____ fossils.

9. Which of the following aspects of detrital sedimentary rocks is most useful in determining source area?
   a. _____ desiccation cracks; b. _____ composition; c. _____ fossils; d. _____ horizontal bedding; e. _____ skeletons of microorganisms.

10. One of the most distinctive features of meandering stream sand bodies is:
    a. _____ point bars; b. _____ alluvial fans; c. _____ outwash; d. _____ ripple marks; e. _____ barrier islands.

11. Submarine fans are made up largely of sediments transported and deposited by:
    a. _____ turbidity currents; b. _____ glacial ice; c. _____ wind; d. _____ water in stream channels; e. _____ waves.

12. Islands composed of sand and separated from the mainland by a lagoon are:
    a. _____ reefs; b. _____ point bars; c. _____ playa lakes; d. _____ barrier islands; e. _____ deltas.

13. In which of the following do foreset beds occur?
    a. _____ alluvial fans; b. _____ point bars; c. _____ floodplains; d. _____ deltas; e. _____ submarine fans.

14. One component of many limestones is micrite, which is:
    a. _____ detrital sediment; b. _____ a calcium sulfide mineral; c. _____ microcrystalline carbonate mud; d. _____ fragments of shells; e. _____ till.

15. Define grain size and sorting, and explain how each is used in environmental analyses.

16. Why are microfossils and trace fossils particularly useful for determining depositional environments?

17. Illustrate three sedimentary structures that can be used to determine paleocurrent directions. Explain how each of these structures forms.

18. How do the deposits of braided and meandering streams differ?

19. What criteria would you use to distinguish desert dunes from dunes on a barrier island?

20. Illustrate and describe the vertical sequence of facies produced by a prograding, stream-dominated delta.

21. How and why do sediments on the inner and outer continental shelf differ?

22. In what ways are limestones and detrital sedimentary rocks similar and dissimilar?

23. Draw a generalized cross section of a carbonate shelf, and show the types of sediments you would expect to find on various parts of the shelf.

24. Why are deep-sea sediments mostly fine grained? What are the sources of these sediments?

Additional Readings

Boggs, S., Jr. 1987. *Principles of sedimentology and stratigraphy.* Columbus, Ohio: Merrill Publishing Co.

Collinson, J. D., and D. B. Thompson. 1989. *Sedimentary structures,* 2d ed. London, England: George Allen & Unwin.

Davis, R. A., Jr. 1992. *Depositional systems: An introduction to sedimentology and stratigraphy.* 2d ed. Englewood Cliffs, N.J.: Prentice-Hall.

Friedman, G. M., J. E. Sanders, and D. C. Kopaska-Merkel. 1992. *Principles of sedimentary deposits.* New York: Macmillan.

Fritz, W. J., and J. N. Moore. 1988. *Basics of physical stratigraphy and sedimentology.* New York: John Wiley & Sons.

Hallam, A. 1981. *Facies interpretation and the stratigraphic record.* San Francisco: W. H. Freeman.

Reading, H. G., ed. 1986. *Sedimentary environments and facies.* Boston: Blackwell Scientific Publications.

Selley, R. C. 1978. *Ancient sedimentary environments.* Ithaca, N.Y.: Cornell University Press.

Walker, R. G., ed. 1984. Facies models. *Geoscience Canada,* reprint series 1. Geological Association of Canada.

CHAPTER 6

Chapter Outline

The Frency botanist-geologist Jean Baptiste Pierre Antoine de Monet de Lamarck
(1744–1829) was the first scientist to propose a formal, widely accepted theory of evolution.
(Photo courtesy of The Granger Collection, New York.)

EVOLUTION

Prologue

On December 27, 1831, Charles Robert Darwin departed from Devonport, England, aboard the H.M.S. *Beagle* as an unpaid naturalist. Nearly five years later the *Beagle* returned to England, and Darwin never ventured far from home again. Nevertheless, the 64,360-kilometer voyage (◆ Fig. 6.1) was the most important experience of his life, an experience that changed his view of nature and ultimately revolutionized all of science.

When Darwin sailed aboard the *Beagle*, he was a little-known recent graduate in theology from Christ's College. In fact, his father, Dr. Robert Darwin, had sent him to Christ's College as a last resort; Charles showed little aptitude for academics, except perhaps for science, and had already withdrawn from medical studies at Edinburgh. Although he completed his theological studies, he was rather indifferent to religion. He nevertheless fully accepted the biblical account of creation as historical fact. Darwin's belief in biblical creation initially endeared him to the *Beagle*'s captain,

◆ **FIGURE 6.1** Route of the H.M.S. *Beagle* during its 1831–1836 voyage. Important localities visited by Darwin are shown.

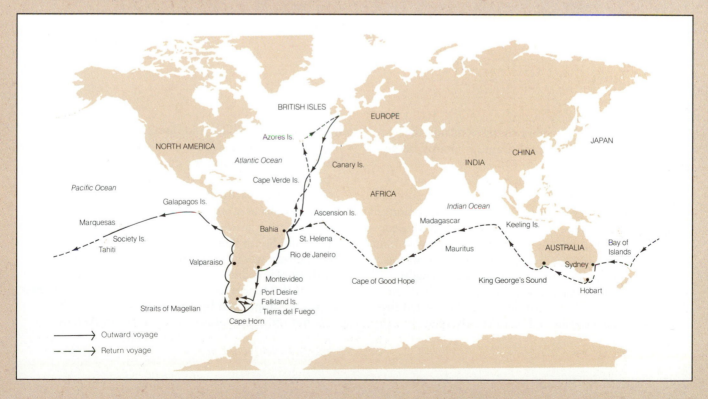

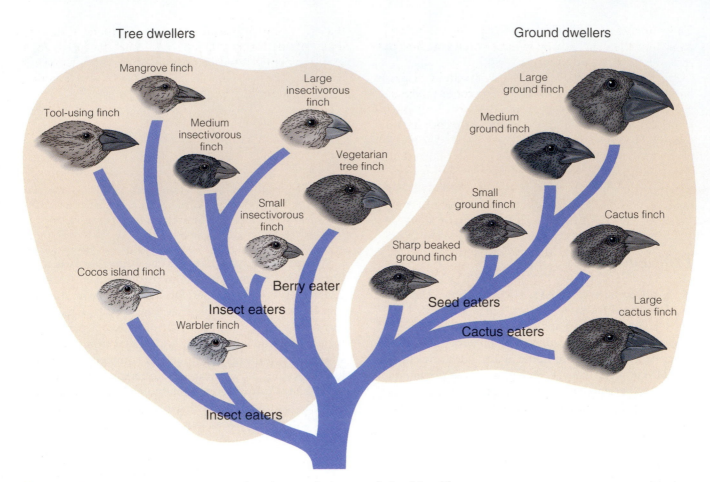

Tree dwellers

Ground dwellers

Mangrove finch

Tool-using finch

Large insectivorous finch

Medium insectivorous finch

Vegetarian tree finch

Small insectivorous finch

Cocos island finch

Berry eater

Insect eaters

Warbler finch

Insect eaters

Large ground finch

Medium ground finch

Small ground finch

Cactus finch

Sharp beaked ground finch

Seed eaters

Large cactus finch

Cactus eaters

◆ **FIGURE 6.2** Darwin's finches arranged to show evolutionary relationships. The six species on the right are ground-dwellers, and the others are adapted for life in trees. Notice that the shapes of the beaks of finches in both groups vary depending on diet.

Robert Fitzroy, but his views changed during the voyage, and his relationship with Captain Fitzroy became strained.

When the voyage began, Darwin was nominally a clergyman with interests in the sciences, especially botany, zoology, and geology. He suffered from prolonged attacks of seasickness, but nevertheless kept detailed notes on his observations; he collected, cataloged, and dissected specimens and eventually became a professional naturalist. The *Beagle* made several lengthy stops in South America where Darwin explored rain forests, experienced an earthquake, and collected fossils. The fossils he collected were clearly related to the living sloth, armadillo, and llama and implied that living species descended from fossil species. Such evidence caused him to question the concept of fixity of species, a concept held by those who accepted biblical creation.

Darwin was particularly fascinated by the plants and animals of oceanic islands. The Cape Verde Islands and the Galapagos Islands are comparable distances west of Africa and South America, respectively (Fig. 6.1). Each island group has its own unique plants and animals, yet these plants and animals most closely resemble those of the nearby continent. The Galapagos, for example, are populated by 13 species of finches (◆ Fig. 6.2), but only one species exists in South America.

These finches are adapted to the different habitats occupied on the mainland by various species such as parrots, flycatchers, and toucans.

Darwin reasoned that the Cape Verdes and Galapagos had received colonists from the nearby continents and that "such colonists would be liable to modification—the principle of inheritance still betraying their original birthplace."[1] The significance of this statement is that it revealed a change in Darwin's view of nature; he no longer accepted the idea of fixity of species. For example, he proposed that an ancestral species of finch had somehow reached the Galapagos Islands from South America and differentiated into the various types he observed (Fig. 6.2). Furthermore, he had read Charles Lyell's *Principles of Geology* during his voyage and accepted uniformitarianism and the great age of the Earth. In short, he had come to view nature as dynamic rather than static.

❖ INTRODUCTION

The term *evolution* comes from the Latin, meaning unrolled. Roman books were written on parchment and rolled on wooden rods, so they were unrolled, or evolved, as they were read. In modern usage, evolution usually refers to change through time. If evolution is defined simply as change through time, it is a pervasive phenomenon; stars evolve, the Earth has evolved and continues to do so, languages and social systems evolve, and life evolves. Change through time is, of course, demonstrable. We know from historical records that languages have changed, and the fossil record reveals that a succession of different life-forms have existed through time.

The biological concept of evolution, most appropriately called *organic evolution*, is concerned with changes in organisms that are inheritable. Evolution so defined is also observable, at least on a small scale. For example, the proportion of hereditary determinants in populations of moths in Great Britain changed in such a way that their color changed from light to dark in several generations. Large-scale evolution is not directly observable but can be inferred from fossil evidence and from studies of comparative anatomy, embryology, and biochemistry of living organisms.

As we discussed in Chapter 1, theories are scientific explanations of natural phenomena that have a large body of supporting evidence. The **theory of evolution** is such an explanation. The central claim of this theory is

that all living organisms are the evolutionary descendants of life-forms that existed during the past. As a scientific theory, it meets the usual criteria for theories; it is a naturalistic explanation, and it can be tested so that its validity can be assessed.

❖ THE EVOLUTION OF AN IDEA

Evolution as a biological concept was seriously considered by some naturalists even before Charles Darwin was born. Indeed, as early as 600 B.C., the ancient Greeks speculated on possible interrelationships among organisms; but overall neither the ancient world nor the medieval gave much thought to evolution. In fact, the prevailing belief well into the eighteenth century was that all important knowledge was contained in the works of Aristotle and the Bible. One did not have to observe nature to understand it, since all the answers were in these two sources, particularly in the first two chapters of Genesis. Literally interpreted, Genesis was taken as the final word on the origin and diversity of life and much of Earth history. According to this view, all species of plants and animals were perfectly fashioned by God during the days of creation and have remained fixed and immutable ever since. To question divine creation in any way was regarded as heresy.

In eighteenth-century Europe, the social and intellectual climate changed, and the absolute authority of the Church in all matters declined. Ironically, the very naturalists who were determined to find physical evidence supporting Genesis as a factual, historical account were responsible for finding more and more evidence that could not be explained by a literal reading of Scripture. This was particularly true of the evidence for a worldwide catastrophic flood. For example, the sedimentary deposits that traditionally had been attributed to the biblical deluge showed clear indications of a noncatastrophic origin. Many naturalists reasoned that this evidence truly reflected the conditions that prevailed when the strata were deposited, for to infer otherwise implied a deception on the part of the Creator, a thesis they could not accept.

It should not be concluded from this short discussion that eighteenth-century naturalists abandoned the Judeo-Christian tradition. As a matter of fact, most did not. Rather, they came to accept Genesis as a symbolic account containing important spiritual messages but not a factual account of creation, a view shared today by most theologians and biblical scholars.

---

[1] C. Darwin, *The Origin of Species* (New York: New American Library, 1958), p. 377.

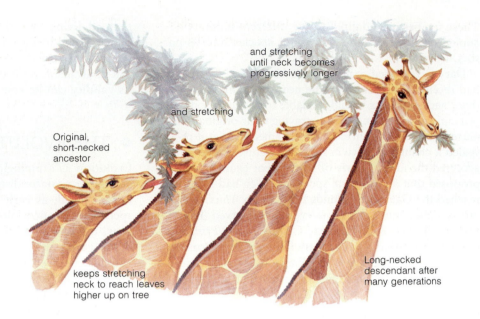

◆ **FIGURE 6.3** According to Lamarck's theory of inheritance of acquired characteristics, ancestral short-necked giraffes stretched their necks to reach leaves higher up on trees, and their descendants were born with longer necks.

Original, short-necked ancestor

and stretching

and stretching until neck becomes progressively longer

keeps stretching neck to reach leaves higher up on tree

Long-necked descendant after many generations

These changing attitudes can be attributed in large part to the Enlightenment, an eighteenth-century philosophical movement that relied on rationalism. In science this meant that one could discover natural laws to explain observations of natural systems; all did not depend on divine revelation. The sciences, particularly biology and chemistry, began to acquire a modern look.

In this changed intellectual atmosphere, the concepts of uniformitarianism and the great age of the Earth were gradually accepted. The French zoologist Georges Cuvier clearly demonstrated that many types of plants and animals had become extinct, and that fossils differed through time. In view of the fossil evidence, as well as observations and studies of living organisms, many naturalists became convinced that species were not immutable. Change from one species to another—that is, evolution—became an acceptable idea. What was lacking, however, was an overall theoretical framework to explain evolution.

Lamarck and the Giraffe's Neck

Erasmus Darwin, a physician, poet, and naturalist and Charles Darwin's grandfather, was the first to propose a process by which gradual evolution could occur. But it was Jean Baptiste de Lamarck (1744–1829), a French botanist-geologist, who was first taken seriously by his colleagues. Pointing out that organisms are adapted to their environments, he proposed that evolution is an adaptive process whereby organisms acquire traits or characteristics during their lifetimes and pass these acquired traits on to their descendants.

According to Lamarck's theory of **inheritance of acquired characteristics** ancestral short-necked giraffes, for example, stretched their necks to browse on leaves of trees (◆ Fig. 6.3). The environmentally imposed characteristic of neck-stretching was thus expressed in future generations as a longer neck. In other words, the capacity for giraffes to produce offspring with longer necks was acquired through the habit of neck-stretching.

Considering the data available in the early 1800s, Lamarck formulated a reasonable theory. Little was known about how traits are passed from parent to offspring, so Lamarck's inheritance mechanism seemed logical and was accepted by many naturalists as a viable mechanism for evolution. In fact, Lamarck's concept of inheritance was not completely refuted until more modern concepts of inheritance were developed. Based on the work of Gregor Mendel in the 1860s, and developments in the early 1900s, we now know that all organisms possess hereditary determinants called *genes* that cannot be modified by any effort of the organism during its lifetime. Numerous attempts have been made to validate Lamarckian inheritance, including one during this century in the former Soviet Union (see Perspective 1.1), but all have failed.

Charles Darwin's Contribution

In 1859 Charles Robert Darwin published *The Origin of Species* in which his ideas on evolution were outlined (♦ Fig. 6.4). Most naturalists almost immediately recognized that evolution provides a unifying theory that explains an otherwise encyclopedic collection of biologic facts. Even most, but not all, churches and theologians eventually accepted the idea of evolution as a naturalistic explanation that does not conflict with the purpose and intent of Scripture.

♦ **FIGURE 6.4** Charles Darwin in 1850. (Photo courtesy of Historical Pictures/Stock Montage.)

While 1859 may mark the beginning of modern evolutionary biology, Darwin had actually formulated his ideas more than 20 years earlier, but aware of the furor his ideas would generate, had been reluctant to publish. As a matter of fact, when the furor did erupt, Darwin, a semi-invalid most of his life, never defended his theories in public; he left their defense to others.

As discussed in the Prologue, Darwin concluded during the voyage of the *Beagle* that species are not fixed. In fact, by the time Darwin returned home in 1836, he had what he thought was clear evidence that species had indeed changed through time, but he had no idea what might bring about change. By 1838, however, his observations of the selection practiced by animal and plant breeders, and a chance reading of Thomas Malthus's essay on population, gave him the elements necessary to formulate his theories.

Animal and plant breeders selected organisms for desirable traits, bred organisms with those traits, and thereby induced a great amount of change. This practice of **artificial selection** yielded a fantastic variety of domestic pigeons, dogs, and plants (♦ Fig. 6.5). If this much change could be produced artificially, was a process that selected among variant types perhaps also acting on natural populations?

Darwin came to appreciate fully the power of selection after reading Malthus's book on population. Malthus argued that far more animals are born than reach maturity, yet the numbers of adult animals of a species remain rather constant. He reasoned that competition for resources resulted in a high infant mortality rate, thus limiting the size of the population. Darwin proposed that a natural process was selecting only a few animals for survival, a process he called **natural selection.**

Natural Selection

For 20 years Darwin kept his ideas on evolution and natural selection largely to himself. But in 1858 he received a letter from Alfred Russell Wallace, a young naturalist working in southern Asia. Wallace had also read Malthus's essay on population and had come to precisely the same conclusion as Darwin. Friends convinced Darwin that his paper and Wallace's should be read at the same session of the Linnaean Society of London. Later some of Darwin's correspondence was published, establishing his scientific priority, and indeed Wallace even insisted that Darwin be given credit in recognition of his earlier discovery and more thorough documentation of the theory.

◆ **FIGURE 6.5** Artificial selection has yielded more than 300 domesticated breeds of pigeons, three of which are shown here. The common rock dove (upper left) is thought to be the parent stock of the domestic breeds. (Photos on left © G. C. Kelley/Photo Researchers, © Tom McHugh/Photo Researchers, on right © Kenneth Fink/ Photo Researchers.)

The Darwin-Wallace theory of natural selection can be summarized as follows:

1. All populations contain heritable variations—size, speed, agility, coloration, and so forth.
2. Some variations are more favorable than others; that is, some variant types have an edge in the competition for resources and avoidance of predators.
3. Not all young survive to reproductive maturity.
4. Those with favorable variations are more likely to survive and pass on their favorable variations.

Evolution by natural selection is then largely a matter of reproductive success, for only those that reproduce can pass on favorable variations. Of course, favorable variations do not guarantee survival for an individual, but within a population, those with favorable variations are more likely to survive and reproduce.

Natural selection is the process proposed by Darwin and Wallace to account for evolution. But is their statement substantially different from Lamarck's evolution by the inheritance of acquired characteristics? In other words, how would Darwin and Wallace explain the giraffe's long neck?

Suppose in an ancestral population of giraffes some individuals had longer necks than most of the others (◆ Fig. 6.6). These long-necked animals could obviously browse higher on trees. The important point here is that some simply have longer necks and therefore enjoy a selective advantage in that they can acquire resources more effectively. Consequently, they are more likely to survive, reproduce, and pass on their favorable variations. Even in the second generation, a few more giraffes might have longer necks, which would give a competitive edge. And so it would go, generation by generation, thereby increasing the proportion of giraffes with longer necks.

❖ MENDEL AND THE BIRTH OF GENETICS

Critics of natural selection were quick to point out that Darwin and Wallace could not account for the origin of variation or explain how it was maintained in populations. They reasoned that should a variant trait arise, it would simply blend with other traits in the population and be lost. At the time, this was a valid criticism, and neither Darwin nor Wallace could effectively answer it. Actually, information that could have given them the answer existed, but it remained in obscurity until 1900.

A monastery in what is now Czechoslovakia may seem to be an unlikely setting for the discovery of the rules of inheritance, and an Austrian monk may seem an unlikely candidate for the "father of genetics." Nevertheless, dur-

ing the 1860s Gregor Mendel, who later became abbot of the monastery, was doing research that answered some of the inheritance problems that plagued Darwin and Wallace. Unfortunately, Mendel's work was published in an obscure journal and went largely unnoticed. Mendel even sent Darwin a copy of his manuscript, but it apparently lay unread on his bookshelf.

Mendel performed a series of controlled experiments with true-breeding strains of garden peas (strains that when self-fertilized always display the same trait, such as a particular flower color). In one experiment, he transferred pollen from white-flowered plants to red-flowered plants, which produced a second generation of all red-flowered plants. But when left to self-fertilize, these plants yielded a third generation with a ratio of red-flowered plants to white-flowered plants of slightly over 3 to 1 (◆ Fig. 6.7).

Mendel concluded from his experiments that traits such as flower color are controlled by a pair of factors, or what we now call **genes.** He also concluded that genes controlling the same trait occur in alternate forms, or **alleles;** that one allele may be dominant over another; and that offspring receive one allele of each pair from each parent. When an organism produces sex cells — pollen and ovules in plants, and sperm and eggs in animals — only one allele

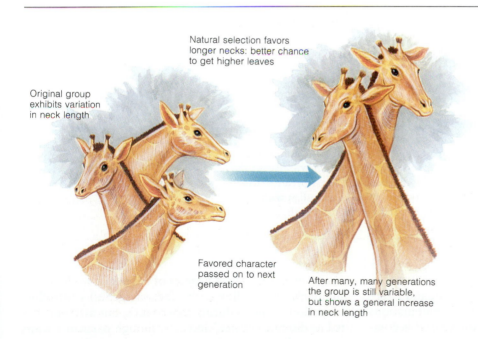

Natural selection favors longer necks: better chance to get higher leaves

Original group exhibits variation in neck length

Favored character passed on to next generation

After many, many generations the group is still variable, but shows a general increase in neck length

◆ **FIGURE 6.6** According to the Darwin-Wallace theory of natural selection, the giraffe's long neck evolved as animals with favorable variations were selected for survival.

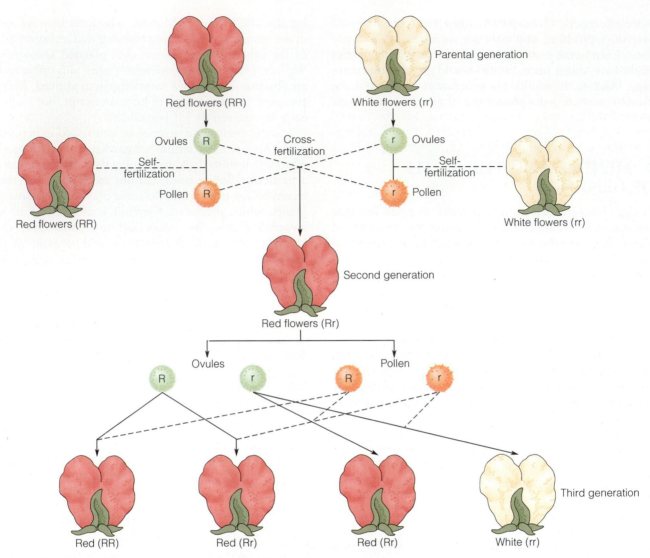

◆ **FIGURE 6.7** In his experiments with flower color in garden peas, Mendel used true-breeding strains. Such plants, shown here as the parental generation, when self-fertilized always yield offspring with the same trait as the parent. However, if the parental generation is cross-fertilized, all plants in the second generation will receive the combination of alleles indicated by Rr; these plants will have red flowers because R is dominant over r. The second generation of plants produces ovules and pollen with the alleles shown and, when left to self-fertilize, produces a third generation with a ratio of three plants with red flowers to one plant with white flowers.

for a trait is present in each sex cell (Fig. 6.7). For example, if R represents the allele for red flower color, and r represents white, the offspring may inherit the combinations of alleles symbolized as RR, Rr, or rr. And since R is dominant over r, only those offspring with the rr combination will have white flowers.

The most important aspects of Mendel's work can be summarized as follows: the factors (genes) controlling traits do not blend during inheritance, but are transmitted as discrete entities; and even though particular traits may not be expressed in each generation, they are not lost. Therefore, some variation in populations is accounted for by alternate expression of genes (alleles), and variation can be maintained.

Even though Mendelian genetics explains much about heredity, we now know the situation is much more complex. For example, our discussion has relied upon a single gene controlling a trait, but, in fact, most traits are controlled by many genes. Nevertheless, Mendel discovered the answers Darwin and Wallace needed, though they went unnoticed until three independent researchers rediscovered them in 1900.

Genes and Chromosomes

The cells of all organisms contain threadlike structures called **chromosomes.** Chromosomes are complex, double-stranded, helical molecules of **deoxyribonucleic acid (DNA)** (◆ Fig. 6.8). Specific segments, or regions, of the DNA molecule are the basic hereditary units, the genes.

The number of chromosomes is specific for a single species but varies among species. For example, the fruit fly *Drosophila* has 8 chromosomes, humans have 46, and horses have 64. However, chromosomes occur in pairs, pairs that carry genes controlling the same characteristics. Remember that the genes on chromosome pairs may occur in different forms, alleles.

In sexually reproducing organisms, the production of eggs and sperm results when parent cells undergo a type of cell division known as **meiosis.** The meiotic process yields cells containing only one chromosome of each pair (◆ Fig. 6.9a). Accordingly, eggs and sperm have only one-half the chromosome number of the parent cell; for example, human eggs and sperm have 23 chromosomes, one of each pair.

During reproduction, a sperm fertilizes an egg, producing a fertilized egg with the full chromosome complement for that species—46 in humans. As Mendel deduced from his garden pea experiments, one-half of the genetic makeup of the fertilized egg comes from each parent. The fertilized egg, however, develops and grows by a cell division process called **mitosis** that does not reduce the chromosome number (Fig. 6.9c).

According to the chromosomal theory of inheritance, chromosomes carrying the genetic determinants, the genes, are passed from one generation to the next. However, changes called **mutations** may occur in genes, and any such change occurring in sex cells is inheritable. We shall have more to say about mutations in the section on "Sources of Variation."

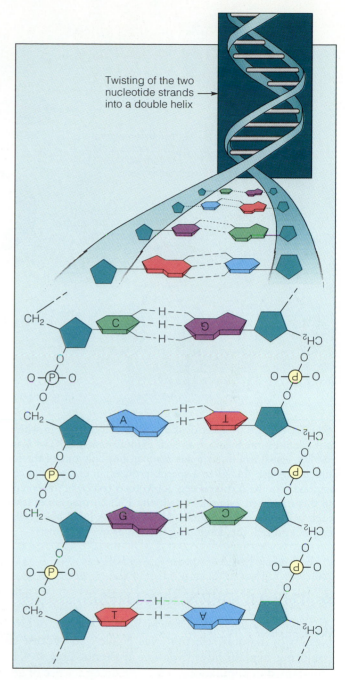

Twisting of the two nucleotide strands into a double helix

◆ **FIGURE 6.8** Chromosomes are double-stranded, helical molecules of deoxyribonucleic acid (DNA) shown here diagrammatically. Specific segments of chromosomes are genes. The two strands of the molecule are joined by hydrogen bonds (H).

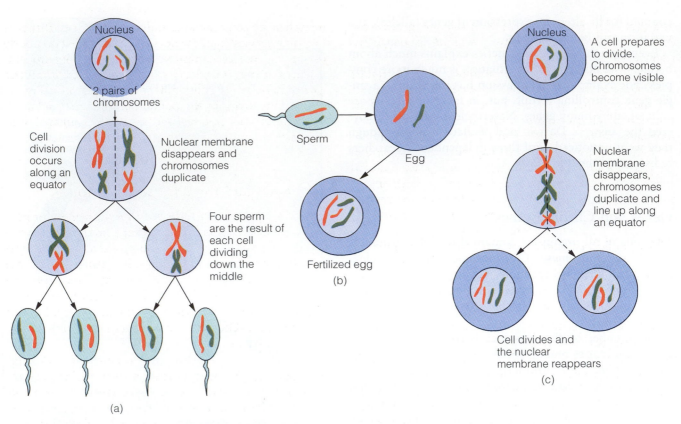

◆ **FIGURE 6.9** (a) Meiosis is a type of cell division in which sex cells are formed; each cell contains one member of each chromosome pair. The formation of sperm cells is shown; eggs form in a similar manner, but only one of the four final cells is a functional egg. (b) When a sperm fertilizes an egg, the full complement of chromosomes is present in the fertilized egg. (c) Mitosis results in the complete duplication of a cell. In this simplified example, a cell with four chromosomes (two pairs) produces two cells, each with four chromosomes. Mitosis occurs in all body cells except sex cells. Once an egg is fertilized, the developing embryo grows by mitosis.

❖ THE MODERN SYNTHESIS

As noted previously, the Darwin-Wallace concept of evolution by natural selection was challenged because it could not explain how variation arose nor how it was maintained in populations. The developing science of genetics partly answered the inheritance problems, but early in this century, geneticists believed that mutation rather than natural selection was the mechanism whereby evolution occurred.

During the 1930s and 1940s, the ideas developed by geneticists, paleontologists, population biologists, and others were brought together to form a **modern synthesis** or neo-Darwinian view of evolution. The chromosome theory of inheritance was incorporated into evolutionary thinking; mutation was seen as one source of variation in populations; Lamarckian concepts of inheritance were completely rejected; and the importance of natural selection was reaffirmed. The modern synthesis also emphasized that evolution is a gradual process, a point that has been challenged by some recent investigators, as will be discussed later.

Sources of Variation

The raw materials for evolution by natural selection are variations within populations. Most variations can be accounted for by sexual reproduction and the reshuffling of alleles from generation to generation. Considering that each of thousands of genes may have several alleles, and that offspring receive one-half of their genes from each parent, the potential for variation is enormous (◆ Fig. 6.10).

New variation may be introduced into a population by a change in genes, a mutation. To understand mutations, we must explore the function of chromosomes. One function of chromosomes is to direct the synthesis of molecules called proteins. During protein synthesis, information in the chromosome structure directs the formation of a protein by selecting the appropriate amino acids in a cell and arranging these amino acids into a sequence. The proteins produced determine the characteristics of an organism. Any change in the information directing protein synthesis is a mutation.

Two additional aspects of mutations are important. First, only mutations that occur in sex cells are inheritable. And second, mutations are random with respect to fitness. That is, there is no evidence of a predetermination of a mutation's effect; it may be harmful (many are), neutral, or beneficial. However, the attributes of harmful versus beneficial can be considered only with respect to the environment.

If a species is well adapted to the environment in which it lives, most mutations would not be particularly useful and perhaps would be harmful. But if the environment changes, what was once a harmful mutation may become beneficial. For example, some plants have developed a tolerance for contaminated soils around mines. Plants of the same species from the normal environment do poorly or die in these contaminated soils, while contaminant-resistant plants do poorly in the normal environment. The mutations for contaminant resistance probably occurred repeatedly in the population, but their adaptive significance was not realized until contaminated soils were present.

Speciation and the Rate of Evolution

In classifying organisms, the term **species** is used for populations of similar individuals that in nature can interbreed and produce fertile offspring. Sheep and goats belong to different species because in nature they do not interbreed, yet in captivity they may hybridize and produce fertile offspring. Obviously, barriers to reproduction are not complete in this case. If reproductive barriers are complete, different species cannot interbreed under any circumstances.

The process of speciation involves a change in the genetic makeup of the population, that is, a change in the frequency of alleles in its **gene pool**. A gene pool is simply all of the genes available in the genetic makeup of a population. There may also be a marked change in form and structure (**morphology**), but morphologic change is

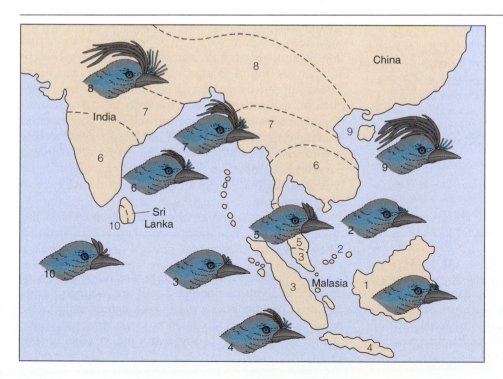

◆ **FIGURE 6.10** Variation in the form of the crest in the drongo (*Dicrurus paradiseus*) of southern Asia.

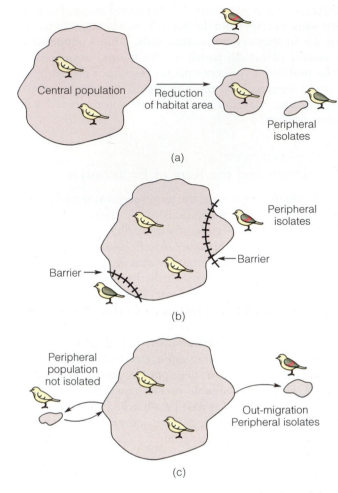

(a)

(b)

Peripheral isolates

Barrier

Barrier

(c)

Peripheral population not isolated

Out-migration Peripheral isolates

◆ **FIGURE 6.11** Allopatric speciation. (a) Reduction of the area occupied by a species may leave small, isolated populations, *peripheral isolates,* at the periphery of the once more extensive range. In this example, members of both peripheral isolates have evolved into new species. (b) Barriers have formed across parts of a central population's range, thereby isolating small populations. (c) Out-migration and the origin of a peripheral isolate.

not necessary in the origin of a new species. In fact, a descendant species may look very much like its ancestral species.

According to the concept of **allopatric speciation,** most species arise when a small part of a population is geographically isolated from its parent population by some kind of barrier. A rising mountain range, a stream, or an invasion of part of a continent by the sea may effectively isolate parts of a once-interbreeding population (◆ Fig.

6.11). Isolation may also occur when a few individuals of a species are accidentally introduced to a remote area.

Once a small population becomes isolated, it no longer shares in the parent population's gene pool, and it is subjected to different selection pressures. Under these conditions, the isolated population evolves independently and eventually may give rise to a reproductively isolated species. Numerous examples of allopatric speciation have been well documented, especially in extremely remote areas such as oceanic islands; the finches of the Galapagos Islands are a good example (Fig. 6.2).

Most investigators agree that allopatric speciation is a common phenomenon, but they disagree about how rapidly new species evolve. Darwin proposed that the origin of a new species is generally a gradual process. That is, the gradual accumulation of minor changes brings about a transition from one species to another, a process commonly called **phyletic gradualism** (◆ Fig. 6.12a). According to this concept, an ancestral species changes gradually and continuously and grades imperceptibly into a descendant species.

Another view, called **punctuated equilibrium,** holds that a species changes little or not at all during most of its history and then evolves rapidly to give rise to a new species. In other words, long periods of equilibrium are occasionally punctuated by short periods of rapid evolution. Accordingly, a new species arises rapidly, perhaps in a few thousand years, but once it has evolved, it remains much the same for the rest of its existence (◆ Fig. 6.12b).

Proponents of punctuated equilibrium claim that the fossil record supports their hypothesis. They argue that few examples of gradual transitions from one species to another are found in the fossil record. In fact, many species appear to have existed for millions of years without noticeable change. According to this view, transitional forms connecting ancestral and descendant species are unlikely to be preserved in the fossil record because species arise rapidly in small, geographically isolated populations. And, as a matter of fact, many species do appear abruptly in the fossil record with no evidence of direct ancestors.

Critics, however, point out that neither Darwin nor those who formulated the modern synthesis insisted that all evolutionary change was gradual and continuous, a view shared by many present-day biologists and paleontologists. Indeed, Darwin allowed for times during which evolutionary change in small populations could be quite rapid. Furthermore, deposition of sediments in most environments is not continuous; thus, the lack of gradual transitions in many cases is simply an artifact of the fossil

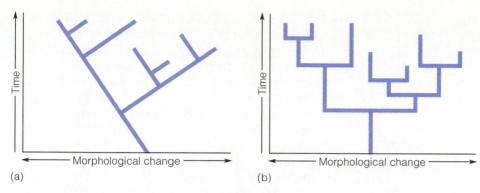

(a)

(b)

◆ **FIGURE 6.12** Comparison of two models for differentiation of organisms from a common ancestor. (a) In the phyletic gradualism model, slow, continuous change occurs as one species evolves into another. (b) According to the punctuated equilibrium model, change occurs rapidly, and new species evolve rapidly. However, little or no change occurs in a species during most of its existence.

record. And finally, despite the incomplete nature of the fossil record, there are a number of examples of gradual transitions from ancestral to descendant species (◆ Fig. 6.13).

Divergent, Convergent, and Parallel Evolution

When an interbreeding population gives rise to diverse descendant types of organisms, the process is referred to as **divergent evolution** (◆ Fig. 6.14). Divergence into numerous related types involves an **adaptive radiation,** which occurs when species of related ancestry exploit different aspects of the environment. At some time in the past, a population of finches, apparently from South America, reached the Galapagos Islands. This ancestral population diversified into the adaptive types of species of finches that Darwin observed (Fig. 6.2).

The fossil record provides many examples of divergence and adaptive radiation. The diversification and adaptive radiation of placental mammals from a shrew-like common ancestor is an excellent example (◆ Fig. 6.15). Following the extinction of dinosaurs and related reptiles at the end of the Mesozoic Era, placental mammals rapidly diversified and eventually gave rise to such adaptive types as whales, bats, elephants, and monkeys.

While divergent evolution leads to organisms that differ markedly from their ancestors, **convergent evolution** and **parallel evolution** are processes whereby similar adaptations arise in different groups (Fig. 6.14). Convergent evolution can be defined as the development of si-

miliar characteristics in distantly related organisms, and parallel evolution as the development of similar characteristics in closely related organisms (Fig. 6.14). The distinction between these two types of evolution depends on the degree of relatedness between the organisms in question, which is not always easy to determine. However, both convergent and parallel evolution occur as a result of different organisms adapting to similar ways of life.

The shells of clams and brachiopods, although not identical, are similar, and they represent an adaptive response to a similar life-style (◆ Fig. 6.16). Since clams

◆ **FIGURE 6.13** Gradual evolution of the small snail *Athleta* showing changes in size, shape, and the development of spines on the shell.

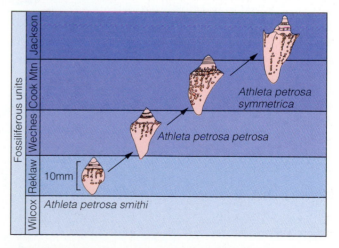

Athleta petrosa symmetrica

Athleta petrosa petrosa

10mm

Athleta petrosa smithi

◆ **FIGURE 6.14** In this example, divergent evolution results as species A diverges and gives rise to two descendant species, B and C. Parallel evolution also involves divergence from a common ancestor, A, but descendant species, B and C, then develop in a similar way as they adapt to a similar life-style. Convergent evolution is much like parallel evolution except that species B and C are distantly related but resemble one another in some aspects because of similar adaptations.

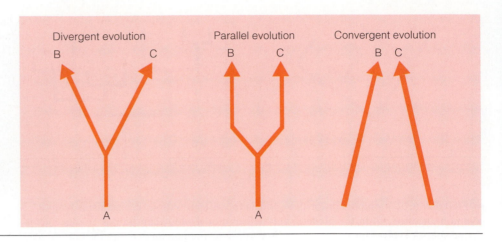

Divergent evolution

Parallel evolution

Convergent evolution

and brachiopods are very distantly related, their similar shells are an example of convergent evolution. Convergent evolution is also well illustrated by the similar adaptations of the Australian marsupial (pouched) mammals and the placental mammals of the other continents (◆ Fig. 6.17). Convergence accounts for the fact that many of the Australian marsupials superficially resemble placental mammals elsewhere.

Another impressive example of convergent evolution is shown by the Cenozoic mammals of North and South America (◆ Fig. 6.18). During most of the Cenozoic, South America was an island continent, and its mamma-

◆ **FIGURE 6.15** Divergent evolution of the placental mammals. An ancestor, probably a small, shrewlike animal, gave rise to diverse descendant groups, each of which adapted to different life-styles.

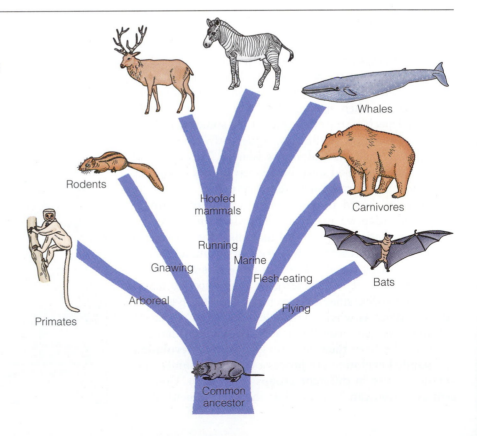

Whales

Rodents

Carnivores

Hoofed
mammals

Running

Gnawing

Marine

Flesh-eating

Bats

Arboreal

Flying

Primates

Common
ancestor

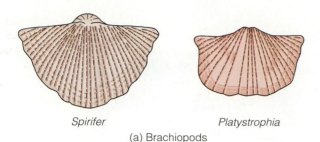

Spirifer   Platystrophia

(a) Brachiopods

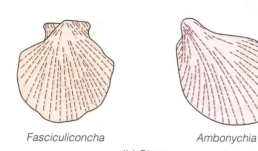

Fasciculiconcha   Ambonychia

(b) Clams

◆ **FIGURE 6.16** (a) Brachiopods and (b) clams are distantly related, but have adapted to a very similar habitat. Convergence is shown by the similarities in their shells; in both groups the shell consists of two parts hinged together. However, the details of the shells of each group differ in a number of aspects.

lian fauna, consisting of both marsupials and placentals, evolved independently. A number of mammals on each continent adapted to similar environments and developed many features in common (Fig. 6.18).

Parallel evolution is also a common phenomenon, but, as previously noted, it is not always easy to distinguish from convergent evolution. For example, jerboas and

◆ **FIGURE 6.17** Convergent evolution among Australian marsupial (pouched) mammals and placental mammals of the other continents. Marsupials and placentals are distantly related, but many adapted in similar ways, giving rise to superficially similar animals.

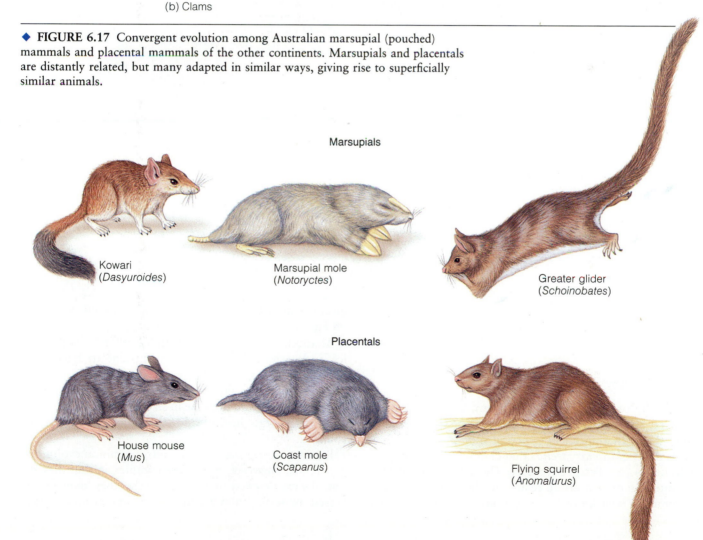

Marsupials

Kowari
(*Dasyuroides*)

Marsupial mole
(*Notoryctes*)

Greater glider
(*Schoinobates*)

Placentals

House mouse
(*Mus*)

Coast mole
(*Scapanus*)

Flying squirrel
(*Anomalurus*)

The Modern Synthesis   155

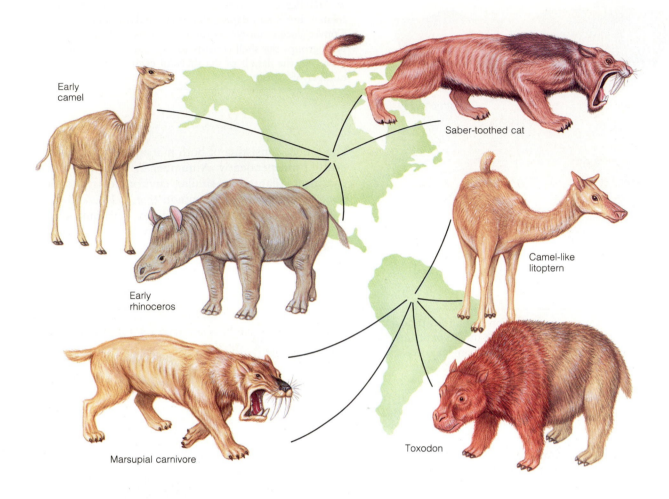

Early
camel

Saber-toothed cat

Early
rhinoceros

Camel-like
litoptern

Marsupial carnivore

Toxodon

◆ **FIGURE 6.18** Before the origin of the Isthmus of Panama a few million years ago, mammals of North and South America evolved independently. Similar adaptations of animals on each continent account for convergence.

kangaroo rats are closely related, but each independently developed similar structures (◆ Fig. 6.19).

Evolutionary Trends

The evolutionary history, or *phylogeny,* of a group can be worked out in some detail if sufficient fossil material is known. Furthermore, constructing such phylogenies reveals the *evolutionary trends* that characterize a group of organisms (see Perspective 6.1). The phylogeny of ammonites, an extinct group related to the living squid and octopus, is well known; one evolutionary trend in this

group was an increasingly complex shell structure (◆ Fig. 6.20).

Titanotheres were mammals that lived in North America and eastern Asia from Early Eocene to Early Oligocene time. A good fossil record records their phylogeny and several evolutionary trends (◆ Fig. 6.21). One trend was simply an increase in size, but titanotheres also developed large horns, and the shape of the skull changed.

Increase in size is one of the most common evolutionary trends. However, trends are complex in that they do not always proceed at the same rate, they may be reversed, or several may occur in the same group. Horses

◆ FIGURE 6.19 Parallel evolution. The kangaroo rat and jerboa are closely related but have independently developed similar features.

Kangaroo rat

Jerboa

◆ FIGURE 6.20 Evolutionary trends in the ammonites. Ammonites with coiled shells evolved from straight-shelled ancestors. Notice that an ammonite's shell is internally divided into chambers, and that the animal lives only in the outermost chamber. The sutures, the lines formed where chamber walls meet the wall of the outer shell, became increasingly more complex as ammonites evolved.

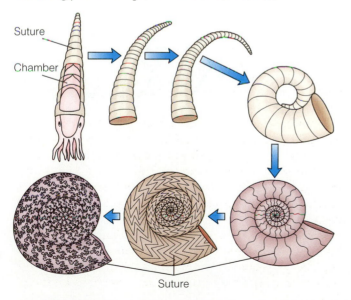

Suture

Chamber

Suture

◆ FIGURE 6.21 Evolutionary trends in the titanotheres, an extinct group of mammals related to modern horses and rhinoceroses. Titanotheres evolved from small ancestors to giants about 2.4 m tall at the shoulder and developed large horns.

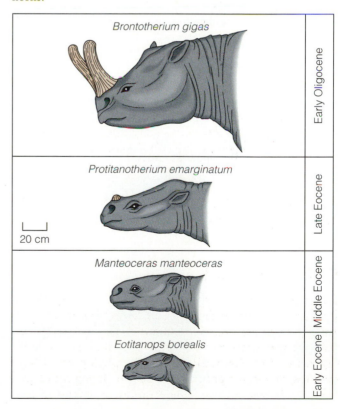

Brontotherium gigas

Early Oligocene

Protitanotherium emarginatum

Late Eocene

20 cm

Manteoceras manteoceras

Middle Eocene

Eotitanops borealis

Early Eocene

CLADISTICS AND CLADOGRAMS

Traditionally, the evolutionary history of many groups of organisms has been shown by *phylogenetic trees* (◆ Fig. 1). Such trees are constructed with the horizontal axis representing anatomical differences between organisms and the vertical axis representing time. These trees show patterns of ancestry and descent based on a wide range of characteristics, although the characteristics used are rarely specified.

Cladistics is derived from the word *clade*, which refers to a group of organisms whose members are more closely related among themselves than they are to other groups. A *cladogram* is a diagram showing these relationships. Cladograms are constructed on the basis of derived characteristics as opposed to primitive characteristics. For example, land-dwelling vertebrate animals (amphibians, reptiles, birds, and mammals) possess bone and paired limbs for locomotion on land. Such characteristics are considered primitive because they are shared by all land-dwelling vertebrates and are therefore of little use in establishing relationships.

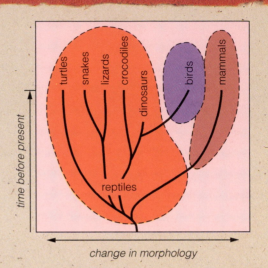

◆ **FIGURE 1** A phylogenetic tree showing the inferred relationships among reptiles, birds, and mammals.

Derived characteristics, on the other hand, are specializations that evolved after the primitive characteristics became established. For example, the amniote egg is a primi-

◆ **FIGURE 2** Cladogram showing the relationships among living reptiles, birds, and mammals. Some of the characteristics used to differentiate the subgroups are indicated.

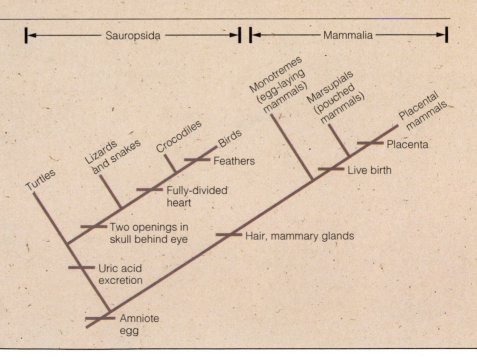

tive characteristic because it is shared by all reptiles, birds, and mammals (◆ Fig. 2). Hair and mammary glands, however, are derived characteristics because they serve to differentiate the mammals from those animals collectively called the Sauropsida (Fig. 2). Live birth is derived and indicates that monotremes (egg-laying mammals) differ from the marsupials (pouched mammals) and placental mammals.

Cladograms usually do not have an absolute time scale associated with them, but they nevertheless show the relative times of appearance of the groups possessing derived characteristics. In other words, they show evolutionary relationships. Figure 2 indicates that all mammals are more closely related to one another than they are to any of the sauropsids, and that crocodiles are more closely related to birds than either is to turtles, snakes, and lizards.

Cladistics and cladograms work well for living organisms, but are somewhat limited when applied to the fossil record. The main problem is recognizing and differentiating primitive characteristics from derived characteristics, especially in groups with poor fossil records. Even in the better known evolutionary lineages, the earliest diversification is usually the most poorly represented by fossils. Nevertheless, cladistics has been applied to a number of groups. The relationships among the Ceratopsidae, a family of horned dinosaurs, have been determined by cladistics (◆ Fig. 3). Notice in Figure 3 that in addition to the family Ceratopsidae, two subfamilies are also shown, the Centrosaurinae and the Chasmosaurinae, both of which were defined by cladistics. So, in addition to showing evolutionary relationships, cladograms can be used to classify organisms. Not all paleontologists and biologists accept cladistics because such analyses depend on traits that are assumed to have arisen solely by descent from a common ancestor. Parallel and convergent evolution are assumed to be of little importance, an assumption that may not be justified in many cases.

◆ **FIGURE 3** (a) Cladogram showing the relationships among the Ceratopsidae, a family of horned dinosaurs. (b) Examples of dinosaurs from each of the subfamilies of the Ceratopsidae.

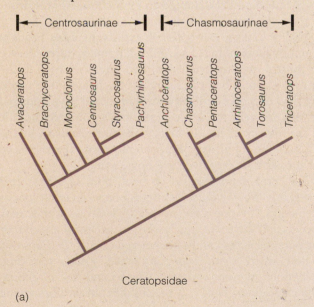

(a)

(b)

evolved from fox-sized ancestors that lived during the Eocene. One trend was an increase in size, but it was not a steady, uniform increase, and, in fact, some horses actually show a reversal in this trend (◆ Fig. 6.22). Horse evolution is also characterized by a reduction in the number of toes, lengthening of the legs and feet, and changes in the teeth and skull. All of these trends are well documented by fossils, but they did not occur at the same rate (Fig. 6.22).

One can view evolutionary trends as a series of adaptations to changing environments or adaptations that occur in response to exploitation of new habitats. The same trends occur repeatedly in the fossil record. Increase in

size is one of the most often repeated trends, but a number of others are known as well. Ammonites, for example, nearly became extinct at the end of the Paleozoic Era, but those that survived into the Mesozoic diversified and gave rise to many adaptive types very similar to those that had existed earlier. The evolutionary significance of this phenomenon is that very similar adaptations have occurred in many groups of organisms in different places throughout the history of life.

Some organisms, however, show little evidence of any evolutionary trends for long periods of time. One good example is the brachiopod *Lingula,* whose shell has not changed significantly since the Ordovician. Such organisms that show little or no change for long periods of time are commonly called **living fossils** (◆ Fig. 6.23). *Lingula* burrows into the muddy sediments of shallow seas. Apparently, once it developed its adaptive strategy, it stayed in an unchanging environment and has changed very little, at least in the appearance of its shell. Of course, *Lingula* may have evolved physiologically or reproductively, but fossils do not reveal changes such as these.

Extinctions

James Audubon (1785–1851) once estimated that a single flock of passenger pigeons numbered more than 1 billion individuals. In 1857 a committee of the Ohio Senate said the passenger pigeon needed no protection. In 1914 the last passenger pigeon died in the Cincinnati Zoo. Steller's sea cow was discovered in 1742 and 27 years later had been hunted to extinction. Rhinoceroses in the wild are now on the verge of extinction.

The preceding examples of extinction were brought about by humankind; and, indeed, humans have a degree of control over their environment unprecedented in the history of life. They can bring about changes that might not have occurred otherwise or that would have occurred at very different rates. However, extinctions are the rule in the history of life. In fact, extinction seems to be the ultimate fate of all species; probably more than 99% of all species that ever lived are now extinct.

The term *extinct* seems so final, and, in a very real sense, it is. But we must distinguish between two types of extinction. In one case a species evolves into a new species that differs so much from its ancestral group that the parent species can be considered extinct. Perhaps this is best called *pseudoextinction* since, strictly speaking, no extinction event occurred. The second case of extinction is terminal. That is, a species dies out without giving

◆ **FIGURE 6.22** Two of the evolutionary trends in horses. Horses increased in size during the Cenozoic but not at a uniform rate, and some horses reversed the trend and became smaller. The present-day horse, *Equus,* has one functional toe on each foot, but its oldest known direct ancestor, *Hyracotherium,* had four-toed forefeet and three-toed back feet. The diagram shows the trend in toe reduction in the forefeet. Forefeet shown are not to scale. Many more types of fossil horses are known in addition to those shown here.

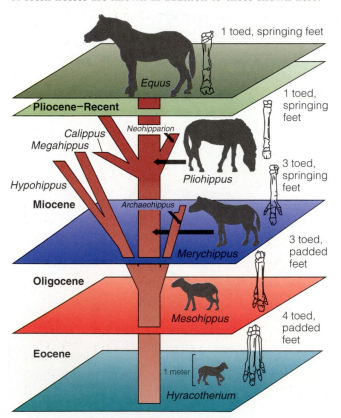

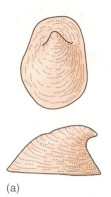

(a)

(b)

rise to anything else. Many examples of terminal extinction are known from the fossil record (◆ Fig. 6.24).

Extinction by one method or another appears to be a continual occurrence in the history of life. But so is the origin of new species that usually quickly exploit the opportunities created by the disappearance of other species. In some cases of extinction, an ecologically equivalent organism may not appear for some time. Ichthyosaurs were Mesozoic marine reptiles that lived like and superficially resembled modern porpoises and dolphins (◆ Fig. 6.25). Yet, whereas ichthyosaurs are unknown after the Mesozoic, porpoises and dolphins did not occupy their niche until some 30 million years later.

At times extinction rates have been greatly accelerated in events called *mass extinctions*. The two greatest crises in the history of life were the mass extinctions at the end of the Paleozoic and Mesozoic eras. In both cases, the diversity of life was sharply reduced. These extinction events, and some of lesser magnitude, will be discussed in greater detail in the following chapters.

❖ EVIDENCE FOR EVOLUTION

When Charles Darwin proposed his theory, he cited such supporting evidence as classification, embryology, comparative anatomy, the geographic distribution of organisms, and, to a limited extent, the fossil record. Darwin had little knowledge of the mechanism of inheritance or biochemistry, however, and molecular biology was completely unknown during his lifetime. Studies in these areas coupled with a better understanding of the fossil record have added to the body of evidence. Today, most

◆ **FIGURE 6.24** Some of the victims of the mass extinction at the end of the Cretaceous Period. One group of small carnivorous dinosaurs probably gave rise to birds, and so live on in their evolutionary descendants; but no other dinosaur group survived the terminal Cretaceous extinction. Among the marine animals, rudists (a type of clam) and ammonites both died out without leaving evolutionary descendants. Extinctions are common phenomena in the history of life. When they occur, they create opportunities for other organisms. For example, mammals coexisted with dinosaurs but were small and not very diverse. When the dinosaurs and their relatives became extinct, however, mammals rapidly exploited the opportunities created by the extinction event.

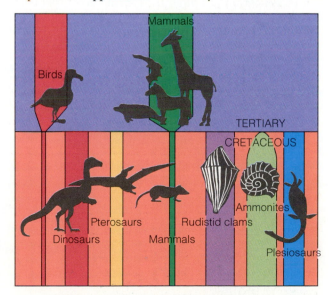

Dolphin

Ichthyosaur

◆ **FIGURE 6.25** Ichthyosaurs were fish-eating, marine reptiles that became extinct at the end of the Mesozoic Era. Following their extinction, no air-breathing animal occupied their niche until the dolphins and porpoises appeared about 30 million years later. Notice that ichthyosaurs and dolphins resemble one another in several features. Both groups adapted to a similar life-style and provide an excellent example of convergent evolution.

scientists accept that the theory of evolution is as well supported by evidence as any major theory.

Classification—A Nested Pattern of Similarities

The Swedish naturalist Carolus Linnaeus (1707–1778) proposed a formal classification of organisms. According to the Linnaean system, an organism is given a two-part name; the coyote, for example, is referred to by its genus (plural, *genera*)) name and its species name, *Canis latrans*.

Table 6.1 shows the basic arrangement of the Linnaean classification although the scheme now used con-

tains more categories. The arrangement is hierarchical. That is, as one proceeds up the list, the categories become more inclusive. The coyote (*Canis latrans*) and the wolf *(Canis lupus)* are members of different species, but they share a large number of characteristics, so both belong to the same genus. The red fox (*Vulpes fulva*) also shares some, but fewer, characteristics with coyotes and wolves, so it, along with other related doglike animals, make up the family Canidae. In turn, all canids share some characteristics with cats, bears, weasels, and raccoons, all of which belong to the order Carnivora. Likewise, all carnivores, rodents, bats, primates, and elephants have hair and mammary glands and constitute the class Mammalia (◆ Fig. 6.26). And so it goes, up to

◆ **TABLE 6.1 The Classification Scheme Now in Use Showing the Hierarchical Arrangement of the Categories**

The boxes include those animals to which the coyote, *Canis latrans*, is most closely related at various levels in the classification scheme (see Figure 6.26).

	Coyote	Wolf	Red fox	Bobcat	Horse	Snapping turtle	Amphioxus	Starfish	White clover
Kingdom	Animalia	Animalia	Animalia	Animalia	Animalia	Animalia	Animalia	Animalia	Plantae
Phylum	Chordata	Chordata	Chordata	Chordata	Chordata	Chordata	Chordata	Eninodermata	Pterophyta
Subphylum	Vertebrata	Vertebrata	Vertebrata	Vertebrata	Vertebrata	Vertebrata	Cephalochordata	Eleutherozoa	
Class	Mammalia	Mammalia	Mammalia	Mammalia	Mammalia	Reptilia	Leptocardii	Asteroidea	Angiospermae
Order	Carnivora	Carnivora	Carnivora	Carnivora	Perissodactyla	Chelonia		Forcipulata	Rosales
Family	Canidae	Canidae	Canidae	Felidae	Equidae	Chelydridae		Asteriidae	Leguminosae
Genus	*Canis*	*Canis*	*Vulpes*	*Lynx*	*Equus*	*Chelydra*	*Branchiostoma*	*Asterias*	*Trifolium*
Species	*latrans*	*lupus*	*vulpes*	*rufus*	*caballus*	*serpentinia*	*virginiae*	*forbesi*	*repens*

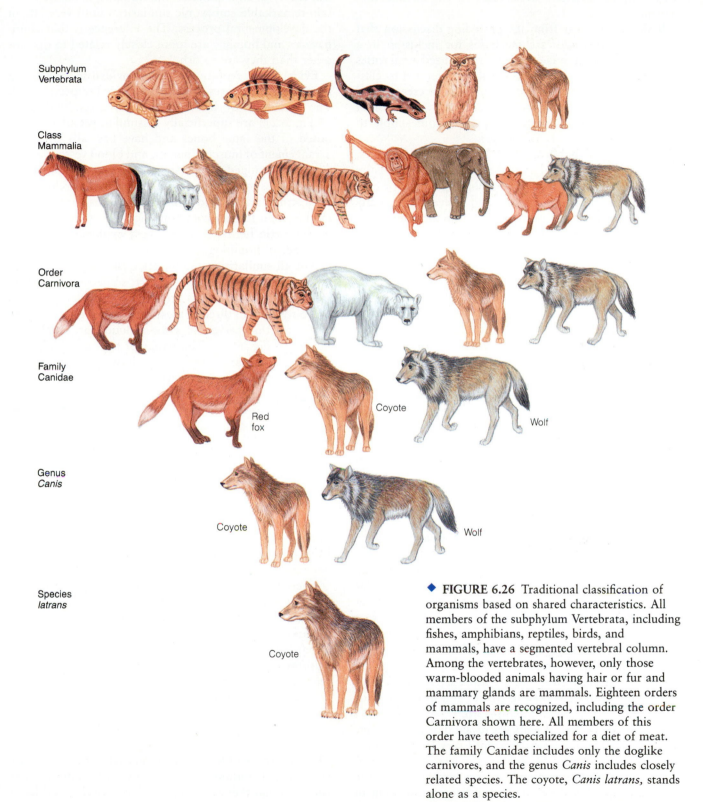

Subphylum
Vertebrata

Class
Mammalia

Order
Carnivora

Family
Canidae

Red
fox

Coyote

Wolf

Genus
Canis

Coyote

Wolf

Species
latrans

Coyote

◆ **FIGURE 6.26** Traditional classification of organisms based on shared characteristics. All members of the subphylum Vertebrata, including fishes, amphibians, reptiles, birds, and mammals, have a segmented vertebral column. Among the vertebrates, however, only those warm-blooded animals having hair or fur and mammary glands are mammals. Eighteen orders of mammals are recognized, including the order Carnivora shown here. All members of this order have teeth specialized for a diet of meat. The family Canidae includes only the doglike carnivores, and the genus *Canis* includes closely related species. The coyote, *Canis latrans*, stands alone as a species.

kingdom, the most inclusive category in the classification scheme.

It should be clear from the preceding discussion that shared characteristics are the bases for inclusion in a specific category. Linnaeus clearly recognized similarities among organisms; however, he simply intended to classify species that he believed were specially created and immutable. He viewed his classification as a reflection of the Great Chain of Being, a sequence from simple to complex, with humans at the apex.

Following the publication of *The Origin of Species,* biologists soon realized that shared characteristics among organisms constitute a strong argument for evolution. In our example, coyotes and wolves share many characteristics because they evolved from a common ancestor in the not too distant past. They also share some characteristics with bears and cats because all members of the order Carnivora had a common ancestor in the more remote past.

Biological Evidence for Evolution

According to evolutionary theory, all life-forms are related, and all living organisms descended with modification from ancestors that lived during the past. If this statement is correct, then there should be evidence of fundamental similarities among all life-forms, and closely related organisms should be more similar to one another than to more distantly related organisms. As a matter of fact, all living things, from bacteria to whales, are composed of the same chemical elements—mostly carbon, nitrogen, hydrogen, and oxygen. Furthermore, the chromosomes of all organisms are composed of DNA, except bacteria, which have RNA; and in all cells, proteins are synthesized in essentially the same way.

Thousands of biochemical tests of numerous organisms have provided compelling evidence for evolutionary relationships. Blood proteins, for example, are similar among all mammals, but they also indicate that among the primates humans are most closely related to the great apes, followed, in order, by the Old World monkeys, the New World monkeys, and the lower primates such as lemurs. Biochemical tests indicate that all birds are related among themselves and are more closely related to turtles and crocodiles than to snakes and lizards, a finding corroborated by the fossil record.

Studies of developing embryos reveal that some organisms are similar to one another until very late in embryonic development, while others are less similar. Fish, chimpanzee, and human embryos are very similar in the earliest stages, but differences soon become apparent in fish embryos. Chimpanzees and humans, in contrast, retain remarkable embryonic similarities until very late in the developmental process. The inference is that chimpanzees and humans are more closely related to one another than they are to fish.

Evidence for evolution is also provided by comparing anatomical structures of organisms (see Perspective 6.2). The forelimbs of humans, whales, dogs, and birds (◆ Fig. 6.27) are superficially dissimilar, yet all are composed of the same bones and have basically the same arrangement of muscles, nerves, and blood vessels; all are similarly arranged with respect to other structures, and all have a similar pattern of embryonic development. Such structures are said to be **homologous.** Descent with modification from a common ancestor—that is, a common genetic heritage—is the best explanation for the existence of homologous organs.

Not all similarities among organisms are evidence of evolutionary relationships; in fact, some similarities are rather superficial. Insects are not closely related to bats and birds, but all three of these animals have wings. Such structures that serve the same function are **analogous organs** (◆ Fig. 6.28). The wings of bats and birds, but not of insects, are homologous, however, and provide evidence for a common ancestry. Insect wings, on the other hand, represent convergence—a similar solution to an adaptive problem, flight—but they differ from bat and bird wings in both structure and embryological development.

Another type of evidence for evolution is provided by observations of small-scale evolution in living organisms (◆ Fig. 6.29, p. 168). We have already mentioned one example, the adaptations of plants to contaminated soils. New insecticides and pesticides must be developed continually as insects and rodents develop resistance to existing ones. Development of antibiotic-resistant strains of bacteria constitutes a continuing problem in medicine.

The important points regarding small-scale evolution are that variation existed, and that some variant types had a better chance to survive and reproduce. The source of variation is irrelevant; it may have already existed, or it may have been introduced by mutations. Regardless of its source, the survival and reproduction of some variant types caused the genetic makeup of the population, its gene pool, to change through time.

Fossils and Evolution

The occurrence of fossil marine animals far from the sea, even high in mountains, led many early naturalists to conclude that they were deposited during the worldwide

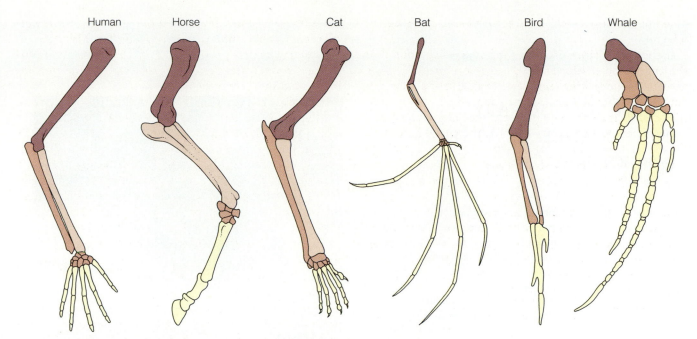

Human   Horse   Cat   Bat   Bird   Whale

◆ **FIGURE 6.27** Homologous organs such as the forelimbs of vertebrates may serve different functions but are composed of the same elements and undergo similar embryological development.

◆ **FIGURE 6.28** The fly's wings serve the same function as wings of birds and bats but have a different structure and different embryological development. Organs that serve the same function are analogous, but the bird and bat wings are also homologous.

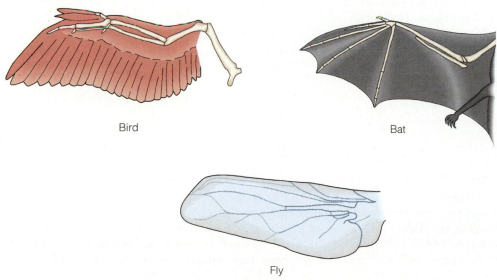

Bird

Bat

Fly

THE EVOLUTIONARY SIGNIFICANCE OF VESTIGIAL STRUCTURES

Probably every living species has **vestigial structures**—nonfunctional or partly functional remnants of organs that were functional in their ancestors. Vestigial structures are leftovers in the evolutionary sense, structures that are probably being lost. For example, one trend in human evolution has been shortening of the jaw, which leaves too little room for the third molars, the so-called wisdom teeth. Some people never develop wisdom teeth at all, but in others, these teeth may not erupt through the gums or may come in crooked and useless. Anyone who has suffered the pain of impacted wisdom teeth can attest to the fact that the human jaw has too little room for these teeth.

Many mammals have dewclaws, which are small, functionless, vestiges of fingers and toes. Dewclaws are particularly obvious in dogs (◆ Fig. 1). The ancestors of dogs were rather flat-footed animals with five functional digits on each foot, but as dogs evolved, they became toe-walkers, and only four digits contact the ground. Consequently, the first digit of each foot, the thumbs and big toes, lost their function and were reduced in size.

Vestigal toes are not restricted to dogs. Indeed, many animals possess such structures. Among the even-toed hoofed mammals, deer and pigs walk on two toes but retain obvious vestiges of two additional toes. Horses have only one functional toe in each foot, the third, which corresponds to the middle finger and middle toe of the human hand and foot. Horses, however, retain remnants of the second and fourth toes as so-called splint-bones, one on each side of the main toe (◆ Fig. 2). These splint-bones are not usually visible because they lie beneath the skin, but a few horses are born with extra toes:

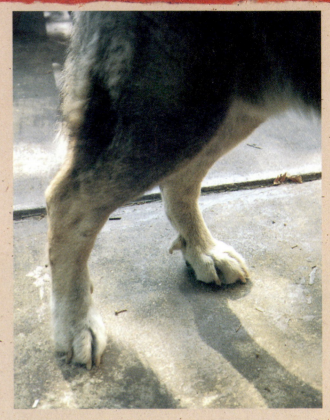

◆ **FIGURE 1** Left hindfoot of a dog showing the dewclaw, which is a vestige of the big toe.

In rare cases, both fore and hind feet may each have two extra digits fairly developed, and all of nearly equal size, thus corresponding to the feet of the extinct [horse] *Protohippus*[1]

[1]Marsh, "Recent Polydactyl Horses," *American Journal of Science* 43 (April 1882): 342.

biblical flood. As early as 1508, Leonardo da Vinci realized that the fossil distribution was not what one would expect in a rising flood. Nevertheless, the flood explanation persisted, and John Woodward (1665–1728) proposed that the fossils had separated out of the floodwater according to their density. In other words, Woodward's hypothesis predicted that the fossils should be arranged in a vertical sequence, with the densest at the bottom followed upward by those of decreasing density. This hypothesis was quickly rejected because field observa-

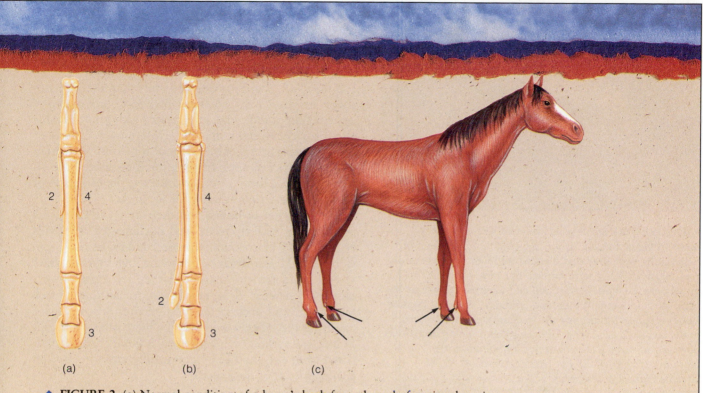

◆ **FIGURE 2** (a) Normal condition of a horse's back foot; the only functional toe is the third, but side splints representing toes 2 and 4 are still present. (b) Horse with an extra toe. (c) Horse with an extra toe on each foot.

Whales adapted to an aquatic life-style and in the process lost the rear limbs possessed by their land-dwelling ancestors. Yet whales still have a remnant of the pelvic girdle, and a few whales have been caught that have rear limbs. Furthermore, one major group of whales have no teeth as adults but as embryos develop a full set of teeth similar to those of their land-dwelling ancestors. These teeth do not erupt through the gums; they are resorbed before birth and thus serve no function at all.

The kiwi, a small flightless bird living only in New Zealand, has tiny remnants of the typical bones of a bird wing.

It seems doubtful that anyone would seriously argue that dogs were specially created with dewclaws or that extra toes in horses really serve some function. Their presence makes sense, however, if they are the remnants of once-functional organs. Charles Darwin had this to say:

They [vestigial organs] have been partially retained by the power of inheritance, and relate to a former state of things.[2]

---

[2]*The Origin of Species* (New York: New American Library, 1958), pp. 419–20.

tions clearly did not support the prediction; fossils of various densities are found throughout the fossil record.

The fossil record does show a sequence, but not one based on density, size, or shape of the fossils. Rather, it shows a sequence of appearances of different life-forms through time (◆ Fig. 6.30). The fact is that older and older fossiliferous strata contain organisms increasingly different from those living today. One-celled organisms appeared before multicelled organisms, plants before animals, and invertebrates before vertebrates. Among the

◆ **FIGURE 6.29** Small-scale evolution of the peppered moths of Great Britain. In pre-industrial England most of these moths were gray and blended well with the lichens on trees. Black varieties were known, but rare, since they stood out against the light background and were easily spotted by birds. With industrialization and pollution, the trees became soot-covered and dark, and the frequency of dark-colored moths increased while that of light-colored moths decreased. However, in rural areas unaffected by pollution, the light-dark frequency did not change. (Photo © M. W. Tweedie/Photo Researchers.)

vertebrates, fish appear first, followed in order by amphibians, reptiles, mammals, and birds (Fig. 6.30). Once established, many primitive groups have persisted to the present, but their order of first appearance is consistent with evolutionary theory.

If the fossil record provided no more than the sequence shown in Figure 6.30, however, it could be interpreted as a succession of independently created groups. But fossils also provide evidence for the evolutionary origins of some of these groups and evidence for evolution within groups. Some examples of such fossil evidence have already been discussed in other contexts, the evolution of ammonites and titanotheres, for example.

The evolution of horses and their living relatives, the rhinoceroses and tapirs, is well documented by fossils (◆ Fig. 6.31). Horses, rhinoceroses, and tapirs may seem to be an odd assortment of animals, but they all share several characteristics that imply they are related by evolutionary descent. If so, one would predict that as each family is traced back in the fossil record, differentiating one from the other would become increasingly difficult. In fact, the earliest members of each family are differentiated from one another by minor differences in size and characteristics of their teeth.

◆ **FIGURE 6.30** The times of appearance of invertebrates and major groups of vertebrate animals.

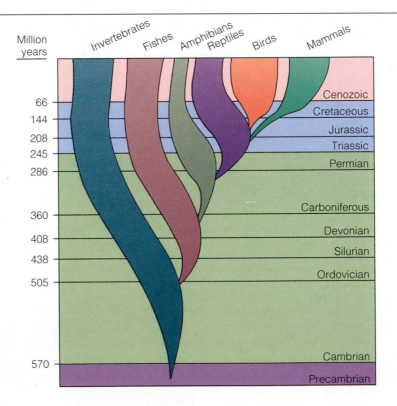

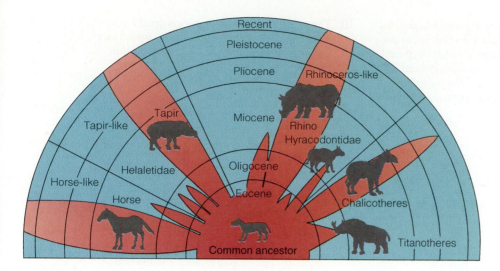

◆ **FIGURE 6.31** The divergence and adaptive radiation of the odd-toed hoofed mammals (order Perissodactyla) which includes horses, rhinoceroses, and tapirs, and their extinct relatives, is well documented by fossils. The evolution of these animals is discussed in more detail in Chapter 19.

Fossils from the Jurassic Solnhofen Limestone of Germany show the first appearance of features we associate with birds. Several fossil specimens showing feather impressions have been discovered, but in almost every other known physical feature, these fossils are more similar to small carnivorous dinosaurs. As a matter of fact, two specimens were classified as dinosaurs until the feather impressions were noticed.

This early birdlike creature, *Archaeopteryx* (ancient feather), illustrates the concept of **mosaic evolution.** All organisms are mosaics of characteristics, some of which are retained from the ancestral condition, while others are more recently evolved. *Archaeopteryx*, for example, retains dinosaurlike teeth, tail, hind limb structure, and brain size but also possesses feathers and a wishbone, characteristics typical of birds.

Many people have the idea that the fossil record provides the main evidence for evolution. Fossils are unquestionably important in evolutionary studies, but one must not overlook the fact that fossils are only one of several lines of evidence that support the concept of evolution. Shared characteristics, biochemistry, comparative anatomy, and small-scale evolution of living organisms are equally important.

Chapter Summary

1. The first formal theory of evolution to be taken seriously was proposed by Jean Baptiste de Lamarck. Inheritance of acquired characteristics was his mechanism for evolution.
2. In 1859 Charles Darwin and Alfred Russell Wallace simultaneously published their views on evolution. They proposed natural selection as the mechanism for evolutionary change.
3. Darwin's observations of variation in natural populations and artificial selection, as well as his reading of Thomas Malthus's essay on population, helped him formulate the idea that natural processes select favorable variants for survival.
4. Gregor Mendel's breeding experiments with garden peas provided some of the answers regarding how variation is maintained and passed on. Mendel's work is the basis for modern genetics.
5. Genes are the hereditary determinants in all organisms. This genetic information is carried in the chromosomes of cells. Only the genes in the chromosomes of sex cells are inheritable.
6. Most variation is maintained in populations by sexual reproduction and mutations.
7. Evolution by natural selection is a two-step process. First, variation must be produced and maintained in interbreeding populations, and second, favorable variants must be selected for survival.
8. An important way in which new species evolve is by allopatric speciation. When a group is isolated from its parent

population, gene flow is restricted or eliminated, and the isolated group is subjected to different selection pressures.

9. Divergent evolution involves an adaptive radiation during which an ancestral stock gives rise to diverse species. The development of similar adaptive types in different groups of organisms results from parallel and convergent evolution.

10. The evidence for evolution includes the way organisms are classified, comparative anatomy and embryology, biochemical similarities, small-scale evolution, and the fossil record.

Important Terms

adaptive radiation
allele
allopatric speciation
analogous organ
artificial selection
chromosomes
convergent evolution
divergent evolution
DNA (deoxyribonucleic acid)

gene
gene pool
homologous organ
inheritance of acquired characteristics
living fossil
meiosis
mitosis
modern synthesis
morphology

mosaic evolution
mutation
natural selection
parallel evolution
phyletic gradualism
punctuated equilibrium
species
theory of evolution
vestigial structure

Review Questions

1. Genes controlling the same trait occur in different forms known as:
   a. _____ homologous organs; b. _____ alleles;
   c. _____ species; d. _____ factors; e. _____ DNA.

2. The type of cell division that produces functional sex cells containing one-half of the chromosomes typical of a species is:
   a. _____ mosaic evolution; b. _____ punctuated equilibrium; c. _____ allopatric speciation; d. _____ meiosis; e. _____ adaptation.

3. The first person to propose a formal theory of evolution was:
   a. _____ Alfred Russell Wallace; b. _____ Jean Baptiste de Lamarck; c. _____ Erasmus Darwin; d. _____ Charles Lyell; e. _____ William Smith.

4. One of the main sources of variation in populations is:
   a. _____ spontaneous generation; b. _____ fixity of species; c. _____ sexual reproduction; d. _____ phyletic gradualism; e. _____ mitosis.

5. According to the concept of punctuated equilibrium:
   a. _____ species change gradually and continually over millions of years; b. _____ most species that ever existed are now extinct; c. _____ the most common evolutionary trend is increase in size; d. _____ new species arise rapidly, perhaps in a few thousands of years.

6. Clams and brachiopods are very distantly related but possess a number of similarities because of:
   a. _____ punctuated equilibrium; b. _____ convergent evolution; c. _____ adaptive divergence; d. _____ parallel adaptation; e. _____ artificial selection.

7. In the biological classification scheme now used, organisms are assigned to genera, families, and orders on the basis of:
   a. _____ type of fossilization; b. _____ shared characteristics; c. _____ time of existence; d. _____ inheritance of acquired characteristics; e. _____ vestigial structures.

8. All of the genes available to an interbreeding population constitute its:
   a. _____ source of mutations; b. _____ chromosome base; c. _____ gene pool; d. _____ phylogeny; e. _____ species resource.

9. Organs containing the same basic structural elements, such as a bird's wing and a dog's forelimb, are said to be:
   a. _____ homologous; b. _____ artificial; c. _____ vestigial; d. _____ mutations; e. _____ mosaics.

10. Genes, the basic units of inheritance, are specific segments of:
    a. _____ sex cells; b. _____ analogous organs; c. _____ fossils; d. _____ species; e. _____ chromosomes.

11. The diversification of a species into two or more descendant species is an example of:
a. _____ the modern synthesis; b. _____ parallel evolution; c. _____ phyletic equilibrium; d. _____ mosaic evolution; e. _____ divergent evolution.

12. In order to be members of a species, organisms must:
a. _____ interbreed and produce fertile offspring; b. _____ breed with other species; c. _____ show rapid evolutionary change; d. _____ exhibit little or no change for several millions of years; e. _____ have a good fossil record.

13. Chromosomes are complex molecules of:
a. _____ drosophila; b. _____ deoxyribonucleic acid; c. _____ carbon dioxide; d. _____ potassium, aluminum, and silicon; e. _____ calcium carbonate.

14. Speciation resulting from isolation of a small part of an interbreeding population is referred to as _____ speciation.
a. _____ adaptive radiation; b. _____ convergent; c. _____ gradual; d. _____ allopatric; e. _____ punctuated.

15. Compare evolution by inheritance of acquired characteristics with evolution by natural selection.

16. How does sexual reproduction maintain variation in interbreeding populations?

17. What observations led Darwin to conclude that natural selection is a mechanism for evolution?

18. What is a mutation? Explain what is meant by harmful, neutral, and beneficial mutations.

19. Explain the process whereby an organism accidentally introduced to a remote island may give rise to a new species.

20. Give examples of divergent and convergent evolution.

21. How does the way in which organisms are classified provide evidence for evolution?

22. Explain how the concept of punctuated equilibrium accounts for the origin of new species. Compare this with phyletic gradualism.

23. What is meant by the term mosaic evolution?

24. How does pseudoextinction differ from terminal extinction?

25. What are the most important aspects of Gregor Mendel's breeding experiments with garden peas?

26. Explain how the fossil record provides evidence for organic evolution.

27. How are adaptive radiation and divergent evolution related?

28. What is a living fossil? Give examples.

Additional Readings

Ayala, F. J. 1978. The mechanisms of evolution. *Scientific American* 239, no. 3: 56–69.

Bajema, C. J. 1985. Charles Darwin and selection as a cause of adaptive evolution, 1837–1859. *The American Biology Teacher* 47: 226–32.

Bingham, R. 1982. On the life of Mr. Darwin. *Science* 82, no. 3: 34–39.

Bowler, P. J. 1990. *Charles Darwin: The man and his influence.* Cambridge, Mass.: Blackwell.

Dobzhansky, T., F. J. Ayala, G. L. Stebbins, and J. W. Valentine. 1977. *Evolution.* San Francisco: W. H. Freeman.

Futuyma, D. J. 1986. *Evolutionary biology.* 2d ed. Sunderland, Mass.: Sinauer Associates.

Grant, P. R. 1991. Natural selection and Darwin's finches. *Scientific American* 265, no. 4: 82–87.

Hoffman, A. 1989. *Arguments on evolution: A paleontologist's perspective.* New York: Oxford University Press.

Holton, N., II. 1968. *The evidence of evolution.* New York: American Heritage Publishing Co.

Mayr, E. 1978. Evolution. *Scientific American* 239, no. 3: 47–55.

CHAPTER 7

Satellite view of the Himalayas, Mount Everest (right center), and the Ganges River plain. The upper portion shows the dry, vegetationless Tibetan Plateau which rises more than 4,300m. The land rises even higher to the snowcapped peaks of the main Himalayas in the center of the photo. The warm, vegetated, and heavily populated Ganges River plain is visible in the lower half of the photo. (Photo courtesy of NASA.)

PLATE TECTONICS: A UNIFYING THEORY

Prologue

A fundamental question in planetary geology has been the nature and extent of volcanic and tectonic activity within the solar system. All of the terrestrial planets—Mercury, Venus, Earth, and Mars—experienced a similar early history that was marked by widespread volcanism and meteorite impacts, both of which helped modify their surfaces. The volcanic and tectonic activity and resultant surface features (other than meteorite craters) of these planets are clearly related to the way in which they transport heat from their interiors to their surfaces.

The Earth appears to be unique in that its surface is broken into a series of individual moving plates. The other three terrestrial planets all have a single, globally continuous plate and thus exhibit fewer types of volcanic and tectonic activity than the Earth.

Mercury's surface is heavily cratered and shows little in the way of primary volcanic structures. Its largest impact basins, however, are filled with what appear to be lava flows similar to the lava plains on the Moon. Mercury also has a global system of lobate scarps that formed when the planet cooled and contracted.

Mars has numerous features that indicate an extensive early period of volcanism. These include *Olympus Mons*, the solar system's largest volcano, lava flows, and uplifted regions believed to have resulted from mantle convection. In addition to volcanic features, Mars also displays abundant evidence of tensional tectonics, including numerous faults and large fault-produced valley structures. Though Mars was tectonically active during the past, there is no evidence that plate tectonics comparable to that on Earth has ever occurred there.

Radar images from the *Magellan* spacecraft, which has been mapping the surface of Venus, reveal a wide variety of spectacular, fascinating, and, in many cases, enigmatic landscapes. Among the features discovered are volcanoes; extensive lava flows and channels; pancake-shaped volcanic features, many of which are surrounded by complex networks of cracks and fractures (◆ Fig. 7.1); intricate networks of fractures and faults, some producing elongated valleys and ridges;

◆ **FIGURE 7.1** This radar image of The Eistla region of Venus made by the *Magellan* spacecraft reveals prominent pancake-shaped volcanic domes 65 km in diameter with broad flat tops less than 1 km in height. A complex network of cracks and fractures surrounds the volcanic features. Geologists think these features result from cooling and the withdrawal of lava.

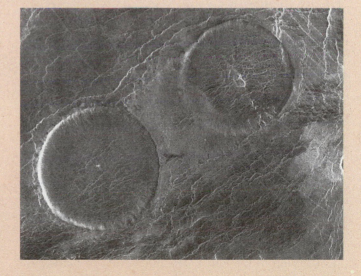

and areas of crumpled crust that superficially resemble mountain ranges formed by compression.

The topographic features of Venus appear to be primarily the result of upwelling and sinking of material within the planet's interior. Blobs of magma rising from the interior could have produced the large, circular, pancake-shaped features and the networks of cracks and fractures that surround them. Furthermore, many of the large elevated regions of the planet such as Maxwell and Aphrodite could have resulted from activity similar to mantle plumes on Earth. As the rising material spread outward near the surface, cooled, and sank back into the mantle, the overlying crust around the highlands would crumple, producing features that appear similar to mountain ranges formed by compression. Though there is no evidence of plate tectonics on Venus, the volcanic and associated features certainly attest to an internally active planet.

While all of the terrestrial planets and the Earth's moon underwent a period of active volcanism during their early history, it appears that only three bodies in the solar system show any signs of present-day volcanism: the Earth, the Jovian moon Io, and perhaps the Neptunian moon Triton.

Io, the innermost of the four large moons of Jupiter, is probably the most volcanically active body yet observed in the solar system. It is brilliantly colored in red, oranges, and yellows resulting from the various sulfur compounds spewed forth by geysers and volcanoes. To date, at least 10 active volcanoes have been discovered on Io.

The Neptunian moon Triton also appears to be volcanically active and to have experienced numerous episodes of deformation. Evidence from images returned by *Voyager 2* indicates that Triton has geysers that are erupting frozen nitrogen crystals and organic compounds.

❖ INTRODUCTION

The decade of the 1960s was a time of geological as well as social and cultural revolution. The ramifications of the newly proposed plate tectonic theory radically changed the way in which geologists viewed the Earth. No longer could it be regarded as an unchanging planet on which continents and ocean basins remained fixed through time. Instead, it was perceived as a dynamically changing planet whose surface consists of constantly moving plates.

Plate tectonics has been the dominant process affecting the evolution of the Earth. The interaction of plates de-

termines the locations of continents, ocean basins, and mountain chains, which in turn affect atmospheric and oceanic circulation patterns that ultimately determine global climates. Furthermore, the physical changes resulting from plate interactions have profoundly influenced the evolution of the Earth's biota.

Geologists now view Earth history in terms of interrelated events that are part of a global panorama of dynamic change through time. For example, the Paleozoic history of the Appalachian Mountains is no longer considered an isolated regional event but rather part of a global interaction of plates that culminated in the creation of a large supercontinent at the end of the Paleozoic.

This chapter provides an overview of plate tectonic theory as it is used to interpret the history of the Earth from the rock and fossil record. This powerful and unifying theory accounts for many seemingly unrelated geologic phenomena, allowing them to be viewed as part of a continuing story rather than as a series of isolated incidents.

❖ EARLY OPINIONS ABOUT CONTINENTAL DRIFT

The earliest maps showing the east coast of South America and the west coast of Africa probably provided people with the first evidence that continents may have once been joined together, then broken apart and moved to their present positions. It was not until 1858, however, that the Frenchman Antonio Snider-Pellegrini suggested in his book, *Creation and Its Mysteries Revealed,* that all of the continents were linked together during the Pennsylvanian Period and later split apart. He based his conclusions on the similarity between plant fossils occurring in coal measures in Europe and North America. Snider-Pellegrini envisioned the rupturing of this single continent as the result of the Great Flood. Some years after Snider-Pellegrini, another Frenchman, Elisée Reclus, discussed the idea of continental movement in his book, *The Earth,* published in 1872. Reclus believed continental drift, not the Great Flood, was responsible for mountain building, volcanism, and earthquake activity.

Also during this time, the Austrian geologist Edward Suess noted the similarity between the Late Paleozoic plant fossils of India, Australia, South Africa, and South America as well as evidence of glaciation in the rock sequences of these southern continents. The plant fossils comprise a unique flora that occurs in the coal layers just above the glacial deposits of these southern continents. This flora is very different from the contemporaneous

(a)

(b)

◆ **FIGURE 7.2** Representative members of the *Glossopteris* flora. Fossils of these plants are found on all five of the Gondwana continents. (a) *Glossopteris* leaves from the Upper Permian Dunedoo Formation and (b) the Upper Permian Illawarra Coal Measures, Australia. (Photos courtesy of Patricia G. Gensel, University of North Carolina.)

coal swamp flora of the northern continents and is collectively known as the ***Glossopteris*** flora after its most conspicuous genus (◆ Fig. 7.2).

In his book, *The Face of the Earth,* published in 1885, Suess coined the term *Gondwanaland* (or **Gondwana**) for the supercontinent composed of the aforementioned southern continents. Gondwana is a locality in India where abundant fossils of the *Glossopteris* flora are found in coal seams. Suess believed these continents were connected by land bridges over which plants and animals migrated. Thus, in his view, the similar fossils of the continents were due to the appearance and disappearance of the connecting land bridges.

In 1910 the American geologist Frank B. Taylor published a pamphlet in which he presented his own theory of continental drift. In it he envisioned the present-day continents as parts of larger polar continents that eventually broke apart and migrated toward the equator. He attributed the breakup of these polar continents to gigantic tidal forces, which he believed were generated when the Moon became a satellite of the Earth about 100 million years ago (we now know that this is incorrect) and thereby slowed the Earth's rotational rate. One of Taylor's most significant contributions was his suggestion that the submarine ridge discovered in the Atlantic Ocean by the *Challenger* expeditions of 1872–1876 might mark the site along which one ancient continent split apart to form the present-day Atlantic Ocean.

Alfred Wegener and Continental Drift

Alfred Wegener (◆ Fig. 7.3) is generally credited with developing the theory of **continental drift**. In his book, *The Origin of Continents and Oceans,* first published in

◆ **FIGURE 7.3** Alfred Wegener proposed the theory of continental drift based on geological, paleontological, and climatological evidence. He is shown here in an expedition hut in Greenland, waiting out the Arctic winter of 1912–1913. (Photo courtesy of the Bildarchiv Preussicher Kulturbesitz.)

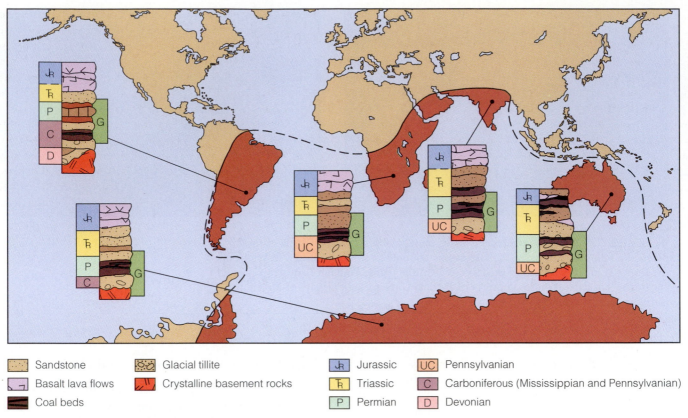

Sandstone	Glacial tillite	Jurassic	Pennsylvanian
Basalt lava flows	Crystalline basement rocks	Triassic	Carboniferous (Mississippian and Pennsylvanian)
Coal beds		Permian	Devonian

◆ **FIGURE 7.4** Marine, nonmarine, and glacial rock sequences of Pennsylvanian to Jurassic age are nearly the same for all Gondwana continents. Such close similarity suggests that they were joined together at one time. The range indicated by G is that of the *Glossopteris* flora.

1915, he portrayed his grand concept of continental movement in a series of maps. He proposed the name **Pangaea** for the supercontinent that resulted from the unification of preexisting continents into a single landmass. According to Wegener, Pangaea began breaking apart during the Cenozoic Era. We now know, however, that this fragmentation began during the Triassic Period.

Wegener amassed a tremendous amount of geological, paleontological, and climatological evidence in support of continental drift. He noted that marine and nonmarine rock sequences of the same age are found on widely separated continents (◆ Fig. 7.4); mountain ranges and glacial deposits match up when continents are united into a single landmass; the shorelines of continents fit together, forming a large supercontinent; many of the same extinct plant and animal groups are found today on widely separated continents, indicating the continents

must have been in proximity at one time (◆ Fig. 7.5); and the rock record indicates that the global climatic belts were not always parallel to the present equator. Wegener argued that this vast amount of evidence from a variety of sources surely indicated that the continents must have been close together at one time in the past.

Additional Support for Continental Drift

With the publication of *The Origin of Continents and Oceans* and its subsequent four editions, some scientists began to take Wegener's unorthodox views seriously. Alexander du Toit, a South African geologist, was one of his more ardent supporters. He further developed Wegener's arguments and introduced more geologic evidence in support of continental drift. In 1937 du Toit published *Our Wandering Continents*, in which he contrasted the glacial deposits of Gondwana with coal deposits of the same age found in the continents of the Northern Hemisphere. To resolve this apparent climatological paradox, du Toit placed the southern continents of Gondwana at or near the South Pole and arranged the northern continents together such that the coal deposits were located at the equator. He named this northern landmass **Laurasia.**

Du Toit also provided additional paleontological support for continental drift by noting that fossils of the Permian freshwater reptile *Mesosaurus* occurred in rocks of the same age in both Brazil and South Africa (Fig. 7.5). Since the physiology of freshwater and marine animals is completely different, most paleontologists find it hard to imagine how a freshwater reptile could have swum across the Atlantic Ocean and then found a freshwater environment nearly identical to its former habitat. Furthermore, if *Mesosaurus* could have swum across the Atlantic Ocean, shouldn't its fossils occur in other localities besides Brazil and South Africa? It is more logical to assume that *Mesosaurus* occupied a large lake that covered portions of both continents when they were united.

In spite of what seemed to be overwhelming evidence presented by Wegener and later by du Toit and others, most geologists simply refused to entertain the idea that continents might have moved in the past. The geologists

◆ **FIGURE 7.5** Some of the plants and animals whose fossils are found today on the widely separated continents of South America, Africa, India, Australia, and Antarctica. These continents were joined together during the Late Paleozoic to form Gondwana, the southern landmass of Pangaea. *Mesosaurus* is a freshwater reptile whose fossils are found in Brazil and Africa; *Glossopteris* is a Permian plant found on all the continents shown; *Lystrosaurus* is an Early Triassic reptile whose fossils are found in Africa, India, and Antarctica; *Cynognathus* is another Early Triassic reptile whose fossils are found in South America and Africa.

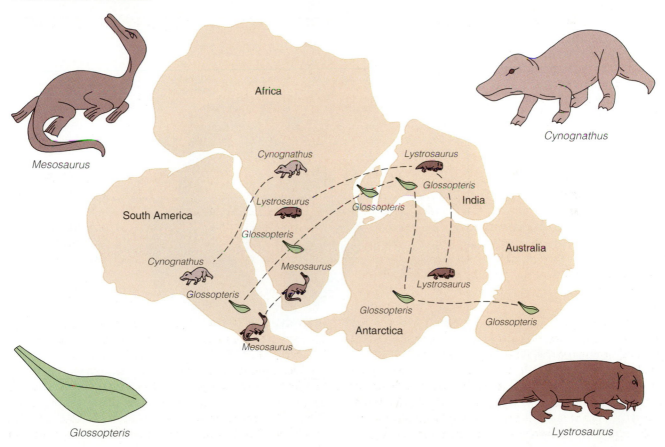

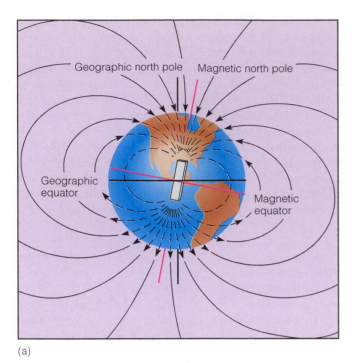

(a)

were not necessarily being obstinate about accepting new ideas; rather they found the proposed mechanisms for continental drift inadequate and unconvincing.

❖ PALEOMAGNETISM AND POLAR WANDERING

Interest in continental drift revived in the 1950s as a result of new evidence from paleomagnetic studies of the Earth. **Paleomagnetism** is the remanent magnetism in ancient rocks recording the direction of the Earth's magnetic poles at the time of the rock's formation. The Earth can be thought of as a giant dipole magnet in which the magnetic poles essentially coincide with the geographic poles (❖ Fig. 7.6). Such an arrangement means that the strength of the magnetic field is not constant, but varies, being weakest at the equator and strongest at the poles. The Earth's magnetic field is believed to result from the different rotation speeds of the outer core and mantle.

When a magma cools, the magnetic iron-bearing minerals align themselves with the Earth's magnetic field, recording both its direction and strength. The tempera-

◆ **FIGURE 7.6** (a) The magnetic field of the Earth has lines of force just like those of a bar magnet. (b) The strength of the magnetic field changes uniformly from the magnetic equator to the magnetic poles. This change in strength causes a dip needle to parallel the Earth's surface only at the magnetic equator, whereas its inclination with respect to the surface increases to 90° at the magnetic poles.

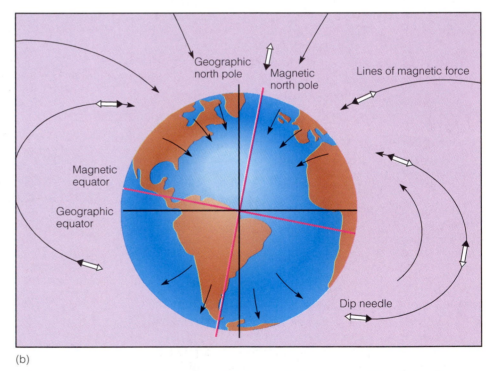

(b)

ture at which iron-bearing minerals gain their magnetization is called the **Curie point.** As long as the rock is not subsequently heated above the Curie point, it will preserve that remanent magnetism. Thus, an ancient lava flow provides a record of the orientation and strength of the Earth's magnetic field at the time the lava flow cooled.

As paleomagnetic research in the 1950s progressed, some unexpected results emerged. When geologists measured the paleomagnetism of recent rocks, they found it was generally consistent with the Earth's current magnetic field. The paleomagnetism of ancient rocks, however, showed different orientations. For example, paleomagnetic studies of Silurian lava flows in North America indicated that the north magnetic pole was located in the western Pacific Ocean at that time, while the paleomagnetic evidence from Permian lava flows indicated a pole in Asia, and that of Cretaceous lava flows pointed to yet another location in northern Asia. When plotted on a map, the paleomagnetic readings of numerous lava flows from all ages in North America trace the apparent movement of the magnetic pole through time (◆ Fig. 7.7). Thus, the paleomagnetic evidence from a single continent could be interpreted in three ways: the continent remained fixed and the north magnetic pole moved; the north magnetic pole stood still and the continent moved; or both the continent and the north magnetic pole moved.

Upon analysis, magnetic minerals from European Silurian and Permian lava flows pointed to a different magnetic pole location than those of the same age from North America (Fig. 7.7). Furthermore, analysis of lava flows from all continents indicated each continent had its own series of magnetic poles. Does this mean there were different north magnetic poles for each continent? That would be highly unlikely and difficult to reconcile with the theory accounting for the Earth's magnetic field. The best explanation for such data is that the magnetic poles have remained at their present locations at the geographic north and south poles and the continents have moved. This interpretation suggests a changing panorama of opening and closing ocean basins with mountain ranges forming along the margins of colliding continents. Furthermore, when the continental margins are fitted together so that the paleomagnetic data point to only one magnetic pole, we find, just as Wegener did, that the rock sequences and glacial deposits match up, and the fossil evidence is consistent with the reconstructed paleogeography.

❖ MAGNETIC REVERSALS AND SEAFLOOR SPREADING

Geologists refer to the Earth's present magnetic field as being normal, that is, with the north and south magnetic poles located approximately at the north and south geographic poles. At numerous times in the geologic past, the Earth's magnetic field has completely reversed. The existence of such **magnetic reversals** was discovered by dating and determining the orientation of the remanent magnetism in lava flows on land (◆ Fig. 7.8). Once their existence was well established, magnetic reversals were also discovered in ocean basalts as part of the extensive mapping of the ocean basins that took place in the 1960s. Although the cause of magnetic reversals is still uncertain, their occurrence in the geologic record is well documented.

In addition to the discovery of magnetic reversals, mapping of the ocean basins also revealed a ridge system 65,000 km long, constituting the most extensive mountain range in the world. Perhaps the best-known part of

---

◆ **FIGURE 7.7** The apparent paths of polar wandering for North America and Europe. The apparent location of the north magnetic pole is shown for different periods on each continent's polar wandering path.

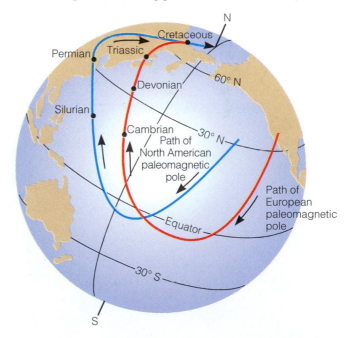

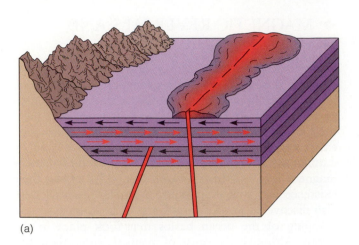

(a)

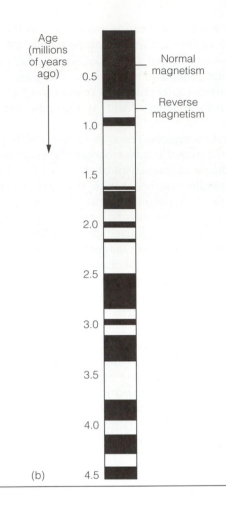

Age
(millions
of years
ago)

0.5 — Normal magnetism

— Reverse magnetism

1.0

1.5

2.0

2.5

3.0

3.5

4.0

4.5

(b)

◆ **FIGURE 7.8** (a) Magnetic reversals recorded in a succession of lava flows are shown diagrammatically by red arrows, whereas the record of normal polarity events is shown by black arrows. The lava flows containing a record of such magnetic-polarity events can be radiometrically dated so that a magnetic time scale as in (b) can be constructed. (b) Magnetic reversals for the last 4.5 million years as determined from lava flows on land. Black bands represent normal magnetism and blue bands represent reverse magnetism.

◆ **FIGURE 7.9** Artistic view of what the Atlantic Ocean basin would look like without water. The major feature is the Mid-Atlantic Ridge. (Photo courtesy of ALCOA.)

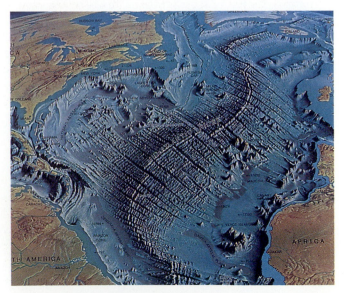

the ridge system is the Mid-Atlantic Ridge, which divides the Atlantic Ocean basin into two nearly equal parts (◆ Fig. 7.9).

In 1962, as a result of the oceanographic research conducted in the 1950s, Harry Hess of Princeton University proposed the theory of **seafloor spreading** to account for continental movement. Hess suggested that continents do not move across oceanic crust, but rather that the continents and oceanic crust move together. He suggested that seafloors separate at oceanic ridges where new crust is formed by upwelling magma. As the magma cools, the newly formed oceanic crust moves laterally away from the ridge. As a mechanism to drive this system, Hess revived the idea of **thermal convection cells** in the mantle; that is, hot magma rises from the mantle, intrudes along rift zone fractures defining oceanic ridges, and thus forms new crust. Cold crust is subducted back into the mantle at deep-sea trenches, where it is heated and recycled, thus completing a thermal convection cell.

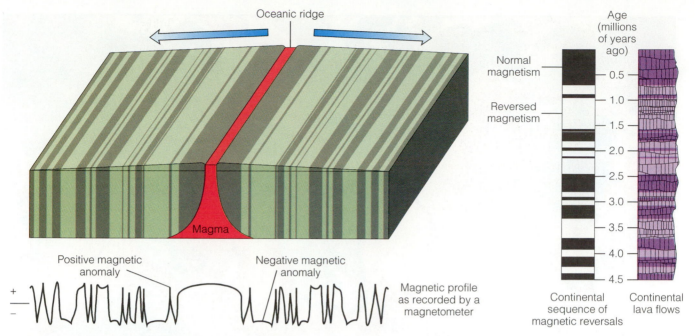

◆ **FIGURE 7.10** The sequence of magnetic anomalies preserved within the oceanic crust on both sides of an oceanic ridge is identical to the sequence of magnetic reversals already known from continental lava flows. Magnetic anomalies are formed when basaltic magma intrudes into oceanic ridges; when the magma cools below the Curie point, it records the Earth's magnetic polarity at the time. Subsequent intrusions split the previously formed crust in half, so that is moves laterally away from the oceanic ridge. Repeated intrusions produce a symmetrical series of magnetic anomalies that reflect periods of normal and reversed polarity. The magnetic anomalies are recorded by a magnetometer, which measures the strength of the magnetic field.

How could Hess's theory be confirmed? Magnetic surveys of the oceanic crust revealed striped **magnetic anomalies** (deviations from the average strength of the Earth's magnetic field) in the rocks that were both parallel to and symmetrical with the oceanic ridges (◆ Fig. 7.10). Furthermore, the pattern of oceanic magnetic anomalies matched the pattern of magnetic reversals already known from studies of continental lava flows (Fig. 7.8). When magma wells up and cools along a ridge summit, it records the Earth's magnetic field at that time as either normal or reversed. As new crust forms at the summit, the previously formed crust moves laterally away from the ridge. These magnetic stripes, representing times of normal or reversed polarity, are parallel to and symmetrical around oceanic ridges (where upwelling magma forms new oceanic crust), conclusively confirming Hess's theory of seafloor spreading.

One of the consequences of the seafloor-spreading theory is its confirmation that ocean basins are geologically young features whose openings and closings are partially responsible for continental movement (◆ Fig. 7.11). Radiometric dating reveals that the oldest oceanic crust is less than 180 million years old, while the oldest continental crust is 3.96 billion years old. Although geologists do not universally accept the idea of thermal convection cells as a driving mechanism for plate movement, most accept that plates are created at oceanic ridges and destroyed at deep-sea trenches, regardless of the driving mechanism involved.

❖ PLATE TECTONICS AND PLATE BOUNDARIES

According to **plate tectonic theory**, the Earth's surface is divided into rigid plates bounded by mid-oceanic ridges, oceanic trenches, faults, and mountain belts (◆ Fig. 7.12). By the late 1960s most geologists accepted the theory of plate tectonics, in part because the evidence

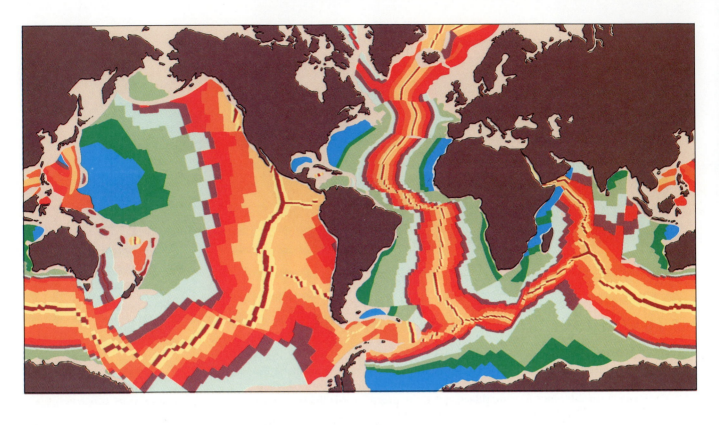

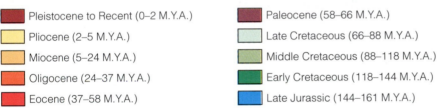

■ Pleistocene to Recent (0–2 M.Y.A.)

■ Pliocene (2–5 M.Y.A.)

■ Miocene (5–24 M.Y.A.)

■ Oligocene (24–37 M.Y.A.)

■ Eocene (37–58 M.Y.A.)

■ Paleocene (58–66 M.Y.A.)

■ Late Cretaceous (66–88 M.Y.A.)

■ Middle Cretaceous (88–118 M.Y.A.)

■ Early Cretaceous (118–144 M.Y.A.)

■ Late Jurassic (144–161 M.Y.A.)

◆ **FIGURE 7.11** The age of the world's ocean basins established from magnetic anomalies demonstrates that the youngest oceanic crust is adjacent to the spreading ridges and that its age increases away from the ridge axis.

was overwhelming, but also because it is a unifying theory that accounts for a variety of apparently unrelated geologic features and events.

Because it is believed that plate tectonics has operated since at least the Proterozoic (see Chapter 10), it is important that we understand how plates move and interact with each other and how ancient plate boundaries are recognized. After all, the movement of plates has had a profound effect on the geologic and biological history of this planet.

Geologists recognize three major types of plate boundaries: *divergent, convergent,* and *transform.* It is along

these boundaries that new plates are formed, consumed, or slide laterally past one another. To understand the implications of plate interactions as they have affected the history of the Earth, geologists must study present plate boundaries.

Divergent Plate Boundaries

Divergent plate boundaries or **spreading ridges** occur where plates are separating and new oceanic lithosphere is forming. These boundaries most commonly occur along oceanic ridges or, less commonly, beneath conti-

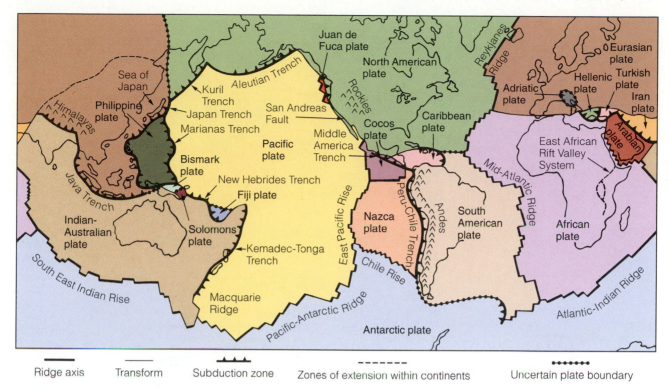

◆ **FIGURE 7.12** The Earth's lithosphere is divided into plates of various sizes that move over the asthenosphere.

nents where the boundary can be recognized by rift valleys. Along a divergent plate boundary at an oceanic

◆ **FIGURE 7.13** Pillow lavas forming along the Mid-Atlantic Ridge. Their distinctive bulbous shape is the result of underwater eruption. (Photo courtesy of Woods Hole Oceanographic Institution.)

ridge, magma wells up and the plates move apart. As the magma cools, new strips of oceanic crust are formed and record the magnetic field of the Earth at that time (Fig. 7.10). Rugged topography, high topographic relief, normal faulting with many associated shallow-focus earthquakes, high heat flow, and basaltic pillow lavas (◆ Fig. 7.13) are common features associated with these oceanic ridges.

Divergent plate boundaries also occur under continents during the early stages of continental breakup (◆ Fig. 7.14). When magma wells up beneath a continent, the crust is initially elevated, stretched, and thinned. Normal faults and a rift valley then begin to form along a central *graben* (the down-dropped fault block) producing shallow-focus earthquakes. During this stage, magma typically intrudes into fractures, forming sills and dikes, as well as flowing onto the graben floor. The East African rift valleys are an excellent example of this stage of continental breakup (◆ Fig. 7.15). As rifting proceeds, the continental crust eventually breaks. If magma continues welling up, the two parts of the continent will move away from each other, as is happening today beneath the Red Sea. As this newly formed narrow

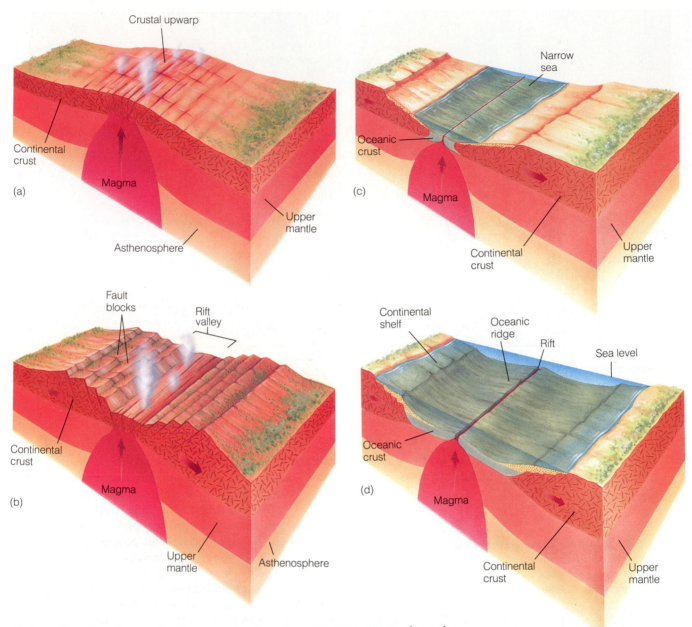

◆ **FIGURE 7.14** History of a divergent plate boundary. (a) Rising magma beneath a continent pushes the crust up, producing numerous cracks and fractures. (b) As the crust is stretched and thinned, rift valleys develop, and lava flows onto the valley floors. (c) Continued spreading further separates the continent until a narrow seaway develops. (d) As spreading continues, an oceanic ridge system forms, and an ocean basin develops and grows.

ocean basin continues to enlarge, it will eventually become an expansive ocean basin such as the Atlantic or Pacific Ocean basins are today.

An Example of Ancient Rifting

What features in the rock record can geologists use to recognize ancient rifting? Associated with regions of con-

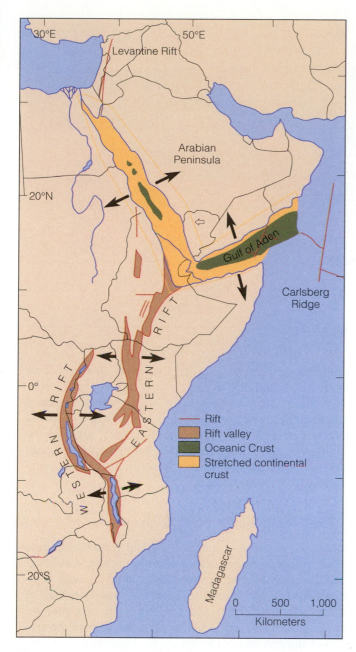

◆ **FIGURE 7.15** The East African rift valley is being formed by the separation of eastern Africa from the rest of the continent along a divergent plate boundary. The Red Sea represents an advanced stage of rifting in which two continental blocks are separated by a narrow sea.

(a)

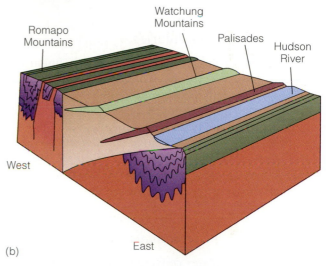

(b)

◆ **FIGURE 7.16** (a) Location of the Triassic fault basins in eastern North America. (b) Cross section of a typical Triassic fault basin, showing the thick accumulation of sediments and the two sills forming the Watchung Mountains and the Palisades.

tinental rifting are normal faults, dikes, sills, lava flows, and thick sedimentary sequences within rift valleys. The Triassic fault basins of the eastern United States are a good example of ancient continental rifting (◆ Fig. 7.16). These fault basins mark the zone of rifting that occurred when North America split apart from Africa. They contain thousands of meters of continental sediment and are riddled with dikes and sills (see Chapter 15).

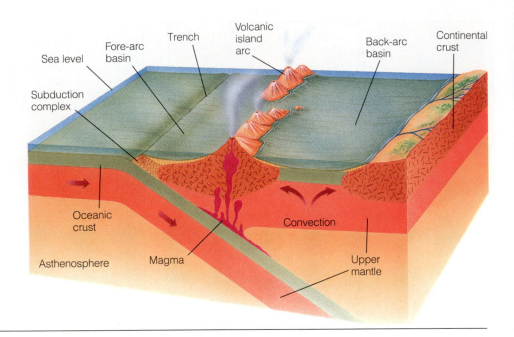

◆ **FIGURE 7.17** Oceanic-oceanic plate boundary. An oceanic trench forms where one oceanic plate is subducted beneath another. On the nonsubducted oceanic plate, a volcanic island arc forms from the rising magma generated from the subducting plate.

Convergent Plate Boundaries

Whereas new crust forms at divergent plate boundaries, old crust is destroyed at many **convergent plate boundaries.** One plate is subducted under another and eventually is resorbed in the asthenosphere. When we talk about convergent plate boundaries, we are really talking about three different types of boundaries: *oceanic-oceanic, oceanic-continental,* and *continental-continental.* The basic processes are the same for all three types of boundaries, but because different types of crust are involved, the results are different.

Oceanic-Oceanic Plate Boundaries

When two oceanic plates converge, one of them is subducted under the other along an **oceanic-oceanic plate boundary** (◆ Fig. 7.17). The subducting plate bends downward to form the outer wall of an oceanic trench. A *subduction complex,* composed of wedge-shaped slices of highly folded and faulted marine sediments and oceanic lithosphere scraped off the descending plate, forms along the inner wall. This subduction complex is elevated as a result of uplift along faults as subduction continues (Fig. 7.17).

As the subducting plate descends into the mantle, it is heated and partially melted, generating a magma commonly of andesitic composition. This magma is less dense than the surrounding mantle rocks and rises to the surface of the nonsubducted plate, forming a curved chain of volcanic islands called a **volcanic island arc.** This arc is nearly parallel to the oceanic trench and is separated from it by a distance of up to several hundred kilometers—the distance depends on the angle of dip of the subducting plate. The Aleutian Islands, Japanese Island chain, and Philippine Island region are good examples of volcanic island arcs resulting from oceanic-oceanic plate convergence.

Oceanic-Continental Plate Boundaries

When oceanic crust is subducted under continental crust along an **oceanic-continental plate boundary,** a subduction complex, consisting of wedge-shaped slices of complexly folded and faulted rocks, forms the inner wall of the trench. Between it and the continent is a *fore-arc basin* containing detrital sediments derived from the erosion of the continents (◆ Fig. 7.18). These sediments are typically flat-lying or only mildly deformed. The andesitic magma generated by subduction rises beneath the continent and either crystallizes as plutons before reaching the surface or erupts at the surface, producing a chain of andesitic volcanoes (also called a volcanic arc). A back-arc basin may be filled with continental detrital sediments, pyroclastic materials, and lava flows, derived from and thickening toward the volcanic arc. An excellent example of an oceanic-continental plate boundary is the Pacific coast of South America where the oceanic Nazca plate is presently being subducted under South

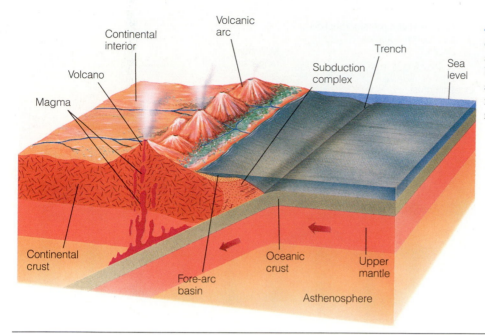

Volcanic
arc

Continental
interior

Volcano

Magma

Trench

Subduction
complex

Sea
level

Continental
crust

Fore-arc
basin

Oceanic
crust

Upper
mantle

Asthenosphere

◆ **FIGURE 7.18** Oceanic-continental plate boundary. When an oceanic plate is subducted beneath a continental plate, an andesitic volcanic mountain range is formed on the continental plate as a result of rising magma.

America. The Peru-Chile Trench marks the site of subduction, and the Andes Mountains are the resulting volcanic mountain chain on the nonsubducting plate (Fig. 7.12). This particular example demonstrates the effect plate tectonics has on our lives. For instance, earthquakes are commonly associated with subduction zones, and the western side of South America is the site of frequent and devastating earthquakes. Furthermore, the southern Andes Mountains act as an effective barrier to moist, easterly blowing Pacific winds, resulting in a desert east of the southern Andes that is virtually uninhabitable.

Continental-Continental Plate Boundaries

Two continents approaching each other will initially be separated by an ocean floor that is being subducted under one continent. The edge of that continent will display the features characteristic of oceanic-continental convergence. As the ocean floor continues to be subducted, the two continents will come closer together until they eventually collide. Because continental lithosphere, which consists of continental crust and the upper mantle, is less dense than oceanic lithosphere (oceanic crust and the upper mantle), it cannot sink into the asthenosphere. Although one continent may partly slide under the other, it cannot be pulled or pushed down into a subduction zone (◆ Fig. 7.19).

When two continents collide, they are welded together along a zone marking the former site of subduction. At

this **continental-continental plate boundary,** an interior mountain belt is formed consisting of thrusted and folded sediments, igneous intrusions, metamorphic rocks, and fragments of oceanic crust. In addition, the entire region is subjected to numerous earthquakes. The Himalayas in central Asia are the result of a continental-continental collision between India and Asia that began about 40 to 50 million years ago and is still continuing.

Ancient Convergent Plate Boundaries

How can former subduction zones be recognized in the rock record? One clue is provided by igneous rocks. The magma that erupts on the Earth's surface, forming island arc volcanoes and continental volcanoes, is of andesitic composition. Another clue can be found in the zone of intensely deformed rocks that occurs between the deep-sea trench where subduction is taking place and the area of igneous activity. Here, sediments and submarine rocks are folded, faulted, and metamorphosed into a chaotic mixture of rocks termed a *melange*.

During subduction, pieces of oceanic lithosphere are sometimes incorporated into the melange and accreted onto the edge of the continent. Such slices of oceanic crust and upper mantle are called **ophiolites** (◆ Fig. 7.20). They consist of a layer of deep-sea sediments that include graywackes (poorly sorted sandstones containing abundant feldspars and rock fragments, usually in a clay-rich matrix), black shales, and cherts. These deep-sea

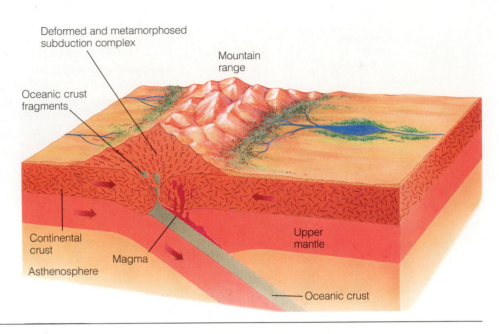

◆ **FIGURE 7.19** Continental -continental plate boundary. When two continental plates converge, neither is subducted because of their great thickness and low and equal densities. As the two plates collide, a mountain range is formed in the interior of a new and larger continent.

sediments are underlain by pillow basalts, a sheeted dike complex, massive gabbro, and layered gabbro, all of which form the oceanic crust. Beneath the gabbro is peridotite, which probably represents the upper mantle. Ophiolites are key features in recognizing plate convergence along a subduction zone (see the Prologue in Chapter 10).

Elongate belts of folded and thrust-faulted marine sediments, andesites, and ophiolites are found in the Appalachians, Alps, Himalayas, and Andes mountains. The combination of such features is good evidence that these mountain ranges resulted from deformation along convergent plate boundaries.

Transform Plate Boundaries

The third type of plate boundary is a **transform boundary.** These occur along transform faults where plates slide laterally past each other, roughly parallel to the direction of plate movement. Although lithosphere is neither created nor destroyed along a transform boundary, the movement between plates results in a zone of intensely shattered rock and numerous shallow-focus earthquakes.

Transform faults are particular types of faults that "transform" or change one type of motion between plates into another type of motion. The majority of transform faults cut oceanic crust and are marked by distinct fracture zones. Transform faults can also extend into continents. One of the best-known transform faults

is the San Andreas fault in California, which separates the Pacific plate from the North American plate (◆ Fig. 7.21). The many earthquakes that affect California are the result of movement along this fault.

◆ **FIGURE 7.20** Ophiolites are sequences of rock on land consisting of deep-sea sediments, oceanic crust, and upper mantle.

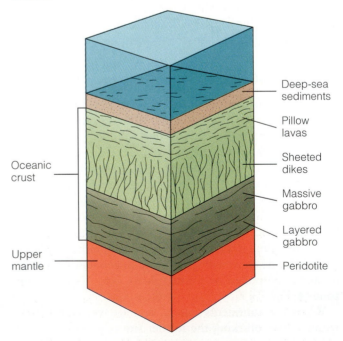

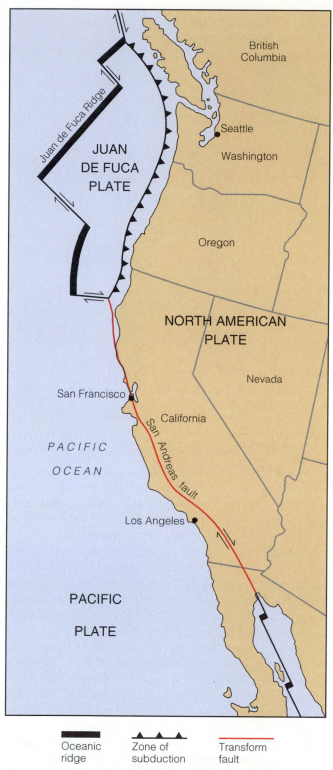

FIGURE 7.21 Transform plate boundary. The San Andreas fault is a transform fault separating the Pacific plate from the North American plate.

Unfortunately, transform faults generally do not leave any characteristic or diagnostic features except for the obvious displacement of the rocks with which they are associated. This displacement is commonly large, on the order of tens to hundreds of kilometers. Such large displacements in ancient rocks can sometimes be related to transform-fault systems.

❖ HOT SPOTS AND MANTLE PLUMES

Before leaving the topic of plate boundaries, we should briefly mention an intraplate feature found beneath both oceanic and continental plates. **Hot spots** are places where stationary columns of magma originating deep within the mantle (mantle plumes) rise through the crust and manifest themselves as volcanoes (❖ Fig. 7.22). Because the mantle plumes remain stationary while the plates move over them, the resulting hot spots leave a trail of extinct, progressively older volcanoes called *aseismic ridges*. Some examples of aseismic ridges and hot spots are the Emperor Seamount–Hawaiian Island chain and Yellowstone National Park in Wyoming.

❖ PLATE MOVEMENT AND MOTION

Rates of plate movement can be calculated in several ways. The least accurate method is to determine the age of the sediments immediately above any portion of the oceanic crust and divide that age by the distance from the spreading ridge. Such calculations give an average rate of movement. A more accurate method of determining both the average rate of movement and the relative motion is by dating the magnetic reversals in the crust of the seafloor (Fig. 7.11). The distance from the axis of an oceanic ridge to any magnetic reversal indicates the width of new seafloor that formed during that time interval. Thus, for a given interval of time, the wider the strip of seafloor, the faster the plate has moved. In this way, not only can the present average rate of movement and relative motion be determined (❖ Fig. 7.23), but the average rate for the past can also be calculated by dividing the distance between reversals by the amount of time elapsed between reversals.

Satellite laser ranging techniques can also be used to determine the average rate of movement as well as the relative motion between any two plates. Laser beams from a station on one plate are bounced off a satellite (in geosynchronous orbit) and returned to a station on a different plate. As the plates move relative to each other, there is an increase in the length of time that the laser

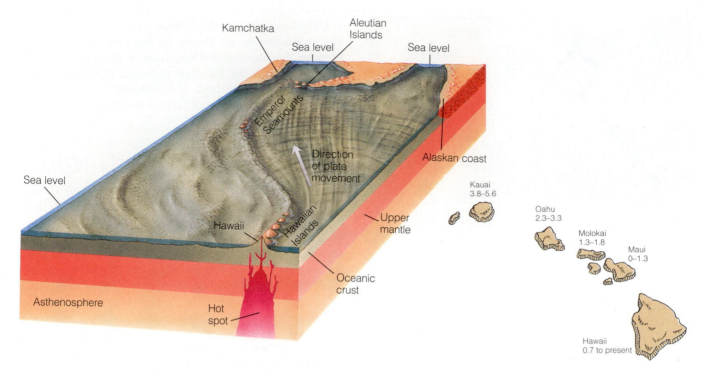

◆ **FIGURE 7.22** The Emperor Seamount–Hawaiian Island chain formed as a result of movement of the Pacific plate over a hot spot. The line of the volcanic islands traces the direction of plate movement. The numbers indicate the age of the islands in millions of years.

beam takes to go from the sending station to the stationary satellite and back to the receiving station. This difference in elapsed time is used to calculate the rate of movement and relative motion between plates.

Plate motions derived from magnetic reversals and satellite laser ranging techniques give only the relative motion of one plate with respect to another. Hot spots allow geologists to determine absolute motion because they provide a fixed reference from which the rate and direction of plate movement can be measured. The Emperor Seamount–Hawaiian Island chain (Fig. 7.22) formed as a result of movement over a hot spot. Thus, the line of the volcanic islands traces the direction of plate movement, and dating the volcanoes enables geologists to determine the rate of movement.

❖ THE DRIVING MECHANISM OF PLATE TECTONICS

A major obstacle to the acceptance of continental drift was the lack of a mechanism to explain continental

movement. When it was shown that continents and ocean floors moved together and oceanic lithosphere formed at spreading ridges by rising magma, most geologists accepted some type of convective heat system as the basic process responsible for plate motion.

Although plate motions can be described in great detail, the question of what exactly drives the plates is still being debated. Two models involving thermal convection cells have been proposed to explain plate movement (◆ Fig. 7.24). In one model, thermal convection cells are restricted to the asthenosphere, whereas in the second model the entire mantle is involved. In both models, spreading ridges mark the ascending limbs of adjacent convection cells, while trenches occur where the convection cells descend back into the Earth's interior. Thus, the locations of spreading ridges and trenches are determined by the convection cells themselves.

Although most geologists agree that the Earth's internal heat plays an important role in plate movement, there are problems with both models. The problem with the first model is explaining the source of heat for the con-

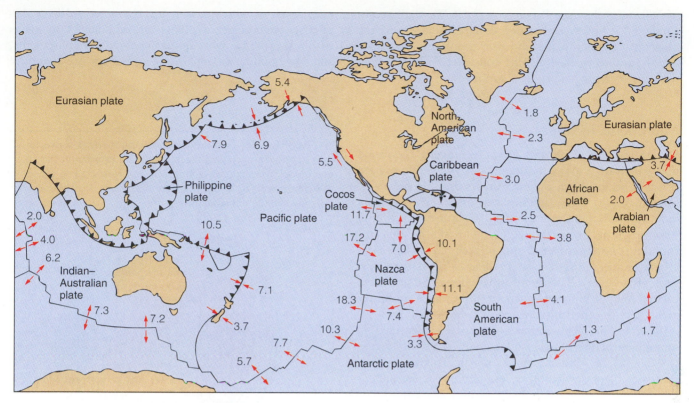

◆ **FIGURE 7.23** This map shows the average rate of movement in centimeters per year and the relative motion of the Earth's plates.

vection cells and why they are restricted to the asthenosphere. The source of heat for the second model comes from the Earth's outer core; however, it is still not known how heat is transferred from the outer core to the mantle, nor how convection can involve both the lower mantle and the asthenosphere.

Even though geologists are fairly certain that some type of convection system is involved in plate movement, a comprehensive theory of plate movement has not as yet been developed, and much still remains to be learned about the Earth's interior.

❖ PLATE TECTONICS AND MOUNTAIN BUILDING

An **orogeny** is an episode of intense rock deformation or mountain building. Orogenies are the consequence of compressive forces related to plate movement. As one plate is subducted under another, sedimentary and vol-

canic rocks are folded and faulted along the plate margin while the more deeply buried rocks are subjected to regional metamorphism. Magma generated within the mantle either rises to the surface to erupt as andesitic volcanoes or cools and crystallizes beneath the Earth's surface, forming intrusive igneous bodies. Typically, most orogenies occur along either oceanic-continental or continental-continental plate boundaries.

As we discussed earlier in this chapter, ophiolites are evidence of ancient convergent plate boundaries. The slivers of ophiolites found in the interiors of such mountain ranges as the Alps, Himalayas, and Urals mark the sites of former ocean basins. The relationship between mountain building and the opening and closing of ocean basins is called the **Wilson cycle** in honor of the Canadian geologist J. Tuzo Wilson, who first suggested that an ancient ocean had closed to form the Appalachian Mountains and then reopened and widened to form the present Atlantic Ocean. According to some geologists, much of the geology of continents can be described in terms of a succession of Wilson cycles (see Perspective 7.1)! We shall see in Chapter 15, however, that new evidence concerning the movement of microplates and their accretion at the margin of continents must also be considered when dealing with the tectonic history of a continent.

Microplate Tectonics

During the late 1970s and 1980s, geologists discovered that portions of many mountain systems are composed of small accreted lithospheric blocks that are clearly of foreign origin. These **microplates** differ completely in their fossil content, stratigraphy, structural trends, and paleomagnetic properties from the rocks of the surrounding mountain system and adjacent crust. These microplates are so different from adjacent rocks that most geologists believe that they formed elsewhere and were carried long distances as parts of other plates until they collided with other microplates or continents.

To date, most microplates have been identified in mountains of the North American Pacific coast region (see Fig. 15.28), but a number of such plates are suspected to be present in other mountain systems, such as the Appalachians (see Chapters 11 and 12). Microplate tectonics is, however, providing a new way of viewing the Earth and of gaining a better understanding of the geologic history of the continents.

❖ PLATE TECTONICS AND THE DISTRIBUTION OF LIFE

The theory of plate tectonics is as revolutionary and far-reaching in its implications for geology as the theory of evolution was for biology when it was proposed. Interestingly, it was the fossil evidence that convinced Wegener, Suess, and du Toit, as well as many other geologists, of the correctness of continental drift. Together, the theories of plate tectonics and evolution have changed the way we view our planet, and we should not be surprised at the intimate association between them. While the relationship between plate tectonic processes and the evolution of life is incredibly complex, paleontological data provide undisputable evidence of the influence of plate movement on the distribution of organisms. (see Fig. 7.5).

The present distribution of plants and animals is not random but is controlled largely by climate and geographic barriers. The world's biota occupy *biotic prov-*

inces, each of which is a region characterized by a distinctive assemblage of plants and animals. Organisms within a province have similar ecological requirements, and the boundaries separating provinces are therefore

◆ **FIGURE 7.24** Two models involving thermal convection cells have been proposed to explain plate movement. (a) In one model, thermal convection cells are restricted to the asthenosphere. (b) In the other model, thermal convection cells involve the entire mantle.

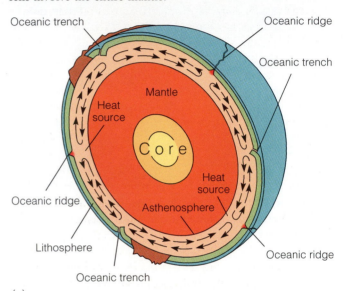

(a)

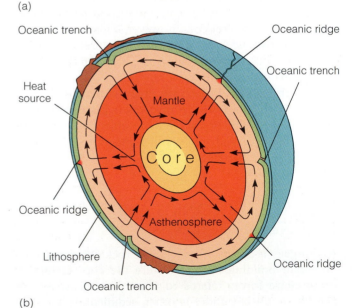

(b)

natural ecological breaks. Climatic or geographic barriers are the most common province boundaries, and these are largely controlled by plate movements.

Because adjacent provinces usually have less than 20% of their species in common, global diversity is a direct reflection of the number of provinces; the more provinces there are in the world, the greater the global diversity. When continents break up, for example, the opportunity for new provinces to form increases, with a resultant increase in diversity. Just the opposite occurs when continents come together. Plate tectonics thus plays an important role in the distribution of organisms and their evolutionary history.

The world's climates result from the complex interaction between wind and ocean currents. These currents are influenced by the number, distribution, topography, and orientation of continents. Temperature is one of the major limiting factors for organisms, and province boundaries often reflect temperature barriers. Because atmospheric and oceanic temperatures decrease from the equator to the poles, most species exhibit a strong climatic zonation. This biotic zonation parallels the world's latitudinal atmospheric and oceanic circulation patterns. Changes in climate thus have a profound effect on the distribution and evolution of organisms.

The distribution of continents and ocean basins not only influences wind and ocean currents, it also affects provinciality by creating physical barriers to, or pathways for, the migration of organisms. Intraplate volcanoes, island arcs, mid-oceanic ridges, mountain ranges, and subduction zones all result from the interaction of plates, and their orientation and distribution strongly influence the number of provinces and hence total global diversity. Thus, provinciality and diversity will be highest when there are numerous small continents spread across many zones of latitude.

When a geographic barrier separates a once-uniform fauna, species may undergo divergence. If conditions on opposite sides of the barrier are sufficiently different, then each species must adapt to the new conditions, migrate, or become extinct. Adaptation to the new environment by various species may involve enough change that new species eventually evolve. The marine invertebrates found on opposite sides of the Isthmus of Panama provide an excellent example of divergence caused by the formation of a geographic barrier. Prior to the rise of this land connection between North and South America, a homogeneous population of bottom-dwelling invertebrates inhabited the shallow seas of the area. After the rise of the Isthmus of Panama by subduction of the Pa-

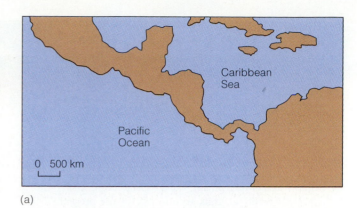

(a)

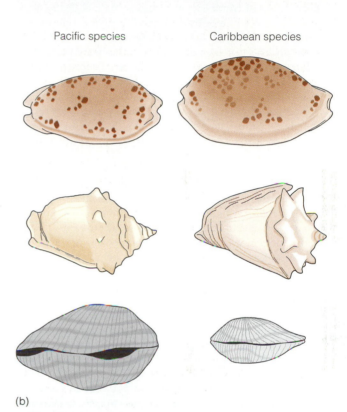

Pacific species   Caribbean species

(b)

◆ **FIGURE 7.25** (a) The Isthmus of Panama forms a barrier that divides a once-uniform fauna. (b) Divergence of molluscan species after the formation of the Isthmus of Panama. Each pair belongs to the same genus but is a different species.

cific plate about 5 million years ago, the original population was divided. In response to the changing environment, new species evolved on opposite sides of the isthmus (◆ Fig. 7.25).

THE SUPERCONTINENT CYCLE

At the end of the Paleozoic Era, all continents were amalgamated into the supercontinent Pangaea. Pangaea began fragmenting during the Triassic Period and continues to do so, thus accounting for the present distribution of continents and oceans. As will be discussed in Chapter 10, it appears that another supercontinent existed at the end of the Proterozoic Eon, and there is some evidence for even earlier supercontinents. Recently, it has been proposed that supercontinents consisting of all or most of the Earth's landmasses form, break up, and re-form in a cycle spanning about 500 million years.

The supercontinent cycle hypothesis is an expansion on the ideas of the Canadian geologist J. Tuzo Wilson. During the early 1970s, Wilson proposed a cycle (now known as the Wilson cycle) that includes continental fragmentation, the opening and closing of an ocean basin, and finally reassembly of the continent. According to the supercontinent cycle hypothesis, heat accumulates beneath a supercontinent because rocks of continents are poor conductors of heat. As a consequence of the heat accumulation, the supercontinent domes upward and fractures. Basaltic magma rising from below fills the fractures. As a basalt-filled fracture widens, it begins subsiding and forms a long narrow ocean such as the present-day Red Sea. Continued rifting eventually forms an expansive ocean basin such as the Atlantic.

According to proponents of the supercontinent cycle, one of the most convincing arguments for their hypothesis is the "surprising regularity" of mountain building caused by compression during continental collisions. Such orogenies occur about every 400 to 500 million years, and each episode of compression-induced orogeny is followed by an episode of rifting about 100 million years later. In other words, a supercontinent fragments and its individual plates disperse following a rifting episode, an interior ocean forms, and then the dispersed fragments reassemble to form another supercontinent (◆ Fig. 1).

In addition to explaining the dispersal and reassembly of supercontinents, the supercontinent cycle hypothesis also explains two distinct types of orogens: *interior orogens* resulting from compression induced by continental collisions and *peripheral orogens* that form at continental margins in response to subduction and igneous activity (Fig. 1). Following the fragmentation of a supercontinent, an interior ocean forms and widens as plates separate. After about 200 million years, however, plate separation ceases as the oceanic crust becomes cooler and denser and begins to subduct at the margins of the interior ocean basin, thus transforming inward-facing passive continental margins into active continental margins (Fig. 1). Rising magma resulting from subduction forms plutons and volcanoes along these continental margins, and an interior orogeny develops when plates converge, causing compression-induced deformation, crustal thickening, and metamorphism (Fig. 1). The present-day Appalachian Mountains of the eastern United States and Canada originally formed as an interior orogen (see Chapters 11 and 12).

Peripheral orogenies are caused by convergence of oceanic plates with the margins of a continent, igneous activity resulting from subduction, and collisions of island arcs with the continent (Fig. 1). Subduction-related volcanism in peripheral orogens is rather continuous, but the collisions of island arcs with the supercontinent are episodic. Peripheral orogenies can develop at any time during the supercontinent cycle. A good example of a presently active peripheral orogeny is the Andes of western South America where the Nazca plate is being subducted beneath the South American plate (see Chapter 17).

◆ **FIGURE 1** The supercontinent cycle. (a) Breakup of a supercontinent and the formation of an interior ocean basin. (b) Subduction along the margins of the interior ocean basin begins approximately 200 million years later, resulting in volcanic activity and deformation along an active oceanic-continental plate boundary. (c) Continental collisions and the formation of a new supercontinent result when all of the oceanic crust of the interior ocean basin is subducted.

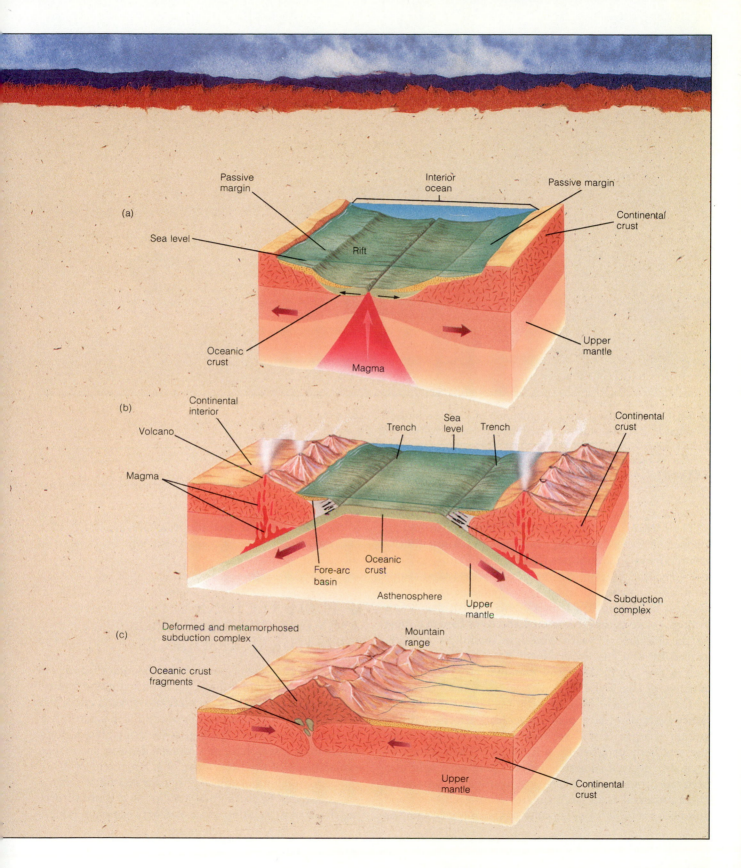

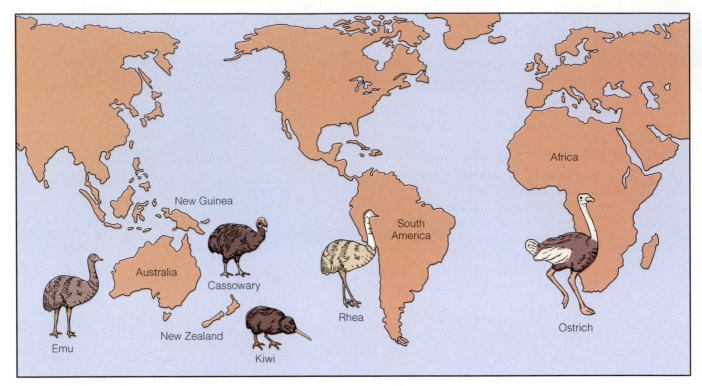

◆ **FIGURE 7.26** Geographic distribution of flightless birds. All except the kiwi are large birds. The ancestors of these birds probably evolved when the Southern Hemisphere continents were assembled into a single landmass, Gondwana. When the continents separated, these ancestral birds were isolated from one another, and each followed its own evolutionary course.

The formation of the Isthmus of Panama also had an impact on the evolution of the North and South American mammalian faunas (see Chapter 19). During most of the Cenozoic Era, South America was an island continent, and its mammalian fauna evolved in isolation from the rest of the world's faunas. When North and South America were connected by the Isthmus of Panama, most of the indigenous South American mammals were replaced by migrants from North America. Surprisingly, only a few South American mammal groups migrated northward (see Fig. 19.33).

The distribution of large flightless birds provides another example of how plate movements affect the distribution of organisms. These birds are now all restricted to the southern continents (◆ Fig. 7.26). Their ancestors probably evolved sometime before Gondwana split into separate continents. When the continents separated, these ancestral birds were isolated from each other and followed their own evolutionary course. Other examples of the role that plate tectonics plays in the distribution and evolution of the Earth's biota will be covered in succeeding chapters.

❖ PLATE TECTONICS AND THE DISTRIBUTION OF NATURAL RESOURCES

In addition to being responsible for the major features of the Earth's crust and influencing the distribution and evolution of the world's biota, plate movements also affect the formation and distribution of the Earth's natural resources. Consequently, geologists are using plate tectonic theory in their search for new mineral deposits and in their explanations of the occurrence of known deposits.

Many metallic mineral deposits are related to igneous and associated hydrothermal activity, so it is not surprising that a close relationship exists between plate boundaries and the occurrence of these valuable deposits. The magma generated by partial melting of a subducting plate rises toward the Earth's surface, and as it cools, it precipitates and concentrates various metallic sulfide ores. Many of the world's major metallic ore deposits, such as the porphyry copper deposits of western North and South America (◆ Fig. 7.27) occur along present-day and ancient convergent plate boundaries.

Divergent plate boundaries also yield valuable resources. Hydrothermal vents are the sites of considerable metallic mineral precipitation (see Perspective 9.2). The island of Cyprus in the Mediterranean is rich in copper and has been supplying all or part of the world's needs for the last 3,000 years.

Studies indicate that minerals of such metals as copper, gold, iron, lead, silver, and zinc are currently forming as sulfides in the Red Sea. The Red Sea is opening as a result of plate divergence and represents the earliest stage in the growth of an ocean basin.

We will discuss natural resources and their formation as part of the geologic history of the Earth in later chapters in this book.

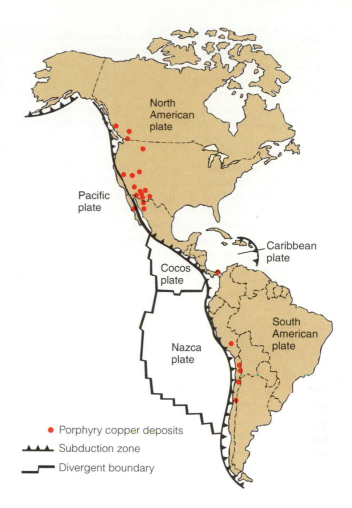

◆ **FIGURE 7.27** Important porphyry copper deposits are located along the west coasts of North and South America.

Chapter Summary

1. The concept of continental movement is not new. The earliest maps showing the similarity between the east coast of South America and the west coast of Africa provided people with the first evidence that continents may once have been united and subsequently drifted away from each other.

2. Alfred Wegener is generally credited with developing the concept of continental drift. He gathered information from many sources to show that the continents were once connected as one supercontinent he called Pangaea. Unfortunately, Wegener could not explain how the continents moved, and most geologists ignored his ideas.

3. The concept of continental drift was revived in the 1950s when paleomagnetic studies of rocks of different ages on the same continent indicated that in the past there apparently had been multiple magnetic north poles instead of

just one as there is today. However, the location of the continents could be rearranged such that the magnetic minerals in the rocks pointed to the same location for the magnetic north pole.

4. Magnetic surveys of the oceanic crust revealed magnetic anomalies in the rocks indicating that the Earth's magnetic field had reversed itself in the past. Because the anomalies are parallel and symmetrical around oceanic ridges, this means that new oceanic lithosphere was forming and the seafloor has been spreading.

5. Radiometric dating reveals the oldest oceanic crust is less than 180 million years old, while the oldest continental crust is 3.96 billion years old. Clearly, the ocean basins are recent geologic features.

6. Plate tectonic theory became widely accepted in the 1970s because of the overwhelming evidence supporting it and

because it provides geologists with a powerful theory for explaining such things as mountain building, global climatic changes, the distribution of the world's biota, and the distribution of some mineral resources.

7. There are three types of plate boundaries: divergent boundaries, where plates move away from each other; convergent boundaries, where one plate is subducted under another plate; and transform boundaries, where two plates slide past each other.

8. Ancient plate boundaries can be recognized by their associated rock assemblages and geologic structures. For divergent boundaries, these may include rift valleys with thick sedimentary sequences and numerous dikes and sills. For convergent boundaries, ophiolites and andesitic rocks are two characteristic features. Transform faults generally do not leave any characteristic or diagnostic features in the rock record.

9. Although a comprehensive theory of plate movement has yet to be developed, geologists believe that some type of convective heat system is involved.

10. The average rate of movement and relative motion of plates can be calculated several ways. These different methods all yield similar average rates of plate movement and indicate that the plates move at different average velocities. Absolute motion of plates can be determined by the movement of plates over mantle plumes.

11. Geologists now realize that plates can grow when microplates collide with the margins of continents.

12. A close relationship exists between the formation of some mineral deposits and plate boundaries. Furthermore, the formation and distribution of the Earth's natural resources are related to plate movements.

Important Terms

continental-continental plate boundary
continental drift
convergent plate boundary
Curie point
divergent plate boundary
Glossopteris flora
Gondwana
hot spot
Laurasia

magnetic anomaly
magnetic reversal
microplate
oceanic-continental plate boundary
oceanic-oceanic plate boundary
ophiolite
orogeny
paleomagnetism

Pangaea
plate tectonic theory
seafloor spreading
spreading ridge
thermal convection cell
transform fault
transform plate boundary
volcanic island arc
Wilson cycle

Review Questions

1. The man who is credited with developing the continental drift hypothesis is:
   a. _____ du Toit; b. _____ Hess; c. _____ Vine;
   d. _____ Wegener; e. _____ Wilson.

2. The name of the supercontinent that formed at the end of the Paleozoic Era is:
   a. _____ Laurasia; b. _____ Gondwana; c. _____ Panthalassa; d. _____ Atlantis; e. _____ Pangaea.

3. Which of the following has been used as evidence for continental drift?
   a. _____ continental fit; b. _____ similarity of fossil plants and animals; c. _____ similarity of rock sequences; d. _____ paleomagnetism; e. _____ all of these.

4. The temperature at which iron-bearing minerals gain their magnetization is the:

   a. _____ eutectic point; b. _____ triple junction;
   c. _____ Curie point; d. _____ magnetic point;
   e. _____ declination point.

5. Deviations from the average strength of the Earth's magnetic field are:
   a. _____ magnetic reversals; b. _____ magnetic anomalies; c. _____ normal polarities; d. _____ reversed polarities; e. _____ none of these.

6. Magnetic surveys of the ocean basins indicate that:
   a. _____ the oceanic crust is youngest adjacent to mid-oceanic ridges; b. _____ the oceanic crust is oldest adjacent to mid-oceanic ridges; c. _____ the oceanic crust is youngest adjacent to the continents; d. _____ the oceanic crust is the same age everywhere; e. _____ answers (b) and (c).

7. Convergent boundaries are areas where:
a. _____ new continental crust is forming; b. _____ new oceanic crust is forming; c. _____ two plates collide; d. _____ two plates move away from each other; e. _____two plates slide laterally past each other.

8. The Aleutian Islands are a good example of a(n) _____ plate boundary.
a. _____ continental-continental; b. _____ oceanic-oceanic; c. _____ oceanic-continental; d. _____ divergent; e. _____ transform.

9. The East African rift valleys are good examples of a(n) _____ plate boundary.
a. _____ continental-continental; b. _____ oceanic-oceanic; c. _____ oceanic-continental; d. _____ divergent; e. _____ transform.

10. Which of the following allows geologists to determine absolute plate motion?
a. _____ hot spots; b. _____ the age of the sediment directly above any portion of the oceanic crust; c. _____ magnetic reversals in the oceanic crust; d. _____ satellite laser ranging techniques; e. _____ all of these.

11. The driving mechanism of plate movement is believed to be:
a. _____ isostasy; b. _____ rotation of the Earth; c. _____ magnetism; d. _____ thermal convection cells; e. _____ none of these.

12. The formation and distribution of many metallic mineral deposits are associated with _____ boundaries.
a. _____ divergent; b. _____ convergent; c. _____ transform; d. _____ answers (a) and (b); e. _____ answers (b) and (c).

13. The most common biotic province boundaries are:
a. _____ geographic barriers; b. _____ biologic barriers; c. _____ climatic barriers; d. _____ answers (a) and (b); e. _____ answers (a) and (c).

14. Small accreted lithospheric blocks that are different from the surrounding rocks are:
a. _____ mantle plumes; b. _____ microplates; c. _____ back-arc basins; d. _____ subduction complexes; e. _____ fore-arc basins.

15. A distinctive assemblage of deep-sea sediments, oceanic crustal rocks, and upper mantle constitutes a:
a. _____ microplate; b. _____ back-arc basin; c. _____ subduction complex; d. _____ fore-arc basin; e. _____ none of these.

16. An episode of intense rock deformation or mountain building is a(n):
a. _____ hot spot; b. _____ orogeny; c. _____ mantle plume; d. _____ Wilson cycle; e. _____ none of these.

17. The relationship between mountain building and the opening and closing of ocean basins is called the:
a. _____ Curie theorem; b. _____ du Toit hypothesis; c. _____ Wilson cycle; d. _____ Hess convection cell; e. _____ none of these.

18. Briefly discuss how rates of movement of plates can be determined.

19. Explain how plate tectonics affects the distribution of organisms.

20. Explain why global diversity increases with an increase in biotic provinces. How does plate movement affect the number of biotic provinces?

21. Briefly discuss how a geologist could use plate tectonic theory to help locate mineral deposits.

22. What evidence convinced Wegener that the continents were once joined together and subsequently broke apart?

23. Briefly discuss microplate tectonics and its contributions to plate tectonic theory.

24. What is the significance of polar wandering in relation to continental drift?

25. Summarize the geologic features characterizing the three different types of plate boundaries.

26. How would you recognize ancient convergent plate boundaries?

27. What are hot spots, and how can they be used to determine the direction and rate of movement of plates?

28. What is the apparent driving mechanism for plate movement?

29. Why is plate tectonics such a powerful unifying theory?

30. What features would an astronaut look for on the Moon or another planet to find out if plate tectonics is currently active or was active in the past?

Additional Readings

Bonatti, E. 1987. The rifting of continents. *Scientific American* 256, no. 3: 96–103.

Brimhall, G. 1991. The genesis of ores. *Scientific American* 264, no. 5: 84–91.

Condie, K. 1989. *Plate tectonics and crustal evolution.* 3d ed. New York: Pergamon Press.

Cox, A., and R. B. Hart. 1986. *Plate tectonics: How it works.* Palo Alto, Calif.: Blackwell Scientific Publishers.

Cromie, W. J. 1989. The roots of midplate volcanism. *Mosaic* 20, no. 4: 19–25.

Gass, I. G. 1982. Ophiolites. *Scientific American* 247, no. 2: 122–31.

Hallam, A. 1972. Continental drift and the fossil record. *Scientific American* 227, no. 5: 56–69.

Jordon, T. H., and J. B. Minster. 1988. Measuring crustal deformation in the American west. *Scientific American* 259, no. 2: 48–59.

Kearey, P., and F. J. Vine. 1990. *Global tectonics*. Palo Alto, Calif: Blackwell Scientific Publishers.

Marshall, L. G. 1988. Land mammals and the great American interchange. *American Scientist* 76, no. 4: 380–88.

Molnar, P. 1986. The structure of mountain ranges. *Scientific American* 255, no. 1: 70–79.

Murphy, J. B., and R. D. Nance. 1992. Mountain belts and the supercontinent cycle. *Scientific American* 266, no. 4: 84–91.

Nance, R. D., T. R. Worsley, and J. B. Moody. 1988. The supercontinent cycle. *Scientific American* 259, no. 1: 72–79.

Vink, G. E., W. J. Morgan, and P. R. Vogt. 1985. The Earth's hot spots. *Scientific American* 252, no. 4: 50–57.

CHAPTER 8

The Andromeda Galaxy, 2.2 million light years away, is the closest major galaxy to our Milky Way. Its full diameter is nearly 150,000 light years and it contains some 300,000,000,000 stars. (Photo courtesy of the Palomar Observatory, California Institute of Technology.)

THE HISTORY OF THE UNIVERSE, SOLAR SYSTEM, AND PLANETS

Prologue

Sometimes the most important scientific discoveries are made quite by accident or while searching for something else. Such was the case for Arno Penzias and Robert Wilson of Bell Telephone Laboratories. What began as a seemingly straightforward research project in the early 1960s resulted in the 1978 Nobel Prize in physics for these two scientists. Penzias and Wilson were originally interested in mapping the radio waves coming from directions away from the main plane of our galaxy. To pick up the extremely faint signals emanating from distant parts of the universe, they needed a very sensitive radio telescope. Such a telescope was available at the Bell Telephone Laboratories at Holmdel, New Jersey.

Penzias and Wilson began their research by calibrating the radio telescope to a frequency at which they did not expect to find many galactic radio waves. After the calibration, they planned to turn to a still lower frequency to map the distant radio waves they were interested in. However, the calibration frequency was far too "noisy," or in the words of Penzias, "The antenna was considerably hotter than expected." This was hardly a promising beginning to a discovery that ranks as one of the most important in astronomy in this century. Penzias and Wilson concluded that something must be wrong with the telescope. One possibility was the two pigeons nesting on it. Even after the pigeons were dislodged, however, the noise persisted.

Early in 1965 Penzias and Wilson reached a point in their observations where they could dismantle the telescope without harming the results of their research. After painstakingly cleaning everything and carefully reassembling it, they found the bothersome noise was still there. "We frankly did not know what to do," lamented Penzias. If the radio noise was not in the telescope, it must be coming from space. If this was the case, then it was spread uniformly throughout the universe. It was always there, day and night, and in every direction the scientists aimed their telescope. The radio noise corresponded to a constant temperature of 2.7° above absolute zero. Absolute zero equals −273°C, the temperature at which all atomic motion stops. The problem was that there was no astronomical explanation for a uniform "noise" permeating the entire universe.

At about the same time Penzias and Wilson were puzzling over their findings, a colleague of theirs recalled reading a paper by Jim Peebles of Princeton University in which he predicted that the universe should have a background temperature of about 10°C above absolute zero. If our present expanding universe began with a "Big Bang," then a relic of that initial high temperature should be pervading all of space. Peeble's calculations pointed to a very low background temperature because of the cooling that has occurred since the formation of the universe.

Based on these calculations, Peebles and another Princeton physicist, Robert Dicke, had just started searching for the background radiation with a radio telescope at Princeton when Penzias contacted Dicke about Peeble's paper. Later in 1965, the two groups published their findings as companion letters in *The Astrophysical Journal*, with Penzias and Wilson referring readers to the "theory" paper of Dicke and Peebles for interpretation of their observations.

After checking Penzias and Wilson's results using other telescopes and different frequencies, astronomers and physicists embraced their findings enthusiastically. What was initially interpreted as some type of problem with the radio telescope turned out to be the fading afterglow of the explosion in which the universe began.

❖ INTRODUCTION

Most people are interested in the question of when, where, and how the universe began as well as the equally intriguing aspect of our place in it. The origin of the universe is of interest not only to scientists but also to theologians and philosophers. How does one comprehend infinity from a finite viewpoint or conceptualize the creation of matter and energy under conditions where the known laws of physics do not apply? Robert Browning wrote, "Ah, but a man's reach should exceed his grasp." In attempting to understand the origin of the universe, our reach must certainly exceed our grasp!

Most scientists think that the universe originated approximately 13 to 20 billion years ago in what is popularly called the "**Big Bang**." At that moment, in a region billions of times smaller than a proton, both space and time were set at zero. Therefore there is no "before the

Big Bang," only what occurred after it. The reason for this is that space and time are unalterably linked to form a space-time continuum as demonstrated by Einstein's theory of relativity. Without space there can be no time.

How do we know the Big Bang took place between 13 and 20 billion years ago? Why couldn't the universe have always existed as we know it today? To answer those questions, we turn our optical and radio telescopes skyward and observe two fundamental phenomena. The first is that the universe is expanding; the second is that it is permeated by background radiation.

That the universe is expanding was first recognized by Edwin Hubble in 1929. By measuring the optical spectra of distant galaxies, Hubble noted that the velocity at which a galaxy moves away from the Earth increases proportionally to its distance from Earth. He observed that the spectral lines (wavelengths of light) of the galaxies are shifted toward the red end of the spectrum; that is, the lines are shifted toward longer wavelengths. Such a redshift would be produced by galaxies receding from each other at tremendous speeds (◆ Fig. 8.1). This is an example of the **Doppler effect**, which is the change in the

---

◆ **FIGURE 8.1** Spectra showing the redshift of the galaxies. Just as the Doppler effect makes a rapidly receding train whistle drop in pitch, the expansion of the universe stretches light waves from receding objects toward the longer wavelengths of the spectrum. This light is said to be *redshifted,* and the amount of the redshift is proportional to the object's distance from us. In this diagram the spectral lines of hydrogen are stretched in proportion to their original wavelengths. All lines in the spectrum of any object, however, exhibit the same ratio of wavelength increase to the original wavelength.

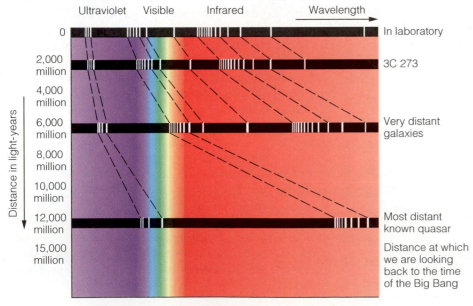

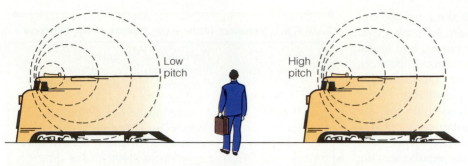

◆ **FIGURE 8.2** The Doppler effect applied to sound waves. The sound waves of an approaching whistle are slightly compressed so that the individual hears a shorter-wavelength, higher-pitched sound. As the whistle passes and recedes from the individual, the sound waves are slightly spread out, and a longer-wavelength, lower-pitched sound is heard.

frequency of a sound, light, or other wave caused by movement of the wave's source relative to the observer (◆ Fig. 8.2).

One way to envision how velocity increases with increasing distance is by reference to the popular analogy of a rising loaf of raisin bread in which the raisins are uniformly distributed throughout the loaf (◆ Fig. 8.3). Suppose that before the dough begins to rise, the raisins are 1 cm apart. After one hour, the dough has risen to the point where the raisins are 2 cm apart. Any raisin is now 2 cm from its nearest neighbor and 4 cm away from the next one, 6 cm from the next, and so on. From the perspective of any raisin, its nearest neighbor has moved away from it at a speed of 1 cm per hour (it originally was 1 cm away and is now 2 cm away), the next raisin over has moved away at a speed of 2 cm/hr (it was 2 cm

◆ **FIGURE 8.3** The motion of raisins in a rising loaf of raisin bread illustrates the relationship that exists between distance and speed and is analogous to an expanding universe. In this diagram, adjacent raisins are located 2 cm apart before the loaf rises. After one hour, any raisin is now 4 cm away from its nearest neighbor and 8 cm away from the next raisin over, and so on. Therefore, from the perspective of any raisin, its nearest neighbor has moved away from it at a speed of 2 cm per hour, and the next raisin over has moved away from it at a speed of 4 cm per hour. In the same way raisins move apart in a rising loaf of bread, galaxies are receding from each other at a rate proportional to the distance between them.

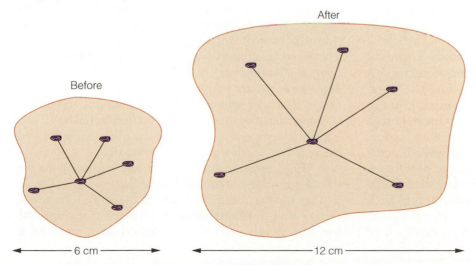

away, and is now 4 cm away), while the next one has moved away at 3 cm/hr (it was originally 3 cm away and is now 6 cm away from the reference raisin), and so on. If the dough continues to rise until the distance between any two raisins is 4 cm, we would find that all the distances between raisin pairs are quadruple what they were to start with. The speeds needed to accomplish this are directly proportional to the distances between any two raisins: the farther away a given raisin is to begin with, the farther it must move to maintain the regular spacing during the expansion, and hence the greater its velocity must be. In the same way that raisins move apart in a rising load of bread, galaxies are receding from each other at a rate proportional to the distance between them, which is exactly what astronomers see when they observe the universe.

A corollary to Hubble's expanding universe is that astronomers can use this expansion rate to calculate how long ago the galaxies were all together in a single point. Such calculations yield an estimated age for the universe of 13 to 20 billion years.

The second important observation providing evidence of the Big Bang was made in 1965 by Arno Penzias and Robert Wilson of Bell Telephone Laboratories when they discovered that there is a pervasive background radiation of 2.7 K—that is, 2.7° above absolute zero (zero degrees Kelvin [0 K] equals −273°C)—everywhere in the universe. This background radiation is believed to be the fading afterglow of the Big Bang.

Based on these two observations (that the universe is expanding and that it has a 2.7 K background radiation), astronomers calculate that the universe, matter, and energy as we know it originated approximately 13 to 20 billion years ago. The universe then is indeed constantly evolving.

❖ THE ORIGIN AND EARLY HISTORY OF THE UNIVERSE

Currently, astrophysicists can reconstruct the history of the universe back to 10^{-43} seconds following the Big Bang (Table 8.1). What was its history before 10^{-43} seconds? No one really knows, since it is impossible at this stage in our knowledge to deal with the infinitely high densities and temperatures that must have been present. Matter as we know it today could not have existed under those conditions, and the universe consisted of pure energy. Many physicists suspect that at the extreme temperatures prevailing before 10^{-43} seconds, the four basic forces—**gravity, electromagnetic force, strong nuclear force,** and **weak nuclear force** (Table 8.2)—were united

◆ ━━━━━━━━━━━━━━━━━━━━━━━━━━━━━━ ◆

TABLE 8.1 Summary of the Early History of the Universe

Big Bang	The origin of the universe.
10^{-43} seconds	Physics starts here, and gravity separates from the other basic forces.
10^{-35} to 10^{-32} seconds	A major inflationary period takes place. The strong force separates, and energy begins to congeal into quarks and electrons and their mirror images, antimatter.
10^{-6} seconds	Quarks bind into protons and neutrons. Matter and antimatter collide with a slight excess of matter left over that comprises the matter in the universe today.
1 second	Electromagnetic and weak nuclear forces separate.
3 minutes	Protons and neutrons fuse into atomic nuclei.
10^5 years	Electrons join with nuclei to make atoms. Photons separate from matter, and the universe bursts forth with light.
10^5 to 10^9 years	The universe becomes clumpy.

into a single unified force. These forces are defined in the table.

By 10^{-43} seconds after the Big Bang, gravity separated from the other basic forces, which remained united, and physics as we know it began. At a temperature estimated at 10^{32} K, the expanding universe at this time is believed to have been only 10^{-28} cm in diameter.

Between 10^{-35} and 10^{-32} seconds after the Big Bang, a major inflationary period took place. The strong force separated out, and energy began to congeal into particles of matter such as *quarks* (one of two fundamental particles; the constituent subparticles of protons and neutrons) and electrons, as well as their mirror images, *antimatter*. Antimatter consists of particles that are opposite to matter in every way except for mass. At the end of this brief inflationary period, the universe was a homogeneous, opaque stew of matter, antimatter, and energy. Its temperature had cooled to 10^{27} K, and it had expanded to about the size of a softball.

By 10^{-6} seconds, the universe had expanded to the size of our solar system and had cooled enough (10^{13} K) so that quarks could bind into protons and neutrons. During that brief time after its formation, the universe was not as symmetrical as it should have been. As it cooled to 10^{13} K, matter and antimatter collided and annihilated each other. Because the universe was not symmetrical, however, a "slight" excess of matter was left over and eventually became our present universe of galaxies, stars, and planets. Without this asymmetry, the universe would instead be an ever expanding and slowly cooling emptiness.

By the time the universe was 1 second old, the electromagnetic and weak nuclear forces had separated. Three minutes after the Big Bang, the temperature had cooled to 10^9 K, and at this temperature protons and neutrons fused to form the nuclei of hydrogen and helium atoms. At about 100,000 years, the temperature had cooled to 3,000 K, and electrons then combined with the previously formed nuclei to form complete atoms of hydrogen and helium. At that time, *photons* (the energetic particles of light) separated from matter, and the universe became transparent as light burst forth for the first time. This liberation of photons is what we observe today as the 3 K background radiation permeating the universe.

Sometime between 100,000 and 1 to 2 billion years after the Big Bang, the universe started to become clumpy. For reasons still not understood, matter began gathering into clouds of different sizes that eventually collapsed to form clusters of galaxies and stars. These galaxies tended to form like beads on a string into superclusters, the largest celestial objects known.

A recent discovery by physicist George Smoot and his team of astrophysicists at the Lawrence Berkeley Laboratory in California appears to have confirmed that the universe underwent a major inflationary period less than a fraction of a second after its formation. Data gathered from sensitive satellite microwave receivers have confirmed the existence of regions where large clusters of galaxies began forming during the early history of the universe. Such regions have long been predicted but never observed. These clusters of galaxies are revealed as variations in cosmic radiation and are crucial in linking the current universe with the Big Bang. These subtle variations in cosmic radiation could not be detected until instruments capable of detecting differences as slight as six parts per million were developed. Confirmation of these preliminary results by other scientists will provide substantive evidence to support the theory of an inflationary universe.

❖ THE CHANGING COMPOSITION OF THE UNIVERSE

As the universe continued expanding and cooling, stars and galaxies formed, and the chemical makeup of the universe changed. Early in its history, the universe was 100% hydrogen and helium, whereas today it is 98% hydrogen and helium by weight. Heavier elements form from lighter elements as a result of fusion reactions in which atomic nuclei combine to form more massive nuclei. Such reactions convert hydrogen to helium and occur in the cores of stars. During their history, stars more massive than our sun may undergo many nuclear reaction steps, in which hydrogen is initially converted to helium, then to carbon, and from carbon to even heavier elements (❖ Fig. 8.4). When these stars die, often explosively, the heavier elements that were formed in their cores are returned to interstellar space and are available for inclusion in new stars. When new stars form, they will have a small amount of these heavier elements. In this way the chemical composition of the galaxies, which are made up of billions of stars, is gradually enhanced in heavier elements.

The chemical composition of the Milky Way Galaxy has thus been changing in the period between the Big Bang and the formation of our solar system. Today, nearly 2% of the total mass of the Milky Way Galaxy is in the form of elements heavier than helium.

❖ THE ORIGIN AND HISTORY OF THE SOLAR SYSTEM

Having looked at the origin and history of the universe, we can now examine how our own solar system formed. In many ways, astronomers know more about the birth,

---

TABLE 8.2 The Four Basic Forces of the Universe

Four forces appear to be responsible for all interactions of matter:

1. **Gravity** is the attraction of one body toward another.
2. The **electromagnetic force** combines electricity and magnetism into the same force and binds atoms into molecules. It also transmits radiation across the various spectra at wavelengths ranging from gamma rays (shortest) to radio waves (longest) through massless particles called *photons*.
3. The **strong nuclear force** binds protons and neutrons together in the nucleus of an atom.
4. The **weak nuclear force** is responsible for the breakdown of an atom's nucleus, producing radioactive decay.

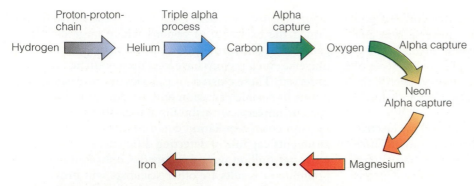

FIGURE 8.4 Stars undergo nuclear reaction steps in which light elements are converted to heavier elements. The first step, the conversion of hydrogen to helium, occurs in all stars. The subsequent conversion to heavier elements depends on the mass of the star. The conversion of helium into carbon involves the triple alpha process. A helium nuclei is called an *alpha particle*, and in the triple alpha process, three helium nuclei combine to form one carbon nucleus. In an *alpha capture*, an alpha particle fuses with the nucleus of an atom, creating a heavier element.

life, and death of distant stars and galaxies than they do about the history of our own solar system. This is because stars are generally too far away for us to observe how planets may be forming around them. The first visual evidence of a stellar disk outside our own solar system was obtained in April 1984 by astronomers at the Las Campañas Observatory in Chile. They observed a huge, disk-shaped cloud of matter surrounding Beta Pictoris, a star twice as large as our Sun in the constellation Pictor, 50 light-years away. This disk-shaped cloud of matter, which may contain planets, provides us with our first telescopic image of what appears to be an evolving solar system (◆ Fig. 8.5). Further study of the cloud should provide astronomers with new insights into the early history of our own solar system.

General Characteristics of the Solar System

Any theory that attempts to explain the origin and history of our solar system must take into account several general characteristics (Table 8.3). All of the planets revolve around the Sun in a counterclockwise direction when viewed from a point high in space above the Earth's North Pole; the orbits around the Sun are nearly circular, and all planetary orbits lie in a common plane, called the *plane of the ecliptic*. Furthermore, all of the planets except Venus and Uranus and nearly all of the planetary moons rotate in a counterclockwise direction, and all of the planets except Uranus and Pluto have axes of rotation that are nearly perpendicular to the plane of the ecliptic (◆ Fig. 8.6).

The nine planets can be divided into two groups based on their chemical and physical properties. The four inner planets—Mercury, Venus, Earth, and Mars—are all small

◆ **FIGURE 8.5** Beta Pictoris, a star about twice the size of our Sun in the constellation Pictor, is surrounded by a disk-shaped cloud of matter that may contain planet-sized objects. This cloud is the first visual evidence of a stellar disk outside our own solar system. (Photo courtesy of NASA.)

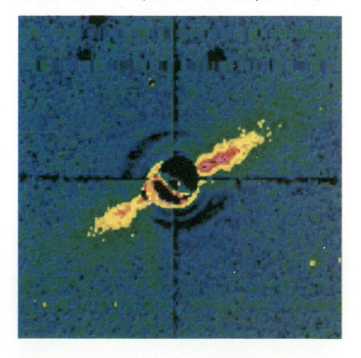

and have high mean densities (Table 8.4), indicating that they are composed of rock and metallic elements. They are known as the **terrestrial planets** because they are similar to *terra,* which is Latin for Earth. The next four planets—Jupiter, Saturn, Uranus, and Neptune—are called the **Jovian planets** because they all resemble Jupiter. The Jovian planets are all large and have low mean densities, indicating they are composed of lightweight gases such as hydrogen and helium, as well as frozen compounds such as ammonia and methane. The outermost planet, Pluto, is small and has a low mean density of slightly more than 2.0 g/cm^3.

The slow rotation of the Sun is another feature that must be accounted for, and it constituted a major problem for many of the early theories of the origin of the solar system. Finally, the nature and distribution of the various types of interplanetary objects such as the asteroid belt, comets, and interplanetary dust must also be explained in any theory of the origin of the solar system.

TABLE 8.3 General Characteristics of the Solar System

1. PLANETARY ORBITS AND ROTATION
   ▶ Planetary and satellite orbits lie in a common plane.
   ▶ Nearly all of the planetary and satellite orbital and spin motions are in the same direction.
   ▶ The rotation axes of nearly all the planets and satellites are roughly perpendicular to the plane of the ecliptic.
2. CHEMICAL AND PHYSICAL PROPERTIES OF THE PLANETS
   ▶ The terrestrial planets are small, have high densities (4.0 to 5.5 g/cm^3), and are composed of rock and metallic elements.
   ▶ The Jovian planets are large, have low densities (0.7 to 1.7 g/cm^3), and are composed of gases and frozen compounds.
3. THE SLOW ROTATION OF THE SUN
4. INTERPLANETARY MATERIAL
   ▶ The existence and location of the asteroid belt.
   ▶ The distribution of interplanetary dust.

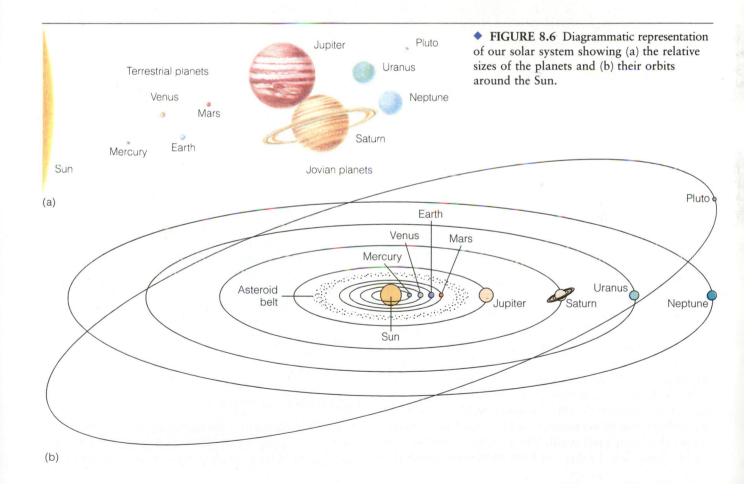

◆ **FIGURE 8.6** Diagrammatic representation of our solar system showing (a) the relative sizes of the planets and (b) their orbits around the Sun.

TABLE 8.4 Characteristics of the Sun, Planets, and Moon

OBJECT	MEAN DISTANCE TO SUN (km × 10⁶)	ORBITAL PERIOD (DAYS)	ROTATIONAL PERIOD (DAYS)	TILT OF AXIS	EQUATORIAL DIAMETER (km)	MASS (kg)	MEAN DENSITY (g/cm³)	NUMBER OF SATELLITES
Sun	—	—	25.5	—	1,391,400	1.99×10^{30}	1.41	—
Terrestrial planets								
Mercury	57.9	88.0	58.7	28°	4,880	3.33×10^{23}	5.43	0
Venus	108.2	224.7	243	3°	12,104	4.87×10^{27}	5.24	0
Earth	149.6	365.3	1	24°	12,760	5.97×10^{24}	5.52	1
Mars	227.9	687.0	1.03	24°	6,787	6.42×10^{23}	3.96	2
Jovian planets								
Jupiter	778.3	4,333	0.41	3°	142,796	1.90×10^{27}	1.33	16
Saturn	1,428.3	10,759	0.43	27°	120,660	5.69×10^{26}	0.69	18
Uranus	2,872.7	30,685	0.72	98°	51,200	8.69×10^{25}	1.27	15
Neptune	4,498.1	60,188	0.67	30°	49,500	1.03×10^{26}	1.76	8
Pluto	5,914.3	90,700	6.39	122°	2,300	1.20×10^{22}	2.03	1
Moon	0.38 (from Earth)	27.3	27.32	7°	3,476	7.35×10^{22}	3.34	—

Evolutionary or Catastrophic Origin of the Solar System?

The various theories proposed to explain the origin and history of the solar system can be divided into two groups: evolutionary and catastrophic. *Evolutionary theories* view the formation of the solar system as part of the normal sequence of events that produced the Sun. *Catastrophic theories* hold that the formation of the Sun was followed by a singular cataclysmic event that disrupted the Sun and formed the planets.

Evolutionary Theories

The first serious evolutionary theory of the origin of the solar system was proposed in 1644 by the French scientist and philosopher René Descartes. He proposed that the solar system formed from some gigantic whirlpool within a universal fluid. Within this vortex, smaller eddies formed the planets and their satellites. Although Descartes never specified the nature of the cosmic material from which the Sun and planets formed, his theory did provide an explanation for why all the planets orbit the Sun in the same direction.

In 1755 the German philosopher Immanuel Kant elaborated on Descartes's idea by using Newton's laws of motion to show that a rotating cloud of gas would flatten into a disk as it contracted. The French mathematician Pierre Simon de Laplace independently proposed the same theory as Kant with some modifications. He stated that as the spinning cloud flattened into a disk, concentric rings of material would form due to rotational forces. These rings would then condense into planets.

The Kant and Laplace theories came to be called the *nebular theory* of the origin of the solar system (◆ Fig. 8.7). This theory had great appeal because it not only explained many of the orbital and rotational features of the planets and satellites, but it also explained how a disk could form from a collapsing ball of interstellar material.

There was, however, one major flaw in the theory. According to the laws of physics, the angular momentum of such a system must remain constant unless some outside force acts on it. This means that if a spinning object shrinks in size, it must increase its rotation rate to compensate for its smaller size, thereby maintaining constant angular momentum (◆ Fig. 8.8). Thus, the Sun, which formed at the center of a collapsing cloud of interstellar material, should have a rapid rotation rate. Instead, it has a rather leisurely 25-day rotation period. Because of this apparent contradiction of the laws of physics, the nebular theory could not be completely accepted until an explanation for the Sun's slow rotation was found.

Catastrophic Theories

Even before Kant proposed his theory, the famous French scientist, Georges Louis de Buffon, suggested the first catastrophic theory for the origin of the solar system in

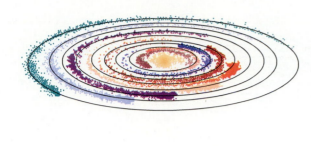

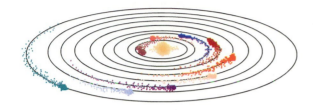

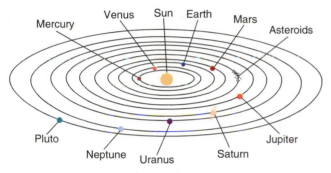

◆ **FIGURE 8.7** The Kant-Laplace nebular theory for the origin of the solar system proposed that the Sun and planets formed from an interstellar cloud that flattened to a rotating disk in which detached rings condensed into the planets.

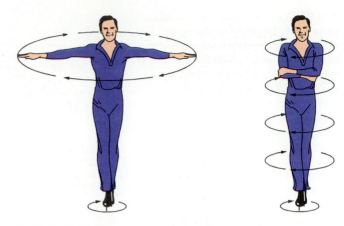

◆ **FIGURE 8.8** A spinning skater illustrates the conservation of angular momentum. As the skater pulls in his arms, he spins at a faster rate. Similarly, as a spinning nebula contracts, its rotational speed increases.

into larger bodies called *planetesimals* that ultimately evolved into the planets and their moons.

The problem with this theory and all similar catastrophic theories is that near collisions of stars are extremely rare events. Furthermore, calculations show that even if a near encounter did occur, the material pulled out of the Sun would be so hot that it would expand and dissipate into space rather than condense into planetary bodies.

◆ **FIGURE 8.9** The encounter theory for the origin of the solar system proposed that a star passed very close to the Sun and pulled away huge filaments of matter that condensed to form planets.

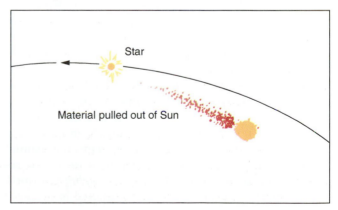

1745. He proposed that a comet passed so close to the Sun that it pulled out gaseous material and dust, which then condensed to form planets. Buffon's idea was largely ignored until the beginning of this century, when the problem of the Sun's slow rotation forced scientists to consider alternatives to the nebular theory.

Probably the best-known catastrophic theory is the *encounter theory* proposed by the English astronomer Forest R. Moulton and the American geologist Thomas C. Chamberlin in 1900 (◆ Fig. 8.9). Their theory was essentially a modification of Buffon's original idea and envisioned another star passing very close to the Sun and pulling away huge filaments of matter that accreted

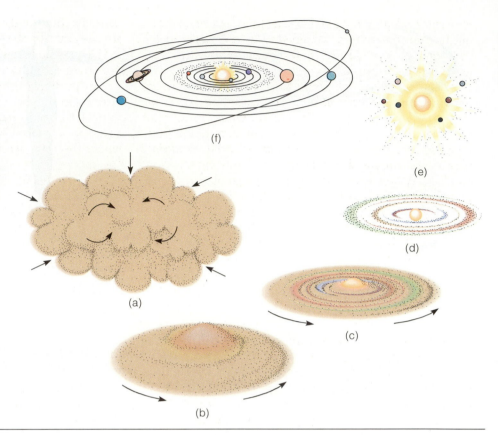

◆ **FIGURE 8.10** The solar nebula theory for the origin of the solar system involves (a) a huge nebula condensing under its own gravitational attraction, then (b) contracting, rotating, and (c) flattening into a disk, with (d) the Sun forming in the center and eddies gathering up material to form planets. As the Sun contracts and begins to visibly shine, (e) intense solar radiation blows away unaccreted gas and dust until finally, (f) the Sun begins burning hydrogen and the planets complete their formation.

Current Theory of the Origin and Early History of the Solar System

When catastrophic theories failed to explain the origin of the solar system, research shifted back to evolutionary models to explain the problem of a slowly rotating Sun. Work by the German physicists C. F. von Weiszacker and Gerard P. Kuiper, and the discovery of solar winds, resulted in the current **solar nebula theory** for the origin of the solar system (◆ Fig. 8.10).

Our solar system was formed about 4.6 billion years ago when interstellar material in a spiral arm of the Milky Way Galaxy condensed and began collapsing. As this cloud gradually collapsed under the influence of gravity, it began to flatten and rotate counterclockwise, with about 90% of its mass concentrated in the central part of the cloud. As rotation and concentration continued, an embryonic Sun, surrounded by a turbulent, rotating cloud of material called a *solar nebula,* formed. The turbulence in this solar nebula resulted in localized eddies forming where condensation of gas and solid particles took place.

As planetesimals formed in these eddies, they all rotated in the same direction around the Sun and in the same direction around their own axes. Something happened, however—perhaps an unusually large collision—that caused Venus to rotate around its axis in the opposite direction. A collision could also explain why Uranus and Pluto do not rotate nearly perpendicular to the plane of the ecliptic. During this early accretionary phase in the solar system's history, collisions between bodies were common, as indicated by the cratering that many planets and satellites exhibit (see Perspective 8.1).

The composition of the various planets can be explained by the fact that every element and compound has a certain temperature-pressure combination at which it condenses from the gaseous phase. In the hot inner portions of the solar nebula, *refractory elements,* which condense at high temperatures, began forming solid particles. It was still too hot in this inner region for *volatile elements* such as hydrogen, helium, ammonia, and methane to condense, so they remained in the gaseous state. In the outer regions of the solar nebula, however, these gases began condensing to form ices.

As condensation took place, gaseous, liquid, and solid particles began accreting into ever-larger masses. The masses became planetesimals and continued to accrete into true planetary bodies with compositions that reflect their distance from the Sun. For example, the terrestrial planets are composed of rock and metallic elements that condense at high temperatures. The Jovian planets, all of which have central rocky cores that are small in comparison to their overall size, are composed mostly of hydrogen, helium, ammonia, and methane, which condense at low temperatures.

While the planets were accreting, material that had been pulled into the center of the nebula also condensed, collapsed, and was heated to several million degrees by gravitational compression. The result was the birth of a star, our Sun.

During the Sun's early history, it emitted a tremendous blast of energy that blew the solar system's unaccreted gases and dust into interstellar space. Such a blast is a normal phase in the evolution of a star and explains why the solar system is so free of extraneous debris. Also during the Sun's early history, its magnetic field interacted with the ionized gases of the solar nebula, slowing down its rotation through a magnetic braking process (◆ Fig. 8.11). The discovery that the Sun's magnetic field exerted a force on the surrounding nebular gas solved the problem of why the Sun has such a slow rotation.

The asteroid belt is the final feature of the solar system that must be explained by the solar nebula theory. Asteroids probably formed as planetesimals in a localized eddy between what eventually became Mars and Jupiter in much the same way as other planetesimals formed the terrestrial planets. The tremendous gravitational field of Jupiter, however, prevented this matter from forming a planet.

The solar nebula theory of the formation of the solar system thus accounts for the similarities in orbits and rotation of the planets and their moons, the differences in composition between the terrestrial and Jovian planets, the slow rotation of the Sun, and the presence of the asteroid belt. Although some details still need to be worked out, the solar nebula theory best explains the features of the solar system and provides a logical explanation for its evolutionary history.

❖ METEORITES—EXTRATERRESTRIAL VISITORS

Meteorites are thought to be pieces of material that originated during the formation of the solar system 4.6 billion years ago. Early in the history of the solar system, a period of heavy meteorite bombardment occurred as the solar system cleared itself of the many pieces of material that had not yet accreted into planetary bodies or moons. Since then, meteorite activity has greatly diminished. Most of the meteorites that currently reach the Earth are probably fragments resulting from collisions of asteroids.

Stones, Irons, and Stony-Irons

Meteorites are classified into three broad groups— *stones, irons, and stony-irons*—based on their proportions of metals and silicate minerals (◆ Fig. 8.12, p. 216). About 93% of all meteorites are composed of iron and magnesium silicate minerals and are thus known as **stones.** Stony meteorites are not all the same and can be divided into three different types.

Ordinary chondrites are the most common type of stone meteorite and are composed of such high-temperature ferromagnesian silicate minerals as olivine and certain pyroxenes (Fig. 8.12b). Age dating reveals they are 4.6 billion years old and thus represent material that existed when the solar system was forming. Most chondrites contain **chondrules,** which are small mineral

◆ **FIGURE 8.11** The slow rotation of the Sun is the result of the interaction of its magnetic force lines with ionized gases of the solar nebula. Thus, the rotation is slowed by a magnetic braking process.

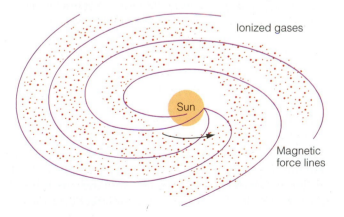

Ionized gases

Sun

Magnetic force lines

METEORITE SHOWERS AND THEIR IMPACT ON PLANETS

A remarkable episode in Earth history occurred soon after the Earth originated and yet left very few traces of its occurrence. When geologists look at the other terrestrial planets and moons they see evidence of a period when meteorite impacts and cratering were far more frequent. The time of this great meteorite shower can be deduced by analyzing the frequency of cratering on the Moon and dating the lunar rocks from different areas. Many of the Moon's large craters are floored by basalt flows, which are younger than the craters and are only lightly to moderately cratered (see Fig. 8.15).

Radiometric dates of samples brought back from the Moon indicate that most of the large craters as well as the highlands are between 3.9 and 4.6 billion years old. The basalts flooring these craters are slightly younger, ranging from 3.2 to 3.8 billion years. The relatively light cratering of these floor basalts indicates that the high intensity of meteorite showering, estimated at more than 1,000 times the present rate, subsided sometime after 4 billion years ago. Analysis of the density of craters on other planets also reveals a period of intense meteorite shower activity between 4.6 and 4.0 billion years ago with only relatively minor meteorite activity since then (see Fig. 8.18).

This period of heavy bombardment took place when the solar system was clearing itself of the many pieces of material that had not yet accreted into planetary bodies or moons. The reason we see little evidence of this activity on Earth is that weathering, erosion, and recycling of crustal rocks have destroyed almost all such evidence.

Since the period of intense meteorite bombardment about 4.0 billion years ago, meteorite impacts on Earth, while not as intense, have still been numerous. Most of these meteorites come from the asteroid belt, where there

◆ **FIGURE 1** Meteor Crater, Arizona, is the result of an Earth-asteroid collision that occurred between 25,000 and 50,000 years ago. It produced a crater 1.2 km in diameter and 180 m deep. (Photo courtesy of D. J. Roddy, USGS.)

are uncounted millions of pieces of rock and small planetoids left over from the origin of the solar system. When these asteroids smash into each other, they send fragments of material into space where they may collide with a planet or moon. The gravitational pull of Jupiter's huge mass also alters the orbits of some asteroids, increasing their chances for collision with planets.

Just as two highways intersect at a crossroad, so, too, do the orbits of Earth and some asteroids, making an eventual collision inevitable. Astronomers have identified at least 40 and estimate there may be as many as 1,000 asteroids larger than a kilometer in diameter whose orbits cross the Earth's. Calculations indicate that we can expect several collisions every million years with asteroids of this size. One of the most famous results of an Earth-asteroid

bodies that formed by rapid cooling when the first solid material of the meteorite was condensing.

Carbonaceous chondrites have the same general composition as ordinary chondrites but also contain about 5% organic compounds (Fig. 8.12c). These organic molecules, which include some amino acids, are not biogen-

ically produced, but may represent chemical precursors of organisms.

Acondrites, the third type of stone meteorite, do not contain chondrules (Fig. 8.12d). Their composition is similar to that of terrestrial basalts, and they are believed to be the result of collisions between larger asteroids.

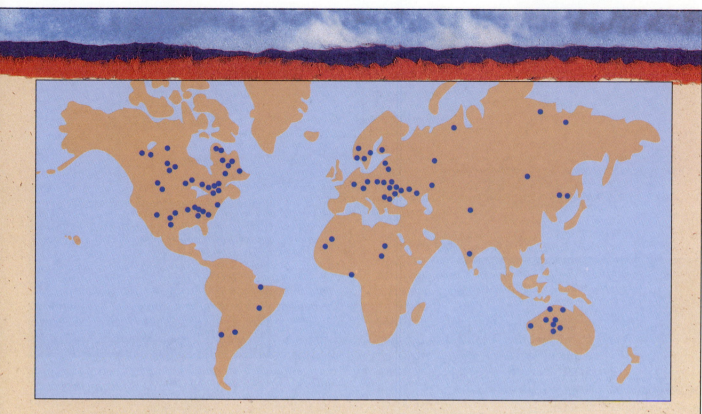

◆ **FIGURE 2** The location of major craters believed to be the result of collisions with massive objects. It should be noted that the distribution is largely a result of our knowledge of these areas and does not purport to be a real distribution of craters.

collision is Meteor Crater, Arizona, which was formed between 25,000 and 50,000 years ago and left a crater 1.2 km in diameter (◆ Fig. 1).

While collisions with Earth are rare, they do happen and can have devastating results. Geologists can identify 116 impact sites on Earth, of which 13 are definitely associated with meteorites; the other 103 are probably impact structures, based on chemical and mineralogical evidence that is distinctive to meteorite impacts (◆ Fig. 2).

A collision with a meteorite about 10 km in diameter is believed to have occurred 65 million years ago. Many scientists think this collision resulted in the extinction of the dinosaurs and all large reptilian groups (see Chapter 16). If such a collision occurred, it would have generated a tremendous amount of dust. This dust would have blocked out the Sun, causing a cessation in photosynthesis and lowering global temperatures, resulting in massive extinctions.

Irons are the second group of meteorites and account for about 6% of all meteorites (Fig. 8.12e). They are composed of a combination of iron and nickel alloys. Their large crystal size and chemical composition indicate that they cooled very slowly in large objects such as asteroids where the hot iron-nickel interior could be in-sulated from the cold of space. Collisions between such slowly cooling asteroids produced the iron meteorites we find today.

Stony-irons, the third group, are composed of nearly equal amounts of iron and nickel and silicate minerals; they make up less than 1% of all meteorites (Fig. 8.12f).

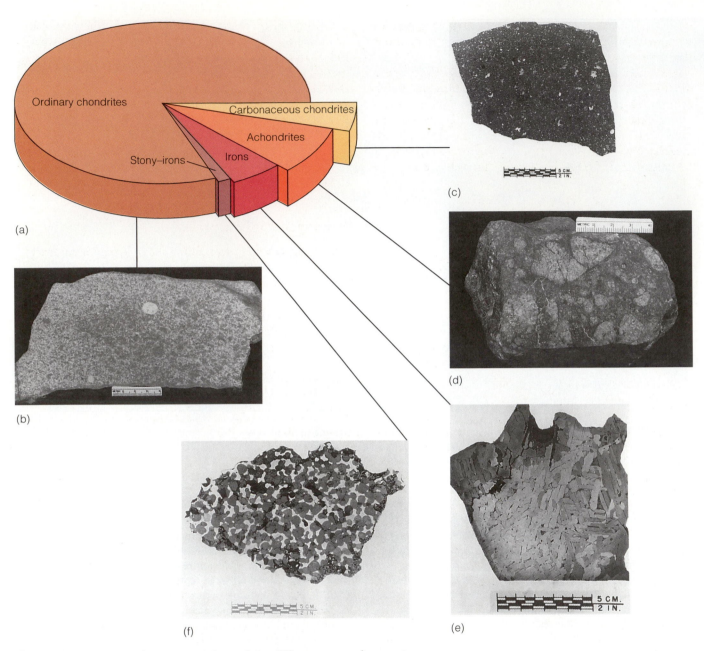

◆ **FIGURE 8.12** (a) Relative proportions of the different types of meteorites.
(b) Polished slab of an ordinary chondrite from Pinto Mountains. (c) Polished slab of a
carbonaceous chondrite from Allende, Mexico. (d) Acondrite from Palo Blanco Creek.
(e) Polished slab of an iron meteorite from Bogou, Upper Volta, Brazil. (f) Polished slab
of a stony-iron from Thiel Mountain, Antarctica. The white minerals are iron-nickel
and the grey minerals are olivine. (Photos c, e, and f courtesy of Brian Mason,
Smithsonian Institution; photos b and d courtesy of Ken Nichols, University of New
Mexico.)

Stony-irons are generally believed to represent fragments from the zone between the metallic and silicate portions of a large differentiated asteroid.

Meteorites are important to us because their age—about 4.6 billion years—and composition provide us with information about the origin and history of the solar system. Furthermore, their collisions with Earth and the other planets have affected the topography of the planets and, although there is no unanimity on this point, perhaps the evolution of life on Earth (see Perspective 8.1).

❖ THE ORIGIN AND DIFFERENTIATION OF THE EARLY EARTH

As matter was accreting in the various turbulent eddies that swirled around the early Sun, one eddy eventually gathered enough material to form the Earth.

The Earth consists of concentric layers of different composition and densities (◆ Fig. 8.13). These density layers are a fundamental feature of the Earth, and presumably this differentiation occurred very early in the Earth's history.

Geologists know that the Earth is 4.6 billion years old. The oldest known rocks, however, are 3.96-billion-year-old metamorphic rocks from Canada. Like the younger crustal rocks, these rocks are composed of relatively light silicate minerals. It appears that a crust, a heavier silicate mantle, and an iron-nickel core were already present 3.96 billion years ago, or 640 million years after the Earth formed. Currently, there are two competing theories of how the core-mantle-crust density layering of the Earth originated (◆ Fig. 8.14). Both theories are consistent with the solar nebula theory of the origin of the solar system.

Homogeneous Accretion

The **homogeneous accretion** theory assumes an early solid Earth of generally uniform composition and density. The iron and nickel of the present core were distributed fairly evenly throughout a larger mass of lighter silicate minerals.

The early Earth is believed to have been rather cool, so the elements and nebular rock fragments accreting to it were solids, rather than gases or liquids. In order for the iron and nickel to concentrate in the core, the interior of

◆ FIGURE 8.13 Cross section of the Earth showing the various layers and their average density. The crust is divided into a continental and oceanic portion. Continental crust is 20 to 90 km thick; oceanic crust is 5 to 10 km thick.

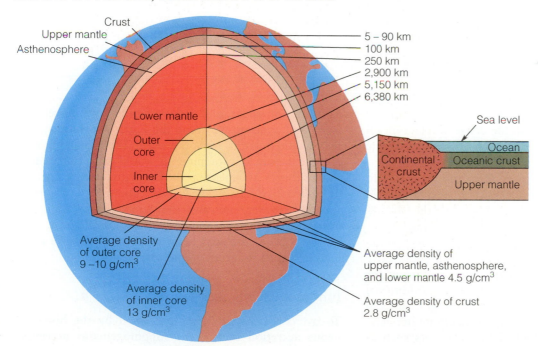

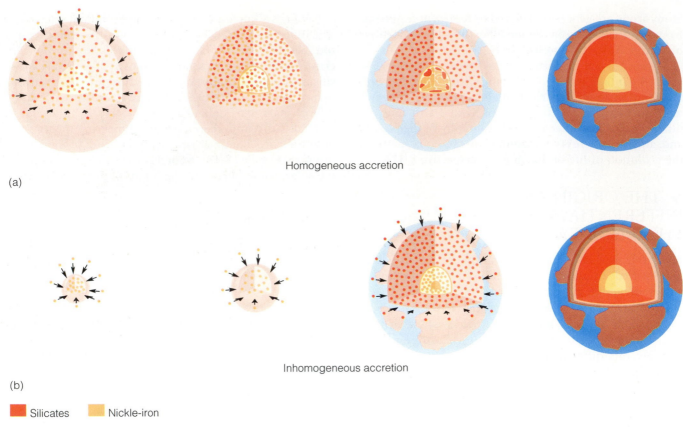

(a)

Homogeneous accretion

(b)

Inhomogeneous accretion

■ Silicates   ■ Nickle-iron

◆ **FIGURE 8.14** Two theories have been proposed to explain the core-mantle-crust zonation of the Earth. (a) According to the homogeneous accretion theory, the early Earth was solid with a generally uniform composition and density throughout. Subsequently, partial melting resulted in iron and nickel settling to the Earth's center and forming the core, while the lighter silicate minerals flowed upward to form the mantle and crust. (b) According to the inhomogeneous accretion theory, the Earth's core, mantle, and crust condensed sequentially from the hot nebular gases.

the early Earth must have been heated enough to become partially melted. At that point, the dense iron and nickel could have sunk through the surrounding lighter silicate minerals and formed the core.

Iron and nickel melt at lower temperatures than silicates and, because they are denser than silicates, would settle to the center of the Earth. Meanwhile the silicates would soften and slowly flow upward, beginning the differentiation of the mantle from the core. The heat would come primarily from the decay of short-lived radioactive isotopes, which were much more abundant during the early history of the Earth than they are today. Additional heat would also have been generated from gravitational compression and from the energy of meteorite impacts.

Many of the easily melted, lighter elements such as calcium, potassium, and sodium would have moved upward and concentrated near the Earth's surface, where they would begin to form the crust. The remaining iron and magnesian silicate minerals would form the mantle.

This theory assumes that the Earth underwent a period of radioactive heating and differentiation. Current research, however, indicates that gravitational accretion energy, even when combined with radiogenic heat, may not have been sufficient to produce the extensive early melting called for from a cold, planetary body.

Inhomogeneous Accretion

To overcome the problems presented by the homogeneous accretion theory, the **inhomogeneous accretion** theory has been proposed. According to this theory, the Earth's core, mantle, and crust condensed sequentially from the hot nebular gases forming the early Earth.

Even though iron and nickel melt at lower temperatures than most silicates, they condense from the gaseous state at a slightly higher temperature. Calculations show that in a cooling cloud of hot nebular gases, iron and nickel would condense first, thus forming the core of the early Earth. As the cloud continued to cool, iron and magnesium silicates would condense to form the mantle, while the lightest and most volatile elements that form the crust would condense last.

Conclusions on Homogeneous versus Inhomogeneous Accretion

Most geologists currently favor the homogeneous accretion theory's explanation of the Earth's internal zonation. There is, however, no conclusive evidence for rejecting the inhomogeneous accretion theory. Both theories are based on the currently accepted solar nebula origin of the solar system, and regardless of which theory, if either, eventually proves correct, geologists do know that the differentiation of the Earth occurred very early in its history. Not only did differentiation lead to the formation of a crust and eventually to continents, but it was probably responsible for the outgassing of light volatile elements from the interior that eventually led to the formation of the oceans and atmosphere.

❖ THE ORIGIN OF THE EARTH-MOON SYSTEM

We probably know more about our Moon than any other celestial object except the Earth (◆ Fig. 8.15). Nevertheless, even though the Moon has been studied for centuries through telescopes and has been sampled directly, many questions remain unanswered.

The Moon is one-fourth the diameter of the Earth, has a low density (3.3 g/cm^3) relative to the terrestrial planets, and exhibits an unusual chemistry in that it is bone-dry, having been largely depleted of most volatile elements (Table 8.4). The Moon orbits the Earth and rotates on its own axis at the same rate, so we always see the same side. Furthermore, the Earth-Moon system is unique among the terrestrial planets. Neither Mercury nor Venus has a moon, and the two small moons of Mars—Phobos and Deimos—are probably captured asteroids.

◆ FIGURE 8.15 The side of the Moon as seen from Earth. The light-colored areas are the lunar highlands, which were heavily cratered by meteorite impacts. The dark-colored areas are maria, which formed when lava flowed out onto the surface. (Photo courtesy of the Lick Observatory.)

The surface of the Moon can be divided into two major parts: the low-lying dark-colored plains, called *maria*, and the light-colored *highlands* (Fig. 8.15). The highlands are the oldest parts of the Moon and are heavily cratered, providing striking evidence of the massive meteorite bombardment that occurred in the solar system more than 4 billion years ago.

Study of the several hundred kilograms of rocks returned by the *Apollo* missions indicates that three kinds of materials dominate the lunar surface: igneous rocks, breccias, and dust. Basalt, a common dark-colored igneous rock on Earth, is one of the several different types of igneous rocks on the Moon and makes up the greater part of the maria. The presence of igneous rocks that are essentially the same as those on Earth shows that magmas similar to those on Earth were generated on the Moon long ago.

The lunar surface is covered with a regolith (or "soil") that is estimated to be 3 to 4 m thick. This gray covering, which is composed of compacted aggregates of rock fragments called breccia, glass spherules, and small particles of dust, is thought to be the result of debris formed by meteorite impacts.

The interior structure of the Moon is quite different from that of the Earth, indicating a different evolutionary history (◆ Fig. 8.16.) The highland crust is thick (65 to 100 km) and comprises about 12% of the Moon's volume. It was formed about 4.4 billion years ago, immediately following the Moon's accretion. The highlands are composed principally of the igneous rock anorthosite, which is made up of light-colored feldspar minerals that are responsible for their white appearance.

A thin covering (1 to 2 km thick) of basaltic lava fills the maria; lava covers about 17% of the lunar surface, mostly on the side facing the Earth. These maria lavas came from partial melting of a thick underlying mantle of silicate composition. Moonquakes occur at a depth of about 1,000 km, but below that depth seismic shear waves apparently are not transmitted. Because shear waves do not travel through liquid, their lack of transmission implies that the innermost mantle may be partially molten. There is increasing evidence that the Moon has a small (600 km to 1,000 km diameter) metallic core comprising 2 to 5% of its volume.

The origin and earliest history of the Moon are still unclear, but the basic stages in its subsequent development are well understood. It formed some 4.6 billion

◆ **FIGURE 8.16** The internal structure of the Moon is different from that of the Earth. The upper mantle is the source for the maria lavas. Moonquakes occur at a depth of 1,000 km. Because seismic shear waves are not transmitted below this depth, it is believed that the innermost mantle is liquid. Below this layer is a small metallic core.

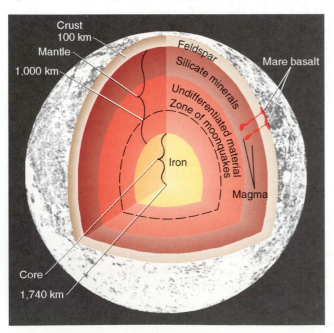

years ago and shortly thereafter was partially or wholly melted, yielding a silicate melt that cooled and crystallized to form the mineral anorthite. Because of the low density of the anorthite crystals and the lack of water in the silicate melt, the thick anorthosite highland crust formed. The remaining silicate melt cooled and crystallized to produce the zoned mantle, while the heavier metallic elements formed the small metallic core.

The formation of the lunar mantle was completed by about 4.4 to 4.3 billion years ago. The maria basalts, derived from partial melting of the upper mantle, were extruded during great lava floods between 3.8 and 3.2 billion years ago.

Numerous models have been proposed for the origin of the Moon, including capture from an independent orbit, formation with the Earth as part of an integrated two-planet system, breaking off from the Earth during accretion, and formation resulting from a collision between the Earth and a large planetesimal These various models are not mutually exclusive, and elements of some occur in others. At this time, scientists cannot agree on a single model, as each has some inherent problems. However, the model that seems to account best for the Moon's particular composition and structure involves an impact by a large planetesimal with a young Earth (◆ Fig. 8.17).

In this model, a giant planetesimal, the size of Mars or larger, crashed into the Earth about 4.6 to 4.4 billion years ago, causing the ejection of a large quantity of hot material that formed the Moon. The material that was ejected was mostly in the liquid and vapor phase and came primarily from the mantle of the colliding planetesimal. As it cooled, the various lunar layers crystallized out in the order we have discussed.

❖ THE PLANETS

A tremendous amount of information about each planet in the solar system has been derived from Earth-based observations and measurements as well as from the numerous space probes launched during the past 30 years. Such information as a planet's size, mass, density, composition, presence of a magnetic field, and atmospheric composition has allowed scientists to formulate hypotheses concerning the origin and history of the planets and their moons.

The Terrestrial Planets

It appears that all of the terrestrial planets had a similar early history during which volcanism and cratering from meteorite impacts were common. After accretion, each planet appears to have undergone differentiation as a

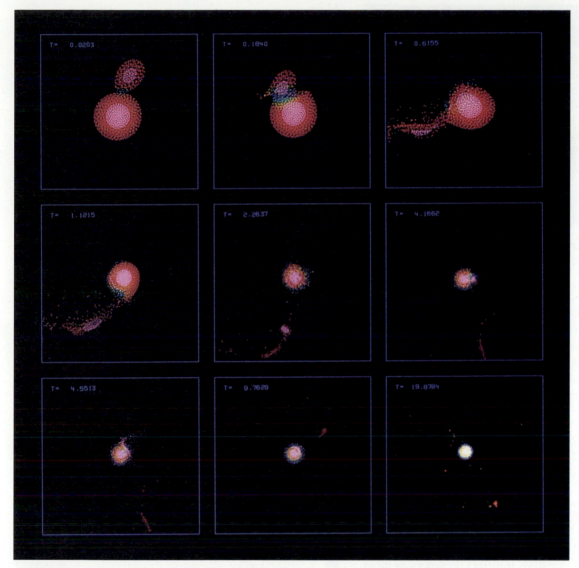

◆ **FIGURE 8.17** According to one hypothesis for the origin of the Moon, a large planetesimal the size of Mars crashed into the Earth 4.6 to 4.4 billion years ago, causing the ejection of a mass of hot material that formed the Moon. This computer simulation shows the formation of the Moon as a result of an Earth-planetesimal collision. (Photo courtesy of W. Benz and W. Slattery, Los Alamos National Laboratories.)

result of heating by radioactive decay. The mass, density, and composition of the planets indicate that each formed a metallic core and a silicate mantle-crust during this phase. Images sent back by the various space probes also clearly show that volcanism and cratering by meteorites continued during the differentiation phase. Volcanic eruptions produced lava flows, and an atmosphere developed on each planet by outgassing.

Mercury

Mercury, the closest planet to the Sun, apparently has changed very little since it was heavily cratered during its early history (◆ Fig. 8.18). Images sent back by *Mariner 10* show a heavily cratered surface with the largest impact basins filled with what appear to be lava flows similar to the lava plains on the Moon. The lava plains are

◆ **FIGURE 8.18** Mercury has a heavily cratered surface that has changed very little since its early history. (Photo courtesy of NASA.)

not deformed, however, indicating that there has been little or no tectonic activity. Another feature of Mercury's surface is a large number of scarps. It is suggested that these scarps formed when Mercury cooled and contracted.

Because Mercury is so small, its gravitational attraction is insufficient to retain atmospheric gases; any atmosphere that it may have held when it formed probably escaped into space very quickly.

Venus

Of all the planets, Venus is the most similar in size and mass to the Earth (Table 8.4). It differs, however, in most other respects. Venus is searingly hot with a surface temperature of 475°C and an oppressively thick atmosphere composed of 96% carbon dioxide and 3.5% nitrogen with traces of sulfur dioxide and sulfuric and hydrochloric acid.

Radar images from orbiting spacecraft as well as from the Venusian surface indicate a wide variety of terrains (◆ Fig. 8.19), some of which are unlike anything seen elsewhere in the solar system (see the Prologue to Chapter 7). Even though no active volcanism has been observed on Venus, the presence of volcanoes, numerous lava flows, areas of crumpled crust, and a network of fractures and faults indicate internal and surface activity has occurred during the past.

Mars

Mars, the red planet, is differentiated, as are all the terrestrial planets, into a metallic core and a silicate mantle and crust (◆ Fig. 8.20). The thin Martian atmosphere consists of 95% carbon dioxide, 2.7% nitrogen, 1.7% argon, and traces of other gases. Mars has distinct seasons during which its polar ice caps of frozen carbon dioxide expand and recede.

Perhaps the most striking aspect of Mars is its surface, many features of which have not yet been satisfactorily explained. Like the surfaces of Mercury and the Moon, the southern hemisphere is heavily cratered, attesting to a period of meteorite bombardment. *Hellas,* a crater with a diameter of 2,000 km, is the largest known impact structure in the solar system and is found in the Martian southern hemisphere.

The northern hemisphere is much different, having large smooth plains, fewer craters, and evidence of extensive volcanism. The largest known volcano in the solar system, *Olympus Mons,* has a basal diameter of 600

◆ **FIGURE 8.19** A nearly complete map of the northern hemisphere of Venus based on radar images beamed back to Earth from the *Magellan* space probe. (Photo courtesy of NASA.)

◆ **FIGURE 8.20** A striking view of Mars is revealed in this mosaic of 102 *Viking* images. The largest canyon known in the solar system, *Valles Marineris,* can be clearly seen in the center of this image, while three of the planet's volcanoes are visible on the left side of the image. (Photo courtesy of JPL/NASA.)

km, rises 27 km above the surrounding plains, and is topped by a huge circular crater 80 km in diameter.

The northern hemisphere is also marked by huge canyons that are essentially parallel to the Martian equator. One of these canyons, *Valles Marineris,* is at least 4,000 km long, 250 km wide, and 7 km deep and is the largest yet discovered in the solar system. If it were present on Earth, it would stretch from San Francisco to New York! It is not yet known how these vast canyons formed, although geologists postulate that they may have started as large rift zones that were subsequently modified by running water and wind erosion. Such hypotheses are based on comparison to rift structures found on Earth and topographic features formed by geologic agents of erosion such as water and wind.

Tremendous wind storms have strongly influenced the surface of Mars and led to dramatic dune formations. Even more stunning than the dunes, however, are the braided channels that appear to be the result of running water. Mars is currently too cold for surface water to exist, yet the channels strongly indicate that there was running water on the planet during the past.

The fresh-looking surfaces of its many volcanoes strongly suggest that Mars was tectonically active during the past and may still be. There is, however, no evidence that plate movement, such as occurs on Earth, has ever occurred.

The Jovian Planets

The Jovian planets are completely unlike any of the terrestrial planets in size or chemical composition (Table 8.4) and have had completely different evolutionary histories. While they all apparently contain a small core in relation to their overall size, the bulk of a Jovian planet is composed of volatile elements and compounds that condense at low temperatures such as hydrogen, helium, methane, and ammonia.

Jupiter

Jupiter is the largest of the Jovian planets (Table 8.4; ◆ Fig. 8.21). With its moons, rings, and radiation belts, it is the most complex and varied planet in the solar system. Jupiter's density is only one-fourth that of Earth, but because it is so large, it has 318 times the mass (Table 8.4). It is an unusual planet in that it emits almost 2.5 times more energy than it receives from the Sun. One explanation is that most of the excess energy is left over from the time of its formation. When Jupiter formed, it heated up because of gravitational contraction (as did all the planets) and is still cooling. Jupiter's massive size insulates its interior, and hence it has cooled very slowly.

Jupiter has a relatively small central core of solid rocky material formed by differentiation. Above this core is a thick zone of liquid metallic hydrogen followed by a thicker layer of liquid hydrogen; above that is a thin layer of clouds. Surrounding Jupiter are a strong magnetic field and an intense radiation belt.

Jupiter has a dense atmosphere of hydrogen, helium, methane, and ammonia, which some believe are the same gases that composed the Earth's first atmosphere. Jupiter's cloudy atmosphere is divided into a series of different colored bands as well as a variety of spots (the Great Red Spot) and other features, all interacting in incredibly complex motions.

◆ **FIGURE 8.21** Jupiter, the largest planet in the solar system, is also the most complex with its moons, rings, and radiation belts. Its cloudy atmosphere displays a regular pattern of bands and spots, the largest of which is the Great Red Spot shown in the lower part of this image. (Photo courtesy of NASA.)

similar, Saturn's atmosphere contains little ammonia because it is farther from the sun and therefore is colder. The cloud layer on Saturn is thicker than on Jupiter, but it lacks the contrast between the different bands. Unlike Jupiter, Saturn has seasons because its axis tilts 27°.

Uranus

Uranus is much smaller than Jupiter, but their densities are about the same (Table 8.4). It is the only planet that lies on its side (◆ Fig. 8.23); that is, its axis of rotation nearly parallels the plane of the ecliptic. Some scientists think that a collision with an Earth-sized body early in its history may have knocked Uranus on its side.

Data gathered by the flyby of *Voyager 2* suggests that Uranus has a water zone beneath its cloud cover. Because the planet's density is greater than if it were composed entirely of hydrogen and helium, it is believed that Uranus must have a dense, rocky core, and this core may be surrounded by a deep global ocean of liquid water.

The atmospheric composition of Uranus is similar to that of Jupiter and Saturn with hydrogen being the dominant gas, followed by helium and some methane. Uranus also has a banded atmosphere and a circulation pattern

Revolving around Jupiter are 16 moons varying greatly in tectonic and geologic activity. Also surrounding Jupiter is a thin, faint ring, a feature shared by all the Jovian planets.

Saturn

Saturn is slightly smaller than Jupiter, about one-third as massive, and about one-half as dense, but has a similar internal structure and atmosphere (Table 8.4; ◆ Fig. 8.22). Saturn, like Jupiter, gives off more energy (2.2 times as much) than it gets from the Sun. Saturn's most conspicuous feature is its ring system, consisting of thousands of rippling, spiraling bands of countless particles.

The composition of Saturn is similar to Jupiter's, but consists of slightly more hydrogen and less helium. Saturn's core is not as dense as Jupiter's, and as in the case of Jupiter, a layer of liquid metallic hydrogen overlies the core, followed by a zone of liquid hydrogen and helium, and, lastly, a layer of clouds.

Even though the atmospheres of Saturn and Jupiter are

◆ **FIGURE 8.22** This image of Saturn was taken by *Voyager 2* from several million kilometers away and shows the ring system of the planet as well as its banded atmosphere. Saturn has an atmosphere similar to that of Jupiter, but has a thicker cloud cover and contains little ammonia. (Photo courtesy of NASA.)

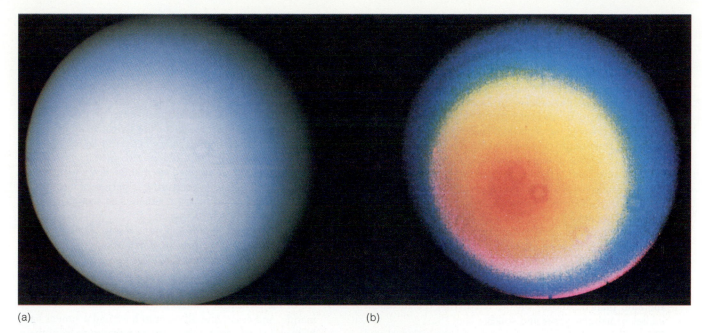

(a)   (b)

◆ **FIGURE 8.23** (a) An image of Uranus taken by *Voyager 2* under ordinary light shows a featureless planet. (b) When color is enhanced by computer processing techniques, Uranus is seen to have zonal flow patterns in its atmosphere.

much like those of Jupiter and Saturn. Surrounding Uranus is a huge corkscrew-shaped magnetic field that stretches for millions of kilometers into space. Uranus has at least nine thin, faint rings and 15 small moons circling it.

Neptune and Pluto

The flyby of *Voyager 2* in August 1989 provided the first detailed look at Neptune and showed it to be a dynamic, stormy planet (◆ Fig. 8.24). Its atmosphere is similar to those of the other Jovian planets, and it exhibits a pattern of zonal winds and giant storm systems comparable to those of Jupiter. The internal structure of Neptune is similar to that of Uranus (Table 8.4), and its atmosphere is composed of hydrogen and helium with some methane.

Pluto is the smallest planet and, strictly speaking, it is not one of the Jovian planets (Table 8.4). Little is known about Pluto, but recent studies indicate it has a rocky core overlain by a mixture of methane gas and ice. It also has a thin, two-layer atmosphere with a clear upper layer overlying a more opaque lower layer. Pluto differs from all the other planets in that it has a highly eccentric orbit that is tilted with respect to the plane of the ecliptic.

◆ **FIGURE 8.24** A shimmering blue planet set against the black backdrop of space, Neptune reveals itself to *Voyager 2*'s instruments during its August 1989 flyby. Shown here is Neptune's turbulent atmosphere with its Great Dark Spot and various wispy clouds. (Photo courtesy of NASA.)

Chapter Summary

1. The universe began with a Big Bang approximately 13 to 20 billion years ago. Astronomers have deduced this from the fact that celestial objects can be seen moving away in what appears to be an ever-expanding universe. Furthermore, the universe has a background radiation of 2.7 K, which represents the cooling remnant of that explosion.

2. During its first 2 billion years, the universe expanded greatly; stars and galaxies formed and combined into larger structures such as clusters and superclusters. During this time the universe became clumpy and enriched in the heavier elements as stars underwent nuclear reactions.

3. In an arm of the Milky Way Galaxy, our own solar system formed from a rotating interstellar cloud of matter about 4.6 billion years ago. As this cloud condensed, it eventually collapsed under the influence of gravity and flattened into a counterclockwise rotating disk. Within this rotating disk, the Sun, planets, and moons accreted from the turbulent eddies of nebular gases and solids.

4. Temperature as a function of distance from the Sun played a major role in the type of planets that evolved. The inner terrestrial planets are composed of rock and metallic elements that condense at high temperatures. The outer Jovian planets plus Pluto, are composed of hydrogen, helium, ammonia, and methane, all of which condense at lower temperatures.

5. During the solar system's early history, the Sun emitted a tremendous blast of energy that swept the solar system free of unaccreted gas and dust. During this time, most of the cratering of the planets and their satellites took place.

6. Meteorites provide vital information about the age and composition of the solar system. The three major groups of meteorites are stones, irons, and stony-irons. Each has a different composition, reflecting a different origin.

7. The Earth formed from one of the swirling eddies of nebular material 4.6 billion years ago, and by at least 3.8 billion years ago, it had differentiated into its present-day structure. It either accreted as a solid body, which then underwent differentiation during a period of heating, or the core, mantle, and crust condensed in sequence from a cooling cloud of nebular gas.

8. The Moon probably formed as a result of a Mars-sized planetesimal crashing into the Earth about 4.6 to 4.4 billion years ago and causing it to eject a large quantity of hot material. As the material cooled, the various lunar layers crystallized, forming a zoned body.

9. All of the terrestrial planets are differentiated into a core, mantle, and crust, and all seem to have had a similar early history during which volcanism and cratering from meteorite impacts were common.

10. The Jovian planets differ from the terrestrial planets in size and chemical composition and followed completely different evolutionary histories. All of the Jovian planets have a small core compared to their overall size; they are mainly composed of volatile elements and compounds that condense at low temperatures, such as hydrogen, helium, methane, and ammonia.

Important Terms

acondrites
Big Bang
carbonaceous chondrites
chondrules
Doppler effect
electromagnetic force
gravity

homogeneous accretion
inhomogeneous accretion
irons
Jovian planets
meteorite
ordinary chondrite
solar nebula theory

stones
stony-irons
strong nuclear force
terrestrial planets
weak nuclear force

Review Questions

1. What two observations lead scientists to conclude that the Big Bang occurred approximately 13 to 20 billion years ago?
a. _____ a steady-state universe and opaque background radiation; b. _____ a steady-state universe and 2.7 K background radiation; c. _____ an expanding universe and opaque background radiation; d. _____ an expanding universe and 2.7 K background radiation; e. _____ a shrinking universe and opaque background radiation.

2. The change in frequency of a sound wave caused by movement of its source relative to an observer is known as the:
a. _____ Curie point; b. _____ Hubble shift; c. _____ Doppler effect; d. _____ Quasar theory; e. _____ none of these.

3. Which of the following is not one of the basic forces in the universe?
a. _____ electromagnetic force; b. _____ gravity; c. _____ strong nuclear force; d. _____ Quasar theory; e. _____ none of these.

4. Within a fraction of a second after the Big Bang, the universe experienced:
a. _____ a period of contraction; b. _____ a period of major expansion; c. _____ a bursting forth of light; d. _____ the separation of gravity from the other forces; e. _____ the formation of atoms.

5. Which of the following statements concerning the solar system is not correct?
a. _____ the orbits of the planets around the Sun are nearly circular; b. _____ all planetary orbits lie in the plane of the ecliptic; c. _____ all of the planets except Venus and Uranus rotate in a counterclockwise direction when viewed from a point in space high above the Earth's North Pole; d. _____ the Sun has a relatively slow rate of rotation; e. _____ all of the planets revolve around the Sun in a clockwise direction when viewed from a point in space high above the Earth's North Pole.

6. The first serious evolutionary theory for the origin of the solar system was proposed by:
a. _____ Kant; b. _____ Descartes; c. _____ Laplace; d. _____ Moulton; e. _____ Chamberlin.

7. The current theory for the origin of the solar system is known as the:
a. _____ nebular theory; b. _____ encounter theory; c. _____ evolutionary theory; d. _____ planetesimal theory; e. _____ solar nebula theory.

8. The most abundant meteorites are:
a. _____ stones; b. _____ acondrites; c. _____ stony-irons; d. _____ irons; e. _____ carbonaceous chondrites.

9. The homogeneous accretion theory states that the:
a. _____ Earth is homogeneous throughout; b. _____ Earth's interior layers condensed sequentially from hot nebular gases; c. _____ Earth's interior layers formed following a period of partial melting of the Earth; d. _____ Earth formed from the collision of hot materials; e. _____ answers (b) and (c).

10. The model that seems to best account for the origin of the Earth's Moon is:
a. _____ capture from an independent orbit; b. _____ impact of a large planetesimal; c. _____ formation with the Earth as part of an integrated two-planet system; d. _____ formation by accretion of nebular material; e. _____ none of these.

11. The terrestrial and Jovian planets differ from each other in:
a. _____ size; b. _____ density; c. _____ chemical composition; d. _____ rings; e. _____ all of these.

12. The terrestrial planet that has changed the least since its formation is:
a. _____ Mercury; b. _____ Venus; c. _____ Earth; d. _____ Mars; e. _____ Pluto.

13. The largest planet in the solar system is:
a. _____ Neptune; b. _____ Uranus; c. _____ Saturn; d. _____ Jupiter; e. _____ Mars.

14. The planet most similar in size and mass to the Earth is:
a. _____ Mercury; b. _____ Venus; c. _____ Mars; d. _____ Neptune; e. _____ Uranus.

15. Which planet shows evidence of plate movements similar to those on Earth?
a. _____ Mercury; b. _____ Venus; c. _____ Mars; d. _____ Neptune; e. _____ none of these.

16. Which planets give off more energy than they receive?
a. _____ Jupiter and Saturn; b. _____ Saturn and Neptune; c. _____ Neptune and Pluto; d. _____ Jupiter and Neptune; e. _____ Saturn and Pluto.

17. Which of the following events did all of the terrestrial planets experience early in their history?
a. _____ accretion; b. _____ differentiation; c. _____ volcanism; d. _____ meteorite impacting; e. _____ all of these.

18. The composition of the universe has been changing since the Big Bang. Yet 98% of its composition by weight still consists of which elements?
a. _____ hydrogen and helium; b. _____ hydrogen and carbon; c. _____ helium and carbon; d. _____ hydrogen and nitrogen; e. _____ helium and nitrogen.

19. Compare the origin and history of the four terrestrial planets, noting the similarities and differences.

20. Compare the origin and history of the four Jovian planets, noting the similarities and differences.

21. Discuss the origin and differentiation of the Earth into its various internal layers.

22. Discuss the origin of the Earth-Moon system.

23. What two fundamental phenomena indicate that the Big Bang occurred 13 to 20 billion years ago?

24. If the oldest terrestrial rocks are 3.96 billion years old, why do geologists think the Earth is 4.6 billion years old?

25. Why is the Doppler effect important to the theory of an expanding universe?

26. Why is an evolutionary theory for the origin of the solar system more appealing than a catastrophic one?

27. How does the solar nebula theory account for the various features of the solar system?

28. What are the three groups of meteorites, and how do they aid geologists in determining the age and composition of the solar system?

29. What other important roles have meteorites played in the history of the solar system?

30. What are the strengths and weaknesses of the homogeneous and inhomogeneous accretion theories for the origin and differentiation of the early Earth?

Additional Readings

Barnes, J. L. Hernquist, and F. Schweizer. 1991. Colliding galaxies. *Scientific American* 265, no. 2.: 40–47.

Binzel, R. P. 1990. Pluto. *Scientific American* 262, no. 6: 50–59.

Binzel, R. P., M. A. Barucci, and M. Fulchignoni. 1991. The origins of the asteroids. *Scientific American* 265, no. 4: 88–95.

Brush, S. G. 1992. How cosmology became a science. *Scientific American* 267, no. 2: 62–71.

Freedman, W. L. 1992. The expansion rate and size of the universe. *Scientific American* 267, no. 5: 54–61.

Grieve, R. A. F. 1990. Impact cratering on the Earth. *Scientific American* 262, no. 4: 66–73.

Hetherington, N. S. 1990. Hubble's cosmology. *American Scientist* 78, no. 2: 142–51.

Horgan, J. 1990. Universal truths. *Scientific American* 263, no. 4: 108–17.

Ingersoll, A. P. 1987. Uranus. *Scientific American* 256, no. 1: 38–45.

Kinoshita, J. 1989. Neptune. *Scientific American* 261, no. 5: 82–91.

Kuhn, K. F. 1991. *In quest of the universe.* St. Paul, Minn.: West Publishing Co.

McSween, H. Y., Jr. 1989. Chondritic meteorites and the formation of planets. *American Scientist* 77, no. 2: 146–53.

Powell, C. S. 1992. The golden age of cosmology. *Scientific American* 267, no. 1: 17–22.

Powell, C. S. 1992. Venus revealed. *Scientific American* 266, no. 1: 16–17.

Saunders, R. S. 1990. The surface of Venus. *Scientific American* 263, no. 6: 60–65.

Snow, T. P. 1991. *The dynamic universe: An introduction to astronomy.* 4th ed. St. Paul, Minn.: West Publishing Co.

Spergel, D. N., and N. G. Turok. 1992. Textures and cosmic structure *Scientific American* 266, no. 3: 52–59.

Taylor, S. R. 1987. The origin of the Moon. *American Scientist* 75, no. 5: 468–77.

Weisskopf, V. F. 1983. The origin of the universe. *American Scientist* 71, no. 5: 473–80.

Wetherill, G. W. 1981. The formation of the Earth from planetesimals. *Scientific American* 244, no. 6: 162–75.

CHAPTER 9

Archaen gneiss deformed by folding and intruded by small granite dikes, Georgian Bay, Ontario, Canada. (Photo courtesy of R. V. Dietrich.)

PRECAMBRIAN HISTORY: THE ARCHEAN EON

Prologue

Imagine a barren, lifeless, waterless, hot planet with a poisonous atmosphere. Volcanoes erupt nearly continuously, meteorites and comets flash through the atmosphere, and cosmic radiation is intense. The planet's crust, which is composed entirely of dark-colored igneous rock, is thin and unstable. Storms form in the turbulent atmosphere, and lightning discharges are common, but no rain falls. Since there is no oxygen in the atmosphere, nothing burns. Rivers and pools of molten rock emit a continuous reddish glow.

This may sound like a science fiction novel, but it is probably a reasonably accurate description of the Earth shortly after it formed (◆ Fig. 9.1). We emphasize "probably" because no record exists for the earliest chapter of Earth history, the interval from 4.6 to 3.8 billion years ago, although one area of rocks 3.96 billion years old is now known in Canada. We can only speculate about what the Earth was like during this time, based on our knowledge of how planets form and about other Earth-like planets.

When the Earth formed, it had a tremendous reservoir of primordial heat, heat generated by colliding particles as the Earth accreted, by compression, and by the decay of short-lived radioactive elements. Many geologists believe that the early Earth was so hot that it was partly or perhaps almost entirely molten. No one knows what the Earth's surface temperature was during its earliest history, but it was almost certainly too hot for liquid water to exist or for any known organism to survive. Volcanism must have been ubiquitous and nearly continuous. Molten rock later solidified to form a thin, discontinuous, dark-colored crust, only to be disrupted by upwelling magmas.

◆ **FIGURE 9.1** The Earth as it is thought to have appeared about 4.6 billion years ago. (Photo courtesy Herb Orth, Life Magazine © Time Warner Inc.; painting by Chesley Bonestell, © The Estate of Chesley Bonestell.)

Archean Eon	Proterozoic Eon	Phanerozoic Eon						
Precambrian		Paleozoic Era						
		Cambrian	Ordovician	Silurian	Devonian	Mississippian	Pennsylvanian	Permian
						Carboniferous		

2,500 M.Y.A.

570 M.Y.A.

Assuming that visitors to the early Earth could tolerate the high temperatures, they would also have to contend with other factors. The atmosphere would be unbreathable by any of today's inhabitants. It probably contained considerable carbon dioxide and water vapor, but little or no oxygen. No ozone layer existed in the upper atmosphere, so our hypothetical visitors would receive a lethal dose of ultraviolet radiation, unless protected, and would be threatened constantly by comet and meteorite impacts. The view of the Moon would have been spectacular because it was much closer to the Earth. However, its gravitational attraction would have caused massive Earth tides. And finally, our visitors would experience a much shorter day because the Earth rotated on its axis in as little as 10 hours.

Eventually, much of the Earth's primordial heat was dissipated into space and its surface cooled. As the Earth cooled, water vapor began to condense, rain fell, and surface water began to accumulate. The bombardment by comets and meteorites slowed. By 3.8 billion years ago, a few small areas of continental crust existed. The atmosphere still lacked oxygen and an ozone layer, but by as much as 3.5 billion years ago, life appeared. Some inconclusive evidence indicates that the most primitive life-forms existed even earlier.

❖ INTRODUCTION

Precambrian is often used informally to refer to both rocks and time. All crustal rocks lying beneath strata of the Cambrian System are called Precambrian. As a geochronologic term, Precambrian includes all geologic time from the Earth's origin 4.6 billion years ago to the beginning of the Phanerozoic Eon 570 million years ago. If all geologic time were represented by a 24-hour day, slightly more than 21 hours of it would be Precambrian (❖ Fig. 9.2). Unfortunately for geologists, not all of this vast interval of time is recorded; only one area of rocks older than 3.8 billion years is known on Earth.

Establishing formal, widely recognized subdivisions of the Precambrian is a difficult task. Precambrian rocks are exposed on all continents, but many have been complexly deformed and altered by metamorphism, and much of Precambrian Earth history is recorded by nonstratified rocks. Thus, the principle of superposition cannot be applied, particularly in older Precambrian rocks, making relative age determinations difficult. In addition, most correlations must be based on radiometric age dates because these rocks contain few biostratigraphically useful fossils.

In 1982, in an effort to standardize Precambrian terminology, the North American Commission on Stratigraphic Nomenclature approved a proposal that recognizes two Precambrian eons, the **Archean** and **Proterozoic** (Table 9.1). This usage has gained wide acceptance in North America and is followed in this book.

The Precambrian subdivisions in Table 9.1 are geochronologic, based on radiometric age dates rather than time-stratigraphic ages. This departs from normal practice in which geologic systems based on stratotypes (see Chapter 4) are the basic time-stratigraphic units. An example will help clarify this point. The Cambrian Period is a geochronologic term, but it corresponds to the Cambrian System, a time-stratigraphic unit based on a body of rock with a stratotype in Wales. By contrast, Precambrian terminology is strictly geochronologic. There are no stratotypes for the subdivisions of the Precambrian.

Phanerozoic Eon										
Mesozoic Era			Cenozoic Era							
Triassic	Jurassic	Cretaceous	Tertiary						Quaternary	
			Paleocene	Eocene	Oligocene	Miocene	Pliocene	Pleistocene		Holocene

66
M.Y.A.

An alternative scheme for designations of Precambrian time was adopted in 1971 by the U.S. Geological Survey (USGS) (Table 9.1). Instead of names, this scheme uses simple letter designations, which are also based on age dates rather than stratotypes. Although this usage is not followed in this book, students will encounter it on recent USGS maps and in some books and articles.

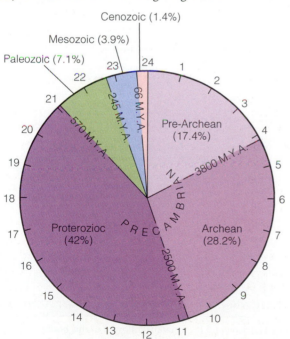

◆ **FIGURE 9.2** Geologic time represented on a 24-hour clock. Precambrian time includes more than 21 hours on this clock, or more than 87% of all geologic time.

❖ PRE-ARCHEAN CRUSTAL EVOLUTION

Geologists know that some continental crust existed at least 3.8 billion years ago because rocks this old are known from several areas, including Minnesota, Greenland, and South Africa. Moreover, many of these are metamorphic, which means they formed from even older rocks. Recently, 3.96-billion-year-old rocks were discovered in Canada. Furthermore, Archean sedimentary rocks in Australia contain detrital zircons dated at 4.2 billion years, indicating that source rocks at least that old were present.

Most geologists agree that some kind of pre-Archean crust probably existed, but because no rocks this old are known, the origin and composition of the earliest crust must be inferred from geochemical considerations. Many investigators think that a crust formed during an early episode of partial melting, magma formation, and rising magmas. Whether this crust was worldwide or more restricted is debatable, as is the cause of the partial melting.

One crustal evolution model relies on internally generated radiogenic heat and decreasing heat production within the Earth through time (◆ Fig. 9.3). An important aspect of this model is the evolution of sialic continental crust. Recall from Chapter 1 that sialic crust contains considerable silicon, oxygen, and aluminum. The earliest crust, however, was probably thin and unstable and was composed of ultramafic igneous rock. Such rock is relatively low in silica (SiO_2) compared with other igneous rocks. According to the model, this early ultramafic crust was disrupted by upwelling basaltic magmas at ridges and consumed at subduction zones (Fig. 9.3a). Thus, due to its high density, which would have made recycling by subduction very likely, the ultramafic crust would have

TABLE 9.1 Two Classification Schemes for Precambrian Rocks and Time in the United States

Time (Millions of Years Ago)	U.S. Geological Survey, 1971	North American Commission on Stratigraphic Nomenclature, 1982*	
	Phanerozoic	Phanerozoic	
500			
	Precambrian Z	Proterozoic	Late Proterozoic
	—— 800 ——		—— 900 ——
1,000	Precambrian Y		Middle Proterozoic
1,500	—— 1,600 ——		—— 1,600 ——
2,000	Precambrian X		Early Proterozoic
2,500	—— 2,500 ——		—— 2,500 ——
	Precambrian W	Archean	Late Archean
3,000			—— 3,000 ——
			Middle Archean
3,400			—— 3,400 ——
3,500			Early Archean (3,800?)
4,000			(Pre-Archean)
4,500			

*This scheme, which was proposed by Harrison and Peterman, is followed in this book.

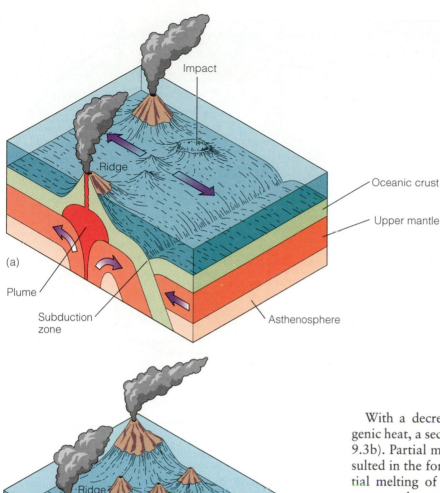

◆ **FIGURE 9.3** Model for origin of pre-Archean crust. The earliest crust may have been composed of ultramafic rock that was disrupted by rising basaltic magmas. (a) Basaltic crust is generated at ridges underlain by mantle plumes. Because of its high density, basaltic crust is consumed at subduction zones and recycled. (b) Andesitic island arcs form at convergent plate margins. Sialic crust grows by collisions of island arcs and intrusions of granitic magmas.

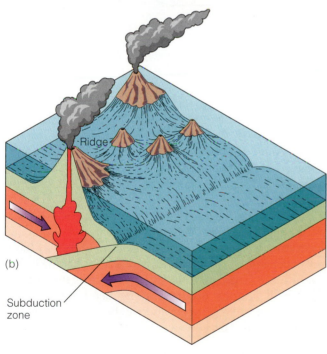

With a decrease in the Earth's production of radiogenic heat, a second stage of crustal evolution began (Fig. 9.3b). Partial melting of earlier-formed basaltic crust resulted in the formation of andesitic island arcs, and partial melting of lower crustal andesites yielded granitic magmas that were emplaced in the earlier-formed crust. Plate motions accompanied by subduction and collisions of island arcs formed several sialic continental nuclei by the Early Archean.

❖ SHIELDS AND CRATONS

Each continent is characterized by a **Precambrian shield** consisting of a vast area of exposed ancient rocks. Continuing outward from the shields are broad platforms of buried Precambrian rocks that underlie much of the continents. The shields and buried platforms are collectively called **cratons** (◆ Fig. 9.4). We can think of the cratons, which are composed of both Archean and Proterozoic rocks, as the ancient nuclei of the continents. The cratons are the relatively stable and immobile parts of the continents and form the foundations upon which Phanerozoic sediments were deposited. These cratons, including their exposed Precambrian shields, have been extremely stable since the beginning of the Phanerozoic Eon. Their

been destroyed. Apparently, only more sialic crust, because of its lower density, is immune to destruction by subduction.

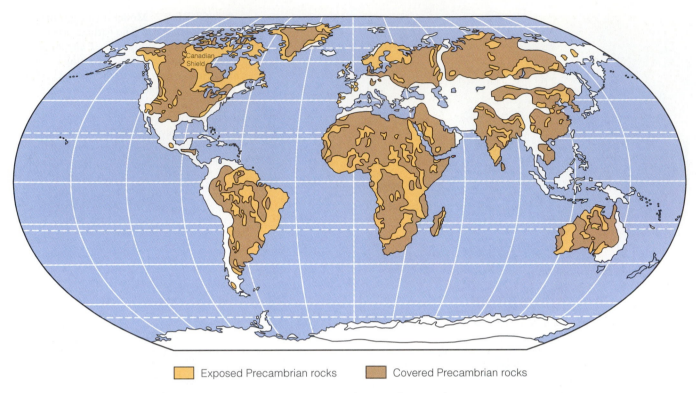

| Exposed Precambrian rocks | Covered Precambrian rocks |

◆ **FIGURE 9.4** Precambrian cratons of the world. The areas of exposed Precambrian rocks are the shields, while the buried Precambrian rocks are the platforms. Shields and platforms collectively make up the cratons.

stability during that time contrasts sharply with their Precambrian history of orogenic activity.

In North America the **Canadian Shield** includes most of northeastern Canada; a large part of Greenland; parts of the Lake Superior region in Minnesota, Wisconsin, and Michigan; and the Adirondack Mountains of New York (Fig. 9.4). In general, the Canadian Shield is a vast area of subdued topography, numerous lakes, and exposed Precambrian rocks, thinly covered in places by Pleistocene glacial deposits. Both Archean and Proterozoic rocks are present, including intrusives, lava flows, various sedimentary rocks, and metamorphic equivalents of all of these (◆ Fig. 9.5).

Beyond the Canadian Shield, exposures of Precambrian rocks are limited to areas of uplift and erosion, as in the Appalachians, the southwestern United States, the Rocky Mountains, and the Black Hills of South Dakota (Fig. 9.4) (see Perspective 9.1). Geophysical evidence and deep drilling demonstrate that Precambrian rocks underlie most of North America.

The geologic history of the Canadian Shield is complex and not fully understood. Nevertheless, we can recognize several smaller cratons within the shield, each of which is delineated on the basis of radiometric ages and structural trends. These cratons and other buried cratons beyond the shield are the subunits that constitute the North American craton. Each of these subunits may have been independent minicontinents that were later assembled into the larger cratonic unit. The amalgamation of these small cratons occurred along deformation belts during the Early Proterozoic and will be considered more fully in Chapter 10.

◆ ARCHEAN ROCKS

Areas underlain by Archean rocks (◆ Fig. 9.6) are characterized by two main types of rock bodies: **greenstone belts** and **granite-gneiss complexes.** By far the most abundant rocks are granites and gneisses. For example,

(a)

(b)

◆ **FIGURE 9.5** Rocks of the Canadian Shield. (a) Outcrop of gneiss, Georgian Bay, Ontario, Canada. (b) Basalt (dark) and granite (light) along the banks of the Chippewa River, Ontario, Canada.

◆ **FIGURE 9.6** Known distribution of Archean rocks. The bold lines outline areas probably underlain by Archean rocks.

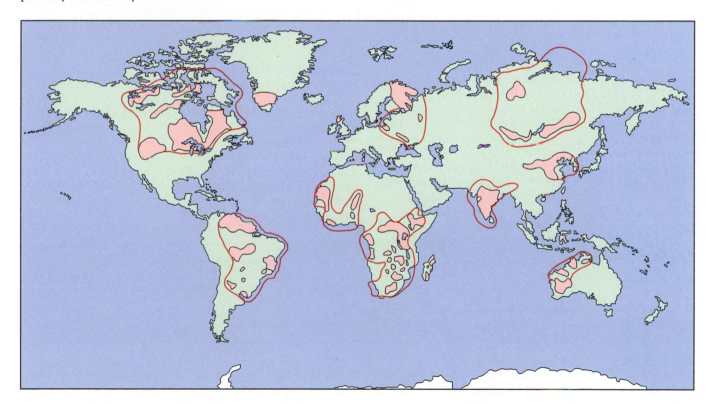

ARCHEAN GOLD AND THE BATTLE OF THE LITTLE BIGHORN

The Black Hills of South Dakota are an eroded domal uplift with a central core of Archean rocks (◆ Fig. 1). In 1876 the Black Hills became the site of a gold rush, which quickly transformed what had been a wilderness into a major mining center. Events leading to this gold rush began in 1874, when Lieutenant Colonel George Armstrong Custer led an army expedition into the Black Hills, the Holy Wilderness of the Sioux Indians. For three months Custer and the thousand soldiers, engineers, and gold miners who accompanied him explored the region, and Custer submitted an official report saying that "gold in satisfactory quantities can be obtained in the Black Hills."[1]

News of the gold in the Black Hills spread rapidly. Many people believed that a gold rush would cure the economic problems that had resulted in the "Panic of 1873," and soon the U.S. government was besieged by thousands of voters demanding that it acquire the Black Hills. According to the treaty of 1868, however, the Black Hills were to belong to the Sioux forever, and they refused to sell their Holy Wilderness. The breakdown of negotiations to purchase the Black Hills led to the Indian War, during which Custer and some 260 of his men were annihilated in June 1876 at the Battle of the Little Bighorn in Montana. Despite this stunning victory, the Sioux could not sustain a war against the U.S. Army, and in September 1876, they were forced to relinquish their claim to the Black Hills.

Miners and settlers began to arrive in the Black Hills in early 1876, and by the following year several mining towns were thriving. During the next 50 years more than $230 million worth of gold was recovered. Indeed, large-scale mining still continues today. In fact, the Homestake Mining Company at Lead, South Dakota, is one of the largest producers of gold in the Western Hemisphere.

Most of the Black Hills' gold comes from the Homestake Formation, an Archean rock unit composed of iron- and carbonate-rich rocks altered to schist by regional metamorphism. The Homestake Formation has been badly deformed, and the gold ores are concentrated along fold axes, especially the axes of synclines. Mining of these ores began at the surface, but now they are being recovered from depths as great as 2.5 km. Huge quantities of rock must be mined and processed, because only about one-third ounce of gold is recovered from each ton of rock. In 1983, the Homestake Mine yielded 16,123 kg of gold valued at more than $198 million.

The time of emplacement of the gold ores in the Homestake Formation was debated for several decades. Some geologists thought they were emplaced during the Tertiary Period when the Homestake Formation was intruded by rhyolite dikes, while others believed they were emplaced during the Proterozoic when intrusions of large granitic bodies occurred. Detailed geochemical studies seem to have resolved the problem: the ores and the rocks of the Homestake Formation formed at the same time by hot spring processes during the Archean, but regional metamorphism later in the Precambrian concentrated the ores along fold axes.

[1] Quoted in T. H. Watkins, *Gold and Silver in the West* (New York: Bonanza Books, 1971), p. 109.

the Rhodesian Province of southern Africa consists of about 83% gneiss and various granitic rocks; the remaining 17% is mostly greenstone belts.

Greenstone Belts

The oldest large, well-preserved greenstone belts are those of South Africa, which date from 3.6 billion years ago. In North America, greenstone belts are most common in the Superior and Slave cratons of the Canadian Shield (◆ Fig. 9.7), and most formed between 2.7 and 2.5 billion years ago.

An idealized greenstone belt consists of three major rock units; the lower and middle units are dominated by volcanic rocks, and the upper unit is sedimentary (◆ Fig. 9.8). Most greenstone belts have a synclinal structure and

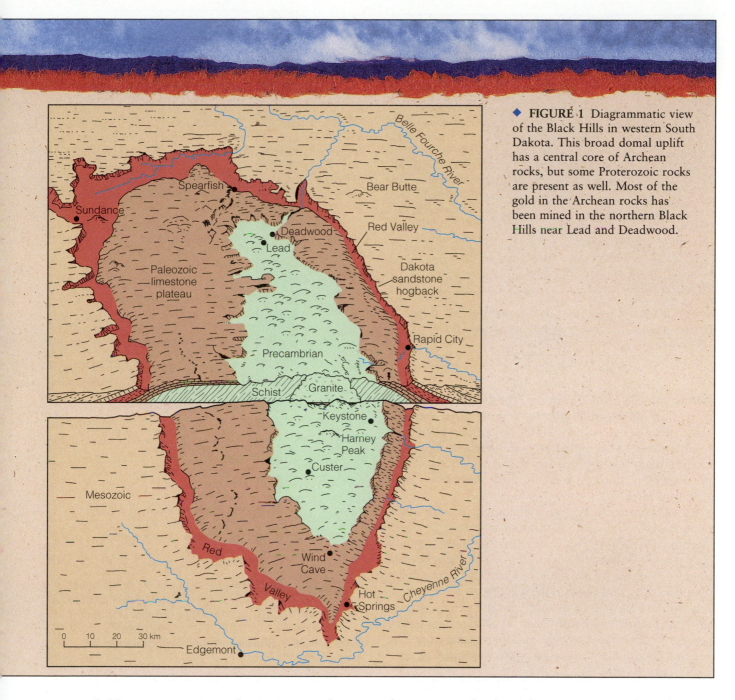

◆ **FIGURE 1** Diagrammatic view of the Black Hills in western South Dakota. This broad domal uplift has a central core of Archean rocks, but some Proterozoic rocks are present as well. Most of the gold in the Archean rocks has been mined in the northern Black Hills near Lead and Deadwood.

are intruded by granitic magmas (◆ Fig. 9.9a), and many are complexly folded and cut by thrust faults. The volcanic rocks of greenstone belts are typically greenish due to the abundance of the mineral chlorite, which formed during low-grade metamorphism.

As the common occurrence of pillow basalts indicates, much of the volcanism responsible for the igneous rocks of greenstone belts was subaqueous (◆ Fig. 9.9b). Shallow water and subaerial eruptions are indicated by pyroclastics, and in some areas large volcanic centers built up above sea level. Perhaps the most interesting igneous rocks in greenstone belts are ultramafic lava flows; such flows are rare in rocks younger than Archean. In order to erupt, an ultramafic magma requires near-surface magma temperatures over 1,600°C; the highest recorded surface magma temperature for recent Hawaiian basalt

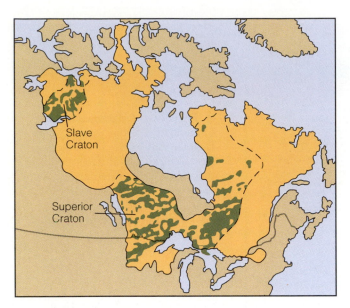

◆ **FIGURE 9.7** Greenstone belts (shown in green) of the Canadian Shield are mostly in the Superior and Slave cratons.

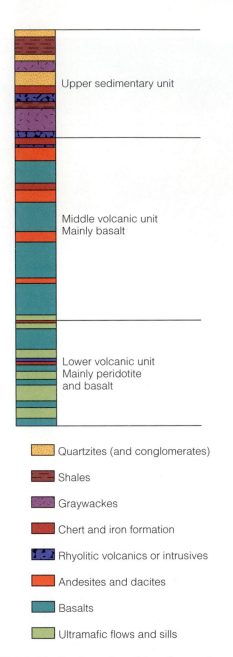

Quartzites (and conglomerates)

Shales

Graywackes

Chert and iron formation

Rhyolitic volcanics or intrusives

Andesites and dacites

Basalts

Ultramafic flows and sills

◆ **FIGURE 9.8** Idealized stratigraphic column of an Archean greenstone belt.

lava flows is 1,350°C. Early in Earth history, however, there was considerably more radiogenic heat; thus the mantle was hotter, perhaps 300°C hotter; and ultramafic magmas could be erupted onto the surface. Since the amount of radiogenic heat has decreased through time, the Earth has cooled; and ultramafic lava flows have ceased to form.

Sedimentary rocks are a minor component in the lower parts of greenstone belts but become increasingly abundant toward the top (Figs. 9.8, 9.9a). The most common sedimentary rocks are successions of *graywacke* and *argillite*. Graywacke is a variety of sandstone containing abundant clay, and those in the greenstone belts are rich in volcanic rock fragments. Argillites are simply slightly metamorphosed mudrocks such as shale. Small-scale graded bedding and cross-bedding indicate that the graywacke-argillite successions were deposited by turbidity currents (see Fig. 5.10). Some of the sedimentary rocks, such as quartz sandstones and shales, in the upper units show clear evidence of shallow-water deposition. These rocks were originally deposited in delta, tidal-flat, barrier-island, and shallow marine-shelf environments (◆ Fig. 9.10).

Other sedimentary rocks occurring in greenstone belts include conglomerate, chert, carbonates, and banded iron formations. Some conglomerates are associated with graywackes and probably represent submarine slumps. Neither chert nor carbonates are very abundant, although there are local exceptions. Banded iron formations occur but are much more common in Proterozoic terranes and will therefore be discussed in the next chapter.

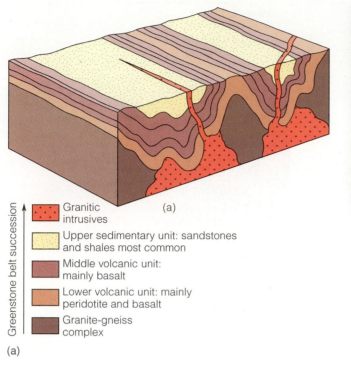

(b)

Granitic
intrusives

Upper sedimentary unit: sandstones
and shales most common

Middle volcanic unit:
mainly basalt

Lower volcanic unit: mainly
peridotite and basalt

Granite-gneiss
complex

Greenstone belt succession

(a)

◆ **FIGURE 9.9** (a) Two adjacent greenstone belts showing their synclinal structure. Older greenstone belts—those more than 2.8 billion years old—have an ultramafic lower unit succeeded upward by a basaltic unit as shown here. In younger greenstone belts, the succession is a basaltic lower unit overlain by an andesite-rhyolite unit. The upper unit in both older and younger greenstone belts consists of sedimentary rocks. (b) Pillow structures in the Ispheming greenstone belt, Marquette, Michigan.

Greenstone Belt Evolution

Most currently popular models for the development of greenstone belts rely on Archean plate movements.

◆ Figure 9.11 shows a model in which greenstone belts develop in **back-arc marginal basins** that subsequently close. Thus, there is an early stage of extension when the

◆ **FIGURE 9.10**   Depositional environments of the Moodies Group, South Africa.

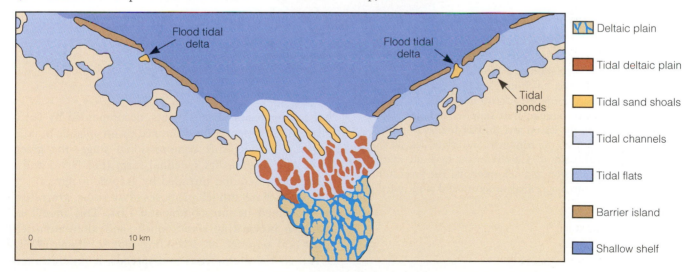

Deltaic plain

Tidal deltaic plain

Tidal sand shoals

Tidal channels

Tidal flats

Barrier island

Shallow shelf

Archean Rocks   **241**

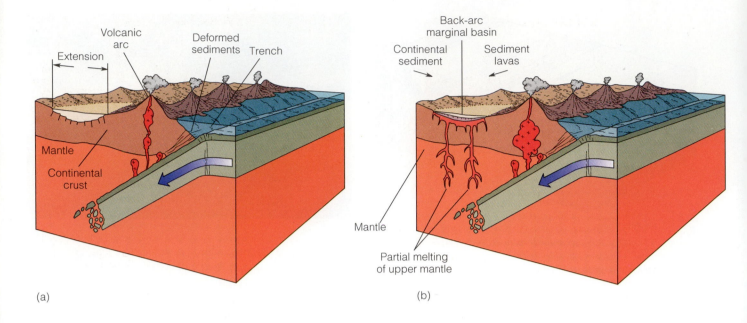

(a)

(b)

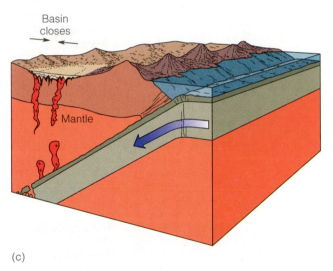

(c)

◆ **FIGURE 9.11** Formation of a greenstone belt in a
back-arc marginal basin. (a) Rifting on the continent side of
a volcanic island arc forms a back-arc marginal basin. Partial
melting of subducted oceanic crust supplies andesitic and
dioritic magmas to the island arc. (b) Basaltic lavas and
sediments derived from the continent and island arc fill the
back-arc marginal basin. (c) Closure of the back-arc marginal
basin causes compression and deformation. The greenstone
belt is deformed into a synclinal structure and is intruded by
granitic magmas.

back-arc marginal basin opens, accompanied by volcanism and sedimentation, followed by an episode of compression. During this compressional stage, the greenstone belt assumes its synclinal form (Fig. 9.9a) and is
metamorphosed and intruded by granitic magmas.

An alternate model proposes that some greenstone
belts formed in **intracontinental rifts.** This model assumes a preexisting sialic (silica- and aluminum-rich)
crust and requires an ascending mantle plume (◆ Fig.
9.12). As the plume rises and spreads, it generates ten-

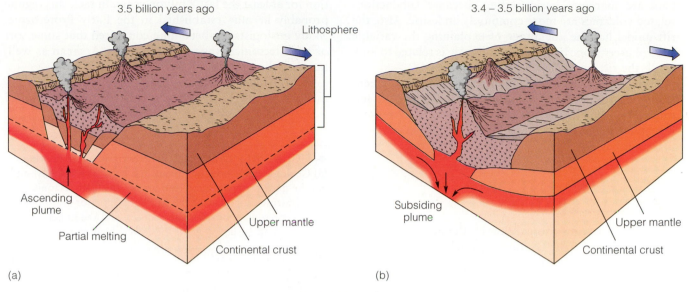

3.5 billion years ago

Lithosphere

Ascending plume

Partial melting

Continental crust

Upper mantle

(a)

3.4 – 3.5 billion years ago

Subsiding plume

Continental crust

Upper mantle

(b)

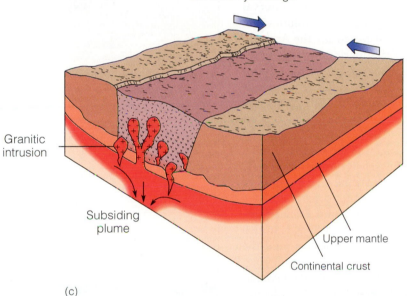

~3.3 billion years ago

Granitic intrusion

Subsiding plume

Upper mantle

Continental crust

(c)

◆ **FIGURE 9.12** Model for the formation of the Barberton greenstone belt of South Africa in an intracontinental rift. (a) An ascending mantle plume causes rifting and volcanism. (b) As the plume subsides, erosion of the rift flanks accounts for deposition of sediments. (c) Closure of the rift causes compression and deformation. Granitic magma intrudes the greenstone belt.

sional forces that cause intracontinental rifting. The plume also serves as the source of the lower and middle volcanic units, and erosion of the rift flanks accounts for the upper sedimentary unit. And finally, there is an episode of subsidence, deformation, low-grade metamorphism, and plutonism (Fig. 9.12).

The rift model has certain appealing aspects. For one, the ultramafic volcanics in some greenstone belts can be more easily accounted for by a mantle plume than by a back-arc marginal basin setting, because subduction-related volcanics are more commonly andesitic. Also, the rift model has the advantage of explaining the variable sizes of greenstone belts because the size is related to how much the rift opens.

Both models may be reasonably used to explain greenstone belts. For belts with ultramafic rocks, the rift setting seems to be the best explanation. But those containing abundant andesites more likely formed in back-arc marginal basins.

❖ DEVELOPMENT OF ARCHEAN CRATONS

We have already mentioned that by the beginning of the Archean several sialic continental nuclei or cratons had formed (Table 9.2). They may have been rather small, however, because rocks older than 3.0 billion years are of limited geographic extent, especially compared with those 3.0 to 2.5 billion years old (this latter interval seems to have been a time of rapid crustal evolution). A plate tectonic model has been proposed for the Archean crustal evolution of the southern Superior craton of Canada (◆ Fig. 9.13). This model incorporates sialic plutonism, greenstone belt formation, and collisions of microcontinents. It accounts for the origin of both greenstone belts and granite-gneiss complexes by the sequential accretion of island arcs (Fig. 9.13).

The events leading to the origin of the southern Superior craton are part of a more extensive orogenic episode that occurred near the end of the Archean. This episode of deformation was responsible for the formation of the Superior and Slave cratons as well as some Archean terranes within other parts of the Canadian Shield. It also affected Archean rocks in Wyoming, Montana, and the Minnesota River Valley.

Deformation during the Late Archean was the last major Archean event in North America. Several sizable cratons had formed, constituting the older parts of the Canadian Shield. These cratons, however, were independent units or microcontinents that were later assembled during the Early Proterozoic to form a craton similar to that outlined by the dashed lines in Figure 9.6. Large parts of the cratons of the other continents had also formed by the Archean.

❖ ARCHEAN PLATE TECTONICS

Undoubtedly, the present tectonic regime of opening and closing oceans has been a primary agent in Earth evolu-

tion for at least the last 2 billion years. In fact, this regime probably became established in the Early Proterozoic. Many geologists are becoming convinced that some sort of plate tectonics was operating in the Archean as well, but they disagree about the details.

Some Archean rocks appear to record plate movements as shown by deformation belts between presumed

◆ **FIGURE 9.13** Origin of the southern Superior craton. (a) Geologic map showing greenstone belts (green areas) and granite-gneiss subprovinces (tan areas). (b) Plate tectonic model for development of the southern Superior craton. The figure represents a north-south section, and the upper diagram is an earlier stage of the lower diagram.

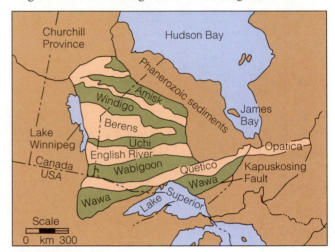

(a)

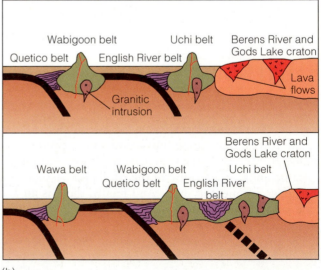

(b)

TABLE 9.2 Chronologic Summary of Events Important in the Archean Development of Cratons (Ages in Thousands of Millions of Years)

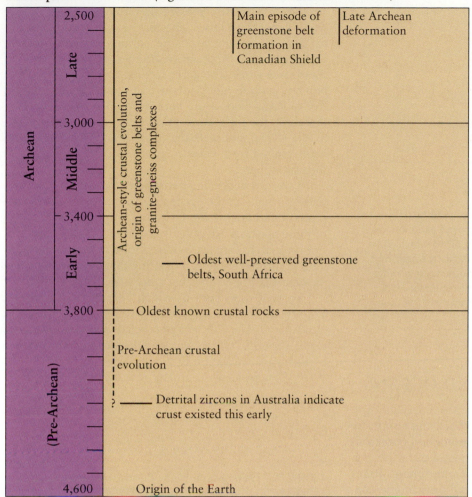

Archean	Late	Archean-style crustal evolution, origin of greenstone belts and granite-gneiss complexes	Main episode of greenstone belt formation in Canadian Shield	Late Archean deformation
	2,500			
	3,000			
	Middle			
	3,400			
	Early	Oldest well-preserved greenstone belts, South Africa		
(Pre-Archean)	3,800	Oldest known crustal rocks		
		Pre-Archean crustal evolution		
		Detrital zircons in Australia indicate crust existed this early		
	4,600	Origin of the Earth		

colliding cratons and island arcs. But ophiolite complexes, which mark younger convergent plate margins, are rare, although Late Archean ophiolites have recently been reported from several areas.

Apparently, the Earth's radiogenic heat production has diminished through time (◆ Fig. 9.14). Thus, in the Archean, when more heat was available, seafloor spreading and plate motions probably occurred faster, and magma was generated more rapidly. Nevertheless, Archean plates seem to have behaved differently from those in the Proterozoic. For example, sedimentary sequences typical of passive continental margins are uncommon in the Archean but quite common in the Proterozoic. Their near-absence in the Archean indicates that

continents with adjacent shelves and slopes were either not present or only poorly developed.

Another factor that favors some kind of Archean plate tectonics is the episode of rapid crustal growth that occurred 3.0 to 2.5 billion years ago. Like continents today, Archean continents probably grew by accretion at convergent plate margins. They probably grew more rapidly since plate motions were faster, thus accounting for an accelerated rate of accretion.

Today most geologists would probably agree that plate tectonics was operative in the Archean. Its details differed, however, from the present style of plate tectonics, which began during the Proterozoic when large, stable cratons were present.

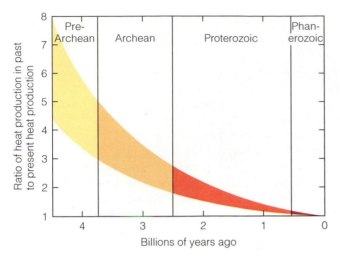

◆ **FIGURE 9.14** Ratio of radiogenic heat production in the past to present heat production. The colored band encloses the ratios according to different models, all of which show an exponential decay of radioactive elements through time. Estimates of heat production for 4 billion years ago range from three to six times the present heat production, whereas heat production at the beginning of the Phanerozoic Eon was only slightly greater than it is now.

❖ THE ATMOSPHERE AND OCEANS

During its earliest history, the Earth was a very inhospitable place. If we could somehow go back and visit it, we would witness a barren, waterless surface and numerous meteorite impacts, be subjected to intense ultraviolet radiation, and be unable to breathe the atmosphere. Today our atmosphere is rich in nitrogen and oxygen and contains important trace amounts of carbon dioxide, water vapor, and other gases (Table 9.3). In the upper atmosphere, ozone (O_3) blocks most of the Sun's ultraviolet radiation.

Since the most abundant elements in the universe are hydrogen and helium, the Earth's earliest atmosphere was probably composed of these gases. If so, gases of such low molecular weight would have escaped into space, since Earth's gravity is insufficient to retain them. Before the Earth had a differentiated core, it lacked a magnetic field and *magnetosphere,* the area around the Earth within which the magnetic field is confined. The absence of a magnetosphere ensured that a strong solar wind, an outflow of ions from the Sun, would sweep away any gases that might otherwise have formed an atmosphere. Once the magnetosphere was established, inter-

nally derived volcanic gases began to accumulate. The process by which atmospheric gases are derived from within the Earth is known as **outgassing** (◆ Fig. 9.15).

Gases emitted by present-day volcanoes are mostly water vapor with lesser amounts of carbon dioxide, sulfur dioxide, carbon monoxide, sulfur, chlorine, nitrogen, and hydrogen. Archean volcanoes probably emitted these same gases. The gases formed an early atmosphere but one notably deficient in free oxygen; and without free oxygen, there could have been no ozone layer. Furthermore, the early atmosphere may have contained ammonia (NH_3) and methane (CH_4), both of which could have resulted from volcanic gases reacting chemically in the atmosphere.

An oxygen-deficient atmosphere appears to have persisted throughout the Archean as indicated by detrital deposits containing pyrite (FeS_2) and uraninite (UO_2). Both of these minerals are quickly oxidized in the presence of free oxygen. Iron in the oxidized state became quite common in the Proterozoic, indicating that at least some free oxygen was present by that time (see Chapter 10).

Two processes, both of which began in the Archean, can account for the introduction of free oxygen into the atmosphere. The first was **photochemical dissociation** of water vapor, in which water molecules were broken up by ultraviolet radiation in the upper atmosphere (◆ Fig. 9.16). This process eventually may have supplied up to

◆━━━━━━━━━━━━━━━━━━━━━━━◆

TABLE 9.3 Composition of the Present-Day Atmosphere

	SYMBOL	PERCENTAGE BY VOLUME*
NONVARIABLE GASES		
Nitrogen	N_2	78.08
Oxygen	O_2	20.95
Argon	Ar	0.93
Neon	Ne	0.002
Others		0.001
VARIABLE GASES AND PARTICULATES		
Water vapor	H_2O	0.1 to 4.0
Carbon dioxide	CO_2	0.034
Ozone	O_3	0.0006
Other gases		Trace
Particulates		Normally trace

*Percentages, except for water vapor, are for dry air.

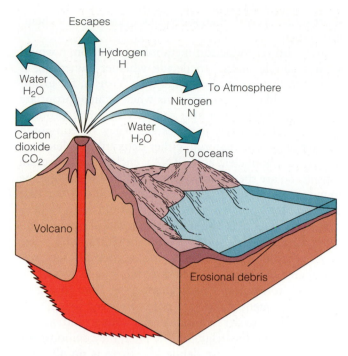

◆ **FIGURE 9.15** Outgassing supplied gases to form an early atmosphere composed of the gases shown. Chemical reactions in the atmosphere also probably yielded methane (CH_4) and ammonia (NH_3).

2% of present-day oxygen levels. At 2% free oxygen, ozone (O_3) will form, creating a barrier against incoming ultraviolet radiation and thus limiting the formation of more free oxygen by photochemical dissociation. Even more important were the activities of photosynthesizing organisms. During **photosynthesis** carbon dioxide and water combine into organic molecules, and oxygen is released as a waste product (Fig. 9.16). Even so, the atmosphere at the end of the Archean may have contained no more than 1% of its present oxygen level.

Recall that the major gas emitted by volcanoes is water vapor (Fig. 9.15), so the early atmosphere was also rich in this compound. Once the Earth cooled sufficiently, water vapor condensed and surface waters began to accumulate. Good evidence exists for oceans in the Early Archean, although their volumes and extent are unknown. We can envision an early hot Earth with considerable volcanic activity and a rapid accumulation of surface waters. Is the volume of oceanic waters still increasing? Perhaps it is; but if so, the rate has slowed considerably because the amount of heat available to generate magmas has decreased (Fig. 9.14). Isotopic studies of present-day volcanic emissions indicate that much of the water vapor emitted is recycled surface water.

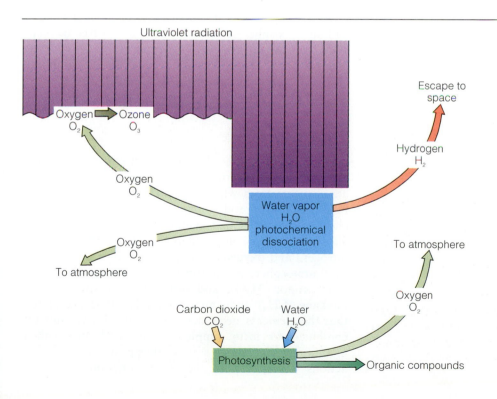

◆ **FIGURE 9.16** The processes of photochemical dissociation and photosynthesis added free oxygen to the atmosphere. Once free oxygen was present, an ozone layer formed in the upper atmosphere and blocked most incoming ultraviolet radiation.

❖ ARCHEAN LIFE

The fossil record reveals that life existed on Earth as much as 3.5 billion years ago. Compared to the present, however, the Archean seems to have been biologically impoverished. Today the Earth's biosphere consists of millions of species of animals, plants, and other organisms, all of which are thought to have evolved from one or a few primordial types. In Chapter 6 we considered the evolutionary processes whereby the diversification of life occurred, but here we are concerned with how life originated in the first place.

First, we must be very clear about what life is; that is, what is living and what is nonliving? Minimally, a living organism must reproduce and practice some kind of metabolism. Reproduction ensures the long-term survival of a group of organisms as a species; metabolism ensures short-term survival of an individual organism as a chemical system.

Using this reproduction-metabolism criterion, it would seem a simple matter to decide whether something is alive. Yet, this is not always the case. Viruses, for example, behave like living organisms when in an appropriate host cell, but outside a host cell they neither reproduce nor manufacture organic molecules. They are composed of a bit of genetic material, either deoxyribonucleic acid (DNA) or ribonucleic acid (RNA), enclosed in a protein capsule. So are they living or nonliving? Biologists can be found on both sides of this question, illustrating that defining life is not as clear-cut as it would seem to be.

Microspheres are simple organic molecules that form spontaneously and show greater organizational complexity than inorganic objects such as rocks. In fact, microspheres can even grow and divide, but in ways that are closer to random chemical processes than to organisms. Consequently, they are not considered living.

One may wonder what bearing viruses and microspheres have on the origin of life. First, they demonstrate that no matter how we define life, there is no absolute criterion for deciding whether or not some things are living. Second, if life originated on Earth by natural processes, it must have passed through a prebiotic stage, perhaps similar to microspheres.

As early as 1924, the great Russian biochemist A. I. Oparin postulated that life originated when the Earth had an atmosphere containing little or no free oxygen. With no oxygen to destroy organic molecules and no ozone layer to block ultraviolet radiation, life could indeed have come into existence from nonliving matter.

The Origin of Life

All investigators agree that two requirements were necessary for the origin of life: (1) a source of the appropriate elements from which organic molecules could have been synthesized and (2) an energy source to promote chemical reactions that synthesized organic molecules. All organisms are composed mostly of carbon, hydrogen, nitrogen, and oxygen, all of which were present in the early atmosphere in the form of carbon dioxide (CO_2), water vapor (H_2O), and nitrogen (N_2) and possibly methane (CH_4) and ammonia (NH_3). It is postulated that the elements necessary for life (C, H, N, and O) combined to form simple organic molecules called **monomers.** The energy sources that promoted these reactions were probably ultraviolet radiation and light-

◆ FIGURE 9.17 Experimental apparatus used by Stanley Miller. Several amino acids characteristic of organisms were artificially synthesized during Miller's experiments.

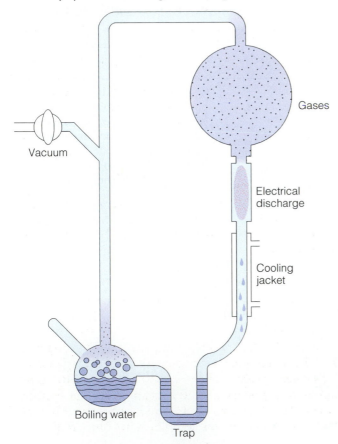

Vacuum

Gases

Electrical discharge

Cooling jacket

Boiling water

Trap

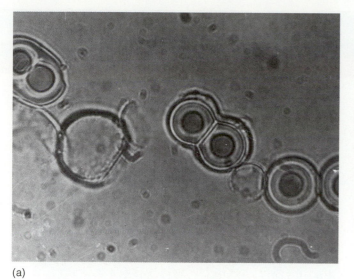

(a)

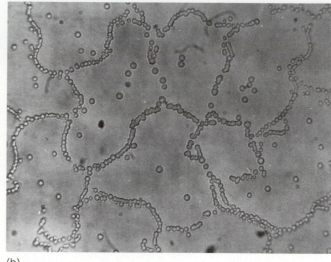

(b)

◆ **FIGURE 9.18** (a) Bacterium-like proteinoid. (b) Proteinoid microspheres. (Photos courtesy of Sidney W. Fox, Distinguished Research Professor, Southern Illinois University at Carbondale.)

ning. Typical monomers characteristic of organisms are amino acids.

Monomers are the basic building blocks of more complex organic molecules, but is it plausible that they originated in the manner postulated? Experimental evidence suggests so. In the 1950s, Stanley Miller synthesized several amino acids by circulating gases approximating the composition of the early atmosphere in a closed glass vessel (◆ Fig. 9.17). This mixture of gases was subjected to an electric spark (simulating lightning), and in a few days the mixture became cloudy. Analysis showed that several amino acids typical of organisms had formed. In more recent experiments, all 20 amino acids common in organisms have been successfully synthesized.

Making monomers in a test tube is one thing, but the molecules of organisms are **polymers** such as proteins and nucleic acids, which consist of linked monomers. Therefore the next question is, how did the process of polymerization occur? This is more difficult to answer, especially if polymerization occurred in an aqueous solution, which usually results in depolymerization. However, Sidney Fox of the University of Florida has synthesized small molecules he calls *proteinoids*, some of which consist of more than 200 amino acid units (◆ Fig. 9.18). He dehydrated concentrated amino acids and found that when heated they spontaneously polymerized to form proteinoids.

At this stage we can refer to these molecules as *protobionts*, which are intermediate between inorganic chemical compounds and living organisms. These protobionts, however, would have been diluted and would have ceased to exist if some kind of outer membrane had not developed. In other words, they had to be self-contained as present-day cells are. Fox's experiments demonstrated that proteinoids will spontaneously aggregate into microspheres (Fig. 9.18), which are bounded by a cell-like membrane and grow and divide much as bacteria do.

Fox's experimental results are interesting, but how can they be related to what may have taken place in the early history of life? Monomers likely formed continuously and in great abundance, accumulated in the oceans, and formed what the British biochemist J. B. S. Haldane characterized as a hot, dilute soup. According to Fox, the amino acids in this hot dilute soup may have washed up onto a beach or perhaps cinder cones, where they were concentrated by evaporation and polymerized by heat. The polymers were then washed back into the sea, where they reacted further.

Not much is known about the next step in the origin of life—the development of a reproductive mechanism. Fox's microspheres divide and are considered by some experts to represent a protoliving system. However, in present-day organisms nucleic acids, either as RNA or DNA, are necessary for reproduction. The problem is

SUBMARINE HYDROTHERMAL VENTS AND THE ORIGIN OF LIFE

Near spreading ridges, seawater seeps down into the oceanic crust through cracks and fissures, is heated by magma, then rises to form hydrothermal vents on the seafloor (◆ Fig. 1). More than 20 years ago, geologists reasoned that a logical implication of plate tectonic theory was the presence of such vents, but none were observed until 1977. The first vents were sighted from the submersible *Alvin* at a depth of about 2,500 m on the Galapagos Rift in the eastern Pacific Ocean basin. Since then, such vents have been observed in several other areas.

Submarine hydrothermal vents are interesting for several reasons. First, the rising hot-water solutions contain large quantities of dissolved mineral matter. When the hot water cools, iron, copper, and zinc sulfides and other minerals precipitate near the vents. Economic geologists are interested in this phenomenon because it may help them understand the origin of some ore deposits.

We are mainly concerned, however, with the biological implications of these vents. Vent organisms (Fig. 1) are the only major communities known that are not dependent on sunlight as a source of energy. Chemosynthetic bacteria that oxidize sulfur compounds from the hot vent waters lie at the base of the food chain. Thus, the ultimate source of energy for the vent communities is radiogenic heat.

Apparently, hydrothermal vents have been present since the Early Archean. Massive sulfide deposits from the Archean appear to have been produced by hydrothermal activity, indicating that vents were present. In fact, if the early Earth really did have more radiogenic heat, more

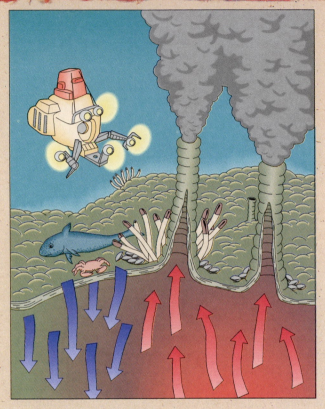

◆ **FIGURE 1** The submersible *Alvin* sheds light on hydrothermal vents at the Galapagos Rift, a branch of the East Pacific Rise. Seawater seeps down through the oceanic crust and becomes heated. The heated seawater rises and builds chimneys on the seafloor. The plume of heated water discharged through the chimney is saturated with dissolved minerals. Chemosynthetic bacteria oxidize sulfur compounds from the hot vent waters and form the base of a food chain for tubeworms, giant clams, anemones, crabs, and several types of fish.

that nucleic acids cannot replicate without enzymes, and enzymes cannot be made without nucleic acids. Or so it seemed until recently.

Recent experimental evidence has demonstrated that small RNA molecules can replicate themselves without the aid of protein enzymes. In view of this evidence, it seems that the first replicating system may have been an RNA molecule. In fact, some researchers propose an early "RNA world" in which these molecules were intermediate between inorganic chemical compounds and the DNA-based molecules of organisms. Just how RNA molecules were naturally synthesized, however,

magmatism, and faster spreading rates, hydrothermal vents should have been more common than they are now.

Recently, investigators at the Oregon State University School of Oceanography have proposed that submarine hydrothermal vents were the sites at which life originated. The researchers point out that the more traditional hypotheses for the origin of life rely upon the synthesis of monomers such as amino acids from atmospheric gases, the formation of polymers, and the origin of some kind of self-replicating system. Recall from our earlier discussion that both monomers and polymers have been synthesized experimentally. The Oregon State investigators claim that

> The physical and chemical conditions of those experiments are found in a natural environment which has been present in the oceans since they first formed, in hot springs associated with submarine volcanism.[1]

The elements and energy source necessary for the synthesis of organic molecules were present in the hydrothermal vents (◆ Fig. 2). In fact, amino acids have been detected in hydrothermal vent solutions. Polymerization may have occurred on the surfaces of clay minerals, a process that has been proposed by other investigators. And finally, protocells would have been discharged with vent fluids and deposited on the seafloor. This scenario is questioned by some who point out that polymerization would be inhibited in this environment.

One of the most intriguing aspects of this hypothesis is that the process may be continuing at present! This does not mean that life is continually originating, since any polymers or protocells would be quickly devoured by existing vent community organisms. Nevertheless, this hypothesis can be tested, at least in part, in a natural environment.

[1]J. B. Corliss, et al., *Oceanologica Acta* 4 (1981): 61.

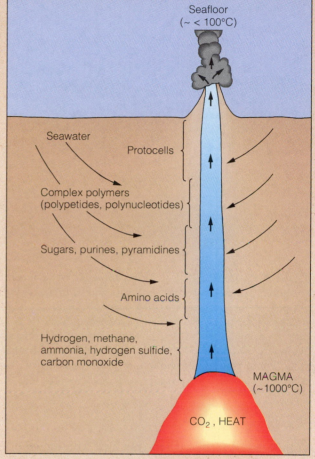

◆ **FIGURE 2** Proposed sequence of events leading to the origin of protocells in a hydrothermal vent system. According to this hypothesis, seawater heated by magma rises through a fracture and cools as it rises. Increasingly complex structures may have formed in the vent system, eventually giving rise to protocells and chemosynthetic bacteria.

remains a mystery, because they cannot easily be synthesized under the conditions that probably prevailed on the early Earth.

It should be apparent from our discussion that much remains to be learned about the origin of life by natural processes. The steps involved and the significance of some experimental results are still debated. A common theme among many investigators, however, is that the earliest organic molecules were synthesized from atmospheric gases. But even this has been questioned by some who propose that life originated in hydrothermal vent systems on the seafloor (see Perspective 9.2).

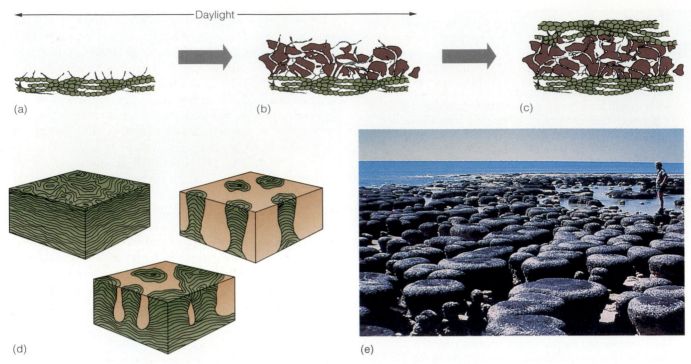

◆ **FIGURE 9.19** (a) through (c). The entrapment of sediment by a mat of cyanobacteria; brown masses are sediment grains. (a) Uncovered mat at beginning of daylight period. (b) Sediment trapping during daylight. (c) Regrowth and sediment binding during darkness. This process is repeated many times and yields the layered structure of stromatolites. (d) Morphological types of stromatolites include irregular mats, columns, and columns linked by mats. (e) Recent stromatolites, Shark Bay, Australia. (Photo courtesy of Phillip E. Playford, Geological Survey of Western Australia.)

The Earliest Organisms

Prior to the mid-1950s, we had very little knowledge of Precambrian life. Investigators had long assumed that the fossils so abundant in Cambrian strata must have had a long earlier history, but no such record was known. Some enigmatic Precambrian fossils had been reported, but these were mostly dismissed as inorganic in origin. In fact, many geologists considered Precambrian rocks to be unfossiliferous.

In the early 1900s, Charles Walcott described layered moundlike structures from the Early Proterozoic Gunflint Iron Formation of Ontario. These structures are now called **stromatolites.** Walcott proposed that they represented reefs constructed by algae, but paleontologists did not demonstrate that stromatolites are the products of organic activity until 1954. Studies of present-day stromatolites show that these structures originate by the entrapment of sediment grains on sticky mats of photo-

synthesizing cyanobacteria or blue-green algae (◆ Fig. 9.19a through c). Morphologically distinct types of stromatolites may form (◆ Fig. 9.19d), all of which are known from Precambrian rocks, some of which contain microfossils.

Currently, the oldest known undisputed stromatolites are from 3.0-billion-year-old rocks in South Africa. Even older probable stromatolites have recently been discovered in the 3.3- to 3.5-billion-year-old Warrawoona Group near North Pole, Australia (◆ Fig. 9.20). Indirect evidence for even more ancient life comes from Archean rocks of western Greenland. These 3.8-billion-year-old rocks contain small carbon spheres that may be of biological origin, but the evidence is not conclusive.

The oldest known fossils are of photosynthesizing organisms, but photosynthesis is a complex metabolic process, and it seems reasonable that it was preceded by an even simpler process. In other words, nonphotosynthesizing organisms must have been present before cyanobacteria appeared.

We have no fossils of these earliest organisms, but they probably resembled tiny bacteria. Since the early atmosphere contained little or no free oxygen, they must have been **anaerobic,** meaning they needed no oxygen. And very likely they were completely dependent on an external source of nutrients. We refer to such organisms as **heterotrophic** as opposed to **autotrophic** organisms, which make their own nutrients, as in photosynthesis. All these early life-forms were unicellular and lacked a cell nucleus. Cells of this type are called **prokaryotic cells.**

We can characterize these earliest organisms as anaerobic, heterotrophic prokaryotes. Their nutrient source was probably adenosine triphosphate (ATP), which was used to drive the energy-requiring reactions in cells. ATP can be synthesized from simple gases and phosphate, so it was probably available in the early Earth environment. The earliest life-forms may have simply acquired their ATP from their surroundings. This situation could not have persisted for long, though, because as more and more cells competed for the same resources, the supply must have diminished. The first organisms to develop a more sophisticated metabolism probably used *fermentation* to meet their energy needs. Fermentation is an anaerobic process in which molecules such as sugars are split, releasing carbon dioxide, alcohol, and energy. In fact, most living prokaryotes ferment.

Other than the origin of life itself, the most significant biological event of the Archean was the development of the autotrophic process of photosynthesis as much as 3.5 billion years ago. These cells were still anaerobic prokaryotes, but as autotrophs they were no longer completely dependent on an external source of preformed organic molecules as a source of nutrients. The Archean microfossils in Figure 9.20 are anaerobic, autotrophic prokaryotes. They belong to the kingdom Monera, which is represented today by bacteria and cyanobacteria or blue-green algae (Appendix B).

◆ **FIGURE 9.20** Photomicrographs (a) and (c) of fossil prokaryotes from the 3.3–3.5 billion-year-old Warrawoona Group, Western Australia. Schematic restorations of (a) and (c) are shown in (b) and (d). (Photos courtesy of J. William Schopf, University of California, Los Angeles.)

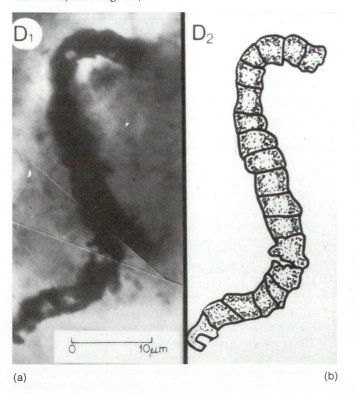

(a) (b)

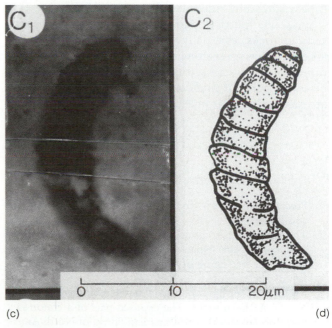

(c) (d)

❖ ARCHEAN MINERAL DEPOSITS

Although a variety of mineral resources occur in Archean rocks, the mineral most commonly associated with them is gold. Archean and Proterozoic rocks near Johannesburg, South Africa, have yielded more than 50% of the world's gold since 1886, much of it from conglomerate layers. A number of gold mining areas also occur within the Superior craton in Ontario, Canada, and the second largest gold mine in the United States is in the Archean Homestake Formation at Lead, South Dakota (see Perspective 9.1).

Gold has been mined for at least 6,000 years. During most of historic time, it has been used for jewelry, ornaments, and ritual objects and has served as a symbol of wealth and as a monetary standard. These traditional uses continue, but gold is now used for many other purposes, including gold plating, electrical circuitry, glass making, and the manufacture of chemicals.

A number of Archean-aged deposits of massive sulfides of zinc, copper, and nickel are known from several areas, including Western Australia, Zimbabwe, and the Abitibi greenstone belt of Ontario, Canada. Many of these deposits probably formed as the result of hydrothermal activity associated with greenstone belt volcanism. Indeed, similar deposits are currently forming adjacent to black smokers, which are types of hydrothermal vents at or near spreading ridges (see Perspective 9.2).

About one-fourth of the world's chrome reserves are in Archean rocks, especially those in Zimbabwe. The deposits occur in greenstone belt rocks and appear to have formed when crystals settled and became concentrated in the lower parts of mafic and ultramafic sills and other intrusive bodies. Chrome is used to produce hard chrome steel and for plating other metals. The United States has very few commercial chrome deposits so it must import most of what it needs.

One area of chrome deposits in the United States is in the Stillwater Complex in Montana. Low-grade ores from this area were mined during both World Wars as well as the Korean War, but they were simply stockpiled; that is, they were not refined for their chrome content. The Stillwater Complex is also a potential source of platinum, as are some other Archean rocks, but most platinum mined today comes from Proterozoic mafic rocks of the Bushveld Complex of South Africa.

Banded iron formations are sedimentary rocks consisting of alternating layers of silica (chert) and iron minerals. About 6% of the world's banded iron formations were deposited during the Archean Eon. Although Archean iron ores are mined in some areas, they are neither as thick nor as extensive as those of the Proterozoic Eon, which constitute the world's major iron deposits (see Chapter 10).

Pegmatites are very coarsely crystalline igneous rocks, commonly associated with plutons. Most are composed of quartz and feldspars and thus correspond to the composition of granite. Archean pegmatites are mostly granitic and of little economic value. Some, such as those of the Herb Lake district in Manitoba, Canada, and the Rhodesian Province in Africa, contain valuable minerals. In addition to minerals of gem quality, Archean pegmatites contain minerals mined for their lithium, beryllium, and rubidium. Furthermore, pegmatites are the only known commercial sources of some of the rare earth elements such as cesium.

Chapter Summary

1. Precambrian time is divided into two eons, the Archean and Proterozoic.

2. The earliest pre-Archean crust may have been composed of ultramafic rocks. None of this crust has been preserved, but by the beginning of the Archean, several sialic continental nuclei existed.

3. Each continent has an ancient, stable craton composed in part of Archean rocks. The exposed part of a craton is a Precambrian shield. The Canadian Shield of North America is made up of several cratons, which are delineated by age dates and structural trends.

4. Archean rocks are predominantly greenstone belts and granite-gneiss complexes. Greenstone belts have a synclinal structure and occur as linear bodies within much more extensive areas of granite and gneiss.

5. Typical greenstone belts can be divided into three major rock sequences. The two lower units are dominated by volcanic rocks, some of which are ultramafic. The upper unit consists mostly of sedimentary rocks.

6. Most Archean sedimentary rocks are preserved in greenstone belts, and the most common rocks are graywacke-argillite assemblages deposited by turbidity currents.

7. The evolution of greenstone belts is controversial, but most models rely on Archean plate motions. Those greenstone belts with ultramafic rocks may have formed in intracon-

tinental rifts, while those rich in andesites probably formed in back-arc marginal basins.

8. During Late Archean deformation, the Superior, Slave, and parts of other cratons of the Canadian Shield were formed. This deformation was the last major Archean event in North America.

9. Plate tectonics similar to that of the present did not begin until the Early Proterozoic, but many geologists think some type of Archean plate tectonics occurred. Archean plates may have moved more rapidly, however, because the Earth possessed more radiogenic heat.

10. The atmosphere and surface waters were derived from internally generated volcanic gases by a process called outgassing. The atmosphere so formed was deficient in free oxygen.

11. Most models for the origin of life require a nonoxidizing atmosphere. Atmospheric gases contained the elements necessary for simple organic molecules. Ultraviolet radiation and lightning probably provided the energy for synthesizing organic molecules.

12. The first naturally synthesized organic molecules were monomers such as amino acids. Monomers formed long chains, which gave rise to more complex molecules called polymers, such as proteins.

13. The method whereby a reproductive mechanism developed is unknown. Some investigators think RNA molecules may have been the first molecules capable of self-replication.

14. The Archean fossil record is very poor. A few localities contain unicellular, prokaryotic bacteria of the kingdom Monera. Stromatolites formed by photosynthesizing bacteria may date from 3.5 billion years ago.

15. The most important Archean mineral resources are gold, chrome, and massive sulfides of zinc, copper, and nickel.

Important Terms

anaerobic
Archean Eon
autotrophic
back-arc marginal basin
Canadian Shield
craton
granite-gneiss complex

greenstone belt
heterotrophic
intracontinental rift
monomer
outgassing
photochemical dissociation
photosynthesis

polymer
Precambrian
Precambrian shield
prokaryotic cell
Proterozoic Eon
stromatolite

Review Questions

1. The lower units in a typical greenstone belt are composed mostly of:
   a. _____ granite; b. _____ volcanic rocks; c. _____ limestone and shale; d. _____ volcanic ash; e. _____ graywacke-argillite.

2. Amino acids are examples of:
   a. _____ photosynthesis; b. _____ heterotrophism; c. _____ cratons; d. _____ stromatolites; e. _____ monomers.

3. Which of the following was probably responsible for adding some free oxygen to the Earth's early atmosphere?
   a. _____ fermentation; b. _____ outgassing; c. _____ carbonization; d. _____ abiotic synthesis; e. _____ photochemical dissociation.

4. The largest exposed area of the North American craton is the:
   a. _____ Wyoming craton; b. _____ American platform; c. _____ Canadian Shield; d. _____ Appalachian Mountains; e. _____ Grand Canyon.

5. The Archean Eon ended _____ billion years ago.
   a. _____ 1.3; b. _____ 2.5; c. _____ 3.8; d. _____ 3.96; e. _____ 1.95.

6. The ancient, stable nuclei of continents are called:
   a. _____ cratons; b. _____ bedrock; c. _____ orogenic belts; d. _____ Proterozoic sedimentary rocks; e. _____ greenstone belts.

7. The most common Archean rocks are:
   a. _____ granite-gneiss complexes; b. _____ greenstone-graywacke associations; c. _____ argillite-limestone facies; d. _____ ultramafic lava flows; e. _____ felsic volcanics.

8. The presence of _____ indicates that much of the greenstone belt volcanism was subaqueous.
   a. _____ pyroclastics; b. _____ argillites; c. _____ andesite; d. _____ pillow structures; e. _____ ultramafic magma.

9. Deposition of graywacke-argillite associations probably resulted from:
a. _____ turbidity currents; b. _____ evaporation; c. _____ volcanism; d. _____ metamorphism; e. _____ organic activity.

10. Greenstone belts containing abundant _____ probably formed in back-arc marginal basins.
a. _____ limestone; b. _____ granite; c. _____ andesite; d. _____ volcanic glass; e. _____ ultramafic lava flows.

11. The process whereby gases are derived from within the Earth and released into the atmosphere by volcanism is:
a. _____ photochemical dissociation; b. _____ oxygen respiration; c. _____ outgassing; d. _____ photosynthesis; e. _____ polymerization.

12. The only known Archean organisms were:
a. _____ prokaryotic cells; b. _____ sponges; c. _____ jellyfish; d. _____ microspheres; e. _____ monomers.

13. Some scientists think that the first self-replicating system may have been a(n):
a. _____ stromatolite; b. _____ RNA molecule; c. _____ bacterium; d. _____ ATP cell; e. _____ proteinoid.

14. Why are Precambrian rocks so difficult to study compared to Phanerozoic rocks?

15. What are Precambrian shields and cratons?

16. Describe the vertical succession of rock types in a typical greenstone belt.

17. Explain how greenstone belts may have formed in back-arc marginal basins and in intracontinental rifts.

18. What sequence of events led to the origin of the southern Superior craton of the Canadian Shield?

19. How does a greater amount of radiogenic heat account for a different style of plate tectonics during the Archean?

20. Explain how photosynthesis and photochemical dissociation of water vapor supplied oxygen to the Archean atmosphere.

21. Summarize the experimental evidence indicating that monomers could have formed by natural processes during the Archean.

22. Describe the earliest life-forms as thoroughly as possible.

23. Although the Archean began 3.8 billion years ago, there is evidence of older rocks. What is this evidence?

24. Explain why ultramafic lava flows are so rare in rocks younger than Archean.

Additional Readings

Condie, K. C. 1981. *Archean greenstone belts*. New York: Elsevier Scientific Publishing.

_____. 1982. *Plate tectonics and crustal evolution*. New York: Pergamon Press.

Dickerson, R. E. 1978. Chemical evolution and the origin of life. *Scientific American* 239, no. 3: 70–85.

Fox, S.W. 1991. Synthesis of life in the lab? Defining a proto-living system. *The Quarterly Review of Biology*, v.66:181–85.

Horgan, J. 1991. In the beginning. *Scientific American* 264, no. 2: 116–25.

McCall, G. J. H., ed. 1977. *The Archean*. Stroudsburg, Penn.: Dowden, Hutchinson & Ross.

Margulis, L. 1982. *Early life*. New York: Van Nostrand Reinhold.

Margulis, L., and L. Olendzenski, eds. 1992. *Environmental evolution: Effects of the origin and evolution of life on planet Earth*. Cambridge, Mass.: MIT Press.

Moorbath, S. 1977. The oldest rocks and the growth of continents. *Scientific American* 236, no. 1: 92–104.

Nisbet, E. G. 1987. *The young Earth: An introduction to Archean geology*. Boston, Mass: Allen & Unwin.

Windley, B. F. 1984. *The evolving continents*. New York: John Wiley & Sons.

CHAPTER 10

Chapter Outline

The Late Proterozoic Jacobsville Sandstone exposed along the shore of Lake Superior at Marquette, Michigan.

PRECAMBRIAN HISTORY: THE PROTEROZOIC EON

Prologue

According to plate tectonic theory, oceanic crust is produced at the axes of oceanic ridges, then moves away from these ridges by seafloor spreading, and is eventually destroyed at subduction zones. This continuous destruction of oceanic crust explains why no oceanic crust older than 180 million years is known except in fragments that were tectonically emplaced on continents. Such fragments consisting of a sequence of upper mantle, oceanic crust, and deep-sea sedimentary rocks are known as **ophiolites.** Ophiolites are one of the primary features used to recognize ancient convergent plate margins.

Probably less than 0.001% of all oceanic crust that ever existed is now preserved as ophiolites. Nevertheless, these small fragments provide geologists with information on the processes by which oceanic crust is generated at oceanic ridges and on the locations of ancient plate collisions. Furthermore, their presence indicates a style of plate tectonics essentially the same as that operating at present. Thus, the oldest ophiolites mark the time when a plate tectonic style comparable to that of the present began.

Ophiolites are well known from Phanerozoic-aged rocks, and a probable 2.65-billion-year-old (Late Archean) ophiolite has recently been reported from Wyoming. The oldest complete ophiolite so far recognized, however, is the 1.96-billion-year-old (Early Proterozoic) Jormua mafic-ultramafic complex of Finland (◆ Fig. 10.1). Even though the Jormua complex is highly deformed, it is similar to younger, well-documented ophiolites. Clearly, the Jormua complex represents a sequence of rocks that can be interpreted as upper mantle, lower oceanic crust, submarine eruptions represented by pillow basalts (◆ Fig. 10.1b), as well as metamorphosed deep-sea clay, oozes, and turbidites.

Asko Kontinen of the Geological Survey of Finland thinks that the Jormua complex formed along a divergent plate margin during an extensional stage in the breakup of an Archean craton. A major ocean basin formed in which deep-sea sediments and turbidites were deposited. A compressional stage occurred about 1.96 billion years ago during which most of the oceanic crust was destroyed by subduction. The Jormua complex, however, was tectonically emplaced on the craton and represents a small fragment of this ancient oceanic crust.

❖ INTRODUCTION

The Proterozoic Eon includes 42% of all geologic time (see Fig. 9.2), yet we review this 1.93-billion-year episode of Earth history in a single chapter. A mere 12% of all geologic time, the Phanerozoic Eon, is the subject of the remainder of this book. This may seem disproportionate, but the study of the Proterozoic is subject to the same problems we encountered in studying Archean history, although to a lesser degree. In short, we know more about the Proterozoic than we do about the Archean, but Phanerozoic history is even better known, hence its greater coverage.

How does the Proterozoic differ from the Archean? The basic difference is the style of crustal evolution. Archean crust-forming processes generated greenstone belts and granite-gneiss complexes that were shaped into cratons; and although these rock assemblages continued to form during the Proterozoic, they did so at a considerably reduced rate. Many Archean rocks have been metamorphosed, although the degree of metamorphism

Archean Eon	Proterozoic Eon	Phanerozoic Eon							

	Paleozoic Era						
Precambrian	Cambrian	Ordovician	Silurian	Devonian	Mississippian	Pennsylvanian	Permian
					Carboniferous		

2,500
M.Y.A.

570
M.Y.A.

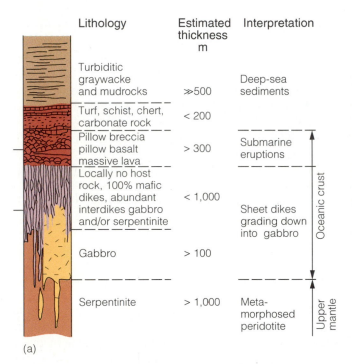

(a)

Lithology	Estimated thickness m	Interpretation
Turbiditic graywacke and mudrocks	≫500	Deep-sea sediments
Turf, schist, chert, carbonate rock	< 200	
Pillow breccia pillow basalt massive lava	> 300	Submarine eruptions
Locally no host rock, 100% mafic dikes, abundant interdikes gabbro and/or serpentinite	< 1,000	Sheet dikes grading down into gabbro
Gabbro	> 100	
Serpentinite	> 1,000	Meta-morphosed peridotite

Oceanic crust

Upper mantle

(b)

(c)

◆ **FIGURE 10.1** (a) Reconstruction of the highly deformed Jormua mafic-ultramafic complex of Finland. This 1.96-billion-year-old sequence of rocks is considered to be the oldest known complete ophiolite sequence. (b) A metamorphosed basaltic pillow lava. The code plate is 12 cm long. (c) Metamorphosed gabbro between mafic dikes. The hammer shaft is 65 cm long. (Photos courtesy of Asko Kontinen, Geological Survey of Finland.)

varies considerably, and some are completely unaltered. In contrast, many Proterozoic rocks are unmetamorphosed and little deformed, and in many areas Protero- zoic rocks are separated from Archean rocks by a profound unconformity. Finally, the Proterozoic is characterized by widespread assemblages of sedimentary

Phanerozoic Eon											
Mesozoic Era			Cenozoic Era								
Triassic	Jurassic	Cretaceous	Tertiary							Quaternary	
			Paleocene	Eocene	Oligocene	Miocene	Pliocene		Pleistocene	Holocene

5
A.

66
M.Y.A.

rocks that are rare in the Archean and by a plate tectonic style that is essentially the same as that of the present.

Changes definitely occurred across the Archean-Proterozoic boundary, but these changes were not synchronous. For example, Archean crustal evolution in the Kaapvaal craton of South Africa was completed nearly 3.0 billion years ago when sedimentary deposits more typical of the Proterozoic began to accumulate. In North America, this same change did not occur until 2.6 billion years ago. Actually, the change in crust-forming processes occurred from 2.95 to 2.45 billion years ago, so placing the Archean-Proterozoic boundary at 2.5 billion years ago is rather arbitrary.

❖ PROTEROZOIC CRUSTAL EVOLUTION

In the last chapter we noted that the Archean cratons assembled through a series of island arc and miniconti-nent collisions. These provided the nuclei around which Proterozoic continental crust accreted, thereby forming much larger cratons. One large landmass, called **Lauren-tia**, consisted mostly of North America and Greenland, parts of northwestern Scotland, and perhaps parts of the Baltic Shield of Scandinavia. This chapter will emphasize the geologic evolution of Laurentia. Table 10.1 summarizes the crust-forming events discussed in the following sections.

Early Proterozoic Evolution of Laurentia

The first major episode in the Proterozoic evolution of Laurentia occurred between 2.0 and 1.8 billion years ago (Table 10.1). During this time, several major **orogens** developed; these are zones of deformed rocks, many of which have been metamorphosed and intruded by plutons. The Archean cratons were sutured along these deformed belts (◆ Fig. 10.2a). In other words, this was an episode of continental growth during which collisions between Archean cratons formed a larger craton, and new crust was accreted at the margins of the cratons so formed. By 1.8 billion years ago, much of what is now Greenland, central Canada, and the north-central United States formed a large craton (Fig. 10.2a).

One Early Proterozoic episode of colliding Archean cratons is recorded in rocks of the Thelon orogen in the northwestern Canadian Shield, where the Slave and Rae cratons collided (Fig. 10.2a.). Here, the geologic record indicates subduction, granitic plutonism, volcanism, intense deformation, and regional metamorphism. The events leading to the formation of the Thelon orogen resulted in the collision and suturing of the Slave and Rae cratons (◆ Fig. 10.2).

Rocks of the Trans-Hudson orogen in northern Saskatchewan record initial rifting, extrusive volcanism, sedimentation, and the formation of oceanic crust, followed by ocean basin closure (◆ Fig. 10.3, p. 264). As the ocean basin closed, a subduction zone formed resulting in island arc collisions, granitic plutonism, intense deformation, and regional metamorphism. The events leading to the formation of the Trans-Hudson orogen resulted in the suturing of the Superior, Wyoming, and Hearne cratons.

Other notable events in the Early Proterozoic evolution of Laurentia include the development of the Wopmay and Penokean orogens (Fig. 10.2a). Rocks of the Wopmay orogen, adjacent to the Slave craton in northwestern Canada, record the oldest completely preserved Wilson cycle. Thus, The Wopmay orogen deserves more detailed discussion.

Several events in the Wopmay orogen can be deduced from studies of the Coronation Supergroup (◆ Fig.

TABLE 10.1 Summary Chart Showing the Ages of Some of the Proterozoic Events Discussed in the Text

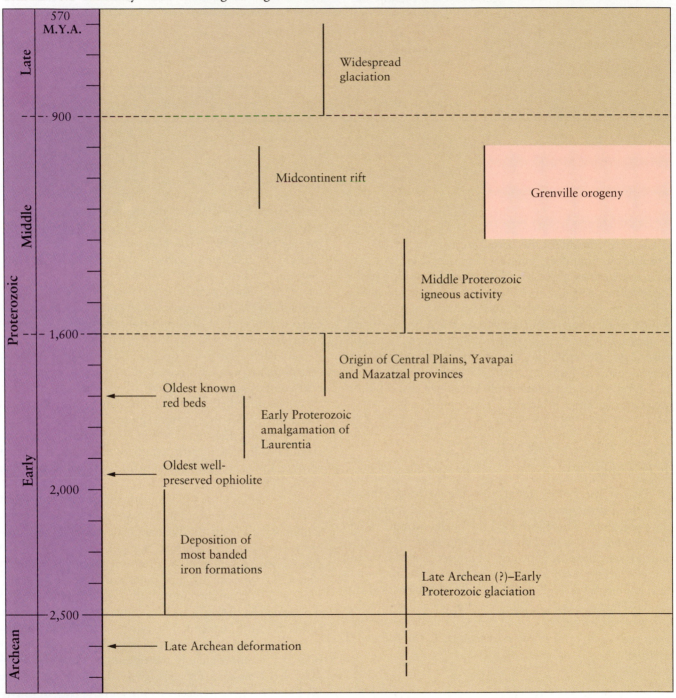

10.4, p. 265). Rifting is indicated by the fault-bounded basins in which turbidites and volcanics of the Akaitcho Group accumulated. Continental rifting established a pas- sive continental margin upon which sediments of the Ep- worth Group were deposited. These sediments are of par- ticular interest because they form a **quartzite-carbonate-**

shale assemblage, a suite of rocks that are characteristic of passive margins and first became abundant and widespread during the Proterozoic.

Conformably overlying the passive-margin rocks are deep-water sediments indicating abrupt deepening of the shelf. Deformation caused by closure of the ocean basin occurred contemporaneously with the deposition of these rocks and resulted in the Coronation Supergroup being thrust eastward over the craton's edge.

South of the Superior craton, the Penokean orogen accounts for the origin of another large segment of Laurentia (Fig. 10.2a). Considerable disagreement exists on the tectonic processes responsible for this orogen. Some geologists think it formed in a manner similar to the

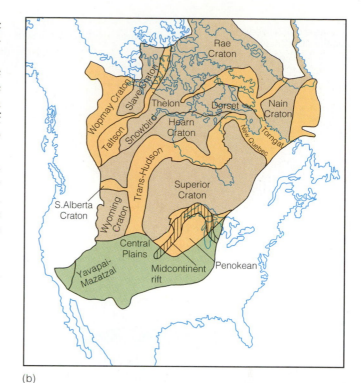

◆ **FIGURE 10.2** Proterozoic evolution of the Laurentian craton. (a) During the Early Proterozoic, the Archean cratons were sutured along deformation belts called *orogens*. (b) Laurentia grew along its southern margin by accretion of the Central Plains, Yavapai, and Mazatzal orogens. (c) A final episode of Protoerozoic accretion occurred during the Grenville orogeny.

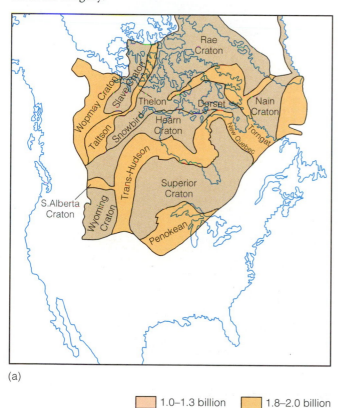

(a)

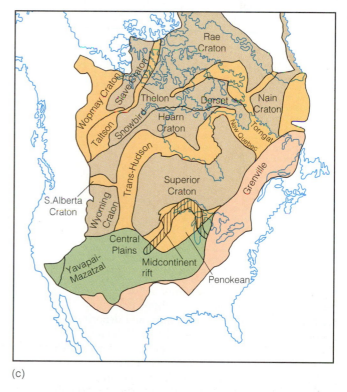

(c)

1.0–1.3 billion		1.8–2.0 billion
1.6–1.8 billion		>2.5 billion

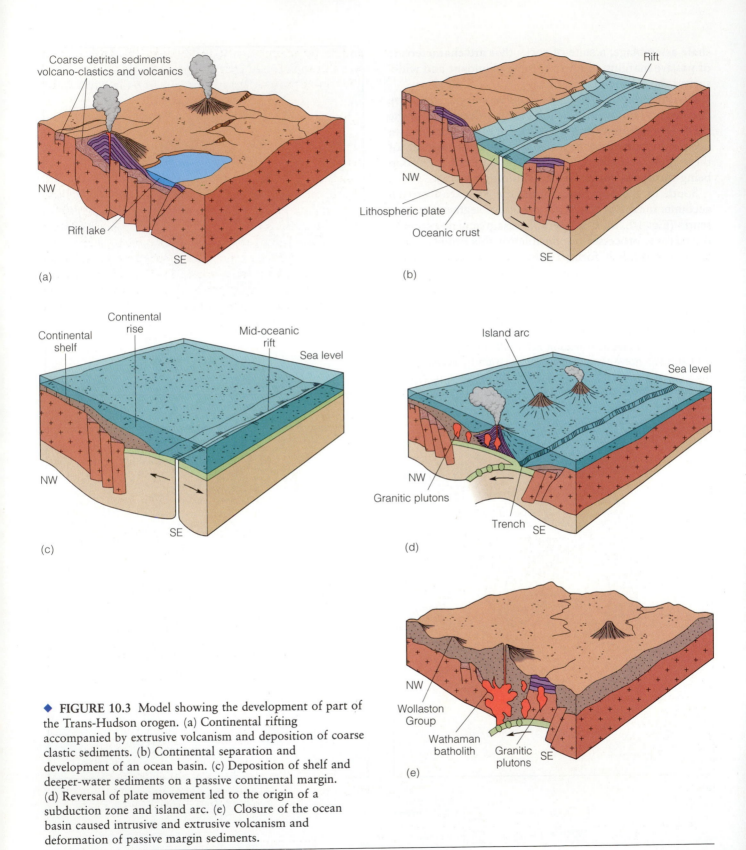

◆ **FIGURE 10.3** Model showing the development of part of the Trans-Hudson orogen. (a) Continental rifting accompanied by extrusive volcanism and deposition of coarse clastic sediments. (b) Continental separation and development of an ocean basin. (c) Deposition of shelf and deeper-water sediments on a passive continental margin. (d) Reversal of plate movement led to the origin of a subduction zone and island arc. (e) Closure of the ocean basin caused intrusive and extrusive volcanism and deformation of passive margin sediments.

Trans-Hudson orogen (Fig. 10.3), but others appeal to intracratonic deformation.

Following the initial episode of amalgamation of Archean cratons, considerable accretion occurred along the southern margin of Laurentia. Between 1.8 and 1.6 billion years ago, growth continued in what is now the southwestern and central United States as successively younger belts were sutured to Laurentia, forming the Central Plains, Yavapai, and Mazatzal orogens (Fig. 10.2b).

The plate tectonic model accounting for the accretion of the Yavapai and Mazatzal orogens contains both extensional and compressional phases. Greenstone belts formed in back-arc marginal basins (see Fig. 9.11), whereas quartzite-carbonate-shale assemblages were deposited in fluvial, tidal, and shallow marine environments on a south-facing continental margin. Continental accretion occurred as northward-migrating island arcs collided with Laurentia, resulting in deformation, metamorphism,

and the emplacement of granitic batholiths. The net effect was the accretion of more than 1,000 km of continental crust along the southern margin of Laurentia (Fig. 10.2b).

Middle Proterozoic Igneous Activity

No major episode in the growth of Laurentia occurred between 1.6 and 1.3 billion years ago. Nevertheless, during this interval extensive igneous activity occurred that was unrelated to orogenic activity (◆ Fig. 10.5). These rocks did not increase the area of Laurentia because they were intruded into or erupted upon already existing crust. These igneous rocks are buried beneath Phanerozoic strata in most areas, in the north-central and southwestern United States, for example. They are exposed in eastern Canada, extend across southern Greenland, and also occur in the Baltic Shield of Scandinavia. Recent investigations show that they are mostly granitic plutons,

◆ **FIGURE 10.4** Rocks of the Coronation Supergroup in the Wopmay orogen of northwestern Canada record the oldest known complete Wilson cycle. See the text for a discussion.

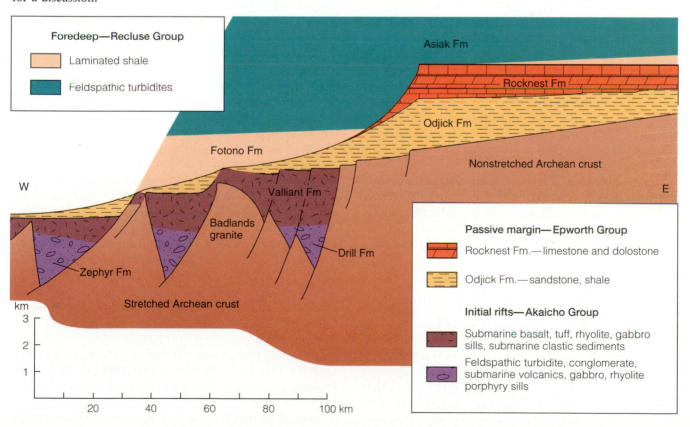

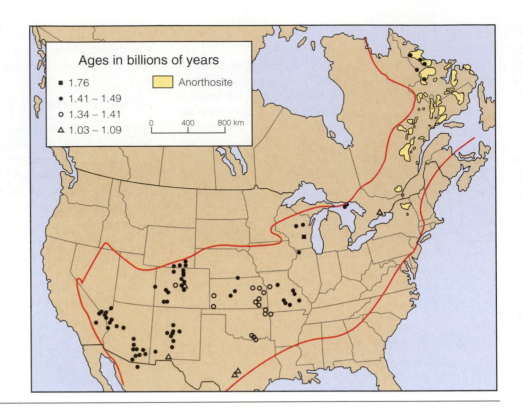

◆ **FIGURE 10.5** Distribution of Middle Proterozoic igneous rock complexes unrelated to orogenic activity. Most of this activity occurred between 1.34 and 1.49 billion years ago.

calderas and their fill, and vast sheets of rhyolite and ash flows, although gabbros and anorthosites (plutonic rocks composed mostly of plagioclase) are known as well.

The origin of these Middle Proterozoic igneous rocks continues to be the subject of debate. Paul Hoffmann of the Geological Survey of Canada postulates that large-scale mantle upwelling, or what he calls a superswell, rose beneath a Proterozoic supercontinent to produce these rocks. According to this hypothesis, the mantle temperature beneath a Proterozoic supercontinent would have been considerably higher than beneath later super-continents because radiogenic heat production within the Earth has decreased (see Fig. 9.14). Accordingly, non-orogenic igneous activity would have occurred following the amalgamation of the first supercontinent.

Middle Proterozoic Orogeny and Rifting

Another major episode in the evolution of Laurentia, the **Grenville orogeny** in the eastern United States and Canada, occurred between 1.3 and 1.0 billion years ago (Fig. 10.2c, Table 10.1). Rocks of the Grenville orogen are extensively exposed in southeastern Canada where they abut the Archean Superior Province (Fig. 10.2c). This belt of deformed rocks extends northeast into Greenland and continues into Scandinavia. It also continues south-west through the area of the present-day Appalachian Mountains (Fig. 10.2c).

The Grenville Supergroup, consisting of sandstones, shales, and carbonates, was highly deformed, metamorphosed, and intruded by various igneous bodies during the formation of the Grenville orogen (◆ Fig. 10.6). Many geologists think the Grenville orogen can be explained by the opening and then closing of an ocean basin. If so, the Grenville Supergroup may represent sediments deposited on a passive continental margin. However, some geologists think that Grenville deformation occurred in an intracontinental setting or was caused by major shearing. Whatever the cause, it represents the final episode of Proterozoic continental accretion of Laurentia.

Contemporaneous with Grenville deformation was an episode of continental rifting in Laurentia. The **Midcontinent rift** is a major feature extending from the Lake Superior basin southwest into Kansas, while a southeasterly trending branch extends through Michigan and into Ohio (◆ Fig. 10.7). It cuts across Archean and Early Proterozoic rocks and, in the east, terminates against the Grenville orogen.

Most of the rift is concealed by younger strata except

in the Lake Superior region where rocks that filled the rift are well exposed. The central part of the rift contains thick accumulations of basaltic lava flows that form extensive lava plateaus. For example, the Portage Lake Volcanics consist of numerous overlapping lava flows forming a volcanic pile several kilometers thick (◆ Fig. 10.8a and d).

The rift formed between 1.2 and 1.0 billion years ago, but the volcanism apparently occurred during a comparatively short interval of only 20 to 30 million years. For the entire rift, the volume of volcanic rocks has been estimated at 300,000 to 1,000,000 km³. This is comparable in volume but not areal extent to the great outpourings of lava during the Cenozoic (see Chapter 17).

Intertonguing with and overlying the volcanics are thick sequences of sedimentary rocks. Near the rift margins coarse-grained rocks such as the Copper Harbor Conglomerate formed large alluvial fans that graded into sandstones and shales of the rift axis (Fig. 10.8a and c).

❖ PROTEROZOIC SUPERCONTINENTS

Thus far this chapter has largely reviewed the geologic evolution of Laurentia, a large landmass composed mostly of North America and Greenland (Fig. 10.2). Paleomagnetic studies and the continuation of the Grenville orogen into Scandinavia suggest that the Baltic Shield may also have been part of or situated close to Laurentia

◆ **FIGURE 10.6** Metamorphosed sedimentary rocks of the Grenville Supergroup near Alexandria, New York. At this locality the Grenville is unconformably overlain by the Upper Cambrian Potsdam Formation. (Photo courtesy of R. V. Dietrich.)

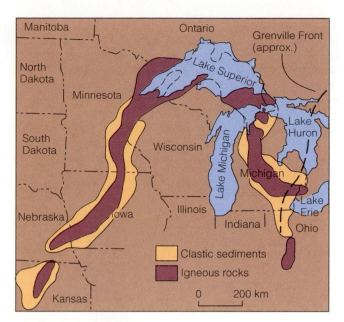

◆ **FIGURE 10.7** Location of the Midcontinent rift. The volcanic and sedimentary rocks that fill the rift are exposed in the Lake Superior region, but elsewhere they are buried beneath younger rocks.

during the Proterozoic (◆ Fig. 10.9). Indeed, some geologists suggest that a collision between the Baltic and Laurentian cratons may have caused some of the early Proterozoic deformation of Laurentia.

But what of Laurasia and Gondwana, the two major components of the Late Paleozoic supercontinent Pangaea? The components that ultimately formed Pangaea were evolving during the Precambrian, but few investigators agree on their size, shape, or associations. One group holds that a single Pangaea-like supercontinent persisted from the Late Archean to the end of the Proterozoic. Other investigators, pointing to uncertainties in the data, maintain that this reconstruction is not justified.

Three or more supercontinents existed in the Late Proterozoic, but their relative locations are unknown. As indicated in ◆ Figure 10.10, East Gondwana consisted of Australia, India, and Antarctica, while West Gondwana included Africa and South America. The third supercontinent, Laurasia, consisted of Laurentia, Europe, and most of Asia.

During the latest Proterozoic, all the Southern Hemisphere continents were strongly deformed. These events may mark the time during which East and West Gondwana were assembled into the larger supercontinent of Gondwana. In fact, some geologists think that Laurentia,

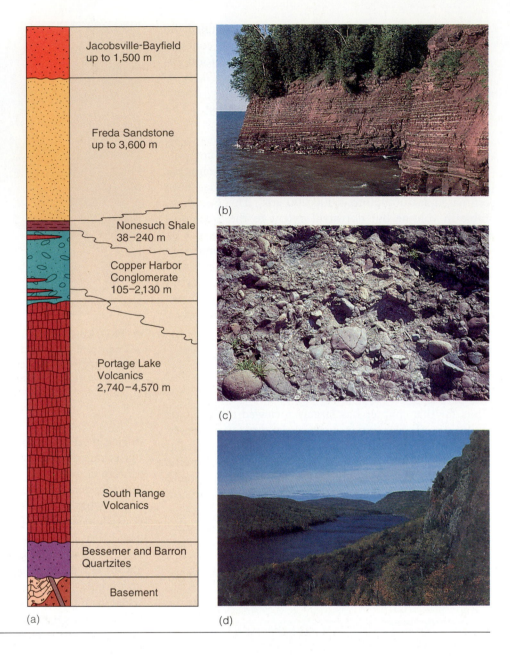

◆ **FIGURE 10.8** Rocks of the Midcontinent rift exposed in the Lake Superior region. (a) Section showing vertical relationships. (b) Freda Sandstone. (Photo courtesy of Albert B. Dickas, University of Wisconsin, Superior.) (c) Copper Harbor Conglomerate. (d) Portage Lake Volcanics. (Photo courtesy of Dan Urbanski, White Pine, Silver City, Michigan.)

Labels in figure (a):
- Jacobsville-Bayfield up to 1,500 m
- Freda Sandstone up to 3,600 m
- Nonesuch Shale 38–240 m
- Copper Harbor Conglomerate 105–2,130 m
- Portage Lake Volcanics 2,740–4,570 m
- South Range Volcanics
- Bessemer and Barron Quartzites
- Basement

(b) (c) (d)

Australia, and Antarctica were connected during the Late Proterozoic, forming a supercontinent (◆ Fig. 10.11).

◆ PROTEROZOIC ROCKS

Some Proterozoic rocks and rock associations deserve special attention. Greenstone belts, for example, are more typical of the Archean, but they continued to form during the Proterozoic, though they differed in detail. Quartzite-carbonate-shale assemblages also merit our attention because they closely resemble sedimentary rock associations of the Phanerozoic. Deposits attributed to two episodes of Proterozoic glaciation are recognized as well.

The onset of a present style of plate tectonics probably occurred in the Early Proterozoic, and the first well-preserved *ophiolites*, which mark convergent plate margins, are of this age (see the Prologue). Banded iron formations and red beds, too, are important Proterozoic rock types, but these are more conveniently discussed in the section on the atmosphere.

Greenstone Belts

Greenstone belts are characteristic of the Archean, and most seem to have formed in the interval between 2.7 and 2.5 billion years ago. They are not, however, restricted to the Archean. Proterozoic greenstone belts occur on several continents, including North America, and at least one has been reported from the Cambrian of Australia. Several greenstone belts formed in the Yavapai and

◆ **FIGURE 10.9** Reconstructions of the Baltic Shield relative to Laurentia for two times during the Proterozoic.

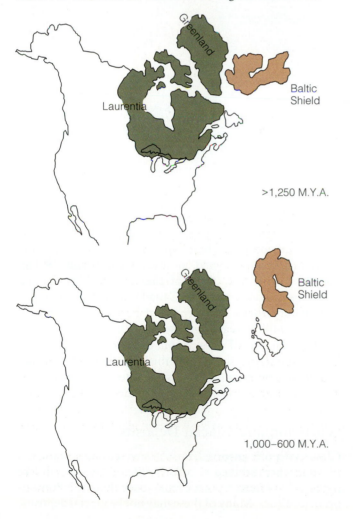

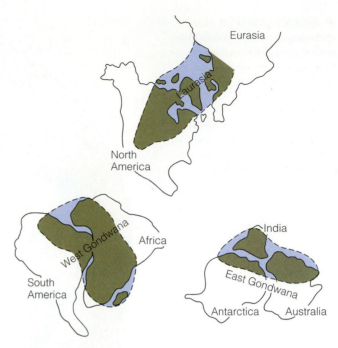

◆ **FIGURE 10.10** Reconstructions of three possible Late Proterozoic supercontinents. Dashed lines indicate areas known or suspected to be underlain by Archean crustal rocks. The relative locations of Laurasia, East, and West Gondwana are unknown.

Mazatzal orogens as crustal accretion occurred south of the Wyoming craton (Fig. 10.2).

Ultramafic rocks are rare in Proterozoic greenstone belts. The near-absence of Proterozoic and younger ultramafics is probably accounted for by the decreasing amount of radiogenic heat produced within the Earth (see Fig. 9.14). Otherwise, Proterozoic and Archean greenstone belts differ very little. Both appear to have formed in similar tectonic settings, intracratonic rifts and back-arc marginal basins (see Figs. 9.11 and 9.12), although the latter setting became more prevalent during the Late Proterozoic.

Quartzite-Carbonate-Shale Assemblages

Fully 60% of known Proterozoic rock successions are composed of *quartzite-carbonate-shale assemblages*. Widespread deposition of this assemblage occurred throughout the Proterozoic along rifted continental margins and in intracratonic basins. The most important aspect of these assemblages is that they represent suites of rocks similar to those of the Phanerozoic. Hence their

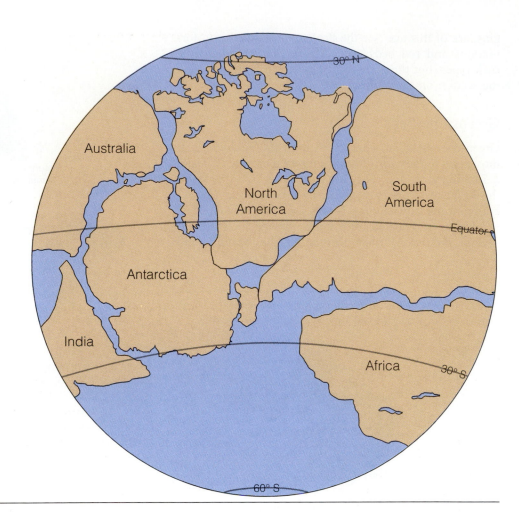

◆ **FIGURE 10.11** Possible arrangement of the continents during the Late Proterozoic.

appearance indicates the widespread distribution of stable cratons with depositional environments much like those of the present.

Early Proterozoic quartzite-carbonate-shale assemblages are widely distributed in the Great Lakes region (◆ Fig. 10.12). Both quartzite and shale are abundant, but conglomerates and graywackes are also present. The quartzites were derived from mature quartz sandstones, and many show wave-formed ripple marks and cross-bedding (◆ Fig. 10.13a). Thick carbonates, mostly dolostone, containing abundant stromatolites, are also present (◆ Fig. 10.13b).

Many of these rocks were deposited on a shallow-water shelf, perhaps a rifted continental margin. Some intermittent deformation took place during the deposition of these rocks, but all were further deformed and intruded by granite batholiths during the Early Proterozoic amalgamation of Laurentia. These rocks are now part of the Penokean orogen (Fig. 10.2).

In the western United States and Canada, quartzite-carbonate-shale assemblages were deposited along the continental margin. These Middle to Late Proterozoic rocks are best preserved in three large basins (◆ Fig. 10.14). In the Belt Basin of the northwestern United States and adjacent parts of Canada, a thick sequence of sedimentary rocks was deposited between 1.45 billion and 850 million years ago (◆ Fig. 10.15). Like the deposits of the Great Lakes region, these rocks are predominantly quartzite, shale, and thick stromatolite-bearing carbonates. Some of the rocks are interpreted as the deposits of large alluvial fans and braided streams.

Glaciation and Glacial Deposits

Glaciers deposit unsorted, unstratified sediment known as *till* that when lithified is called *tillite*. Tillites or till-like deposits have been reported from more than 300 Precambrian localities. Many of these may not be glacial deposits,

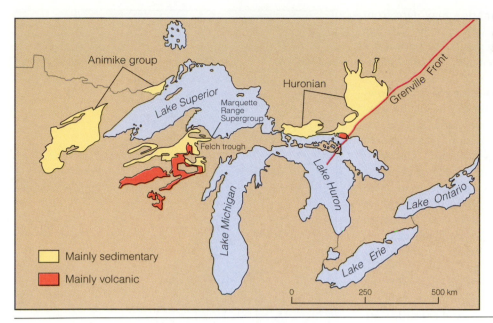

◆ **FIGURE 10.12** The geographic distribution of Early Proterozoic rocks in the Great Lakes region.

but some almost certainly are. In fact, two major episodes of Proterozoic glaciation have been recognized—one occurred during the Early Proterozoic and the other during the Late Proterozoic (Table 10.1).

But how can we be sure there were any Proterozoic glaciers? Tillite is simply a type of conglomerate, but other processes, such as *mass wasting,* can produce very similar deposits. The deposits can be distinguished by their areal extent: mass-wasting deposits are very lim-ited, whereas true tillites cover much larger areas. In addition, a suspected tillite can be established as a glacial deposit by its association with a striated bedrock pavement (◆ Fig. 10.16a) and with finely laminated argillites (varved sediments) that contain large clasts dropped from floating ice (◆ Fig. 10.16b).

Early Proterozoic glacial deposits are known from Canada, the United States, Australia, and South Africa. An example is the Huronian Supergroup of Ontario

---

◆ **FIGURE 10.13** Early Proterozoic sedimentary rocks of the Great Lakes region. (a) Wave-formed ripple marks in the Mesnard Quartzite. The crests of the ripples point toward the observer. (b) Outcrop of the Kona Dolomite showing stromatolites.

(a)

(b)

where the tillite of the Bruce Formation may date from 2.7 billion years ago, making it Late Archean in age (◆ Fig. 10.17). Tillites of the Gowganda Formation are widespread, overlie a striated pavement, and are associated with varved deposits.

Similar deposits thought to be the same age as the Gowganda tillite also occur in Michigan, the Medicine Bow Mountains of Wyoming, and Quebec, Canada. This distribution indicates that North America may have had an extensive Early Proterozoic ice sheet centered southwest of Hudson Bay (◆ Fig. 10.18). Tillites of about the same age also occur on other continents; the dating of these deposits is not precise enough to determine whether there was a single, widespread glacial episode or a number of glacial events at different times in different places.

Widespread Late Proterozoic glaciation occurred between 900 and 600 million years ago (Table 10.1). Tillites and other glacial deposits have been recognized on all continents except Antarctica. Glaciation was not continuous during this entire interval but was episodic with four major glacial periods recognized. ◆ Figure 10.19 shows the approximate distribution of these Late Proterozoic glaciers, but we should emphasize that they are approximate, because the geographic extent of glacial ice is unknown. In addition, the glaciers shown in Figure 10.19 were not all present at the same time. Nevertheless, the most extensive glaciation in Earth history was during the latest Proterozoic when ice sheets seem to have been present even in near-equatorial areas.

❖ THE EVOLVING ATMOSPHERE

In the last chapter, we noted that many geologists are convinced that the Archean atmosphere contained little or no free oxygen. In other words, the atmosphere was not strongly oxidizing as it is now. The late Preston Cloud, a specialist on the early history of life, estimated that the free oxygen content of the atmosphere increased from about 1% to 10% through the Proterozoic. It was not until about 400 million years ago that oxygen reached its present level. Most of the atmospheric oxygen was released as a waste product by photosynthesizing cyanobacteria (see Fig. 9.16). As indicated by fossil stromatolites, cyanobacteria became common about 2.3 billion years ago.

Banded Iron Formations

Banded iron formations (BIFs) are sedimentary rocks consisting of alternating thin layers of silica (chert) and

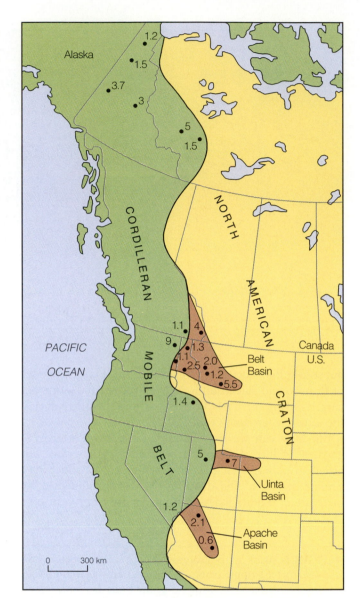

◆ **FIGURE 10.14** Late Proterozoic basins of sedimentation in the western United States and Canada. The numbers indicate the approximate thickness of Late Proterozoic rocks in kilometers.

iron minerals (◆ Fig. 10.20, p. 276). The iron in these formations is mostly iron oxide (hematite and magnetite), but iron silicate, iron carbonate, and iron sulfide also occur.

Archean BIFs are relatively small lenticular (lens-shaped) bodies measuring a few kilometers across and a few meters thick; most appear to have been deposited in greenstone belts. Proterozoic BIFs are much more common; they are typically hundreds of meters thick and can

◆ **FIGURE 10.15** Rocks of the Late Proterozoic Belt Supergroup in Glacier National Park, Montana. The present topography resulted from erosion by glaciers during the Pleistocene and Holocene epochs.

◆ **FIGURE 10.16** (a) Evidence for Proterozoic glaciation in Norway. The Bigganjarga tillite overlies a striated bedrock surface on sandstone of the Veidnesbotn Formation. (Photo courtesy of Anna Siedlecka, Geological Survey of Norway.) (b) Finely laminated argillite with clasts dropped from floating ice in the Gowganda Formation. (Photo courtesy of the Geological Survey of Canada.)

(a)

(b)

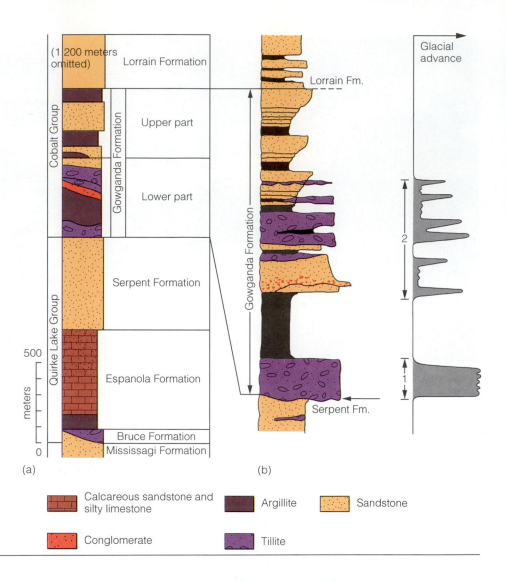

◆ **FIGURE 10.17** Precambrian glacial deposits. (a) Partial section of the Huronian Supergroup, Ontario, showing tillites of the Bruce and Gowganda formations. The Bruce Formation may represent Late Archean glacial deposits. (b) Expanded view of the Early Proterozoic Gowganda Formation. Two major glacial advances are recognized.

be traced for hundreds of kilometers. Most were deposited in shallow marine waters.

BIFs are found throughout the geologic column, but most formed during three main episodes of deposition (Table 10.2). These three depositional periods are bounded by major orogenies representing periods of increased global plate tectonic activity. The BIFs accumulated during the intervening quiescent intervals. The period from 2.5 to 2.0 billion years represents a unique time in Earth history, a time during which 92% of the Earth's BIFs formed (Table 10.2). Also remarkable are the consistent layering and extent of BIFs in shallow marine environments and their chemical purity; BIFs are composed mostly of iron oxides and silicon dioxide as chert.

Iron is a highly reactive element. In the presence of oxygen, it combines to form rustlike oxides that are not readily soluble in water. In the absence of free oxygen, however, iron is easily taken into solution and can accumulate in large quantities in the world's oceans. As we mentioned in Chapter 9, the Archean atmosphere was deficient in free oxygen, thus, it seems likely that little oxygen was dissolved in seawater. The increase in abundance of stromatolites about 2.3 billion years ago resulted in an increase in free oxygen in the oceans, since oxygen is a metabolic waste product of photosynthesizing cyanobacteria that form stromatolites. Apparently, this introduction of free oxygen into the world's oceans helped cause the precipitation of dissolved iron and silica and thus the formation of the BIFs.

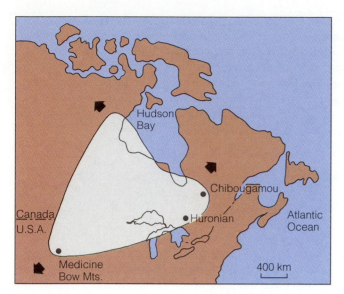

◆ **FIGURE 10.18** Glacial deposits of about the same age in Ontario and Quebec, Canada, and in Michigan and Wyoming suggest that an Early Proterozoic ice sheet was centered southwest of Hudson Bay.

◆ ——————————————————————————————— ◆

TABLE 10.2 Times of Deposition of Banded Iron Formations

TIME OF DEPOSITION (BILLIONS OF YEARS)	PERCENTAGE OF TOTAL BIF DEPOSITED
3.5–3.0	6%
2.5–2.0	92%
1.0–0.5	2%

The current model favored for BIF precipitation involves a Precambrian ocean with an upper oxygenated layer overlying a large volume of oxygen-deficient water that contained reduced iron and silica. Upwelling of these deep, iron- and silica-rich waters onto newly created shallow continental shelves resulted in large-scale precipitation of iron oxides and silica as BIFs (◆ Fig. 10.21). A likely major source for this iron and silica was submarine volcanism, although weathering of the continents probably supplied iron as well. Recent research has shown that hydrothermal vents near spreading ridges (see Perspective 9.2) remove large quantities of iron and silica from the oceanic crust. The hydrothermally derived iron and silica combined with free oxygen released by photosynthesizing organisms and triggered the precipitation of massive quantities of BIFs between 2.5 and 2.0 billion years ago.

Red Beds

Continental **red beds**—red sandstones and shales—first appeared about 1.8 billion years ago, following the deposition of the Proterozoic banded iron formations. The color of red beds is caused by the presence of ferric oxide, usually as the mineral hematite (Fe_2O_3), which forms under oxidizing conditions. These deposits become increasingly abundant through the Proterozoic and are quite common in the Phanerozoic.

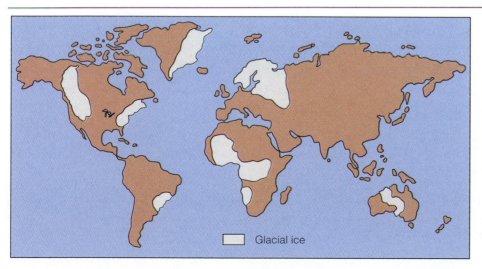

◆ **FIGURE 10.19** Major Late Proterozoic glacial centers shown on a map with the continents in their present positions. Extent of ice centers is hypothetical and approximate.

Glacial ice

◆ **FIGURE 10.20** Precambrian banded iron formations. Early Proterozoic BIF at Ishpeming, Michigan. At this location the rocks are brilliantly colored alternating layers of red chert and silver iron minerals.

The appearance of continental red beds indicates an oxidizing atmosphere. But as we stated earlier, the Proterozoic atmosphere contained little free oxygen, perhaps as little as 1 to 2% in the early part of that eon. Is this sufficient to account for oxidized iron in sediments? Probably not; other attributes of this atmosphere must be considered.

Before abundant free oxygen was present, no ozone (O_3) layer existed in the upper atmosphere. Accordingly,

as free oxygen (O_2) was released to the atmosphere as a waste product during photosynthesis, ultraviolet radiation converted some of it into O and O_3, both of which oxidize surface materials more efficiently than O_2. Once an ozone layer became established in the upper atmosphere, most of the ultraviolet radiation failed to penetrate to the surface, and O_2 became the primary agent for oxidation of surface deposits.

❖ PROTEROZOIC LIFE

In the last chapter, we noted that Archean fossils are not very common, and all are varieties of bacteria—unicellular prokaryotes. The Early Proterozoic is likewise characterized by these organisms. Bacteria and cyanobacteria were the only life-forms present. Apparently little diversification had occurred up to this point in life history. In fact, cells more advanced than prokaryotic cells did not appear until about 1.4 billion years ago. Even the well-known, 1.8- to 2.1-billion-year-old Gunflint Iron Formation of Ontario, Canada, with 12 species of microorganisms contains only bacteria and cyanobacteria (◆ Fig. 10.22).

Actually, the lack of organic diversity during the Archean and Early Proterozoic is not too surprising because prokaryotic cells reproduce asexually. Recall our discussion in Chapter 6 on the sources of variation in populations. Most variation results from sexual reproduction, during which the offspring receives half of its

◆ **FIGURE 10.21** Depositional model for the origin of banded iron formations.

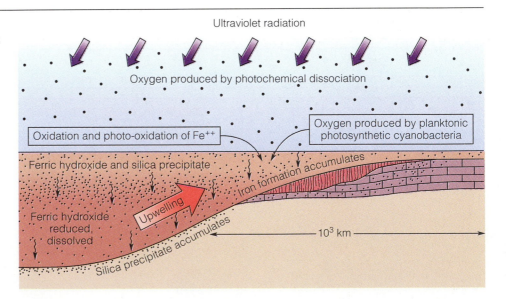

genetic makeup from each parent. Mutations may introduce new variations into any population, but the effects of mutations in prokaryotes are limited.

Consider this. Suppose a beneficial mutation occurred in a bacterium. How could this mutation spread throughout the population? It could not, since the bacterium with the mutation does not share its genes with other bacteria.[1] By contrast, a beneficial mutation in a sexually reproducing organism could spread quickly and widely in an interbreeding population.

Before the appearance of cells capable of sexual reproduction, evolution was a comparatively slow process, accounting for the low organic diversity during the Archean and Early Proterozoic. This situation did not persist, however. By the Middle Proterozoic, cells appeared that reproduced sexually, and the tempo of evolution picked up markedly.

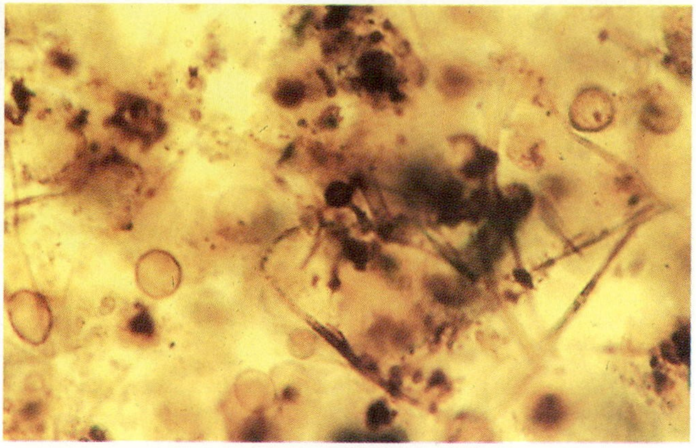

A New Type of Cell Appears

The appearance of **eukaryotic cells** marks one of the most important events in the history of life. Eukaryotic cells are considerably more organizationally complex than prokaryotic cells and have a membrane-bounded nucleus that contains the genetic material (◆ Fig. 10.23). Organ-

------

[1] Bacteria usually reproduce by binary fission and give rise to two cells having the same genetic makeup. Under some conditions, bacteria engage in conjugation during which some genetic material is transferred from one cell to another.

◆ **FIGURE 10.22** Photomicrographs of spheroidal and filamentous microfossils from stromatolitic chert of the Gunflint Iron Formation, Ontario.

isms composed of eukaryotic cells are **eukaryotes** whereas those with prokaryotic cells are **prokaryotes.** Most eukaryotes are multicellular, although those of the kingdom Protoctista are unicellular (Appendix B). In marked contrast to prokaryotes, most eukaryotes reproduce sexually and most are aerobic; thus, eukaryotes could not have appeared until some free oxygen was present in the atmosphere.

A number of Proterozoic fossil localities have yielded unicellular eukaryotes, but the oldest one appears to be

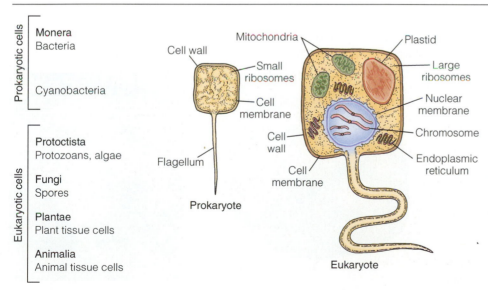

◆ **FIGURE 10.23** Prokaryotic and eukaryotic cells. Note that eukaryotes have a cell nucleus containing the genetic material and several organelles such as mitochondria and plastids.

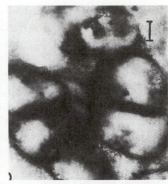

◆ **FIGURE 10.24** Microfossils from the Beck Springs Dolomite, California. These fossils may be the oldest known eukaryotic cells. (Photos courtesy of Preston Cloud, University of California, Santa Barbara.)

in the 1.2- to 1.4-billion-year-old Beck Springs Dolomite of southeastern California (◆ Fig. 10.24, Table 10.3). Fossils that appear to be unicellular algae from the 1.0-billion-year-old Bitter Springs Formation of Australia show evidence of both mitosis and meiosis, processes used only by eukaryotic cells. Although the fossils of the Beck Springs Dolomite are commonly cited as the first known eukaryotes, this is a bit uncertain, because it is difficult to be sure that microscopic fossils are eukaryotes rather than prokaryotes.

The fossil evidence for the appearance of eukaryotic cells comes from size and relative complexity. Eukaryotic cells are larger, commonly much larger, than prokaryotic cells (Fig. 10.23). Cells larger than 60 microns appear in abundance about 1.4 billion years ago. As for relative complexity, prokaryotic cells are typically simple spher-

ical or platelike structures. Proterozoic fossils of branched filaments, flask-shaped organisms, and some containing what appear to be internal, membrane-bounded structures all indicate a eukaryotic level of organization. Theories of how eukaryotic cells originated from prokaryotic cells are considered in Perspective 10.1.

Additional evidence for eukaryotes in the Proterozoic comes from a fossil group called *acritarchs*. These hollow fossils are probably the cysts of planktonic algae (◆ Fig. 10.25a and b), which became quite abundant during the Late Proterozoic but first appeared about 1.4 billion years ago. Numerous microfossils with vase-shaped skeletons have been recovered from Late Proterozoic rocks of the Grand Canyon (◆ Fig. 10.25c). These have been tentatively identified as cysts of some kind of algae.

◆ **FIGURE 10.25** These common Late Proterozoic microfossils are thought to represent eukaryotic organisms. (a) and (b) Acritarchs are probably the cysts of algae. (Photos courtesy of Andrew H. Knoll, Harvard University.) (c) Vase-shaped microfossil, probably a cyst of some kind of algae. (Photo courtesy of Bonnie Bloeser.)

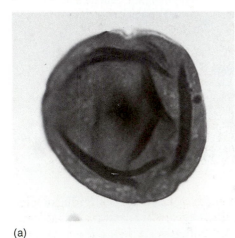

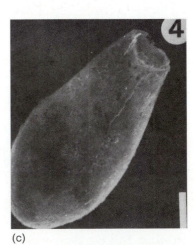

(a)   (b)   (c)

TABLE 10.3 Summary Chart for Proterozoic Fossils Discussed in the Text. Dashed vertical lines indicate uncertainties in age ranges.

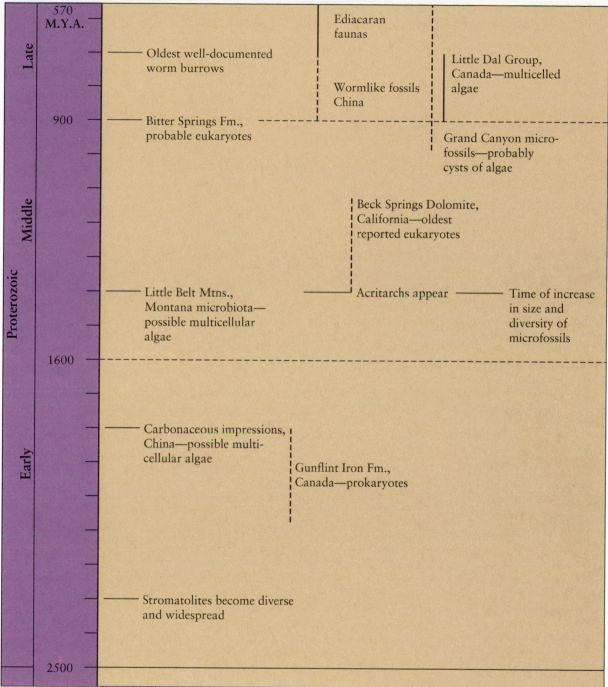

The precise time when eukaryotes appeared is debatable, but undoubtedly they were present by the Late Proterozoic. Perhaps they appeared as early as 1.4 billion years ago, the maximum age of the Beck Springs Dolomite. Most authorities agree that eukaryotes had appeared by at least 1.0 billion years ago. During the Late

THE ORIGIN OF EUKARYOTIC CELLS

Stratigraphic position and radiometric ages tell us that eukaryotic cells were present by the Middle Proterozoic. But although we know when eukaryotes appeared, the fossil record reveals nothing about how they evolved from prokaryotic ancestors. The study of living microorganisms, however, gives us some idea of how this may have occurred.

A currently popular theory among evolutionary biologists holds that eukaryotic cells formed from several prokaryotic cells that had established a symbiotic relationship. Symbiosis, the living together of two or more dissimilar organisms, is quite common among organisms today. It may take the form of parasitism in which one organism lives at the expense of another, or the two may coexist with mutual benefit. For example, lichens, once considered to be plants, are actually symbiotic fungi and algae.

In a symbiotic relationship, each symbiont must be capable of metabolism and reproduction, but the degree of dependence in some symbiotic relationships is such that one symbiont cannot live independently. Many parasites, for example, cannot exist outside a host organism. This may have been the case with Proterozoic prokaryotes: two or more prokaryotes may have entered into a symbiotic relationship (◆ Fig. 1), and the symbionts became increasingly interdependent until the unit could exist only as a whole.

Supporting evidence for the symbiosis theory comes from the study of living eukaryotic cells. For example, eukaryotic cells contain internal structures called organelles that have their own complements of genetic material. Although these organelles, such as plastids and mitochondria, cannot exist independently today, they apparently were once capable of reproduction as free-living organisms.

Another way of evaluating the symbiosis theory is to look at protein synthesis. Prokaryotic cells synthesize proteins, but can be thought of as a single system, whereas eukaryotes are a combination of protein-synthesizing systems; that is, some of the organelles within eukaryotes such as mitochondria and plastids are capable of protein synthesis. These organelles, with their own genetic material and protein-synthesizing capabilities, are thought to have been free-living bacteria

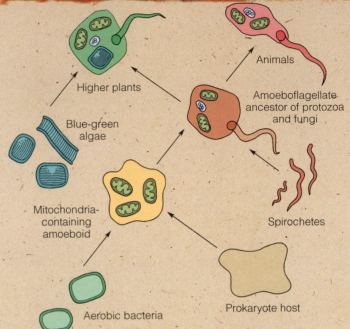

◆ **FIGURE 1** Symbiosis theory for the origin of eukaryotic cells. An aerobic bacterium and a larger host of the kingdom Monera united to form a mitochondria-containing amoeboid. An amoeboflagellate was formed by a union of the amoeboid and a bacterium of the spirochete group; this amoeboflagellate was the direct ancestor of two kingdoms—Fungi and Animalia. Another kingdom, Plantae, was founded when this amoeboflagellate formed a union with blue-green algae (cyanobacteria) that became plastids.

that entered into a symbiotic relationship. With time, the interdependence of the various units grew until life was possible only as an integrated whole (Fig. 1).

More recently, another theory for the origin of eukaryotic cells has been proposed. According to this theory, the sequence of events leading to the origin of eukaryotes began when early prokaryotes lost the ability to synthesize an essential component of cell walls. One solution to this problem was to develop an internal skeleton consisting of microtubules and microfilaments (◆ Fig. 2a). This cytoskeleton, as it is called, allowed for

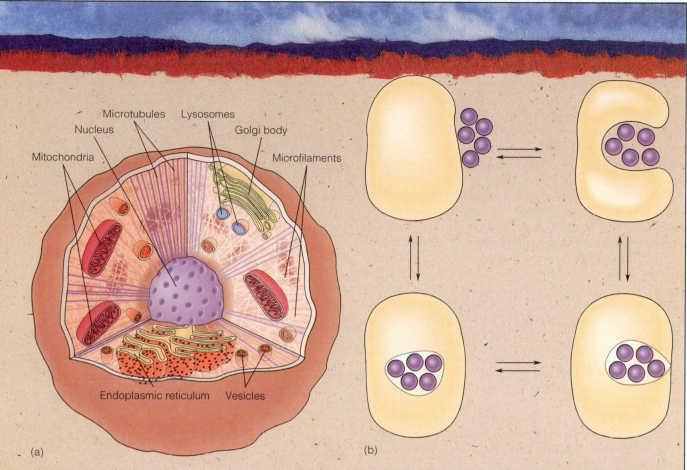

◆ **FIGURE 2** (a) Diagrammatic view of a cell showing the cytoskeleton consisting of microtubules and microfilaments. (b) According to one theory for the origin of eukaryotic cells, cells acquired such structures as a nucleus, mitochondria, and plastids by infolding and enveloping materials.

an outer fluid cell membrane that infolded so that material coming into the cell could be enveloped. Such infolding is thought to have resulted in the origin of such intracellular structures as the cell nucleus (◆ Fig. 2b). According to this theory, the first eukaryotic cell was anaerobic; later it acquired a free-living aerobic organism by symbiosis, thus accounting for the origin of aerobic eukaryotes.

Although the fossil record does not record the acquisition of organelles or symbiosis, living eukaryotes can give some idea of what the first eukaryotes may have been like. The present-day giant amoeba *Pelomyxa*, which lives in the mud of ponds, lacks mitochondria. Two types and hundreds of individual bacteria, however, have a symbiotic relationship with *Pelomyxa* and perform the same function as mitochondria.

Pelomyxa provides evidence for the symbiotic theory for the origin of eukaryotes. However, recent studies of *Giardia*, a single-celled eukaryote, seem to indicate that eukaryotes may have acquired their internal membrane-bounded organelles by infolding of the cell wall. Although *Giardia* is a eukaryote, it shares many characteristics with prokaryotes and is capable of acquiring materials from outside by infolding of the cell wall.

Proterozoic, eukaryotes show considerable diversification whereas prokaryotes exhibit little change.

Multicellular Organisms

Multicellular organisms are not only composed of many cells, often billions, but also have cells specialized to perform specific functions such as reproduction and respiration. We know from fossils that multicellular organisms appeared in the Late Proterozoic, but we have no fossil evidence of the transition from their unicellular ancestors.

The study of present-day organisms gives some clues as to how this transition may have occurred. Perhaps some unicellular organism divided and formed a group of cells that did not disperse but remained together as a colony. The cells in some colonies may have become somewhat specialized, similar to the situation in the living *colonial organisms* (◆ Fig. 10.26). Further specialization might have led to simple multicellular organisms such as sponges that consist of cells specialized for reproduction, respiration, and food gathering. In other words, specialization of cells might have led to the development of organs with specific functions.

Is there any particular advantage to being multicellular? After all, until the Late Proterozoic, life seems to have thrived even though all known organisms were unicellular. In fact, unicellular organisms are quite successful at what they do, but what they do is limited. For example, they cannot become very large, because as size increases, proportionally less of the cell is exposed to the external environment in relation to its volume. In other words, as size increases, the amount of surface area compared to volume decreases; consequently, the process of transferring materials from the external environment into the cell becomes less efficient. Multicellular organisms live longer, since cells can be replaced; and more offspring can be produced, because some cells are specialized for reproduction. And finally, cells have increased functional efficiency as they become specialized into organs.

◆ **FIGURE 10.27** Carbonaceous impressions in Proterozoic rocks in the Little Belt Mountains, Montana. These may be impressions of multicellular algae, but this is uncertain. (Photo courtesy of Robert Horodyski, Tulane University.)

Multicellular Algae

Well-preserved carbonaceous impressions of multicellular algae are known from rocks 800 million years old. And probable algae are preserved in 1,000- to 700-million-year-old rocks in Spitzbergen, China, India, and the Little Dal Group of northwestern Canada (Table 10.3). The size, composition, and general shape of these impressions indicate photosynthesizing eukaryotes, probably planktonic algae. In fact, some organic sheets in the Little Dal Group closely resemble the living green alga known as sea lettuce.

Even older rocks contain carbonaceous impressions that may be multicellular algae, but this is uncertain. For example, the 1.4-billion-year-old microbiota from the Little Belt Mountains of Montana contains filaments and spherical forms of uncertain affinities (◆ Fig. 10.27, Table 10.3). Carbonaceous macroscopic filaments from 1.8-billion-year-old rocks of China are also suggestive of multicellular algae.

◆ **FIGURE 10.26** Unicellular versus multicellular organisms. Although *Gonium* consists of as few as four cells, all cells are alike and can produce a new colony. *Volvox* has some cells specialized to perform different functions and has thus crossed the threshold that separates unicellular from multicellular organisms.

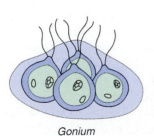

Gonium

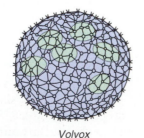

Volvox

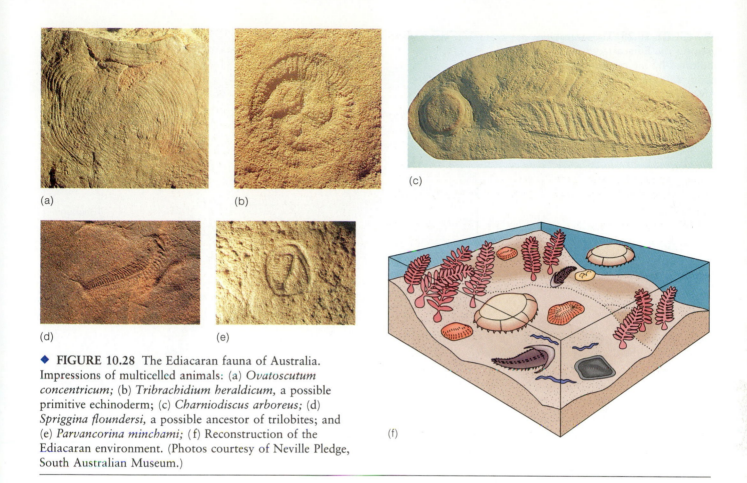

◆ **FIGURE 10.28** The Ediacaran fauna of Australia. Impressions of multicelled animals: (a) *Ovatoscutum concentricum;* (b) *Tribrachidium heraldicum,* a possible primitive echinoderm; (c) *Charniodiscus arboreus;* (d) *Spriggina floundersi,* a possible ancestor of trilobites; and (e) *Parvancorina minchami;* (f) Reconstruction of the Ediacaran environment. (Photos courtesy of Neville Pledge, South Australian Museum.)

The Ediacaran Fauna

In 1947, the Australian geologist R. C. Sprigg discovered impressions of soft-bodied animals in rocks of the Edi-acara Hills of South Australia. Additional discoveries by geologists and amateur paleontologists turned up what appeared to be impressions of algae and various animals, many of which bear no resemblance to living organisms. These discoveries have provided some of the evidence that has partly resolved one of the great mysteries in the history of life.

Before the discovery of these Ediacaran fossils, geologists were perplexed by the apparent absence of animal fossils in strata older than the Cambrian. The shelly faunas of the Cambrian Period seemed to have appeared abruptly in the fossil record. In fact, the evidence for life before the Cambrian was so sparse that all pre-Paleozoic time was once referred to as the *Azoic,* meaning without life. Geologists assumed that the Cambrian shelly faunas must have been preceded by more primitive animals, but since none were known, a worldwide unconformity, the Lipalian interval, was proposed to account for the absence of such fossil-bearing rocks.

The rock unit in which the Ediacara Hills fossils were discovered, the Pound Quartzite, was initially thought to be Cambrian. However, a joint investigation by the South Australian Museum and the University of Adelaide demonstrated that the fossil-bearing strata lie more than 150 m below the oldest recognized Cambrian strata. Eventually, it became clear that what had been discovered was a unique assemblage of soft-bodied animals preserved as molds and casts on the undersides of sandstone layers (◆ Fig. 10.28). Martin Glaessner of the University of Adelaide believes that these Ediacaran animals lived in a nearshore, shallow marine environment.

Some investigators, including Glaessner, are of the opinion that at least three present-day invertebrate phyla

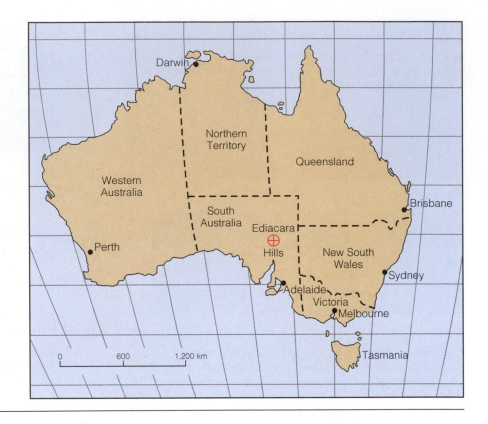

◆ **FIGURE 10.29** The stratotype area for the Ediacarian System, indicated by the cross in a circle, is about 480 km north of Adelaide, Australia.

are represented: jellyfish and sea pens (phylum Cnidaria), segmented worms (phylum Annelida), and primitive members of the phylum Arthropoda, the phylum that includes insects, spiders, and crabs. One wormlike Ediacaran fossil, *Spriggina,* has been cited as a possible ancestor of trilobites, and another, *Tribachidium,* may be a primitive echinoderm (Fig. 10.28b and c).

Researchers disagree, however, on exactly what these Ediacaran animals were and how they should be classified. Adolph Seilacher of Tübingen, Germany, for example, thinks that they represent an early evolutionary radiation quite distinct from the ancestry of the present-day invertebrate phyla. In other words, he believes that the Ediacaran animals are not members of the phyla noted above, and that they are not ancestral to the complex invertebrates that appeared in abundance during the Cambrian Period. In fact, Seilacher thinks they represent an evolutionary "dead end." He also believes that these animals fed by passive absorption because there is no conclusive evidence of a mouth in the preserved specimens.

Ediacara-type faunas are now known on all continents except Antarctica. Collectively, these **Ediacaran faunas,** as they are commonly called, lived between 570 and 670 million years ago. These animals were widespread during this time, but their fossils are rare. Their scarcity should come as no surprise, however, since all lacked durable skeletons.

The discovery of pre-Paleozoic fossil-bearing strata has prompted Preston Cloud, formerly of the University of California at Santa Barbara, and Glaessner to propose a new geologic period and system, the Ediacarian. According to this proposal, the Ediacarian Period constitutes the first period of the Paleozoic Era, and the stratotype for the Ediacarian System is in South Australia (◆ Fig. 10.29). Thus, strata of the Ediacarian System were deposited during the Ediacarian Period, which began 670 million years ago when multicelled organisms appeared in the fossil record and ended 570 million years ago when shelly faunas of the Cambrian Period first appeared.

Cloud and Glaessner's proposal is a reasonable one and has been accepted by many geologists. Many other geologists, however, prefer the more traditional time scale, and in this book we consider the Ediacaran fauna to be latest Proterozoic in age.

Other Proterozoic Animal Fossils

Although scarce, there is some evidence for pre-Ediacaran animals. For example, a jellyfish-like impression is known from rocks 2,000 m below the Pound Quartzite. And in many areas burrows, presumably made by worms, occur in rocks at least 700 million years old.

Wormlike fossils associated with fossil algae were recently reported from 700- to 900-million-year-old strata in China (◆ Fig. 10.30, Table 10.3). Perhaps these are worms, but both their biological affinities and their age have been questioned. For the present, we can consider these wormlike fossils as persuasive but not conclusive evidence for pre-Ediacaran animals.

All known Proterozoic animals were soft-bodied; that is, they lacked the durable exoskeletons that characterize many Phanerozoic invertebrates. There is some evidence, however, that the earliest stages of skeletonization occurred during the Late Proterozoic. For example, some Ediacaran animals may have had chitin, possibly a chitinous carapace, and others may have possessed some calcareous skeletal elements.

◆ **FIGURE 10.30** Wormlike body fossils from the Late Proterozoic of China. (Photos courtesy of Sun Weiguo, Nanjing Institute of Geology and Palaeontology, Academia Sinica, Nanjing, People's Republic of China.)

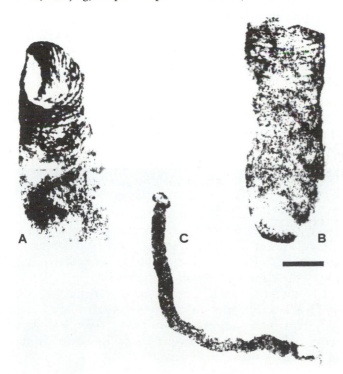

By the latest Proterozoic, several skeletonized animals probably existed. Evidence for this conclusion comes from minute scraps of shell-like material and denticles from larger animals, and spicules, presumably from sponges. Durable skeletons of chitin (a complex organic substance), silica, and calcium carbonate, however, began appearing in abundance at the beginning of the Phanerozoic Eon 570 million years ago.

❖ PROTEROZOIC MINERAL DEPOSITS

The most notable mineral deposits of Proterozoic age are banded iron formations (BIFs). BIFs are present in all Precambrian cratons (◆ Fig. 10.31); as noted earlier, 92% of all BIFs were deposited during the Late Proterozoic, although a few Archean and Phanerozoic examples are known. These deposits constitute the world's major iron ores. The largest producers of iron ores include Brazil, Australia, China, the Ukraine, Sweden, South Africa, Canada, and the United States. Even though the United States is a major producer, it must still import about 30% of the iron ore used, mostly from Canada and Venezuela.

In North America, most of the large iron mines are in the Lake Superior region. Huge deposits of BIFs in Ontario, Canada, and adjacent states have been mined extensively for decades. In 1988, Minnesota was the leading producer of iron ore in the United States, accounting for 72% of the total production, while Michigan produced 25% of the total.

The richest ores—those containing up to 70% iron—in the Lake Superior region were depleted by the time of World War II. Continued mining has been possible because a method was developed to separate the iron ore from unusable rock of lower-grade ores and then shape the iron into pellets (◆ Fig. 10.32). These pellets contain about 65% iron and are easily shipped via the Great Lakes to the steel-producing centers.

The Sudbury mining district in Ontario, Canada, is an important area of nickel and platinum production. Nickel is essential in the production of nickel alloys such as stainless steel and Monel metal (nickel plus copper), which are valued for their strength and resistance to corrosion and heat. The United States must import more than 50% of all nickel used; most of these imports come from the Sudbury mining district.

In addition to its economic importance, the Sudbury Basin, an elliptical area measuring more than 59 by 27

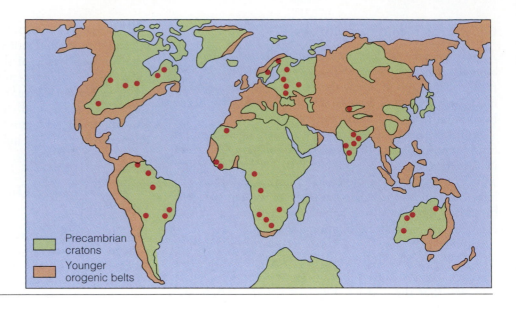

Precambrian cratons

Younger orogenic belts

km, is interesting from the geological perspective. One hypothesis for the concentration of ores is that they were mobilized from metal-rich rocks beneath the basin following a high-velocity meteorite impact.

Some platinum for jewelry, surgical instruments, and chemical and electrical equipment is also exported to the United States from Canada, but the major exporter is South Africa. The Bushveld Complex of South Africa is a layered complex of igneous rocks from which both platinum and chromite, the only ore of chromium, are mined. Much of the chromium used in the United States is imported from South Africa; it is used mostly in the manufacture of stainless steel.

Economically recoverable oil and gas have been discovered in Proterozoic rocks in China and Siberia, arousing some interest in the Midcontinent rift as a potential source of hydrocarbons. So far, considerable land has been leased for exploration, and numerous geophysical studies have been done. However, even though some rock units within the rift are known to contain petroleum, no producing oil or gas wells are currently operating.

A number of Proterozoic pegmatites are important economically. The Dunton pegmatite in Maine, whose age is generally considered to be Late Proterozoic, has yielded magnificent gem-quality specimens of tourmaline and other minerals (◆ Fig. 10.33). Other pegmatites are

◆ **FIGURE 10.32** Iron pellets from a mine in northern Michigan. The pellets are about 1 cm in diameter.

◆ **FIGURE 10.33** Tourmaline from the Dunton mine in Maine.

mined for gemstones as well as for tin; industrial minerals, such as feldspars, micas, and quartz; and minerals containing such elements as cesium, rubidium, lithium, and beryllium.

More than 20,000 pegmatites have been identified in the country rocks adjacent to the Harney Peak Granite in the Black Hills of South Dakota. These pegmatites formed about 1.7 billion years ago when the granite was emplaced as a complex of dikes and sills. A few of these have been mined for gemstones, tin, lithium, and micas. In addition, some of the world's largest known mineral crystals have been discovered in these pegmatites.

Chapter Summary

1. The crust-forming processes that characterized the Archean continued into the Proterozoic but at a considerably reduced rate.

2. Archean cratons served as nuclei about which Proterozoic crust accreted. One large landmass that formed by this process is called Laurentia. It consisted mostly of North America and Greenland.

3. The major events in the Proterozoic evolution of Laurentia were an Early Proterozoic episode of amalgamation of cratons, Middle Proterozoic igneous activity, and the Middle Proterozoic Grenville orogeny and Midcontinent rift.

4. By the Late Proterozoic, there were probably three or more supercontinents. Laurasia consisted of Laurentia and what is now Europe and most of Asia. East and West Gondwana may have united in the latest Proterozoic to form Gondwana.

5. Greenstone belts formed during the Proterozoic but at a reduced rate, and they differ in detail from their Archean counterparts.

6. Ophiolite sequences, which mark convergent plate margins, are first well documented from the Early Proterozoic. A present-day style of plate tectonics seems to have been established during the Early Proterozoic.

7. Quartzite-carbonate-shale assemblages are known from the Late Archean but become common in Proterozoic rocks. These rock assemblages were deposited on passive continental margins and in intracratonic basins.

8. Widespread glaciation occurred during the Early and Late Proterozoic.

9. The atmosphere became progressively richer in free oxygen through the Proterozoic. Photosynthesizing cyanobacteria were largely responsible for oxygenation of the atmosphere, although photochemical dissociation of water vapor contributed some oxygen.

10. During the period from 2.5 to 2.0 billion years ago, most of the world's iron ores were deposited as banded iron formations.

11. The first continental red beds were deposited about 1.8 billion years ago. The widespread occurrence of oxidized iron in sedimentary rocks indicates an oxidizing atmosphere.

12. Unicellular prokaryotes are the only known life-forms from the Early Proterozoic. Eukaryotic cells first appeared during the Middle Proterozoic.

13. The oldest fossils of multicellular organisms are carbonaceous impressions, probably of algae, in rocks between 1 billion and 700 million years old.

14. The Late Proterozoic Ediacaran faunas include the oldest well-documented animal fossils other than burrows. Animals were widespread at this time, but all were soft-bodied, so fossils are not common.

15. Most of the world's iron ore production is from Proterozoic banded iron formations. Other important resources include nickel, platinum, and a variety of materials mined from pegmatites.

Important Terms

banded iron formation (BIF)
Ediacaran faunas
eukaryotic cell
Grenville orogeny

Laurentia
Midcontinent rift
multicellular organism
ophiolite

orogen
quartzite-carbonate-shale assemblage
red beds

Review Questions

1. The oldest well-documented fossil animals occur in the:
   a. _____ Lake Superior fauna; b. _____ Ediacaran fauna; c. _____ Llano fauna; d. _____ Wopmay fauna; e. _____ Bitter Springs Formation.

2. The Proterozoic Eon began _____ billion years ago.
   a. _____ 570; b. _____ 4.6; c. _____ 3.8; d. _____
   2.5; e. _____ 1.9.
3. The large landmass consisting mostly of North America
   and Greenland that formed during the Proterozoic was:
   a. _____ Gondwana; b. _____ Hudsonia; c. _____
   Laurentia; d. _____ the Trans-Hudson orogen;
   e. _____ the Wyoming craton.
4. Widespread Middle Proterozoic igneous rocks consist
   mostly of:
   a. _____ ultramafic lava flows and andesite; b. _____
   ash flow deposits and plutons; c. _____ colliding island
   arcs; d. _____ greenstone belts; e. _____ sedimentary
   and metamorphic rocks.
5. The last major North American orogen to form during
   the Proterozoic was the _____ orogen.
   a. _____ Kenoran; b. _____ Wopmay; c. _____
   Grenville; d. _____ Colorado; e. _____ Belt
   Supergroup.
6. Thick accumulations of basaltic lava flows forming
   extensive lava plateaus occur in the:
   a. _____ Gowganda Formation; b. _____ Midcontinent
   rift; c. _____ Penokean orogen; d. _____ Grenville
   orogen; e. _____ Wyoming craton.
7. Archean and Proterozoic greenstone belt rocks are
   similar except in the latter there is little or no:
   a. _____ ultramafic rock; b. _____ basalt; c. _____
   graywacke; d. _____ argillite; e. _____ granite.
8. By far the most common associations of Proterozoic
   rocks are:
   a. _____ basalt-andesite–ash fall; b. _____
   sandstone-granite-basalt; c. _____ granite-andesite-tillite;
   d. _____ banded iron formation–tillite-andesite; e. _____
   quartzite-carbonate-shale.
9. Tillite(s) is(are):
   a. _____ structures in limestone formed by bacteria;
   b. _____ a type of iron-rich rock; c. _____ lithified
   glacial deposits; d. _____ alternating dark and light
   layers formed in lakes; e. _____ striated bedrock
   pavement.
10. Most banded iron formations (BIFs) were deposited
    during the:
    a. _____ Late Proterozoic; b. _____ Middle Archean;
    c. _____ Early Proterozoic; d. _____ Trans-Hudson
    orogen; e. _____ Midcontinent rift.
11. The presence of red beds during the Proterozoic
    indicates:
    a. _____ widespread glaciation; b. _____ that the
    atmosphere contained some free oxygen; c. _____ that
    animals had appeared; d. _____ a chemically reducing
    atmosphere; e. _____ that carbonate deposition was
    becoming increasingly common.
12. Probably the most important event in the evolution of
    life during the Proterozoic was the appearance of:
    a. _____ photosynthesizing organisms; b. _____ algae;
    c. _____ the first autotroph; d. _____ eukaryotic cells;
    e. _____ ophiolites.
13. The evidence for the origin of eukaryotic cells comes
    from the study of:
    a. _____ fossils; b. _____ chemicals preserved in
    Archean rocks; c. _____ tracks and trails of trilobites;
    d. _____ present-day organisms; e. _____ Proterozoic
    glacial deposits.
14. Summarize the major difference between the Archean
    and Proterozoic.
15. Discuss the events leading to the acretion of continental
    crust during the Early Proterzoic.
16. Discuss the Middle Proterozoic episode of igneous activity
    and its importance in the evolution of Laurentia.
17. What is the Midcontinent rift, and what kinds of rocks
    does it contain?
18. Discuss the significance of ophiolites and
    quartzite-carbonate-shale assemblages in establishing
    that the Proterozoic was characterized by a modern style
    of plate tectonics.
19. What kind of evidence would confirm that a suspected
    Proterozoic tillite was actually a glacial deposit?
20. How are BIFs thought to have been deposited?
21. The atmosphere seems to have been chemically oxidizing
    by 1.8 billion years ago. What evidence supports this
    statement?
22. What evidence indicates that eukaryotic cells appeared
    between 1.4 and 1.0 billion yhears ago?
23. Explain how a symbiotic relationship among Proterozoic
    prokaryotes may have given rise to eukaryotes.
24. Since the transition from unicellular to multicellular
    organisms is not recorded by fossils, how can we
    account for this event?
25. Briefly review the evidence for the presence of animals in
    the Late Proterozoic.

◆ ━━━ ◆

Additional Readings

Condie, K. C. 1989. *Plate tectonics and crustal evolution.*
3rd ed. New York: Pergamon Press.

Glaessner, M. F. 1984. *The dawn of animal life.* New York:
Cambridge University Press.

Hambrey, M. 1992. Secrets of a tropical ice age. *New Scientist* 133, no. 1806: 42–49.

Hoffmann, P. F. 1988. United plates of America, the birth of
a craton: Early Proterozoic assembly and growth of Lau-

rentia. *Annual Review of Planetary Sciences* 16: 543–603.

Kabnick, K. S., and D. A. Peattie. 1991. *Giardia:* A missing link between prokaryotes and eukaryotes. *American Scientist* 79, no. 1: 34–43.

Knoll, A. H. 1991. End of the Proterozoic Eon. *Scientific American* 265, no. 4: 64–73.

Kroner, A., ed. 1987. *Proterozoic lithospheric evolution.* Geodynamics Series, v. 17. Washington, D.C.: American Geophysical Union.

Margulis, L. 1982. *Early life.* New York: Van Nostrand Reinhold.

Margulis, L., and L. Olendzenski, eds. 1992. *Environmental evolution: Effects of the origin of life on planet Earth.* Cambridge, Mass.: MIT Press.

Mendaris, L. G., Jr., C. W. Byers, D. M. Michelson, and W. C. Shanks. 1983. *Proterozoic geology: Selected papers from an International Symposium.* The Geological Society of America, Memoir 161.

Schopf, J. W. 1978. The evolution of the earliest cells. *Scientific American* 239, no. 3: 111–37.

Schopf, J. W., ed. 1992. *Major events in the history of life.* Boston, Mass.: Jones and Bartlett Publishers.

Schofp, J.W., and C. Klein, eds., 1992. *The Proterozoic biosphere.* New York: Cambridge University Press.

Windley, B. F. 1984. *The evolving continents.* New York: John Wiley & Sons.

CHAPTER 11

Major John Wesley Powell, who led the first geologic expedition down the Grand Canyon in Arizona. (Photo courtesy of the National Park Service, U.S. Department of the Interior.)

GEOLOGY OF THE EARLY PALEOZOIC ERA

Prologue

"The Grand Canyon is the one great sight which every American should see," declared President Theodore Roosevelt. "We must do nothing to mar its grandeur." And so, in 1908, he named the Grand Canyon a national monument to protect it from exploitation. In 1919 the Grand Canyon National Monument was upgraded to a national park primarily because both its scenery and the geology exposed in the canyon are unparalleled.

When people visit the Grand Canyon, many are astonished by the seemingly limitless time represented by the rocks exposed in the walls. For most people, staring down 1.5 km at the rocks in the canyon is their only exposure to the concept of geologic time.

Major John Wesley Powell was the first geologist to explore the Grand Canyon region. Major Powell, a Civil War veteran who lost his right arm in the battle of Shiloh, led a group of hardy explorers down the uncharted Colorado River through the Grand Canyon in 1869. Without any maps or other information, Powell and his group ran the many rapids of the Colorado River in fragile wooden boats, hastily recording what they saw. Powell wrote in his diary that "all about me are interesting geologic records. The book is open and I read as I run."

From this initial reconnaissance, Powell led a second expedition down the Colorado River in 1871. This second trip included a photographer, a surveyor, and three topographers. This expedition made detailed topographic and geologic maps of the Grand Canyon area as well as the first photographic record of the region.

Probably no one has contributed as much to the understanding of the Grand Canyon as Major Powell. In recognition of his contributions, the Powell Memorial was erected on the South Rim of the Grand Canyon in 1969 to commemorate the hundredth anniversary of his first expedition.

When we stand on the rim and look down into the Grand Canyon, we are really looking far back in time, all the way back to the early history of our planet. More than one billion years of history are recorded in the rocks of the Grand Canyon, ranging from mountain-building episodes to periods of transgressions and regressions of shallow seas.

The oldest rocks exposed in the Grand Canyon record two major mountain-building episodes during the Proterozoic Eon. The first episode, represented by the Vishnu and Brahma schists, records a time of uplift, deformation, and metamorphism. This mountain range was eroded to a rather subdued landscape and was followed by deposition of approximately 4,000 m of sediments and lava flows of the Grand Canyon Supergroup. These rocks and the underlying Vishnu and Brahma schists were uplifted and formed a second Proterozoic mountain range. This mountain range was also eroded to a nearly flat surface by the end of the Proterozoic Eon.

The first sea of the Paleozoic Era transgressed over the region during the Cambrian Period, depositing sandstones, siltstones, and limestones. A major unconformity separates the Cambrian rocks from the Mississippi limestones exposed as the cliff-forming Redwall Limestone. Another unconformity separates the Redwall Limestone from the overlying Permian Kaibab Limestone, which forms the rim of the Grand Canyon.

The Grand Canyon in all its grandeur is a most appropriate place to start our discussion of the Paleozoic history of North America.

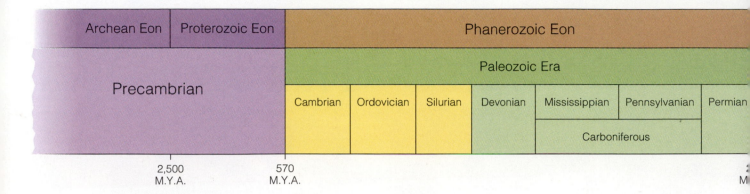

Archean Eon	Proterozoic Eon	Phanerozoic Eon						
Precambrian		Paleozoic Era						
		Cambrian	Ordovician	Silurian	Devonian	Mississippian	Pennsylvanian	Permian
						Carboniferous		

2,500
M.Y.A.

570
M.Y.A.

❖ INTRODUCTION

Having reviewed the geologic history of the Archean and Proterozoic eons, we now turn our attention to the Phanerozoic Eon, comprising the remaining 12% of geologic time. First, however, let us briefly examine how the Archean and Proterozoic differ from the Phanerozoic.

Recall that during the Archean, crust-forming processes generated greenstone belts and granite-gneiss complexes, which were shaped into cratons. The most common Archean sedimentary rocks are graywackes and argillites, typical of tectonically active regions. While a modern style of plate tectonics did not begin until the Early Proterozoic, many geologists believe that the Late Archean style of plate tectonics involved numerous collisions of microcontinents, resulting in rapid crustal growth. The Proterozoic can be characterized as a time during which the sedimentary-rock assemblages and plate tectonic style were like those of the Phanerozoic. During the Proterozoic, plate movements, extensive plutonism, and regional metamorphism thickened, stabilized, and increased the area of continental crust.

At the beginning of the Phanerozoic, there were six major continental landmasses, four of which straddled the paleoequator. Plate movements during the Phanerozoic created a changing panorama of continents and ocean basins whose positions affected atmospheric and oceanic circulation patterns and created new environments for habitation by the rapidly evolving biota.

The Paleozoic history of most continents involves major mountain-building activity along the continental borders and numerous shallow-water marine transgressions and regressions over their interiors. These transgressions and regressions were caused by global changes in sea level probably related to plate activity and glaciation.

The following chapters present the geologic history of North America in terms of those major transgressions and regressions rather than a period-by-period chronology. While we will focus on North American geologic history, we will endeavor to place those events in a global context.

❖ CONTINENTAL ARCHITECTURE: CRATONS AND MOBILE BELTS

During the Precambrian, continental accretion and orogenic activity led to the formation of sizable continents. At least three large continents existed during the Late Proterozoic (see Fig. 10.10), and some geologists believe that these landmasses later collided to form a single Pangaea-like supercontinent (see Fig. 10.11). This supercontinent began breaking apart sometime during the latest Proterozoic. By the beginning of the Paleozoic Era, six major continents were present. Each continent can be divided into two major components: a *craton* and one or more *mobile belts*.

Cratons are the relatively stable and immobile parts of continents and form the foundation upon which Phanerozoic sediments were deposited. Cratons typically consist of two parts: a shield and a platform (❖ Fig. 11.1).

Shields are the exposed portion of the crystalline basement rocks of a continent and are composed of Precambrian metamorphic and igneous rocks. The metamorphic and igneous rocks reveal a history of extensive orogenic activity during the Precambrian (see Chapters 9 and 10). During the Phanerozoic, however, shields were extremely stable and formed the foundation of the continents.

Extending outward from the shields are buried Precambrian rocks that constitute a platform, another part of the craton. Overlying the platform are flat-lying or

Phanerozoic Eon										
Mesozoic Era			Cenozoic Era							
Triassic	Jurassic	Cretaceous	Tertiary						Quaternary	
			Paleocene	Eocene	Oligocene	Miocene	Pliocene		Pleistocene	Holocene

45
Y.A.

66
M.Y.A.

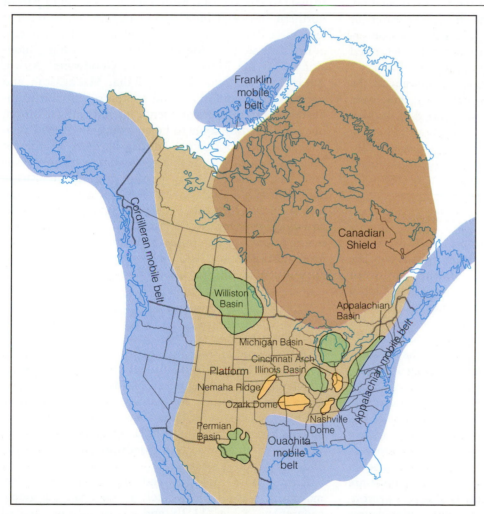

◆ **FIGURE 11.1** The mobile belts and major cratonic structures of North America that formed during the Paleozoic Era.

gently dipping Phanerozoic sedimentary rocks. Phanerozoic rocks deposited on the platform include detrital and chemical sedimentary rocks that were deposited in widespread shallow seas that transgressed and regressed over the craton. These seas, called **epeiric seas,** were a common feature of most Paleozoic cratonic histories. Changes in sea-level caused primarily by continental glaciation as well as by plate movement were responsible for the advance and retreat of the seas.

While most of the Paleozoic platform rocks are still essentially flat-lying, in some places they were gently folded into regional arches, domes, and basins (Fig. 11.1). In many cases some of these structures stood out as low islands during the Paleozoic Era and supplied sediments to the surrounding epeiric seas.

Mobile belts are elongated areas of mountain-building activity. They occur along the margins of continents where sediments are deposited in the relatively shallow waters of the continental shelf and the deeper waters at the base of the continental slope. During plate convergence along these margins, the sediments are deformed and intruded by magma, creating mountain ranges.

Four mobile belts formed around the margin of the North American craton during the Paleozoic; these were the **Franklin, Cordilleran, Ouachita,** and **Appalachian mobile belts** (Fig. 11.1). Each was the site of mountain building in response to compressional forces along a convergent plate boundary and formed such mountain ranges as the Appalachians and Ouachitas.

❖ PALEOZOIC PALEOGEOGRAPHY

One of the major lessons plate tectonic theory teaches us is that the Earth's geography is constantly changing. The present-day configuration of the continents and ocean basins is merely a snapshot in time. As the plates move about the Earth, the location of continents and ocean basins is constantly changing and being modified. One of the goals of historical geology is to provide paleogeographic reconstructions of the world for the geologic past. By synthesizing all of the pertinent paleoclimatic, paleomagnetic, paleontologic, sedimentologic, stratigraphic, and tectonic data available, geologists can prepare paleogeographic maps of what the world looked like at a particular time in the geologic past (see Perspective 11.1). For example, the paleogeographic history of the Mesozoic and Cenozoic eras is generally well known because magnetic anomaly patterns in oceanic crust are well preserved. These patterns allow geologists to reconstruct the positions of the continents with a high degree of certainty for this time interval.

The paleogeographic history of the Paleozoic Era is not as precisely known, however, in part because the magnetic anomaly patterns preserved in the oceanic crust were destroyed when much of the Paleozoic oceanic crust

was subducted during the formation of Pangaea. Paleozoic paleogeographic reconstructions are therefore based primarily on structural relationships, climate-sensitive sediments such as red beds, evaporites, and coals, as well as the distribution of plants and animals.

Recall that by the beginning of the Paleozoic, six major continents were present. In addition to these large landmasses, geologists have also identified numerous small microcontinents and island arcs associated with various microplates that were present during the Paleozoic. We will be primarily concerned, however, with the history of the six major continents and their relationship to each other. The six major Paleozoic continents are **Baltica** (Russia west of the Ural Mountains and the major part of northern Europe), **China** (a complex area consisting of at least three Paleozoic continents that were not widely separated and are here considered to include China, Indochina, and the Malay Peninsula), **Gondwana** (Africa, Antarctica, Australia, Florida, India, Madagascar, and parts of the Middle East and southern Europe), **Kazakhstania** (a triangular continent centered on Kazakhstan, but considered by some to be an extension of the Paleozoic Siberian continent), **Laurentia** (most of present North America, Greenland, northwestern Ireland, Scotland, and part of eastern Russia), and **Siberia** (Russia east of the Ural Mountains and Asia north of Kazakhstan and south of Mongolia). The paleogeographic reconstructions that follow (◆ Fig. 11.2) are based on the methods used to determine and interpret the location, geographic features, and environmental conditions on the paleocontinents (see Chapter 5 and Perspective 11.1).

Early Paleozoic Global History

In contrast to today's global geography, the Cambrian world consisted of six major continents dispersed around the globe at low tropical latitudes (◆ Fig. 11.2a). Water circulated freely among ocean basins, and the polar regions were apparently ice-free. By the Late Cambrian, epeiric seas had covered large areas of Laurentia, Baltica, Siberia, Kazakhstania, and China, while major highlands were present in northeastern Gondwana, eastern Siberia, and central Kazakhstania.

During the Ordovician and Silurian periods, plate movement played a major role in the changing global geography (◆ Fig. 11.2b and c). Gondwana moved southward during the Ordovician and began to cross the South Pole as indicated by Upper Ordovician tillites found today in the Sahara Desert. In contrast to the passive continental margin Laurentia exhibited during the Cambrian, an active convergent plate boundary formed

PALEOGEOGRAPHIC RECONSTRUCTIONS AND MAPS

The key to any reconstruction of world paleogeography is the correct positioning of the continents in terms of latitude and longitude as well as orientation of the paleocontinent relative to the paleonorth pole. The main criteria used for paleogeographic reconstructions are paleomagnetism, biogeography, tectonic patterns, and climatology.

Paleomagnetism provides the only source of quantitative data on the orientations of the continents. For the Paleozoic Era, however, the paleomagnetic data are often inconsistent and contradictory due to secondary magnetizations acquired through the effects of metamorphism or weathering.

The distribution of faunas and floras provides a useful check on the latitudes determined by paleomagnetism and can provide additional limits on longitudinal separation of continents. As is well known, the distribution of plants and animals is controlled by both climatic and geographic barriers. Such information can be used to position continents and ocean basins in a way that accounts for the biogeographic patterns indicated by fossil evidence.

Tectonic activity is indicated by deformed sediments associated with andesitic volcanics and ophiolites. Such features allow geologists to recognize ancient mountain ranges and zones of subduction. These mountain ranges may subsequently have been separated by plate movement, so the identification of large, continuous mountain ranges provides important information about continental positions in the geologic past.

Climate-sensitive sedimentary rocks are used to interpret past climatic conditions. Desert dunes are typically well sorted and cross-bedded on a large scale, and associated with other deposits, they indicate an arid environment. Coals form in freshwater swamps where climatic conditions promote abundant plant growth. Evaporites result when evaporation exceeds precipitation, such as in desert regions or along hot, dry shorelines. Tillites result from glacial activity and indicate cold, wet environments.

Paleogeographic features can be determined by associations of sedimentary rocks and sedimentary structures. For example, large-scale cross-beds may indicate aeolian or windblown conditions such as in deserts. Delta complexes and deep-sea fans have characteristic internal features and three-dimensional forms that can be recognized in the rock record, just as coal and associated deposits usually follow a particular sequence. These features can be used to interpret such geographic features as lakes, streams, swamps, and shallow and deep marine areas.

Former mountain ranges can be recognized by folded and faulted sedimentary rocks associated with metamorphic and igneous rocks. We have already mentioned the association of andesites and ophiolites as evidence of former mountain building.

By combining all relevant geologic, paleontologic, and climatologic information, geologists can construct paleogeographic maps (see Figs. 5.32 and 11.2). Such maps are simply interpretations of the geography of an area for a particular time in the geologic past. The majority of paleogeographic maps show the distribution of land and sea, probable climatic regimes, and such geographic features as mountain ranges, swamps, and glaciers.

---

along its eastern margin during the Ordovician as indicated by the Late Ordovician Taconic orogeny that occurred in New England. During the Silurian, Baltica moved northwestward relative to Laurentia and collided with it during the Late Silurian as indicated by the Caledonian orogeny. Following this orogeny, the southern part of the Iapetus Ocean still remained open between Laurentia and Gondwana (Fig. 11.2c). Siberia and Kazakhstania moved from a southern equatorial position during the Cambrian to north temperate latitudes by the end of the Silurian Period.

With this plate tectonic overview in mind, we will now focus our attention on North America (Laurentia) and the role it played in the Early Paleozoic geologic history of the world.

❖ EARLY PALEOZOIC EVOLUTION OF NORTH AMERICA

Following the tectonism that occurred in North America during the Late Proterozoic (see Chapter 10), the first 100 million years of the Paleozoic Era was a time of

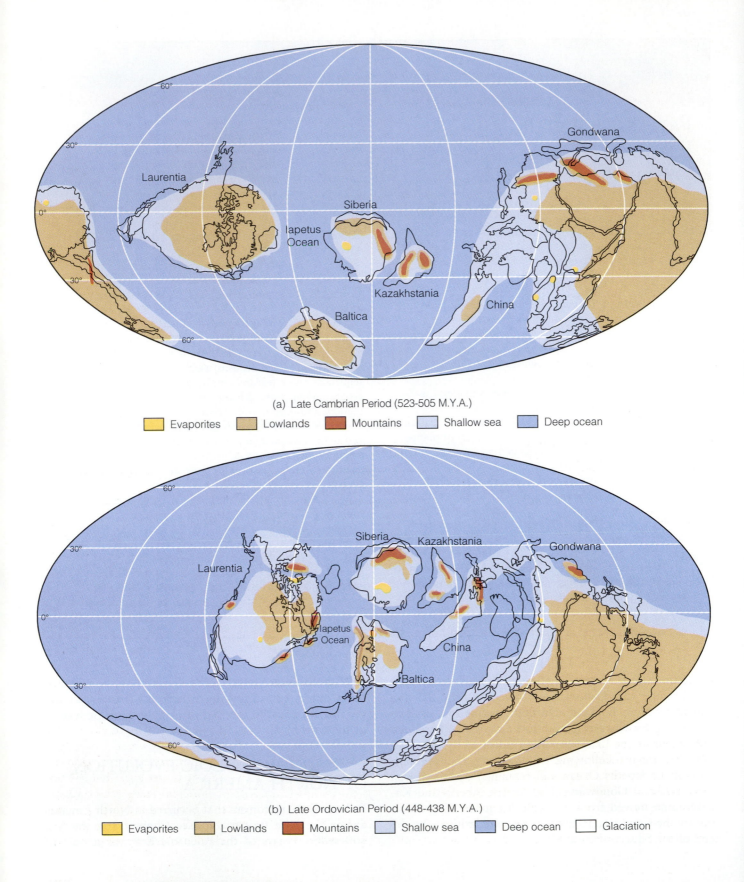

(a) Late Cambrian Period (523-505 M.Y.A.)

☐ Evaporites   ☐ Lowlands   ☐ Mountains   ☐ Shallow sea   ☐ Deep ocean

(b) Late Ordovician Period (448-438 M.Y.A.)

☐ Evaporites   ☐ Lowlands   ☐ Mountains   ☐ Shallow sea   ☐ Deep ocean   ☐ Glaciation

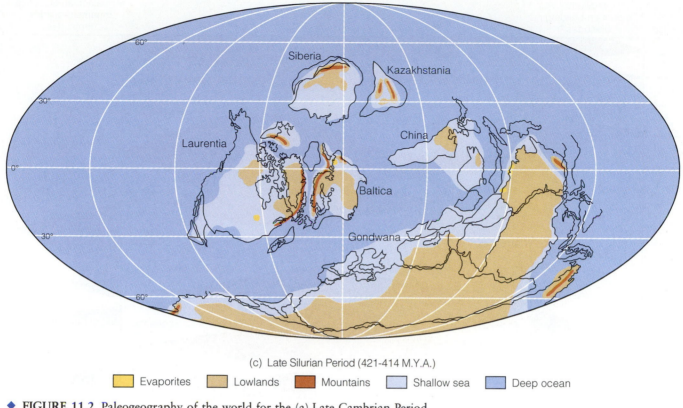

(c) Late Silurian Period (421-414 M.Y.A.)

☐ Evaporites   ☐ Lowlands   ☐ Mountains   ☐ Shallow sea   ☐ Deep ocean

◆ **FIGURE 11.2** Paleogeography of the world for the (a) Late Cambrian Period,
(b) Late Ordovician Period, and (c) Late Silurian Period.

relative stability for the North American craton and its
continental margins. During this time the continental
margins were generally passive, and the craton experienced a major transgression, regression, and renewed
transgression. During the next 70 to 80 million years,
epeiric seas continued to advance and retreat over the
craton. The eastern margin of North America, however,
changed from a passive continental margin to an active
convergent plate boundary as Laurentia and Baltica
moved toward each other (Fig. 11.2b and c). It is therefore convenient to divide the history of the North American craton into two parts, one part dealing with the
relatively stable continental interior over which epeiric
seas transgressed and regressed, and the other with the
mobile belts where mountain building occurred.

In 1963 the American geologist Laurence L. Sloss proposed that the sedimentary-rock record of North America could be subdivided into six cratonic sequences. A
cratonic sequence is a large-scale (greater than supergroup) lithostratigraphic unit representing a major
transgressive-regressive cycle bounded by cratonwide un-

conformities (◆ Fig. 11.3). The transgressive phase,
which is usually covered by younger sediments, commonly is well preserved, while the regressive phase of
each sequence is marked by an unconformity. Where
rocks of the appropriate age are preserved, each of the
six unconformities can be shown to extend across the
various sedimentary basins of the North American craton and into the mobile belts along the cratonic margin.

Geologists have also recognized major unconformity
bounded sequences in cratonic areas outside North
America. Such global transgressive and regressive cycles
are caused by sea level changes and are believed to result
from major tectonic and glacial events.

The realization that rock units can be divided into
cratonic sequences and that these sequences can be further subdivided and correlated provides the foundation
for an important new concept in geology that allows
high-resolution analysis of time and facies relationships
within sedimentary rocks. **Sequence stratigraphy** is the
study of rock relationships within a time-stratigraphic
framework of related facies bounded by erosional or

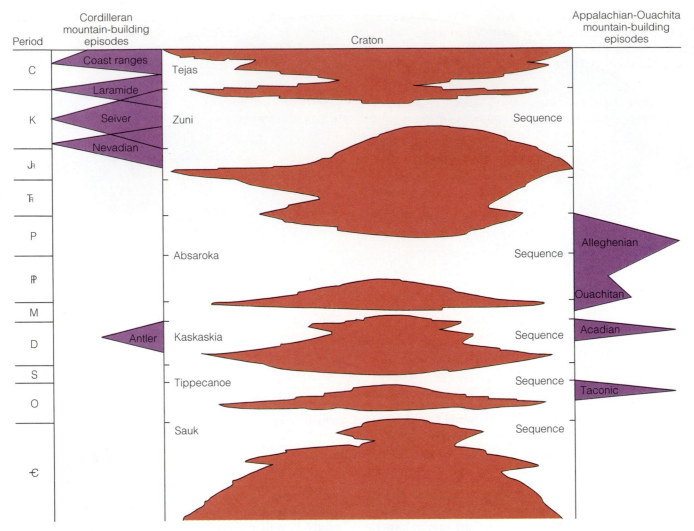

FIGURE 11.3 Cratonic sequences of North America. The white areas represent sequences of rocks that are separated by large scale unconformities shown as brown areas. The major Cordilleran orogenies are shown on the left side of the figure, and the major Appalachian orogenies are shown on the right side.

nondepositional surfaces. The basic unit of sequence stratigraphy is the *sequence,* which is a succession of rocks bounded by unconformities and their equivalent conformable strata. Sequence boundaries form as a result of the relative drop in sea level. Sequence stratigraphy is becoming an important tool in geology because it allows geologists to subdivide sedimentary rocks into related units that are bounded by time-stratigraphically significant boundaries. Geologists are using sequence stratigraphy for high-resolution correlation and mapping as well as interpreting and predicting depositional environments.

❖ THE SAUK SEQUENCE

Rocks of the **Sauk sequence** record the first major transgression onto the North American craton (Fig. 11.3). During the Late Proterozoic and Early Cambrian, deposition of marine sediments was limited to the passive shelf areas of the Appalachian and Cordilleran borders of the craton. The craton itself was above sea level and experiencing extensive weathering and erosion as streams flowed across a barren and plantless landscape. Because North America was located in a tropical climate at this time and there is no evidence of any terrestrial vegetation, weathering and erosion of the exposed Precambrian basement rocks must have proceeded at a very rapid rate. During the Middle Cambrian, the transgressive phase of the Sauk began with epeiric seas encroaching over the craton (see Perspective 11.2). By the Late Cambrian, the Sauk Sea had covered most of North America, leaving only a portion of the Canadian Shield and a few large islands above sea level (◆ Fig. 11.4).

◆ **FIGURE 11.4** Paleogeography of North America during the Cambrian Period. Note the position of the Cambrian paleoequator. During this time North America straddled the equator as indicated in Figure 11.2a.

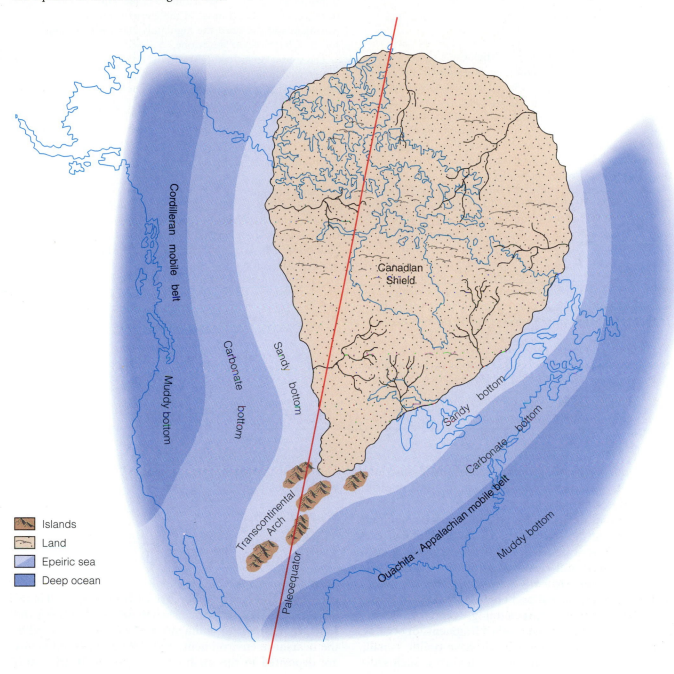

Islands
Land
Epeiric sea
Deep ocean

Canadian Shield

Cordilleran mobile belt

Muddy bottom

Carbonate bottom

Sandy bottom

Sandy bottom

Carbonate bottom

Muddy bottom

Transcontinental Arch

Paleoequator

Ouachita - Appalachian mobile belt

PICTURED ROCKS NATIONAL LAKESHORE

Exposed along the south shore of Lake Superior between Au Sable Point and Munising in Michigan's Upper Peninsula is the beautiful and imposing wavecut sandstone called Pictured Rocks cliffs (◆ Fig. 1). The rocks exposed in this area, part of which is designated a national lakeshore, comprise the Upper Cambrian Munising Formation, which is divided into two members: the lower Chapel Rock Sandstone and the upper Miner's Castle Sandstone (Fig. 1). The Munising Formation unconformably overlies the Upper Proterozoic Jacobsville Sandstone and is unconformably overlain by the Middle Ordovician Au Train Formation. The reddish brown, coarse-grained Jacobsville Sandstone was deposited in streams and lakes over an irregular erosion surface (Fig. 1). Following deposition, the Jacobsville was slightly uplifted and tilted.

As we have discussed previously, the Sauk Sea covered most of North America by the Late Cambrian Period, with only a part of the Canadian Shield and a few large islands remaining above sea level (Fig. 11.4). By the Late Cambrian, the transgressing Sauk Sea reached the Michigan area. During the first phase of this transgression, the Chapel Rock Sandstone was deposited. The principal source area for this unit was the Northern Michigan highlands, an area that corresponds to the present Upper Peninsula. Following deposition of the Chapel Rock Sandstone, the Sauk Sea retreated from the area.

The Miner's Castle Sandstone represents a second transgression of the Sauk Sea in the area. This second transgression covered most of the Upper Peninsula of Michigan and drowned the highlands that were the source for the Chapel Rock Sandstone.

The source area for the Miner's Castle Sandstone was the Precambrian Canadian Shield area to the north and northeast. The Miner's Castle Sandstone contains rounder, better sorted, and more abundant quartz grains than the Chapel Rock Sandstone, indicating a different source area. A major unconformity separates the Miner's Castle Sandstone from the overlying Middle Ordovician Au Train Formation.

One of the most prominent features of Pictured Rocks National Lakeshore is Miner's Castle, a wavecut projection along the shoreline (Fig. 1). The lower sandstone unit at water level is the Chapel Rock Sandstone, while the rest of the feature is composed of the Miner's Castle Sandstone. The two turrets of the castle formed as sea stacks during a time following the Pleistocene when the water level of Lake Superior was much higher.

These islands, collectively referred to as the **Transcontinental Arch,** extended from New Mexico to Minnesota and the Lake Superior region.

The sediments deposited on both the craton and along the shelf area of the craton margin show abundant evidence of shallow-water deposition. The only difference between the shelf and craton deposits is that the shelf deposits are thicker. In both areas, the sands are generally clean and well sorted and commonly contain ripple marks and small-scale cross-bedding. Many of the carbonates are bioclastic (composed of fragments of organic remains), contain stromatolites, or have oolitic (small, spherical calcium carbonate grains) textures. Such sedimentary structures and textures are evidence of shallow-water deposition.

The Cambrian of the Grand Canyon Region: A Transgressive Facies Model

Recall from Chapter 4 that sediments become increasingly finer the farther away from land one goes. Therefore, in a stable environment where sea level remains the same, coarse detrital sediments are typically deposited in the nearshore environment, and finer-grained sediments are deposited in the offshore environment. Carbonates

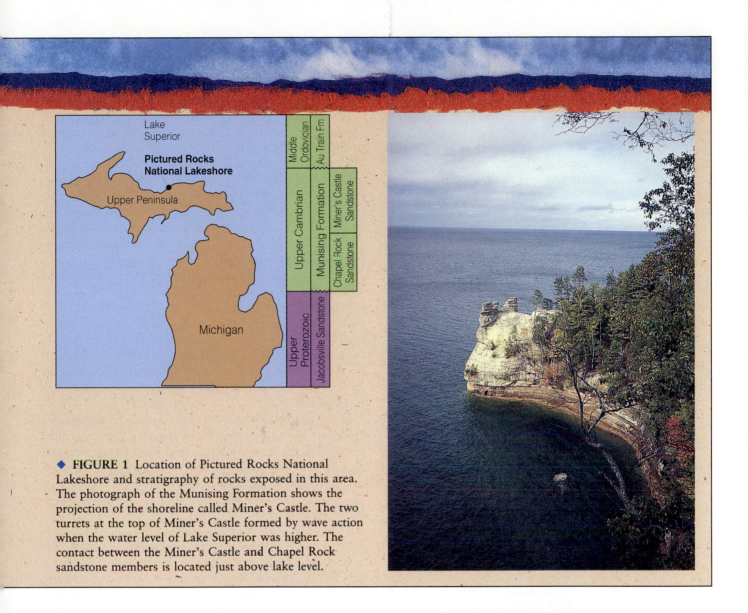

◆ **FIGURE 1** Location of Pictured Rocks National Lakeshore and stratigraphy of rocks exposed in this area. The photograph of the Munising Formation shows the projection of the shoreline called Miner's Castle. The two turrets at the top of Miner's Castle formed by wave action when the water level of Lake Superior was higher. The contact between the Miner's Castle and Chapel Rock sandstone members is located just above lake level.

form farthest from land in the area beyond the reach of detrital sediments. During a transgression, these facies (sediments that represent a particular environment) migrate in a landward direction (see Fig. 4.12).

The Cambrian rocks of the Grand Canyon region (see the Prologue) provide an excellent model for illustrating the sedimentation patterns of a transgressing sea. The Grand Canyon region occupied the passive shelf and western margin of the craton during Sauk time. During the Late Proterozoic and Early Cambrian, most of the craton was above sea level as the Sauk Sea was still largely restricted to the margins of the craton(continental shelves and slopes). In the Grand Canyon region, the Tapeats Sandstone represents the basal transgressive shoreline deposits that accumulated as marine waters transgressed across the shelf and just onto the western margin of the craton during the Early Cambrian (◆ Fig. 11.5). These sediments are clean, well-sorted sands of the type one would find on a beach today. As the transgression continued into the Middle Cambrian, muds of the Bright Angel Shale were deposited over the Tapeats Sandstone. By the Late Cambrian, the Sauk Sea had transgressed so far onto the craton that in the Grand Canyon region, carbonates of the Muav Limestone were being deposited over the Bright Angel Shale. This vertical succession of sandstone (Tapeats), shale (Bright Angel), and limestone (Muav) forms a typical transgressive sequence and represents a progressive migration of offshore facies toward the craton through time (Fig. 11.5).

Cambrian rocks of the Grand Canyon region also illustrate how many formations are time-transgressive; that is, their age is not the same every place they are found. Mapping and correlations based on faunal evidence indicate that deposition of the Mauv Limestone had already started on the shelf before deposition of the Tapeats Sandstone was completed on the craton. Faunal analysis of the Bright Angel Shale indicates that it is Early Cambrian in age in California and Middle Cambrian in age in the Grand Canyon region, thus illustrating the time-transgressive nature of formations and facies.

This same facies relationship also occurred elsewhere on the craton as the seas encroached from the Appalachian and Ouachita mobile belts onto the craton interior (◆ Fig. 11.6). Carbonate deposition dominated on the craton as the Sauk transgression continued during the Early Ordovician, and the islands of the Transcontinental Arch were soon covered by the advancing Sauk Sea. By the end of Sauk time, the majority of the craton was submerged beneath a warm, equatorial epeiric sea (Fig. 11.2a).

❖ THE TIPPECANOE SEQUENCE

As the Sauk Sea regressed from the craton during the Early Ordovician, it revealed a landscape of low relief. The rocks exposed were predominantly limestones and dolostones that experienced deep (in some places up to 50 m) and extensive erosion because North America was still located in a tropical environment (◆ Fig. 11.7). The resulting cratonwide unconformity marks the boundary between the Sauk and Tippecanoe sequences.

Like the Sauk sequence, deposition of the **Tippecanoe sequence** began with a major transgression onto the craton. This transgressing sea deposited clean quartz sands over most of the craton. The best known of the Tippecanoe basal sandstones is the St. Peter Sandstone, an almost pure quartz sandstone used in manufacturing glass. It occurs throughout much of the midcontinent and resulted from numerous cycles of weathering and erosion of Proterozoic and Cambrian sandstones deposited during the Sauk transgression (◆ Fig. 11.8).

The Tippecanoe basal sandstones were followed by widespread carbonate deposition (Fig. 11.7). The limestones were generally the result of deposition by calcium carbonate–secreting organisms such as corals, brachiopods, stromatoporoids, and bryozoans. In addition to the limestones, there were also many dolostones. Most of the dolostones formed as magnesium was substituted for some of the calcium in calcite and, in the process, converted the limestones into dolostones.

In the eastern portion of the craton, the carbonates grade laterally into shales. These shales mark the farthest extent of detrital sediments derived from weathering and erosion of the highlands formed during the Taconic orogeny, a tectonic event we will discuss later.

Tippecanoe Reefs and Evaporites

Widespread reefs and evaporite deposits are major features of the Tippecanoe sequence. Remember that during the Ordovician and Silurian periods, Laurentia was moving northward, yet was still located within tropical latitudes (Fig. 11.2b and c). Most of the North American craton during that time was covered by a warm epeiric sea. In the present-day Great Lakes region, basins surrounded by large reefs developed. The evaporation of seawater in these basins resulted in the precipitation of thick evaporite mineral deposits such as gypsum and halite. Before discussing Silurian reefs and evaporites, we will briefly examine present-day reefs to gain a better understanding of reef environments in general.

General Characteristics of Modern Organic Reefs

Organic reefs are limestone structures constructed by living organisms, some of which contribute skeletal mate-

◆ **FIGURE 11.5** Cross section of Cambrian strata exposed in the Grand Canyon region illustrating the transgressive nature of the three formations.

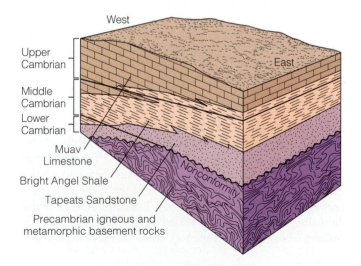

West

East

Upper Cambrian

Middle Cambrian

Lower Cambrian

Muav Limestone

Bright Angel Shale

Tapeats Sandstone

Nonconformity

Precambrian igneous and metamorphic basement rocks

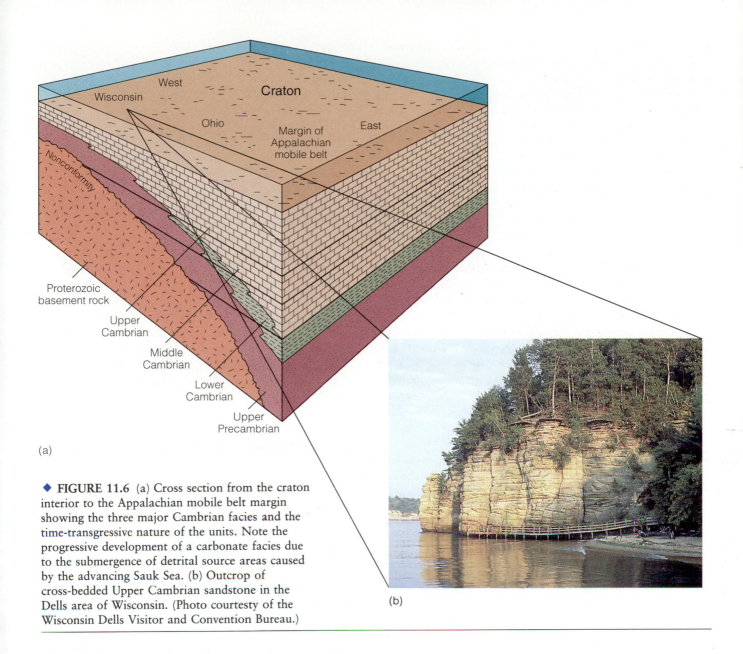

(a)

◆ **FIGURE 11.6** (a) Cross section from the craton interior to the Appalachian mobile belt margin showing the three major Cambrian facies and the time-transgressive nature of the units. Note the progressive development of a carbonate facies due to the submergence of detrital source areas caused by the advancing Sauk Sea. (b) Outcrop of cross-bedded Upper Cambrian sandstone in the Dells area of Wisconsin. (Photo courtesy of the Wisconsin Dells Visitor and Convention Bureau.)

(b)

rials to the reef framework (◆ Fig. 11.9, p. 306). Today corals and calcareous algae are the most prominent reef-builders, but in the geologic past, other organisms played a major role. Regardless of the organisms dominating reef communities, reefs appear to have occupied the same ecological niche in the geologic past that they do today. Because of the ecological requirements of reef-building organisms, reefs today are confined to a narrow latitudinal belt between 30 degrees north and south of the equator. Corals, the major reef-building organisms today, require warm, clear, shallow water of normal salinity for optimal growth.

The size and shape of a reef are largely the result of the interaction between the reef-building organisms, the bottom topography, wind and wave action, and subsidence of the seafloor. Reefs also alter the area around them by forming barriers to water circulation or wave action.

Reefs are commonly long, linear masses that form a barrier between a shallow platform on one side and a comparatively deep marine basin on the other side. Such reefs are known as *barrier reefs* (Fig. 11.9). Reefs create and maintain a steep seaward front that absorbs incoming wave energy. As skeletal material breaks off from the reef front, it accumulates along a fore-reef slope. The

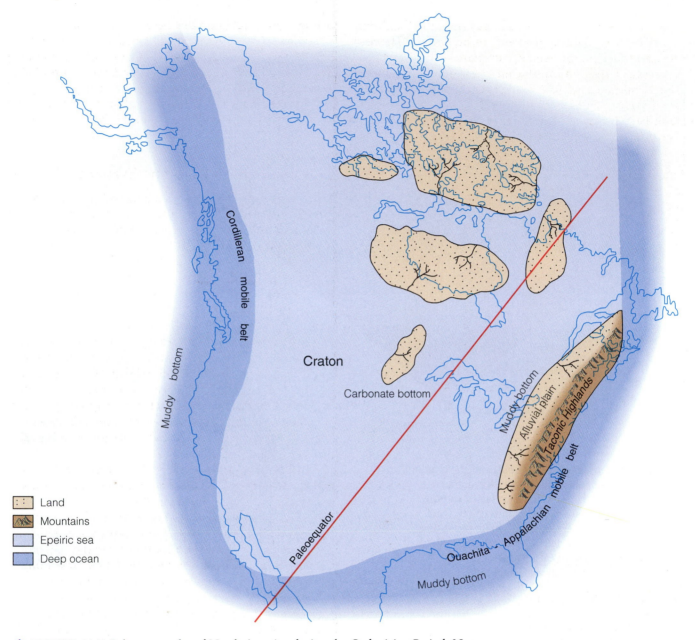

◆ FIGURE 11.7 Paleogeography of North America during the Ordovician Period. Note that the position of the equator has changed, indicating North America was rotating in a counterclockwise direction.

reef barrier itself is porous and composed of reef-building organisms. The lagoon area is a low-energy, quiet-water zone where fragile, sediment-trapping organisms thrive. The lagoon area can also become the site of evaporitic deposits when circulation to the open sea is cut off.

Modern examples of barrier reefs are the Florida Keys, Bahama Islands, and Great Barrier Reef of Australia. Other types of reefs include circular fringing reefs that develop around islands, Pacific atolls that are built on submerged volcanic peaks, and small patch reefs, several meters in diameter, that form in a variety of settings.

Reefs have been common features since the Cambrian and have been built by a variety of organisms. The first skeletal builders of reeflike structures were archaeocyathids. These conical-shaped organisms lived during the Cambrian and had double, perforated, calcareous shell walls. Archaeocyathids built small mounds that have been found on all continents except South America (see Fig. 13.9). Beginning in the Middle Ordovician, stromatoporoid-coral reefs became common in the low latitudes, and similar reefs remained so throughout the rest of the Phanerozoic Eon. The burst of reef building seen in the Late Ordovician through Devonian probably occurred in response to evolutionary changes triggered by the appearance of extensive carbonate seafloors and platforms beyond the influence of detrital sediments.

◆ **FIGURE 11.8** (a) The transgression of the Tippecanoe Sea resulted in the deposition of the St. Peter Sandstone (Middle Ordovician) over a large area of the craton. (b) Outcrop of St. Peter Sandstone in Governor Dodge State Park, Wisconsin. (Photo courtesy of the Wisconsin Department of Natural Resources.)

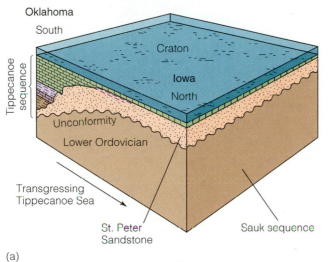

(a)

(b)

Silurian Organic Reefs and Evaporite Facies

The Middle Silurian rocks (Tippecanoe sequence) of the present-day Great Lakes region are world famous for their reef and evaporite deposits and have been extensively studied (◆ Fig. 11.10). The most famous structure in the region is the Michigan Basin. It is a broad, circular basin surrounded by large barrier reefs. No doubt these reefs contributed to increasingly restricted circulation and the precipitation of Upper Silurian evaporites within the basin (◆ Fig. 11.11, p. 308).

Within the rapidly subsiding interior of the basin, other types of reefs are found. *Pinnacle reefs* are tall, spindly structures up to 100 m high. They reflect the rapid upward growth needed to maintain themselves near sea level during subsidence of the basin (Fig. 11.11). In addition to the pinnacle reefs, bedded carbonates and thick sequences of salt and anhydrite are also found in the Michigan Basin.

As the Tippecanoe Sea gradually regressed from the craton during the Late Silurian, precipitation of evaporite minerals occurred in the Appalachian, Ohio, and Michigan basins. In the Michigan Basin alone, approximately 1,500 m of sediments were deposited, nearly half of which are halite and anhydrite. How did such thick sequences of evaporites accumulate? One possibility is that a drop in sea level occurred so that the tops of the barrier reefs were as high as or above seal level, thus preventing the influx of new seawater into the basin. Evaporation of the basinal seawater would result in the precipitation of salts. A second possibility is that the reefs grew upward so close to sea level that they formed a sill or barrier that eliminated interior circulation (◆ Fig. 11.12, p. 309).

Because North America was still near the equator during the Silurian Period (Fig. 11.2c), temperatures were probably high. As circulation to the Michigan Basin was restricted, seawater within the basin evaporated, forming a brine. Because the brine was heavy, it concentrated near the bottom, and minerals were precipitated onto the basin floor. Some seawater flowed in over the sill and through channels cut in the barrier reefs, but this replenishment only added new seawater that later became concentrated as brine. In this way, the brine in the basin became increasingly concentrated until the salts could no longer be held in solution and therefore precipitated to form evaporite minerals.

The order and type of salts that precipitate from seawater depend on their solubility, the original concentration of seawater, and local conditions of the basin. In general, salts precipitate in order beginning with the least soluble and ending with the most soluble. There-

fore, calcium carbonate usually precipitates out first, followed by gypsum,* and lastly halite. Many lateral shifts and interfingering of the limestone, anhydrite, and halite facies may occur, however, due to variations in the amount of seawater entering the basin and changing geologic conditions.

---

*Recall from Chapter 5 that gypsum ($CaSO_4 \cdot 2H_2O$) is the common sulfate precipitated from seawater, but when deeply buried, gypsum loses its water and is converted to anhydrite ($CaSO_4$).

◆ FIGURE 11.9 (a) Modern reef community showing the various reef-building organisms. (b) Diagrammatic block diagram of a modern reef showing the various environments within the reef complex. (Photo courtesy of L. J. Lipke, Amoco Production Company.)

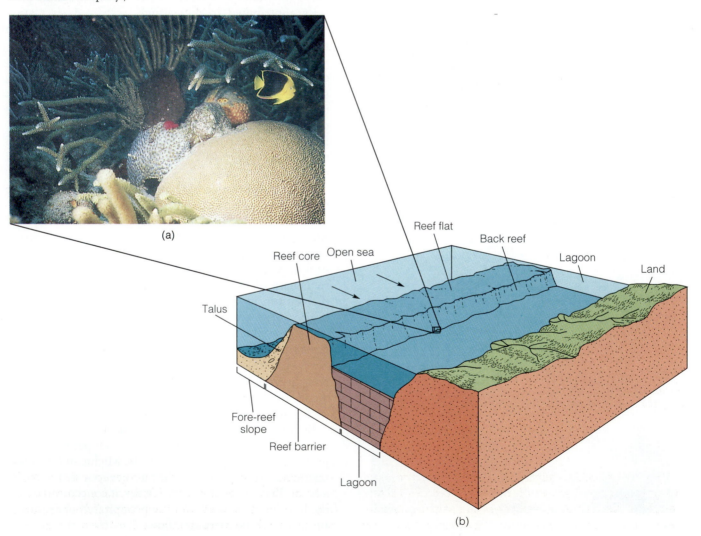

(a)

(b)

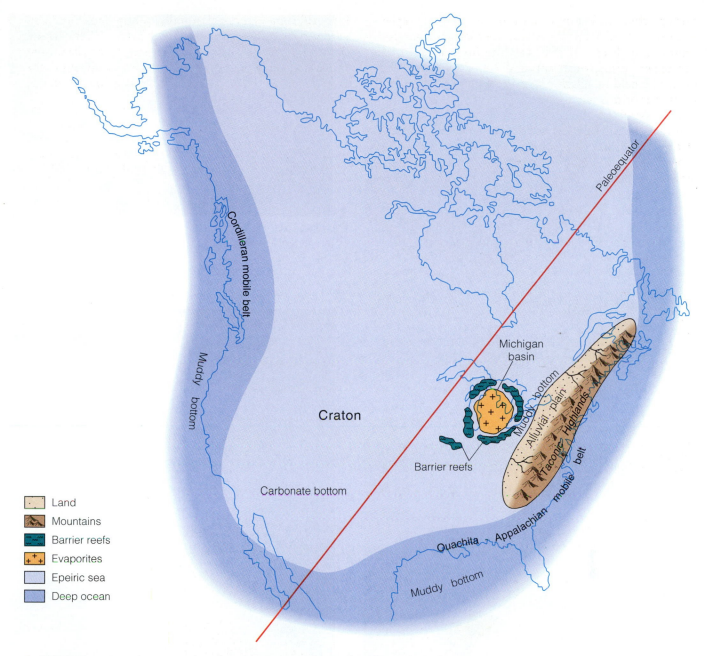

Legend:
- Land
- Mountains
- Barrier reefs
- Evaporites
- Epeiric sea
- Deep ocean

Cordilleran mobile belt

Muddy bottom

Craton

Carbonate bottom

Michigan basin

Barrier reefs

Muddy bottom

Alluvial plain

Taconic Highlands

Appalachian mobile belt

Ouachita

Muddy bottom

Paleoequator

◆ **FIGURE 11.10** Paleogeography of North America during the Silurian Period. Note the development of reefs in the Michigan, Ohio, and Indiana-Illinois-Kentucky areas.

Thus, the periodic evaporation of seawater proposed by this model could account for the observed vertical and lateral distribution of evaporites in the Michigan Basin. Associated with those evaporites, however, are pinnacle reefs, and the organisms that constructed those reefs could not have lived in such a highly saline environment (Fig. 11.11). How, then, can such contradictory features be explained? Numerous models have been proposed,

Laminar
stromatoporoid
Barrier
reef
Anhydrite
Halite
Evaporite
Carbonate
Pinnacle
reef
Meters
0
100
(a)
Stromatoporoid
Barrier reef
Stromatolites
Algal
Coral algal
Crinoidal
Laminar
stromatoporoid
Clinton Fm.
Niagara Fm.
(d)
(b)
(c)

◆ **FIGURE 11.11** (a) Generalized cross section of the northern Michigan Basin during the Silurian Period. (b) Stromatoporoid barrier-reef facies. (c) Evaporite facies. (d) Carbonate facies. (Photos courtesy of Sue Monroe).

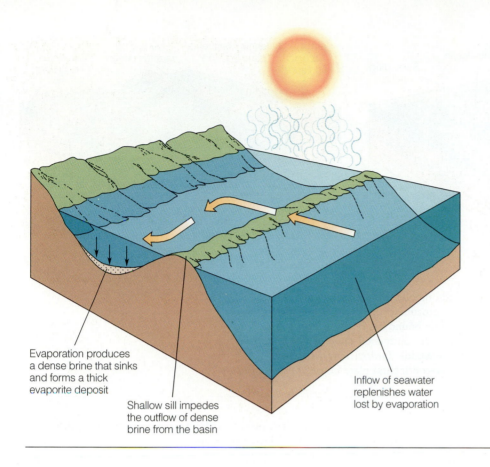

Evaporation produces a dense brine that sinks and forms a thick evaporite deposit

Shallow sill impedes the outflow of dense brine from the basin

Inflow of seawater replenishes water lost by evaporation

◆ **FIGURE 11.12** Silled basin model for evaporite sedimentation by direct precipitation from seawater. Vertical scale is greatly exaggerated.

ranging from cessation of reef growth followed by evaporite deposition, to alternation of reef growth and evaporite deposition. Although the Michigan Basin has been studied extensively for years, no model yet proposed completely explains the genesis and relationship of its various reef, carbonate, and evaporite facies.

The End of the Tippecanoe Sequence

By the Early Devonian, the regressing Tippecanoe Sea had retreated to the craton margin exposing an extensive lowland topography. During this regression, marine deposition was initially restricted to a few interconnected cratonic basins and, finally by the end of the Tippecanoe, to only the mobile belts surrounding the craton.

During the Early Devonian as the Tippecanoe Sea regressed, the craton experienced mild deformation resulting in the formation of many domes, arches, and basins. These structures were mostly eroded during the time the craton was exposed so that they were eventually covered by deposits from the encroaching Kaskaskia Sea, producing a regional unconformity that forms the boundary between the two sequences.

❖ THE APPALACHIAN MOBILE BELT AND THE TACONIC OROGENY

Having examined the Sauk and Tippecanoe geologic history of the craton, we turn our attention to the Appalachian mobile belt, where the first Phanerozoic orogeny began during the Middle Ordovician.

Throughout Sauk time, the Appalachian region was a broad, passive, continental margin. Sedimentation was closely balanced by subsidence as thick, shallow marine sands were succeeded by extensive carbonate deposits. During this time, the **Iapetus Ocean** was widening as a result of movement along a divergent plate boundary (◆ Fig. 11.13a).

Beginning with the subduction of the Iapetus plate beneath Laurentia (an oceanic-continental convergent plate boundary—see Chapter 7), the Appalachian mobile belt

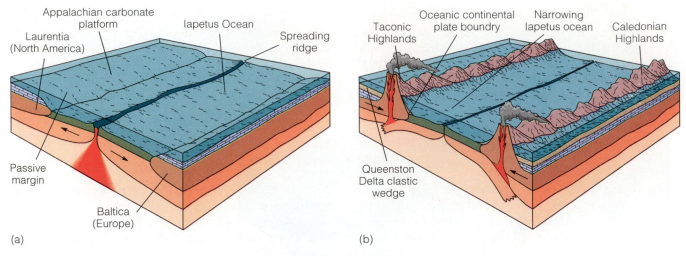

(a)

(b)

◆ **FIGURE 11.13** Evolution of the Appalachian mobile belt from the Late Proterozoic to the Late Ordovician. (a) During the Late Proterozoic to the Early Ordovician, the Iapetus Ocean was opening along a divergent plate boundary. Both the east coast of Laurentia and the west coast of Baltica were passive continental margins where large carbonate platforms existed. (b) Beginning in the Middle Ordovician, the passive margins of Laurentia and Baltica became oceanic-continental plate boundaries resulting in orogenic activity.

was born (◆ Fig. 11.13b). The resulting **Taconic orogeny,** named after the present-day Taconic Mountains of eastern New York, central Massachusetts, and Vermont, was the first of several orogenies to affect the Appalachian region.

Obviously, any record of mountain building is complex, and we cannot go into all the details or subdivisions in a book such as this. We can, however, provide an overview of the event and explain its underlying causes in a plate tectonic framework as well as describe the evidence used by geologists to reconstruct such events.

The Appalachian mobile belt can be divided into two depositional environments. The first is the extensive, shallow-water carbonate platform that formed the broad eastern continental shelf and stretched from Newfoundland to Alabama. It formed during the Sauk Sea transgression onto the craton when carbonates were deposited in a large, shallow sea. The shallow-water depth on the platform is indicated by stromatolites, desiccation cracks, and other sedimentary structures.

Carbonate deposition ceased along the east coast during the Middle Ordovician and was replaced by deepwater deposits characterized by thinly bedded black shales, graded beds, coarse sandstones, graywackes, and associated volcanics. This suite of sediments marks the onset of mountain building, in this case, the Taconic orogeny. The subduction of the Iapetus plate beneath

Laurentia resulted in volcanism and downwarping of the carbonate platform, forming an area where sediments accumulated (Fig. 11.13). Throughout the Appalachian mobile belt, facies patterns, paleocurrents, and sedimentary structures all indicate that these deposits were derived from the east, where the Taconic Highlands and associated volcanoes were rising.

Additional structural, stratigraphic, petrologic, and sedimentologic evidence has provided much information on the timing and origin of this orogeny. For example, at many locations within the Taconic belt, pronounced angular unconformities occur where steeply dipping Lower Ordovician rocks are overlain by gently dipping or horizontal Silurian and younger rocks.

Other evidence includes volcanic activity in the form of deep-sea lava flows, volcanic ash layers, and intrusive bodies in the area from present-day Georgia to Newfoundland. These igneous rocks show a clustering of radiometric ages between 440 to 480 million years ago. In addition, regional metamorphism coincides with the radiometric dates.

The final piece of evidence for the Taconic orogeny is the development of a large **clastic wedge,** an extensive accumulation of mostly detrital sediments that are deposited adjacent to an uplifted area. These deposits are thickest and coarsest nearest the highland area and become thinner and finer grained away from the source area, eventually grading into the carbonate cratonic facies (◆ Fig. 11.14). The clastic wedge resulting from the erosion of the Taconic Highlands is referred to as the **Queenston Delta.** Careful mapping and correlation of these deposits indicate that more than 600,000 km³ of rock were eroded from the Taconic Highlands. Based on this figure, geologists estimate the Taconic Highlands were at least 4,000 m high.

The Taconic orogeny marked the first pulse of mountain building in the Appalachian mobile belt and was a response to the subduction taking place beneath the east coast of Laurentia. As the Iapetus Ocean narrowed and closed, another orogeny occurred in Europe during the Silurian. The *Caledonian orogeny* was essentially a mirror image of the Taconic orogeny and the Acadian orogeny (see Chapter 12) and was part of the global mountain-building episode that occurred during the Paleozoic Era. Even though the Caledonian orogeny occurred during Tippecanoe time, we will discuss it in the next chapter because it was intimately related to the Acadian orogeny.

❖ EARLY PALEOZOIC MINERAL RESOURCES

Early Paleozoic age rocks contain a variety of resources including sand and gravel for construction, building stone, and limestone used in the manufacture of cement. Important sources of industrial or silica sand are the Upper Cambrian Jordan Sandstone of Minnesota and Wisconsin, the Lower Silurian Tuscarora Sandstone in Pennsylvania and Virginia, and the Middle Ordovician St. Peter Sandstone. The latter, the basal sandstone of the Tippecanoe sequence (Fig. 11.8), occurs in several states, but the best-known area of production is in La Salle County, Illinois. Silica sand has a variety of uses including the manufacture of glass, refractory bricks for blast furnaces, and molds for casting iron, aluminum, and copper alloys. Some silica sands, called hydraulic fracturing sands, are pumped into wells to fracture oil- or gas-bearing rocks and provide permeable passageways for the oil or gas to migrate to the well.

Thick deposits of Silurian evaporites, mostly rock salt (NaCl) and rock gypsum (CaSO$_4$ · H$_2$O) altered to rock anhydrite (CaSO$_4$), underlie parts of Michigan, Ohio, New York, and adjacent areas in Ontario, Canada. These rocks are important sources of various salts. In addition, barrier and pinnacle reefs in carbonate rocks associated with these evaporites are the reservoirs for oil and gas in Michigan and Ohio. Ordovician rocks in these states as well as Oklahoma and Texas have also yielded hydrocarbons.

The host rocks for deposits of lead and zinc in southeast Missouri are Cambrian dolostones, although some Ordovician rocks contain these metals as well. These deposits have been mined since 1720 but have been largely depleted. Now most lead and zinc mined in Missouri come from Mississippian-aged sedimentary rocks.

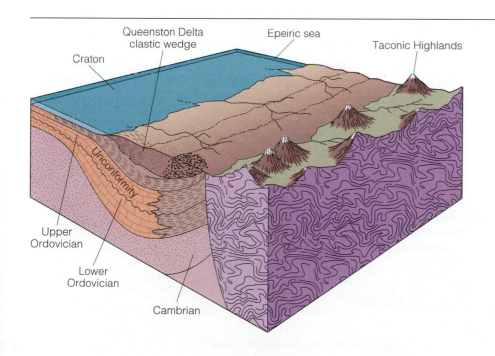

◆ **FIGURE 11.14.** Reconstruction of the Taconic Highlands and Queenston Delta clastic wedge. The clastic wedge consists of thick coarse-grained detrital sediments nearest the highlands and thins laterally into finer-grained sediments on the craton.

The Silurian Clinton Formation crops out from Alabama north to New York, and equivalent rocks are found in Newfoundland. This formation has been mined for iron in many places. In the United States, the richest ores and most extensive mining occurred near Birmingham, Alabama, but only a small amount of ore is currently produced in that area. One of the common ores in the Clinton Formation consists of hematite oolites in a matrix of hematite and calcite (◆ Fig. 11.15).

◆ **FIGURE 11.15** Oolitic hematite. (Photo courtesy of Sue Monroe.)

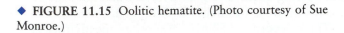

Chapter Summary

Table 11.1 provides a summary of the geologic history of the North American craton and mobile belts as well as global events and sea level changes for the Early Paleozoic.

1. Six major continents existed at the beginning of the Paleozoic Era; four of them were located near the paleoequator.
2. While Laurentia was moving northward, Gondwana moved to a south polar location, as indicated by tillite deposits.
3. Most continents consist of two major components: a relatively stable craton over which epeiric seas transgressed and regressed, surrounded by mobile belts in which mountain building took place.
4. The geologic history of North America can be divided into cratonic sequences that reflect cratonwide transgressions and regressions. The Sauk and Tippecanoe sequences were deposited during the latest Proterozoic to Early Devonian.
5. The Sauk Sea was the first major transgression onto the craton. At its peak it covered the craton except for a series of large, northeast-southwest trending islands called the Transcontinental Arch.
6. The Tippecanoe sequence began with deposition of an extensive sandstone over the exposed and eroded Sauk landscape.
7. During Tippecanoe time, extensive carbonate deposition occurred. In addition, large barrier reefs enclosed basins, resulting in evaporite deposition within these basins.
8. The eastern edge of North America was a stable carbonate platform during Sauk time. During Tippecanoe time an oceanic-continental convergent plate boundary formed, resulting in the Taconic orogeny, the first of several orogenies to affect the Appalachian mobile belt.
9. The newly formed Taconic Highlands shed sediments into the western epeiric sea, producing a clastic wedge called the Queenston Delta.
10. Early Paleozoic age rocks contain a variety of resources including building stone, limestone for cement, hydrocarbons, evaporites, and iron ore.

Important Terms

Appalachian mobile belt
Baltica
China
clastic wedge
Cordilleran mobile belt
craton
cratonic sequence
epeiric sea

Franklin mobile belt
Gondwana
Iapetus Ocean
Kazakhstania
Laurentia
mobile belt
organic reef
Ouachita mobile belt

Queenston Delta
Sauk sequence
sequence stratigraphy
Siberia
Taconic orogeny
Tippecanoe sequence
Transcontinental Arch

Review Questions

1. An elongated area marking the site of former mountain building is a:
   a. _____ craton; b. _____ platform; c. _____ shield;
   d. _____ epeiric sea; e. _____ mobile belt.

2. Which of the following was not a Paleozoic continent?
   a. _____ Gondwana; b. _____ Baltica; c. _____ Kazakhstania; d. _____ Eurasia; e. _____ Laurentia.

3. A major transgressive-regressive cycle bounded by cratonwide unconformities is (a)n:
   a. _____ cratonic sequence; b. _____ epeiric sea;
   c. _____ orogeny; d. _____ biostratigraphic unit;
   e. _____ cyclothem.

4. During the Sauk sequence, the only area above sea level besides a portion of the Canadian Shield was the:
   a. _____ Appalachian mobile belt; b. _____ Transcontinental Arch; c. _____ Taconic Highlands;
   d. _____ Queenston Delta; e. _____ cratonic margin.

5. Which famous formation forms the basal sandstones of the Tippecanoe sequence?
   a. _____ Tapeats; b. _____ Oriskany; c. _____ St. Peter; d. _____ Clinton; e. _____ Jacobsville.

6. The predominant cratonic lithologies of the Tippecanoe sequence are:
   a. _____ coals; b. _____ sandstones and shales;
   c. _____ graywackes and cherts; d. _____ volcanics;
   e. _____ evaporites and reef carbonates.

7. During the Early Paleozoic Era, orogenic activity was confined to which mobile belt?
   a. _____ Cordilleran; b. _____ Ouachita; c. _____ Appalachian; d. _____ Franklin; e. _____ answers (a) and (b).

8. In which area of North America were reefs abundant during the Tippecanoe sequence?
   a. _____ Cincinnati Arch; b. _____ Michigan Basin;
   c. _____ Franklin mobile belt; d. _____ Williston Basin; e. _____ Transcontinental Arch.

9. Which orogeny took place along the western margin of Baltica during the Silurian Period?
   a. _____ Caledonian; b. _____ Acadian; c. _____ Taconic; d. _____ Antler; e. _____ Alleghenian.

10. The Taconic orogeny resulted from what type of plate movement?
    a. _____ oceanic-oceanic convergent; b. _____ oceanic-continental convergent; c. _____ continental-continental convergent; d. _____ divergent;
    e. _____ transform.

11. The Paleozoic ocean separating Laurentia from Siberia and Baltica was the:
    a. _____ Panthalassa; b. _____ Tethys; c. _____ Caledonian; d. _____ Iapetus; e. _____ Kaskaskia.

12. Which was the first major trangressive sequence onto the North American craton?
    a. _____ Sauk; b. _____ Tippecanoe; c. _____ Kaskaskia; d. _____ Absaroka; e. _____ Zuni.

13. One of the major sources of Early Paleozoic age iron ore is in which formation?
    a. _____ St. Peter; b. _____ Tuscarora; c. _____ Jordan; d. _____ Oriskany; e. _____ Clinton.

14. During which sequence was the eastern margin of Laurentia a passive plate margin?
    a. _____ Sauk; b. _____ Tippecanoe; c. _____ Kaskaskia; d. _____ Absaroka; e. _____ Zuni.

15. What is the order of precipitation of seawater salts from the least soluble to the most soluble?
    a. _____ halite, gypsum, carbonate; b. _____ gypsum, carbonate, halite; c. _____ carbonate, halite, gypsum;
    d. _____ halite, carbonate, gypsum; e. _____ carbonate, gypsum, halite.

16. Discuss how the Cambrian rocks of the Grand Canyon illustrate the sedimentation patterns of a transgressive sea.

17. Draw a diagrammatic cross section of a present-day reef, labeling and defining the various environments within the reef complex.

18. Briefly discuss the various natural resources that are found in Early Paleozoic age rocks.

19. Briefly discuss the Early Paleozoic geologic history of the world.

20. Summarize the differences between the Archean and Proterozoic eons and the Phanerozoic Eon.

21. How can geologists determine the locations of continents during the Paleozoic Era?

22. Why are cratonic sequences a convenient way to study the geologic history of the Paleozoic Era?

23. Where was the Transcontinental Arch? What evidence indicates that it existed?

24. Why did greater carbonate deposition occur on the craton during the Ordovician and Silurian periods than during the Cambrian Period?

25. Discuss how evaporites of the Michigan Basin may have formed during the Silurian Period.

26. What were the major differences between the Appalachian, Ouachita, and Cordilleran mobile belts during the Early Paleozoic?

27. What evidence in the rock record indicates that the Taconic orogeny occurred?

28. What evidence indicates that the Iapetus Ocean began closing during the Middle Ordovician?

TABLE 11.1   Summary of Early Paleozoic Geologic Events

Age (Millions of Years)	Geologic Period	Sequence	Relative Changes in Sea Level		Cordilleran Mobile Belt
			Rising	Falling	
408 —	Silurian	Tippecanoe			
438 —	Ordovician	Tippecanoe		Present sea level ←	
505 —	Cambrian	Sauk			
570 —					

TABLE 11.1   **Summary of Early Paleozoic Geologic Events** (continued)

Craton	Ouachita Mobile Belt	Appalachian Mobile Belt	Major Events Outside North America
Extensive barrier reefs and evaporites common.			Caledonian orogeny
Queenston Delta clastic wedge. Transgression of Tippecanoe Sea. Regression exposing large areas to erosion.		Taconic orogeny	Continental glaciation in Southern Hemisphere.
Canadian Shield and Transcontinental Arch only areas above sea level. Transgression of Sauk Sea.			

Additional Readings

Bally, A. W., and A. R. Palmer, eds. 1989. *The geology of North America: An overview.* The Geology of North America. vol. A. Boulder, Colo.: Geological Society of America.

Bambach, R. K., C. R. Scotese, and A. M. Ziegler. 1980. Before Pangaea: The geographies of the Paleozoic world. *American Scientist 68, no. 1:* 26–38.

Catacosinos, P. A., and P. A. Daniels, Jr., eds. 1991. *Early sedimentary evolution of the Michigan Basin.* Geological Society of America Special Paper 256. Boulder, Colo.: Geological Society of America.

Dallmeyer, R. D., ed. 1989. *Terranes in the Circum-Atlantic Paleozoic orogens.* Geological Society of America Special Paper 230. Boulder, Colo.: Geological Society of America.

Harwood, D. S., and M. M. Miller, eds. 1990. *Paleozoic and Early Mesozoic paleogeographic relations: Sierra Nevada, Klamath Mountains, and related terranes.* Geological Society of America Special Paper 255. Boulder, Colo.: Geological Society of America.

Hatcher, R. D., Jr., W. A. Thomas, and G. W. Viele, eds. 1989. *The Appalachian-Ouachita orogen in the United States.* The Geology of North America. vol. F-2. Boulder, Colo.: Geological Society of America.

McKerrow, W. S., and C. R. Scotese, eds. 1990. *Palaeozoic palaeogeography and biogeography.* The Geological Society Memoir No. 12. London, England: Geological Society of London.

Moullade, M., and A. E. M. Nairn, eds. 1991. *The Phanerozoic geology of the world I: The Palaeozoic.* New York: Elsevier Science Publishing Co.

Sloss, L. L., ed. 1988. *Sedimentary cover—North American Craton: U.S..* The Geology of North America. vol. D-2. Boulder, Colo.: Geological Society of America.

Stewart, J. H., C. H. Stevens, and A. E. Fritsche, eds. 1977. *Paleozoic paleogeography of the western United States.* Pacific Coast Paleogeography Symposium 1. Tulsa, Okla.: Society of Economic Paleontologists and Mineralogists.

Wilgus, C. K., B. S. Hastings, C. G. St. C. Kendall, H. Posamentier, C. A. Ross, and J. Van Wagoner, eds. 1988. *Sea-level changes: An integrated approach.* Society of Economic Paleontologists and Mineralogists Special Publication No. 42. Tulsa, Okla.: Society of Economic Paleontologists and Mineralogists.

CHAPTER 12

Chapter Outline

Seedless vascular plant fossils of *Neuropteris* in iron-carbonate concretions from Pennsylvanian-aged deposits, Mazon Creek locality, Will County, Illinois. (Photo courtesy of the Field Museum of Natural History, Chicago.)

GEOLOGY OF THE LATE PALEOZOIC ERA

Prologue

The unusual state of preservation of the Pennsylvanian-aged Mazon Creek biota of northeastern Illinois provides us with significant insights about the soft-part anatomy of organisms rarely preserved in the fossil record. The biota is divided into marine and nonmarine components and contains the only known or oldest fossil representatives of several major animal groups.

The Mazon Creek fossils occur in spheroidal- to elliptical-shaped iron-carbonate concretions ranging from 1 to 30 cm long. Rapid burial and the formation of concretions around the organisms were primarily responsible for the excellent preservation of the fossils. Not only are the hard parts of organisms preserved, but also impressions and carbonaceous films of the soft-bodied animals and plants as well. More than 320 species of animals and 350 species of plants have been described from this classic fossil assemblage.

The environment in which these plants and animals lived was a large delta where sluggish southward-flowing rivers emptied into a subtropical epeiric sea that covered most of the present state of Illinois. Two major habitats are represented: a swampy forested lowland of the subaerial delta and the shallow-marine environment of the actively prograding delta.

A diverse marine fauna lived in the warm, shallow waters of the delta front and included cnadarians, mollusks, echinoderms, arthropods, worms, and fish. From this fauna have come the only known fossils of several animal groups, including the lamprey, whose gills and liver can be discerned in the impressions.

More than 350 species of plants lived in the swampy lowlands surrounding the delta. Almost all of the plants were seedless vascular plants, typical of the kinds that comprised the Pennsylvanian coal-forming swamps of North America. Also found in the swampy lowlands were numerous insects, including millipedes and centipedes as well as spiders and other animals such as scorpions and amphibians. In the ponds, lakes, and rivers were many fish, shrimp, and ostracodes.

Local farmers and townspeople began collecting the Mazon Creek biota in the mid-1800s. Their efforts were soon followed by scientific studies by such famous geologists and paleontologists as J. D. Dana, L. Lesquereux, and E. D. Cope. Study of the Mazon Creek fossils has been uneven through the years, with much of it concentrating on the descriptions of the many plant and animal species recovered. Today research is still largely concerned with describing the complete assemblage, but researchers are also interested in determining the phylogenetic relationships of the organisms and the paleoecologic significance of the assemblage.

❖ INTRODUCTION

The Late Paleozoic Era, composed of the Devonian, Carboniferous (Mississippian and Pennsylvanian in North American terminology), and Permian periods, witnessed the formation of the supercontinent Pangaea from the various landmasses that were present when the era began. Collisions between continents along convergent plate boundaries led to the assembly of Pangaea with the result that by the end of the Permian, the world's geography consisted of one large landmass surrounded by a global ocean. During the Late Paleozoic, the various continents experienced a variety of climatic conditions as indicated by such sedimentary rocks as coals, evaporites,

Archean Eon	Proterozoic Eon	Phanerozoic Eon						
Precambrian		Paleozoic Era						
		Cambrian	Ordovician	Silurian	Devonian	Mississippian	Pennsylvanian	Permian
						Carboniferous		

2,500 M.Y.A.   570 M.Y.A.

and tillites. Major glacial-interglacial intervals occurred over much of Gondwana as it continued moving over the South Pole during the Late Mississippian to Early Permian. The growth and retreat of these glaciers profoundly affected the world's biota and also contributed to global changes in sea level. The mountain building that resulted from plate convergence also strongly influenced oceanic and atmospheric circulation patterns. By the end of the Paleozoic, widespread arid and semiarid conditions prevailed over much of Pangaea.

❖ LATE PALEOZOIC PALEOGEOGRAPHY

The Late Paleozoic was a time of continental collisions, mountain building, fluctuating sea levels, and varied climates. These events resulted from plate interactions and major continental ice-sheet advances and retreats. We will initially examine the changing paleogeography of the Late Paleozoic and then will place North American events into this global framework.

The Devonian Period

Laurentia and Baltica collided along a convergent plate boundary to form the larger continent of **Laurasia** during the Silurian. This collision, which closed the northern Iapetus Ocean, is marked by the **Caledonian orogeny.** During the Devonian, as the southern Iapetus Ocean narrowed between Laurasia and Gondwana, mountain building continued along the eastern margin of Laurasia with the **Acadian orogeny** (◆ Fig. 12.1a). The erosion of the resulting highlands provided vast amounts of reddish fluvial sediments that covered large areas of northern Europe (Old Red Sandstone) and eastern North America (the Catskill Delta). Other Devonian tectonic events,

probably related to the collision of Laurentia and Baltica, include the Cordilleran **Antler orogeny,** the **Ellesmere orogeny** along the northern margin of Laurentia (which may reflect the collision of Laurentia with Siberia), and the change from a passive continental margin to an active convergent plate boundary in the Uralian mobile belt of eastern Baltica. The distribution of reefs, evaporites, and red beds, as well as the existence of similar floras throughout the world, suggests a rather uniform global climate during the Devonian Period.

The Carboniferous Period

During the Carboniferous Period, southern Gondwana moved over the South Pole, resulting in extensive continental glaciation (◆ Figs. 12.1b and 12.2a). The advance and retreat of these glaciers produced global changes in sea level that affected sedimentation patterns on the cratons. As Gondwana continued moving northward, it first collided with Laurasia during the Early Carboniferous and continued suturing with it throughout the rest of the Carboniferous (Figs. 12.1b and 12.2a). Because Gondwana rotated clockwise relative to Laurasia, deformation generally progressed in a northeast-to-southwest direction along the Hercynian, Appalachian, and Ouachita mobile belts of the two continents. The final phase of collision between Gondwana and Laurasia is indicated by the Ouachita Mountains of Oklahoma, which were formed by thrusting during the Late Carboniferous and Early Permian.

Elsewhere, Siberia collided with Kazakhstania and moved toward the Uralian margin of Laurasia (Baltica), colliding with it during the Early Permian. It has recently been suggested that the northwestern margin of China collided with the southwestern margin of Siberia during the Late Carboniferous. Thus, by the end of the Carbonif-

Phanerozoic Eon									
Mesozoic Era			Cenozoic Era						
Triassic	Jurassic	Cretaceous	Tertiary					Quaternary	
			Paleocene	Eocene	Oligocene	Miocene	Pliocene	Pleistocene	Holocene

245
M.Y.A.

66
M.Y.A.

erous, the various continental landmasses were fairly close together as Pangaea began taking shape.

The Carboniferous coal basins of eastern North America, western Europe, and the Donets Basin of Ukraine all lay in the equatorial zone, where rainfall was high and temperatures were consistently warm. The absence of strong seasonal growth-rings in fossil plants from these coal basins is indicative of such a climate. The fossil plants found in the coals of Siberia and China, however, show well-developed growth-rings, signifying seasonal growth with abundant rainfall and distinct seasons such as occur in the temperature zones (latitudes 40 degrees to 60 degrees north).

Glacial conditions and the movement of large continental ice sheets in the high southern latitudes are indicated by widespread tillites and glacial striations in southern Gondwana (see Fig. 7.4). These ice sheets spread toward the equator and, at their maximum growth, extended well into the middle temperate latitudes.

The Permian Period

The assembly of Pangaea was essentially concluded during the Permian with the completion of many of the continental collisions that began during the Carboniferous (◆ Fig. 12.2b). Although geologists generally agree on the configuration and location of the western half of the supercontinent, there is no consensus on the number or configuration of the various terranes and continental blocks that composed the eastern half of Pangaea. Regardless of the exact configuration of the eastern portion, geologists know that the supercontinent was surrounded by various subduction zones and moved steadily northward during the Permian. Furthermore, an enormous single ocean, **Panthalassa,** surrounded Pangaea and spanned the Earth from pole to pole. Waters of this ocean probably circulated more freely than at present, resulting in more equable water temperatures.

The formation of a single large landmass had climatic consequences for the terrestrial environment as well. Terrestrial Permian sediments indicate that arid and semi-arid conditions were widespread over Pangaea. The mountain ranges produced by the **Hercynian, Alleghenian,** and **Ouachita orogenies** were high enough to create rain shadows that blocked the moist, subtropical, easterly winds—much as the southern Andes Mountains do in western South America today. This produced very dry conditions in North America and Europe, as evident from the extensive Permian evaporites found in western North America, central Europe, and parts of Russia. Permian coals, indicative of abundant rainfall, were mostly limited to the northern temperate belts (latitude 40 degrees to 60 degrees north), while the last remnants of the Carboniferous ice sheets retreated to the mountainous regions of eastern Australia.

❖ LATE PALEOZOIC HISTORY OF NORTH AMERICA

The Late Paleozoic cratonic history of North America included periods of extensive shallow-marine carbonate deposition and large coal-forming swamps as well as dry, evaporite-forming terrestrial conditions. Cratonic events largely resulted from changes in sea level due to Gondwanan glaciation and tectonic events related to the assembly of Pangaea. Mountain building that began with the Ordovician Taconic orogeny continued with the Caledonian, Acadian, Alleghenian, and Ouachitia orogenies. These orogenies were part of the global tectonic process that resulted in the assembly of Pangaea by the end of the Paleozoic Era.

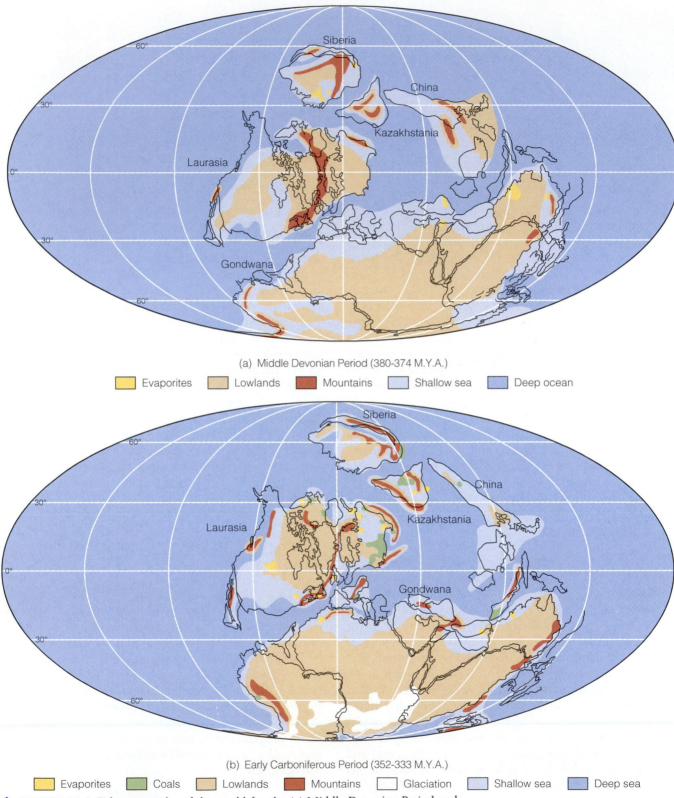

(a) Middle Devonian Period (380-374 M.Y.A.)

Evaporites   Lowlands   Mountains   Shallow sea   Deep ocean

(b) Early Carboniferous Period (352-333 M.Y.A.)

Evaporites   Coals   Lowlands   Mountains   Glaciation   Shallow sea   Deep sea

◆ **FIGURE 12.1** Paleogeography of the world for the (a) Middle Devonian Period and (b) Early Carboniferous Period.

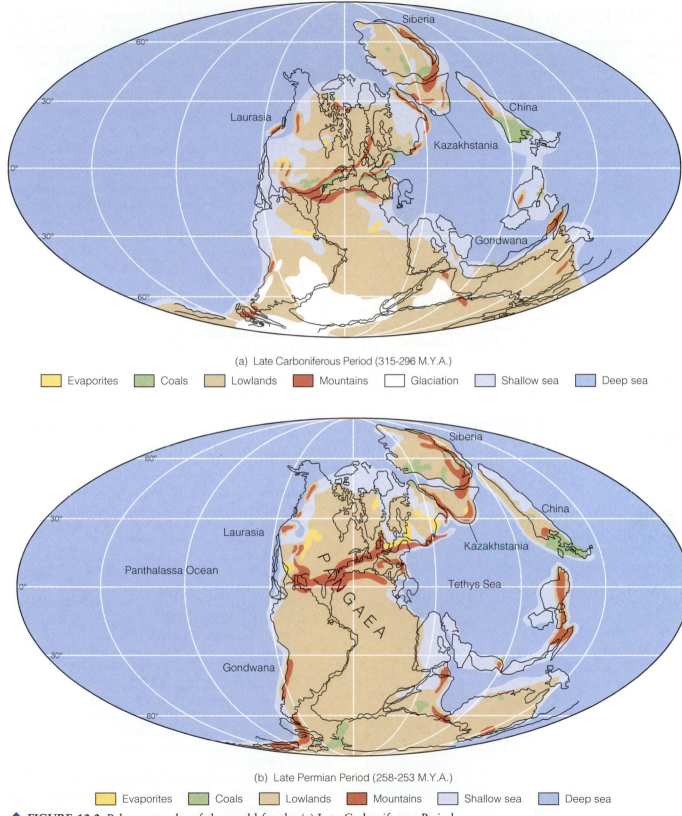

(a) Late Carboniferous Period (315-296 M.Y.A.)

☐ Evaporites ☐ Coals ☐ Lowlands ☐ Mountains ☐ Glaciation ☐ Shallow sea ☐ Deep sea

(b) Late Permian Period (258-253 M.Y.A.)

☐ Evaporites ☐ Coals ☐ Lowlands ☐ Mountains ☐ Shallow sea ☐ Deep sea

◆ **FIGURE 12.2** Paleogeography of the world for the (a) Late Carboniferous Period and (b) Late Permian Period.

❖ THE KASKASKIA SEQUENCE

The boundary between the Tippecanoe sequence and the overlying **Kaskaskia sequence** is marked by a major unconformity. As the Kaskaskia Sea transgressed over the low relief landscape of the craton during the middle Early Devonian, the majority of the basal beds deposited consisted of clean, mature, quartz sandstones. A good example is the Oriskany Sandstone of New York and Pennsylvania and its lateral equivalents (◆ Fig. 12.3). The Oriskany Sandstone, like the basal Tippecanoe St. Peter Sandstone, is an important glass sand as well as a good gas-reservoir rock.

The source areas for the basal Kaskaskia sandstones were primarily the eroding highlands of the Appalachian mobile belt area (◆ Fig. 12.4), exhumed Cambrian and Ordovician sandstones cropping out along the flanks of the Ozark Dome, and exposures of the Canadian Shield in the Wisconsin area. The lack of similar sands in the Silurian carbonate beds below the Tippecanoe-Kaskaskia unconformity indicates that the source areas of the basal Kaskaskia detrital rocks were submerged when the Tippecanoe sequence was deposited. Stratigraphic studies indicate that these source areas were uplifted and the Tippecanoe carbonates removed by erosion prior to the Kaskaskia transgression. Kaskaskian basal rocks elsewhere on the craton consist of carbonates that are frequently difficult to differentiate from the underlying Tippecanoe carbonates unless they are fossiliferous.

Except for widespread Late Devonian and Early Mississippian black shales, the majority of Kaskaskian rocks are carbonates, including reefs, and associated evaporite deposits. In many other parts of the world, such as southern England, Belgium, central Europe, Australia, and Russia, the Middle and early Late Devonian epochs were times of major reef building (see Perspective 12.1)

Reef Development in Western Canada

The Middle and Late Devonian reefs of western Canada contain large reserves of petroleum and have therefore been widely studied from outcrops and in the subsurface (◆ Fig. 12.5). These reefs began forming as the Kaskaskia Sea transgressed southward into western Canada. By the end of the Middle Devonian, they had coalesced into a large barrier-reef system that restricted the flow of oceanic water into the back-reef platform, thus creating conditions for evaporite precipitation (Fig. 12.5). In the back reef, up to 300 m of evaporites were precipitated in much the same way as in the Michigan Basin during the Silurian (see Fig. 11.11). More than half of the world's potash, which is used in fertilizers, comes from these Devonian evaporites. By the middle of the Late Devonian, reef growth stopped in the western Canada region, although nonreef carbonate deposition continued.

Black Shales

In North America, many areas of carbonate-evaporite deposition gave way to a greater proportion of shales and coarser detrital rocks beginning in the Middle Devonian and continuing into the Late Devonian. This change to detrital deposition resulted from the formation of new source areas brought on by the mountain-building activity associated with the Acadian orogeny in North America (Fig. 12.4).

As the Devonian Period ended, a conspicuous change in sedimentation occurred over the craton with the appearance of widespread black shales. In the eastern United States, these black shales are commonly called the Chattanooga Shale, but are known by a variety of local names elsewhere (for example, New Albany Shale and Antrim Shale). Although these black shales are best developed from the cratonic margins along the Appalachian mobile belt to the Mississippi Valley, correlatives can also be found in many western states and in western Canada (◆ Fig. 12.6).

The Late Devonian–Early Mississippian black shales of North America are typically noncalcareous, thinly bedded, and usually less than 10 m thick (Fig. 12.6). If

◆ **FIGURE 12.3** Extent of the basal units of the Kaskaskia sequence in the eastern and north-central United States.

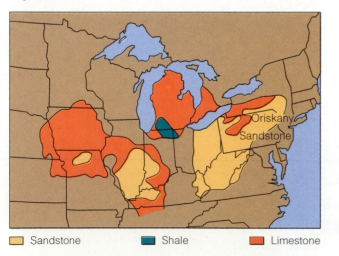

Oriskany Sandstone

☐ Sandstone   ☐ Shale   ☐ Limestone

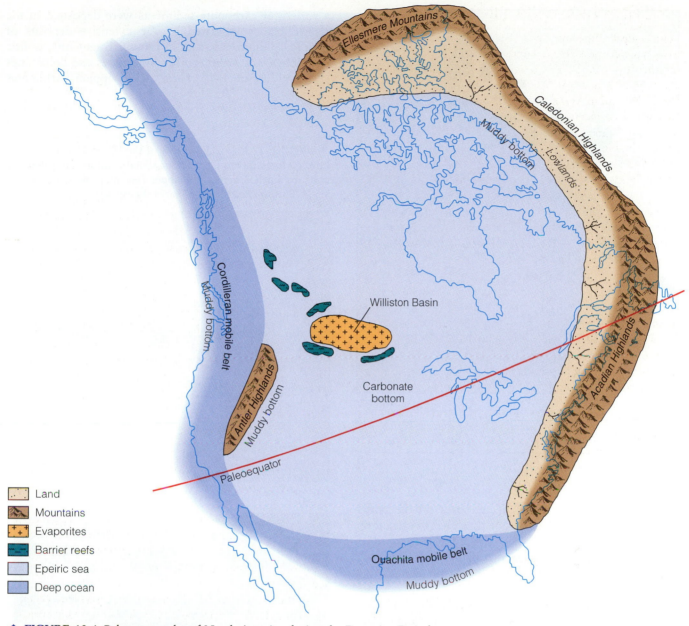

LEGEND

- Land
- Mountains
- Evaporites
- Barrier reefs
- Epeiric sea
- Deep ocean

◆ **FIGURE 12.4** Paleogeography of North America during the Devonian Period.

they are rich in land-plant remains, they are commonly termed *carbonaceous*. If marine productivity was high, the shales are *bituminous* and contain mostly degraded algal material. Fossils are typically rare in these black shales, but some Upper Devonian black shales do contain rich conodont faunas with large numbers of individuals. Because most black shales lack body fossils, they are difficult to date and to correlate. In places where they can be dated, usually by conodonts (pelagic animals), acri-

tarchs (pelagic algae), or plant spores, the lower beds are Late Devonian, and the upper beds are Early Mississippian in age.

Although the origin of these extensive black shales is still being debated, the essential features required to produce them include undisturbed anaerobic bottom water, a reduced supply of coarser detrital sediment, and high organic productivity in the overlying oxygenated waters. High productivity in the surface waters leads to a shower

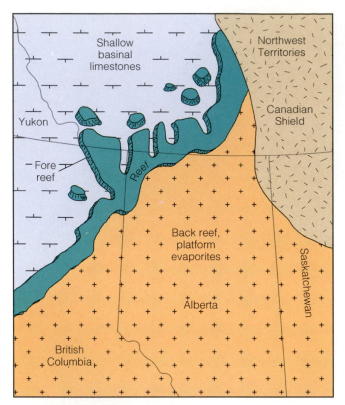

♦ **FIGURE 12.5** Reconstruction of the extensive Devonian Reef complex of western Canada. These extensive reefs controlled the regional facies of the Devonian epeiric seas.

a variety of carbonate sediments were deposited in the epeiric sea as indicated by the extensive deposits of crinoidal limestones (rich in crinoid fragments), oolitic limestones, and various other limestones and dolostones (♦ Fig. 12.9, p. 331). These Mississippian carbonates

♦ **FIGURE 12.6** (a) The extent of the Upper Devonian to Lower Mississippian Chattanooga Shale and its equivalent units in North America. (b) Upper Devonian New Albany Shale, Button Mold Knob Quarry, Kentucky.

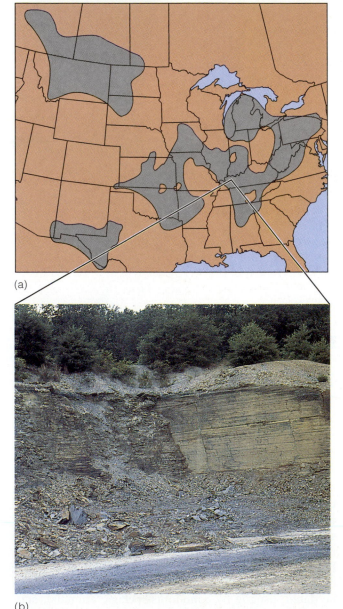

(a)

(b)

of organic material, which decomposes on the undisturbed substrate and depletes the dissolved oxygen at the sediment-water interface. All of the models proposed for the depositional environments of black shales involve a quiet, anaerobic environment beneath an aerated one (♦ Fig. 12.7).

The wide extent of such apparently shallow-water black shales in North America remains puzzling. Nonetheless, these shales are rich in uranium and are an important source rock of oil and gas in the Appalachian region.

The Late Kaskaskia—A Return to Extensive Carbonate Deposition

Following deposition of the widespread Late Devonian–Early Mississippian black shales, carbonate sedimentation on the craton dominated the remainder of the Mississippian Period (♦ Fig. 12.8, p. 330). During this time,

display cross-bedding, ripple marks, and well-sorted fossil fragments, all of which are indicative of a shallow-water environment. Analogous features can be observed on the present-day Bahama Banks. In addition, numerous small organic reefs occurred throughout the craton during the Mississippian. These were all much smaller than the large barrier-reef complexes that dominated the earlier Paleozoic seas.

During the Late Mississippian regression of the Kaskaskia Sea from the craton, carbonate deposition was replaced by vast quantities of detrital sediments. The resulting sandstones, particularly in the Illinois Basin, have been studied in great detail because they are excellent petroleum reservoirs. Prior to the end of the Mississippian, the Kaskaskia Sea had retreated to the craton margin, once again exposing the craton to widespread weathering and erosion that resulted in a cratonwide unconformity when the Absaroka Sea began transgressing back over the craton.

❖ THE ABSAROKA SEQUENCE

The **Absaroka sequence** includes uppermost Mississippian through Lower Jurassic rocks. In this chapter, however, we will only be concerned with the Paleozoic rocks of the Absaroka sequence. The extensive unconformity separating the Kaskaskia and Absaroka sequences essentially divides the strata into the North American Mississippian and Pennsylvanian systems. These two systems are equivalent to the European Lower and Upper Carboniferous systems, respectively. The rocks of the Absaroka sequence are not only different from those of the Kaskaskia sequence, but they are also the result of quite different tectonic regimes affecting the North American craton.

The lowermost sediments of the Absaroka sequence are confined to the margins of the craton. These deposits are generally thickest in the east and southeast, near the emerging highlands of the Appalachian and Ouachita mobile belts, and thin westward onto the craton. The lithologies also reveal lateral changes from nonmarine detrital rocks and coals in the east, through transitional marine-nonmarine beds, to largely marine detrital rocks and limestones farther west (◆ Fig. 12.10, p. 331).

Cyclothems

One of the characteristic features of Pennsylvanian rocks is their cyclical pattern of alternating marine and nonmarine strata. Such rhythmically repetitive sedimentary

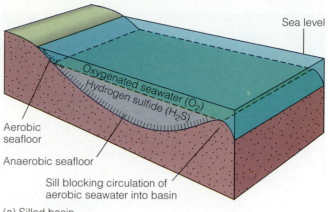

(a) Silled basin

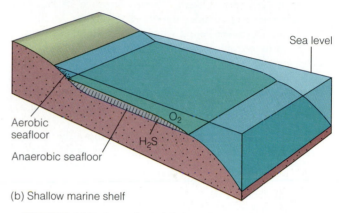

(b) Shallow marine shelf

◆ **FIGURE 12.7** Several models have been proposed to explain the formation of widespread Late Devonian–Early Mississippian black shales. All of the models involve a layer of quiet anaerobic seawater below an aerated layer. (a) Silled basin environment. (b) Shallow-marine shelf environment.

sequences are called **cyclothems.** They result from repeated alternations of marine and nonmarine environments, usually in areas of low relief. Though seemingly simple, cyclothems reflect a delicate interplay between nonmarine deltaic and shallow-marine interdeltaic and shelf environments.

For purposes of illustration, we can look at a typical coal-bearing cyclothem from the Illinois Basin (◆ Fig. 12.11, p. 332). Such a cyclothem contains nonmarine units, capped by a coal and overlain by marine units. Figure 12.11c shows the depositional environments that produced the cyclothem. The initial units represent deltaic and fluvial deposits. Above them is an underclay that frequently contains root casts from the plants and trees that comprise the overlying coal. The coal bed results from accumulations of plant material and is overlain by

THE CANNING BASIN, AUSTRALIA—A DEVONIAN GREAT BARRIER REEF

One of the largest and most spectacularly exposed fossil-reef complexes in the world is the Great Barrier Reef of the Canning Basin, Western Australia (◆ Fig. 1). This barrier-reef complex developed during the Middle and Late Devonian Period when the Canning Basin was covered by a tropical epeiric sea (Fig. 12.1a). The reefs are now exposed as limestone ridges that extend for some 350 km along the northern margin of the Canning Basin, but they probably continued around the present northern coastal region to join with similar reefs exposed in the Bonaparte Basin to the east.

The limestone reefs rise 50 to 100 m above the surrounding plains looking much the same as they did when the area was covered by the Devonian epeiric sea. The shales and other soft sediments deposited in the open ocean in front of the reef complex (see Fig. 11.9) have been eroded away, leaving the resistant limestone reefs standing as ridges.

The reefs themselves were constructed primarily by calcareous algae, stromatoporoids, and tabulate and

◆ **FIGURE 1** Aerial View of Windjana Gorge showing the Devonian Great Barrier Reef exposed as a limestone ridge.

rugose corals, which also were the main components of the other major reef complexes in the world at that time. An interesting feature of these Canning Basin reefs is the contribution of column-shaped stromatolites which are found in the reef, back-reef, and marginal-slope areas of the reef complex. Stromatolites are an unusual component because they ceased to be abundant by the end of the Proterozoic Eon. Throughout the Phanerozoic Eon,

marine units of alternating limestones and shales, usually with an abundant marine invertebrate fauna. The marine cycle ends with an erosion surface. A new cyclothem begins with a nonmarine deltaic sandstone.

Cyclothems represent transgressive and regressive sequences with an erosional surface separating one cyclothem from another. Thus, an idealized cyclothem passes upward from fluvial-deltaic deposits, through coals, to detrital shallow-water marine sediments, and finally to limestones typical of an open marine environment. It should be noted that Figure 12.11 represents ideal conditions. Frequently, all beds are not preserved due to abrupt changes from marine to nonmarine conditions or to removal of some units by erosion.

Such places as the Mississippi delta, the Florida Everglades, and the Dutch lowlands represent modern coal-forming environments similar to those that existed during the Pennsylvanian (◆ Fig. 12.12, p. 333). By studying these

modern analogues, geologists can make reasonable deductions about conditions that existed in the geologic past.

The coal swamps that existed during the Pennsylvanian Period must have been large lowland areas neighboring the sea. In such cases, a very slight rise in sea level would have flooded large areas, while slight drops would have exposed large areas, resulting in alternating marine and nonmarine environments. The same result could have been caused by rising sea level and progradation of a large delta, such as occurs today in Louisiana.

Such regularity and cyclicity in sedimentation over a large area requires an explanation, and the origin of the Pennsylvanian cyclothems has long been debated. In most cases, local cyclothems of limited extent can be explained by rapid but slight changes in sea level in a swamp-delta complex of low relief near the sea or by localized crustal movement. Explaining widespread cyclothems is more difficult.

stromatolites typically formed only in areas generally inhospitable to animals.

The outcrop along Windjana Gorge beautifully reveals the various features and facies of the Devonian Great Barrier Reef complex (◆ Fig. 2). The reef core consists of unbedded limestones composed predominantly of calcareous algae, stromatoporoids, and corals. The back-reef facies (see Fig. 11.9) is bedded and makes up the major part of the total reef complex environment. A diverse and abundant fauna of calcareous algae, stromatoporoids, various corals, some bivalves, gastropods, cephalopods, brachiopods, and crinoids lived in this lagoonal area behind the reef core.

In front of the reef core was the steep fore-reef slope (see Fig. 11.9) that supported some organisms, including algae, sponges, and stromatoporoids. This facies contains considerable reef talus, an accumulation of debris eroded by waves from the reef front. The ocean-basin deposits contain the fossils of mainly nektonic and planktonic organisms such as fish, radiolarians, cephalopods, and conodonts.

Near the end of the Late Devonian, nearly all the reef-building organisms as well as much of the associated fauna of the Canning Basin Great Barrier Reef became extinct. As we will discuss in Chapter 13, few massive tabulate-rugose-stromatoporoid reefs are known from latest Devonian or younger rocks.

◆ **FIGURE 2** Outcrop of the Devonian Great Barrier Reef along Windjana Gorge. The talus of the fore-reef area can be seen on the left side of the picture sloping away from the reef core, which is unbedded. To the right of the reef core is the back-reef facies, which is horizontally bedded. (Photos courtesy of Geoffrey Playford, The University of Queensland, Brisbane, Australia.)

The hypothesis currently favored by most geologists is a rise and fall of sea level related to advances and retreats of Gondwanan continental glaciers. When the Gondwanan ice sheets advanced, sea level dropped, and when they melted, sea level rose. Late Paleozoic cyclothem activity on all of the cratons closely corresponds to Gondwanan glacial-interglacial cycles.

Cratonic Uplift—The Ancestral Rockies

As we discussed in Chapter 11, cratons are stable areas, and when they do experience deformation, it is usually only mild and results in structures of great dimensions. The Pennsylvanian Period, however, was a time of unusually severe deformation of the craton and resulted in uplifts of sufficient magnitude to expose Precambrian basement rocks. In addition to newly formed highlands and basins, many previously formed arches and domes, such as the Cincinnati Arch, Nashville Dome, and Ozark Dome, were also reactivated (see Fig. 11.1).

During the Late Absaroka, the area of greatest deformation occurred in the southwestern part of the North American craton where a series of fault-bounded uplifted blocks formed the **Ancestral Rockies** (◆ Fig. 12.13a, p. 334). These mountain ranges had diverse geologic histories and were not all elevated at the same time. The group of ranges along the Texas-Oklahoma border and Texas Panhandle constitute the Oklahoma uplift; those across northeastern Arizona, the Kaibab-Defiance-Zuni uplift; and those in the Colorado, Utah, and New Mexico area, namely, the Front Range—Pedernal and Uncompahgre uplifts, are referred to collectively as the Ancestral Rocky Mountains. Uplift of these mountains, some of which were elevated more than 2 km along near-vertical faults, resulted in the erosion of the overlying Paleozoic sediments and exposure of the Precambrian igneous and

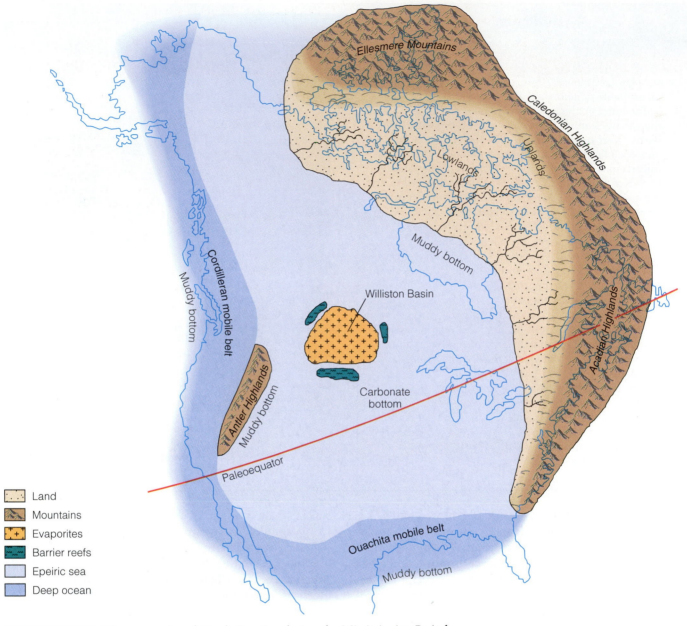

Land

Mountains

Evaporites

Barrier reefs

Epeiric sea

Deep ocean

◆ **FIGURE 12.8** Paleogeography of North America during the Mississippian Period.

metamorphic basement rocks (◆ Fig. 12.13b). As the mountains eroded, tremendous quantities of coarse, red arkosic sand and conglomerate were deposited in the surrounding basins (Fig. 12.13b). Currently, these sediments are preserved in many areas including the rocks of the Garden of the Gods near Colorado Springs (◆ Fig. 12.14, p. 335) and at the Red Rocks Amphitheatre near Morrison, Colorado.

Intracratonic mountain ranges are unusual, and their cause has long been debated. It is now believed that the collision of Gondwana with Laurussia (Fig. 12.2a) produced great stresses in the southwestern region of the North American craton. These crustal stresses were relieved by faulting which resulted in uplift of cratonic blocks and downwarp of adjacent basins, forming a series of ranges and basins.

◆ **FIGURE 12.9** Mississippian limestones exposed near Bowling Green, Kentucky.

The Late Absaroka—More Evaporite Deposits and Reefs

The various basins bordering the Ancestral Rockies contain extensive marine and nonmarine sediments and thus provide an excellent stratigraphic record for deciphering the geologic history of this region. The Paradox Basin,

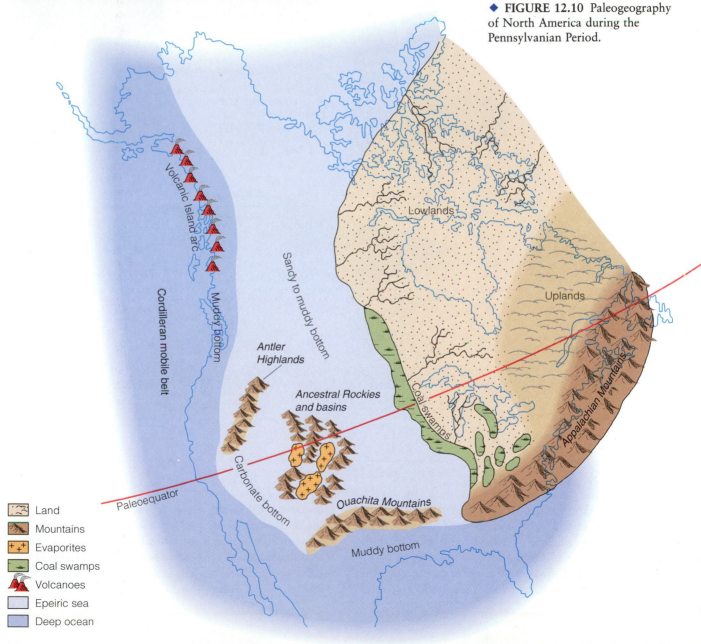

◆ **FIGURE 12.10** Paleogeography of North America during the Pennsylvanian Period.

Lowlands

Uplands

Volcanic Island arc

Cordilleran mobile belt

Muddy bottom

Sandy to muddy bottom

Antler Highlands

Ancestral Rockies and basins

Coal swamps

Appalachian Mountains

Carbonate bottom

Ouachita Mountains

Paleoequator

Muddy bottom

Land

Mountains

Evaporites

Coal swamps

Volcanoes

Epeiric sea

Deep ocean

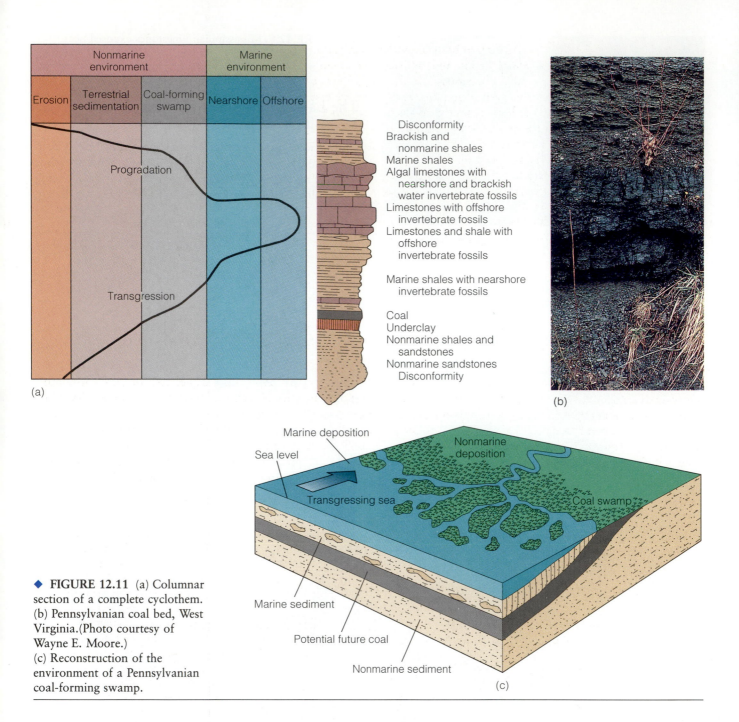

Disconformity
Brackish and
  nonmarine shales
Marine shales
Algal limestones with
  nearshore and brackish
  water invertebrate fossils
Limestones with offshore
  invertebrate fossils
Limestones and shale with
  offshore
  invertebrate fossils

Marine shales with nearshore
  invertebrate fossils

Coal
Underclay
Nonmarine shales and
  sandstones
Nonmarine sandstones
Disconformity

(a)

(b)

Marine deposition
Sea level
Nonmarine
deposition
Transgressing sea
Coal swamp
Marine sediment
Potential future coal
Nonmarine sediment
(c)

◆ **FIGURE 12.11** (a) Columnar section of a complete cyclothem. (b) Pennsylvanian coal bed, West Virginia.(Photo courtesy of Wayne E. Moore.) (c) Reconstruction of the environment of a Pennsylvanian coal-forming swamp.

located between the Uncompahgre uplift and the Kaibab-Defiance-Zuni uplift, has been extensively studied by petroleum geologists and thus serves as a good example of the geologic history of the region (Fig. 12.13a).

The Four Corners region (the area where the four states of Arizona, New Mexico, Colorado, and Utah come together) was covered during the Early Pennsylvanian by the transgressing Absaroka Sea. These marine sediments, consisting mostly of shales, were deposited on a well-developed Middle Mississippian karst topography. Access to the Paradox Basin along its western margin became limited during the Middle Pennsylvanian resulting in thick, cyclical deposits of gypsum, anhydrite, and salt in the central basin area. Fossiliferous and oolitic limestones were deposited around the perimeter of the basin, while patch and barrier reefs grew along the western margin, further restricting the flow of marine waters into the central basin. It is in the reef facies of the Paradox Basin that oil and gas discoveries have been made. Sediments eroded from the rising Uncompahgre highlands completely filled the Paradox Basin with arkosic sediments during the Late Pennsylvanian.

While the Paradox Basin was filling with sediment during the Late Pennsylvanian, the Absaroka Sea slowly began retreating from the craton. During the Early Permian, the Absaroka Sea occupied a narrow region from Nebraska through West Texas (◆ Fig. 12.15, p. 336). By the Middle Permian, the sea had retreated to West Texas and southern New Mexico. The thick evaporite deposits in Kansas and Oklahoma provide evidence of the restricted nature of the Absaroka Sea during the Early and Middle Permian and its southwestward retreat from the central craton.

During the Middle and Late Permian, the Absaroka Sea was restricted to West Texas and southern New Mexico, forming an interrelated complex of lagoonal, reef, and open-shelf environments (◆ Fig. 12.16, p. 337). Three basins separated by two submerged platforms formed in this area during the Permian. Massive reefs grew around the basin margins (◆ Fig. 12.17, p. 337),

while limestones, evaporites, and red beds were deposited in the lagoonal areas behind the reefs. As the barrier reefs grew and the passageways between the basins became more restricted, Late Permian evaporites gradually filled the individual basins with deposits up to 600 m thick.

Spectacular deposits representing the geologic history of this region can be seen today in the Guadalupe Mountains of Texas and New Mexico where the Capitan Limestone forms the caprock of these mountains (◆ Fig. 12.18, p. 337). These reefs have been extensively studied because of the tremendous oil production that comes from this region.

By the end of the Permian Period, the Absaroka Sea had retreated from the craton, leaving continental red beds over most of the southwestern and eastern region.

❖ HISTORY OF THE LATE PALEOZOIC MOBILE BELTS

Having examined the history of the craton for the Kaskaskia and Absaroka sequences, we now turn our attention to the orogenic activity in the mobile belts. The mountain building that occurred during this time had a profound influence on the climate and sedimentary history of the craton. In addition, it was part of the global tectonic regime that sutured the continents together, forming Pangaea by the end of the Paleozoic Era.

◆ FIGURE 12.12 The Okefenokee Swamp, Florida, is a modern example of a coal-forming environment, similar to those that existed during the Pennsylvanian Period. It contains a variety of angiosperm plants and trees that may eventually be converted into coal under the proper geologic conditions. (Photo © Patricia Caulfied/Photo Researchers, Inc.)

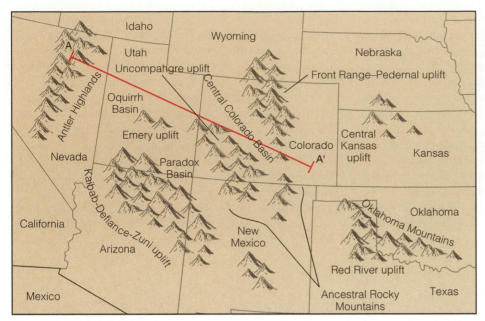

(a)

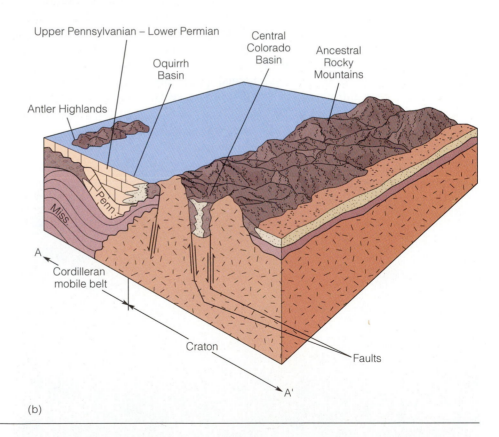

◆ **FIGURE 12.13** (a) Location of the principal Pennsylvanian highland areas and basins of the southwestern part of the craton. (b) Cross section of the Ancestral Rockies, which were elevated by faulting during the Pennsylvanian Period. Erosion of these mountains produced coarse, red-colored sediments that were deposited in the adjacent basins.

(b)

Cordilleran Mobile Belt

During the Late Proterozoic and Early Paleozoic, the Cordilleran area was a passive continental margin along which extensive continental shelf sediments were deposited (see Figs. 10.14 and 11.5). Thick sections of marine sediments graded laterally into thin cratonic units as the Sauk Sea transgressed onto the craton. Beginning in the Middle Paleozoic, an island arc (called the *Klamath Arc*) formed on the western margin of the craton. A collision between this eastward-moving island arc and the western border of the craton occurred during the Late Devonian and Early Mississippian resulting in a highland area. Evidence for this collision can be seen along the Roberts Mountain thrust fault in central Nevada where deep-water continental slope and rise deposits were thrust eastward more than 150 km over shallow-water continental shelf carbonates. This orogenic event, called the *Antler orogeny*, was caused by subduction and resulted in the closing of the narrow ocean basin that separated the Klamath island arc system from the craton (◆ Fig. 12.19). Erosion of the Antler Highlands produced large quantities of sediment that were deposited to the east in the epeiric sea covering the craton and to the west in the deep sea. The Antler orogeny was the first in a series of orogenic events to affect the Cordilleran mobile belt. During the Mesozoic and Cenozoic, this area was the site of major tectonic activity by both oceanic-continental convergence and accretion of various terranes.

Ouachita Mobile Belt

The Ouachita mobile belt extends for approximately 2,100 km from the subsurface of Mississippi to the Marathon region of Texas. Approximately 80% of the former mobile belt is buried beneath a Mesozoic and Cenozoic sedimentary cover. The two major exposed areas in this region are the Ouachita Mountains of Oklahoma and Arkansas and the Marathon Mountains of Texas. Based on extensive study of the subsurface geology and the Ouachita and Marathon mountains, geologists have learned that this region had a very complex geologic history (◆ Fig. 12.20, p. 338).

During the Late Proterozoic to Early Mississippian, shallow-water detrital and carbonate sediments were slowly deposited on a broad continental shelf, while in the deeper-water portion of the adjoining mobile belt, bedded cherts and shales were also slowly accumulating (Fig. 12.20a). Beginning in the Mississippian Period, the rate of sedimentation increased dramatically as the region changed from a passive continental margin to an active convergent plate boundary (Fig. 12.20b). Increasing rates of sedimentation as well as the addition of volcanic tuffs interbedded with graywackes and black shales indicate the development of a volcanic arc–subduction zone to the south. Rapid deposition of sediments continued into the Pennsylvanian with the formation of a clastic wedge that thickened to the south. As much as 16,000 m of Mississippian and Pennsylvanian-

◆ **FIGURE 12.14** These spectacular exposures are the erosional remnants of Pennsylvanian arkoses that were later uplifted during the Laramide orogeny and are now exposed at Red Rock Canyon, Garden of the Gods, near Colorado Springs, Colorado.

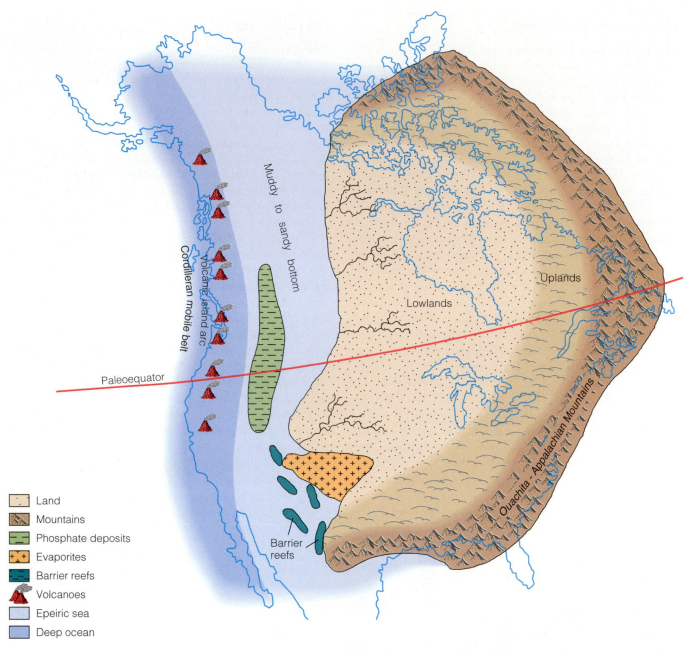

Land

Mountains

Phosphate deposits

Evaporites

Barrier reefs

Volcanoes

Epeiric sea

Deep ocean

Muddy to sandy bottom

Cordilleran mobile belt

Volcanic island arc

Paleoequator

Uplands

Lowlands

Barrier reefs

Ouachita - Appalachian Mountains

◆ **FIGURE 12.15** Paleogeography of North America during the Permian Period.

aged rocks crop out in the Ouachita Mountains, attesting to the rapid rate of sedimentation during this time. The formation of a clastic wedge marks the beginning of uplift of the area and formation of a mountain range.

Thrusting of sediments continued throughout the Pennsylvanian and Early Permian as a result of the compressive forces generated along the zone of subduction as Gondwana collided with Laurasia (◆ Fig. 12.20c). The

collision of Gondwana and Laurasia is marked by the formation of a large mountain range, most of which was eroded during the Mesozoic Era. Only the rejuvenated Ouachita and Marathon mountains remain of this once lofty mountain range.

The Ouachita deformation was part of the general worldwide tectonic activity that occurred when Gondwana united with Laurasia. The Hercynian, Appala-

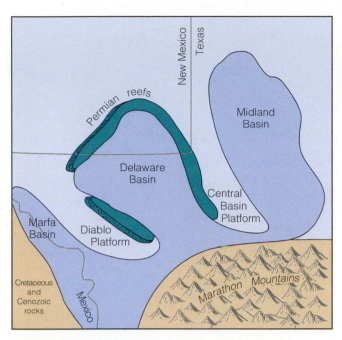

◆ **FIGURE 12.16** Location of the West Texas Permian basins and surrounding reefs.

◆ **FIGURE 12.17** A reconstruction of the Middle Permian Capitan Limestone reef environment. Shown are brachiopods, corals, bryozoans, and large glass sponges. (Photo courtesy of Rubin's Studio of Photography, The Petroleum Museum, Midland, Texas.)

chian, and Ouachita mobile belts were continuous and marked the southern boundary of Laurasia (Fig. 12.2). The tectonic activity that resulted in the uplift in the Ouachita mobile belt was very complex and involved not

◆ **FIGURE 12.18** The prominent light-colored Capitan Limestone forms the caprock of the Guadalupe Mountains. The Capitan Limestone is rich in fossil corals and associated reef organisms. (Photo courtesy of Bill Cornell, The University of Texas at El Paso.)

◆ **FIGURE 12.19** Reconstruction of the Cordilleran mobile belt during the Early Mississippian, showing the effects of the Antler orogeny.

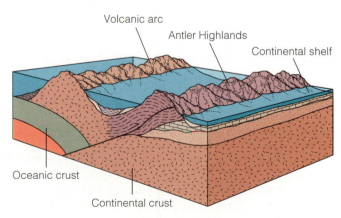

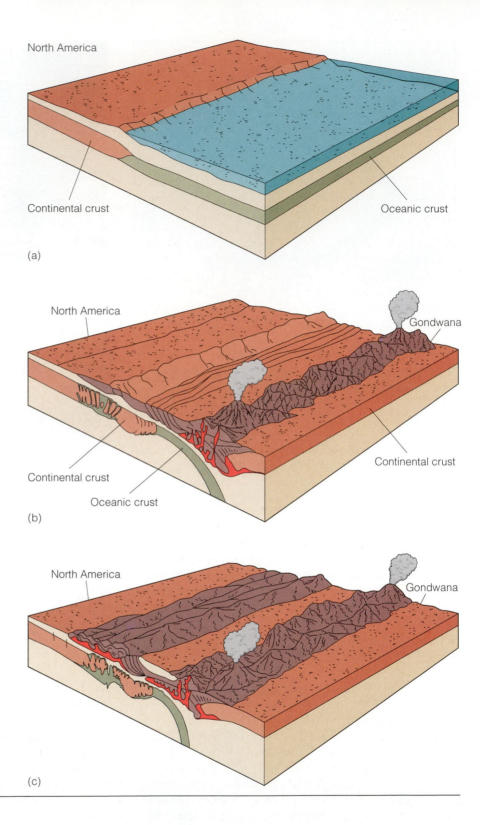

◆ **FIGURE 12.20** Plate tectonic model for deformation of the Ouachita mobile belt. (a) Depositional environment prior to the beginning of orogenic activity. (b) Incipient continental collision between North America and Gondwana began during the Mississippian to Pennsylvanian. (c) Continental collision continued during the Pennsylvanian Period.

only the collision of Laurasia and Gondwana, but several microplates between the continents that eventually became part of Central America. The compressive forces impinging on the Ouachita mobile belt also affected the craton by causing broad uplift of the southwestern part of North America as previously discussed.

Appalachian Mobile Belt

Caledonian Orogeny

The Caledonian mobile belt extends along the western border of Baltica and includes the present-day countries of Scotland, Ireland, and Norway (see Fig. 11.2c). During the Middle Ordovician, subduction along the boundary between the Iapetus plate and Baltica (Europe) began, forming a mirror image of the convergent plate boundary off the east coast of Laurentia (North America).

The culmination of the *Caledonian orogeny* (mentioned earlier) occurred during the Late Silurian and Early Devonian with the formation of a mountain range along the margin of Baltica. Red-colored sediments deposited along the front of the Caledonian highlands formed a large clastic wedge. These deposits are known as the Old Red Sandstone.

Acadian Orogeny

The third Paleozoic orogeny to affect Laurentia and Baltica began during the Late Silurian and concluded at the end of the Devonian Period. The *Acadian orogeny* (mentioned earlier) affected the Appalachian mobile belt from Newfoundland to Pennsylvania as sedimentary rocks were folded and thrust against the craton.

As with the preceding Taconic and Caledonian orogenies, the Acadian orogeny occurred along an oceanic-continental convergent plate boundary. As the northern Iapetus Ocean continued to close during the Devonian, the plate carrying Baltica finally collided with Laurentia, forming a continental-continental convergent plate boundary along the zone of collision (Fig. 12.1a).

As the greater metamorphic and igneous activity indicates, the Acadian orogeny was more intense and of longer duration than the Taconic orogeny. Radiometric dates from the metamorphic and igneous rocks associated with the Acadian orogeny cluster between 360 and 410 million years ago. Furthermore, just as with the Taconic orogeny, deep-water sediments were folded and thrust northwestward, producing profound angular un-

conformities separating Upper Silurian from Mississippian rocks.

Weathering and erosion of the Acadian Highlands produced a thick clastic wedge called the **Catskill Delta,** named for the Catskill Mountains in upstate New York where it is well exposed. The Catskill Delta, composed of red, coarse conglomerates, sandstones, and shales, contains nearly three times as much sediment as the Queenston Delta (see Fig. 11.14).

The Devonian rocks of New York are among the best studied on the continent. A cross section of the Devonian strata clearly reflects an eastern source (Acadian Highlands) for the Catskill facies (◆ Fig. 12.21). These clastic rocks can be traced from eastern Pennsylvania, where the coarse clastics are approximately 3 km thick, to Ohio, where the deltaic facies are only about 100 m thick and consist of cratonic shales and carbonates.

The red beds of the Catskill Delta derive their color from the hematite found in the sediments. Plant fossils and oxidation of the hematite indicate the beds were deposited in a subaerial environment. Toward the west, the red beds grade laterally into gray sandstones and shales containing fossil tree trunks, which indicate a swamp or marsh environment.

The Old Red Sandstone

The red beds of the Catskill Delta have a European counterpart in the Devonian Old Red Sandstone of the British Isles (Fig. 12.21). The Old Red Sandstone was a Devonian clastic wedge that grew eastward from the Caledonian Highlands onto the Baltica craton. The Old Red Sandstone, just like its North American Catskill counterpart, contains numerous fossils of freshwater fish, early amphibians, and land plants.

By the end of the Devonian Period, Baltica and Laurentia were sutured together, forming Laurasia (Fig. 12.21). The red beds of the Catskill Delta can be traced north, through Canada and Greenland, to the Old Red Sandstone of the British Isles and into northern Europe. These beds were deposited in similar environments along the flanks of developing mountain chains formed by tectonic forces at convergent plate boundaries.

Geologists now believe that the Taconic, Caledonian, and Acadian orogenies were all part of the same major orogenic event related to the closing of the Iapetus Ocean (see Figs. 11.13 and 12.21). This event began with paired oceanic-continental convergent plate boundaries during the Taconic and Caledonian orogenies and culminated with a continental-continental convergent plate boundary during the Acadian orogeny as Laurentia and Baltica

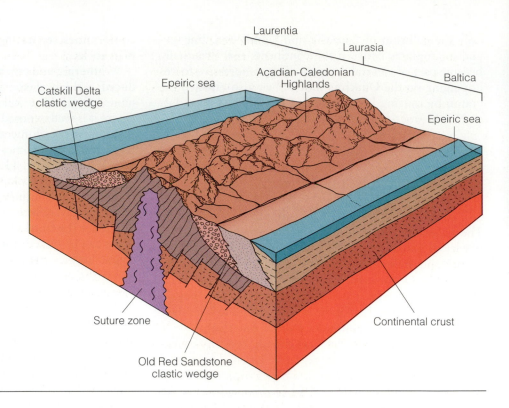

◆ **FIGURE 12.21** Cross Section showing the area of collision between Laurentia and Baltica. Note the bilateral symmetry of the Catskill Delta clastic wedge and the Old Red Sandstone and their relationship to the Acadian and Caledonian Highlands.

became sutured. Following this, the Hercynian-Alleghenian orogeny began, followed by orogenic activity in the Ouachita mobile belt.

Hercynian-Alleghenian Orogeny

The Hercynian mobile belt of southern Europe and the Appalachian and Ouachita mobile belts of North America mark the zone along which Europe (part of Laurasia) collided with Gondwana (Fig. 12.1). While Gondwana and southern Laurasia collided during the Pennsylvanian and Permian in the area of the Ouachita mobile belt, eastern Laurasia (Europe and southeastern North America) joined together with Gondwana (Africa) as part of the *Hercynian-Alleghenian orogeny* (Fig. 12.2).

Initial contact along the Hercynian mobile belt between eastern Laurasia and Gondwana began during the Mississippian Period. The greatest deformation occurred during the Pennsylvanian and Permian periods. This event is referred to as the *Hercynian orogeny* (mentioned earlier). The central and southern parts of the Appalachian mobile belt (from New York to Alabama) were folded and thrust toward the craton as eastern Laurasia and Gondwana were sutured. This event in North America is referred to as the *Alleghenian orogeny* (also mentioned earlier).

These three Late Paleozoic orogenies (Hercynian, Alleghenian, and Ouachita) represent the final joining of Laurasia and Gondwana into the supercontinent Pangaea during the Permian.

❖ PALEOZOIC MICROPLATES AND THE FORMATION OF PANGAEA

We have generally presented the geologic history of the mobile belts surrounding the Paleozoic continents in terms of subduction along convergent plate boundaries. It is becoming increasingly clear, however, that accretion along the continental margins is more complicated than the somewhat simple, large-scale plate interactions that we have described. Geologists now recognize that numerous terranes or microplates existed during the Paleozoic and were involved in the orogenic events that occurred during that time. In this chapter and the previous one, we have only been concerned with the six major Paleozoic continents. A careful examination of the Paleozoic global paleogeographic maps (Figs. 11.2, 12.1, and 12.2) shows that there were numerous microplates, and their location and role during the formation of Pangaea must be taken into account.

We presented the history of the Ouachita, Appalachian, and Hercynian mobile belts in terms of the closing of the Iapetus Ocean along zones of subduction that resulted in the collision of Laurasia and Gondwana and the formation of Pangaea. Geologists now know from detailed geologic and paleontologic studies that numerous microplates were also involved in the construction of Pangaea. For example, the small continent of Avalonia is composed of some coastal parts of New England, southern New Brunswick, much of Nova Scotia, the Avalon Peninsula of eastern Newfoundland, southeastern Ireland, Wales, England, and parts of Belgium and northern France. This microplate existed as a separate continent during the Ordovician and collided with Baltica during the Silurian and Laurentia during the Devonian (see Figs. 11.2 and 12.1).

Florida and parts of the eastern seaboard of North America make up the Piedmont microplate that was part of the larger Gondwana continent. This microplate became sutured to Laurasia during the Pennsylvanian Period. Numerous microplates occupied the region between Gondwana and Laurasia that eventually became part of Central America during the Pennsylvanian collision between these continents.

Thus, while the basic history of the formation of Pangaea during the Paleozoic remains the same, geologists now realize that microplates also played an important role. Furthermore, the recognition of terranes within mobile belts helps explain some previously anomalous geologic situations.

❖ LATE PALEOZOIC MINERAL RESOURCES

Late Paleozoic-aged rocks contain a variety of mineral resources including energy resources and metallic and nonmetallic mineral deposits. Petroleum and natural gas are recovered in commercial quantities from rocks ranging from the Devonian through Permian. For example, Devonian-aged rocks in the Michigan Basin, Illinois Basin, and the Williston Basin of Montana, South Dakota, and adjacent parts of Alberta, Canada, have yielded considerable amounts of hydrocarbons. Permian reefs and

other strata in the western United States, particularly Texas, have also been productive rocks.

Although Permian-aged coal beds are known from several areas including Asia, Africa, and Australia, much of the coal in North America and Europe is Pennsylvanian (Late Carboniferous). Large areas in the Appalachian region and the midwestern United States are underlain by vast coal deposits (◆ Fig. 12.22). These coal deposits formed from the lush vegetation such as the nonflowering seed-bearing plants called *lycopsids* that were the dominant plants in Pennsylvanian coal swamps (see Fig. 14.26).

Much of this coal is characterized as bituminous coal, which contains about 80% carbon. It is a dense, black coal that has been so thoroughly altered that plant remains can be seen only rarely. Bituminous coal is used to make *coke*, a hard, gray substance made up of the fused ash of bituminous coal. Coke is used to fire blast furnaces during the production of steel.

Some of the Pennsylvanian coal from North America is called *anthracite*, a metamorphic type of coal containing up to 98% carbon. Most anthracite is in the Appalachian region (Fig. 12.22). It is an especially desirable type of coal because it burns with a smokeless flame and it yields more heat per unit volume than other types of coal. Unfortunately, it is the least common type of coal, so much of the coal used in the United States is bituminous.

A variety of Late Paleozoic evaporite deposits are important nonmetallic mineral resources. The Zechstein evaporites of Europe extend from Great Britain across the North Sea and into Denmark, the Netherlands, Germany, and eastern Poland and Lithuania. In addition to the evaporites themselves, Zechstein deposits form the caprock for the large reservoirs of the gas fields of the Netherlands and part of the North Sea region. Furthermore, the *Kupferschiefer*, the chief copper deposits of Germany since 1150 A.D., are associated with Zechstein evaporites.

Other important evaporite mineral resources include those of the Permian Delaware Basin of West Texas and New Mexico, and Devonian evaporites in the Elk Point basin of Canada. In Michigan, gypsum is mined and used in the construction of wallboard.

In 1989, about two-thirds of the silica sand mined in the United States came from east of the Mississippi River, and much of this came from Upper Paleozoic rocks. For example, the Devonian Ridgeley Formation is mined in West Virginia, Maryland, and Pennsylvania, and the Devonian Sylvania Sandstone is mined near Detroit, Michigan. Recall from Chapter 11 that silica sand is used in

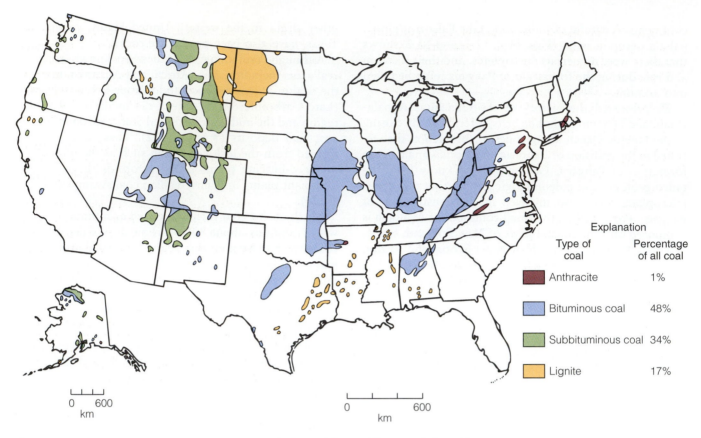

◆ **Figure 12.22** Distribution of coal deposits in the United States. The age of the coals in the midwestern states and the Appalachian region are mostly Pennsylvanian, whereas those in the west are mostly Cretaceous and Tertiary.

the manufacture of glass; for refractory bricks in blast furnaces; for molds for casting aluminum, iron, and copper alloys; and for a variety of other uses.

Late Paleozoic limestones from many areas in North America are used in the manufacture of cement. For example, mining of limestone in Michigan supported production of cement valued at more than $270 million in 1989. Limestone is also mined and used in blast furnaces when steel is produced.

Metallic mineral resources including tin, copper, gold, and silver are also known from Late Paleozoic-aged rocks, especially those that have been deformed during mountain building. Although the precise origin of the Missouri lead and zinc deposits remains unresolved, much of the ores of these metals come from Mississippian rocks. In fact, mines in Missouri accounted for about 89% of all 1989 domestic production of lead ores, and about 18% of all zinc ores.

Chapter Summary

Table 12.1 provides a summary of the geologic events occurring on the North American craton and mobile belts and shows how they relate to global events and changes in sea level during the Late Paleozoic.

1. During the Late Paleozoic, Baltica and Laurentia collided, forming Laurasia. Siberia and Kazakhstania collided and finally were sutured to Laurasia. Gondwana moved over the South Pole and experienced several glacial-interglacial peri-

ods, resulting in global changes in sea level and transgressions and regressions along the low-lying craton margins.

2. Laurasia and Gondwana underwent a series of collisions beginning in the Carboniferous. During the Permian, the formation of Pangaea was completed. Surrounding the supercontinent was a global ocean, Panthalassa.

3. Much of the history of the North American craton can be deciphered from the rocks of the Kaskaskia and Absaroka sequences.

4. The basal beds of the Kaskaskia sequence deposited over the exposed Tippecanoe surface consisted of either mature sandstones, derived from the eroding Taconic Highlands, or carbonate rocks.

5. Most of the Kaskaskia sequence was dominated by carbonates and associated evaporites. The Devonian Period was a time of major reef building in western Canada, southern England, Belgium, Australia, and Russia.

6. A persistent and widespread black shale, the Chattanooga Shale and its equivalents, was deposited over a large area of the craton during the Late Devonian and Early Mississippian.

7. The Mississippian Period was dominated for the most part by carbonate deposition.

8. Transgressions and regressions over the low-lying craton resulted in cyclothems and the formation of coals during the Pennsylvanian Period.

9. Cratonic mountain building occurred during the Pennsylvanian Period, and thick nonmarine detrital rocks and evaporites were deposited in the intervening basins.

10. By the Early Permian, the Absaroka Sea occupied a narrow zone of the south-central craton. Here, several large reefs and associated evaporites developed. By the end of the Permian Period, the Absaroka Sea had retreated from the craton.

11. The Cordilleran mobile belt was the site of a minor Devonian orogeny called the Antler orogeny during which deep-water sediments were thrust eastward over shallow-water sediments.

12. Mountain building occurred in the Ouachita mobile belt during the Pennsylvanian and Early Permian. This tectonic activity was partly responsible for the cratonic uplift that occurred in the southwest, producing the Ancestral Rockies.

13. The Caledonian, Acadian, Hercynian, and Alleghenian orogenies were all part of the global tectonic activity resulting from the assembly of Pangaea.

14. During the Paleozoic Era, numerous microplates existed and played an important role in the formation of Pangaea.

15. Late Paleozoic–aged rocks contain a variety of mineral resources including petroleum, coal, evaporites, silica sand, lead, zinc, and other metallic deposits.

Important Terms

Absaroka sequence
Acadian orogeny
Alleghenian orogeny
Ancestral Rockies
Antler orogeny

Caledonian orogeny
Catskill Delta
cyclothem
Ellesmere orogeny
Hercynian orogeny

Kaskaskia sequence
Laurasia
Ouachita orogeny
Panthalassa Ocean

Review Questions

1. The first Paleozoic orogeny to occur in the Cordilleran mobile belt was the:
   a. _____ Acadian; b. _____ Alleghenian; c. _____ Antler; d. _____ Caledonian; e. _____ Ellesmere.

2. Extensive continental glaciation of the Gondwana continent occurred during which period?
   a. _____ Devonian; b. _____ Silurian; c. _____ Carboniferous; d. _____ Cambrian; e. _____ Permian.

3. Major reef development in western Canada occurred in which sequence?
   a. _____ Sauk; b. _____ Tippecanoe; c. _____ Absaroka; d. _____ Zuni; e. _____ Kaskaskia.

4. Extensive cratonic black shales were deposited during which two periods?

a. _____ Late Silurian–Early Devonian; b. _____ Late Devonian–Early Mississippian; c. _____ Late Mississippian–Early Pennsylvanian; d. _____ Late Pennsylvanian–Early Permian; e. _____ Late Permian–Early Triassic.

5. Which of the following is not a requirement for the formation of extensive black shales?
   a. _____ undisturbed anaerobic bottom water; b. _____ reduced supply of detrital sediment; c. _____ high organic productivity in the overlying oxygenated waters; d. _____ low-energy depositional site; e. _____ turbulent anaerobic bottom water.

6. Which of the following was not a source area for the basal Kaskaskia sandstones?

(Continued on page 346)

TABLE 12.1   Summary of Late Paleozoic Geologic Events

Age (Millions of Years)	Geologic Period		Sequence	Relative Changes in Sea Level		Cordilleran Mobile Belt
				Rising	Falling	
245	Permian		Absaroka			
286	Carboniferous	Pennsylvanian	Absaroka			
320	Carboniferous	Mississippian	Kaskaskia			
360	Devonian		Kaskaskia			Antler orogeny
	Devonian		Tippecanoe	Present sea level		Klamath Arc
408			Tippecanoe			

TABLE 12.1   Summary of Late Paleozoic Geologic Events (continued)

Craton	Ouachita Mobile Belt	Appalachian Mobile Belt	Major Events Outside North America
Deserts, evaporites, and continental red beds in southwestern United States. Extensive reefs in Texas area.		Formation of Pangaea	
		Allegheny orogeny	Hercynian orogeny
Coal swamps common.	Ouachita orogeny		
Formation of Ancestral Rockies.			Continental glaciation in Southern Hemisphere.
Transgression of Absaroka Sea.			
Widespread black shales and limestones.			
Widespread black shales. Catskill Delta clastic wedge.		Acadian orogeny	Old Red Sandstone clastic wedge in British Isles.
Extensive barrier-reef formation in Western Canada. Transgression of Kaskaskia Sea.			Caledonian orogeny

a. _____ Transcontinental Arch; b. _____ Appalachian mobile belt highlands; c. _____ margins of the Ozark Dome; d. _____ Canadian Shield in Wisconsin; e. _____ answers (b) and (c).

7. Rhythmically repetitive sedimentary sequences are:
a. _____ tillites; b. _____ cyclothems; c. _____ orogenies; d. _____ reefs; e. _____ evaporites.

8. Uplift in the southwestern part of the craton during the Late Absaroka resulted in which mountainous region?
a. _____ Ancestral Rockies; b. _____ Antler Highlands; c. _____ Appalachians; d. _____ Marathon; e. _____ none of these.

9. During which period did the Ouachita mobile belt change from a passive plate margin to an active plate margin?
a. _____ Silurian; b. _____ Devonian; c. _____ Mississippian; d. _____ Pennsylvanian; e. _____ Permian.

10. Weathering of which highlands or mountains produced the Catskill Delta?
a. _____ Acadian; b. _____ Alleghenian; c. _____ Antler; d. _____ Appalachian; e. _____ Caledonian.

11. The European counterpart to the Devonian Catskill Delta is the:
a. _____ Phosphoria Formation; b. _____ Oriskany Sandstone; c. _____ Capitan Limestone; d. _____ Old Red Sandstone; e. _____ none of these.

12. What type of plate boundary resulted from the Alleghenian orogeny?
a. _____ divergent; b. _____ oceanic-oceanic convergent; c. _____ oceanic-continental convergent; d. _____ continental-continental convergent; e. _____ transform.

13. Which orogeny was not part of the closing of the Iapetus Ocean?
a. _____ Acadian; b. _____ Alleghenian; c. _____ Antler; d. _____ Caledonian; e. _____ Taconic.

14. During the Late Kaskaskia, what type of deposition predominated on the craton?
a. _____ black shales; b. _____ carbonates; c. _____ volcanics; d. _____ metamorphics; e. _____ cherts and graywackes.

15. What is the main economic deposit of a cyclothem?
a. _____ evaporites; b. _____ carbonates; c. _____ gravels; d. _____ ores; e. _____ coals.

16. What are some Late Paleozoic economic deposits and how did they form?

17. Discuss how plate movement during the Late Paleozoic affected worldwide weather patterns.

18. Compare the formation of the Oriskany Sandstone with that of the St. Peter Sandstone.

19. Discuss the formation of the Permian reefs.

20. Provide a geologic history for the Iapetus Ocean during the Paleozoic Era.

21. What are cyclothems? How do they form? Why are they economically important?

22. What was the relationship between the Ouachita orogeny and the cratonic uplifts that occurred on the craton during the Pennsylvanian Period?

23. What is the evidence for glaciation on Gondwana during the Late Paleozoic?

24. What is the evidence for arid conditions in Laurentia during the Permian?

25. What do the various models proposed for the deposition of the Late Devonian–Early Mississippian black shales have in common? How do they differ? What is the evidence for a shallow-water depositional environment?

26. How are the Caledonian, Acadian, Ouachita, Hercynian, and Alleghenian orogenies related to modern concepts of plate tectonics?

27. Compare the Taconic, Caledonian, and Acadian orogenies in terms of the tectonic forces that caused them and the sedimentary features that resulted.

28. How does the origin of evaporite deposits of the Kaskaskia sequence compare with the origin of evaporites of the Tippecanoe sequence?

29. How did the sedimentary environments of the Kaskaskia sequence differ from those of the Absaroka sequence in North America?

30. How did the formation of Pangaea and Panthalassa affect the world's climate at the end of the Paleozoic Era?

Additional Readings

Bally, A. W., and A. R. Palmer, eds. 1989. *The geology of North America: An overview.* The Geology of North America. vol. A. Boulder, Colo.: Geological Society of America.

Bambach, R. K., C. R. Scotese, and A. M. Ziegler. 1980. Before Pangaea: The geographies of the Paleozoic world. *American Scientist* 68, no. 1: 26–38.

Dallmeyer, R. D., ed. 1989. *Terranes in the Circum-Atlantic Paleozoic orogens.* Geological Society of America Special Paper 230. Boulder, Colo.: Geological Society of America.

Dineley, D. L. 1984. *Aspects of a stratigraphic system—The Devonian.* New York: John Wiley & Sons.

Fouch, T. D., and E. R. Magathan, eds. 1980. *Paleozoic paleogeography of the west-central United States.* Rocky

Mountain Paleogeography Symposium 1. Denver, Colo.: Rocky Mountain Section, Society of Economic Paleontologists and Mineralogists.

Harwood, D. S., and M. M. Miller, eds. 1990. *Paleozoic and Early Mesozoic paleogeographic relations: Sierra Nevada, Klamath Mountains, and related terranes*. Geological Society of America Special Paper 255. Boulder, Colo.: Geological Society of America.

Hatcher R. D., Jr., W. A. Thomas, and G. W. Viele, eds. 1989. *The Appalachian-Ouachita orogen in the United States*. The Geology of North America. vol. F-2. Boulder, Colo.: Geological Society of America.

McKerrow, W. S., and C. R. Scotese, eds. 1990. *Palaeozoic palaeogeography and biogeography*. The Geological Society Memoir No. 12. London, England: Geological Society of London.

Moullade, M., and A. E. M. Nairn, eds. 1991. *The Phanerozoic geology of the world I: The Paleozoic*. New York: Elsevier Science Publishing Co.

Sloss, L. L., ed. 1988. *Sedimentary cover—North American Craton: U.S.* The Geology of North America. vol. D-2. Boulder, Colo.: Geological Society of America.

Stewart, J. H., C. H. Stevens, and A. E. Fritsche, eds. 1977. *Paleozoic paleogeography of the western United States*. Pacific Coast Paleogeography Symposium 1. Tulsa, Okla.: Society of Economic Paleontologists and Mineralogists.

Woodrow, D. L., and W. D. Sevan. 1985. The Catskill Delta. *Geological Society of America Special Paper* 201: 1–246.

Ziegler, P. A. 1989. *Evolution of Laurussia: A study in Late Palaeozoic plate tectonics*. Boston, Mass.: Kluwer Academic Publishers.

CHAPTER 13

Diorama of the environment and biota of the Phyllopod bed of the Burgess Shale. In the background is the vertical wall of a submarine escarpment with algae growing on it. The large cylindrical ribbed organisms on the muddy bottom in the foreground are sponges. (Photo courtesy of the Smithsonian Institution, Transparency No. 86-13471A.)

LIFE OF THE PALEOZOIC ERA: INVERTEBRATES

Prologue

On August 30 and 31, 1909, near the end of the summer field season, Charles D. Walcott, geologist and head of the Smithsonian Institution, was searching for fossils along a trail on Burgess Ridge between Mount Field and Mount Wapta, near Field, British Columbia, Canada. On the west slope of this ridge, he discovered the first soft-bodied fossils from the Burgess Shale, a discovery of immense importance in deciphering the early history of life. During the following week, Walcott and his collecting party split open numerous blocks of shale, many of which yielded the impressions of a number of soft-bodied organisms beautifully preserved on bedding planes. Walcott returned to the site the following summer and located the shale stratum that was the source of his fossil-bearing rocks in the steep slope above the trail. He quarried the site and shipped back thousands of fossil specimens to the United States National Museum of Natural History, where he later cataloged and studied them.

The importance of Walcott's discovery is not that it was another collection of well-preserved Cambrian fossils, but rather that it allowed geologists a rare glimpse into a world previously almost unknown—that of the soft-bodied animals that lived some 530 million years ago. The beautifully preserved fossils from the Burgess Shale present a much more complete picture of a Middle Cambrian community than deposits containing only fossils of the hard parts of organisms. Specifically, the Burgess Shale contains species of trilobites, sponges, brachiopods, mollusks, and echinoderms, all of which have hard parts and are characteristic of Cambrian faunas throughout the world. But in addition to the diverse skeletonized fauna, a large and varied fossil assemblage of soft-bodied animals is also present. In fact, 60% of the total fossil assemblage is composed of soft-bodied animals, which usually are not preserved. In all, more than 100 genera of animals, at least 60 of which were soft-bodied and preserved as impressions, have been recovered from the Burgess Shale. This proportion of soft-bodied animals to those with hard parts is comparable to modern marine communities.

The Burgess Shale fauna reveals the evolutionary stage that marine life had reached by the Middle Cambrian: highly advanced organisms comprised complex communities as diverse in structure and adaptation as many modern marine communities. The Burgess Shale fauna makes speculation about the early evolution of multicelled organisms difficult because it shows how little we know about early marine life. In fact, some of the organisms preserved may not represent any known phyla at all.

What conditions led to the remarkable preservation of the Burgess Shale fauna? When it was deposited, the Burgess Shale was located at the base of a steep submarine escarpment. The animals whose exquisitely preserved fossil remains are found in the Burgess Shale lived in and on mud banks that formed along the top of this escarpment. Periodically, this unstable area would slump and slide down the escarpment as a turbidity flow. At the base, the mud and animals carried with it were deposited in a deep-water anoxic environment devoid of life. In such an environment, bacterial degradation did not destroy the buried animals, and they were compressed by the weight of the overlying sediments, eventually resulting in their preservation as carbonaceous impressions.

❖ INTRODUCTION

In the following two chapters, we examine the history of Paleozoic life as a series of interconnected biologic and geologic events in which the underlying principles and

Archean Eon	Proterozoic Eon	Phanerozoic Eon				
Precambrian		Paleozoic Era				
		Cambrian	Ordovician	Silurian	Devonian	Mississippian	Pennsylvanian	Permian
						Carboniferous	

2,500
M.Y.A.

570
M.Y.A.

245
M.Y.A.

processes of evolution and plate tectonics played a major role. The opening and closing of ocean basins, transgressions and regressions of epeiric seas, and the changing positions of the continents had a profound effect on the evolution of the marine and terrestrial communities.

A time of tremendous biologic change began with the appearance of skeletonized animals near the Precambrian-Cambrian boundary. Following this event, marine invertebrates began a period of adaptive radiation and evolution during which the Paleozoic marine invertebrate community greatly diversified. Indeed, the history of the Paleozoic marine invertebrate community was one of diversifications and extinctions.

Vertebrates also evolved during the Paleozoic. The earliest fossil records of vertebrates are of fish, which evolved during the Cambrian. One group of fish was ancestral to the first land animals, the amphibians, which evolved during the Devonian. Reptiles evolved from a group of amphibians during the Pennsylvanian Period and were the dominant vertebrate animals on land by the end of the Paleozoic Era.

Plants preceded animals onto the land. Both plants and animals had to solve the same basic problems in making the transition from water to land. The method of reproduction proved to be the major barrier to expansion into new environments for both groups. With the evolution of the seed in plants and the amniote egg in animals, this limitation was removed, and both groups were able to move into all terrestrial environments.

The end of the Paleozoic Era was a time of major extinctions. The marine invertebrate community was greatly decimated, and many amphibians and reptiles on land also became extinct.

❖ THE FIRST SHELLED FOSSILS

Several important questions that arise regarding the early history of life relate to the acquisition of a mineralized exoskeleton and its adaptive significance, the composition of the exoskeleton, and the rapid radiation of skeletonized animals starting near the beginning of the Cambrian Period. Early geologists observed that the remains of skeletonized animals appeared rather abruptly in the fossil record. Charles Darwin addressed this problem in *On the Origin of Species* and observed that without a

◆ FIGURE 13.1 The first shelled fossils. These small calcium carbonate tubes presumably housed wormlike suspension-feeding organisms. They are found associated with Ediacaran faunas from Namibia and China. (a) *Cloudina*. (b) *Sinotubulites*.

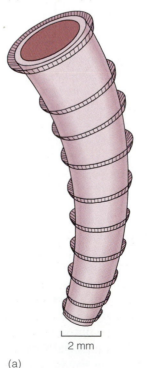

2 mm

(a)

1 mm

(b)

Phanerozoic Eon										
Mesozoic Era			Cenozoic Era							
Triassic	Jurassic	Cretaceous	Tertiary						Quaternary	
			Paleocene	Eocene	Oligocene	Miocene	Pliocene		Pleistocene	Holocene

245
M.Y.A.

66
M.Y.A.

convincing explanation, such an event was difficult to reconcile with his newly expounded evolutionary theory.

Most scientists recognize that the variety and complexity of Cambrian life suggest that multicelled organisms must have had a Precambrian history during which they lacked hard parts and thus did not leave a fossil record. As we discussed in Chapter 10, the oldest apparently multicelled organisms, a possible alga, are found in the 1.4-billion-year-old Belt Supergroup of Montana (see Fig. 10.27), and recently wormlike fossils have been reported from rocks 700 to 900 million years old in China (see Fig. 10.30).

Impressions of the first unequivocally multicelled animals belong to the widely distributed Ediacaran fauna and are found in rocks between 570 to 670 million years old (see Fig. 10.28). Associated with Ediacaran faunas in Namibia and southern China are the first shelled fossils (◆ Fig. 13.1). These are small calcium carbonate tubes, presumably housing wormlike suspension-feeding organisms. In addition, small organic tube-shaped fossils, also presumably housing wormlike suspension-feeding animals occur with the calcareous tubes. By the latest part of the Proterozoic, several skeletonized animals had made their appearance, yet durable skeletons of chitin, silica, and calcium carbonate did not begin to appear in abundance until the beginning of the Phanerozoic Eon 570 million years ago.

❖ THE CAMBRIAN PERIOD AND THE EMERGENCE OF A SHELLY FAUNA

The Early Cambrian was characterized by a low-diversity shelly fauna consisting of animals that used both calcium carbonate and calcium phosphate to construct their skeletons. They included small worm tubes, mollusks and echinoderms, archaeocyathids, and brachiopods (◆ Fig. 13.2). It is likely that this fauna was yet another "experiment," like the Ediacaran fauna of the Proterozoic Eon, but this experiment was very successful. By the Middle Cambrian, a large number of the major groups of invertebrate animals had evolved. Many, such as the brachiopods, are still around today, while others, including the archaeocyathids and trilobites, are extinct.

◆ **FIGURE 13.2** Three small Lower Cambrian shelly fossils. (a) A conical sclerite (a piece of the armor covering) of *Lapworthella* from Australia. (b) *Archaeooides,* an enigmatic spherical fossil from the Mackenzie Mountains, Northwest Territories, Canada. (c) The tube of an anabaritid from the Mackenzie Mountains, Northwest Territories, Canada. (Photos courtesy of Simon Conway Morris and Stefan Bengtson, University of Cambridge, England.)

(a)

(b)

(c)

The Cambrian Period and the Emergence of a Shelly Fauna   351

The Cambrian Period was also a time during which new body plans evolved (◆ Fig. 13.3), and animals moved into new niches. As might be expected, the Cambrian witnessed a higher percentage of such experiments than any other period of geologic history.

The emergence of so many different organisms at this juncture in the Earth's history resulted in the development of new relationships and community structures as organisms underwent tremendous evolutionary change and filled previously unoccupied niches.

❖ THE ACQUISITION AND SIGNIFICANCE OF HARD PARTS

A striking aspect of the Early Cambrian fauna is that many animals already had fully developed features; that is, their anatomies indicate an extended period of evolution before the evolution of hard parts. Furthermore, the

	Cambrian			Ordovician	
	Early	Middle	Late	Early	Middle
Arthropoda (Trilobites)					→
Brachiopoda					→
Echinodermata					→
Mollusca (Gastropoda)					→
Porifera					→
Mollusca (Bivalvia)					→
Mollusca (Cephalopoda)					→
Protozoa					→
Bryozoa					→
Cnidaria (Tabulate corals)					→
Cnidaria (Rugose corals)					→

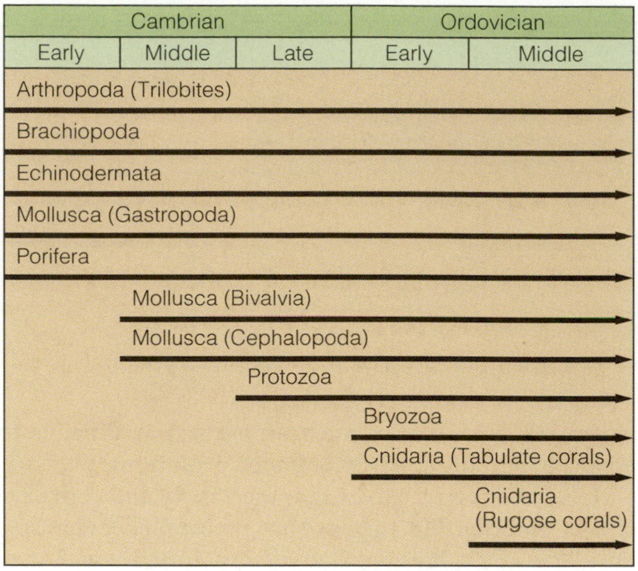

◆ FIGURE 13.4 The first recorded occurrence of selected members of the major marine invertebrate phyla.

◆ FIGURE 13.3 *Helicoplacus*, a primitive echinoderm that became extinct 20 million years after its first appearance about 510 million years ago. Such an organism (a representative of one of several short-lived echinoderm classes) illustrates the "experimental" nature of the Cambrian invertebrate fauna. (Photo by Porter M. Kier, courtesy of J. Wyatt Durham, University of California, Berkeley.)

major skeletonized animal groups did not all evolve at once, but rather evolved throughout the Cambrian and Ordovician periods (◆ Fig. 13.4). A preskeletonized period of evolution occurred during which members of a phylum evolved for an unknown period of time as soft-bodied organisms.

But why did invertebrates initially acquire skeletons and why did such acquisitions occur over an extended period of time? Various explanations have been proposed, but none is completely satisfactory or universally accepted. Many geologists long believed that hard parts appeared rather suddenly in the fossil record as a result of changes in the chemistry of the Late Precambrian oceans. According to this hypothesis, the Precambrian oceans were deficient in calcium, carbonate, and phosphate ions, which are the materials found in mineralized skeletons. Without sufficient quantities of these ions, invertebrates could not easily construct a skeleton. This explanation is now generally rejected because numerous Late Precambrian and Early Cambrian carbonate rocks are known from around the world as well as the earliest known phosphate deposits. The occurrence of these rocks indicates that sufficient calcium, carbonate, and phosphate were available for skeleton construction.

Another explanation is that mineralized skeletons evolved as a response to invertebrates' need to eliminate mineral matter from their metabolic systems. One way to do this is to secrete a buildup of excess ions as a solid. In this way a mineralized skeleton may have evolved as a means of eliminating high levels of calcium and phosphate ions. The evolution of a skeleton would then prove to be advantageous to the organism in a variety of ways that we will discuss shortly. Because we now know that the acquisition of hard parts by the various invertebrate phyla occurred over an extended period of time and was not a sudden event, this explanation is appealing to many paleontologists.

The formation of an exoskeleton confers many advantages on an organism: (1) It provides protection against ultraviolet radiation, allowing animals to move more eas-

ily into shallower waters. (2) It helps prevent drying out in an intertidal environment. (3) It provides protection against predators. Previous assumptions that predators in Cambrian communities were not important have proved to be wrong. Recent evidence including actual fossils of predators and specimens of damaged prey, as well as antipredatory adaptations in some animals, indicates that the impact of predation was great (◆ Fig. 13.5). With predators playing an important role in the Cambrian marine ecosystem, any mechanism or feature that protected an animal would certainly be advantageous and confer an adaptive advantage to the organism. (4) A supporting skeleton, whether an exo- or endoskeleton, allows animals to increase their size. (5) It also provides attachment sites for development of strong muscles, thus increasing locomotor efficiency in mobile animals.

Currently, there is no clear answer as to why marine invertebrates acquired mineralized skeletons over an extended period of time. Hard parts undoubtedly originated because of a variety of biologic and environmental factors rather than a single one. Whatever the reason, the acquisition of a mineralized skeleton was a major evolutionary innovation allowing invertebrates to occupy a wide variety of marine habitats successfully.

◆ FIGURE 13.5 (a) Reconstruction of *Anomalocaris,* a predator from the Early and Middle Cambrian Period. It was about 45 cm long and probably fed on trilobites. Its gripping appendages presumably carried food to its mouth.
(b) Wounds to the body of the trilobite *Olenellus robsonensis* (arrow). The wounds have healed, demonstrating that they occurred when the animal was alive and were not inflicted on an empty shell. (Photo courtesy of the Geological Survey of Canada, Photo 202109-5.)

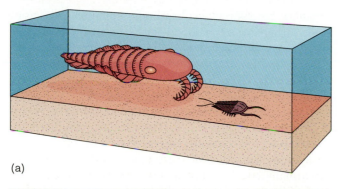

(a)

(b)

❖ PALEOZOIC INVERTEBRATE MARINE LIFE

Having considered the origin, differentiation, and evolution of the Precambrian-Cambrian marine biota, we now examine the changes that occurred in the marine invertebrate community during the Paleozoic Era. Rather than focusing on the history of each invertebrate phylum (Table 13.1), we will survey the evolution of the marine invertebrate communities through time, concentrating on the major features and the changes that took place. To do that, we need to briefly examine the nature and structure of living marine communities so that we can make a reasonable interpretation of the fossil record.

The Present Marine Ecosystem

In analyzing the present-day marine ecosystem, we must look at where organisms live and how they get around, as well as how they feed (◆ Fig. 13.6). Organisms that live in the water column above the seafloor are called *pelagic.* They can be divided into two main groups: the floaters, or **plankton,** and the swimmers, or **nekton.**

TABLE 13.1 The Major Invertebrate Groups and Their Stratigraphic Ranges

Phylum Protozoa	Cambrian-Recent	**Phylum Mollusca**	Cambrian-Recent
Class Sarcodina	Cambrian-Recent	Class Monoplacophora	Cambrian-Recent
Order Foraminifera	Cambrian-Recent	Class Gastropoda	Cambrian-Recent
Order Radiolaria	Cambrian-Recent	Class Bivalvia	Cambrian-Recent
		Class Cephalopoda	Cambrian-Recent
Phylum Porifera	Cambrian-Recent		
		Phylum Annelida	Precambrian-Recent
Phylum Archaeocyatha	Cambrian		
		Phylum Arthropoda	Cambrian-Recent
Phylum Cnidaria	Cambrian-Recent	Class Trilobita	Cambrian-Permian
Class Anthozoa	Ordovician-Recent	Class Crustacea	Cambrian-Recent
Order Tabulata	Ordovician-Permian	Class Insecta	Silurian-Recent
Order Rugosa	Ordovician-Permian		
Order Scleractinia	Triassic-Recent	**Phylum Echinodermata**	Cambrian-Recent
Class Hydrozoa	Cambrian-Recent	Class Blastoidea	Ordovician-Permian
Order Stromatoporoida	Cambrian-Cretaceous	Class Crinoidea	Cambrian-Recent
		Class Echinoidea	Ordovician-Recent
Phylum Bryozoa	Ordovician-Recent	Class Asteroidea	Ordovician-Recent
Phylum Brachiopoda	Cambrian-Recent	**Phylum Hemichordata**	Cambrian-Recent
Class Inarticulata	Cambrian-Recent	Class Graptolithina	Cambrian-Mississippian
Class Articulata	Cambrian-Recent		

Plankton are mostly passive and go where the current carries them. Plant plankton such as diatoms, dinoflagellates, and various algae, are called *phytoplankton* and are mostly microscopic. Animal plankton are called *zooplankton* and are also mostly microscopic. Examples of zooplankton include foraminifera, radiolarians, and jellyfish. The nekton are swimmers and are mainly vertebrates such as fish; the invertebrate nekton include cephalopods.

Organisms that live on or in the seafloor make up the **benthos.** They can be characterized as *epifauna* (animals) or *epiflora* (plants), for those that live on the seafloor, and the *infauna,* which are animals living in and moving through the sediments. The benthos can be further divided into those organisms that stay in one place, called *sessile,* and those that move around on or in the seafloor, called *mobile.*

The feeding strategies of organisms are also important in terms of their relationships with other organisms in the marine ecosystem. There are basically four feeding groups: **suspension-feeding** animals remove or consume microscopic plants and animals as well as dissolved nutrients from the water; **herbivores** are plant eaters; **carnivore-scavengers** are meat eaters; and **sediment-deposit feeders** ingest sediment and extract the nutrients from it.

We can define an organism's place in the marine ecosystem by where it lives and how it eats. For example, an articulate brachiopod is a benthonic, epifaunal suspension feeder, whereas a cephalopod is a nektonic carnivore.

An ecosystem includes several **trophic levels** which are tiers of food production and consumption within a feeding hierarchy. The feeding hierarchy and hence energy flow in an ecosystem comprise a food web of complex interrelationships among the producers, consumers, and decomposers (◆ Fig. 13.7). The **primary producers,** or **autotrophs,** are those organisms that manufacture their own food. Virtually all marine primary producers are phytoplankton. Feeding on the primary producers are the primary consumers, which are mostly suspension feeders. Secondary consumers feed on the primary consumers, and thus are predators, while tertiary consumers, which are also predators, feed on the secondary consumers. In addition to the producers and consumers, there are also transformers and decomposers. These are bacteria that break down the dead organisms that have not been consumed into organic compounds that are then recycled.

When we look at the marine realm today, we see a complex organization of organisms interrelated by trophic interactions and affected by changes in the physical environment. When one part of the system changes, the whole structure changes, sometimes almost insignificantly, other times catastrophically.

As we examine the evolution of the Paleozoic marine ecosystem, keep in mind how geologic and evolutionary changes can have a significant impact on the composition and structure of the ecosystem. For example, the major transgressions onto the craton opened up vast areas of shallow seas that could be inhabited. The movement of continents affected oceanic circulation patterns as well as causing environmental changes as the continents and their epeiric seas moved through different climatic zones.

Cambrian Marine Community

Although almost all the major invertebrate phyla appeared in the fossil record during the Cambrian Period, many were represented by only a few species. While trace fossils are common, and echinoderms diverse, trilobites, brachiopods, and archaeocyathids comprised the majority of Cambrian skeletonized life (◆ Fig. 13.8).

Trilobites were by far the most conspicuous element of the Cambrian marine invertebrate community and made up about half of the total fauna. Trilobites were benthonic mobile sediment-deposit feeders that crawled or swam along the seafloor. They first appeared in the Early Cambrian, reached their peak in the Late Cambrian, and then suffered mass extinctions near the end of the Cambrian from which they never fully recovered. As yet no consensus exists on what caused the trilobite extinctions,

◆ **FIGURE 13.6** Where and how animals and plants live in the marine ecosystem.
Plankton: (a) jellyfish. Nekton: (b) fish and (c) cephalopod. Benthos: (d) through (k).
Sessile epiflora: (d) seaweed. Sessile epifauna: (g) bivalve, (i) coral, and (j) crinoid.
Mobile epifauna: (k) starfish and (h) gastropod. Infauna: (e) worm and (f) bivalve.
Suspension feeders: (g) bivalve, (i) coral, and (j) crinoid. Herbivores: (h) gastropod.
Carnivores-scavengers: (k) starfish. Sediment-deposit feeders: (e) worm.

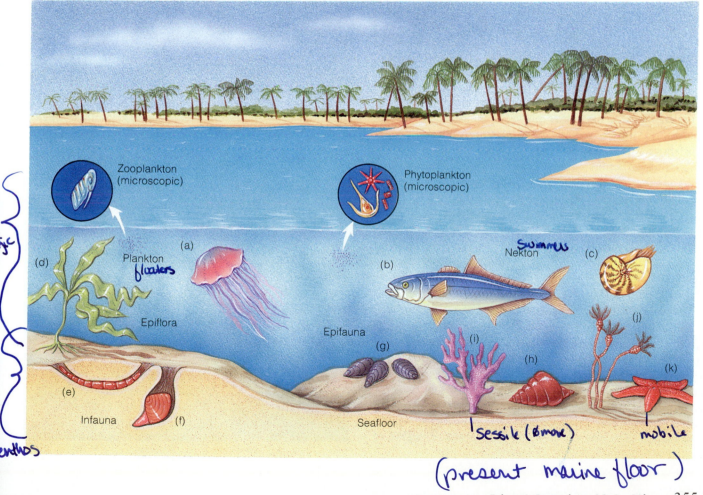

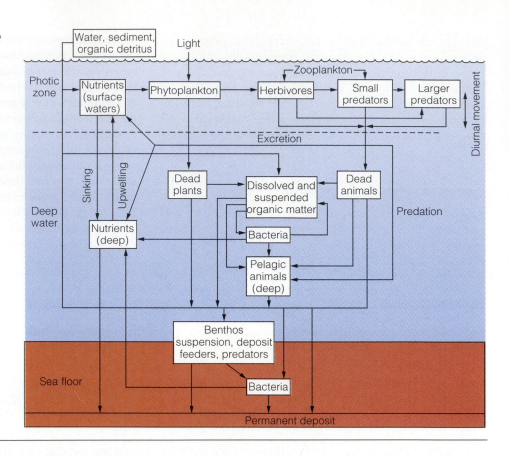

◆ **FIGURE 13.7** Marine food web showing the relationships among the producers, consumers, and decomposers.

but it has been suggested that a sudden, temporary cooling of the seas may have played a major role. As supporting evidence, the trilobites living in the cool, deeper offshore waters escaped the mass extinctions.

Trilobite faunas can be grouped into distinct faunal realms that characterize the paleogeography of the time. Furthermore, trilobites are excellent guide fossils. Those of the Early Cambrian were cosmopolitan in distribution, while those of the Middle and Late Cambrian were more provincial.

Cambrian **brachiopods** were mostly primitive types called *inarticulates*. They secreted a chitinophosphate shell, composed of the organic compound chitin combined with calcium phosphate. Inarticulate brachiopods also lacked a tooth-and-socket-arrangement along the hinge line. The *articulate* brachiopods, which have a tooth-and-socket arrangement, were also present but did not become abundant until the Ordovician Period.

The third major group of Cambrian organisms were the **archaeocyathids** (◆ Fig. 13.9). These organisms were benthonic sessile suspension feeders that constructed reeflike structures. The rest of the Cambrian fauna consisted of

representatives of the other major phyla, including many organisms that were short-lived evolutionary experiments.

◆ **FIGURE 13.8** Reconstruction of a Cambrian marine community. Floating jellyfish, swimming arthropods, benthonic sponges, and scavenging trilobites are shown. (Photo courtesy of the Carnegie Museum of Natural History.)

◆ FIGURE 13.9 Reconstruction of a Cambrian reeflike structure built by archaeocyathids.

The Burgess Shale Biota

No discussion of Cambrian life would be complete without mentioning one of the best examples of a preserved soft-bodied fauna and flora, the Burgess Shale biota. As the Sauk Sea transgressed from the Cordilleran shelf onto the western edge of the craton, Early Cambrian sands were covered by a Middle Cambrian black, anoxic mud that allowed a diverse soft-bodied benthonic community to be preserved. As we discussed in the Prologue, these fossils were discovered in 1909 by Charles D. Walcott near Field, British Columbia. They represent one of the most significant fossil finds of the century because they consist of impressions of soft-bodied animals and plants (◆ Fig. 13.10), which are very rarely preserved in the fossil record. This discovery, therefore, allows us a valuable glimpse of rarely preserved organisms as well as the soft-part anatomy of many extinct groups.

In recent years, the reconstruction, classification, and interpretation of many of the Burgess Shale fossils have undergone a major change that has led to new theories and explanations of the Cambrian explosion of life. Recall that during the Late Proterozoic multicellular organisms evolved, and shortly thereafter animals with hard

◆ FIGURE 13.10 Some of the fossil animals preserved in the Burgess Shale.
(a) *Burgessia bella*, an arthropod. (b) *Waptia fieldensis*, another arthropod.
(c) *Burgessochaeta setigera*, an annelid worm. (d) *Marrella splendens*, an arthropod that is the most abundant fossil animal in the Phyllopod bed of the Burgess Shale. (Photos courtesy of the Smithsonian Institution.)

(a)

(b)

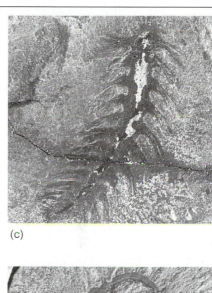

(c)

(d)

parts made their first appearance. These were followed by an explosion of invertebrate phyla during the Cambrian, some of which are now extinct. These Cambrian phyla represent the root stock and basic body plans from which all present-day invertebrates evolved. The question that paleontologists are now hotly debating is how many phyla arose during the Cambrian, and at the center of that debate are the Burgess Shale fossils. For years, most paleontologists placed the bulk of the Burgess Shale organisms into existing phyla, with only a few assigned to phyla that are now extinct. Thus, the phyla of the Cambrian world were viewed as being essentially the same in number as the phyla of the present-day world, but with fewer species in each phylum. According to this view, the history of life has been simply a gradual increase in the diversity of species within each phylum through time. The number of basic body plans has therefore remained more-or-less constant since the initial radiation of multicellular organisms.

This view has recently been challenged by Harvard paleobiologist Stephen Jay Gould in his 1989 book *Wonderful Life*. According to Gould, the initial explosion of varied life-forms in the Cambrian was promptly followed by a short period of experimentation and then extinction of many phyla. The richness and diversity of modern life-forms are the result of repeated variations of the basic body plans that survived the Cambrian extinctions. In other words, life was much more diverse in terms of phyla during the Cambrian than it is today. The reason members of the Burgess Shale biota look so strange to us is that no living organisms possess their basic body plan, and therefore many of them have been placed into new phyla.

Recent discoveries of age-equivalent fossils at other localities has led some paleontologists to reassign some of the Burgess Shale specimens back into extant phyla, thereby reducing the diversity of the Burgess Shale fauna. If these reassignments to known phyla prove to be correct, then no massive extinction event followed the Cambrian explosion, and life has gradually increased in diversity through time. Currently, there is no clear answer to this debate, and the outcome will probably be decided as more fossil discoveries are made.

Ordovician Marine Community

A major transgression that began during the Middle Ordovician (Tippecanoe sequence) resulted in the most widespread inundation of the craton known. This vast

◆ **FIGURE 13.11** Reconstruction of a Middle Ordovician seafloor fauna. Cephalopods, crinoids, colonial corals, graptolites, trilobites, and brachiopods are shown. (Photo courtesy of the Field Museum of Natural History, Chicago, #Geo80820c.)

epeiric sea, which experienced a uniformly warm climate during this time, opened numerous new marine habitats that were soon filled by a variety of organisms.

Not only did sedimentation patterns change dramatically from the Cambrian to the Ordovician, but the fauna underwent equally striking changes in composition. Whereas the Cambrian invertebrate community was dominated by three groups—trilobites, brachiopods, and archaeocyathids—the Ordovician was characterized by the adaptive radiation of many other animal phyla, (such as bryozoans and corals), with a consequent dramatic increase in the diversity of the total shelly fauna. The Ordovician invertebrate community, however, was dominated by epifaunal benthonic sessile suspension feeders (◆ Fig. 13.11). The Ordovician was also a time of increased diversity and abundance of the **acritarchs** (organic-walled phytoplankton of unknown affinity), which were the major phytoplankton group of the Paleozoic Era and the primary food source of the suspension feeders (◆ Fig. 13.12).

Whereas during the Cambrian, archaeocyathids were the main builders of reeflike structures, bryozoans, stromatoporoids, and tabulate and rugose corals assumed that role beginning in the Middle Ordovician. Many of these reefs were small patch reefs similar in size to those of the Cambrian but of a different composition, whereas others were quite large. As with present-day reefs, Ordovician reefs exhibited a high diversity of organisms and were dominated by suspension feeders (◆ Fig. 13.13).

Finally, three Ordovician fossil groups have proved to be particularly useful for biostratigraphic correlation — the *articulate brachiopods, graptolites,* and *conodonts* (◆ Fig. 13.14). The articulate brachiopods, present since the Cambrian, began a period of major diversification in the shallow-water marine environment during the Ordovician (Fig. 13.14a). They became a conspicuous element of the invertebrate fauna during the Ordovician and in succeeding Paleozoic periods.

Most **graptolites** were planktonic animals carried about by ocean currents. Because most graptolites were planktonic, and most individual species existed for less than a million years, graptolites are excellent guide fossils. They were especially abundant during the Ordovician and Silurian periods. Due to the fragile nature of their organic skeleton, graptolites are most commonly found in black shales (Fig. 13.14b).

Conodonts are a group of well-known small toothlike fossils composed of the mineral apatite (calcium phosphate), the same mineral that composes bone (Fig. 13.14c). Although conodonts have been known for more than 130 years, their affinity has been the subject of debate until the discovery of the conodont animal in 1983 (◆ Fig. 13.15). Several specimens of carbonized impressions of the conodont animal from Lower Carboniferous rocks of Scotland reveal that it is a member of a group of primitive jawless animals assigned to the phylum Chordata. Study of the specimens indicates that the conodont animal was probably an elongate swimming organism that swam by moving its segmented body in waves as it moved through the water. The conodont elements themselves are located just behind the mouth and probably served as a feeding apparatus. The wide distribution and short stratigraphic range of individual conodont species make them excellent fossils for biostratigraphic zonation. Conodonts are easily recovered from limestones by dissolving the rock in acid — a process that does not attack the calcium phosphate material of which the conodonts are composed.

The end of the Ordovician was a time of mass extinctions in the marine realm. More than 100 families of marine invertebrates did not survive into the Silurian, and in North America alone, approximately one-half of the brachiopods and bryozoans died out. What caused such an event? Many geologists believe these extinctions were the result of the extensive glaciation that occurred in Gondwana at the end of the Ordovician Period (see Chapter 11).

Mass extinctions, those geologically rapid events in which an unusually high percentage of the fauna and/or flora becomes extinct, have occurred throughout geologic time (at or near the end of the Ordovician, Devonian, Permian, and Cretaceous periods) and currently are the focus of much research and debate (see Perspective 13.1).

◆ **FIGURE 13.12** Acritarchs from the Upper Ordovician Sylvan Formation, Oklahoma. Acritarchs are organic-walled phytoplankton and were the primary food source of suspension feeders during the Paleozoic Era.

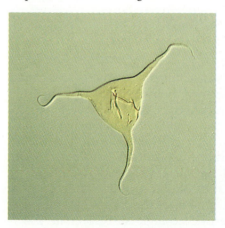
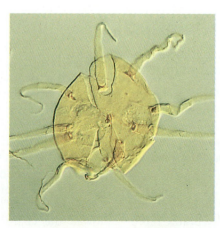
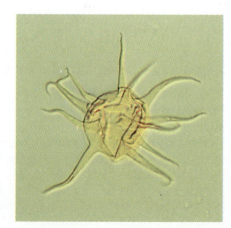

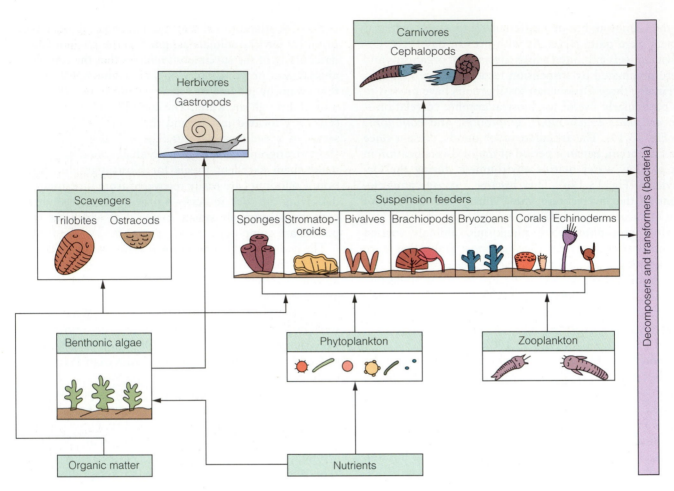

◆ **FIGURE 13.13** Trophic analysis of an Ordovician reef community showing the relationships among the various organisms of the community. The phytoplankton and benthonic algae are primary producers, occupying the lowest trophic level. The primary consumers are the suspension feeders, which make up the majority of the community. The highest trophic level is occupied by the carnivores, which in this community are the cephalopods.

Silurian and Devonian Marine Communities

The mass extinction at the end of the Ordovician was followed by rediversification and recovery of many of the decimated groups. Brachiopods, bryozoans, gastropods, bivalves, corals, crinoids, and graptolites were just some of the groups that rediversified again beginning during the Silurian.

As we discussed in Chapters 11 and 12, the Silurian and Devonian were times of major reef building. While most of the Silurian radiations of invertebrates represented repopulating of niches, organic reef-builders di-

versified in new ways, building massive reefs larger than any produced during the Cambrian or Ordovician. This repopulation was probably due in part to renewed transgressions over the craton, and although a major drop in sea level occurred at the end of the Silurian, the Middle Paleozoic sea level was generally high (see Table 11.1).

The Silurian and Devonian reefs were dominated by tabulate and colonial rugose corals and stromatoporoids (◆ Fig. 13.16). The growth of these reefs generally followed the characteristic ecologic succession in which small, sticklike tabulate and rugose corals initially colo-

(a)

(b)

(c)

◆ **FIGURE 13.14** Representative brachiopods, graptolites, and conodonts. (a) Brachiopods. (b) Graptolites: *Phyllograptus angustifolius,* Norway. (Photos courtesy of Sue Monroe.) (c) Conodonts: *Cahabagnathus sweeti,* Copenhagen Formation (Middle Ordovician), Monitor Range, Nevada (left); *Phragmodus flexuosus,* Lenoir Limestone (Middle Ordovician), Friendsville, Tennessee (right); *Scolopodussp,* Shingle Limestone, Single Pass, Nevada (middle). (Photo courtesy of Stig M. Bergstrom, Ohio State University.)

nized the subtidal seafloor. Following this, an intermediate stage began in which broad, moundlike tabulate and rugose corals dominated, forming low mounds. The final mature stage was reached when algae and stromatoporoids formed a ridge on the seaward side such that a quiet water zone and lagoon formed behind it. In the lagoon

◆ **FIGURE 13.15** The conodont animal preserved as a carbonized impression in the Lower Carboniferous Granton Shrimp Bed in Edinburgh, Scotland. The animal measures about 40 mm long and 2 mm wide. (Photo by J.K. Ingham, supplied courtesy of R.J. Aldridge, University of Leicester, Leicester, England.)

◆ **FIGURE 13.16** Reconstruction of a Middle Devonian reef from the Great Lakes area. It represents the mature stage of an ecological succession. Shown are corals, ammonoids, trilobites, and brachiopods. (Photo courtesy of the Field Museum of Natural History, Chicago, #Geo80821c.)

EXTINCTIONS: CYCLICAL OR RANDOM?

Throughout geologic history, various plant and animal species have become extinct. In fact, extinction is a common feature of the fossil record, and the rate of extinction through time has fluctuated only slightly. Just as new species evolve, others become extinct. There have, however, been brief intervals in the past during which mass extinctions have eliminated large numbers of species. Extinctions of this magnitude could only occur due to radical changes in the environment on a regional or global scale. The cause of these mass extinctions has been the subject of study and debate by geologists for many years.

When we look at the different mass extinction events that have occurred throughout the geologic past, several common themes stand out. The first is that mass extinctions have affected life both in the sea and on land. Second, tropical organisms, particularly in the marine realm, apparently are more affected than organisms from the temperate and high latitude regions. Third, some animal groups repeatedly experience mass extinctions. During the first mass extinction event, such groups are severely affected but not wiped out. Following the initial crisis, the survivors diversify, only to

have their numbers reduced further by another mass extinction. Three marine invertebrate groups in particular display this characteristic: the trilobites, graptolites, and ammonoids. Each of these groups experienced high rates of extinction, followed by high rates of speciation. The fourth theme of mass extinctions is the apparent periodicity displayed during the Phanerozoic Eon. It has been proposed that mass extinctions have occurred approximately every 26 million years.

When we look at the mass extinctions for the last 650 million years, we see that the first event involved only the acritarchs. Several extinction events occurred during the Cambrian, and these affected only marine invertebrates, particularly trilobites. Three other marine mass extinctions took place during the Paleozoic Era: one at the end of the Ordovician, involving many invertebrates; one near the end of the Devonian, affecting the major barrier reef–building organisms as well as the primitive armored fish; and the most severe at the end of the Permian, when between 70 to 90% of the marine species became extinct. On land, a group of reptiles called pelycosaurs also became extinct at the end of the Permian.

The Mesozoic Era experienced several mass extinctions, the most devastating occurring at the end of the Cretaceous, when all large animals, including dinosaurs and seagoing animals such as plesiosaurs and ichthyosaurs, became extinct. Many scientists believe the terminal Cretaceous

---

area, a wide variety of invertebrates flourished, including brachiopods, crinoids, bryozoans, and mollusks, among others. While the fauna of these Silurian and Devonian reefs was somewhat different from that of earlier reefs and reeflike structures, the general composition and structure are the same as in present-day reefs.

The Silurian and Devonian periods were also the time when *eurypterids* (arthropods with scorpionlike bodies and impressive pincers) were abundant, especially in brackish and freshwater habitats (◆ Fig. 13.17). *Ammonoids,* a subclass of the cephalopods, evolved from nautiloids during the Early Devonian and rapidly diversified. With their distinctive suture patterns, short stratigraphic ranges, and widespread distribution, ammonoids are excellent guide fossils for the Devonian through Cretaceous periods (◆ Fig. 13.18).

Another mass extinction occurred near the end of the Devonian. Geologists divide the Late Devonian Epoch

into two ages—Frasnian and Famennian. This second great mass extinction occurred in Late Frasnian to Early Famennian time and resulted in a worldwide near-total collapse of the massive reef communities. On land, however, the seedless vascular plants were seemingly unaffected, although the diversity of freshwater fish was greatly reduced.

Thus, extinctions were most extensive in the marine realm, particularly in the reef and pelagic communities. Although the massive tabulate-rugose-stromatoporoid reefs declined during the Frasnian, they became virtually extinct by the end of the Famennian Age. Approximately 80% of the Frasnian brachiopod genera did not survive into the Famennian Age. Ammonoids experienced a similar decline, and many gastropods and bryozoans also became extinct during this time.

The acritarchs, the main food source of the suspension feeders, underwent a concurrent dramatic decline in

extinction event was caused by a meteorite impact, although that theory is still being vigorously debated.

Several mass extinction events occurred during the Cenozoic Era. The most severe was near the end of the Eocene Epoch.

Though many scientists think of the marine mass extinctions as sudden events from a geological perspective, they were rather gradual from a human perspective, occurring over hundreds of thousands or even millions of years. Furthermore, many geologists believe that climatic changes, rather than some catastrophe, were primarily responsible for the extinctions, particularly in the marine realm. Evidence of glacial episodes or other signs of climatic change have been correlated with the extinction events recorded in the fossil record.

One of the more controversial issues to emerge from the mass extinction debate is whether they occur periodically. Several scientists have proposed that extinctions occur regularly through time. For example, David Raup and John Sepkoski of the University of Chicago have suggested a periodicity of about 26 million years for mass extinction events since the end of the Paleozoic Era. According to them, the most recent mass extinction occurred nearly 11 million years ago. They based their conclusions on the geologic ranges of marine invertebrate families. By graphing all family-level extinctions of marine invertebrates since the Middle Permian, and treating each family extinction as if it occurred at the end of a geologic age, they found that their extinction peaks best fit a 26-million-year periodicity. Furthermore, some of the peaks, such as the one representing the end of the Cretaceous, coincided with recognized mass extinctions.

The idea of periodicity has remained controversial, in part because the dates used for the geologic ages are imperfect and vary among different researchers. For instance, estimates for the end of the Oxfordian Epoch of the Jurassic Period range from 140 to 156 million years ago. Furthermore, some geologists question whether the plotted peaks represent anything more than what would occur randomly.

What Raup and Sepkoski's data show is not that the major extinction events occur precisely every 26 million years, but rather that the occurrence is closer to 26-million-year intervals than would be expected for a strictly random pattern. It has been suggested that some type of recurring extraterrestrial event, such as a meteorite impact can account for the apparent periodicity (see Chapter 16).

At this point, there does not seem to be sufficient evidence to attribute every major extinction event to an extraterrestrial cause. We have acquired a tremendous amount of knowledge about extinctions during the past decade and can only hope that our picture of the evolution and extinction of the Earth's biota will become clearer during the next decade.

abundance and diversity, perhaps contributing in part to the extinctions of many suspension-feeding invertebrates at this time.

The demise of the Middle Paleozoic reef communities serves to highlight the geographic aspects of the Late Devonian mass extinction event. The tropical groups were most severely affected; in contrast, the polar communities were seemingly little affected. Apparently, an episode of global cooling was largely responsible for the extinctions near the end of the Devonian. During such a cooling, the disappearance of tropical conditions would have had a severe effect on reef and other warm-water organisms. Cool-water species, on the other hand, could have simply migrated toward the equator. While cooling temperatures certainly played an important role in the Late Devonian extinctions, the closing of the Iapetus Ocean and the orogenic events of this time undoubtedly also played a role by reducing the area of shallow shelf environments where many marine invertebrates lived.

Carboniferous and Permian Marine Communities

The Carboniferous invertebrate marine community responded to the Late Devonian extinctions in much the same way the Silurian invertebrate marine community responded to the Late Ordovician extinctions—that is, by renewed adaptive radiation and rediversification. The brachiopods and ammonoids quickly recovered and again assumed important ecologic roles, while other groups, such as the lacy bryozoans and crinoids, reached their greatest diversity during the Carboniferous. With the decline of the stromatoporoids and the tabulate and rugose corals, large organic reefs like those existing earlier in the Paleozoic virtually disappeared and were re-

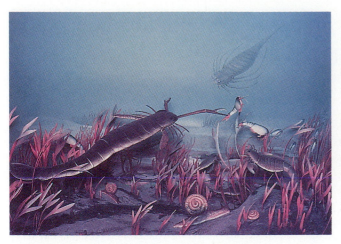

◆ **FIGURE 13.17** Reconstruction of a Silurian brackish-marine bottom scene near Buffalo, New York. Shown are algae, eurypterids, worms, and shrimp. (Photo courtesy of the Field Musuem of Natural History, Chicago, #Geo80819c.)

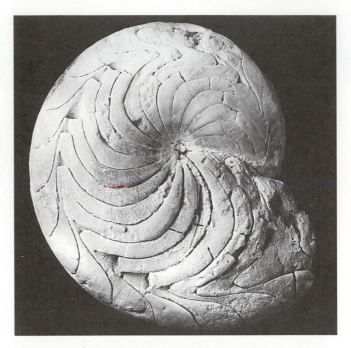

◆ **FIGURE 13.18** *Imitoceras rotatorium* (DeKoninck), a goniatitic ammonoid cephalopod from the Lower Mississippian Rockford Limestone, near Rockford, Indiana. The distinctive suture pattern, short stratigraphic range, and wide distribution make ammonoids excellent guide fossils. (Photo courtesy of the Smithsonian Institution.)

◆ **FIGURE 13.19** Marine life during the Mississippian based on an Upper Mississippian fossil site at Crawfordville, Indiana. Invertebrate animals shown include crinoids, blastoids, lacy bryozoans, and small corals. (Photo © American Museum of Natural History, Trans. #K10257.)

placed by small patch reefs. These reefs were dominated by crinoids, blastoids, lacy bryozoans, brachiopods, and calcareous algae and flourished during the Late Paleozoic (◆ Fig. 13.19). In addition, bryozoans, crinoids, and fusulinids (spindle-shaped foraminifera) contributed large amounts of skeletal debris to the formation of the vast bedded limestones that constitute the majority of Mississippian sedimentary rocks.

The Permian invertebrate marine faunas resembled those of the Carboniferous. Due to the restricted size of the shallow seas on the craton and the reduced shelf space along the continental margins, however, they were more restricted in their distribution (see Fig. 12.15). The spiny and odd-shaped productids dominated the brachiopod assemblage and constituted an important part of the reef complexes that formed in the Texas region during the Permian (◆ Fig. 13.20). The fusulinids (◆ Fig. 13.21), which first appeared during the Late Mississippian and greatly diversified during the Pennsylvanian, experienced a further diversification during the Permian; more than 5,000 species are known from the Permian Period alone. Because of their abundance, diversity, and worldwide occurrence, fusulinids are important guide fossils for Pennsylvanian and Permian strata. Bryozoans, sponges, and some types of calcareous algae also were common elements of the Permian invertebrate fauna.

◆ **FIGURE 13.20** Reconstruction of a Permian patch-reef community from the Glass Mountains of West Texas. Shown are algae, productid brachiopods, cephalopods, and corals. (Photo © American Museum of Natural History, trans. #K10269.)

The Permian Marine Invertebrate Extinction Event

The greatest recorded mass extinction event to affect the marine invertebrate community occurred at the end of the Permian Period (◆ Fig. 13.22). Before the Permian ended, roughly one-half of all marine invertebrate families and perhaps 90% of all marine invertebrate species became extinct. Fusulinids, rugose and tabulate corals, two bryozoan orders, and two brachiopod orders as well as trilobites and blastoids did not survive the end of the Permian. All of these groups had been very successful during the Paleozoic Era, although the trilobites had decreased in abundance during the Early Paleozoic.

What caused such a crisis for the marine invertebrates? Many hypotheses have been proposed, but no completely satisfactory answer has been found. Two currently discussed hypotheses are (1) a reduction of living area related to widespread regression of the seas and suturing of the continents when Pangaea formed and (2) decreased

◆ **FIGURE 13.21** Fusulinids are large benthonic foraminifera that are excellent guide fossils for the Pennsylvanian and Permian periods. (a) *Leptotriticites tumidus*, Early Permian, Kansas. (b) *Triticites*, Pennsylvanian, Kansas. (Photos courtesy of G. A. Sanderson, Amoco Production Co.)

(a)

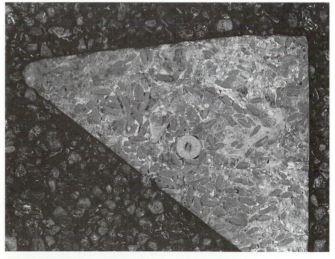

(b)

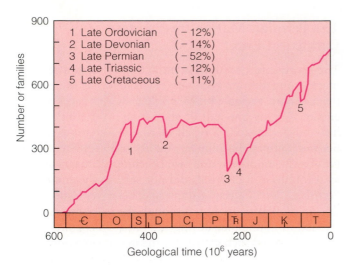

1 Late Ordovician	(−12%)
2 Late Devonian	(−14%)
3 Late Permian	(−52%)
4 Late Triassic	(−12%)
5 Late Cretaceous	(−11%)

◆ **FIGURE 13.22** Phanerozoic diversity for marine invertebrate and vertebrate families. Note the three episodes of Paleozoic mass extinctions, with the greatest occurring at the end of the Permian Period.

ocean salinity due to widespread arid climates. As we discussed earlier, there was a trend toward continental convergence during the Paleozoic, culminating in the formation of Pangaea by the end of the Permian. Such continental convergence resulted in regression of the epeiric seas from the cratons and reduction of the shallow-water shelf area surrounding each continent. Decreased ocean salinity would affect those organisms with narrow salinity tolerances, including most marine invertebrates. Extensive marginal marine evaporite deposits were formed during the Permian Period, and it is hypothesized that removal of the salts from the ocean to form these deposits lowered ocean salinity to lethal levels for many invertebrates. Other calculations, however, show that not enough salts could have been removed by this method to have significantly lowered the salinity level of the entire Panthalassic Ocean.

The Permian mass extinctions were probably caused by a combination of many interrelated geologic and biologic factors. In any case, the surviving marine invertebrate faunas of the Early Triassic were of very low diversity and were widely distributed around the world. This wide distribution indicates that either the groups were tolerant of a wide temperature range, or the latitudinal temperature gradient was very gentle, enabling species of normal temperature tolerance to spread far and wide.

Chapter Summary

Table 13.2 summarizes the major evolutionary and geologic events of the Paleozoic Era and shows their relationships to each other.

1. Soft-bodied multicelled organisms presumably had a long Precambrian history during which they lacked hard parts. Hard parts appeared in different invertebrate groups during the Cambrian and Ordovician periods and provided such advantages as protection against predators and support for muscles, enabling organisms to grow large and increase locomotor efficiency. Hard parts probably evolved as a result of several factors rather than just one.

2. Marine organisms are classified as plankton if they are floaters, nekton if they swim, and benthos if they live on or in the seafloor.

3. Marine organisms can be divided into four basic feeding groups: suspension feeders, which remove or consume microscopic plants and animals as well as dissolved nutrients from water; herbivores, which are plant eaters; carnivore-scavengers, which are meat eaters; and sediment-deposit feeders, which ingest sediment and extract nutrients from it.

4. The marine ecosystem consists of various trophic levels of food production and consumption. At the base are primary producers, upon which all other organisms are dependent.

Feeding on the primary producers are the primary consumers, which in turn can be fed upon by higher levels of consumers. The decomposers are bacteria that break down the complex organic compounds of dead organisms and recycle them within the ecosystem.

5. The Cambrian invertebrate community was dominated by three major groups—the trilobites, brachiopods, and archaeocyathids. Little specialization existed among the invertebrates, and most phyla were represented by only a few species.

6. The Middle Cambrian Burgess Shale contains one of the finest examples of a well-preserved soft-bodied biota in the world. The significance of this biota in terms of its impact on the history of life is currently being debated.

7. The Ordovician marine invertebrate community marked the beginning of the dominance by the shelly fauna and the start of large-scale reef building. The end of the Ordovician Period was a time of major extinctions for many of the invertebrate phyla.

8. The Silurian and Devonian periods were times of diverse faunas dominated by reefs, while the Caroniferous and Permian periods saw a great decline in invertebrate diversity.

9. A major extinction occurred at the end of the Paleozoic Era, affecting the invertebrates as well as the vertebrates.

Important Terms

acritarch	conodont	primary producer
archaeocyathid	graptolite	sediment-deposit feeder
benthos	herbivore	suspension feeder
brachiopod	nekton	trilobite
carnivore-scavenger	plankton	trophic level

Review Questions

1. The first shelled fossils were:
   a. _____ trilobites; b. _____ small calcium carbonate tubes; c. _____ archaeocyathids; d. _____ brachiopods; e. _____ mollusks.

2. The age of the Burgess Shale fauna is:
   a. _____ Cambrian; b. _____ Ordovician; c. _____ Silurian; d. _____ Devonian; e. _____ Mississippian.

3. Organisms living on or in the seafloor are:
   a. _____ epifauna; b. _____ epiflora; c. _____ infauna; d. _____ benthos; e. _____ all of these.

4. The most abundant elements of the Cambrian marine invertebrate community were:
   a. _____ trilobites; b. _____ brachiopods; c. _____ mollusks; d. _____ archaeocyathids; e. _____ echinoderms.

5. An exoskeleton is advantageous because it:
   a. _____ provides protection against predators; b. _____ prevents drying out in an intertidal environment; c. _____ provides attachment sites for development of strong muscles; d. _____ provides protection against ultraviolet radiation; e. _____ all of these.

6. The first organisms to construct reeflike structures were:
   a. _____ corals; b. _____ bryozoans; c. _____ archaeocyathids; d. _____ sponges; e. _____ mollusks.

7. The major phytoplankton group of the Paleozoic Era and the primary food source of the suspension feeders were:
   a. _____ dinoflagellates; b. _____ coccolithophorids; c. _____ diatoms; d. _____ acritarchs; e. _____ graptolites.

8. The greatest recorded mass extinction event to affect the marine invertebrate community occurred at the end of which period?
   a. _____ Cambrian; b. _____ Ordovician; c. _____ Silurian; d. _____ Devonian; e. _____ Permian.

9. Which two periods experienced the greatest reef-building activity during the Paleozoic?
   a. _____ Cambrian-Ordovician; b. _____ Ordovician-Silurian; c. _____ Silurian-Devonian; d. _____ Devonian-Mississippian; e. _____ Mississippian-Permian.

10. Which group was particularly abundant during the Silurian and Devonian periods, especially in brackish and freshwater habitats?
    a. _____ ammonoids; b. _____ conodonts; c. _____ graptolites; d. _____ stromatoporoids; e. _____ eurypterids.

11. What type of invertebrates dominated the Ordovician invertebrate community?
    a. _____ epifaunal benthonic sessile suspension feeders; b. _____ infaunal benthonic sessile suspension feeders; c. _____ epifaunal benthonic mobile suspension feeders; d. _____ infaunal nektonic carnivores; e. _____ epifloral planktonic primary producers.

12. The Early Cambrian was characterized by a:
    a. _____ high-diversity soft-bodied fauna; b. _____ high-diversity shelly fauna; c. _____ low-diversity shelly fauna; d. _____ low-diversity soft-bodied fauna; e. _____ none of these.

13. Fossils of the earliest multicelled animals are found in what age deposits?
    a. _____ Early Precambrian; b. _____ Late Precambrian; c. _____ Early Cambrian; d. _____ Middle Cambrian; e. _____ Late Cambrian.

14. A _____ is an example of an epifaunal benthonic suspension feeder.
    a. _____ trilobite; b. _____ cephalopod; c. _____ graptolite; d. _____ articulate brachiopod; e. _____ gastropod.

15. Which group of invertebrates are excellent guide fossils for the Pennsylvanian and Permian periods?
    a. _____ trilobites; b. _____ gastropods; c. _____ sponges; d. _____ bivalves; e. _____ fusulinids.

16. The second great mass extinction occurred near the end of the Devonian Period and resulted in a global near-total collapse of which invertebrate groups?
    a. _____ corals and stromatoporoids; b. _____ cephalopods and gastropods; c. _____ conodonts and graptolites; d. _____ fusulinids and radiolarians; e. _____ brachiopods and blastoids.

17. The fossils of the Burgess Shale are significant because they provide a rare glimpse of _____ .

(continued on page 370)

TABLE 13.2   Major Evolutionary and Geologic Events of the Paleozoic Era

Age (Millions of Years)	Geologic Period		Invertebrates	Vertebrates
245				
	Permian		Largest mass extinction event to affect the invertebrates.	Acanthodians, placoderms, and pelycosaurs become extinct. Therapsids and pelycosaurs the most abundant reptiles.
286	Carboniferous	Pennsylvanian	Fusulinids diversify.	Reptiles evolve. Amphibians abundant and diverse.
320		Mississippian	Crinoids, lacy bryozans, blastoids become abundant. Renewed adaptive radiation following extinctions of many reef-builders.	
360	Devonian		Extinctions of many reef-building invertebrates near end of Devonian. Reef building continues. Eurypterids abundant.	Amphibians evolve. All major groups of fish present—Age of Fish.
408	Silurian		Major reef building. Diversity of invertebrates remains high.	Ostracoderms common. Acanthodians, the first jawed fish, evolve.
438	Ordovician		Extinctions of a variety of marine invertebrates near end of Ordovician. Major adaptive radiation of all invertebrate groups. Suspension feeders dominant.	Ostracoderms diversify.
505	Cambrian		Many trilobites become extinct near end of Cambrian. Trilobites, brachiopods, and archaeocyathids are most abundant.	Earliest vertebrates—jawless fish called ostracoderms.
570				

Plants	Major Geologic Events
Gymnosperms diverse and abundant.	Formation of Pangaea. Alleghenian orogeny. Hercynian orogeny.
Coal swamps with flora of seedless vascular plants and gymnosperms.	Coal-forming swamps common. Formation of Ancestral Rockies. Continental glaciation in Gondwana.
Gymnosperms appear (may have evolved during Late Devonian).	Ouachita orogeny.
First seeds evolve. Seedless vascular plants diversify.	Widespread deposition of black shale. Antler orogeny. Acadian orogeny.
Early land plants—seedless vascular plants.	Caledonian orogeny. Extensive barrier reefs and evaporites.
Plants move to land?	Continental glaciation in Gondwana. Taconic orogeny.
	First Phanerozoic transgression (Sauk) onto North American craton.

a. _____ the first shelled animals; b. _____ the soft-part anatomy of extinct groups; c. _____ soft-bodied animals; d. _____ answers (a) and (b); e. _____ answers (b) and (c).

18. Discuss the significance of the appearance of the first shelled animals and possible causes for the acquisition of a mineralized exoskeleton.

19. Draw a marine food web that shows the relationships among the producers, consumers, and decomposers.

20. Discuss the ecological succession of a typical Devonian reef.

21. Discuss the major differences between the Cambrian marine community and the Ordovician marine community.

Additional Readings

Cowen, R. 1989. *History of life*. Palo Alto, Calif.: Blackwell Scientific Publications.

Donovan, S. K., ed. 1989. *Mass extinctions*. New York: Columbia University Press.

Gould, S. J. 1989. *Wonderful life*. New York: W. W. Norton.

Lane, N. G. 1992. *Life of the past*. 3d ed. Columbus, Ohio: Charles E. Merrell Publishing Co.

Levinton, J.S. 1992. The big bang of animal evolution. *Scientific American* 267, no. 5: 84–93.

McMenamin, M. A. S. 1987. The emergence of animals. *Scientific American* 256, no. 4: 94–102.

McMenamin, M. A. S., and D. L. Schulte McMenamin. 1990. *The emergence of animals: The Cambrian breakthrough*. New York: Columbia University Press.

Reid, M. 1992. Ghosts of the Burgess Shale. *Earth* 1, no. 5: 38–45.

Raup, D. M., and J. J. Sepkoski. 1984. Periodicity of extinctions in the geologic past. *Proceedings of the National Academy of Sciences*, 81: 801–5.

Stanley, S. M. 1987. *Extinction*. New York: Scientific American Books.

Stearn, C. W., and R. L. Carroll. 1989. *Paleontology: The record of life*. New York: John Wiley & Sons.

Valentine, J. W., and E. M. Moores. 1974. Plate tectonics and the history of life in the ocean. *Scientific American* 230, no. 4: 80–89.

Whittington, H. B. 1985. *The Burgess Shale*. New Haven, Conn.: Yale University Press.

CHAPTER 14

Carboniferous sandstone casts of *Stigmaria* tree stumps are preserved in Glasgow's Victoria Park, Scotland. These casts were formed when an influx of sand filled the interior of the rotting trunks. The bark was later altered to thin coal layers. (Photo courtesy of NAGT.)

LIFE OF THE PALEOZOIC ERA: VERTEBRATES AND PLANTS

Prologue

Humans have always been intrigued with finding the earliest representative of a particular group of organisms, whether of our own species or the simplest one-celled organism. Recently, two discoveries have yielded the oldest complete fossil fish known and the oldest known amphibians in North America. These separate discoveries have contributed important information for interpreting the history of fish and amphibians.

During a fossil-collecting expedition in southern Bolivia, a team of international paleontologists recovered the remains of at least 30 remarkably well-preserved jawless armored fish from Lower Ordovician rocks, 470 million years old. At least 10 of the specimens were virtually complete (◆ Fig. 14.1). The oldest known vertebrate fossils are phosphatic plate and scale fragments of the jawless armored fish *Anatolepis*. These fragments are Late Cambrian (510 million years) and Early Ordovician (470 million years) in age and come from widely scattered areas in North America and the island of Spitzbergen. Far more complete remains of a similar fish *(Arandaspis)* are known from Lower Ordovician rocks in Australia; in these specimens, however, the bone itself is not preserved, but is represented by well-preserved molds in the rock. Thus, the finds in Bolivia represent the oldest complete fossil fish. Upon viewing the Bolivian specimens, David K. Elliott, vertebrate paleontologist at Northern Arizona University, Flagstaff, Arizona, said, "This is one of the most

(a)

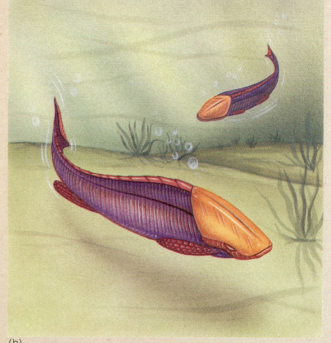

(b)

◆ **FIGURE 14.1** (a) Fossil specimen of the Bolivian Early Odovician armored jawless fish *Sacabambaspis*. (Photo courtesy of Pierre-Yves Gagnier, Institute de Paleontologie, Paris.) (b) A reconstruction of its mode of life.

Archean Eon	Proterozoic Eon	Phanerozoic Eon						
Precambrian		Paleozoic Era						
		Cambrian	Ordovician	Silurian	Devonian	Mississippian	Pennsylvanian	Permian
						Carboniferous		

2,500 M.Y.A.   570 M.Y.A.   2... M.Y...

exciting and important discoveries in lower-vertebrate studies in the last 50 years."

Reconstructions of the fish indicate they lived in shallow marine waters and were probably poor swimmers. They are different enough from the Late Cambrian–Early Ordovician North American, Spitzbergen, and Australian fish to be placed in a new genus, *Sacabambaspis*, named after a village near the fossil discovery. The fossils indicate that fish were widespread and abundant at least 470 million years ago and must, therefore, have had a longer history than paleontologists had previously assumed.

Another important fossil find was reported in June 1988, when geologists in Iowa announced the discovery of the oldest *tetrapod* (four-legged vertebrate animal) fossil site in North America. The fossils were recovered from the Mississippian (335-million-year-old) St. Louis Formation, near Delta in southeastern Iowa. Several hundred tetrapod fossils have been collected, representing at least two amphibian and eight fish species.

Only 22 other fossil sites of comparable age (Late Devonian and Mississippian) are known, and most of those have yielded only fragmentary material. The Delta site represents the oldest amphibian fossils to be found in midcontinental North America.

The two amphibian species include the earliest temnospondyl (animals 1 to 2 m long, resembling salamanders) and a species informally called *protoanthracosaur* (◆ Fig. 14.2). The temnospondyls had well-developed legs for walking on land, yet probably spent most of their time in the water.

The discovery of these fossils provides important information about the morphology and phylogenetic relationships among the earliest tetrapods. In addition, it enables paleontologists to better understand the relationship between the earliest amphibians and their fish ancestors.

❖ INTRODUCTION

In the previous chapter, we examined the Paleozoic history of invertebrates, beginning with the acquisition of hard parts and concluding with the massive Permian extinction event, in which upwards of 90% of all invertebrate species became extinct. In this chapter we examine the Paleozoic evolutionary history of plants and **vertebrates**, those animals with a segmented vertebral column.

◆ **FIGURE 14.2** Side view of a fossil protoanthracosaur skull from the Mississippian St. Louis Formation near Delta, Iowa. (Photo reprinted courtesy of *Nature*, v. 333, p. 770, © 1992 MacMillan Magazines Ltd.)

Phanerozoic Eon							
Mesozoic Era			Cenozoic Era				
Triassic	Jurassic	Cretaceous	Tertiary				Quaternary	
			Paleocene	Eocene	Oligocene	Miocene	Pliocene	Pleistocene	Holocene

45
M.Y.A.

66
M.Y.A.

Vertebrates first evolved in the sea, and their earliest fossil record consists of phosphatic plate and scale fragments from Late Cambrian–aged rocks. By the end of the Devonian, all of the major fish groups had evolved. During the Devonian, one group of fish evolved into amphibians, beginning the colonization of land by these animals. Before the end of the Paleozoic, reptiles had evolved from amphibians and had become the dominant group of terrestrial vertebrates.

Plants preceded animals onto the land, probably sometime during the Ordovician. In making the transition from water to land, both plants and animals had to solve the same basic problems. For both groups, the method of reproduction proved to be the major barrier to expansion into the various terrestrial environments. With the evolution of the seed in plants and the amniote egg in animals, this limitation was removed, and both groups were able to expand into all the terrestrial habitats.

The end of the Paleozoic Era was a time of major extinctions. Not only was the marine invertebrate community greatly decimated, but many amphibians and reptiles also became extinct.

❖ VERTEBRATE EVOLUTION

A **chordate** is an animal that has, at least during part of its life cycle, a notochord, a dorsal hollow nerve cord, and gill slits (◆ Fig. 14.3). Vertebrates, which are animals with backbones, are simply a subphylum of chordates.

The ancestors and early members of the phylum Chordata were soft-bodied organisms that left few fossils. Consequently, we know very little about the early evolutionary history of the chordates or vertebrates. Surprisingly, a very close relationship exists between echinoderms and chordates. They may even have shared a common ancestor, because the development of the embryo is the same in both groups and differs completely from other invertebrates (◆ Fig. 14.4). Furthermore, the biochemistry of muscle activity, blood proteins, and the larval stages are very similar in both echinoderms and chordates.

The evolutionary pathway to vertebrates may have begun with a sessile suspension-feeding animal with exposed cilia on its arms (◆ Fig. 14.5). Subsequently, these organisms evolved into nektonic gilled animals, perhaps looking somewhat like *Amphioxus* (see Fig. 14.3). With

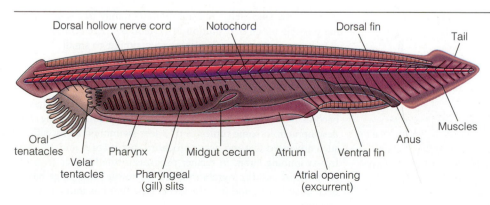

◆ FIGURE 14.3 The structure of the lancelet *Amphioxus* illustrates the three characteristics of a chordate: a notochord, a dorsal hollow nerve cord, and gill slits.

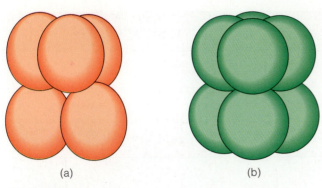

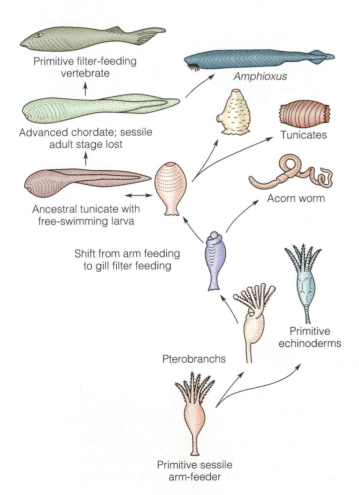

(a)   **(b)**

◆ **FIGURE 14.4** (a) Arrangement of cells resulting from spiral cleavage. In this arrangement, cells in successive rows are nested between each other. Spiral cleavage is characteristic of all invertebrates except the echinoderms. (b) Arrangement of cells resulting from radial cleavage is characteristic of chordates and echinoderms. In this configuration, cells are directly above each other.

Primitive filter-feeding vertebrate

Amphioxus

Advanced chordate; sessile adult stage lost

Tunicates

Ancestral tunicate with free-swimming larva

Acorn worm

Shift from arm feeding to gill filter feeding

Primitive echinoderms

Pterobranchs

Primitive sessile arm-feeder

the modification of the notochord to vertebrae, the first true vertebrates evolved.

❖ FISH

The most primitive vertebrates are the fishes, and the oldest fish remains are found in the Upper Cambrian Deadwood Formation in northeastern Wyoming (◆ Fig. 14.6). Here phosphatic scales and plates of *Anatolepis,* a primitive member of the class Agnatha (jawless fish) have been recovered from marine sediments. All known Cambrian and Ordovician fossil fish have been found in shallow, nearshore marine deposits, while the earliest nonmarine fish remains have been found in Silurian strata. This does not prove that fish originated in the oceans, but it does lend strong support to the idea.

◆ **FIGURE 14.6** A fragment of a plate from *Anatolepis* cf. *A. heintzi* from the Upper Cambrian Deadwood Formation of Wyoming. *Anatolepis* is the oldest known fish. (Photo courtesy of John E. Repetski, U.S.G.S.)

◆ **FIGURE 14.5** (left) A diagrammatic family tree suggesting the possible mode of evolution of vertebrates. The echinoderms may have arisen from forms somewhat similar to small pterobranchs; the acorn worm may have evolved from pterobranch descendants that had evolved a gill-feeding system but were somewhat more advanced in other regards. Tunicates represent a stage in which, in the adult, the gill apparatus has become highly evolved. The important point, though, is the development in some tunicates of a free-swimming larva with such advanced features as a notochord, nerve cord, and free-swimming habit. In further progress to the lancelet *Amphioxus* and the vertebrates, the old sessile adult stage has been abandoned, and it is the larval type that has initiated the advance.

As a group, fish range from the Late Cambrian to the present (◆ Fig. 14.7). The oldest and most primitive of the class Agnatha are the **ostracoderms,** whose name means "bony skin" (Table 14.1). These are armored jawless fish that first evolved during the Late Cambrian, reached their zenith during the Silurian and Devonian, and then became extinct.

The majority of ostracoderms lived on the sea bottom.

◆ **FIGURE 14.7** Geologic ranges of the major fish groups.

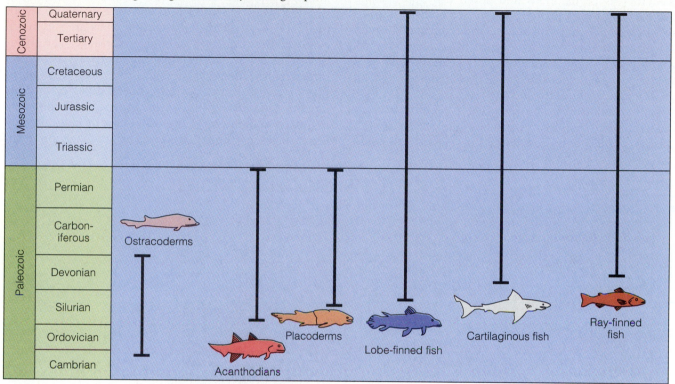

TABLE 14.1 Brief Classification of Fish Showing the Groups Referrred to in the Text

CLASSIFICATION	GEOLOGIC RANGE	LIVING EXAMPLE

Class Agnatha (jawless fish)
  Early members of the class are called ostracoderms
	Late Cambrian–Recent	Lamprey, hagfish
		  No living ostracoderms

Class Acanthodii (the first fish with jaws)	Early Silurian–Permian	None
Class Placodermi (armored jawed fish)	Late Silurian–Permian	None
Class Chondrichthyes (cartilagenous fish)	Devonian–Recent	Sharks, rays, skates
Class Osteichthyes (bony fish)	Devonian–Recent	Tuna, perch, bass, pike, catfish, trout, salmon, lungfish, *Latimeria*
Subclass Actinopterygii (ray-finned fish)		Tuna, perch, bass, pike, catfish, trout, salmon
Subclass Sarcopterygii (lobe-finned fish)		Lungfish, *Latimeria*
Order Dipnoi		Lungfish
Order Crossopterygii		*Latimeria*
Suborder Rhipidistia		None

Hemicyclaspis is a good example of a bottom-dwelling ostracoderm (◆ Fig. 14.8a). It had a flattened underside, vertical scales, eyes on the top of its head, and a mouth on the bottom. The vertical scales allowed the fish to wiggle sideways, propelling itself along the seafloor, while the eyes on the top of its head allowed it to see such predators as cephalopods and jawed fish approaching from above. While moving along the sea bottom, it probably sucked up small bits of food and sediments through its jawless mouth. Another type of ostracoderm, represented by *Pteraspis,* was more elongated and probably an active swimmer, although it also seemingly fed on small pieces of food it could suck up.

The evolution of jaws was a major evolutionary advance among primitive vertebrates. While their jawless ancestors could only feed on detritus, jawed fish could eat plants and also become active predators, thus opening many new ecological niches. The evolution of the vertebrate jaw is an excellent example of evolutionary opportunism. Various studies suggest that the jaw originally evolved from the first three gill arches of jawless fish. In present-day fish, water coming in through the mouth is pumped backward and passed over the gills where oxygen and carbon dioxide are exchanged with the blood. Because the gills are soft, they are supported by gill arches composed of bone or cartilage. The evolution of the jaw may thus have been related to respiration rather than feeding (◆ Fig. 14.9). By evolving joints in the forward gill arches, jawless fish could open their mouths wider. Every time a fish opened and closed its mouth, it would pump more water past the gills, thereby increasing the oxygen intake. The modification from

◆ **FIGURE 14.8** Reconstruction of a Devonian seafloor showing (a) an ostracoderm *(Hemicyclaspis)*, (b) a placoderm *(Bothriolepis)*, (c) an acanthodian *(Parexus)* and (d) a ray-finned fish *(Cheirolepis)*.

rigid to hinged forward gill arches enabled fish to increase both their food consumption and oxygen intake, and the evolution of the jaw as a feeding structure rapidly followed.

The remains of the first jawed fish, called **acanthodians** (Fig. 14.8c, Table 14.1), are found in Lower Silurian nonmarine rocks. Acanthodians are an enigmatic group of fish characterized by large spines, scales covering much of the body, jaws, teeth, and reduced bony armor. Their relationship to other fish has not been well established. None of the known fossil ostracoderms are likely ancestors, and differences in the tail, spines, and teeth separate them from members of the bony fish. The acanthodians were most abundant during the Devonian, declined in importance through the Carboniferous, and became extinct during the Permian.

While we do not know how the acanthodians were related to other, more complex fish, we do know that during the Devonian a major adaptive radiation of jawed fish occurred. **Placoderms** (Table 14.1), whose name means "plate-skinned," arose in the Late Silurian and, like acanthodians, reached their peak of abundance and diversity during the Devonian. Placoderms were heavily armored jawed fish that lived in both fresh water and the ocean. The placoderms exhibited considerable variety, including small bottom-dwellers called *antiarchs* (Fig. 14.8b) as well as *anthrodires*, which were the major predators of the Devonian seas (◆ Fig. 14.10a). Arthrodires are best represented by *Dunkleosteus,* a Late Devonian fish that lived in the mid-continental North American epeiric seas (Fig. 14.10a). It was by far the largest fish of the time, attaining a length of more than 12 m. It had a heavily armored head and shoulder region, a huge jaw lined with razor-sharp bony teeth, and a flexible tail, all features consistent with its status as a ferocious predator.

In addition to the abundant acanthodians, placoderms, and ostracoderms, other fish groups, such as the cartilaginous and bony fish, also evolved during the Devonian Period. It is small wonder, then, that the Devonian is informally called the "Age of Fish," since all major fish groups were present during this time period.

The **cartilaginous fish,** class Chrondrichthyes (Table 14.1), represented today by sharks, rays, and skates, first evolved during the Middle Devonian, and by the Late Devonian, primitive marine sharks such as *Cladoselache* were quite abundant (◆ Fig. 14.10b). Cartilaginous fishes have never been as numerous nor as diverse as their cousins, the bony fishes, but they were, and still are, important members of the marine vertebrate fauna.

Along with the cartilaginous fish, the **bony fish,** class Osteichthyes (Table 14.1), also first evolved during the Devonian. Because bony fish are the most varied and numerous of all the fishes, and because the amphibians evolved from them, their evolutionary history is particularly important. There are two groups of bony fish: the common **ray-finned fish** (◆ Figure 14.10d) and the less familiar **lobe-finned fish** (Table 14.1).

The term *ray-finned* refers to the way the fins are supported by thin bones that spread away from the body (◆ Fig. 14.11a). From a modest freshwater beginning during the Devonian, the ray-finned fish, which include most of the familiar fish such as trout, bass, perch, salmon, and tuna, rapidly diversified to dominate the Mesozoic and Cenozoic seas.

Present-day lobe-finned fish are characterized by muscular fins. The fins do not have radiating bones, but rather articulating bones with the fin attached to the body by a fleshy shaft (◆ Fig. 14.11b). Two major groups of lobe-finned fish are recognized: the *lung fish* and *crossopterygians* (Table 14.1). The lung fish were

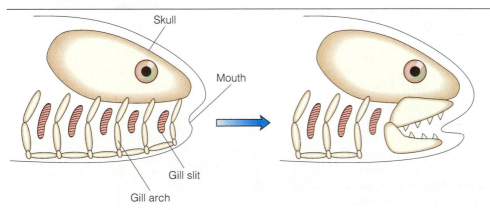

◆ **FIGURE 14.9** The evolution of the vertebrate jaw is believed to have occurred from the modification of the first two or three anterior gill arches. This theory is based on the comparative anatomy of living vertebrates.

Skull

Mouth

Gill slit

Gill arch

◆ FIGURE 14.10 Recreation of a Late Devonian marine scene from the midcontinent of North America. (a) The giant placoderm *Dunkleosteus* (length more than 12 m) is pursuing (b) the shark *Cladoselache* (length up to 1.2 m). Also shown are (c) the bottom-dwelling placoderm *Bothriolepis* and (d) the swimming ray-finned fish *Cheirolepis*, both of which attained a length of 40–50 cm.

fairly abundant during the Devonian, but today only three freshwater genera exist, one each in South America, Africa, and Australia. Their present-day distribution presumably reflects the Mesozoic breakup of Gondwana.

Studies of present-day lung fish indicate that lungs evolved from saclike bodies on the ventral side of the esophagus. These saclike bodies enlarged and improved their capacity for oxygen extraction, eventually evolving into lungs. When the lakes or streams in which lung fish live become stagnant and dry up, they breathe at the surface or burrow into the sediment to prevent dehydration. When the water is well oxygenated, however, lung fish rely upon gill respiration.

The **crossopterygians** are a second group of lobe-finned fish and a most important group because it was from them that the amphibians evolved. During the Devonian, two separate branches of crossopterygians evolved. One led to the amphibians, which we will discuss shortly, while the other invaded the sea. This latter group, the *coelacanths,* were thought to have become extinct at the end of the Cretaceous. In 1938, however, fisherman caught a coelacanth in the deep waters off Madagascar (see Fig. 6.23), and since then several dozen more have been caught.

◆ FIGURE 14.11 Arrangement of fin bones for (a) a typical ray-finned fish and (b) a lobe-finned fish. The muscles extend into the fin of the lobe-finned fish, allowing greater flexibility of movement than for the ray-finned fish.

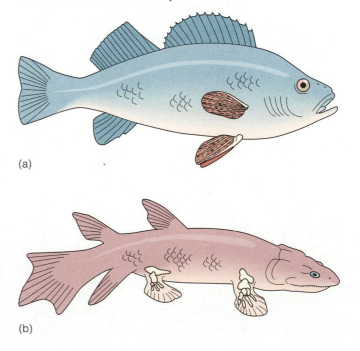

(a)

(b)

The group of crossopterygians that is ancestral to amphibians are called *rhipidistians* (Table 14.1). These fish, attaining lengths of over 2 m, were the dominant freshwater predators of the Late Paleozoic. *Eusthenopteron*, a good example of a rhipidistian crossopterygian, had an elongate body that enabled it to move swiftly in the water, as well as paired muscular fins that could be used for locomotion on land (◆ Fig. 14.12). The structural similarity between crossopterygian fish and the earliest amphibians is striking and one of the better documented transitions from one major group to another (◆ Fig. 14.13).

Before discussing this transition and the evolution of amphibians, it would be useful to place the evolutionary history of Paleozoic fish in the larger context of Paleozoic evolutionary events. Certainly, the evolution and diversification of jawed fish as well as eurypterids and ammonoids had a profound effect on the marine ecosystem. Previously, defenseless organisms either evolved defensive mechanisms or suffered great losses, possibly even extinction. Recall from Chapter 13 that trilobites experienced major extinctions at the end of the Cambrian, recovered slightly during the Ordovician, then declined greatly from the Silurian to their ultimate demise at the end of the Permian. Perhaps their lightly calcified external covering made them easy prey for the rapidly evolving jawed fish and cephalopods. Ostracoderms, although armored, would also have been easy prey for the swifter jawed fishes.

◆ **FIGURE 14.12** *Eusthenopteron,* a member of the rhipidistian crossopterygians. The crossopterygians are the group from which the amphibians are believed to have evolved. *Eusthenopteron* had an elongate body and paired fins that could be used for moving about on land.

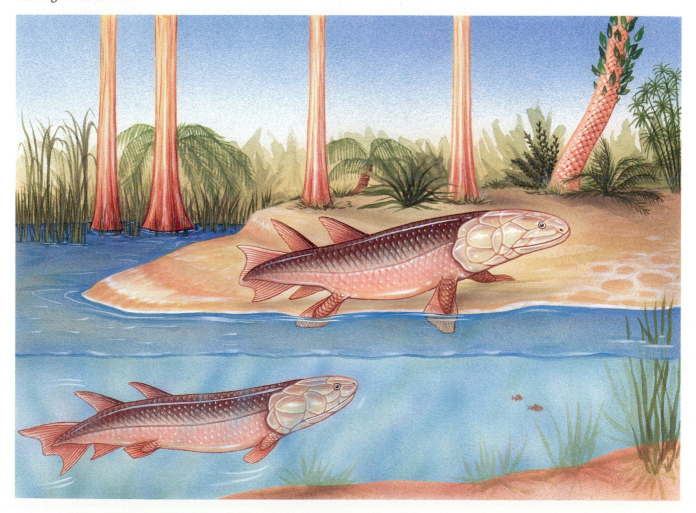

Ostracoderms became extinct by the end of the Devonian, a time that coincides with the rapid evolution of jawed fish. The placoderms also became extinct by the end of the De-

vonian, while the acanthodians decreased in abundance after the Devonian and became extinct by the end of the Paleozoic Era. On the other hand, cartilaginous and ray-finned bony fish expanded during the Late Paleozoic, as did the ammonoid cephalopods, the other major predator of the Late Paleozoic seas.

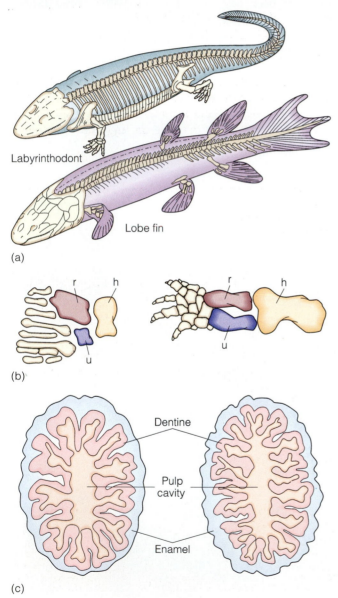

◆ **FIGURE 14.13** Similarities between the crossopterygian lobe-finned fish and the labyrinthodont amphibians. (a) Skeletal similarity. (b) Comparison of the limb bones of a crossopterygian (left) and amphibian (right); colors identifies the bones (u = ulna, shown in blue, r = radius, mauve, h = humerus, gold) that the two groups have in common. (c) Comparison of tooth cross sections shows the complex and distinctive structure found in both the crossopterygians (left) and amphibians (right).

❖ AMPHIBIANS—VERTEBRATES INVADE THE LAND

Although amphibians were the first vertebrates to live on land, they were not the first land-living organisms. Land plants, which probably evolved from green algae, first evolved during the Ordovician. Furthermore, insects, millipedes, spiders, and even snails invaded the land before amphibians (see Perspective 14.1).

The transition from water to land required that several barriers be surmounted. The most critical for animals were desiccation, reproduction, the effects of gravity, and the extraction of oxygen from the atmosphere by lungs rather than from water by gills. These problems were partly solved by the crossopterygians; they already had a backbone and limbs that could be used for walking and lungs that could extract oxygen (Fig. 14.13).

The earliest amphibian fossils are found in the Upper Devonian Old Red Sandstone of eastern Greenland. These amphibians had streamlined bodies, long tails, and fins. In addition, they had four legs, a strong backbone, a rib cage, and pelvic and pectoral girdles, all of which were structural adaptations for walking on land (◆ Fig. 14.14). The earliest amphibians appear to have had many characteristics that were inherited from the crossopterygians with little modification (Fig. 14.13).

Because amphibians did not evolve until the Late Devonian, they were a minor element of the Devonian terrestrial ecosystem. Like other groups that moved into new and previously unoccupied niches, however, amphibians underwent rapid adaptive radiation and became abundant during the Carboniferous and Early Permian. The Late Paleozoic amphibians did not at all resemble the familiar frogs, toads, newts, and salamanders that make up the modern amphibian fauna. Rather they displayed a broad spectrum of sizes, shapes, and modes of life (◆ Fig. 14.15). One group of amphibians were the **labyrinthodonts,** so named for the labyrinthine wrinkling and folding of the chewing surface of their teeth (Fig. 14.13c). Most labyrinthodonts were large animals, as much as 2 m in length. These typically sluggish creatures lived in swamps and streams, eating fish, vegetation, insects, and other small amphibians (Fig. 14.15).

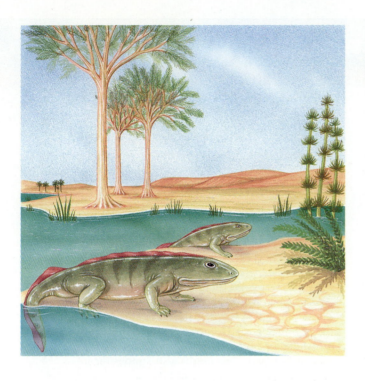

◆ **FIGURE 14.14** Reconstruction of a Late Devonian landscape in the eastern part of Greenland. Shown is *Ichthyostega,* an amphibian that grew to a length of about 1 m. The flora of the time was diverse, consisting of a variety of small and large seedless vascular plants.

Labyrinthodonts were very abundant during the Carboniferous when swampy conditions were widespread, but soon declined in abundance during the Permian, perhaps in response to changing climatic conditions. Only a few species survived into the Triassic.

❖ EVOLUTION OF THE REPTILES—THE LAND IS CONQUERED

Amphibians were limited in colonizing the land because they had to return to water to lay their gelatinous eggs. The evolution of the **amniote egg** (◆ Fig. 14.16) freed reptiles from this constraint. In such an egg, the developing

◆ **FIGURE 14.15** Reconstruction of a Carboniferous coal swamp. The varied amphibian fauna of the time is shown, including the large labyrinthodont amphibian *Eryops* (foreground), the larval *Branchiosaurus* (center), and the serpent-like *Dolichosoma* (background).

A Late Ordovician Invasion of the Land

Recently, the question of when the land was first colonized has been much discussed. There is excellent evidence that the land was colonized by vascular plants during the Silurian Period. Fragments of plant remains have been found in Middle Silurian rocks indicating primitive plants had invaded the land by that time. These early plants, like their Devonian descendants, had to live near bodies of water due to their reproductive requirements.

While Silurian plant fossil evidence is represented by a handful of geographically scattered fossils, the spore evidence for a Silurian land colonization is much more abundant and diverse. This may be due in part to the greater chance for preservation of spores, which have a resistant organic covering. In the North Atlantic region, Silurian vascular plant megafossil evidence is based primarily on specimens of *Cooksonia* (Fig. 14.22). Yet at least 14 spore genera are described for the Silurian Period.

Discoveries of probable vascular plant megafossils and characteristic spores indicate to many paleontologists that the evolution of vascular plants occurred well before the Middle Silurian. Sheets of cuticlelike cells—that is, the cells that cover the surface of modern land plants—and tetrahedral clusters that closely resemble the spore tetrahedrals of primitive land plants have been reported from Middle to Upper Ordovician rocks from western Libya by Jane Gray and her co-workers at the University of Oregon (◆ Fig. 1).

The interpretation of these spores has been controversial and not completely accepted by all paleontologists. If they are in fact from land plants, this means that land plants had a long pre-Silurian record. And if land plants existed during the Ordovician, when did animals invade the land?

The first vertebrate animals made the transition from water to land during the Late Devonian. Arthropods, however, including scorpions and flightless insects, evolved at least by the Early Devonian, based on their fossil representatives found in the Lower Devonian Rhynie Chert of Scotland. This unit is also famous for its beautifully preserved fossil plant flora.

The discovery of fossil burrows within a buried soil of the Upper Ordovician Juanita Formation in central Pennsylvania represents the oldest reported nonmarine trace fossils. While no hard parts or traces of organisms are associated with the burrows, their size, shape, and arrangement are consistent with a bilaterally symmetrical burrower that was resistant to

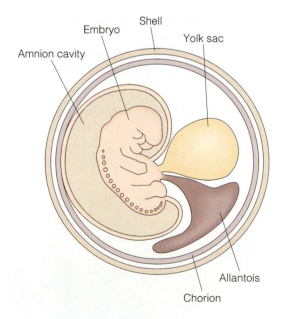

◆ **FIGURE 14.16** The amniote egg. In an amniote egg, the embryo is surrounded by a liquid sac (amnion cavity) and provided with a food source (yolk sac) and waste sac (allantois). The evolution of the amniote egg freed reptiles from having to return to the water for reproduction and allowed them to inhabit all parts of the land.

embryo is surrounded by a liquid-filled sac called the *amnion* and provided with both a yolk, or food sac, and an allantois, or waste sac. In this way the emerging reptile is, in essence, a miniature adult, bypassing the need for a larval stage in the water. The evolution of the amniote egg allowed vertebrates to colonize all parts of the land because they no longer had to return to the water as part of their reproductive cycle. The oldest known structure thought to be a reptilian egg comes from Lower Permian rocks of Texas, although some paleontologists question its authenticity. It is, however, generally acknowledged that the amniote egg evolved at the latest by the Early Pennsylvanian when the earliest reptile fossils are found.

desiccation. It is hypothesized that these burrows may have been made by millipedes. Millipedes first appear in marine rocks of Early Silurian age, and some present-day millipedes are active burrowers.

The existence of sizable burrowing organisms on dry land during the Late Ordovician implies that some type of terrestrial vegetation existed for them to feed on. Burrowing animals could conceivably have fed on soil algae, now believed to have been in existence since the Late Precambrian. The discovery of possible primitive land plant spores from Middle to Upper Ordovician rocks means that these plants could have supported large populations of burrowing herbaceous arthropods as well as litter organisms. The discovery of Late Ordovician trace fossils in paleosoils adds to our increasing knowledge of Early Paleozoic terrestrial ecosystems.

Many of the differences between amphibians and reptiles are physiological and are not preserved in the fossil record. Nevertheless, amphibians and reptiles differ sufficiently in skull structure, jawbones, ear location, and limb and vertebral construction to suggest that reptiles evolved from labyrinthodont ancestors by at least the early Pennsylvanian.

The oldest known reptiles are from the Lower Pennsylvanian Joggins Formation in Nova Scotia, Canada. Here, remains of *Hylonomus* are found in the sediments filling in tree trunks (◆ Fig. 14.17). These earliest reptiles were small and agile and fed largely on grubs and insects. They belonged to the group of reptiles known as **captorhinomorphs,** the group from which all other reptiles evolved (◆ Fig. 14.18). During the Permian Period, reptiles diversified and began displacing many amphibians. The success of the reptiles is due partly to their advanced method of reproduction and their more advanced jaws and teeth as well as to their ability to move rapidly on land.

◆ **FIGURE 14.17** Reconstruction and skeleton of the oldest known reptile, *Hylonomus lyelli* from the Pennsylvanian Period. Fossils of this animal have been collected from sediments that filled tree stumps. *Hylonomus lyelli* was about 30 cm long.

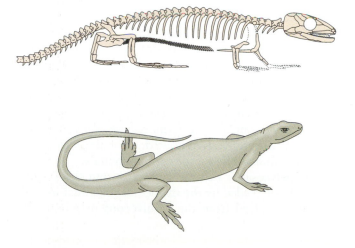

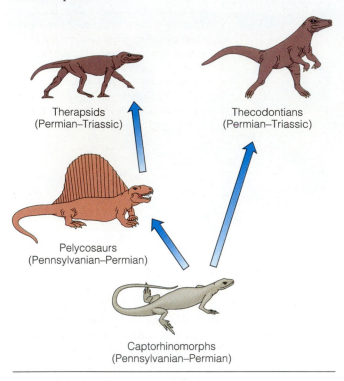

Therapsids
(Permian–Triassic)

Thecodontians
(Permian–Triassic)

Pelycosaurs
(Pennsylvanian–Permian)

Captorhinomorphs
(Pennsylvanian–Permian)

The **pelycosaurs,** or finback reptiles, evolved from the captorhinomorphs during the Pennsylvanian and were the dominant reptile group by the Early Permian. They evolved into a diverse assemblage of herbivores, exemplified by *Edaphosaurus,* and carnivores such as *Dimetrodon* (◆ Fig. 14.19). An interesting feature of the pelycosaurs is their sail. It was formed by vertebral spines that, in life, were covered with skin. The sail has been variously explained as a type of sexual display, a means of protection, and a display to look more ferocious, but current consensus seems to be that the sail served as some type of thermoregulatory device, raising the reptile's temperature by catching the sun's rays or cooling it by facing the wind. Because pelycosaurs are considered to be the group from which mammal-like reptiles evolved, it is interesting that they may have had some sort of body-temperature control. Furthermore, the teeth of *Dimetrodon* were slightly differentiated, a condition characteristic of mammal-like reptiles and mammals.

The pelycosaurs became extinct during the Permian and were succeeded by the **therapsids,** mammal-like reptiles that evolved from the carnivorous pelycosaur lineage and rapidly diversified into herbivorous and carnivorous lineages (◆ Fig. 14.20). Therapsids were small to medium-sized animals displaying the beginnings of many mammalian features; fewer bones in the skull due to fusion of many of the small skull bones; enlargement of the lower jawbone; differentiation of the teeth for various functions such as nipping, tearing, and chewing food; and a more vertical position of the legs for greater flexibility, as opposed to the way the legs sprawled out to the side in primitive reptiles.

Furthermore, many paleontologists believe therapsids were endothermic, or warm-blooded, enabling them to maintain a constant internal body temperature. This characteristic would have allowed them to expand into a variety of habitats, and indeed the Permian rocks do indicate a wide latitudinal distribution.

As the Paleozoic Era came to an end, the therapsids constituted about 90% of the known reptile genera and occupied a wide range of ecological niches. The mass extinctions that decimated the marine fauna at the close of the Paleozoic had an even greater effect on the terrestrial population. By the end of the Permian, about 50% of the invertebrate marine families were extinct compared with 75% of the amphibians and 80% of the reptile families. Plants, on the other hand, apparently did not experience as great a turnover as the invertebrates and animals.

❖ PLANT EVOLUTION

When plants made the transition from water to land, they had to solve the same problems that animals did: desiccation, support, and the effects of gravity. Plants did so by evolving a variety of structural adaptations that were fundamental to the subsequent radiations and diversification that occurred during the Silurian, Devonian, and later periods (Table 14.2). Most experts agree that the ancestors of land plants first evolved in a marine environment, then moved into a freshwater environment and finally onto land. In this way, the differences in osmotic pressures between salt and fresh water were overcome while the plant was still in the water.

What was the land surface like before the plants made the transition to land? During the Early Paleozoic, the land was probably covered by scattered cyanobacterial microbial mats, which were followed later by various coccoid and filamentous green algae. Microfossil evidence indicates that sometime around the Middle Ordovician the terrestrial vegetation changed. Spores recov-

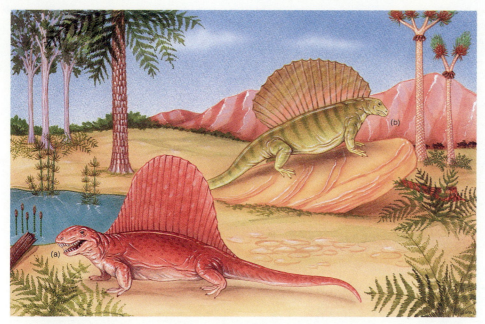

◆ **FIGURE 14.19** Most pelycosaurs, or finback reptiles have a characteristic sail on their back. One hypothesis explains the sail as a type of thermoregulatory device. Other hypotheses are that it was a type of sexual display or a device to make the reptile look more intimidating. Shown here are (a) the carnivore *Dimetrodon* and (b) the herbivore *Edaphosaurus*.

◆ **FIGURE 14.20** Reconstruction of a Late Permian scene in southern Africa showing various therapsids including *Dicynodon* (left foreground) and *Moschops* (right). Many paleontologists believe therapsids were endothermic and may have had a covering of fur as shown here.

ered from Middle to Upper Ordovician rocks from western Libya by Jane Gray and her co-workers at the University of Oregon suggest that land plants evolved well before the earliest known Middle Silurian fossil plant evidence indicates (see Perspective 14.1). Though many paleontologists now accept the microfossil evidence for a pre-Silurian land flora, some still believe the spores were produced by aquatic algae instead of land plants.

The higher land plants are composed of two major groups, the nonvascular and vascular plants. Most land plants are **vascular,** meaning they have a tissue system of specialized cells for the movement of water and nutrients. The **nonvascular** plants, such as bryophytes (liver-

TABLE 14.2 Major Events in the Evolution of Land Plants. The Devonian Period was a time of rapid evolution for the land plants. Major events were the appearance of leaves, heterospory, secondary growth, and the emergence of seeds.

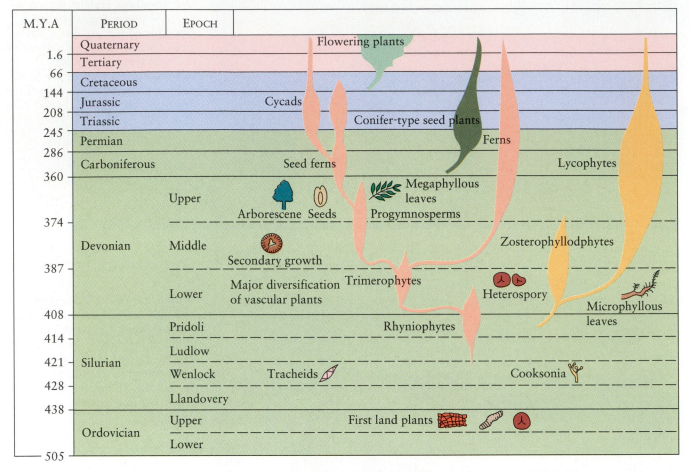

worts, hornworts, and mosses) and fungi, do not have these specialized cells and are typically small and usually live in low, moist areas.

The earliest land plants from the Middle to Late Ordovician were probably small and bryophyte-like in their overall organization (but not necessarily related to bryophytes). The evolution of vascular tissue in plants was an important step as it allowed for the transport of food and water. The ancestor of terrestrial vascular plants was probably some type of green alga. While no fossil record of the transition from green algae to terrestrial vascular plants exists, comparison of their physiology reveals a strong link. Primitive *seedless vascular plants* (discussed later in this chapter) such as ferns resemble green algae in their pigmentation, important metabolic enzymes, and two-phase reproductive cycle (◆ Fig. 14.21). Furthermore, the green algae are one of the few plant groups to have made the transition from marine to fresh water.

The evolution of terrestrial vascular plants from an aquatic, probably green algal ancestry was accompanied by various modifications that allowed them to occupy this new and harsh environment. In addition to the primary function of transporting water and nutrients throughout a plant, vascular tissue also provides some support for the plant body. Additional strength is derived from the organic compounds *lignin* and *cellulose,* which are found throughout a plant's walls.

The problem of desiccation was circumvented by the evolution of *cutin,* an organic compound found in the outer-wall layers of plants. Cutin also provides additional resistance to oxidation, the effects of ultraviolet light, and the entry of parasites.

Roots evolved in response to the need to collect water and nutrients from the soil and to help anchor the plant in the ground. The evolution of *leaves* from tiny outgrowths on the stem or from branch systems provided

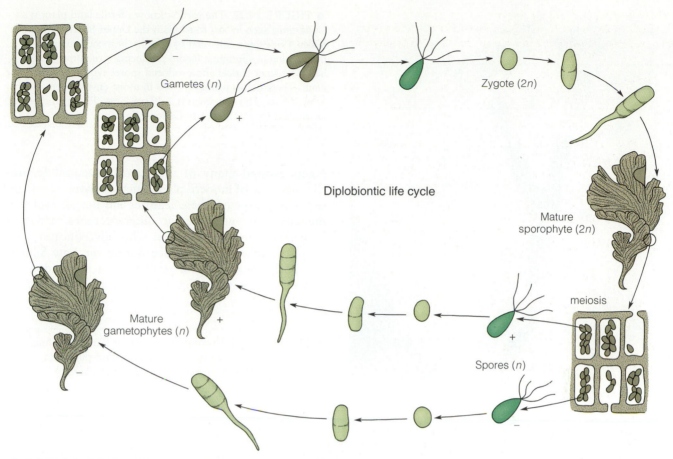

Gametes (n)

Diplobiontic life cycle

Zygote (2n)

Mature
sporophyte (2n)

meiosis

Mature
gametophytes (n)

Spores (n)

◆ **FIGURE 14.21** The diplobiontic life cycle of the green algae *Ulva*. Here the zygote grows to form a multicellular sporophyte plant that produces spores by meiosis. These develop into a mature gametophyte plant that produces gametes that fuse to produce a zygote, which in turn grows into the sporophyte plant. This alternation of generations is the same method of reproduction found in seedless vascular plants and is evidence that the vascular plants evolved from a green algal ancestor.

plants with an efficient light-gathering system for photosynthesis.

Silurian and Devonian Floras

The earliest known vascular land plants are small Y-shaped stems assigned to the genus *Cooksonia* from the Middle Silurian of Wales and Ireland. Together with Upper Silurian and Lower Devonian species from Scotland, New York State, former Czechoslovakia, and the Commonwealth of Independent States, these earliest plants were small, simple, leafless stalks with a spore-producing structure at the tip (◆ Fig. 14.22); they are known as **seedless vascular plants** because they did not produce a seed. They also did not have a true root sys-

tem. A *rhizome*, the underground part of the stem, transferred water from the soil to the plant and anchored the plant to the ground. The sedimentary rocks in which these plant fossils are found indicate that they lived in low, wet, marshy, freshwater environments.

An interesting parallel can be seen between seedless vascular plants and amphibians. When they made the transition from water to land, they had to overcome the problems such a transition involved. Both groups, while very successful, nevertheless required a source of water in order to reproduce. In the case of amphibians, their gelatinous egg had to remain moist, while the seedless vascular plants required water for the sperm to travel through to reach the egg.

From this simple beginning, the seedless vascular

◆ **FIGURE 14.22** The earliest known fertile land plant was *Cooksonia*, seen in this fossil from the Upper Silurian of South Wales. *Cooksonia* consisted of upright, branched stems terminating in sporangia (spore-producing structures). It also had a resistant cuticle and produced spores typical of a vascular plant. These plants probably lived in moist environments such as mud flats. This specimen is 1.49 cm long. (Photo courtesy of Dianne Edwards, University College, England.)

plants evolved many of the major structural features characteristic of modern plants such as leaves, roots, and secondary growth. These features did not all evolve simultaneously but rather at different times, a pattern known as *mosaic evolution*. This diversification and adaptive radiation took place during the Late Silurian and Early Devonian and resulted in a tremendous increase in diversity (◆ Fig. 14.23). From the end of the

◆ **FIGURE 14.23** Reconstruction of an Early Devonian landscape showing some of the earliest land plants. (a) *Dawsonites.* (b) *Protolepidodendron.* (c) *Bucheria.*

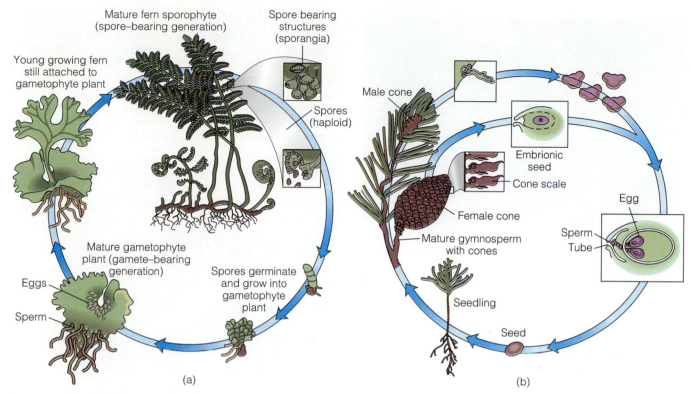

(a)

(b)

◆ **FIGURE 14.24** (a) Generalized life history of a seedless vascular plant. The mature sporophyte plant produces spores, which upon germination grow into small gametophyte plants that produce sperm and eggs. The fertilized eggs grow into the spore-producing mature plant, and the sporophyte-gametophyte life cycle begins again. (b) Generalized life history of a gymnosperm plant. The mature plant bears both male cones that produce sperm-bearing pollen grains and female cones that contain embryonic seeds. Pollen grains are transported to the female cones by the wind. Fertilization occurs when the sperm moves through a moist tube growing from the pollen grain and unites with the embryronic seed which then grows into a cone-bearing mature plant.

Early Devonian to the end of the Devonian, the number of plant genera remained about the same, yet the composition of the flora changed. Whereas the Early Devonian landscape was dominated by relatively small, low-growing, bog-dwelling types of plants, the Late Devonian witnessed forests of large tree-sized plants up to 10 m tall.

In addition to the diverse seedless vascular plant flora of the Late Devonian, another significant floral event took place. The evolution of the seed at this time liberated land plants from their dependence on moist conditions and allowed them to spread over all parts of the land.

Seedless vascular plants require moisture for successful fertilization because the sperm must travel to the egg on the surface of the gamete-bearing plant (gametophyte) to produce a successful spore-generating plant (sporophyte). Without moisture, the sperm would dry out before reaching the egg (◆ Fig. 14.24a). In the seed method of reproduction, the spores are not released to the environment as they are in the seedless vascular plants, but are retained on the spore-bearing plant, where they grow into the male and female forms of the gamete-bearing generation. In the case of the **gymnosperms**, or flowerless seed plants, these are male and female cones (◆ Fig. 14.24b). The male cone produces pollen, which contains the sperm and has a waxy coating to prevent desiccation, while the egg, or embryonic seed, is contained in the female cone. After fertilization, the seed then develops into a mature, cone-bearing plant. In this way the need for a moist environment

Plant Evolution   391

(a)

(b)

(d)

◆ **FIGURE 14.25** (a) Specimen and (b) reconstruction of the Early Devonian plant *Chaleuria cirrosa* from New Brunswick, Canada. This plant was heterosporous, meaning that it produced two sizes of spores. (c) The larger spores, which are thought to have produced female plants, range in size from 60 to 180 μm in diameter. (d) The smaller spores, which are believed to have produced male plants, have a size range of 30–40 μm in diameter. (Photos and reconstruction courtesy of Patricia G. Gensel, University of North Carolina.)

for the gametophyte generation is solved. The significance of this development is that seed plants, like reptiles, were no longer restricted to wet areas, but were free to migrate into previously unoccupied dry environments. As will be discussed in Chapter 16, the evolution of the flowering plants (angiosperms) did not occur until the Cretaceous Period.

Before seed plants evolved, an intermediate evolutionary step was necessary. This was the development of *heterospory*, whereby a species produces two types of spores: a large one (megaspore) that gives rise to the female gamete-bearing plant and a small one (microspore) that produces the male gamete-bearing plant. Prior to heterospory, plants were *homosporous;* that is, they produced spores that were all the same size and, upon germinating, grew into gametophyte plants that produced both male and female gametes.

The earliest evidence of heterospory is found in the Early Devonian plant *Chaleuria cirrosa*, which produced spores of two distinct sizes (◆ Fig. 14.25). The appearance of heterospory was followed several million years later by the emergence of progymnosperms—Middle and Late Devonian plants with fernlike reproductive habits and a gymnosperm anatomy—which gave rise in the Late Devonian to such other gymnosperm groups as the seed

ferns and conifer-type seed plants. While the seedless vascular plants dominated the flora of the Carboniferous coal-forming swamps, the gymnosperms made up an important element of the Late Paleozoic flora, particularly in the nonswampy areas.

Late Carboniferous and Permian Floras

As discussed earlier, the rocks of the Pennsylvanian Period (Late Carboniferous) are the major source of the world's coal. Coal results from the alteration of plants living in low, swampy areas. The geologic and geographic conditions of the Pennsylvanian were ideal for the growth of seedless vascular plants, and consequently these coal swamps had a very diverse flora (◆ Fig. 14.26).

◆ **FIGURE 14.26** Reconstruction of a Pennsylvanian coal swamp with characteristic vegetation. The amphibian is *Eogyrinus*.

◆ **FIGURE 14.27** Living sphenopsids include the horsetail *Equisetum*.

It is evident from the fossil record that while the Early Carboniferous flora was similar to its Late Devonian counterpart, a great deal of evolutionary experimentation was occurring that would lead to the highly successful Late Paleozoic flora of the coal swamps and adjacent habitats. Among the seedless vascular plants, the *lycopsids* and *sphenopsids* were the most important coal-forming groups of the Pennsylvanian Period.

The lycopsids were present during the Devonian, chiefly as small plants, but by the Pennsylvanian, they were the dominant element of the coal swamps, achieving heights up to 30 m in such genera as *Lepidodendron* and *Sigillaria*. The Pennsylvanian lycopsid trees are interesting because they lacked branches except at their top. The leaves were elongate and similar to the individual palm leaf of today. As the trees grew, the leaves were replaced from the top, leaving prominent and characteristic rows or spirals of scars on the trunk. Today, the lycopsids are represented by small temperate-forest ground pines.

The sphenopsids were the other important coal-forming plant group and are characterized by being jointed and having horizontal underground stem-bearing roots. Many of these plants, such as *Calamites*, averaged

5 to 6 m tall. Living sphenopsids include the horsetail (*Equisetum*) and scouring rushes (◆ Fig. 14.27). Small seedless vascular plants and seed ferns formed a thick undergrowth or ground cover beneath these treelike plants.

Not all plants were restricted to the coal-forming swamps. Among those plants occupying higher and drier ground were some of the *cordaites*, a group of tall gymnosperm trees that grew up to 50 m and probably formed vast forests (◆ Fig. 14.28). Another important non-swamp dweller was *Glossopteris*, the famous plant so abundant in Gondwana (see Fig. 7.2), whose distribution is cited as critical evidence that the continents have moved through time.

The floras that were abundant during the Pennsylvanian persisted into the Permian, but due to climatic and geologic changes resulting from tectonic events (see Chapter 12), they declined in abundance and importance. By the end of the Permian, the cordaites became extinct, while the lycopsids and sphenopsids were reduced to mostly small, creeping forms. The gymnosperms, whose life-style was more suited to the warmer and drier Permian climates, diversified and came to dominate the Permian, Triassic, and Jurassic landscapes.

◆ **FIGURE 14.28** A cordaite forest from the Late Carboniferous. Cordaites were a group of gymnosperm trees that grew up to 50 m tall.

Chapter Summary

Table 14.3 summarizes the major evolutionary and geologic events of the Paleozoic Era and shows their relationships to each other.

1. Chordates are characterized by a notochord, dorsal hollow nerve cord, and gill slits. The earliest chordates were soft-bodied organisms that were rarely fossilized. Vertebrates are a subphylum of the chordates.
2. The fish are the earliest known vertebrates. Their first fossil occurrence is in Upper Cambrian rocks. They have had a long and varied history, including jawless and jawed armored forms (ostracoderms and placoderms), cartilaginous forms, and bony forms. The crossopterygians, a group of lobe-finned fish, gave rise to the amphibians.
3. The link between the crossopterygians and the earliest amphibians is very convincing and includes a close similiarity

of bone and tooth structures. The transition from fish to amphibians occurred during the Devonian. During the Carboniferous, the labyrinthodont amphibians were the dominant terrestrial vertebrate animals.
4. The earliest fossil record of reptiles is from the Late Carboniferous. The evolution of an amniote egg was the critical factor in the reptiles' ability to colonize all parts of the land.
5. The pelycosaurs were the dominant reptile group during the Early Permian, while the therapsids dominated the landscape for the rest of the Permian Period.
6. Plants had to overcome the same basic problems as the animals, namely desiccation, reproduction, and gravity in making the transition from water to land.
7. The earliest fossil record of land plants is from Middle to Upper Ordovician rocks. These plants were probably small and bryophyte-like in their overall organization.

(continued on page 398)

TABLE 14.3   Major Evolutionary and Geologic Events of the Paleozoic Era

Age (Millions of Years)	Geologic Period		Invertebrates	Vertebrates
245 —	Permian		Largest mass extinction event to affect the invertebrates.	Acanthodians, placoderms, and pelycosaurs become extinct. Therapsids and pelycosaurs the most abundant reptiles.
286 —	Carboniferous	Pennsylvanian	Fusulinids diversify.	Reptiles evolve. Amphibians abundant and diverse.
320 —		Mississippian	Crinoids, lacy bryozans, blastoids become abundant. Renewed adaptive radiation following extinctions of many reef-builders.	
360 —	Devonian		Extinctions of many reef-building invertebrates near end of Devonian. Reef building continues. Eurypterids abundant.	Amphibians evolve. All major groups of fish present—Age of Fish.
408 —	Silurian		Major reef building. Diversity of invertebrates remains high.	Ostracoderms common. Acanthodians, the first jawed fish, evolve.
438 —	Ordovician		Extinctions of a variety of marine invertebrates near end of Ordovician. Major adaptive radiation of all invertebrate groups. Suspension feeders dominant.	Ostracoderms diversify.
505 —	Cambrian		Many trilobites become extinct near end of Cambrian. Trilobites, brachiopods, and archaeocyathids are most abundant.	Earliest vertebrates—jawless fish called ostracoderms.
570 —				

TABLE 14.3 Major Evolutionary and Geologic Events of the Paleozoic Era (continued)

Plants	Major Geologic Events
Gymnosperms diverse and abundant.	Formation of Pangaea. Alleghenian orogeny. Hercynian orogeny.
Coal swamps with flora of seedless vascular plants and gymnosperms.	Coal-forming swamps common. Formation of Ancestral Rockies. Continental glaciation in Gondwana.
Gymnosperms appear (may have evolved during Late Devonian).	Ouachita orogeny.
First seeds evolve. Seedless vascular plants diversify.	Widespread deposition of black shale. Antler orogeny. Acadian orogeny.
Early land plants—seedless vascular plants.	Caledonian orogeny. Extensive barrier reefs and evaporites.
Plants move to land?	
	Continental glaciation in Gondwana. Taconic orogeny.
	First Phanerozoic transgression (Sauk) onto North American craton.

8. The evolution of vascular tissue was an important event in plant evolution as it allowed food and water to be transported throughout the plant and provided the plant with additional support.

9. The ancestor of terrestrial vascular plants was probably some type of green alga based on such similarities as pigmentation, metabolic enzymes, and a two-phase reproductive cycle.

10. The earliest seedless vascular plants were small, leafless stalks with spore-producing structures on their tips. From this simple beginning, plants evolved many of the major structural features characteristic of modern plants.

11. By the end of the Devonian Period, forests with tree-sized plants up to 10 m had evolved. The Late Devonian also witnessed the evolution of the flowerless seed plants (gymnosperms) whose reproductive style freed them from having to stay near water.

12. The Carboniferous Period was a time of vast coal swamps, where conditions were ideal for the seedless vascular plants. With the onset of more arid conditions during the Permian, the gymnosperms became the dominant element of the world's flora.

Important Terms

acanthodian
amniote egg
bony fish
captorhinomorph
cartilaginous fish
chordate
crossopterygian

gymnosperm
labyrinthodont
lobe-finned fish
nonvascular
ostracoderm
pelycosaur

placoderm
ray-finned fish
seedless vascular plant
therapsid
vascular
vertebrate

Review Questions

1. Which of the following must an organism possess during at least part of its life cycle to be classified as a chordate?
   a. _____ vertebrae, dorsal hollow nerve cord, gill slits;
   b. _____ notochord, ventral solid nerve cord, lungs;
   c. _____ notochord, dorsal hollow nerve cord, gill slits;
   d. _____ vertebrae, dorsal hollow nerve cord, lungs;
   e. _____ notochord, dorsal solid nerve cord, lungs.

2. Based on similarity of embryo development, which invertebrate phylum is most closely allied with the chordates?
   a. _____ Arthropoda; b. _____ Annelida; c. _____ Porifera; d. _____ Echinodermata; e. _____ Mollusca.

3. The oldest fish remains are of what age?
   a. _____ Cambrian; b. _____ Ordovician; c. _____ Silurian; d. _____ Devonian; e. _____ Mississippian.

4. Jawless armored fish are:
   a. _____ ostracoderms; b. _____ placoderms; c. _____ cartilaginous; d. _____ bony; e. _____ none of these.

5. The "Age of Fish" refers to which period?
   a. _____ Cambrian; b. _____ Ordovician; c. _____

Silurian; d. _____ Devonian; e. _____ Mississippian.

6. Placoderms are:
   a. _____ jawless fish; b. _____ primitive amphibians; c. _____ advanced reptiles; d. _____ plants; e. _____ jawed armored fish.

7. The first jawed fish were:
   a. _____ ostracoderms; b. _____ acanthodians; c. _____ placoderms; d. _____ cartilaginous; e. _____ lobe-finned.

8. The ancestors of amphibians belong to which group?
   a. _____ placoderms; b. _____ ostracoderms; c. _____ acanthodians; d. _____ crossopterygians; e. _____ antiarchs.

9. The earliest amphibian fossils are found in what age rocks?
   a. _____ Silurian; b. _____ Devonian; c. _____ Mississippian; d. _____ Pennsylvanian; e. _____ Permian.

10. Labyrinthodonts are:
    a. _____ reptiles; b. _____ fish; c. _____ amphibians; d. _____ plants; e. _____ none of these.

11. In which period were amphibians most abundant?
a. _____ Silurian; b. _____ Devonian; c. _____ Mississippian; d. _____ Pennsylvanian; e. _____ Permian.

12. The most significant evolutionary change that allowed reptiles to colonize all parts of the land was:
a. _____ endothermy; b. _____ origin of limbs capable of supporting the animals on land; c. _____ evolution of the amniote egg; d. _____ evolution of a watertight skin; e. _____ evolution of tear ducts.

13. The earliest reptiles were:
a. _____ captorhinomorphs; b. _____ labyrinthodonts; c. _____ acanthodians; d. _____ pelycosaurs; e. _____ therapsids

14. The finback reptiles *Dimetrodon* and *Edaphosaurus* were:
a. _____ therapsids; b. _____ pelycosaurs; c. _____ labyrinthodonts; d. _____ acanthodians; e. _____ captorhinomorphs.

15. The dominant reptile group during the Permian, and the one that gave rise to the mammals, was the:
a. _____ labyrinthodonts; b. _____ captorhinomorphs; c. _____ therapsids; d. _____ pelycosaurs; e. _____ acanthodians.

16. Which problems did plants have to contend with to make the successful transition from water to land?
a. _____ desiccation; b. _____ photosynthesis; c. _____ gravity; d. _____ answers (a) and (b); e. _____ answers (a) and (c).

17. Most paleontologists now believe that plants made the transition from water onto land during which period?
a. _____ Cambrian; b. _____ Ordovician; c. _____ Silurian; d. _____ Devonian; e. _____ Mississippian.

18. Which algal group was the probable ancestor of vascular plants?
a. _____ red; b. _____ blue-green; c. _____ green; d. _____ brown; e. _____ yellow.

19. Which plant group was the first to successfully invade the land?
a. _____ seedless vascular; b. _____ gymnosperms; c. _____ naked seed bearing; d. _____ angiosperms; e. _____ flowering.

20. During which period were the seedless vascular plants most abundant?
a. _____ Silurian; b. _____ Devonian; c. _____ Mississippian; d. _____ Pennsylvanian; e. _____ Permian.

21. Which plant group occupied the higher and drier grounds during the Late Paleozoic?
a. _____ cordaites; b. _____ lycopsids; c. _____ sphenopsids; d. _____ glossopterids; e. _____ answers (b) and (c).

22. The first plant group that did not require a wet area for part of its life cycle was the:
a. _____ seedless vascular plants; b. _____ gymnosperms; c. _____ angiosperms; d. _____ flowering plants; e. _____ labyrinthodonts.

23. Discuss the Paleozoic evolutionary history of the amphibians.

24. Discuss the significance and advantages of the pelycosaur sail.

25. Discuss the evidence for an Ordovician faunal and floral colonization of the land.

26. Outline the evolutionary history of fish.

27. Describe the problems that had to be overcome before organisms could inhabit the land.

28. Why were the reptiles so much more successful at extending their habitat than the amphibians?

29. In what ways are therapsids more mammal-like than pelycosaurs?

30. What are the major differences between seedless vascular plants and gymnosperms? What is the significance of these differences in terms of exploiting the terrestrial environment?

Additional Readings

Bolt, J. R., R. M. McKay, B. J. Witzke, and M. P. McAdams. 1988. A new Lower Carboniferous tetrapod locality in Iowa. *Nature* 333, no. 6175: 768–70.

Carroll, R. L. 1988. *Vertebrate paleontology and evolution.* New York: W. H. Freeman.

Colbert, E. H., and M. Morales. 1991. *Evolution of the vertebrates.* 4th ed. New York: John Wiley & Sons.

Cowen, R. 1989. *History of life.* Palo Alto, Calif.: Blackwell Scientific Publications.

Gensel, P. G., and H. N. Andrews. 1987. The evolution of early land plants. *American Scientist* 75, no. 5: 478–89.

Gray, J., and W. Shear. 1992. Early life on land. *American Scientist* 80, no. 5: 444–56.

Lane, N. G. 1992. *Life of the past.* 3d ed. Columbus, Ohio: Charles E. Merrell Publishing Co.

Schopf, J. W., ed. 1992. *Major events in the history of life.* Boston: Jones and Bartlett.

Thomas, B., and R. Spicer. 1987. *Evolution and palaeobiology of land plants.* Portland, Oregon: Dioscorides Press.

Thomson, K. W. 1991. Where did tetrapods come from? *American Scientist* 79, no. 6: 488–90.

White, M. E. 1986. *The greening of Gondwana.* Frenchs Forest, NSW, Australia: Reed Books Limited.

CHAPTER 15

Chapter Outline

Ownes Valley and eastern scarp along the Sierra Nevada, California. (Photo © Peter Kresan.)

GEOLOGY OF THE MESOZOIC ERA

Prologue

Approximately 150 to 210 million years after the emplacement of massive plutons created the Sierra Nevada (Nevadan orogeny), gold was discovered at Sutter's Mill on the South Fork of the American River at Coloma, California. On January 24, 1848, James Marshall, a carpenter building a sawmill for John Sutter, found bits of the glittering metal in the mill's tailrace. Soon, settlements throughout the state were completely abandoned as word of the chance for instant riches spread throughout California. Within a year after the news of the gold discovery reached the East Coast, the Sutter's Mill area was swarming with more than 80,000 prospectors, all hoping to make their fortune. In all, at least 250,000 gold seekers prospected the Sutter's Mill area, and though most were Americans, they came from all over the world, even as far away as China. Most of them thought the gold was simply waiting to be taken.

Of course, no one gave any thought to the consequences of so many people converging on the Sutter's Mill area, all intent on making easy money. In reality, only a small percentage of prospectors ever hit it big or were even moderately successful. The rest barely eked out a living until they eventually abandoned the dream and went home.

Although some prospectors dug $30,000 worth of gold dust a week out of a single claim and gold was found practically on the surface of the ground, most of this easy gold was recovered very early during the gold rush. Most prospectors made only a living wage working their claims. Nevertheless, during the five years from 1848 to 1853 that constituted the gold rush proper, more than $200 million in gold was extracted.

Would-be prospectors could follow three basic routes to the gold fields. The most popular was the overland journey by wagon train from the East to California—a route fraught with peril, including the threat of starvation, disease, and the crossing of the Sierra Nevada. Those who took this route and survived finally arrived at Sutter's Mill and dispersed from there to prospect. The Panama route to California began with a voyage by ship to Chagres on the Isthmus of Panama. Here the potential prospectors and their goods were transported to Panama City by boat and mules and then traveled by ship to California. For the most part, the prospectors who took this route to California were young, vigorous men without wives and children. They faced the usual hardships of steamy jungles, disease, and overcrowded ships, many of which were unseaworthy. The third route, by ship around the southern tip of South America, was the longest, but it avoided the overcrowding encountered on the Panama route and the perils of the overland journey. Those who went by ship arrived at San Francisco, where many decided to stay after seeing that there were opportunities to make money in this new town.

The earliest prospectors came from the West Coast and were, for the most part, honest and hardworking. Those who arrived from elsewhere in 1849 and later were also mostly honest and hardworking, but, unfortunately, some were thieves, outlaws, and murderers. Life in the mining camps was extremely hard and expensive. Frequently, the shopowners and traders made more money than the prospectors did. Gambling halls and saloons sprang up all over since the men had little to do with their time except gamble and drink. Family life was virtually nonexistent as women comprised only 2% of the population.

Archean Eon	Proterozoic Eon	Phanerozoic Eon						
Precambrian		Paleozoic Era						
		Cambrian	Ordovician	Silurian	Devonian	Mississippian	Pennsylvanian	Permian
						Carboniferous		

2,500
M.Y.A.

570
M.Y.A.

245
M.Y.

The gold these prospectors sought was mostly in the form of placer deposits (◆ Fig. 15.1). Gold is almost always associated with felsic igneous rocks, where it forms from mineral-rich solutions moving through cracks and fractures of the igneous body. Weathering of the gold-bearing igneous rocks and mechanical separation of minerals by density during stream transport formed placer deposits. Many prospectors searched for the mother lode, but all of the gold recovered came from placers. One of the most common methods of mining these deposits was by panning. A prospector dipped a shallow pan into a stream bed, swirled the material around, and poured off the lighter material. The gold, being about six times heavier than most sand grains and rock chips, concentrated on the bottom of the pan and could be picked out.

◆ FIGURE 15.1 "How the California mines are worked" from a letter-sheet of the early 1850s. Shown on the left is the way dry or hill digging was done. The engraving on the right shows how gold was recovered from the river beds. The quartz mill in the center was the means by which quartz rock containing gold was pulverized so that the minute particles of gold could be gathered and amalgamated with quicksilver. (Photo courtesy of the Huntington Library.)

❖ INTRODUCTION

The dawn of the Mesozoic ushered in a new era in Earth history. The major geologic event was the breakup of Pangaea, which affected oceanic and climatic circulation patterns and influenced the evolution of the terrestrial and marine biotas. Though the Mesozoic Era is popularly known as the Age of Reptiles, it also was the time during which birds, mammals, and angiosperms (flowering plants) first evolved (see Chapter 16).

The Mesozoic Era (245 to 66 million years ago) is divided into three periods beginning with the Triassic (37 million years), followed by the Jurassic (64 million years), and finally the Cretaceous (78 million years). The stratotypes for the systems from which these periods derive their names are in the Hercynian Mountains of central Germany (Triassic), the Jura Mountains of northwest Switzerland (Jurassic), and the Paris Basin of northern France (Cretaceous).

Because most of the Mesozoic geologic history of North America involves the continental margins, we fo-

Phanerozoic Eon											
Mesozoic Era			Cenozoic Era								
Triassic	Jurassic	Cretaceous	Tertiary						Quaternary		
			Paleocene	Eocene	Oligocene	Miocene	Pliocene		Pleistocene	Holocene	

66
M.Y.A.

cus our attention on the eastern, gulf, and western coastal regions. The transgressions and regressions of the final two epeiric seas and their depositional sequences (Absaroka and Zuni) will be incorporated into the geologic history of each of these regions.

❖ THE BREAKUP OF PANGAEA

Just as the formation of Pangaea influenced geologic and biologic events during the Paleozoic, the breakup of this supercontinent profoundly affected geologic and biologic events during the Mesozoic. The movement of continents affected the global climatic and oceanic regimes as well as the climates of the individual continents. Populations became isolated or were brought into contact with other populations, leading to evolutionary changes in the biota. So great was the effect of this breakup on the world, that it forms the central theme of this chapter.

Pangaea's breakup can be documented by the extensional structures (fault-block basins or rift valleys) that formed along the newly created continental margins. Modern-day analogues for the different stages of continental rifting can be found in various parts of the world such as the East African Rift valleys (see Fig. 7.15). As a continent begins to rift, fault-block basins are formed (see Figure 7.14). Associated with these tensional structures are basaltic dikes, sills, and lava flows, which can be radiometrically dated to determine the time of rifting. As the continents continue separating, and basaltic oceanic crust forms between them (see Fig. 7.14), rates and direction of movement can be determined by radiometric and paleomagnetic dating of the oceanic crust (see Chapter 7).

During the initial phase of plate fragmentation, structures called **triple junctions** may form where three plates join. Though some are stable—such as the Galapagos

triple junction at the boundary of the Pacific, Cocos, and Nazca plates—and have maintained the same configuration for a long time (see Fig. 1.10), most triple junctions are unstable and migrate laterally as plate boundaries change position.

The type of triple junction in which we are most interested involves three ridges, usually formed initially by a mantle plume. When a mantle plume rises beneath a continent, the crust is stretched and typically fractures in a three-pronged pattern (❖ Fig. 15.2). The crust, however, usually separates along two of the three fractures, thus leaving the third fracture inactive. This third fracture is called a failed arm, or **aulacogen,** and eventually becomes filled with sediment. Some of the world's major river systems, including the lower Niger, Mississippi, and Amazon, apparently occupy segments of failed rifts.

A modern example of a triple junction with an aulacogen is the triple junction at the East African Rift valleys, Red Sea, and Gulf of Aden (see Fig. 17.5). A Mesozoic example of the same type of triple junction involved the separation of South America from Africa during the Cretaceous Period (❖ Fig. 15.3). Here, two arms separated to form the South Atlantic, while the third arm was abandoned and remains as the Benue Trough of Africa in which the Niger River flows.

The Fragmenting of Pangaea

Geologic, paleontologic, and paleomagnetic data indicate that the fragmentation of Pangaea occurred in four general stages. The first stage involved the rifting between Laurasia and Gondwana during the Late Triassic. By the end of the Triassic, the expanding Atlantic Ocean had separated North America from Africa (❖ Fig. 15.4). This was followed by the rifting of North America from

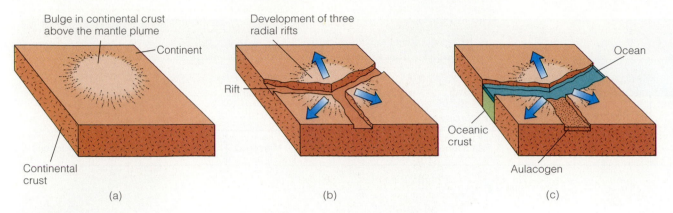

◆ FIGURE 15.2 Formation of a triple junction and an aulacogen. (a) A bulge forms over a rising mantle plume. (b) The outward radial flow from the mantle plume produces three radial rifts. (c) The continent is divided into two parts as new oceanic crust forms between the separating continents. The third rift becomes an aulacogen, or failed arm, and eventually fills with sediment.

South America sometime during the Late Triassic and Early Jurassic.

Separation of the continents allowed water from the Tethys Sea to flow into the expanding central Atlantic Ocean, while Pacific Ocean waters flowed into the newly formed Gulf of Mexico, which at that time was little more than a restricted bay (◆ Fig. 15.5). During that time, these areas were located in the low tropical latitudes where high temperatures and high rates of evaporation were ideal for the formation of thick evaporite deposits.

The second stage in Pangaea's breakup involved rifting and movement of the various Gondwana continents during the Late Triassic and Jurassic periods. As early as the Late Triassic, Antarctica and Australia, which remained sutured together, separated from South America and Africa, while India split away from all four Gondwana continents and began moving northward (Fig. 15.4).

The third stage of breakup began during the Late Jurassic, when South America and Africa began separating (◆ Fig. 15.6a). The rifting and subsequent separation of these two continents formed a narrow basin where thick evaporite deposits accumulated from the evaporation of southern ocean waters (Fig. 15.5). During this stage, the

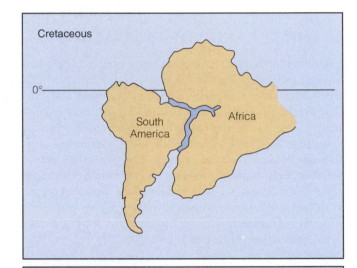

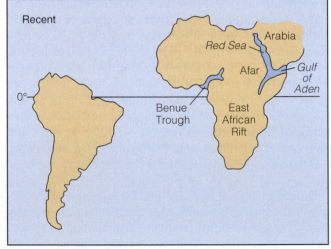

◆ FIGURE 15.3 The Benue Trough is an aulacogen that formed as a result of continental rifting between Africa and South America during the Cretaceous Period. The East African Rift valley may also be an aulacogen that formed when Africa and Arabia began separating 25 million years ago.

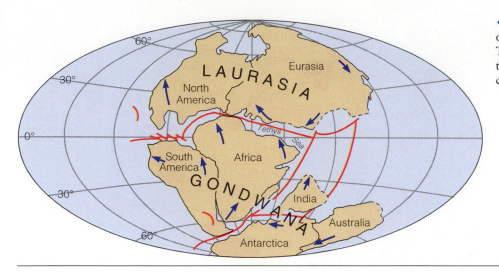

◆ **FIGURE 15.4** Paleogeography of the world during the Late Triassic Period. Blue arrows show the direction of movement for the continents.

eastern end of the Tethys Sea began closing as a result of the clockwise rotation of Laurasia and the northward movement of Africa. This narrow Late Jurassic and Cretaceous seaway between Africa and Europe was the forerunner of the present Mediterranean Sea.

◆ **FIGURE 15.5** Evaporites accumulated in shallow basins as Pangaea broke apart during the Early Mesozoic. Water from the Tethys Sea flowed into the central Atlantic Ocean, while water from the Pacific Ocean flowed into the newly formed Gulf of Mexico. Marine water from the south flowed into the southern Atlantic Ocean.

By the end of the Cretaceous, Australia and Antarctica had separated, India had nearly reached the equator, South America and Africa were widely separated, and the eastern side of what is now Greenland had begun separating from Europe (◆ Fig. 15.6b).

A global rise in sea level during the Cretaceous resulted in worldwide transgressions onto the continents. These transgressions were caused by higher heat flow along the oceanic ridges due to increased rifting and the consequent expansion of oceanic crust (see Fig. 4.15). By the Middle Cretaceous, sea level probably was as high as at any time since the Ordovician, and approximately one-third of the present land area was inundated by epeiric seas.

The final stage in the breakup of Pangaea occurred during the Cenozoic. During this stage, Australia continued moving northward, and Greenland completely separated from Europe and rifted from North America to form a separate landmass.

The Effects of the Breakup of Pangaea on Global Climates and Ocean Circulation Patterns

By the end of the Permian Period, Pangaea extended from pole to pole, covered about one-fourth of the Earth's surface, and was surrounded by Panthalassa, a global ocean that encompassed about 300 degrees of longitude (see Fig. 12.2b). Such a configuration exerted tremendous influence on the world's climate and resulted in generally arid conditions over large parts of Pangaea's interior.

◆ **FIGURE 15.6** Paleogeography of the world during (a) the Jurassic Period and (b) the Cretaceous Period.

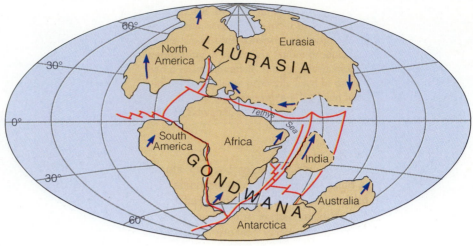

(a) Jurassic Period (208–144 M.Y.A)

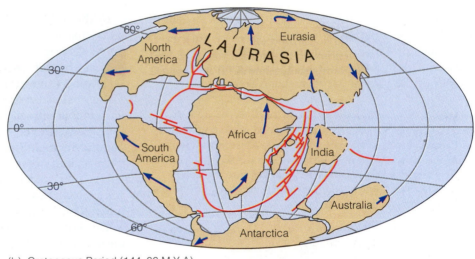

(b) Cretaceous Period (144–66 M.Y.A)

The world's climates result from the complex interaction between wind and ocean currents and the location and topography of the continents. In general, dry climates occur on large landmasses in areas remote from sources of moisture and where barriers to moist air, such as mountain ranges, exist. Wet climates occur near large bodies of water or where winds can carry moist air over land.

Past climatic conditions can be inferred from the distribution of climate-sensitive deposits such as evaporites, red beds, desert dunes, and coals. Evaporites occur where evaporation exceeds precipitation. Desert dunes and red beds are characteristic deposits of arid regions, and although both may form locally in humid regions, they are most widespread in dry areas. Coal forms in both warm and cool humid climates. Vegetation that is eventually converted into coal requires at least a good seasonal water supply; thus, coal deposits are indicative of humid conditions.

Widespread Triassic evaporites, red beds, and desert dunes in the low and middle latitudes of North and South America, Europe, and Africa indicate dry climates in those regions (◆ Fig. 15.7). Coal deposits formed mainly in the high latitudes, indicating humid condi-

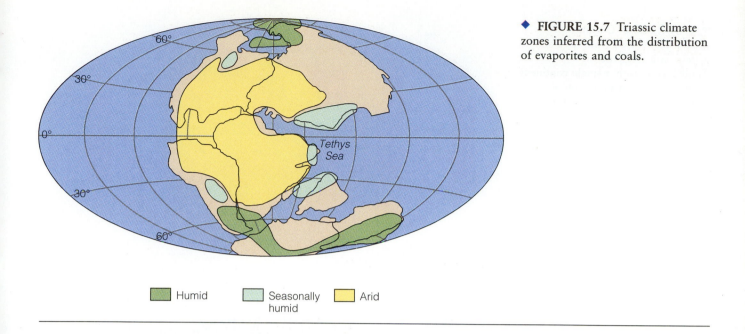

◆ **FIGURE 15.7** Triassic climate zones inferred from the distribution of evaporites and coals.

Humid Seasonally humid Arid

tions. These high-latitude coals are analogous to today's Scottish peat bogs or Canadian muskeg. The lands bordering the Tethys Sea were probably dominated by seasonal monsoon rains resulting from the warm moist winds and warm oceanic currents impinging against the east-facing coast of Pangaea (Fig. 15.7).

The temperature gradient between the tropics and the poles also affects oceanic and atmospheric circulation. The greater the temperature difference between the tropics and the poles, the steeper the temperature gradient, and the faster the circulation of the oceans and atmosphere. Oceans absorb about 90% of the solar radiation they receive, while continents absorb only about 50%, even less if they are snow covered. The rest of the solar radiation is reflected back into space. Therefore, areas dominated by seas are warmer than those dominated by continents. By knowing the distribution of continents and ocean basins, geologists can calculate the average annual temperature for any region on Earth. From this information a temperature gradient can be determined. For example, the temperature gradient of the Northern Hemisphere is currently 41°C, but is calculated to have been 20°C during the Triassic. This means that the present-day circulation of the atmosphere and oceans is faster than it was during the Triassic.

The breakup of Pangaea during the Late Triassic caused the global temperature gradient to increase because the Northern Hemisphere continents moved further northward, displacing higher-latitude ocean waters. Due to the steeper global temperature gradient caused by a decrease in temperature in the high latitudes and the changing positions of the continents, oceanic and atmospheric circulation patterns greatly accelerated during the Mesozoic (◆ Fig. 15.8). Though the temperature gradient and seasonality on land were increasing during the Jurassic and Cretaceous, the middle- and higher-latitude oceans were still quite warm, because warm waters from the Tethys Sea were circulating to the higher latitudes. This resulted in a relatively equable worldwide climate to the end of the Cretaceous.

❖ THE MESOZOIC HISTORY OF NORTH AMERICA

Having examined the global aspects of the breakup of Pangaea, we can now relate those events to the Mesozoic history of North America (◆ Figs. 15.9, 15.10, and 15.11, pp. 409–411).

In terms of tectonism and sedimentation, the beginning of the Mesozoic Era was essentially the same as the preceding Permian Period in North America (see Fig. 12.15). Terrestrial sedimentation continued over much of the craton, while block faulting and igneous activity began in the Appalachian region as North America and Africa began separating. The newly forming Gulf of

◆ **FIGURE 15.8** Oceanic circulation evolved from (a) a simple pattern in a single ocean (Panthalassa) with a single continent (Pangaea) to (b) a more complex pattern in the newly formed oceans of the Cretaceous Period.

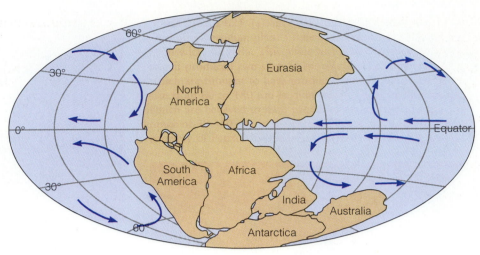

(a) Triassic Period

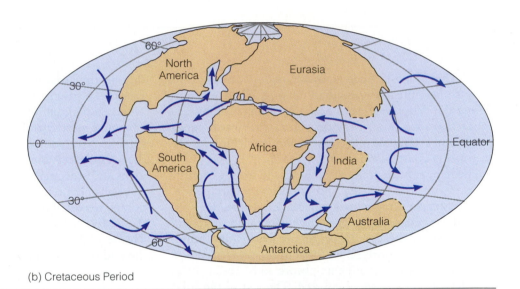

(b) Cretaceous Period

Mexico experienced extensive evaporite deposition during the Late Triassic and Jurassic as North America separated from South America (Fig. 15.5). Thousands of meters of sediments were deposited in the Atlantic and Gulf Coastal region during the Cretaceous as a result of the worldwide rise in sea level.

Marine deposition was continuous over much of the North American Cordillera. A volcanic island arc system that formed off the western edge of the craton during the Permian was sutured to North America sometime later during the Permian or Triassic. This event is referred to as the *Sonoma orogeny* and will be discussed later in the chapter. During the Jurassic, the entire Cordilleran area was involved in a series of major mountain-building

episodes that resulted in the formation of the Sierra Nevada, the Rocky Mountains, and other lesser mountain ranges. While each orogenic episode has its own name, the entire mountain-building event is simply called the *Cordilleran orogeny* (also discussed later in this chapter). With this simplified overview of the Mesozoic history of North America in mind, we will now examine the specific regions of the continent.

❖ CONTINENTAL INTERIOR

Recall that the history of the North American craton can be divided into unconformity-bound sequences reflecting advances and retreats of epeiric seas over the craton (see

Fig. 11.3). While these transgressions and regressions played a major role in the Paleozoic geologic history of the continent, they were not as important during the Mesozoic. During the Mesozoic, most of the continental interior was well above sea level and did not experience epeiric sea inundation. Consequently, the two Mesozoic cratonic sequences, the Absaroka (Late Mississippian to Early Jurassic) and **Zuni** (Early Jurassic to Early Paleocene) (Fig. 11.3), are incorporated as part of the his-tory of the three continental margin regions of North America.

❖ EASTERN COASTAL REGION

During the Early and Middle Triassic, coarse detrital sediments derived from the erosion of the recently uplifted Appalachians (Alleghenian orogeny) filled the various intermontane basins and spread over the surrounding areas. As

◆ **FIGURE 15.9** Paleogeography of North America during the Triassic Period.

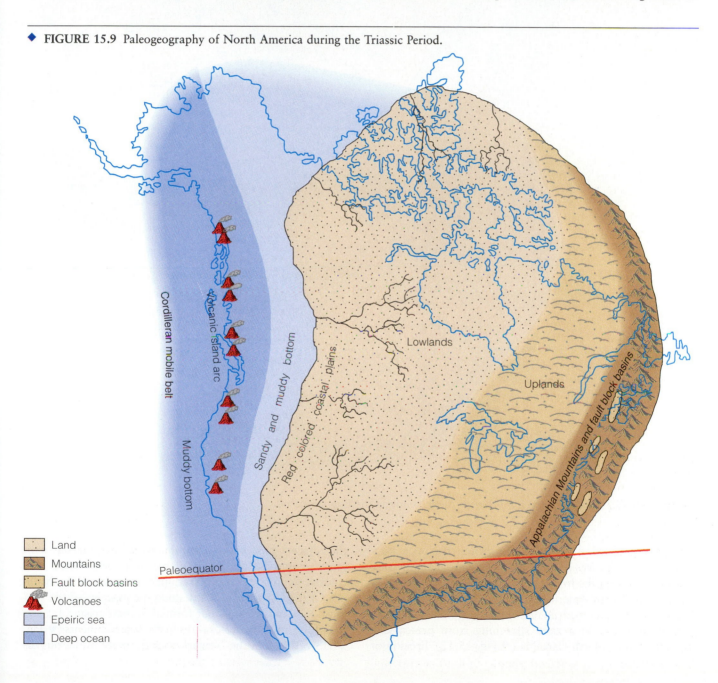

Land
Mountains
Fault block basins
Volcanoes
Epeiric sea
Deep ocean

Cordilleran mobile belt
Volcanic island arc
Muddy bottom
Sandy and muddy bottom
Red colored coastal plains
Lowlands
Uplands
Appalachian Mountains and fault block basins
Paleoequator

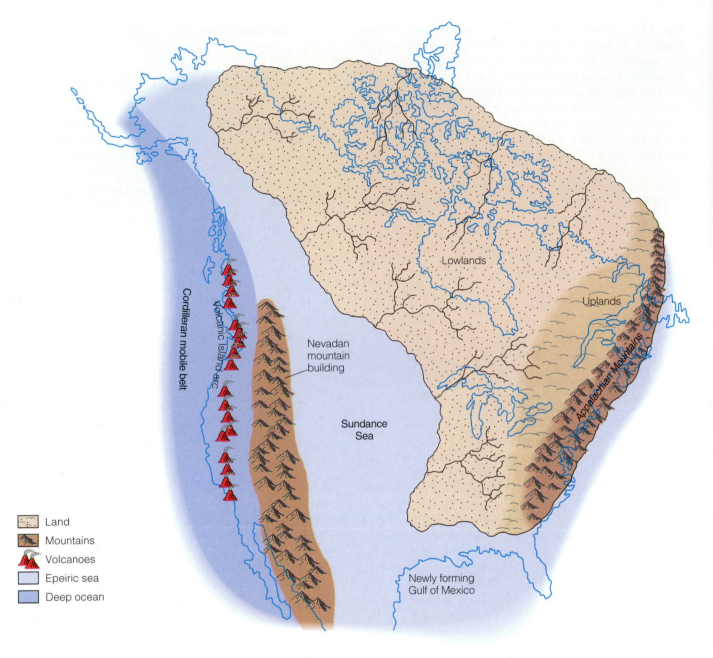

Land

Mountains

Volcanoes

Epeiric sea

Deep ocean

Cordilleran mobile belt

Volcanic Island arc

Nevadan mountain building

Sundance Sea

Lowlands

Uplands

Appalachian Mountains

Newly forming Gulf of Mexico

◆ **FIGURE 15.10** Paleogeography of North America during the Jurassic Period.

erosion continued during the Mesozoic, this once lofty mountain system was reduced to a low-lying plain. During the Late Triassic, the first stage in the breakup of Pangaea began with North America separating from Africa. Fault-block basins developed in response to upwelling magma beneath Pangaea in a zone stretching from present-day Nova Scotia to North Carolina (◆ Fig. 15.12). Erosion of the adjacent fault-block mountains filled these basins with great quantities (up to 6,000 m) of poorly sorted red-colored nonmarine detrital sediments known as the *Newark Group*. Reptiles roamed along the margins of the various lakes and streams that formed in these basins, leaving their footprints and trackways in the soft sediments (◆ Fig. 15.13). Although the Newark rocks contain numerous di-

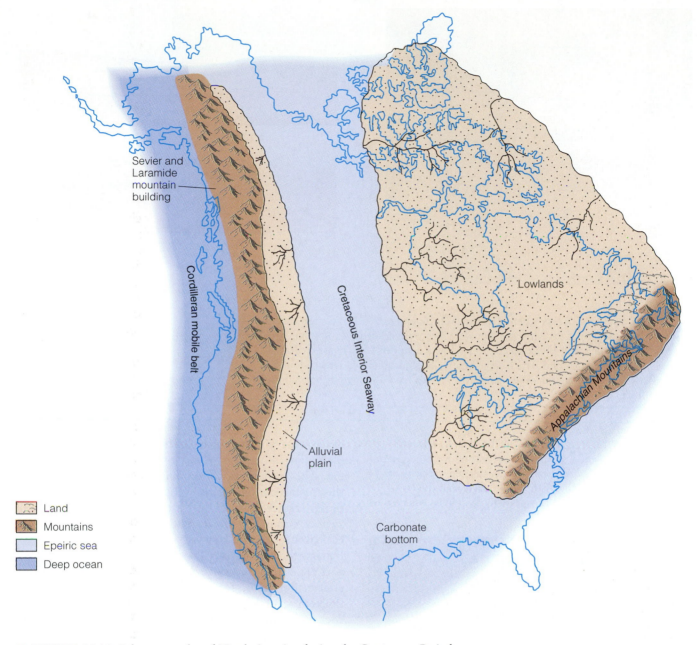

Labels in figure:
Sevier and Laramide mountain building

Cordilleran mobile belt

Cretaceous Interior Seaway

Alluvial plain

Lowlands

Appalachian Mountains

Carbonate bottom

Legend:
- Land
- Mountains
- Epeiric sea
- Deep ocean

◆ **FIGURE 15.11** Paleogeography of North America during the Cretaceous Period.

nosaur footprints, they are almost completely devoid of dinosaur bones.

Concurrent with sedimentation in the fault-block basins were extensive lava flows that blanketed the basin floors as well as intrusions of numerous dikes and sills. The most famous intrusion is the prominent Palisades sill along the Hudson River in the New York–New Jersey

area (◆ Fig. 15.14). Radiometric dates obtained from the Palisades sill indicate that it was intruded about 200 million years ago. While the Newark Group is considered Late Triassic in age, recent spore and pollen research indicates that deposition in some areas began during the Early Jurassic.

As the Atlantic Ocean grew, rifting ceased along the

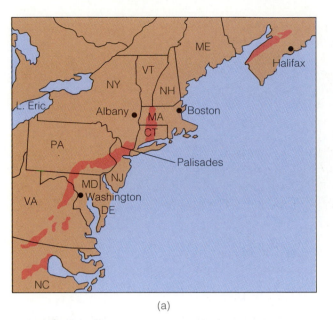

(a)

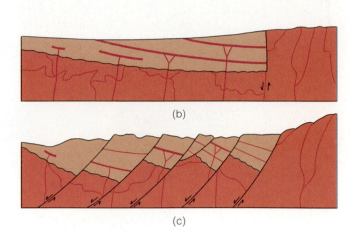

(b)

(c)

◆ **FIGURE 15.12** (a) Areas where Triassic fault-block basin deposits crop out in eastern North America. (b) After the Appalachians were eroded to a low-lying plain by the Middle Triassic, fault-block basins formed as a result of Late Triassic rifting between North America and Africa. (c) These valleys accumulated tremendous thicknesses of sediments and were themselves broken by a complex of normal faults during rifting.

◆ **FIGURE 15.13** Reptile tracks in the Triassic Newark Group were uncovered during the excavation for a new state building in Hartford, Connecticut. Because the tracks were so spectacular, the building site was moved, and the excavation was designated as a state park. (Photo courtesy of Dinosaur State Park.)

eastern margin of North America, and this once active plate margin became a passive, trailing continental margin. The fault-block mountains that were produced by this rifting continued to erode during the Jurassic and Early Cretaceous until only a broad, low-lying erosional surface remained. The sediments resulting from erosion contributed to the growing eastern continental shelf. During the Cretaceous Period, the Appalachian region was reelevated and once again shed sediments onto the continental shelf, forming a gently dipping, seaward-thickening wedge of rocks up to 3,000 m thick. These rocks are currently exposed in a belt extending from Long Island, New York, to Georgia.

❖ GULF COASTAL REGION

The Gulf Coastal region was above sea level until the Late Triassic (Fig. 15.9). As North America separated from South America during the Late Triassic, however, the Gulf of Mexico began to form (Fig. 15.10). With oceanic waters flowing into this newly formed, shallow, restricted basin, conditions were ideal for evaporite for-

◆ **FIGURE 15.14** Palisades of the Hudson River. This sill was one of many that were intruded into the Newark sediments during the Late Triassic rifting that marked the separation of North America from Africa. (Photo courtesy of Palisades Interstate Park Commission.)

mation. More than 1,000 m of evaporites were precipitated at this time, and most geologists believe that these Jurassic evaporites are the source for the Tertiary salt domes found today in the Gulf of Mexico and southern Louisiana (see Fig. 17.38). The history of these salt domes and their associated petroleum accumulations will be discussed in Chapter 17.

By the Late Jurassic, circulation in the Gulf of Mexico was less restricted, and evaporite deposition ended. Normal marine conditions returned to the area with alternating transgressing and regressing seas, resulting in the deposition of sandstones, shales, and limestones. These sedimentary rocks were later covered and deeply buried by great thicknesses of Cretaceous and Cenozoic sediments.

During the Cretaceous, the Gulf Coastal region, like the rest of the continental margin, was inundated by transgressing seas. The relatively low-lying Gulf Coastal area was invaded by northward-moving marine waters during the Early Cretaceous. This can be seen in facies relationships in which nearshore sandstones are overlain by finer sediments characteristic of deeper waters. Following an extensive regression at the end of the Early Cretaceous, a major transgression began during which a wide seaway extended from the Arctic Ocean to the Gulf of Mexico (Fig. 15.11). Cretaceous sediments that were deposited in the Gulf Coastal region formed a seaward-thickening wedge (◆ Fig. 15.15).

Reefs were widespread in the Gulf Coastal region during the Cretaceous. Bivalves called *rudists* were the main constituent of many of these reefs (see Fig. 16.5). Because of their high porosity and permeability, rudistoid reefs make excellent petroleum reservoirs. A good example of a Cretaceous reef complex occurs in Texas (◆ Fig. 15.16). Here the reef trend had a strong influence on the carbonate platform deposition of the region. The facies patterns of these carbonate rocks are as complex as those found in the major barrier-reef systems of the Paleozoic Era.

❖ WESTERN REGION

Triassic Tectonics

With the exception of the Late Devonian–Early Mississippian Antler orogeny (see Fig. 12.19), the Cordilleran region of North America experienced little tectonism during the Paleozoic. During the Permian, however, an island arc and ocean basin analogous to the present-day Japanese Islands and Sea of Japan formed off the western North American craton (Fig. 15.9). This was followed by subduction of an oceanic plate beneath the island arc and the thrusting of oceanic and island arc rocks eastward against the craton margin (◆ Fig. 15.17). This event ini-

◆ **FIGURE 15.15** Outcrop of the Cretaceous Demopolis Chalk (Selma Group) along the banks of the Tombigbee River near Demopolis, Alabama. (Photo courtesy of Mirza A. Beg, Geological Survey of Alabama.)

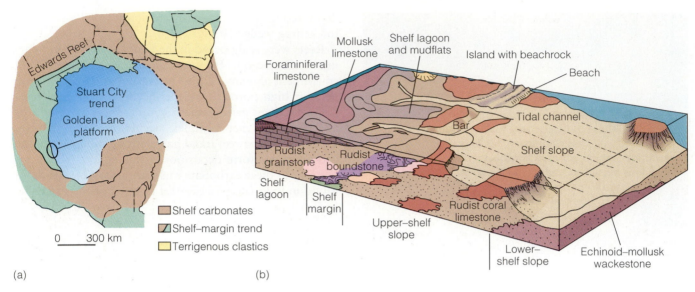

(a)

(b)

◆ **FIGURE 15.16** (a) Early Cretaceous shelf-margin facies around the Gulf of Mexico Basin. The reef trend is shown as a black line. (b) Reconstruction of the depositional environment and facies changes across the Stuart City reef trend, South Texas.

tiated the **Sonoma orogeny** at or near the Permian-Triassic boundary.

Following the Late Paleozoic–Early Mesozoic destruction of the volcanic island arc during the Sonoma orog-

eny, the western margin of North America became an oceanic-continental convergent plate boundary similar to the one that now exists along the western margin of South America. During the Late Triassic, a steeply dip-

◆ **FIGURE 15.17** Tectonic activity that culminated in the Permian-Triassic Sonoma orogeny in western North America. The Sonoma orogeny was the result of a collision between the southwestern margin of North America and an island arc system.

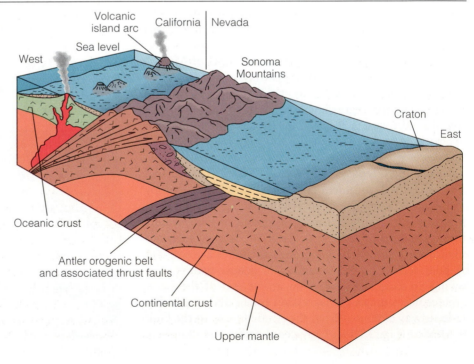

◆ **FIGURE 15.18** (a) Stratigraphic section of Triassic formations in the western United States. (b) View of East Entrance of Zion Canyon, Zion National Park, Utah. The light-colored massive rocks are the Jurassic Navajo Sandstone, while the slope-forming rocks below the Navajo are the Upper Triassic Kayenta Formation.

ping subduction zone developed along the western margin of North America in response to the westward movement of North America over the Pacific plate as the Atlantic Ocean basin opened. This newly created oceanic-continental plate boundary controlled Cordilleran tectonics for the rest of the Mesozoic Era and for most of the Cenozoic Era; this subduction zone marks the beginning of the modern circum-Pacific orogenic system.

Triassic Sedimentation

Concurrent with the tectonism occurring in the Cordilleran mobile belt, Early Triassic sedimentation on the western continental shelf consisted of shallow-water marine sandstones, shales, and limestones. During the Middle and Late Triassic, the western shallow seas regressed further west, exposing large areas of former seafloor to erosion. Marginal marine and nonmarine Triassic rocks, particularly red beds, contribute to the spectacular and colorful scenery of such areas as the Painted Desert in western North America (see Perspective 15.1).

The Lower Triassic Moenkopi Formation of the southwestern United States consists of a succession of brick-red and chocolate-colored mudstones (◆ Fig. 15.18). Such sedimentary structures as desiccation cracks and ripple marks, as well as fossil amphibians and reptiles and their tracks, indicate deposition in a variety of continental environments, including stream channels, floodplains, and fresh and brackish water ponds. Thin tongues of marine limestones indicate three brief incursions of the sea. Local beds with gypsum and halite crystal casts attest to a rather arid climate.

Unconformably overlying the Moenkopi is the Upper Triassic Shinarump Conglomerate, a persistent unit generally less than 50 m thick that was derived from the erosion of newly emergent areas in Arizona, western Colorado, and Idaho (Fig. 15.18).

Above the Shinarump are the multicolored shales, siltstones, and sandstones of the Chinle Formation (Fig. 15.18). This Upper Triassic formation is widely exposed over the Colorado Plateau and is probably most famous for its petrified wood, spectacularly exposed in Petrified Forest National Park, Arizona (see Perspective 15.1). Fossil ferns occur here, but the park is best known

PETRIFIED FOREST NATIONAL PARK

Petrified Forest National Park is located in eastern Arizona about 42 km east of Holbrook. The park consists of two sections: the Painted Desert, which is north of Interstate 40, and the Petrified Forest, which is south of the Interstate.

The Painted Desert is a brilliantly colored landscape whose colors and hues change constantly throughout the day. The multicolored rocks of the Triassic Chinle Formation have been weathered and eroded to form a badlands topography of numerous gullies, valleys, ridges, mounds, and mesas. The Chinle Formation is composed predominantly of different-colored shale beds. These shales and associated volcanic ash layers are easily weathered and eroded. Interbedded locally with the shales are lenses of conglomerates, sandstones, and limestones,

which are more resistant to weathering and erosion than the shales and form resistant ledges.

The Petrified Forest was originally set aside as a national monument to protect the large number of petrified logs that lay exposed in what is now the southern part of the park (◆ Fig.1). When the transcontinental railroad constructed a coaling and watering stop in Adamana, Arizona, passengers were encouraged to take excursions to "Chalcedony Park," as the area was then called, to see the petrified forests. In a short time, collectors and souvenir hunters hauled off tons of petrified wood, quartz crystals, and Indian relics. It was not until a huge rock crusher was built to crush the logs for the manufacture of abrasives that the area was declared a national monument and the petrified forests preserved and protected.

During the Triassic Period, the climate of the area was much wetter than today, with many rivers, streams, and lakes. About 40 different fossil plant species have been identified from the Chinle Formation. These include numerous seedless vascular plants such as rushes and

---

for its abundant and beautifully preserved logs of gymnosperms, especially conifers and plants called *cycads* (see Fig. 16.12a). Fossilization resulted from the silicification of the plant tissues. Weathering of volcanic ash beds interbedded with fluvial and deltaic Chinle sediments provided most of the silica for silicification. Some trees were preserved in place, but most were transported during floods and deposited on sandbars and on floodplains, where fossilization occurred. After burial, silica-rich groundwater percolated through the sediments and silicified the wood.

Though best known for its petrified wood, the Chinle Formation has also yielded fossils of labyrinthodont amphibians, phytosaurs, and small dinosaurs (see Chapter 16 for a discussion of the latter two animal groups).

The Wingate Sandstone, a desert dune deposit, and the Kayenta Formation, a stream and lake deposit, overlie the Chinle Formation. These two formations are well exposed in southwestern Utah and complete the Triassic stratigraphic succession in the Southwest (Fig. 15.18).

Jurassic and Cretaceous Tectonics

The Mesozoic geologic history of the North American Cordilleran mobile belt is very complex, involving the

eastward subduction of the oceanic Pacific plate under the continental North American plate. Activity along this oceanic-continental convergent plate boundary resulted in an eastward movement of deformation. This orogenic activity progressively affected the trench and continental slope, the continental shelf, and the cratonic margin, causing a thickening of the continental crust.

Two subduction zones, dipping in opposite directions from each other, formed off the west coast of North America during the Middle and early Late Jurassic (◆ Fig. 15.19). The more westerly subduction zone was eliminated by the westward-moving North American plate, which overrode the oceanic Pacific plate. The Franciscan Group, an unusual rock formation consisting of a chaotic mixture of rocks, then began accumulating in the remaining submarine trench (◆ Fig. 15.20a).

The Franciscan Group, which is up to 7,000 m thick, consists mostly of graywacke, but also contains breccias, siltstones, black shales, cherts, pillow basalts, volcanic breccias, and greenstones (◆ Fig. 15.20b). Fossils are extremely rare in the Franciscan, but radiolarians, clams, ammonites, and foraminifera indicate an age of Late Jurassic and Cretaceous. The various rock types suggest that continental-shelf, slope, and deep-sea environments were brought together in a submarine trench when

◆ **FIGURE 1** Petrified Forest National Park, Arizona. All of the logs here are *Araucarioxylon*, which is the most abundant tree in the park. The petrified logs have been weathered from the Chinle Formation and are mostly in the position in which they were buried some 200 million years ago. (Photo © Stephen J. Kraseman/Photo Researchers.)

ferns as well as gymnosperms such as cycads and conifers. Such plants thrive in floodplains and marshes. Most of the logs are conifers and belong to the genus *Araucarioxylon*. Some of these trees were more than 60 m tall and up to 4 m in diameter. Apparently, most of the conifers grew on higher ground or riverbanks. Although many trees were buried in place, most appear to have been uprooted and transported by raging streams during times of flooding. Burial of the logs was rapid, and groundwater saturated with silica from the ash of nearby volcanic eruptions quickly permineralized the trees.

Deposition continued in the Colorado Plateau region during the Jurassic and Cretaceous, further burying the Chinle Formation. During the Laramide orogeny, the Colorado Plateau area was uplifted and eroded, exposing the Chinle Formation. Since the Chinle is mostly shales, it was easily eroded, leaving the more resistant petrified logs and log fragments exposed on the surface—much as we see them today.

North America overrode the subducting Pacific oceanic crust. This subduction resulted in metamorphism at many localities, indicated by blueschist facies rocks that occur throughout the Franciscan Group.

East of the Franciscan Group and currently separated from it by a major thrust fault is the Great Valley Group. It consists of more than 16,000 m of Cretaceous conglomerates, sandstones, siltstones, and shales. These sediments were deposited on the continental shelf and slope in a fore-arc basin setting at the same time the Franciscan deposits were accumulating in the submarine trench (Fig. 15.20).

The general term **Cordilleran orogeny** is applied to the mountain-building activity that began during the Jurassic and continued into the Cenozoic (◆ Fig. 15.21, p. 420). Similar to the Paleozoic orogenic activity in the Appalachian mobile belt, the Cordilleran orogeny consisted of a series of individual mountain-building events that occurred in different regions at different times. Most of this Cordilleran orogenic activity is related to the continued westward movement of the North American plate.

The first phase of the Cordilleran orogeny, the **Nevadan orogeny** (Fig. 15.21), began during the Late Jurassic and continued into the Cretaceous as large volumes of granite-granodiorite magma were generated at depth be-

neath the western edge of North America. These granitic masses ascended as huge batholiths that are now recognized as the Sierra Nevada, Southern California, Idaho, and Coast Range batholiths (◆ Fig. 15.22, p. 420).

By the Late Cretaceous, most of the volcanic and plutonic activity had migrated eastward into Nevada and Idaho. This migration was probably caused by a change from high-angle to low-angle subduction, which resulted in the subducting oceanic plate reaching its melting depth farther east (◆ Fig. 15.23, p. 421). Thrusting occurred progressively further east so that by the Late Cretaceous, it extended all the way to the Idaho-Washington border.

The second phase of the Cordilleran orogeny, the **Sevier orogeny,** was mostly a Cretaceous event (Fig. 15.22). As subduction of the Pacific plate beneath the North American plate continued, compressive forces generated along this convergent plate boundary were transmitted eastward, resulting in numerous overlapping, low-angle thrust faults (◆ Fig. 15.24a, p. 421). This thrusting produced generally north-south–trending mountain ranges consisting of blocks of Paleozoic shelf and slope strata. Though the term *Sevier* is usually applied to orogenic events occurring from southern California to Utah, the same deformational style was also occurring within a tectonic belt stretching from Montana to western Canada.

◆ **FIGURE 15.19** Interpretation of the tectonic evolution of the Sierra Nevada during the Mesozoic Era.

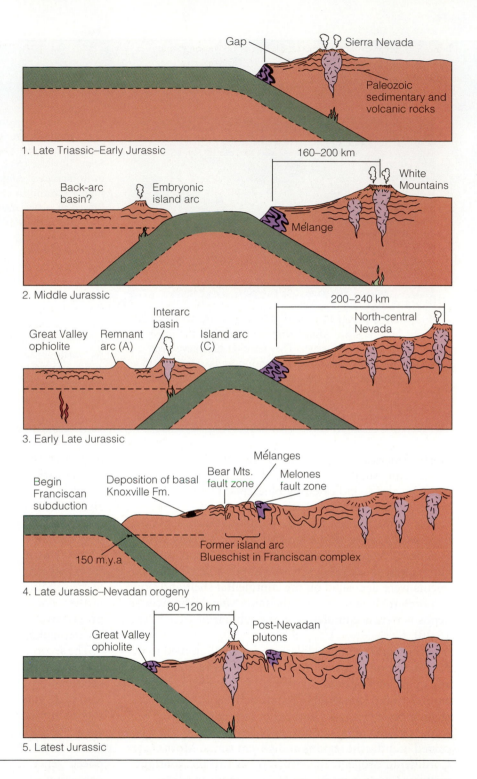

Gap

Sierra Nevada

Paleozoic sedimentary and volcanic rocks

1. Late Triassic–Early Jurassic

160–200 km

Back-arc basin?

Embryonic island arc

White Mountains

Mélange

2. Middle Jurassic

200–240 km

Great Valley ophiolite

Remnant arc (A)

Interarc basin

Island arc (C)

North-central Nevada

3. Early Late Jurassic

Mélanges

Bear Mts. fault zone

Deposition of basal Knoxville Fm.

Melones fault zone

Begin Franciscan subduction

150 m.y.a

Former island arc
Blueschist in Franciscan complex

4. Late Jurassic–Nevadan orogeny

80–120 km

Great Valley ophiolite

Post-Nevadan plutons

5. Latest Jurassic

During the Late Cretaceous to Early Cenozoic, the final pulse of the Cordilleran orogeny occurred (Fig. 15.21). The **Laramide orogeny** developed east of the Sevier orogenic belt in the present-day Rocky Mountain areas of New Mexico, Colorado, and Wyoming. The Laramide orogeny differed considerably in tectonic style

◆ **FIGURE 15.20** (a) Reconstruction of the depositional environment of the Franciscan Group during the Late Jurassic and Cretaceous periods. (b) Exposures of the Franciscan Group along the central California coast.

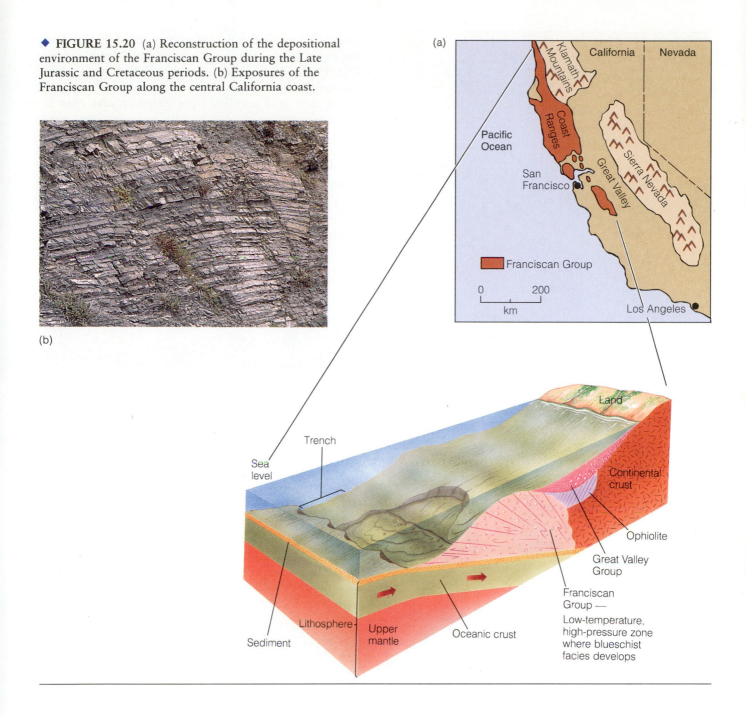

from the Nevadan and Sevier orogenies. Though thrust faulting still occurred in the northern Rocky Mountain region, anticlines, domes, and basins were the more characteristic style of deformation in the middle and southern Rockies. Most of the features of the present-day Rocky Mountains resulted from the Cenozoic phase of the Laramide orogeny, and for that reason, it will be discussed in Chapter 17.

Jurassic and Cretaceous Sedimentation

Having examined the Jurassic and Cretaceous orogenic events of western North America, we now turn our attention to how the orogenic activity affected sedimentation patterns. The Early Jurassic deposits in a large part of the western region consist mostly of clean, cross-bedded sandstones indicative of windblown deposits.

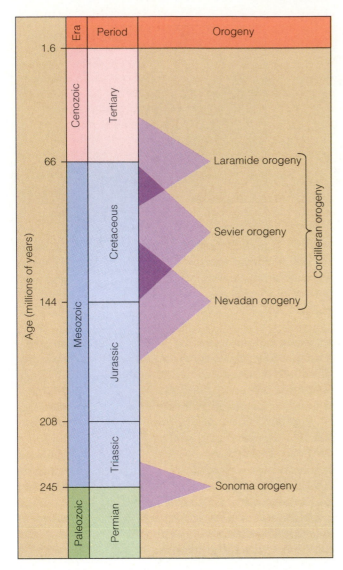

◆ **FIGURE 15.21** Mesozoic orogenies occurring in the Cordilleran mobile belt.

Middle Jurassic when a wide seaway called the **Sundance Sea** twice flooded the interior of western North America (Fig. 15.10). The resulting deposits, called the Sundance Formation, were largely derived from tectonic highlands to the west that paralleled the shoreline and were the result of intrusive igneous activity and associated volcanism that began during the Triassic.

During the Late Jurassic, the folding and thrust faulting that began as part of the Nevadan orogeny in Nevada, Utah, and Idaho formed a large mountain chain

◆ **FIGURE 15.22** Location of Jurassic and Cretaceous batholiths in western North America.

The thickest and most prominent of these is the Navajo Sandstone, a widespread cross-bedded sandstone that accumulated along the southwestern margin of the craton. The sandstone's most distinguishing feature is its large-scale cross-beds, some of which are more than 25 m high (◆ Fig. 15.25). The upper part of the Navajo contains smaller cross-beds as well as dinosaur and crocodilian fossils. Based on its sedimentary structures, facies relationships and fossil remains, the Navajo Sandstone probably represents a coastal dune environment.

Marine conditions returned to the area during the

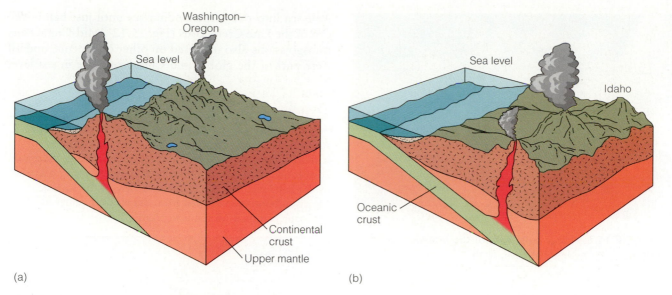

(a)

(b)

◆ **FIGURE 15.23** A possible cause for the eastward migration of igneous activity in the Cordilleran region during the Cretaceous Period was a change from (a) high-angle to (b) low-angle subduction. As the subducting plate moved downward at a lower angle, the depth of melting moved farther to the east.

◆ **FIGURE 15.24** (a) Reconstruction showing the associated tectonic features of the Late Cretaceous Sevier orogeny due to subduction of the Pacific plate under the North American plate. (b) The Keystone thrust fault, a major fault in the Sevier overthrust belt is exposed west of Las Vegas, Nevada. The sharp boundary between the light-colored Mesozoic rocks and the overlying dark-colored Paleozoic rocks marks the trace of the Keystone thrust fault. (Photo © John Shelton.)

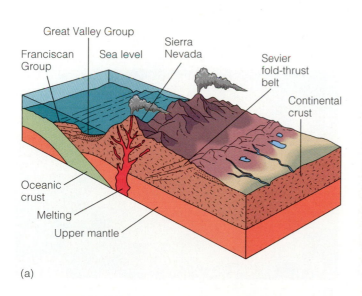

(a)

(b)

◆ **FIGURE 15.25** Large cross-beds of the Jurassic Navajo Sandstone in Zion National Park, Utah.

paralleling the coastline (◆ Fig. 15.26a). As the mountain chain grew and shed sediments eastward, the Sundance Sea retreated northward (Fig. 15.26a). A large part of the area formerly occupied by the Sundance Sea was then covered by multicolored sandstones, mudstones, shales, and occasional lenses of conglomerates that comprise the world-famous Morrison Formation (◆ Fig. 15.26b).

The Morrison Formation contains the world's richest assemblage of Jurassic dinosaur remains. Although most of the dinosaur skeletons are broken up, as many as 50 individuals have been found together in a small area. Such a concentration indicates that the skeletons were brought together during times of flooding and deposited on sandbars in stream channels. Soils in the Morrison indicate that the climate was seasonably dry. The presence of turtle, crocodile, and fish fossils also indicates that the lakes at the time were probably very shallow and saline.

Although most major museums have either complete dinosaur skeletons or at least bones from the Morrison Formation, the best place to see the bones still embedded in the rocks is the visitors' center at Dinosaur National Monument near Vernal, Utah (see Perspective 15.2).

Shortly before the end of the Early Cretaceous, Arctic waters spread southward over the craton, forming a large inland sea in the Cordilleran foreland basin area. By the beginning of the Late Cretaceous, this incursion joined the northward-transgressing waters from the Gulf area to create an enormous inland sea called the **Cretaceous Interior Seaway** that occupied the area east of the Sevier orogenic belt. Extending from the Gulf of Mexico to the Arctic Ocean, and more than 1,500 km wide at its maximum extent, this seaway effectively divided North

America into two large landmasses until just before the end of the Late Cretaceous (Fig. 15.12). Mid-Cretaceous transgressions also occurred on other continents, and all were part of the global mid-Cretaceous rise in sea level

◆ **FIGURE 15.26** (a) Paleogeography of western North America during the Late Jurassic. As the Sundance Sea withdrew from the western interior, the nonmarine Morrison Formation accumulated in part of the area formerly occupied by the Sundance Sea. (b) Panoramic view of the Jurassic Morrison Formation as seen from the visitors' center at Dinosaur National Monument, Utah.

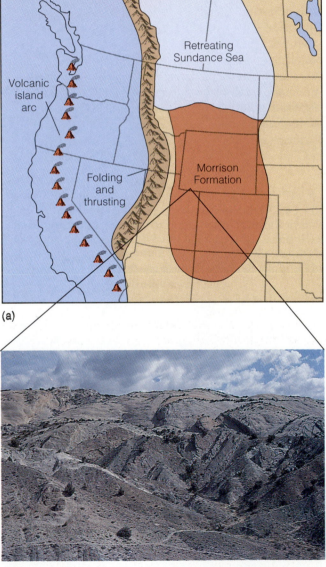

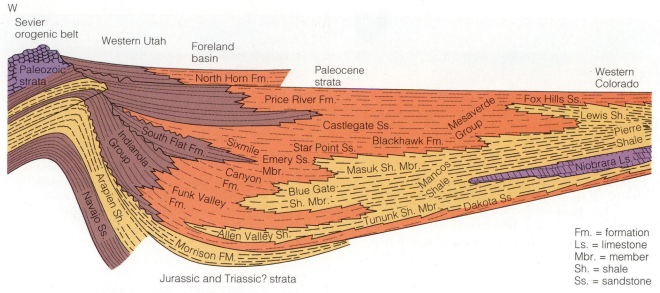

W

Sevier orogenic belt

Western Utah

Foreland basin

Paleocene strata

Western Colorado

Paleozoic strata

North Horn Fm.

Price River Fm.

Castlegate Ss.

Fox Hills Ss.

Lewis Sh.

Mesaverde Group

Blackhawk Fm.

Pierre Shale

Indianola Group

South Flat Fm.

Sixmile

Star Point Ss.

Emery Ss.

Mbr.

Masuk Sh. Mbr.

Mancos Shale

Niobrara Ls.

Canyon Fm.

Blue Gate Sh. Mbr.

Funk Valley Fm.

Arapien Sh.

Navajo Ss.

Tununk Sh. Mbr.

Dakota Ss.

Allen Valley Sh.

Morrison FM.

Jurassic and Triassic? strata

Fm. = formation
Ls. = limestone
Mbr. = member
Sh. = shale
Ss. = sandstone

◆ **FIGURE 15.27** This restored west-east cross section of Cretaceous facies of the western Cretaceous Interior Seaway shows their relationship to the Sevier orogenic belt.

that resulted from accelerated seafloor spreading as Pangaea continued to fragment.

Cretaceous deposits less than 100 m thick indicate that the eastern margin of the Cretaceous Interior Seaway subsided slowly and received little sediment from the emergent, low relief craton to the east. The western shoreline, however, shifted back and forth, primarily in response to fluctuations in the supply of sediment from the Cordilleran Sevier orogenic belt to the West. The facies relationships show lateral changes from conglomerate and coarse sandstone adjacent to the mountain belt through finer sandstones, siltstones, shales, and even limestones and chalks in the east (◆ Fig. 15.27). During times of particularly active thrusting and uplift, these coarse clastic wedges of gravel and sand prograded even further east.

As the Mesozoic Era ended, the Cretaceous Interior Seaway withdrew from the craton. During this regression, marine waters retreated to the north and south, and marginal marine and continental deposition formed widespread coal-bearing deposits on the coastal plain.

❖ MICROPLATE TECTONICS AND THE GROWTH OF WESTERN NORTH AMERICA

In the preceding sections, we discussed orogenies along convergent plate boundaries resulting in continental ac-

cretion. Much of the material accreted to continents during such events is simply eroded older continental crust, but a significant amount of new material is added to continents as well—igneous rocks that formed as a consequence of subduction and partial melting, for example. While subduction is the predominant influence on the tectonic history in many regions of orogenesis, other processes are also involved in mountain building and continental accretion, especially the accretion of microplates.

During the late 1970s and 1980s, geologists discovered that portions of many mountain systems are composed of small accreted lithospheric blocks that are clearly of foreign origin. These **microplates** differ completely in their fossil content, stratigraphy, structural trends, and paleomagnetic properties from the rocks of the surrounding mountain system and adjacent craton. In fact, these microplates are so different from adjacent rocks that most geologists think they formed elsewhere and were carried great distances as parts of other plates until they collided with other microplates or continents.

Geologic evidence indicates that more than 25% of the entire Pacific coast from Alaska to Baja California consists of accreted microplates. The accreting microplates are composed of volcanic island arcs, oceanic ridges, seamounts, and small fragments of continents that were scraped off and accreted to the continent's margin as the oceanic plate with which they were carried was subducted under the continent. It is estimated that more than 100 different-sized microplates have been added to

DINOSAUR NATIONAL MONUMENT

In 1909, Earl Douglass of the Carnegie Museum discovered "Dinosaur Ledge," a sandstone unit in the Upper Jurassic Morrison Formation that contained numerous dinosaur bones. After 13 years of excavating the layer, the Carnegie Museum had removed parts of 300 dinosaur specimens, two dozen of which were sufficiently complete to be reassembled (♦ Fig. 1). Ten different dinosaur species were represented as well as many other reptiles. It was by far the finest collection of dinosaur remains in the world. In the years that followed, the Smithsonian Institution and the University of Utah also worked this rich quarry and recovered more fossil specimens. Even more bones remained buried in an untouched part of the dipping sandstone ledge, and all that was needed to reveal them was the removal of the overlying layers of shale and siltstone.

Recognizing the scientific importance of this unit, the dinosaur quarry and 80 acres surrounding it were designated a national monument by President Wilson on October 4, 1915. Less than a year later, it was included in the newly created National Park System. In 1938, Dinosaur National Monument was further expanded to 200,000 acres.

As far back as 1915, Earl Douglass envisioned an exhibit in which the dinosaur bones would be exposed in relief in the tilted sandstone bed exactly where they came to rest 140 million years ago. In a letter to Dr. Charles Walcott, secretary of the Smithsonian Institution, Douglass wrote: "I hope that the Government, for the benefit of science and the people, will uncover a large area, leave the bones and skeletons in relief, and house them in. It would make one of the most astounding and instructive sights imaginable." Not until 1953, however, did work begin on a truly unique museum constructed around the still-buried sandstone ledge. The sediment overlying the remaining tilted sandstone ledge was removed, and the dinosaur bones were carefully exposed in bas relief. This quarry wall now forms the north wall of the visitors' center at Dinosaur National Monument (♦ Fig. 2). The structure was completed and opened to the public in 1958.

♦ **FIGURE 1** *Camarasaurus lentus* is one of the best preserved and complete skeletons recovered from the Morrison Formation by the Carnegie Museum at Dinosaur National Monument. (Photo courtesy of the Carnegie Museum of Natural History.)

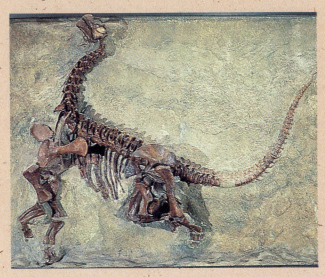

the western margin of North America during the last 200 million years (♦ Fig. 15.28).

The basic plate tectonic reconstruction of orogenies and continental accretion remains unchanged, but the details of such reconstructions are decidedly different in view of microplate tectonics. For example, growth along active continental margins is faster than along passive continental margins because of the accretion of microplates. Furthermore, these accreted microplates are often new additions to a continent, rather than reworked older continental material.

So far, most microplates have been identified in mountains of the North American Pacific Coast region, but a number of such plates are suspected to be present in other mountain systems as well. They are more difficult to recognize in older mountain systems, such as the Appalachians, however, because of greater deformation and erosion. Thus, microplate tectonics provides a new way

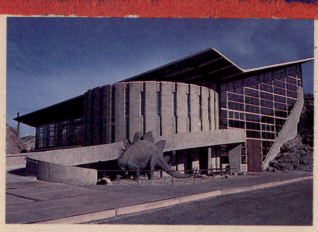

(a)

(b)

◆ **FIGURE 2** (a) Visitors' center, Dinosaur National Monument. (b) North wall of visitors' center showing dinosaur bones in bas relief, just as they were deposited 140 million years ago.

What was the landscape like during the Jurassic when dinosaurs roamed the area that is now Dinosaur National Monument? The land for miles around the present-day quarry was a low-lying desert during the Early Jurassic. This is indicated by the large cross-bedded dune sands of the Navajo Sandstone. Following deposition of the Navajo, a shallow sea transgressed from the west, depositing sandstones, siltstones, and limestones. During the Late Jurassic, the area from Mexico to Canada and from central Utah to the Mississippi River was above sea level. To the west was a mountain chain. Streams flowing from these mountains deposited sand and silt in the adjacent plains. Small lakes were numerous and some swamps were present. These stream, lake, and swamp deposits of the Jurassic coastal plain comprise the Morrison Formation.

Semitropical conditions prevailed during the Late Jurassic in the area, and forests of ginkos, cycads, and tree-ferns covered the land. In this setting pterosaurs glided through the air while dinosaurs roamed the landscape below. Crocodiles, turtles, and small mammals were also present. Most of the bones of the dinosaurs and other animals were deposited in the stream beds. During Cretaceous through Eocene time, the area was uplifted during the Laramide orogeny, and later the Green and Yampa rivers cut deep canyons in the area, exposing the rocks that make up Dinosaur National Monument.

of viewing the Earth and of gaining a better understanding of the geologic history of the continents.

❖ MESOZOIC MINERAL RESOURCES

Although much of the coal in North America is Pennsylvanian or Tertiary in age, important Mesozioc coals occur in the Rocky Mountains states. These are mostly lignite and bituminous coals, but some local anthracites

occur as well. Particularly widespread in western North American are coals of Cretaceous age. Mesozoic coals are also known from Alberta and British Columbia, Canada, as well as from Australia, Russia, and China.

Large concentrations of petroleum occur in many areas of the world, but more than 50% of all proven reserves are in the Persian Gulf region (◆ Fig. 15.29). During the Mesozoic, what is now the Gulf region was a broad passive continental margin extending eastward

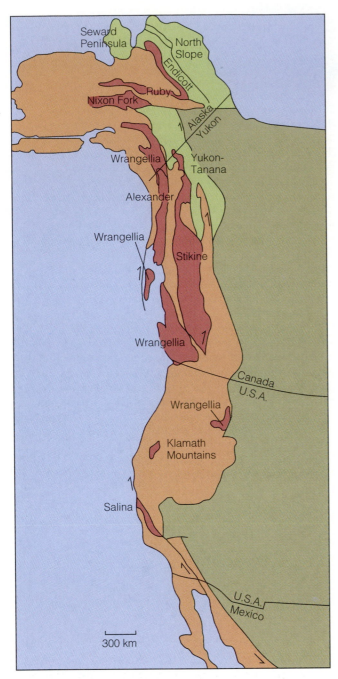

◆ **FIGURE 15.28** Some of the accreted lithospheric blocks called microplates that form the western margin of the North American Craton. The dark brown blocks probably originated as parts of continents other than North America. The light green blocks are possibly displaced parts of North America. The North American craton is shown in dark green.

from Africa. This continental margin lay near the equator where countless microorganisms lived in the surface waters, particularly during the Cretaceous Period when most of the petroleum formed. The remains of these organisms accumulated with the bottom sediments and were buried, beginning the complex processes of oil generation and formation of source beds. Several transgressions and regressions occurred during which some of the reservoir rocks formed as extensive, thick regressive sandstones, oolitic limestones, algal reef limestones, and reefs composed of the shells of clams. Overlying the reservoir rocks are cap rocks that include widespread shale and evaporite units.

Similar conditions existed in what is now the Gulf Coast region of the United States and Central America. Here petroleum and natural gas also formed on a broad shelf over which transgressions and regressions occurred. In this region, the hydrocarbons are largely in reservoir rocks that were deposited as distributary channels on deltas and as barrier-island and beach sands. Some of these hydrocarbons are associated with structures formed adjacent to rising salt domes. The salt, called the Louann Salt, initially formed in a long, narrow sea when North America separated from Europe and North Africa during the fragmentation of Pangaea (Fig. 15.5).

The richest uranium ores in the United States are widespread in the Colorado Plateau area of Colorado and adjoining parts of Wyoming, Utah, Arizona, and New Mexico. These ores, consisting of fairly pure masses of a complex potassium-, uranium-, vanadium-bearing mineral called *carnotite*, are associated with plant remains in sandstones that were deposited in ancient stream channels. Some petrified trees also contain large quantitites of uranium.

As noted in Chapter 10, Proterozoic banded iron formations are the main sources of iron ores. There are, however, important exceptions. For example, the Jurassic-aged "Minette" iron ores of Western Europe, composed of oolitic limeonite and hematite, are important ores in France, Germany, Belgium, and Luxembourg. In Great Britain, low-grade iron ores of Jurassic age consist of oolitic siderite, which is an iron carbonate. And in Spain, Cretaceous rocks are the host rocks for iron minerals.

South Africa, the world's leading producer of gem-quality diamonds and among the leaders in industrial diamond production, mines these minerals from conical igneous intrusions called kimberlite pipes. Kimberlite pipes are composed of dark gray or blue igneous rock known as kimberlite. Diamonds, which form at great

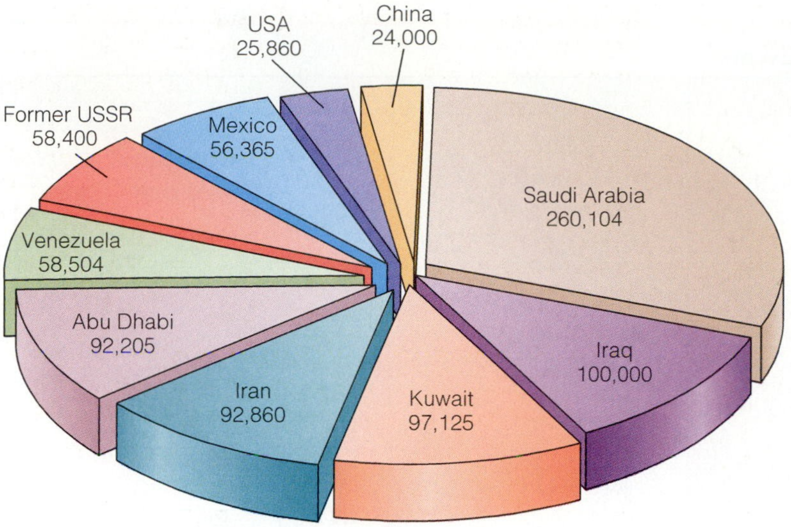
USA
25,860
China
24,000
Former USSR
58,400
Mexico
56,365
Saudi Arabia
260,104
Venezuela
58,504
Abu Dhabi
92,205
Iraq
100,000
Iran
92,860
Kuwait
97,125

◆ **FIGURE 15.29** The top 10 countries in proven oil reserves in 1989. Numbers indicate millions of barrels.

depth where pressure and temperature are high, are brought to the surface during the explosive volcanism that forms kimberlite pipes. Although kimberlite pipes have formed throughout geologic time, the most intense episode of such activity in South Africa and adjacent countries was during the Cretaceous Period. Emplacement of Triassic and Jurassic diamond-bearing kimberlites also occurred in Siberia.

In the Prologue we noted that the mother lode or source for the placer deposits mined during the California gold rush is in Jurassic-aged intrusive rocks of the Sierra Nevada. Gold placers are also known in Cretaceous-aged conglomerates of the Klamath Mountains of California and Oregon.

Porphyry copper was originally named for copper deposits in the western United States mined from porphyritic granodiorite, but the term now applies to large, low-grade copper deposits disseminated in a variety of rocks. These prophyry copper deposits are an excellent example of the relationship between convergent plate boundaries and the distribution, concentration, and exploitation of valuable metallic ores. Magma generated by partial melting of a subducting plate rises toward the surface, and as it cools, it precipitates and concentrates various metallic ores. The world's largest copper deposits were formed during the Mesozoic and Tertiary in a belt along the western margins of North and South America (see Fig. 7.27).

Chapter Summary

Table 15.1 provides a summary of the geologic history of North America as well as global events and sea level changes during the Mesozoic.

1. The breakup of Pangaea can be divided into four stages.
   a. The first stage involved the separation of North America from Africa during the Late Triassic, followed by the separation of North America from South America.
   b. The second stage involved the separation of Antarctica, India, and Australia from South America and Africa during the Jurassic. During this stage, India broke away from the still-united Antarctica and Australia landmass.
   c. During the third stage, South America separated from Africa, while Europe and Africa began to converge.
   d. In the last stage, Greenland separated from North America and Europe.
2. The breakup of Pangaea influenced global climatic and atmospheric circulation patterns. While the temperature gradient from the tropics to the poles gradually increased during the Mesozoic, overall global temperatures remained equable.
3. An increased rate of seafloor spreading during the Cretaceous Period caused sea level to rise and transgressions to occur.

(continued on page 430)

TABLE 15.1   Summary of Mesozoic Geologic Events

Age (Millions of Years)	Geologic Period	Sequence	Relative Changes in Sea Level Rising — Falling	Cordilleran Mobile Belt	
66	Cretaceous	Zuni		Jurassic and Cretaceous tectonism controlled by eastward subduction of the Pacific plate beneath North America and accretion of microplates.	Laramide orogeny
					Sevier orogeny
144			← Present sea level		Nevadan orogeny
	Jurassic				
208		Absaroka			
	Triassic			Subduction zone develops as a result of westward movement of North America.	
245				Sonoma orogeny	

Cordilleran orogeny (label spanning Nevadan, Sevier, and Laramide orogeny)

TABLE 15.1 Summary of Mesozoic Geologic Events (continued)

North American Interior	Gulf Coastal Region	Eastern Coastal Region	Global Plate Tectonic Events
	Major Late Cretaceous transgression. Reefs particularly abundant.	Appalachian region uplifted.	South America and Africa are widely separated. Greenland begins separating from Europe.
	Regression at end of Early Cretaceous. Early Cretaceous transgression and marine sedimentation.	Erosion of fault-block mountains formed during the Late Triassic to Early Jurassic.	
	Sandstones, shales, and limestones are deposited in transgressing and regressing seas. Thick evaporites are deposited in newly formed Gulf of Mexico.		South America and Africa begin separating in the Late Jurassic.
		Fault-block mountains and basins develop in eastern North America.	
		Deposition of Newark Group; lava flows, sills, and dikes.	Breakup of Pangaea begins with rifting between Laurasia and Gondwana.
	Gulf of Mexico begins forming during Late Triassic.		Supercontinent Pangaea still in existence.

4. Except for incursions along the continental margin and two major transgressions (the Sundance Sea and the Cretaceous Interior Seaway), the North American craton was above sea level during the Mesozoic Era.

5. The Eastern Coastal Plain was the initial site of the separation of North America from Africa that began during the Late Triassic. During the Cretaceous Period, it was inundated by marine transgressions.

6. The Gulf Coastal region was the site of major evaporite accumulation during the Jurassic as North America rifted from South America. During the Cretaceous, it was inundated by a transgressing sea, which, at its maximum, connected with a sea transgressing from the north to create the Cretaceous Interior Seaway.

7. Mesozoic rocks of the western region of North America were deposited in a variety of continental and marine environments. One of the major controls of sediment distribution patterns was tectonism.

8. Western North America was affected by four interrelated orogenies: the Sonoma, Nevadan, Sevier, and Laramide. Each involved batholithic intrusions as well as eastward thrust faulting and folding.

9. The cause of the Sonoma, Nevadan, Sevier, and Laramide orogenies was the changing angle of subduction of the oceanic Pacific plate under the continental North American plate. The timing, rate, and to some degree the direction of plate movement was related to seafloor spreading and the opening of the Atlantic Ocean.

10. Orogenic activity associated with the oceanic-continental convergent plate boundary in the Cordilleran mobile belt explains the structural features of the western margin of North America. It is believed, however, that more than 25% of the North American western margin originated from the accretion of microplates.

Important Terms

aulacogen
Cordilleran orogeny
Cretaceous Interior Seaway
Laramide orogeny

microplate
Nevadan orogeny
Sevier orogeny
Sonoma orogeny

Sundance Sea
triple junction
Zuni sequence

Review Questions

1. The failed rift of a triple junction is a(n):
   a. _____ trough; b. _____ terrane; c. _____ sill;
   d. _____ aulacogen; e. _____ none of these.

2. Evidence for the breakup of Pangaea includes:
   a. _____ rift valleys; b. _____ dikes; c. _____ sills;
   d. _____ great quantities of poorly sorted nonmarine detrital sediments; e. _____ all of these.

3. Which cratonic sequence was deposited during Early Jurassic to Early Paleocene?
   a. _____ Tippecanoe; b. _____ Absaroka; c. _____ Zuni; d. _____ Sauk; e. _____ Kaskaskia.

4. The time of greatest post-Paleozoic inundation of the craton occurred during which geologic period?
   a. _____ Triassic; b. _____ Jurassic; c. _____ Cretaceous; d. _____ Paleogene; e. _____ Neogene.

5. Which formation or group filled the Eastern Coastal region Triassic fault-block basins?
   a. _____ Newark; b. _____ Chinle; c. _____ Moenkopi; d. _____ Navajo; e. _____ Shinarump.

6. The first Mesozoic orogeny in the Cordilleran region was the:
   a. _____ Antler; b. _____ Laramide; c. _____ Nevadan; d. _____ Sevier; e. _____ Sonoma.

7. Which group of invertebrates were the main constituent of many of the extensive Cretaceous carbonate reefs?
   a. _____ corals; b. _____ ammonites; c. _____ rudists; d. _____ echinoids; e. _____ gastropods.

8. What type of climates dominated the Triassic low and middle latitudes?
   a. _____ hot and humid; b. _____ cool and humid; c. _____ cold and dry; d. _____ warm and dry; e. _____ glacial.

9. The breakup of Pangaea began with initial Triassic rifting between which two continental landmasses?
   a. _____ South America and Africa; b. _____ Laurasia and Gondwana; c. _____ North America and Eurasia; d. _____ Antarctica and India; e. _____ India and Australia.

10. The Jurassic formation or group famous for dinosaur fossils is the:
    a. _____ Navajo; b. _____ Chinle; c. _____ Sundance; d. _____ Morrison; e. _____ Franciscan.

11. The orogeny responsible for the present-day Rocky Mountains is the:
    a. _____ Sevier; b. _____ Sonoma; c. _____ Nevadan; d. _____ Antler; e. _____ Laramide.

12. The first major seaway to flood North America was the:
a. _____ Sundance; b. _____ Newark; c. _____ Zuni;
d. _____ Cordilleran; e. _____ Cretaceous Interior
Seaway.

13. The Mesozoic tectonic history of the North American
Cordilleran region is very complex and involves:
a. _____ oceanic-continental convergence; b. _____
continental-continental convergence; c. _____
microplate accretion; d. _____ answers (a) and (b);
e. _____ answers (a) and (c).

14. A possible cause for the eastward migration of igneous
activity in the Cordilleran region during the Cretaceous
was a change from:
a. _____ oceanic-oceanic convergence to
oceanic-continental convergence; b. _____ high-angle to
low-angle subduction; c. _____ divergent to convergent
plate margin activity; d. _____ divergent plate margin
activity to subduction; e. _____ subduction to divergent
plate margin activity.

15. The formation or group responsible for the spectacular
scenery of the Painted Desert and Petrified Forest is the:
a. _____ Chinle; b. _____ Wingate; c. _____
Franciscan; d. _____ Navajo; e. _____ Morrison.

16. The age of the thick evaporite deposits of the Gulf
Coastal region that form the Tertiary salt domes is:
a. _____ Permian; b. _____ Triassic; c. _____ Jurassic;
d. _____ Cretaceous; e. _____ Tertiary.

17. The Sierra Nevada, Southern California, Idaho, and
Coast Range batholiths formed as a result of which
orogeny?
a. _____ Sonoma; b. _____ Nevadan; c. _____ Sevier;
d. _____ Laramide; e. _____ none of these.

18. Which formation or group consists of a chaotic mixture of
rocks brought together in a submarine trench when North
America overrode the subducting Pacific oceanic plate?
a. _____ Morrison; b. _____ Newark; c. _____
Navajo; d. _____ Chinle; e. _____ Franciscan.

19. Discuss the geological evidence for the breakup of Pangaea.

20. Discuss the depositional environments, tectonic setting,
and depositional processes of the Triassic Newark Group
in the Eastern Coastal region.

21. Discuss the depositional environments, tectonic setting,
and depositional processes of the Triassic Moenkopi,
Shinarump, and Chinle formations in the Cordilleran
region.

22. Provide a general global history of the breakup of
Pangaea.

23. Compare the tectonics of the Sonoma and Antler
orogenies.

24. How did the breakup of Pangaea affect oceanic and
climatic circulation patterns?

25. How did the Mesozoic rifting that took place on the east
coast of North America affect the tectonics in the
Cordilleran mobile belt?

26. Discuss the tectonics of the Cordilleran mobile belt
during the Mesozoic Era.

27. Using a diagram, explain how increased seafloor
spreading can cause a rise in sea level along the
continental margins.

28. The American West is blessed with spectacular scenery.
What geologic conditions led to this beauty?

29. Compare the tectonic setting and depositional
environment of the Gulf of Mexico evaporites with the
evaporite sequences of the Paleozoic Era.

30. How does microplate tectonics change our
interpretations of the geologic history of the western
margin of North America, and how does it relate to the
Mesozoic orogenies that took place in that area?

Additional Readings

Bally, A. W., and A. R. Palmer, eds. 1989. *The geology of North America: An overview.* The Geology of North America. vol. A. Boulder, Colo.: Geological Society of America.

Ben-Avraham, Z. 1981. The movement of continents. *American Scientist* 69, no. 3: 291–99.

Bonatti, E. 1987. The rifting of continents. *Scientific American* 256, no. 3: 96–103.

Harwood, D. S., and M. M. Miller, eds. 1990. *Paleozoic and Early Mesozoic paleogeographic relations; Sierra Nevada, Klamath Mountains, and related terranes.* Geological Society of America Special Paper 255. Boulder, Colo.: Geological Society of America.

Howell, D. G. 1989. *Tectonics of suspect terranes.* New York: Chapman and Hall.

Jones, D. L., A. Cox, P. Coney, and M. Beck. 1982. The growth of western North America. *Scientific American* 247, no. 5: 70–128.

Murray, G. E. 1961. *Geology of the Atlantic and Gulf Coastal Province of North America.* New York: Harper and Row.

Nations, J. D., and J. G. Eaton, eds. 1991. *Stratigraphy, depositional environments, and sedimentary tectonics of the western margin, Cretaceous Western Interior Seaway.* Geological Society of America Special Paper 260. Boulder, Colo.: Geological Society of America.

Salvador, A., ed. 1991. *The Gulf of Mexico Basin.* The Geology of North America. vol. J. Boulder, Colo.: Geological Society of America.

CHAPTER 16

Restoration of the oldest known dinosaur, *Staurikosaurus,* from the Late Triassic of Argentina and Brazil. It measured about 2 m long. (Photo © 1990 Mark Hallett, All Rights Reserved.)

LIFE OF THE MESOZOIC ERA

Prologue

About 80 million years ago, in what is now northern Montana, a wide coastal plain with rivers, swamps, and marshes sloped gently eastward to the sea. Unlike the semiarid climate of today, this area was subtropical, probably much as southern Louisiana is now. Parts of the region were covered by dense vegetation including bald cypress, redwoods, and broad-leaf trees. Like the flora, the fauna was also varied. Fish, amphibians, turtles, and crocodiles lived in the streams and swamps, flying reptiles soared in the skies, and small primitive mammals scurried about.

Several types of dinosaurs lived in this area as well. Duck-billed dinosaurs were particularly abundant, but they shared their habitat with armored dinosaurs and herds of large, single-horned dinosaurs called *Monoclonius*. The large predator *Albertosaurus*, which looked much like *Tyrannosaurus*, and the small carnivore *Troodon* preyed upon the old, the weak, and unprotected juveniles.

Studies conducted by Princeton University and Montana State University under the direction of paleontologist John R. Horner have demonstrated that at least three dinosaur species used this area as a nesting ground. Numerous eggs, some unhatched, and a number of nests have been recovered. The eggs are of three types; the largest were laid by a duck-billed dinosaur, the genus *Maiasaura* (meaning "good mother lizard") (◆ Fig. 16.1). These eggs are about 20 cm long, and the newborn measured 30 to 35 cm.

Eight maiasaur nests have been found at one site; the nests measure 2 m in diameter and are spaced about 7 m apart, or about the average length of an adult maiasaur. Each bowl-shaped nest contained 20 to 25 eggs laid in a circular pattern. It seems that these dinosaurs

◆ **FIGURE 16.1** Scene from the Late Cretaceous of Northern Montana. A female *Maiasaura* leads her young to a feeding area. (Photo © Douglas Henderson.)

nested in colonies much as some living birds do. Some nests contain the remains of juveniles up to 1 m long, which is considerably greater than their length at the time of hatching. This evidence suggests that the young remained in the nest area for some time after hatching, and it further implies that they received parental care (Fig. 16.1). It seems likely that adults protected and fed the young.

These discoveries seem to provide considerable evidence for colonial nesting behavior. Additionally, smaller dinosaurs called *hypsilophodonts*, which laid intermediate-sized eggs, also in circular clutches, used the same nesting area repeatedly (◆ Fig. 16.2). Their nests occur in three superposed rock layers in a vertical sequence of strata 3 m thick, indicating that they

Archean Eon	Proterozoic Eon	Phanerozoic Eon						
Precambrian		Paleozoic Era						
		Cambrian	Ordovician	Silurian	Devonian	Mississippian	Pennsylvanian	Permian
						Carboniferous		

2,500
M.Y.A.

570
M.Y.A.

245
M.Y.A.

◆ **FIGURE 16.2** Hypsilophodont eggs from Egg Mountain, Montana. (Photo courtesy of the Museum of the Rockies.)

nested in the same area again and again. Hypsiloph-odont young did not remain in the nest as the maia-saur young did, but they did stay in the immediate nest area. Both hypsilophodont and maiasaur young stayed in the company of adults long after hatching. The smallest eggs found in this area were laid in two parallel rows, but the identity of the egg-layer is unknown.

❖ INTRODUCTION

The Mesozoic Era is commonly referred to as the "Age of Reptiles," alluding to the fact that reptiles were the most diverse and abundant land-dwelling vertebrate animals. This is perhaps the most interesting chapter in the history of life because the reptiles included the dinosaurs and their relatives, such as flying reptiles and marine reptiles. The Mesozoic diversification of reptiles was an important evolutionary event, but other equally important events also occurred. For example, mammals evolved

from the mammal-like reptiles during the Triassic, while birds evolved from reptiles, probably small carnivorous dinosaurs, during the Jurassic.

Vast changes occurred in land-plant communities too. When the Mesozoic Era began, the dominant land plants were holdovers from the Paleozoic, but during the Cretaceous, the first flowering plants evolved. These plants diversified rapidly and soon became the most diverse and abundant land plants.

The Mesozoic was also a time of resurgence of marine invertebrates. Diversity among marine invertebrates had been drastically reduced as a result of the Permian extinction event, but survivors rapidly diversified, giving rise to increasingly complex marine invertebrate communities. Ammonites and planktonic microorganisms called *foraminifera* were particularly abundant and diverse.

The breakup of Pangaea that began during the Triassic continued throughout the Mesozoic. Nevertheless, the proximity of continents and mild Mesozoic climates allowed land animals and plants to occupy extensive geographic areas. As the fragmentation of Pangaea continued, however, some continents, Australia and South America especially, became isolated, and their faunas evolved independently.

Another mass extinction event occurred at the end of the Mesozoic. Once again organic diversity declined markedly as the dinosaurs and several other reptiles and some marine invertebrates died out. Because dinosaurs were victims of this extinction event, it has received more publicity than any other event although the Permian extinctions were of greater impact (see Chapter 13).

❖ MARINE INVERTEBRATES

Following the wave of extinctions that occurred at the end of the Paleozoic, the Mesozoic was a time when the marine invertebrates repopulated the seas. As the Atlan-

Mesozoic Era			Cenozoic Era					
Triassic	Jurassic	Cretaceous	Tertiary				Quaternary	
			Paleocene	Eocene	Oligocene	Miocene	Pliocene	Pleistocene	Holocene

5
.A.

66
M.Y.A.

◆ **FIGURE 16.3** Scleractinian corals evolved during the Triassic and proliferated in the warm, clear, shallow marine waters of the Mesozoic Era. Most living corals are scleractinians, represented here by the so-called staghorn coral. (Photo courtesy of Sue Monroe.)

tic and Indian oceans formed, the epeiric seas once again inundated the low-lying continental margins, and invertebrate marine life returned with a flourish. Gone were many Paleozoic invertebrates such as the fusulinid foraminifera, blastoids, rugose corals, and trilobites.

The Early Triassic invertebrate marine fauna, though widely distributed, was not very diverse. By the Late Triassic, the seas were once again richly populated with invertebrates. The mollusks became increasingly diverse and abundant throughout the Mesozoic. The brachiopods, however, never completely recovered from their near extinction at the end of the Paleozoic, and though they diversified during the Triassic and Jurassic, they declined in numbers during the Cretaceous, remaining a minor invertebrate phylum during the Cenozoic. In areas of warm, relatively clear, shallow marine waters, corals

again proliferated. These corals were of a new and familiar type, the *scleractinians* (◆ Fig. 16.3). Echinoids, which were rare during the Paleozoic, greatly diversified during the Mesozoic (◆ Fig. 16.4).

One of the major differences between Paleozoic and Mesozoic marine invertebrate communities was the increased abundance and diversity of burrowing organisms. Except for a few animals with hard parts, such as the inarticulate brachiopod *Lingula,* almost all Paleozoic burrowers were soft-bodied animals such as worms. The

◆ **FIGURE 16.4** Echinoids, or sea urchins, were rare during the Paleozoic but became diverse and abundant during the Mesozoic. (Photo courtesy of Sue Monroe.)

◆ **FIGURE 16.5** Two different genera of Cretaceous reef-building rudistid bivalves.

bivalves and echinoids, which were epifaunal elements during the Paleozoic, evolved various means of entering the infaunal habitats. This trend toward an infaunal existence may reflect an adaptive response to increasing predation from the rapidly evolving fish and cephalopods.

The end of the Mesozoic witnessed another major extinction event, this time involving mostly the large reptiles. The only extinctions of consequence in the marine invertebrate fauna were those affecting the ammonoid cephalopods, planktonic foraminifera and a group of reef-forming *rudist* bivalves (◆ Fig. 16.5).

Beginning with the primary producers, we will now examine some of the major Mesozoic marine plant and invertebrate groups. The *coccolithophores* are an important group of living phytoplankton (◆ Fig. 16.6a). Their remains, which are constructed of microscopic calcareous plates, collect in tremendous numbers on the ocean floor when the organisms die. Coccolithophore remains comprise many Cretaceous chalk beds such as the White Cliffs of Dover in England and the Demopolis Chalk of the southwestern United States. Coccolithophores first appeared during the Jurassic and diversified tremendously during the Cretaceous. They continue to be abundant today and are one of the major primary producers supporting suspension feeders in present-day oceans. *Diatoms* first evolved during the Cretaceous, but were more important as primary producers during the Cenozoic. Diatoms construct their shells out of silica. They are presently most abundant in cooler waters (◆ Fig. 16.6b). *Dinoflagellates* were common during the Mesozoic and today are the major primary producers in warm waters. Their remains are easily preserved as organic cysts (◆ Fig. 16.6c).

The mollusks, as previously noted, were the major invertebrate phylum of the Mesozoic. The mollusks include six classes, only three of which—the gastropods, bivalves, and cephalopods—are significant members of

the marine invertebrate fauna. The gastropods increased in abundance and diversity during the Mesozoic, becoming most abundant during the Cretaceous, when carnivorous forms appeared. It was the bivalves and cephalopods, however, that dominated the invertebrate community of the Mesozoic.

Mesozoic bivalves diversified to inhabit many epifaunal and infaunal niches. Oysters and other clams became particularly diverse and abundant epifaunal suspension feeders and, in spite of a reduction at the end of the Cretaceous, continued to be important throughout the Cenozoic to the present (◆ Fig. 16.7). The reef-forming rudists were a significant group of Mesozoic bivalves (Fig. 16.5). These epifaunal bivalves were geologically short-lived and so are excellent guide fossils for the Late

◆ **FIGURE 16.6** (a) *Calcidiscus macintyrei,* a Miocene coccolith from the Gulf of Mexico (left); *Discoaster variabilis,* a Pliocene-Miocene coccolith from the Gulf of Mexico (right). (b) *Actinoptychus senarius,* an Upper Miocene centric diatom from Java (left); *Cocconeis pellucida,* an Upper Miocene pinnate diatom from Java (right). (c) *Deflandrea spinulosa,* an Eocene dinoflagellate from Alabama (left); *Spiniferites mirabilis,* a Neogene dinoflagellate from the Gulf of Mexico (right). (Scanning electron photomicrographs courtesy of *a,* Merton E. Hill; *b,* John Barron, United States Geological Survey; *c,* John H. Wrenn, Amoco Production Co.)

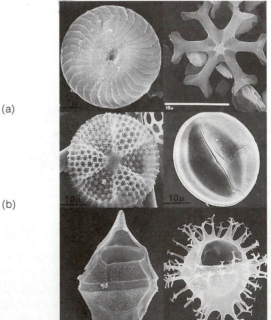

(a)

(b)

(c)

◆ **FIGURE 16.7** Bivalves, represented here by two Cretaceous forms, were particularly diverse and abundant during the Mesozoic. (Photo courtesy of Sue Monroe.)

Jurassic through Cretaceous. Rudists are also significant because they formed large tropical reefs, displacing corals as the main reef-builders. Bivalves also expanded into the infaunal niche during the Mesozoic. By burrowing into the sediment, they escaped predation from cephalopods and fish.

◆ **FIGURE 16.8** Cephalopods were an important Mesozoic invertebrate group. The ammonites, which are characterized by extremely complex suture patterns, were particularly abundant and diverse during the Jurassic and Cretaceous and are excellent guide fossils for those periods. This specimen, *Scaphites preventricosus,* is from the Upper Cretaceous Colorado Formation, Toole County, Montana, and shows the complex suture pattern characteristic of ammonites. (Photo courtesy of Smithsonian Institution, Photo No. 106677.)

Cephalopods were one of the most important Mesozoic invertebrate groups. Their rapid evolution and nektonic life-style make them excellent guide fossils (◆ Fig. 16.8). Recall that two orders of cephalopods arose during the Paleozoic: the Nautiloidea, with simple sutures, and the Ammonoidea, with wrinkled sutures. The Ammonoidea are divided into three groups: the goniatites, ceratites, and ammonites. The ammonites, which are characterized by extremely complex suture patterns, were present during all three Mesozoic periods but were most prolific during the Jurassic and Cretaceous. Their tremendous success was a reflection of their ability to adapt to a wide range of marine environments. While most ammonites were coiled, some attaining diameters of 2m, others were uncoiled and led a near benthonic existence.

In spite of their successful adaptations, the ammonites became extinct at the end of the Cretaceous; explanations for their demise have ranged from competition with fish to extinction caused by a meteorite impact. Although the ammonites became extinct at or near the end of the Cretaceous, two other groups of cephalopods survived into the Cenozoic—the nautiloids and the belemnoids, a group of squidlike cephalopods that was highly successful during the Jurassic and Cretaceous (◆ Fig. 16.9).

◆ **FIGURE 16.9** Belemnoid cephalopods were abundant during the Jurassic and Cretaceous periods. (Photo courtesy of Sue Monroe.)

Marine Invertebrates   **437**

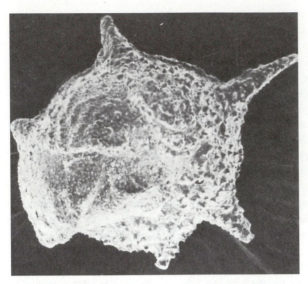

◆ **FIGURE 16.10** Planktonic foraminifera, represented here by *Globotruncana calcarata* from the Cretaceous Pecan Gap Chalk of Texas, became diverse during the Jurassic and Cretaceous, but were affected by the extinction event at the end of the Cretaceous. (Photo courtesy of B. A. Masters.)

Although the mollusks dominated the Mesozoic, other phyla were also important. The stalked echinoderms, which were abundant during the Paleozoic, became minor members of the Mesozoic marine invertebrate community. However, the echinoids, which were exclusively epifaunal during the Paleozoic, branched out into the infaunal habitat and became very diverse and abundant (Fig. 16.4). Bryozoans, although rare in Triassic strata, diversified and expanded during the Jurassic and Cretaceous.

As is true today, where shallow marine waters were warm and clear, coral reefs proliferated. Mesozoic corals belong to the order Scleractinia (Fig. 16.3). Scleractinians use aragonite rather than calcite in constructing their skeletons, and most today have a symbiotic relationship with certain dinoflagellates. Because dinoflagellates require sufficient sunlight to function, they can live only in shallow waters. Whether scleractinian corals evolved from the rugose order or from an as yet unknown soft-bodied group that left no fossil record is still unresolved.

Lastly, the foraminifera underwent an explosive radiation during the Jurassic and Cretaceous that continued to the present. The planktonic forms (◆ Fig. 16.10) in particular underwent rapid diversification, but they were also affected by the extinction event at the end of the

Cretaceous. Most of the planktonic genera became extinct at the end of the Cretaceous, and only a few genera survived into the Cenozoic.

In general terms, we can think of the Mesozoic as a time of increasing complexity of the marine invertebrate community as it evolved from a simple one with low diversity and short food chains at the beginning of the Triassic to a highly complex community with interrelated food chains near the end of the Cretaceous. This evolutionary history reflects the change in geologic conditions influenced by plate tectonic activity that we discussed in Chapter 15.

❖ FISHES AND AMPHIBIANS

The cartilaginous fishes, which includes sharks and their relatives, increased in abundance through the Mesozoic, but even so they never came close to matching the diversity of the bony fishes. Nevertheless, sharks were, and still are, important elements of the marine fauna.

Among the bony fishes, the lung fishes and crossopterygians are represented by few Mesozoic genera and species. In fact, the crossopterygians declined and were almost extinct by the close of the era; only one living species is known (see Fig. 6.23), and the group has no known Cenozoic fossil record.

The remaining bony fishes belong to three groups, which for convenience we can call primitive, intermediate, and advanced (◆ Fig. 16.11). Superficially, these fishes look much alike, but important changes occurred as one group replaced another. By the end of the Cretaceous, the advanced group, the teleosts, had become the dominant marine and freshwater fishes.

A few labyrinthodont amphibians persisted into the Mesozoic but died out by the end of the Triassic. Since the Pennsylvanian, the time of their greatest diversity, amphibians have made up only a small part of the total vertebrate fauna. Frogs and salamanders appeared during the Mesozoic, but their fossil records are poor.

❖ PRIMARY PRODUCERS ON LAND—PLANTS

As a prelude to our discussion of Mesozoic land animals, we will consider land-plant communities. After all, plants as photosynthesizers, lie at the base of the food chain. In other words, they are the primary producers on land, while the animals, as consumers, are dependent upon plants.

Triassic and Jurassic land-plant communities, like

those of the Late Paleozoic, were composed of seedless vascular plants such as ferns, horsetail rushes, and club mosses and various gymnosperms. Among the gymnosperms, however, the large seed ferns became extinct by the end of the Triassic. The *ginkgos* remained abundant throughout the Mesozoic Era; ginkgos nearly disappeared during the Cenozoic and only one species survives today. *Conifers* also persisted from the Late Paleozoic, continued to diversify, and now are the most common land plants at high elevations and at the margins of the Arctic. A new type of gymnosperm, the *cycads,* evolved during the Triassic. Cycads superficially resemble palm trees, and several varieties still exist in tropical and subtropical areas (◆ Fig. 16.12a).

Marked changes in the composition of land-plant communities occurred during the Cretaceous. The long

(a)

◆ **FIGURE 16.11** Three stages in the evolution of bony fishes. (a) The primitive chondrosteans and (b) the intermediate holosteans are represented by few living species. (c) In contrast, the advanced teleosts include most species of living bony fishes.

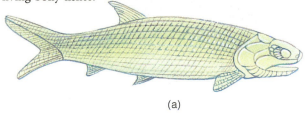

(a)

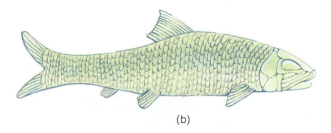

(b)

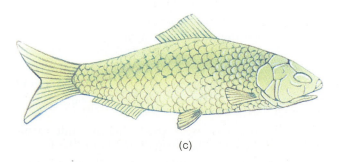

(c)

(b)   (c)

◆ **FIGURE 16.12** (a) Living cycads, southern California. (b) and (c) Fossil angiosperms from the Lower Cretaceous Potomac Group of the eastern United States. (b) *Sapindopsis,* Cecil County, Maryland. (c) *Aralia* from New Jersey. (Photos b and c courtesy of Leo J. Hickey, Yale University.)

---

dominance of seedless plants and gymnosperms ended as many were replaced by **angiosperms,** or flowering plants (◆ Fig. 16.12b and c). The earliest angiosperms probably evolved from a specialized group of seed ferns. In any case, since the angiosperms evolved, they have adapted to nearly every terrestrial habitat from high mountains to low deserts. Some have even adapted to shallow coastal waters, and a few varieties are carnivorous.

Several factors account for the phenomenal success of flowering plants, but chief among them is their method of reproduction (◆ Fig. 16.13). Two developments were particularly important: the evolution of flowers, which attract animal pollinators, especially insects, and the evolution of enclosed seeds. Angiosperm seeds are dispersed

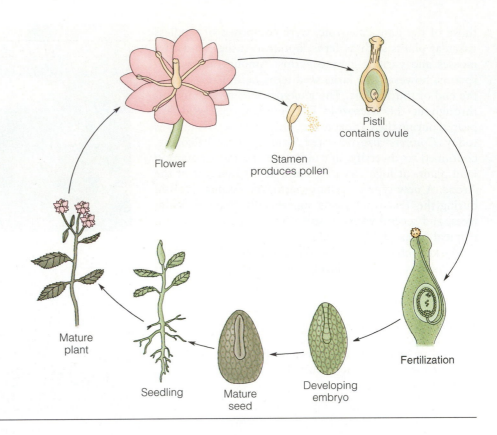

◆ **FIGURE 16.13** The reproductive cycle in angiosperms.

Flower

Stamen produces pollen

Pistil contains ovule

Fertilization

Developing embryo

Mature seed

Seedling

Mature plant

widely in several ways—blown by the wind, carried in a fleshy fruit, or carried in burrs that become attached to animal fur.

Seedless plants and gymnosperms are important and still flourish in many environments; in fact, many botanists regard the ferns and conifers as emerging groups. Nevertheless, a measure of the angiosperms' success is that today they account for more than 90 percent of all land-plant species.

❖ REPTILES

The evolutionary relationships among the groups of fossil and living reptiles are summarized in ◆ Figure 16.14. Reptile diversification began during Pennsylvanian time with the evolution of the captorhinomorphs, apparently the first animals to lay amniote eggs (see Chapter 14). From this basic stock of so-called *stem reptiles* all other reptiles evolved. Birds and mammals, too, have their ancestors among the reptiles, so they are a part of this major evolutionary diversification.

Recall from Chapter 14 that the captorhinomorphs

and pelycosaurs were the dominant land vertebrates of the Pennsylvanian and Permian periods. Here we continue our story of reptile diversification with a group called *thecodontians*.

Thecodontians and the Ancestry of Dinosaurs

Thecodontian is an informal term for a variety of Late Permian and Triassic reptiles characterized by teeth set into individual sockets as are those of crocodiles, dinosaurs, mammal-like reptiles, and mammals. Compared with most other groups of reptiles, thecodontians had a comparatively brief history, existing for only 45 to 50 million years. Furthermore, their adaptive radiation was rather limited since it resulted in the origin of few genera and species. Thecodontians are nevertheless important because they included the ancestors of dinosaurs, flying reptiles, and three of the five orders of living reptiles (Fig. 16.14).

Some thecodontians were small, lightly built carnivores (◆ Fig. 16.15). These small predators had well-developed forelimbs and moved primarily on all four

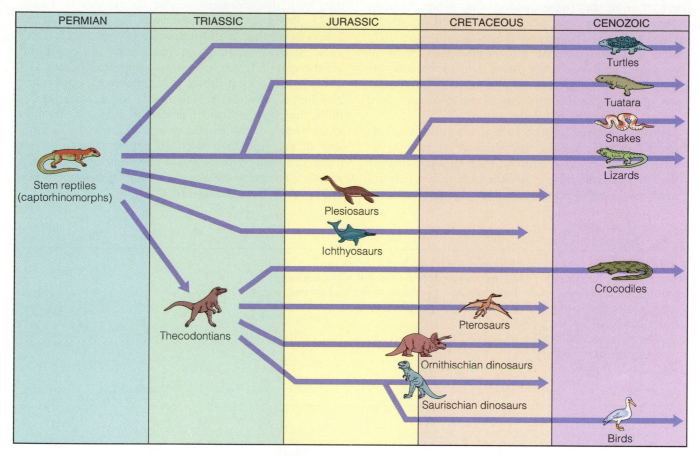

◆ **FIGURE 16.14** Relationships among fossil and living reptiles and birds.

◆ **FIGURE 16.15** Scene from the Early Triassic of South Africa showing the small bipedal thecodontian *Euparkeria*. The vegetation consists of ginkgos, cycads, and horsetails.

limbs; that is, they were **quadrupedal.** When running, however, they rose onto their hind limbs and moved in a **bipedal** fashion. The largest thecodontians were fully quadrupedal and covered with bony armor. Some of these were herbivores, while others—the phytosaurs, for example—were predators (◆ Fig. 16.16). Phytosaurs were crocodile-like in appearance and probably lived much as existing crocodiles do.

Dinosaurs evolved from thecodontians during the Late Triassic, but their specific thecodontian ancestor is debated. The traditional interpretation is that two distinct orders of dinosaurs were established when they first appeared, each of which may have had an independent or-igin from thecodontians similar to the one in Figure 16.15. Pelvic structure is the basis for recognizing the orders **Saurischia** and **Ornithischia** (◆ Fig. 16.17, Table 16.1). Saurischian dinosaurs had a lizard-like pelvis and are therefore referred to as lizard-hipped dinosaurs. Ornithischians had a bird-like pelvis; hence they are called bird-hipped dinosaurs. However, some investigators think that both orders of dinosaurs shared a common ancestor among the thecodontians and that the subdivision of the dinosaurs based on pelvic structure is not justified. In this book we follow the traditional approach and thus recognize the classification of dinosaurs shown in Figure 16.17 and Table 16.1.

◆ **FIGURE 16.16** Armored, quadrupedal thecodontians of the Triassic. *Desmatosuchus*, on the right, had a formidable appearance, but was a herbivore possessing weak jaws and teeth. The phytosaur *Rutiodon*, in the background, was a large carnivore that lived much as present-day crocodiles do.

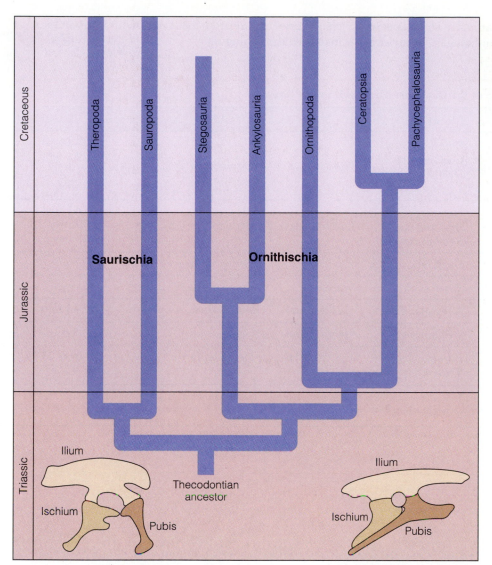

◆ **FIGURE 16.17** Origin of and inferred relationships among dinosaurs. Both ornithischian and saurischian dinosaurs evolved from thecodontians, but each may have had an independent origin from that group. The pelvis of each order of dinosaurs is shown.

Dinosaurs

More than any other type of animal, the dinosaurs have inspired awe and have thoroughly captured the public imagination. As often as not, unfortunately, their popularization in cartoons, movies, and many books has led to misunderstandings about dinosaurs in general. It is true that many were large, indeed the largest animals ever to live on land. But not all were large. In fact, dinosaurs varied from giants to those no larger than a chicken (◆ Fig. 16.18, Table 16.1).

A common but erroneous perception of dinosaurs is that they were poorly adapted animals that had trouble surviving. True, they became extinct, but to consider this a failure is to ignore the fact that for more than 140 million years they were the dominant land vertebrates. During their existence, they diversified into numerous types and adapted to a wide variety of environments. Eventually, the dinosaurs did die out, an event that then enabled mammals to become the predominant land vertebrates.

Nor were dinosaurs the lethargic beasts often portrayed in various media. Recent evidence indicates that at least some dinosaurs may have been very active and possibly warm-blooded. It also appears that some species cared for their young long after hatching, a behavioral characteristic most often associated with birds and mammals (see the Prologue).

TABLE 16.1 Summary Chart of Some of the Dinosaurs Mentioned in the Text

ORDER	SUBORDER	GENUS	LOCOMOTION	FEEDING	LENGTH*	WEIGHT*
Saurischia	Theropoda	*Compsognathus* *Coelophysis* *Deinonychus* *Tyrannosaurus*	Bipedal	Carnivorous	60 cm 3 m 2.4–4 m 14 m	2–3 kg 29.5 kg 75 kg 7 tons
	Sauropoda	*Diplodocus* *Apatosaurus* *Brachiosaurus†*	Quadrupedal	Herbivorous	27 m 21 m 23–27m	10.6 tons 30 tons 77 tons
Ornithischia	Ornithopoda	Duck-bills	Bipedal	Herbivorous	‡	‡
	Pachycephalosauria	*Stegoceras*	Bipedal	Herbivorous	2.5 m	55 kg
	Ankylosauria	*Ankylosaurus*	Quadrupedal	Herbivorous	6 m	2–3 tons
	Stegosauria	*Stegosaurus*	Quadrupedal	Herbivorous	9 m	1.8 tons
	Ceratopsia	*Psittacosaurus* *Triceratops*	Quadrupedal	Herbivorous	.8–1.5 m 9 m	Up to 23 kg 5.4 tons

*Lengths and weights mostly from D. Lambert, *A Field Guide to Dinosaurs* (New York: Avon Books, 1983). Tons are metric tons (1 metric ton = 2,204 lb).
†Partial remains of what appear to be even larger brachiosaurids have been discovered. For these, lengths greater than 30 m and weights up to 136 metric tons have been claimed.
‡Duck-billed dinosaurs varied considerably in size: lengths ranged from 3.6 to 15 m, and some larger species weighed more than 4 metric tons.

◆ **FIGURE 16.18** *Compsognathus* is one of the smallest known dinosaurs. It was about 60 cm long and weighed 2 to 3 kg. Lizard bones have been found inside the rib cage of *Compsognathus* indicating that this small carnivore caught such prey.

◆ **FIGURE 16.19** Scene from the Late Cretaceous showing the ankylosaur *Euoplocephalus* (foreground), the large theropod *Tyrannosaurus*, and the ceratopsian *Triceratops* (right).

Saurischian Dinosaurs

Two groups of saurischians are recognized, theropods and sauropods (Fig. 16.17, Table 16.1). Theropods were carnivorous bipeds that ranged in size from tiny, *Compsognathus* (Fig. 16.18) to *Tyrannosaurus* (◆ Fig. 16.19), the largest terrestrial carnivore known. A partic-ularly interesting theropod was *Deinonychus,* which means terrible claw. It was a predator with large, sickle-like claws on its hind feet that it probably used to kill its prey (◆ Fig. 16.20).

Included among the sauropods were the giant, quadru-pedal herbivores such as *Apatosaurus, Diplodocus,* and

◆ **FIGURE 16.20** *Deinonychus* ("terrible claw") from the Early Cretaceous of Montana is shown here using its sickle-like claws to kill its prey. *Deinonychus* was a comparatively small theropod that measured 2.4 to 4 m long. (Photo © John Agnew, courtesy of the Academy of Natural Sciences, Philadelphia.)

◆ **FIGURE 16.21** (below) Late Jurassic scene showing *Brachiosaurus,* the largest of the sauropods, and the plated ornithischian dinosaur *Stegosaurus.*

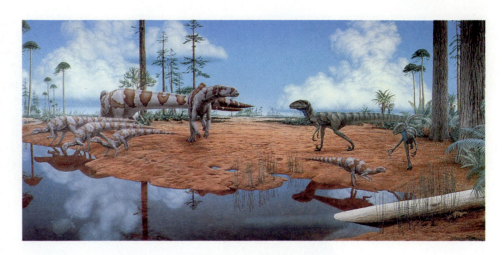

Brachiosaurus (◆ Fig. 16.21), the largest known land animals of any kind. Recent discoveries of fossils in Colorado, New Mexico, and elsewhere suggest that even larger sauropods existed (Table 16.1) (see Perspective 16.1). Evidence from fossil trackways indicates that sauropods moved in herds. They depended on their size and herding behavior rather than speed as their primary protection from predators.

Ornithischian Dinosaurs

The great diversity of ornithischians is manifested by the fact that five distinct groups are recognized: ornithopods, pachycephalosaurs, ankylosaurs, stegosaurs, and ceratopsians (Fig. 16.17, Table 16.1). Ornithopods include the duck-billed dinosaurs, which had flattened, bill-like mouths (◆ Fig. 16.22). These dinosaurs were particularly varied and abundant during the Cretaceous, and some species were characterized by head crests (Fig. 16.22). Such crests may have functioned as a means of species recognition, as display devices to attract mates, or as resonating chambers to amplify bellowing. Some duck-billed dinosaurs practiced colonial nesting and care of the young (see the Prologue). All ornithopod genera were herbivores and primarily bipedal, but their well-developed forelimbs allowed them to walk in a quadrupedal fashion, too.

◆ **FIGURE 16.22** Three duck-billed dinosaurs from the Late Cretaceous of western North America. *Anatosaurus* (back, right) and two of the several duck-billed dinosaurs with head crests, *Corythosaurus* (left) and *Parasaurolophus* (foreground). Notice that duck-billed dinosaurs had well-developed forelimbs and were capable of both bipedal and quadrupedal locomotion.

◆ **FIGURE 16.23** Two Late Cretaceous pachycephalosaurs, *Stegoceras* (foreground) and *Pachycephalosaurus*.

The pachycephalosaurs constitute a most peculiar group of ornithischian dinosaurs. The most distinctive feature of these bipedal herbivores is the dome-shaped skull that resulted from thickening of the bones (◆ Fig. 16.23). According to one hypothesis, these domed skulls were used in intraspecific butting contests for dominance and mates. The few known genera of pachycephalosaurs lived mostly during the Late Cretaceous.

Ankylosaurs were heavily armored, quadrupedal her-bivores, and some were quite large (Fig. 16.19, Table 16.2). Bony armor protected the back, flanks, and top of the head, and the tail ended in a large, bony clublike growth. No doubt a blow delivered by the powerful tail could seriously injure an attacking predator.

The stegosaurs, represented by the familiar genus *Stegosaurus* (Fig. 16.21, Table 16.1), were quadrupedal herbivores with bony spikes on the tail, which were undoubtedly used for defense, and bony plates on the back. The exact arrangement of these plates is debated but many paleontologists believe they functioned as a device to absorb and dissipate heat.

The final group of ornithischian dinosaurs is the ceratopsian or horned dinosaurs. A rather good fossil record indicates that large, Late Cretaceous genera such as *Triceratops* evolved from small, Early Cretaceous ancestors (Fig. 16.19). The later ceratopsians were characterized by huge heads, a large bony frill over the top of the neck, and a large horn or horns on the skull. Fossil trackways indicate that these large, quadrupedal herbivores moved in herds.

We conclude our discussion of dinosaurs with a few general comments to put them in proper perspective. Dinosaurs are fascinating animals, but they are not part of our experience, so we tend to overlook some other, equally fascinating aspects of the present-day animal world. For example, as large as some dinosaurs were, none are known to have been as large as living blue whales. The massive armor of the ankylosaurs is awesome, but we may fail to realize that the most completely armored vertebrate animal that has ever lived is alive today—the turtle. The carnivorous dinosaurs are commonly portrayed as ferocious, aggressive beasts, and this is probably fairly accurate. Nevertheless, carnivores such as *Tyrannosaurus* probably behaved much as large carnivores living today. They went for the easy kill, preying upon the old, the weak, and the young, or simply dined on carrion when it was available.

Warm-Blooded Dinosaurs?

All living reptiles are **ectotherms,** that is, cold-blooded animals whose body temperature varies in response to the outside temperature. **Endotherms,** warm-blooded animals, such as birds and mammals, are capable of maintaining a rather constant body temperature regardless of the outside temperature. Some investigators think that dinosaurs, or at least some dinosaurs, were endotherms.

Proponents of dinosaur endothermy note that dinosaur bones are penetrated by numerous passageways that, when the animals were living, contained blood vessels. Bones of endotherms typically have this structure, but considerably fewer of these passageways are found in bones of ectotherms. Living crocodiles and turtles have this so-called endothermic bone structure, yet they are ectotherms. And in some small mammals the bone structure is more typical of ectotherms, yet we know that they are capable of maintaining a constant body temperature. It may be that bone structure is more related to body size and growth patterns than to endothermy.

Because endotherms have high metabolic rates, they must eat more than ectotherms of comparable size (◆ Fig. 16.24). Consequently, endothermic predators require large prey populations. They would therefore constitute a much smaller proportion of the total animal population than their prey. In contrast, the proportion of ectothermic predators to their prey population is much greater. Where data are sufficient to allow an estimate, dinosaur predators appear to have made up 3 to 5 percent of the total population. These figures are comparable to present-day mammalian populations. However, a number of uncertainties about the composition of fossil communities make this argument for endothermy unconvincing to many paleontologists.

Living endotherms have a large brain in relation to body size. A relatively large brain is not necessary for endothermy, but endothermy does seem to be a prerequisite for having a large brain because a complex nervous system requires a rather constant body temperature. Some dinosaurs, particularly the small carnivores, did have a large brain in relation to their body, but many did not. That the small carnivorous ones usually had a large brain seems to be a good argument for endothermy, but there is an even more compelling argument. The relationship of birds to small carnivores such as *Compsognathus* (Fig. 16.18) implies that these small dinosaurs were

◆ **FIGURE 16.24** Evidence cited for endothermy in dinosaurs. If dinosaurs were endotherms, a large population of prey animals would have been necessary to support a low proportion of carnivores. This illustration shows the proportion of herbivores to carnivores for dinosaurs in Alberta, Canada. The proportions shown are consistent with populations of living endotherms, but some paleontologists question the accuracy of these data.

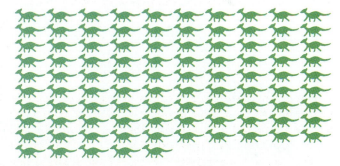

Herbivorous

Carnivorous

THE LARGEST DINOSAURS

The largest living land animal, the African elephant, is about 4 m tall and weighs about 6 metric tons. In fact, living elephants weigh as much as, or more than, many dinosaurs. Nevertheless, it has been estimated that "more than half of the dinosaurs weighed more than 2 metric tons—a size attained by only about 2 percent of modern mammals."[1] Thus, many dinosaurs were very large

[1]J. L. Marx, "Warm-Blooded Dinosaurs: Evidence Pro and Con," *Science* 199 (1978): 1425.

◆ **FIGURE 1** It was once thought that the large dinosaurs, such as these sauropods, were too heavy to walk on dry land and must have been semiaquatic. Evidence from fossil footprints indicates that sauropods were quite capable of walking on land. Furthermore, if they had submerged themselves, as in this illustration, breathing would have been difficult because water pressure would prevent the chest from expanding.

animals, but just how large is "very large"? Certainly, the giants among the dinosaurs were the sauropods, especially the *brachiosaurids* such as *Brachiosaurus* (Fig. 16.21); *Brachiosaurus* is estimated to have been 23 to 27 m long, and may have weighed 77 metric tons!

It was once believed that the giant sauropods were too heavy to walk on land, hence they were commonly depicted partly or completely submerged in swamps and lakes (◆ Fig. 1). However, recent studies indicate that the sauropods were quite capable of walking on dry land, although they no doubt visited swamps and lakes (where the chances of preservation of their bones was greater). Some paleontologists now think that many sauropods could even rear up on their hind legs to browse high up on trees.

Just how large the largest dinosaur was is unknown. Partial remains from Colorado and New Mexico indicate that some were larger than *Brachiosaurus*. These giants among giants are known by such common names as supersaurus, ultrasaurus, and seismosaurus. Ultrasaurus may have been more than 30 m long and possibly weighed 136 metric tons (◆ Fig. 2).

It may seem odd that ultrasaurus's estimated length is not much greater than that of *Brachiosaurus*, yet its estimated weight is more than 1.7 times as much. The reason is that the weight of an animal, or any object, increases by a factor of 8 if the linear dimensions are doubled. In other words, if the linear dimensions increase

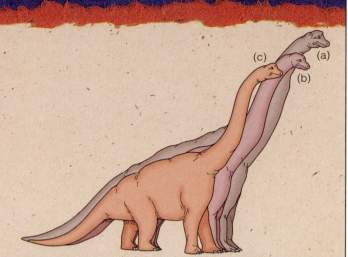

◆ **FIGURE 2** (a) Ultrasaurus and (b) supersaurus compared with (c) *Brachiosaurus*. Ultrasaurus was probably 20 to 25 percent larger than *Brachiosaurus*, but its size is estimated from partial remains.

by a factor of 2, the weight increases eightfold. Thus, since ultrasaurus was perhaps 20 to 25 percent larger than *Brachiosaurus*, it probably weighed more than 1.7 times as much.

endothermic or at least trending in that direction. The large sauropods were probably not endothermic, but nevertheless may have been able to maintain their body temperatures within narrow limits as endotherms do. A large animal heats up and cools down slowly because it has a small surface area compared to its volume. With proportionately less surface area to allow heat loss, sauropods probably retained body heat more efficiently than smaller dinosaurs.

One further point on endothermy in dinosaurs is that the flying reptiles, the *pterosaurs*, evolved from thecodontians as did the dinosaurs (Fig. 16.14). At least one species of pterosaur had hair or hairlike feathers. This is interesting because an insulating covering of hair or feathers is known only in endotherms. Furthermore, the physiology of active flight requires endothermy. Such evidence indicates that perhaps both thecodontians and dinosaurs were endothermic.

Obviously, considerable disagreement exists on dinosaur endothermy. In general, a fairly good case can be made for endothermic, small, carnivorous dinosaurs and pterosaurs, but for the others the question is still open.

Flying Reptiles

The first vertebrate animals to fly are called **pterosaurs** (◆ Fig. 16.25). Pterosaurs evolved from thecodontians during the Triassic and were abundant until their extinction at the end of the Mesozoic. Pterosaur flight adaptations include a wing membrane supported by an elongate fourth finger (Fig. 16.25), light hollow bones, and development of those parts of the brain associated with muscular coordination and sight. Size varied considerably. Some of the early species ranged from sparrow to robin size, while one Cretaceous pterosaur from Texas had a

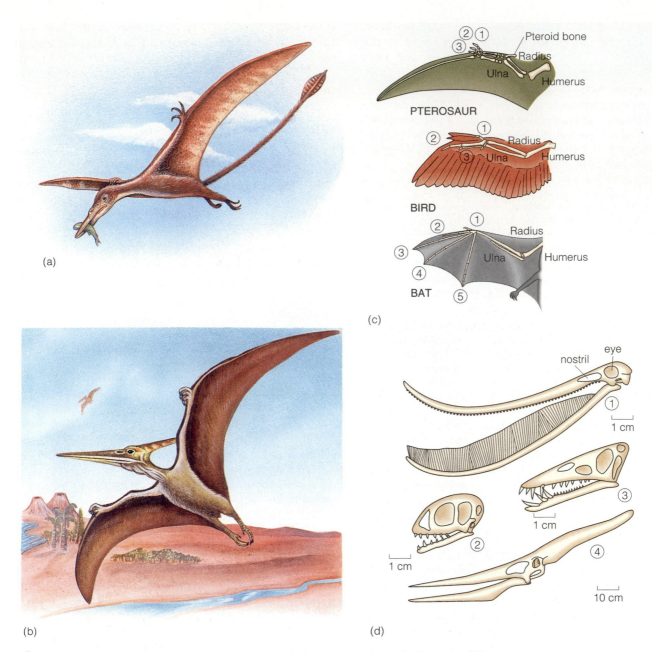

(a)

(c)

PTEROSAUR

Pteroid bone
Radius
Ulna
Humerus

BIRD

Radius
Ulna
Humerus

BAT

Radius
Ulna
Humerus

(b)

(d)

nostril   eye

1 cm

1 cm

1 cm

10 cm

◆ **FIGURE 16.25** (a) Long-tailed and (b) short-tailed pterosaurs from the Jurassic of Europe. *Pteranodon* (b) was a Cretaceous pterosaur with a wingspan of more than 6 m. (c) Comparison of the wings of a pterosaur, a bird, and a bat. In the pterosaur the fourth finger provides the support for the wing, in the bird the second finger provides support, and in the bat the second through fifth finger support the wing. (d) Skulls of pterosaurs: (1) had baleenlike teeth and may have strained plankton from seawater; (2) was a tiny pterosaur that may have eaten insects with its peglike teeth; and (3) and (4) probably ate fish since fish bones have been found within their rib cages.

wingspan of at least 12 m. At least one pterosaur had a coat of hair or hairlike feathers (♦ Fig. 16.26). That this pterosaur had a hair- or feather-covered body and was a flier strongly suggests that it, and perhaps all pterosaurs, were endotherms.

Were pterosaurs, particularly the larger ones, active, wing-flapping fliers or simply gliders? Experiments with scale models indicate that *Pteranodon* (Fig. 16.25b) could actively fly, but once airborne probably took advantage of thermal updrafts to stay aloft and ranged far out over the open sea, mostly by soaring. Perhaps much like the present-day frigate bird, *Pteranodon* glided close to the surface and plucked fish from the water with its long beak. Most paleontologists agree that the small pterosaurs were wing-flapping fliers.

The function of the large crest on the head of *Pteranodon* is uncertain. It may have functioned aerodynamically in maneuvering or braking, or it may have served as a counterbalance for the long beak. Some specimens seem to lack this feature, indicating that it may have been a sexual characteristic of some sort. If so, whether males or females had crests is unknown.

Marine Reptiles

For most people **ichthyosaurs** are probably the most familiar of the Mesozoic marine reptiles (♦ Fig. 16.27). These animals were about 3 m long and were completely aquatic. Aquatic adaptations included a streamlined, somewhat fishlike body, a powerful tail for propulsion, and flipperlike forelimbs for maneuvering. The numerous sharp teeth indicate that ichthyosaurs were fish eaters.

Ichthyosaurs were so thoroughly aquatic that it is doubtful they could venture onto land at all. This poses a reproductive problem since reptile eggs will not survive if laid in water. Female ichthyosaurs probably retained the eggs in their bodies and gave birth to live young. Some fossils with young ichthyosaurs within the body cavity support this interpretation.

A second group of Mesozoic marine reptiles, the **plesiosaurs** occurred in two varieties, short necked and long necked (Fig. 16.27). Most plesiosaurs were between 3.6 and 6 m long, but one species from Antarctica measures 15 m. Long-necked plesiosaurs were heavy-bodied animals with mouthfuls of sharp teeth and limbs specialized into oarlike paddles. They probably rowed themselves through the water and may have used their long necks in snakelike fashion to capture fish. Plesiosaurs probably

came ashore to lay their eggs. No doubt they were clumsy on land but no more so than living walruses or seals. Another group of Mesozoic marine reptiles, the mosasaurs, existed only during the Cretaceous (see Perspective 16.2).

Crocodiles, Turtles, Lizards, and Snakes

By Jurassic time, crocodiles had replaced phytosaurs (Fig. 16.16) as the dominant freshwater predators. All crocodiles are amphibious, spending much of their time in water, but they are also well equipped for walking on land. Overall, crocodile evolution has been conservative, involving changes mostly in size from a meter or so in Jurassic forms to 15 m in some Cretaceous species.

Turtles, too, have been evolutionarily conservative since their appearance during the Triassic. The most

---

♦ **FIGURE 16.26** *Sordes pilosus*, whose name means "hairy devil," was a small Jurassic pterosaur with an insulating coat of hair or hairlike feathers.

◆ **FIGURE 16.27** Mesozoic marine reptiles. (a) a long-necked plesiosaur and (b) Ichthyosaurs.

(a)

remarkable feature of turtles is their heavy, bony armor; turtles are more thoroughly armored than any other vertebrate animal, living or fossil. Turtle ancestry is uncertain. One Permian animal had eight broadly expanded ribs, which may represent the first stages in the development of turtle armor.

Lizards and snakes are closely related to one another, and, in fact, lizards were ancestral to snakes. The limbless condition in snakes (some lizards are limbless, too) and skull modifications that allow snakes to open their mouths very wide are the main differences between these two groups. Lizards are known from Upper Permian strata, but they did not become abundant until the Late Cretaceous. Snakes first appear during the Cretaceous, but the families to which most living snakes belong differentiated since the Early Miocene. Although their Cretaceous and Early Tertiary fossil record is poor, one Early Cretaceous genus from Israel appears to show characteristics intermediate between snakes and their lizard ancestors.

❖ BIRDS

In Chapter 6 we discussed the origin of birds to illustrate how fossils provide evidence for evolution. We pointed out that *Archaeopteryx* (from Jurassic strata in Germany) has such avian features as a wishbone and feathers (◆ Fig. 16.28a), but otherwise it more closely resembles the small carnivorous dinosaurs.

Until recently, *Archaeopteryx* was the only known pre-Cretaceous bird, but the discovery of fossils of two crow-sized individuals called *Protoavis* has perhaps changed that situation. These fossils are from Triassic rocks, so they predate *Archaeopteryx,* and some investigators think they were more birdlike than *Archaeopteryx. Protoavis* had hollow bones and the breastbone structure of birds, but because no impressions of feathers were found on these specimens, some investigators believe they may simply have been small carnivorous dinosaurs. If *Protoavis* proves to be a bird rather than a reptile, it would imply that birds evolved earlier than originally thought.

(b)

Few Mesozoic birds other than *Archaeopteryx* and *Protoavis* are known, but two have recently been discovered. One specimen, from China, is slightly younger than *Archaeopteryx* and possesses both primitive and advanced characteristics. For example, it retains abdominal ribs similar to those of *Archaeopteryx* and the theropod dinosaurs, but it has a reduced tail typical of present-day birds (◆ Fig. 16.28b). Another specimen, this one from Spain, is 20 to 30 million years younger than *Archaeopteryx;* it, too, is a mix of primitive and advanced characteristics, but it does appear to lack abdominal ribs.

Two other Late Cretaceous birds are known from rocks in Kansas. One bird, *Hesperornis,* was a swimming bird with powerful kicking feet and vestigial wings (◆ Fig. 16.29). *Ichthyornis* was an active flyer with well-developed wings, who seems to have lived much like terns. Neither of these two genera had the long-tail characteristic of *Archaeopteryx,* but both seem to have had teeth. Teeth in birds were completely lost by the end of the Mesozoic, since no known Cenozoic bird had teeth.

❖ FROM REPTILE TO MAMMAL

Therapsids, or the advanced mammal-like reptiles, were briefly described in Chapter 14. These reptiles diversified into numerous species of herbivores and carnivores, and during the Permian they were the dominant terrestrial vertebrates. One particular group of carnivorous therapsids called **cynodonts** was the most mammal-like of all and by the Late Triassic gave rise to the class Mammalia (Fig. 16.14).

Therapsids and the Origin of Mammals

The transition from cynodonts to mammals is well documented by fossils and is so gradational that classification of some fossils as either reptile or mammal is difficult. We can easily recognize living mammals as those warm-blooded animals with hair or fur that have mammary glands and, except for the platypus and spiny anteater, give birth to live young (◆ Fig. 16.30).

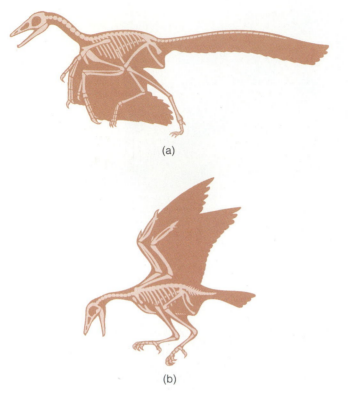

(a)

(b)

◆ **FIGURE 16.28** (a) *Archaeopteryx*, a Jurassic age bird from Germany, has feathers and a wishbone and is therefore classified as a bird. In almost all other anatomical features, however, it more closely resembles theropod dinosaurs. (b) This sparrow-sized bird, which was recently found in China, has a shortened tail, a characteristic more typical of birds.

Obviously, most of these criteria for recognizing living mammals are inadequate for classifying fossils. For them, we must use skeletal structure only. Several skeletal modifications characterize the transition from mammal-like reptiles to mammals, but distinctions between the two groups are based largely on details of the middle ear, the lower jaw, and the teeth (Table 16.2).

Reptiles have only one small bone in the middle ear—the stapes—while mammals have three—the incus, the malleus, and the stapes. Also, the lower jaw of a mammal is composed of a single bone called the *dentary*, but a reptile's jaw is composed of several bones (◆ Fig. 16.31, p. 460). In addition, a reptile's jaw is hinged to the skull at a contact between the articular and quadrate bones, while in mammals the dentary contacts the squamosal bone of the skull (Fig. 16.31).

During the transition from cynodonts to mammals, the quadrate and articular bones that had formed the joint between the jaw and skull in reptiles were modified into the incus and malleus of the mammalian middle ear (Fig. 16.31). Fossils clearly document the progressive enlargement of the dentary until it became the only element in the mammalian jaw (◆ Fig. 16.32, p. 460). Likewise, a progressive change from the reptile to mammal jaw joint is documented by fossil evidence. In fact, some of the most advanced cynodonts were truly transitional because they had a compound jaw joint consisting of (1) the articular and quadrate bones typical of reptiles and (2) the dentary and squamosal bones as in mammals (Table 16.2).

In Chapter 6 we noted that the study of embryos provides evidence for evolution. Opossum embryos clearly show that the middle ear bones of mammals were orig-

◆ **FIGURE 16.29** A Cretaceous bird from Kansas. Restoration of *Hesperornis*, a toothed, aquatic bird with vestigial wings. (Photo courtesy of the Canadian Museum of Nature, Ottawa, Canada; painting by Eleanor M. Kish.)

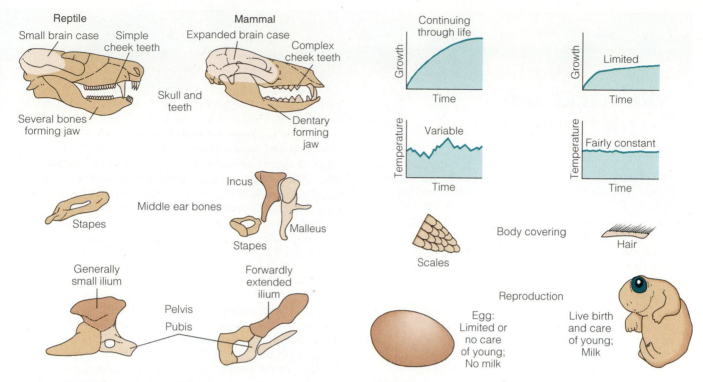

◆ **FIGURE 16.30** Some contrasting characteristics of reptiles and mammals. Most living mammals can be easily identified by these characteristics. The platypus and spiny anteater, however, both lay eggs, but in most other respects are mammals.

TABLE 16.2 Summary Chart Showing Some Characteristics and How They Changed during the Transition from Reptiles to Mammals

	TYPICAL REPTILE	CYNODONT	MAMMAL
Lower Jaw	Dentary and several other bones.	Dentary enlarged, other bones reduced.	Dentary bone only, except in earliest mammals.
Jaw-Skull Joint	Articular-quadrate	Articular-quadrate; some advanced cynodonts had both the reptile jaw-skull joint and the mammalian jaw-skull joint.	Dentary-squamosal
Middle Ear Bones	Stapes	Stapes	Stapes, incus, malleus
Secondary Palate	Absent	Partially developed	Well developed
Teeth	No differentiation	Some differentiation	Fully differentiated
Cold-versus Warm-Blooded	Cold-blooded	Probably warm-blooded	Warm-blooded

MOSASAURS AND PALEOPATHOLOGY

The mosasaurs were a group of Late Cretaceous marine lizards related to the present-day Komodo dragon or monitor lizard. Mosasaurs ranged from small species only 2.5 m long to giants such as *Tylosaurus,* measuring more than 9 m long (◆ Fig. 1). Although mosasaurs had a worldwide distribution, they are best known from the Niobrara Chalk of Kansas, which was deposited in the Cretaceous Interior Seaway (see Fig. 15.11).

Mosasaur limbs resembled paddles and were probably used mostly for maneuvering, while their long tail provided propulsion. All were predators, and the preserved stomach contents of one specimen include the diving bird *Hesperornis* (Fig. 16.29), a marine bony fish, a shark, and part of a smaller mosasaur. Other specimens show that mosasaurs also ate several varieties of invertebrates including ammonoids. One genus had crushing teeth and is thought to have eaten bivalves.

Many mosasaur bones from the Niobrara Chalk show evidence of shark tooth marks, but in most cases it cannot be determined whether sharks preyed on live mosasaurs or simply fed on carcasses. In one case, however, a mosasaur shows a condition known as infectious spondylitis (an inflammation caused by a microorganism), which caused seven vertebrae to fuse together. The tip of a shark tooth found within these fused bones clearly indicates that this individual had been attacked by a shark and survived.

The bones of some mosasaurs show the pathological condition known as avascular necrosis, an affliction characterized by zones of dead bone resulting from too little blood being supplied to a part of the skeleton. This condition is well known in humans but had not previously been detected in any other animal. In humans there are only three known causes of avascular necrosis: exposure to radiation, bismuth poisoning, and decompression syndrome, or what is more commonly called the bends. In humans and presumably in mosasaurs too, decompression syndrome results when a diver surfaces too rapidly and nitrogen bubbles in the blood disrupt the blood supply. No evidence has been found to indicate that avascular necrosis in mosasaurs was caused by either radiation or any kind of poisoning.

Evidence for avascular necrosis occurs in the vertebrae of mosasaurs worldwide. Not all genera were affected, however, and in those showing signs of this condition, the incidence varied. For example, *Clidastes* is thought to have been a near-surface dweller, and no specimen examined showed evidence of avascular necrosis. Based on studies of its anatomy, *Platecarpus* is thought to have been a deep diver, and these animals showed the highest incidence of the condition. All tylosaurs also suffered from avascular necrosis, but fewer vertebrae were affected than in *Platecarpus.*

Avascular necrosis is not restricted to mosasaurs; it has also been detected in several varieties of Mesozoic and Cenozoic marine turtles and is known in some living turtles. Other Mesozoic marine reptiles, however, show no evidence of this syndrome, even though some ichthyosaurs were probably deep divers. Apparently, ichthyosaurs and plesiosaurs evolved features to avoid this condition whereas mosasaurs did not. Although many mosasaurs show signs of avascular necrosis, and none possessed any modifications for deep diving, these conditions had nothing to do with their extinction at the end of the Cretaceous Period. Whatever caused the extinction of dinosaurs, flying reptiles, and other marine reptiles also was responsible for the extinction of mosasaurs.

---

inally part of the jaw. In fact, even when opossums are born, the middle ear elements are still attached to the dentary (Fig. 16.31d), but as they develop further, these elements migrate to the middle ear, and a typical mammal jaw joint develops.

Several other aspects of cynodonts also indicate they were ancestral to mammals. Their teeth were somewhat differentiated into distinct types in order to perform specific functions. In mammals the teeth are fully differentiated into incisors, canines, and chewing teeth, but typical reptiles do not have differentiated teeth. Another mammalian feature, the secondary palate, was partially developed in advanced cynodonts (◆ Fig. 16.33). This secondary palate is a bony shelf above the mouth that separates the nasal passages from the mouth cavity. It is an adaptation for eating and breathing at the same time, a necessary requirement for endotherms with their high demands for oxygen.

◆ **FIGURE 1** *Tylosaurus* was a large mosasaur from the Late Cretaceous. It measured up to 9 m long.

❖ MESOZOIC MAMMALS

Even though mammals appeared during the Late Trias-sic, their diversity remained low during the rest of the Mesozoic (◆ Fig. 16.34, p. 462; Table 16.3). A few Me-sozoic mammals may have been as large as raccoons, but most were quite small—about the size of shrews or rats. Their fossil record, which consists of isolated teeth, jaw fragments, and fragmentary bones, is not particularly good, although a few skulls and even some fairly com-plete skeletons are known. Nevertheless, these fossils are important because two-thirds of mammalian history is recorded by Mesozoic fossils.

The first mammals retained several reptilian character-istics, but had mammalian features as well. For example, the Triassic triconodonts (◆ Fig. 16.35, p. 462) had the

◆ **FIGURE 16.31** (a) The skull of an opossum showing the typical mammalian dentary-squamosal jaw joint. (b) The skull of a cynodont shows the articular-quadrate jaw joint of reptiles. (c) Enlarged view of an adult opossum's middle ear bones. (d) View of the inside of a young opossum's jaw showing that the elements of the middle ear are attached to the dentary during early development. This is the same arrangement of bones that is found in the adults of the ancestral mammals.

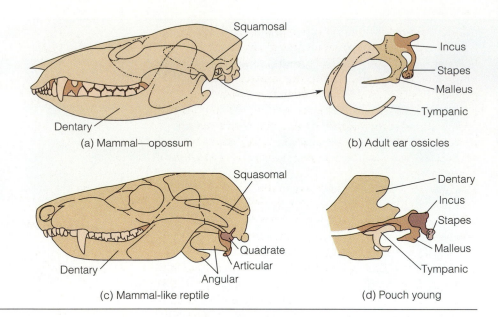

(a) Mammal—opossum

(b) Adult ear ossicles

(c) Mammal-like reptile

(d) Pouch young

◆ **FIGURE 16.32** During the transition from reptiles to mammals, the dentary bone (brown) became progressively larger until it became the only bone in the lower jaw.

Pelycosaur
(Primitive mammal-like reptile)

Primitive therapsid

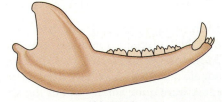

Advanced therapsid, a cynodont

Mammal

fully differentiated teeth typical of mammals, but they had both the reptile and mammal types of jaw joints. The earliest symmetrodonts retained several reptilian bones in the lower jaw, but by the Cretaceous only the dentary was present. In short, some mammalian features evolved more rapidly than others, thereby accounting for animals that possessed characteristics of both reptiles and mammals. Recall from Chapter 6 that evolution in different features of the same organism at such varying rates is called *mosaic evolution*.

Figure 16.34 shows that early mammals diverged into two distinct branches. One branch includes the triconodonts and their probable evolutionary descendants, the **monotremes,** or egg-laying mammals. Living monotremes are the platypus and spiny anteater of the Australian region. Also included in this branch are the *multituberculates,* the first mammalian herbivores (◆ Fig. 16.36); all other Mesozoic mammals probably preyed on insects, worms, and grubs. Multituberculates seem to have lived much like present-day rodents and, in fact, were replaced by rodents during the Early Cenozoic.

The second evolutionary branch shown in Figure 16.34 includes the symmetrodonts and the **eupantotheres** and their descendants—the **marsupial** (pouched) **mammals** and the **placental mammals.** All living mammals except monotremes have ancestries that can be traced back through this branch.

TABLE 16.3 Classification of Mesozoic Mammals

CLASS MAMMALIA	LIVING TYPES
Subclass Prototheria—Egg-laying mammals	
Order Docodonta	
Order Tricondonta	
Order Monotremata	Platypus, spiny anteater
Sublcass Allotheria	
Order Multituberculata	
Subclass Theria—Mammals that give birth to live young	
Order Symmetrodonta	
Order Eupantotheria	
Order Marsupialia	Opossum, kangaroo, wombat
Order Creodonta	
Order Condylartha	
Order Insectivora	Shrew, mole, hedgehog

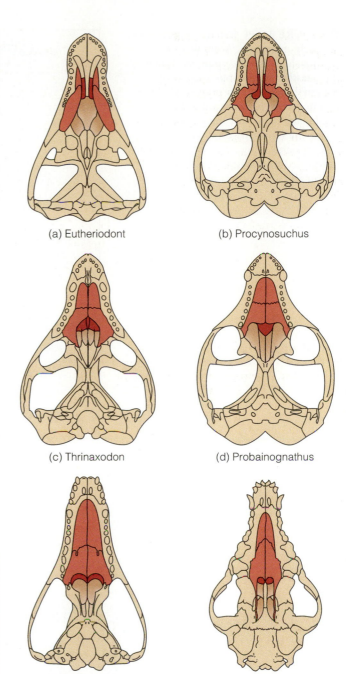

(a) Eutheriodont

(b) Procynosuchus

(c) Thrinaxodon

(d) Probainognathus

(e) Morganucodon

(f) Modern mammal (dog)

◆ **FIGURE 16.33** Views of the bottoms of skulls of (a) an early therapsid (b through d) three cynodonts, (e) an early mammal, and (f) a living mammal, showing the progressive development of the bony secondary palate (brown).

◆ **FIGURE 16.34** Relationships among the early mammals. A number of relationships are uncertain, as shown by the question marks, but apparently mammalian evolution proceeded along two branches. One branch led to the egg-laying mammals, the monotremes, which are represented by the living platypus and spiny anteater. The other branch led to marsupial and placental mammals.

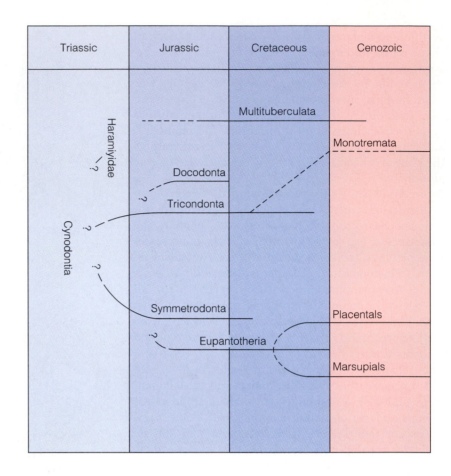

Triassic	Jurassic	Cretaceous	Cenozoic

Haramiyidae

Multituberculata

Monotremata

Docodonta

Tricondonta

Cynodontia

Symmetrodonta

Placentals

Eupantotheria

Marsupials

◆ **FIGURE 16.35** One of the earliest mammals, the triconodont *Triconodon.*

Eupantotheres were shrew-sized animals with a poor fossil record, but details of their teeth indicate they were ancestral to both marsupial and placental mammals. Furthermore, one eupantothere fossil possesses bones projecting from the pelvis that could have supported a pouch, so in this respect they were similar to living marsupials. The divergence of marsupials and placentals from a common ancestor probably occurred during the Early Cretaceous, but undoubtedly both were present by the Late Cretaceous. The earliest known placental mammals were members of the order *Insectivora* (◆ Fig. 16.37), an order represented today by shrews, moles, and hedgehogs.

❖ MESOZOIC CLIMATES AND PALEOGEOGRAPHY

The present continental positions and climatic patterns largely restrict the distribution of organisms. Land plants and animals have little opportunity to colonize distant areas because of physical barriers (especially the ocean basins) and climatic barriers. Mesozoic barriers to migration were apparently not as effective as they are now, because some Mesozoic organisms are known from areas that are now widely separated.

Fragmentation of the supercontinent Pangaea began by the Late Triassic and continues to the present, but during much of the Mesozoic, close connections existed

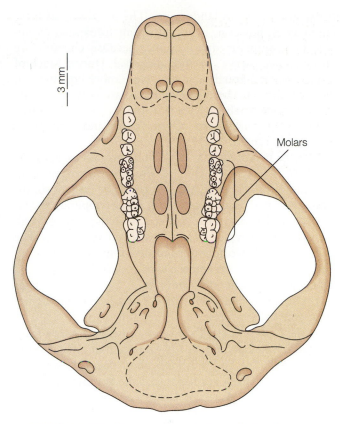

Molars

◆ **FIGURE 16.36** Bottom view of the skull of a Cretaceous multituberculate from Mongolia. Multituberculates are so called because of the numerous rounded projections or tubercules on their molars.

◆ **FIGURE 16.37** The oldest known placental mammals were members of the order Insectivora such as those in this scene from the Late Cretaceous. These animals probably fed on insects, worms, and grubs.

between the various landmasses. The proximity of these landmasses, however, is not sufficient to explain Mesozoic biogeographic distributions, because climates are also effective barriers to wide dispersal. During much of the Mesozoic, though, climates were more equable and lacked the strong north and south zonation characteristic of the present. In short, Mesozoic plants and animals had greater opportunities to occupy much more extensive geographic ranges.

Pangaea persisted as a single unit through most of the Triassic. The Triassic climate was warm-temperate to tropical, although some areas, such as the present southwestern United States, were arid. Mild temperatures extended 50 degrees north and south of the equator, and even the polar regions may have been temperate. The fauna was truly worldwide in its distribution. Phytosaurs (Fig. 16.16) were present in North America, Europe, and Madagascar. Some dinosaurs had continuous ranges across Laurasia and Gondwana. The peculiar gliding lizards were in New Jersey and England.

By the Late Jurassic, Laurasia had become partly fragmented by the opening North Atlantic, but a connection still existed. The South Atlantic had begun to open so that a long, narrow sea separated the southern parts of Africa and South America. Otherwise the southern continents were still close together.

The mild Triassic climate persisted into the Jurassic. Ferns, whose living relatives are now restricted to the tropics of southeast Asia, are known from areas as far as 63° south latitude and 75° north latitude. Dinosaurs roamed widely across Laurasia and Gondwana. Specimens from the Morrison Formation in western North America and those in the Tendagura beds of eastern Africa are quite similar. For example, the giant dinosaur *Brachiosaurus* (Fig. 16.21) is known from both areas. Stegosaurs (Fig. 16.21) and some families of carnivorous dinosaurs lived throughout Laurasia and in Africa.

By the Late Cretaceous, the North Atlantic had opened further, and Africa and South America were completely separated. South America remained an island continent until late in the Cenozoic. Its fauna, evolving in isolation, became increasingly different from faunas of the other continents. Marsupial mammals reached Australia from South America via Antarctica, but the South American connection was eventually severed. Placentals, other than bats and a few rodents, never reached Australia. This explains why the marsupials continue to dominate the continent's fauna even today.

Cretaceous climates were more strongly zoned by latitude, but they remained warm and equable until the close of that period. Climates then became more seasonal and cooler, a trend that persisted into the Cenozoic. Dinosaur and mammal fossils demonstrate that interchange was still possible, especially between the various components of Laurasia.

As one might expect, the paleontology of Antarctica is poorly known. Marine reptiles, including plesiosaurs 15 m long have been found on Seymour Island near the tip of the Antarctic Peninsula, but since they were marine reptiles, their presence tells us little about continental connections. Nevertheless, Antarctica remained close to South America well into Cenozoic time, as indicated by the fact that a land-dwelling marsupial of Eocene age has been found there. An Early Tertiary coal bed on Seymour Island indicates that the climate was temperate.

❖ MASS EXTINCTIONS—A CRISIS IN THE HISTORY OF LIFE

Extinctions have occurred continuously throughout the history of life. This so-called *background extinction* differs in degree from **mass extinctions**, which are times of accelerated extinction rates. One such mass extinction event occurred at the close of the Mesozoic, an event second in magnitude only to the extinctions at the end of the Paleozoic (see Chapter 13). Casualities of the Mesozoic extinction event included dinosaurs, flying reptiles, marine reptiles, and several kinds of marine invertebrates. Among the latter were the ammonites, which had been so abundant through the Mesozoic, the rudistid clams, and some planktonic organisms.

Numerous ideas have been proposed to explain Mesozoic extinctions, but most have been dismissed as improbable or untestable. A new proposal was made recently based on a discovery at the Cretaceous-Tertiary boundary in Italy—a clay layer 2.5 cm thick, with an abnormally high concentration of the platinum group element iridium (◆ Fig. 16.38). Since this discovery, high iridium concentrations have been identified at many other Cretaceous-Tertiary boundary sites. The significance of this discovery lies in the fact that iridium is rare in crustal rocks but occurs in much higher concentrations in some meteorites. Several investigators proposed a meteorite impact to explain this iridium anomaly and further postulated that the impact of a large meteorite, perhaps 10 km in diameter, set in motion a chain of events that led to extinctions (◆ Fig. 16.39). Some Cretaceous-Tertiary boundary sites also contain soot and shock-metamorphosed quartz grains, both of which are cited as further evidence of an impact event. The meteorite-impact scenario goes something like this. Upon impact, about 60 times the mass of the meteorite was blasted from the Earth's crust high into the atmosphere, and the heat generated at impact started raging fires that added more particulate matter to the atmosphere. Sunlight was blocked for several months, causing a temporary cessation of photosynthesis; food chains collapsed, and extinctions followed. In addition, with sunlight greatly diminished, the Earth's surface temperatures were drastically reduced and could have added to the biologic stress.

The iridium anomaly is real, but its origin and significance are debatable. We know very little about the distribution of iridium in crustal rocks or how it may be distributed and concentrated. Some geologists suggest that the iridium was derived from within the Earth by volcanism, but it is not conclusively supported by evidence.

Some investigators now claim that they have found the probable impact site centered on the town of Chicxulub on the Yucatán Peninsula of Mexico. The structure lies beneath layers of sedimentary rock, so it has been detected only in drill holes and by geophysical work. For example, magnetic data indicate that the structure is symmetrical and measures 180 km in diameter. Furthermore, it appears to be the right age. Evidence supporting the conclusion that the Chicxulub structure is an impact crater includes shocked quartz, what appear to be the deposits of huge waves, and tektites, which are small pieces of rock that were melted during the proposed impact and hurled into the atmosphere. Although an impact origin for this structure is gaining acceptance, some geologists think it is some kind of volcanic feature.

Even if a meteorite did hit the Earth, did it lead to these extinctions? If so, both terrestrial and marine extinctions must have occurred at the same time. To date, strict time equivalence between terrestrial and marine extinctions has not been demonstrated. The selective nature of the extinctions is also a problem. In the terrestrial realm, large animals were the most drastically affected, but not all dinosaurs were large, and crocodiles, close relatives of dinosaurs, were unaffected. Likewise, tropical plants seem to have suffered no ill effects from the supposed lower temperatures. This is puzzling because they have little resistance to cold. Some paleontologists think that dinosaurs, some marine invertebrates, and many plants were already on the decline and headed for extinction before the end of the Cretaceous. A meteorite impact, if one actually occurred, may have simply hastened the process. There is even some evidence indicating that dinosaurs survived into the Early Cenozoic, several tens of thousands of years after the proposed impact.

Investigators at the University of Chicago have proposed that the Mesozoic extinction event was only one of several to occur at 26-million-year intervals during the last 250 million years (◆ Fig. 16.40). One possible cause of these cyclic extinctions is periodic meteorite showers. Some investigators have suggested that a companion star to the Sun with a highly eccentric orbit could provide a

◆ FIGURE 16.38 Stratigraphy of the Cretaceous-Tertiary section in Italy. The 2.5 cm thick boundary clay shows an abundance of iridium. Even though iridium is present in crustal rocks in quantities measured in parts per billion (ppb), its concentration increases markedly at the Cretaceous-Tertiary boundary.

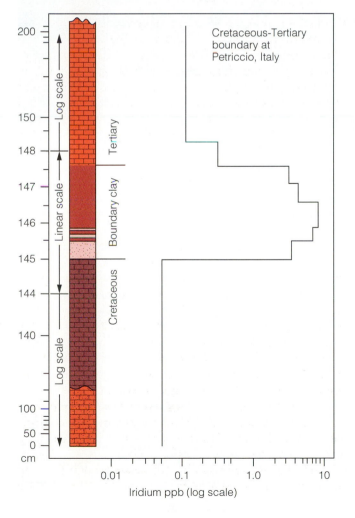

◆ **FIGURE 16.39** Proposed meteorite impact at the end of Cretaceous. (Photo courtesy of the Canadian Museum of Nature, Ottawa, Canada; painting by Eleanor M. Kish.)

◆ **FIGURE 16.40** Record of extinctions for 9,773 genera of marine fossil animals. According to the calculations of investigators at the University of Chicago, mass extinctions have occurred at 26-million-year intervals shown by the peaks on the graph.

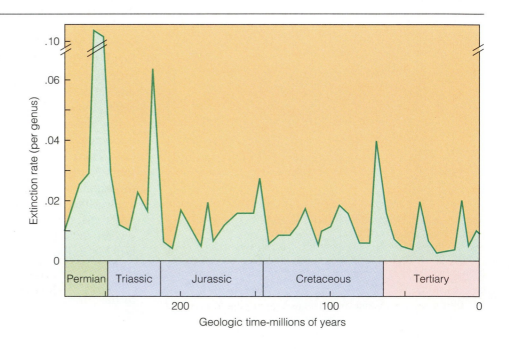

mechanism for periodic meteor showers. In this scenario, when this star is close to the Sun, it perturbs cometary orbits, thereby causing terrestrial impacts and mass extinctions.

Since attributing one mass extinction to a meteorite impact has generated controversy, one can imagine just how controversial this new proposal is. The evidence for cyclic mass extinctions shown in Figure 16.40 seems compelling, but not all paleontologists are convinced. The critics of cyclic mass extinctions think that some minor extinction events have been overemphasized. Furthermore, they point out that if mass extinctions are caused by periodic meteorite impacts, iridium anomalies corresponding with extinction events should be present, but few such anomalies have been identified.

In the final analysis, we have no widely accepted explanation for Mesozoic extinctions. We do know that vast shallow seas occupied large parts of the continents during the Cretaceous, and that by the latest Cretaceous they had largely withdrawn. We also know that the mild, equable climates of the Mesozoic became harsher and more seasonal by the end of the Mesozoic. Changes such as these seem adequate to many paleontologists to explain Mesozoic mass extinctions. But the fact remains that this extinction event was very selective, and no explanation accounts for all aspects of this crisis in the history of life.

Chapter Summary

Table 16.4 is a summary of biological and physical events for the Mesozoic Era.

1. Among the marine invertebrates, survivors of the Permian extinction event diversified and gave rise to increasingly complex Mesozoic marine invertebrate communities.

2. Some of the most abundant and diverse invertebrates were ammonites and foraminifera.

3. Triassic and Jurassic land-plant communities were composed of seedless plants and gymnosperms. Angiosperms, or flowering plants, appeared during the Early Cretaceous, diversified rapidly, and soon became the dominant land plants.

4. The Triassic thecodontian reptiles included crocodile-like predators, armored herbivores, and small bipedal carnivores that were ancestral to dinosaurs.

5. Dinosaurs appeared during the Late Triassic, but were most abundant and diverse during the Jurassic and Cretaceous. Based on pelvic structure, two distinct orders of dinosaurs are recognized—Saurischia (lizard-hipped) and Ornithischia (bird-hipped).

6. The carnivores and the giant quadrupeds were saurischian dinosaurs. The ornithischians, some of which were heavily armored, were more diverse than the saurischians. All ornithischians were herbivores.

7. Bone structure, brain size, and predator-prey relationships have been cited as evidence for endothermy in dinosaurs. Considerable disagreement on dinosaur endothermy exists, but a fairly good case can be made for endothermic, small, carnivorous dinosaurs, because they are closely related to birds.

8. Pterosaurs were the first flying vertebrate animals. Small pterosaurs were probably active, wing-flapping fliers, while large ones may have depended more on thermal updrafts and soaring to stay aloft. At least one pterosaur species had hair or feathers, so it was very likely endothermic.

9. The fish-eating, porpoise-like ichthyosaurs were thoroughly adapted to an aquatic life. Female ichthyosaurs probably retained eggs within their bodies and gave birth to live young. Plesiosaurs were heavy-bodied marine reptiles that probably came ashore to lay eggs.

10. During the Jurassic, crocodiles replaced phytosaurs as the dominant freshwater predators. Turtles and lizards were present through most of the Mesozoic. Snakes appeared during the Cretaceous.

11. Birds probably evolved from small carnivorous dinosaurs. The oldest known bird, *Archaeopteryx,* appeared during the Jurassic, but few other Mesozoic birds are known. Recent finds of *Protoavis* in Triassic rocks may represent a bird older than *Archaeopteryx.*

12. Among the mammal-like reptiles, the cynodonts show a transformation from the reptilian to the mammalian condition. The earliest mammals evolved during the Late Triassic, but they are difficult to distinguish from advanced cynodonts. Details of the teeth, the middle ear, and lower jaw are used to distinguish the two.

13. Several types of Mesozoic mammals existed, but all were small, and their diversity was low. A group of Mesozoic mammals called eupantotheres gave rise to both marsupials and placentals during the Cretaceous.

14. The same types of some Mesozoic animals are found as fossils in areas that are now widely separated. Because the continents were close together during much of the Mesozoic and climates were mild even at high latitudes, animals and plants dispersed very widely.

(continued on page 470)

TABLE 16.4 Summary of Biological and Physical Events for the Mesozoic Era

Age (Millions of Years)	Geologic Period	Invertebrates	Vertebrates
66 —			
	Cretaceous	Continued diversification of ammonites and belemnoids. Rudist become major reef-builders. Extinction of ammonites, rudists, and most planktonic foraminifera at end of Creataceous.	Extinctions of dinosaurs, flying reptiles, and marine reptiles. Placental and marsupial mammals diverge.
144 —			*GREATEST DIVERSITY OF DINOSAURS*
	Jurassic	Ammonites and belemnoid cephalopods increase in diversity. Scleractinian coral reefs common. Appearance of rudist bivalves.	First birds (may have evolved in Late Triassic). Time of giant sauropod dinosaurs.
208 —			
	Triassic	The seas are repopulated by invertebrates that survived the Permian extinction event. Bivalves and echinoids expand into the infaunal niche.	Mammals evolve from cynodonts. Cynodonts become extinct. Thecodontians give rise to dinosaurs. Flying reptiles and marine reptiles evolved.
245			

Plants	Climate	Plate Tectonics
Angiosperms evolve and diversify rapidly. Seedless plants and gymnosperms still common but less varied and abundant.	North-south zonation of climates more marked, but remains equable. Climate becomes more seasonal and cooler at end of Cretaceous.	Further fragmentation of Pangaea. South America and Africa have separated. Australia separated from South America but remains connected to Antarctica. North Atlantic continues to open.
Seedless vascular plants and gymnosperms only.	Much like Triassic. Ferns with living relatives restricted to tropics live at high latitudes, indicating mild climates.	Fragmentation of Pangaea continues, but close connections exist among all continents.
Land flora of seedless vascular plants and gymnosperms as in Late Paleozoic.	Warm-temperate to tropical. Mild temperatures extend to high latitudes; polar regions may have been temperate. Local areas of aridity.	Fragmentation of Pangaea begins in Late Triassic.

15. Mesozoic mass extinctions account for the disappearance of dinosaurs, several other groups of reptiles, and a number of marine invertebrates. Paleontologists do not agree on the cause or causes of Mesozoic extinctions. One hypothesis holds that the extinctions were caused by the impact of a large meteorite with the Earth. Many paleontologists reject the meteorite proposal and claim that withdrawal of epeiric seas and climatic changes can account for this extinction event.

◆————————————————————————————◆

Important Terms

angiosperm
bipedal
cynodont
ectotherm
endotherm
eupantothere

ichthyosaur
marsupial mammal
mass extinction
monotreme
ornithischian
placental mammal

plesiosaur
pterosaur
quadrupedal
saurischian
thecondontian
therapsid

◆————————————————————————————◆

Review Questions

1. The mammal-like reptiles that were most like mammals were the:
   a. _____ cynodonts; b. _____ monotremes; c. _____ symmetrodonts; d. _____ thecodontians; e. _____ bipeds.

2. Sauropod dinosaurs were:
   a. _____ carnivorous; b. _____ fast-running bipeds; c. _____ the largest dinosaurs; d. _____ descendants of therapsids; e. _____ particularly varied and abundant during the Early Triassic.

3. The remains of _____comprise the Cretaceous chalk beds exposed in the White Cliffs of Dover, England.
   a. _____ foraminifera; b. _____ echinoids; c. _____ scleractinian corals; d. _____ coccolithophores, e. _____ rudist bivalves.

4. An important Mesozoic event in the history of land plants was the:
   a. _____ extinction of cycads; b. _____ origin of ferns; c. _____ first appearance of angiosperms; d. _____ dominance of ginkgos; e. _____ all of these.

5. All carnivorous dinosaurs are classified as:
   a. _____ phytosaurs; b. _____ theropods; c. _____ *Euparkeria*; d. _____ marsupials; e. _____ pachycephalosaurs.

6. The first vertebrate animals to fly were the:
   a. _____ bats; b. _____ insects; c. _____ mosasaurs; d. _____ pterosaurs; e. _____ ankylosaurs.

7. Which of the following is a marine reptile?
   a. _____ ichthyosaur; b. _____ ultrasaur; c. _____ brachyosaur; d. _____ triconodont; e. _____ multituberculate.

8. Which statement is correct?
   a. _____ *Archaeopteryx* had feathers on its wings, but the rest of its body was covered with scales; b. _____ *Archaeopteryx* had reptile-like teeth; c. _____ *Protoavis* specimens have clear impressions of feathers; d. _____ all Mesozoic birds had long tails; e. _____ the oldest fossil birds come from Cretaceous rocks.

9. A typical reptile's jaw-skull joint is between which of these pairs of bones?
   a. _____ squamosal-secondary palate; b. _____ articular-quadrate; c. _____ ethmoid-occipital; d. _____ ilium-dentary; e. _____ stapes-incus.

10. The group of mammals directly ancestral to placentals and marsupials was probably the:
    a. _____ Eupantotheria; b. _____ Haramyidae; c. _____ Docodonta; d. _____ Insectivora; e. _____ Multituberculata.

11. Animals that walk on two legs are characterized as:
    a. _____ aquatic; b. _____ land dwellers; c. _____ bipeds; d. _____ therapsids; e. _____ warm-blooded.

12. Because of their rapid evolution and nektonic life-style, the _____are excellent Mesozoic guide fossils.
    a. _____ bivalves; b. _____ cephalopods; c. _____ cynodonts; d. _____ angiosperms; e. _____ trilobites.

13. The probable ancestors of dinosaurs and flying reptiles were:
    a. _____ gymnosperms; b. _____ phytosaurs; c. _____ ammonoids; d. _____ thecodontians; e. _____ ceratopsians.

14. Which of the following is a mammalian trait?
    a. _____ lower jaw composed of several bones;
    b. _____ articular-quadrate jaw-skull joint; c. _____
    middle ear with stapes, incus, and malleus; d. _____
    partially developed secondary palate; e. _____
    sprawling stance.
15. Briefly outline the major changes in the composition of
    Mesozoic marine invertebrate communities.
16. Discuss the changing aspects of land-plant communities
    during the Mesozoic.
17. What are the names of the two dinosaur orders, and
    how are they differentiated from one another?
18. Discuss several ways in which herbivorous dinosaurs
    protected themselves from predators.
19. What evidence indicates that some dinosaurs nested in
    colonies and used the same nesting sites repeatedly?
20. Briefly discuss the adaptations for flight seen in
    pterosaurs. Is there any evidence for endothermy in
    pterosaurs? If so, what?
21. In what ways was *Archaeopteryx* similar to and different
    from small carnivorous dinosaurs?
22. What skeletal features of cynodonts indicate that they
    were endotherms?
23. Explain what is meant when we say the earliest
    mammals show a mosaic evolution.
24. How does the breakup of Pangaea help us better
    understand the distribution of Mesozoic fossil plants and
    animals?
25. Briefly summarize the evidence for and against the
    proposal that a meteorite impact caused Mesozoic mass
    extinctions.

Additional Readings

Bakker, R. T. 1986. *The dinosaur heresies*. New York:
    William Morrow and Company.

Benton, M. 1991. The myth of Mesozoic cannibals. *New Sci-
    entist* 132, no. 1790: 40–44.

Donovan, S. K., ed. 1989. *Mass extinctions*. New York: Co-
    lumbia University Press.

Hopson, J. A. 1987. The mammal-like reptiles: A study of
    transitional fossils. *The American Biology Teacher* 49,
    no. 1: 16–26.

Horner, J. R. 1984. The nesting behavior of dinosaurs.
    Scientific American 250, no. 4: 130–37.

Hsü, H. J. 1986. *The great dying*. New York: Harcourt Brace
    Jovanovich.

Lambert, D. 1983. *A field guide to dinosaurs*. New York:
    Avon Books.

Langston, W. 1981. Pterosaurs. *Scientific American* 244,
    no. 1: 122–36.

Martin, L. D., and B. M. Rothschild. 1989. Paleopathology
    and diving mosasaurs. *American Scientist* 77, no. 5:
    460–67.

McGowan, C. 1991. *Dinosaurs, spitfires, and sea dragons*.
    Cambridge, Mass.: Harvard University Press.

Monastersky, R. 1992. Closing in on the killer. *Science News*
    141, no. 5: 72–75.

Russell, D. A. 1988. *An odyssey in time: The dinosaurs of
    North America*. Toronto: Ont.: University of Toronto
    Press.

Savage, R. J. G. 1986. *Mammalian evolution: An illustrated
    guide*. New York: Facts on File Publications.

Stewart, W. N. 1983. *Paleobotany and the evolution of
    plants*. New York: Cambridge University Press.

Weishampel, D. B. P. Dodston, and H. Osmóska, eds. 1990.
    The Dinosauria. Berkeley, Calif.: University of California
    Press.

Wellenhofer, P. 1990. *Archaeopteryx*. *Scientific American*
    262, no. 5: 70–77.

Wilford, J. N. 1985. *The riddle of the dinosaurs*. New York:
    Alfred A. Knopf.

CHAPTER 17

Shiprock is a volcanic neck 550 m high in northwest New Mexico. It formed about 27 million years ago during the Oligocene Epoch. (Photo courtesty of Sue Monroe.)

CENOZOIC GEOLOGIC HISTORY: TERTIARY PERIOD

Prologue

During the Eocene Epoch, several small plutons were emplaced in northeastern Wyoming. The best known of these, Devil's Tower, was established as our first national monument by President Theodore Roosevelt in 1906. The tower rises nearly 260 m above its base and stands more than 390 m above the floodplain of the nearby Belle Fourche River (◆ Fig. 17.1). Visible from 48 km away, it served as a landmark for early travelers in the area.

The Cheyenne and Lakota Sioux Indians called the tower Mateo Tepee, which means "Grizzly Bear Lodge". It was also called the "Bad God's Tower," and reportedly, "Devil's Tower" was a translation of this phrase. According to one Indian legend, the tower formed when the Great Spirit caused it to rise up from the ground, carrying with it several Indian children who were trying to escape from a gigantic grizzly bear. Another legend tells of six brothers and a woman who were also being pursued by a grizzly bear. The youngest brother carried a small rock, and when he sang a song, the rock grew to the present size of Devil's Tower. In both legends, the bear's attempts to reach the Indians, left deep scratches in the tower's rocks (◆ Fig. 17.2).

◆ **FIGURE 17.2** The origin of Devil's Tower according to Cheyenne legend. (Photo of painting by Herbert Collins, courtesy of Devil's Tower National Monument.)

◆ **FIGURE 17.1** Devil's Tower, Wyoming. The rubble at the base of the tower is an accumulation of collapsed columns.

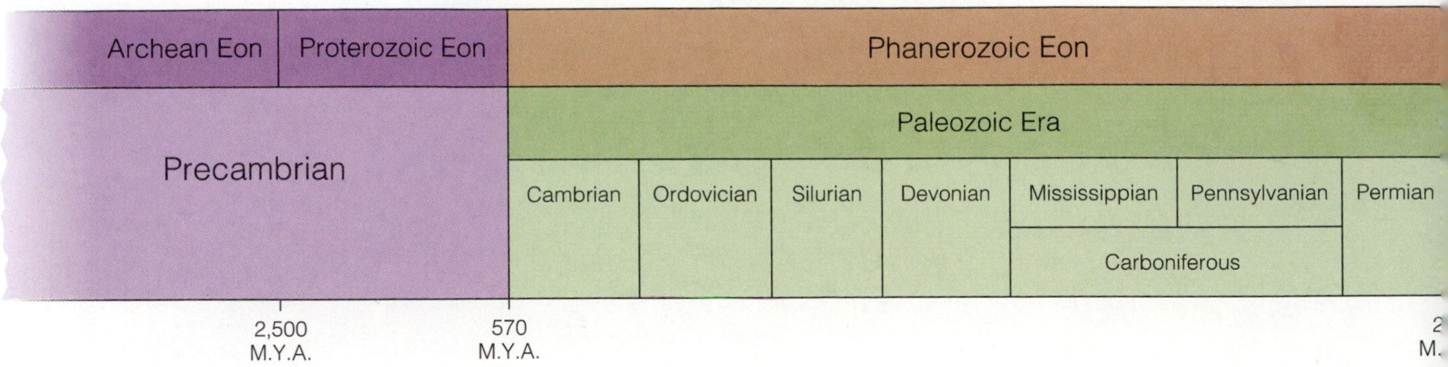

Archean Eon	Proterozoic Eon	Phanerozoic Eon						
Precambrian		Paleozoic Era						
		Cambrian	Ordovician	Silurian	Devonian	Mississippian	Pennsylvanian	Permian
						Carboniferous		

2,500 M.Y.A. 570 M.Y.A. 2... M.

Geologists have a less dramatic explanation for the tower's origin. The near vertical striations (the bear's scratch marks) are simply the lines formed by the intersection of columnar joints. Columnar joints form in magma or lava as it cools and contracts, thus forming columns. Many of the columns are six-sided, but four-, five-, and seven-sided columns occur as well. The larger columns measure about 2.5 m across, and the pile of rubble at the tower's base is an accumulation of collapsed columns (Fig. 17.1).

Geologists agree that Devil's Tower was emplaced as a small pluton, and subsequent erosion exposed it in its present form. The type of pluton and the extent of its modification by erosion are debated, however. Some geologists think Devil's Tower is the eroded remnant of a more extensive sill, or laccolith, while others think it is simply the magma that solidified in the neck of a volcano. Many similar, though less spectacular, small intrusions are found throughout the northern Black Hills of South Dakota. One of these, Bear Butte, is of particular religious and historical significance to the Cheyenne and Lakota Sioux.

❖ INTRODUCTION

Traditionally, the Cenozoic has been divided into two periods, the **Tertiary** (66 to 1.6 million years ago) and the **Quaternary** (1.6 million years ago to the present). Both periods are further divided into epochs (◆ Fig. 17.3). The terms Tertiary and Quaternary are widely used among geologists, but a different scheme for designations of Cenozoic time is becoming increasingly popular. This scheme also recognizes two periods, the *Paleogene* and *Neogene,* but they do not correspond directly to the Tertiary and Quaternary periods (Fig. 17.3). In this book we will follow the traditional usage.

◆ **FIGURE 17.3** The geologic time scale for the Cenozoic Era.

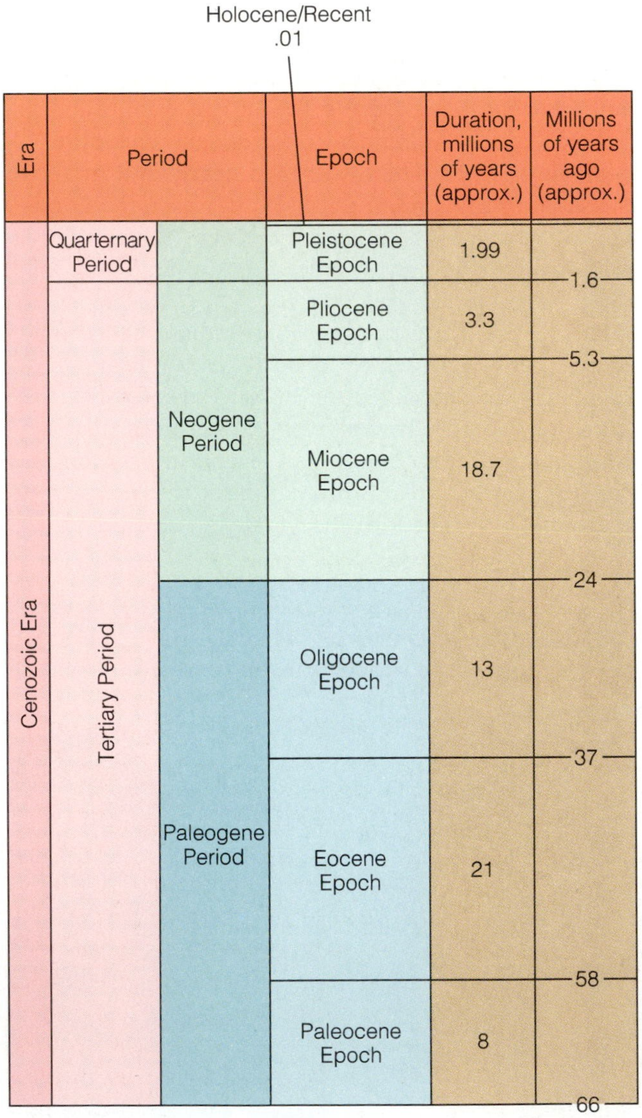

Holocene/Recent .01

Era	Period	Epoch	Duration, millions of years (approx.)	Millions of years ago (approx.)
Cenozoic Era	Quaternary Period	Pleistocene Epoch	1.99	
				1.6
	Tertiary Period — Neogene Period	Pliocene Epoch	3.3	
				5.3
		Miocene Epoch	18.7	
				24
	Tertiary Period — Paleogene Period	Oligocene Epoch	13	
				37
		Eocene Epoch	21	
				58
		Paleocene Epoch	8	
				66

474   Chapter 17   Cenozoic Geologic History: Tertiary Period

Phanerozoic Eon											
Mesozoic Era			Cenozoic Era								
Triassic	Jurassic	Cretaceous	Tertiary							Quaternary	
			Paleocene	Eocene	Oligocene	Miocene	Pliocene		Pleistocene	Holocene	

45
Y.A.

66
M.Y.A.

In Chapter 9 we noted that Precambrian time accounts for more than 87% of all geologic time. Thus, if all geologic time were represented by a 24-hour day, 21 hours would be the duration of the Precambrian. At 66 million years, the Cenozoic is brief, accounting for only about 20 minutes of our 24-hour day. Nevertheless, 66 million years is certainly long enough for significant evolution of the Earth and its biosphere to have occurred.

Many of the features of the Earth have long histories, but the present distribution of land and sea and the topographic expression of continents and their landforms are all the end products of Cenozoic processes. The Appalachian Mountain region, for example, began its evolution during the Precambrian, but its present expression is largely the product of Cenozoic uplift and erosion. In short, the present distinctive aspect of the Earth developed very recently in the context of geologic time.

❖ CENOZOIC PLATE TECTONICS — AN OVERVIEW

The Late Triassic fragmentation of the supercontinent Pangaea (see Fig. 15.4) began an episode of plate motions that continues even now. As the Americas moved westward, the Atlantic Ocean basin opened until it attained its present dimensions (◆ Fig. 17.4).

Spreading ridges such as the Mid-Atlantic Ridge and East Pacific Rise were established, and it is along these features that new oceanic crust is continuously generated. The age distribution of oceanic crust in the Pacific, however, is decidedly asymmetric because much of the crust in the eastern Pacific has been consumed at subduction zones along the western Americas (see Fig. 7.11).

Another important event was the northward movement of the Indian plate and its collision with southern Asia (Fig. 17.4). India's long journey began during the Cretaceous when it separated from Gondwana and moved progressively north. In doing so, India, along with the northward-moving African plate, caused the closure of the Tethys Sea (Fig. 17.4). During the Early Tertiary, Australia separated from Antarctica and moved north to its present position.

The breakup of Pangaea and the drift of its various fragments account for the present geographic distribution of continents and oceans (Fig. 17.4). But continental rifting is not restricted to the Late Triassic. A triple junction is currently located at the junction of the East African Rift System, the rift in the Red Sea, and the rift in the Gulf of Aden (◆ Fig. 17.5). Rifting in this region began during the Late Tertiary and continues at present.

Rifting in East Africa seems to be in its early stage, since the continental crust has not yet stretched and thinned enough for oceanic crust to form from below. In the Red Sea region, rifting was preceded by vast eruptions of basalt. In the succeeding rifting stage, a long, narrow sea formed, and by the Late Pliocene, oceanic crust began forming along the rift axis (Fig. 17.5b). Rifting began even earlier in the Gulf of Aden than in the Red Sea. By the Late Miocene, the continental crust had stretched and thinned, and upwelling basaltic magma was sufficient to form oceanic crust.

Studies of this Late Cenozoic rifting event in Africa and adjacent regions are important for two reasons. First, they allow us to better understand the Late Triassic rifting of Pangaea and the origin of the Atlantic Ocean basin. Second, all three branches of these rifts remain active today, although rifting in East Africa is very slow. As a consequence of rifting in the Red Sea and Gulf of Aden, the Arabian plate has separated from Africa. Arabia's northward motion is responsible for a major north-south trending shear zone, the Dead Sea fault, and much of the present tectonic activity in the Middle East (Fig. 17.5a).

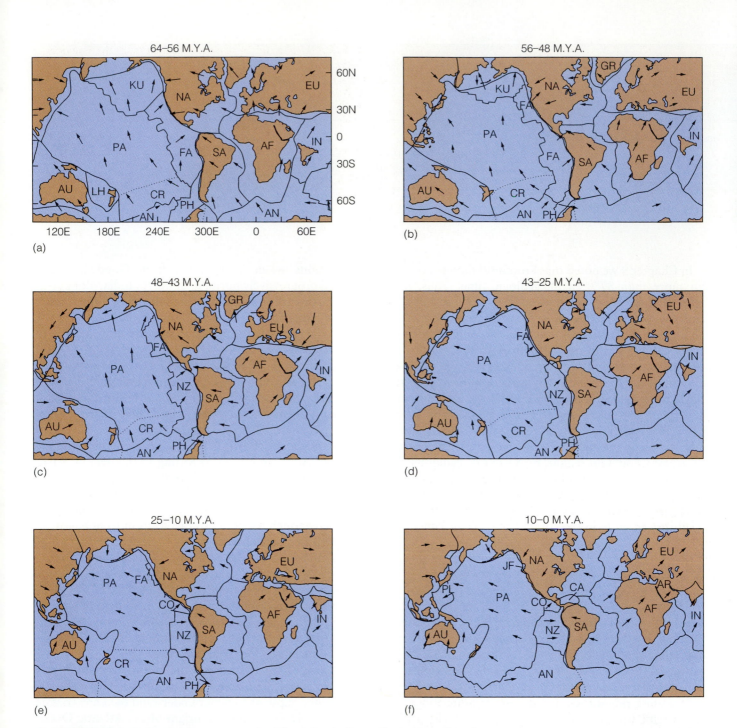

64–56 M.Y.A.

KU, NA, EU, PA, FA, SA, AF, IN, AU, LH, CR, PH, AN

60N, 30N, 0, 30S, 60S

120E, 180E, 240E, 300E, 0, 60E

(a)

56–48 M.Y.A.

GR, KU, NA, EU, PA, FA, SA, AF, IN, AU, CR, AN, PH

(b)

48–43 M.Y.A.

GR, NA, EU, FA, PA, AF, IN, NZ, SA, AU, CR, AN, PH

(c)

43–25 M.Y.A.

EU, NA, FA, PA, AF, IN, NZ, SA, AU, CR, AN, PH

(d)

25–10 M.Y.A.

NA, EU, PA, FA, CO, AF, NZ, SA, IN, AU, CR, AN, PH

(e)

10–0 M.Y.A.

JF, NA, EU, PL, PA, CA, CO, AR, AF, NZ, SA, IN, AU, AN

(f)

◆ **FIGURE 17.4** Cenozoic plate movements. The Atlantic Ocean basin opened as the Americas moved westward from Europe and Africa, while the Pacific Ocean basin decreased in size. The Farallon plate was mostly consumed beneath the Americas; the Nazca, Cocos, and Juan de Fuca plates are remnants of this plate. Africa moved northward and partly closed the Tethys Sea. India also moved northward and collided with Asia. Australia moved northward to its present position. Abbreviations for plates: AF, African; AN, Antarctic; AR, Arabian; AU, Australian; CA, Caribbean; CO, Cocos; CR, Chatham Rise; EU, Eurasian; FA, Farallon; GR, Greenland; IN, Indian; JF, Juan de Fuca; KU, Kula; LH, Lord Howe; NA, North American; NZ, Nazca; PA, Pacific; PL, Philippine; PH, Phoenix; SA, South American.

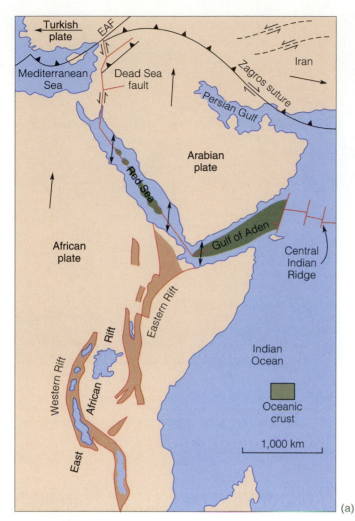

◆ **FIGURE 17.5** (a) A triple junction exists at the junction of the East African Rift System and the rifts in the Red Sea and Gulf of Aden. Oceanic crust began forming in the Gulf of Aden about 10 million years ago. Rifting began later in the Red Sea, and oceanic crust is now forming. In East Africa, the continental crust has not yet stretched and thinned enough for oceanic crust to form from below. Much of the tectonic activity in the Middle East is caused by the Arabian plate moving northward against Eurasia. (b) A schematic cross section of the Red Sea. Rifting that began during the Miocene Epoch has formed a long, narrow sea.

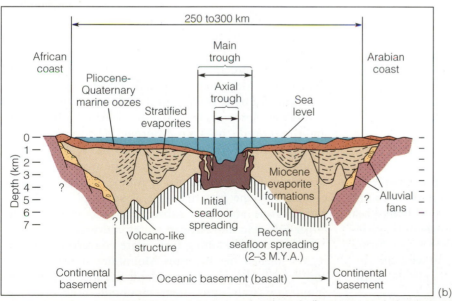

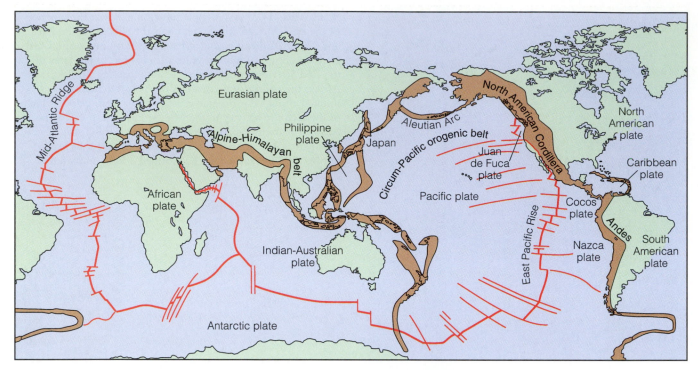

◆ **FIGURE 17.6** Areas of Cenozoic orogenic activity. Orogenesis occurred mainly in two major belts—the Alpine-Himalayan belt and the circum-Pacific belt.

❖ CENOZOIC OROGENIC BELTS

Cenozoic orogenic activity was largely concentrated in two major zones or belts, the **Alpine-Himalayan belt** and the **circum-Pacific belt** (◆ Fig. 17.6). Each belt is composed of a number of *orogens*, or zones of deformed rocks, many of which have been metamorphosed and intruded by plutons. In many of these orogens, deformation dates well back into the Mesozoic but continued into the Cenozoic, and some, such as the Himalayan orogen, are orogenically active today.

The Alpine-Himalayan Orogenic Belt

The Alpine-Himalayan orogenic belt includes the mountainous regions of the Mediterranean and extends eastward through the Middle East and India and into southeast Asia (Fig. 17.6). Remember that the Tethys Sea separated much of Gondwana from Eurasia during Mesozoic time (Fig. 17.4). Plate motions that began during the Mesozoic culminated in the Cenozoic with the closure of the Tethys Sea as Africa and India moved northward and collided with Eurasia.

The Alps

The **Alpine orogeny** produced a zone of deformation in southern Europe extending from the Atlantic Ocean eastward to Greece and Turkey. Concurrent deformation also occurred south of the Mediterranean basin along the northwest coast of Africa (Fig. 17.6). Unraveling the complexities of this deformational event has been a long and arduous task, but the broad picture is now becoming clear, even though many details are still poorly understood.

Although events leading to the Alpine orogeny began during the Mesozoic, major deformation took place during the Eocene to Late Miocene. Deformation was caused by the northward movements of the African and Arabian plates against Eurasia, but the story is not quite so simple. The complexities of Alpine geology are compounded by the fact that several small plates collided with Europe and were subsequently deformed as the larger Eurasian and African plates converged.

Deformation resulting from plate convergence in this region formed the Pyrenees Mountains between Spain and France, the Alps of mainland Europe, and the Apennines of Italy, and other mountain ranges (◆ Fig.

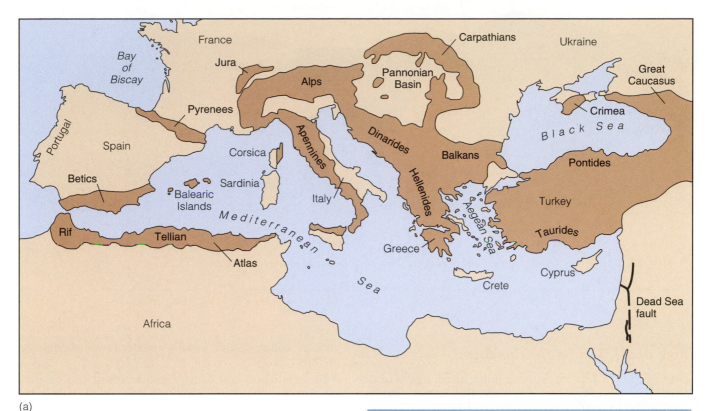

(a)

◆ **FIGURE 17.7** (a) The Alpine system of Europe. (b) A view of the Alps near Grindelwald, Switzerland. A horn peak is visible on the skyline, and the valley contains a glacier. (Photo courtesy of R. V. Dietrich.)

17.7). The compressional forces generated by plate collisions produced complex thrust faults and large overturned folds called *nappes* (◆ Fig. 17.8). Another result of plate convergence in this region was the formation of an isolated sea in the Mediterranean basin, which had formerly been a part of the Tethys Sea (Fig. 17.4) (See the Prologue to Chapter 5).

The Atlas Mountains of northwestern Africa (Fig. 17.7) also formed as the African plate collided with Eurasia. Farther east in the Mediterranean basin, Africa is still forcing oceanic lithosphere northward beneath Greece and Turkey (◆ Fig. 17.9). Active volcanoes in Italy and seismic activity in much of southern Europe and the Middle East indicate that the Mediterranean basin remains geologically active. In 1990, for example, an earthquake measuring 7.3 on the Richter scale killed 40,000 people in Iran, and an earthquake in Turkey in 1992 resulted in at least 570 fatalities.

(b)

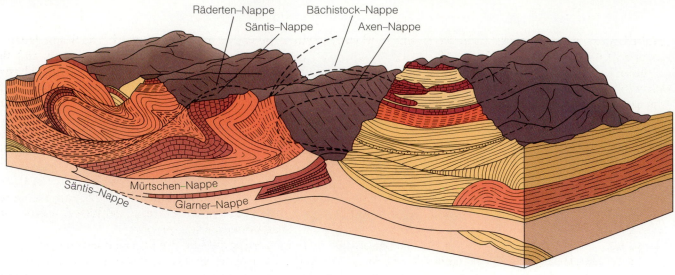

Räderten–Nappe   Bächistock–Nappe
Säntis–Nappe   Axen–Nappe

Säntis–Nappe   Mürtschen–Nappe
Glarner–Nappe

◆ **FIGURE 17.8** Cross section of the European Alps showing large, faulted overturned folds called nappes.

Roof of the World—The Himalayas

During Early Cretaceous time, India broke away from Gondwana and began moving north (Fig. 17.4). At the same time, a subduction zone formed along the south-facing margin of Asia where oceanic lithosphere was consumed (◆ Fig. 17.10a). Partial melting of this descending oceanic lithosphere generated magma, which rose to form a volcanic chain and large granitic intrusions in what is now Tibet. India eventually approached this chain and destroyed it as it collided with Asia to form a **collision orogen.** As a result, two continental plates became sutured—India and Asia were now one.

The exact time of India's collision with Asia is uncertain, but sometime between 40 and 50 million years ago India's northward drift rate decreased abruptly, from between 15 and 20 cm per year to about 5 cm per year. Because continental lithosphere is not dense enough to be subducted, this decrease in rate seems to mark the time of collision and India's resistance to subduction (◆ Fig. 17.10b).

Because of its low density and resistance to subduction, the leading margin of India was underthrust beneath Asia, causing crustal thickening, thrusting, and uplift. Sedimentary rocks that were deposited in the sea south of Asia were thrust northward into Tibet, and two major thrust faults carried Paleozoic and Mesozoic rocks of Asian origin onto the Indian plate (◆ Fig 17.10c). Rocks that were deposited in the shallow seas along India's northern margin now form the higher parts of the Himalayas.

As the Himalayas were uplifted, they were also eroded, but at a rate insufficient to match the uplift. Much of the debris shed from the rising mountains was transported to the south and deposited as a vast blanket on the Ganges Plain and as huge submarine fans in the Arabian Sea and in the Bay of Bengal (◆ Fig. 17.11, p. 484). Continued movement on the Main Boundary fault has folded and

◆ **FIGURE 17.9** In the eastern Mediterranean, Africa is approaching Eurasia, and oceanic lithosphere is being subducted beneath Turkey.

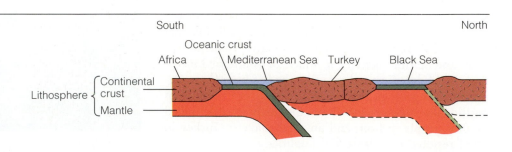

South   North

Oceanic crust
Africa   Mediterranean Sea   Turkey   Black Sea

Lithosphere { Continental crust / Mantle

faulted rocks of the Ganges Plain along the southern margin of the Himalayas (◆ Fig. 17.10d).

Since its collision with Asia, India has been underthrust about 2,000 km beneath Asia! Currently, India continues moving north at about 5 cm per year. In other words, the Himalayas are still forming.

The Circum-Pacific Orogenic Belt

The Pacific plate is being consumed at subduction zones along the western and northern margins of the Pacific Ocean basin (Fig. 17.4). This process has continued throughout the Cenozoic, giving rise to orogens in the Aleutians, the Philippines, Japan, and several other areas in the southwestern Pacific Ocean basin (Fig. 17.6).

Orogens of the western and northern Pacific are **arc orogens** characterized by subduction of oceanic lithosphere, deformation, and igneous activity. Japan, for example, is bounded on the east by the Japan Trench,

where the Pacific plate is subducted. The Sea of Japan, a **back-arc marginal basin,** lies between Japan and the mainland of Asia. The origin of back-arc marginal basins is not fully understood, but according to one theory, Japan was once part of mainland Asia and was separated as back-arc spreading occurred (◆ Fig. 17.12, p. 484). When separation began sometime during the Cretaceous, Japan moved eastward over the Pacific plate, and oceanic crust formed in the Sea of Japan. Japan's geology is complex, and much of its deformation predates the Cenozoic. Nevertheless, considerable deformation, metamorphism, and volcanism occurred during the Cenozoic and, in fact, continues to the present.

Spreading at the East Pacific Rise is carrying the Cocos and Nazca plates eastward, where they are being subducted beneath Central and South America, respectively (Fig. 17.4). Volcanism and seismic activity indicate that the orogens in Central and South America remain active. We are reminded of this activity by events such as the

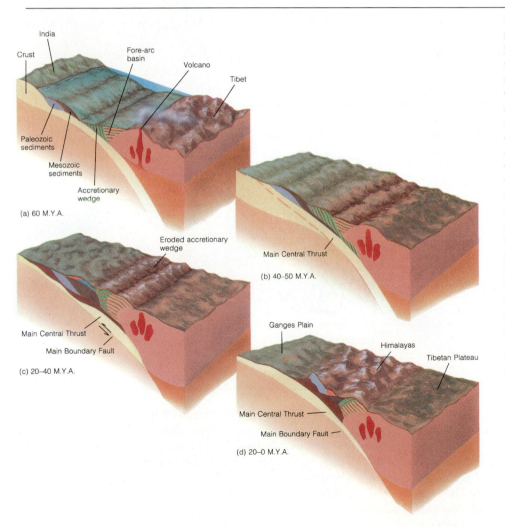

◆ **FIGURE 17.10** Simplified cross sections showing the collision of India with Asia and the origin of the Himalayas. (a) The northern margin of India before its collision with Asia. Oceanic lithosphere was subducted beneath southern Tibet as India approached Asia. (b) About 40 to 50 million years ago, India collided with Asia, but since India was too light to be subducted, it was underthrust beneath Asia. (c) Convergence continues accompanied by thrusting of rocks of Asian origin onto the Indian subcontinent. (d) Since about 10 million years ago, India has moved beneath Asia along the Main Boundary fault. Shallow marine sedimentary rocks that were deposited along India's northern margin now form the higher parts of the Himalayas. Sediment eroded from the Himalayas has been deposited on the Ganges Plain.

THE PERILS OF LIVING NEAR A CONVERGENT PLATE MARGIN

The active state of the Central and South American orogens was demonstrated by two tragic events in 1985.

An earthquake of 8.1 magnitude on the Richter scale claimed 9,000 lives in Mexico City, and a mudflow engulfed Armero, Colombia, and several villages, claiming 23,000 lives. In addition, several tens of thousands of people were injured or left homeless, and property damages of several billions of dollars resulted.

Even though earthquakes and mudflows are very different geologic phenomena, both of these events were related to convergent plate margin activities. The earthquake that struck Mexico City, the most populous

◆ **FIGURE 1** Most of the earthquake and igneous activity around the margins of the Pacific results from plate convergence as illustrated in the two insets.

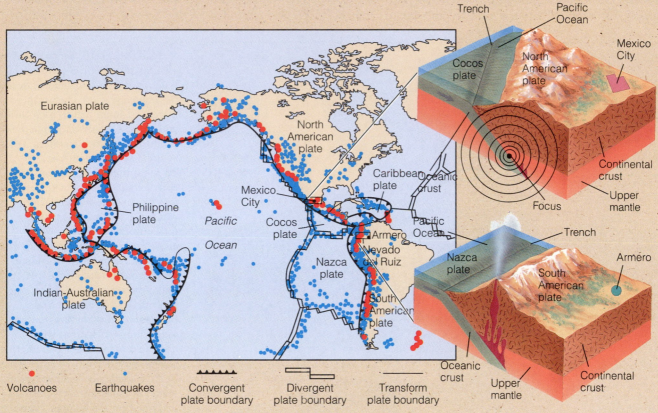

1985 earthquake that struck Mexico City and the tragic volcanic mudflows in Colombia in the same year (see Perspective 17.1).

In South America, the highest mountains in the Western Hemisphere, the Andes (Fig. 17.6), resulted from Mesozoic-Cenozoic plate convergence. The formation of the present-day Andes resulted from crustal thickening as sedimentary rocks were deformed, uplifted, and intruded by Mesozoic-Cenozoic granitic plutons. In addition, a major thrust fault carries the mountains over the Brazilian Shield, further increasing the crustal thickness of South America (◆ Fig. 17.13). Even though the Andes

◆ **FIGURE 2** Earthquake damage in Mexico City. One floor piled on another when this building collapsed. (Photo courtesy of M. Celebi, USGS.)

urban area in the world, resulted from subduction of the Cocos plate at the Middle American Trench (◆ Fig. 1). Sudden movement of the Cocos plate beneath Central America generated seismic waves that traveled outward in all directions. The violent shaking experienced in Mexico City and elsewhere was caused by these seismic waves.

Much of the damage in Mexico City (◆ Fig. 2) occurred because the city rests upon an unstable foundation. The underlying strata are mostly layers of sand, silt, and mud that accumulated in a large lake. In fact, Mexico City was built on the site once occupied by the Aztec capital city of Tenochtitlan, which was located on an island in this ancient lake. Unfortunately, unconsolidated sediments such as those beneath Mexico City amplify the shaking during earthquakes. Consequently, buildings on such unstable foundations are commonly more heavily damaged than those built on solid bedrock foundations.

Less than two months after the Mexico City earthquake, Colombia experienced its greatest recorded natural disaster. The Nazca plate is being subducted beneath South America (Fig. 1), and Nevado del Ruiz is one of several active volcanoes where rising magma from the partially melted descending plate is erupted. A rather minor eruption of Nevado del Ruiz partially melted the glacial ice on the mountain, and the meltwaters rushed down the valleys, picking up sediment as they went, and finally became viscous mudflows.

The city of Armero, Colombia, lies in the valley of the Lagunilla River, one of the river valleys inundated by the mudflows. Of the 23,000 inhabitants of Armero, about 20,000 died, and most of the city was destroyed. Another 3,000 people were killed in other nearby valleys.

Natural disasters are not restricted to convergent plate margins—large earthquakes occasionally shake areas in continental interiors. Nevertheless, events such as these are common at convergent plate margins. Central and South America have experienced similar geologic upheavals in the past and can expect continuing activity of this type in the future.

continue to be compressed at their margins, parts of the high Andes are subjected to tensional forces, resulting in downward movement of large blocks along normal faults.

Following its separation from Africa during the Mesozoic, South America was an island continent until the Late Tertiary. The land connection between North and South America was formed as a result of subduction at the Middle America Trench, together with arc magmatism (◆ Fig. 17.14). We shall have more to say about this land connection between the Americas in Chapter 18 since the connection played an important role in the intercontinental migrations and extinctions of animals.

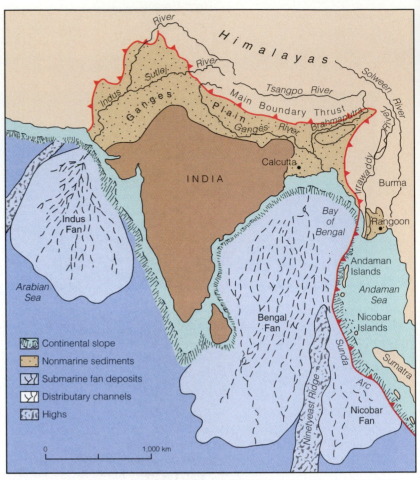

◆ **FIGURE 17.11** Sediment eroded from the Himalayas has been deposited as a vast blanket on the Ganges Plain and as large submarine fans in the Arabian Sea and the Bay of Bengal.

◆ **FIGURE 17.12** The back-arc marginal basin occupied by the Sea of Japan is thought to have formed by back-arc spreading as shown in this model.

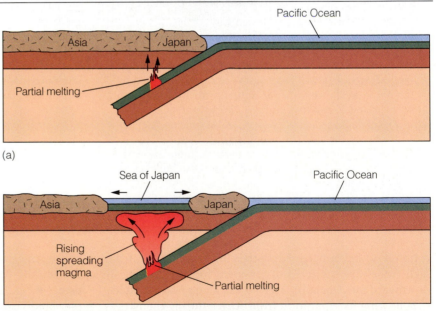

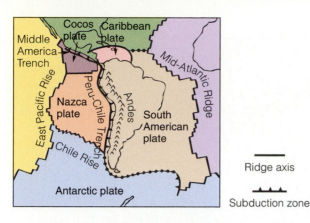

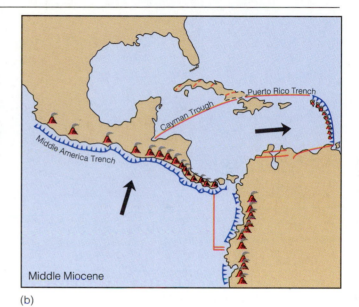

◆ **FIGURE 17.13** The Andes of South America formed as a result of convergence of the Nazca and South American plates.

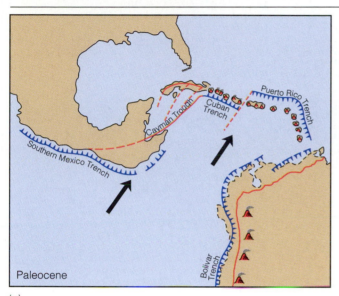

(a)

(b)

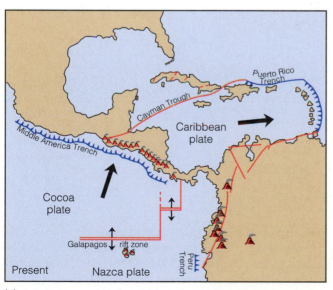

(c)

◆ **FIGURE 17.14** Origin of the land connection between North and South America. (a) During the Paleocene, the Pacific plate was subducted at the Cuban and Puerto Rico trenches. (b) and (c) From the Early Oligocene to the present, however, subduction of the Pacific plate has occurred at the Middle America Trench, resulting in arc magmatism and the origin of the land connection between North and South America.

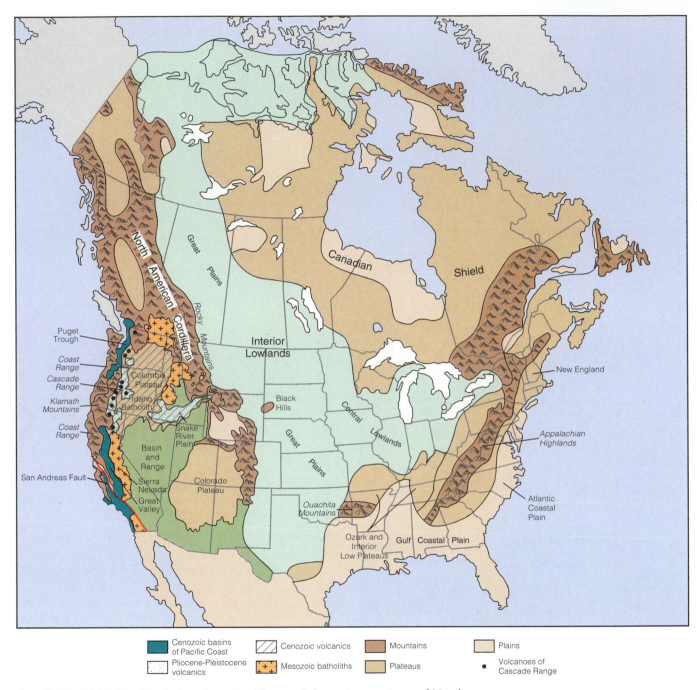

♦ FIGURE 17.15 The North American Cordillera and the major provinces of North America discussed in the text.

❖ THE NORTH AMERICAN CORDILLERA

Bounded on the east by lowlands, and on the west by the Pacific Ocean, the **North American Cordillera** is a com-

plex mountainous region extending from Alaska into Mexico (Fig. 17.6). In the western United States, the Cordillera widens to about 1,200 km (♦ Fig. 17.15); according to William R. Dickinson of Stanford University, "Perhaps no segment of the circum-Pacific [orogenic belt] has

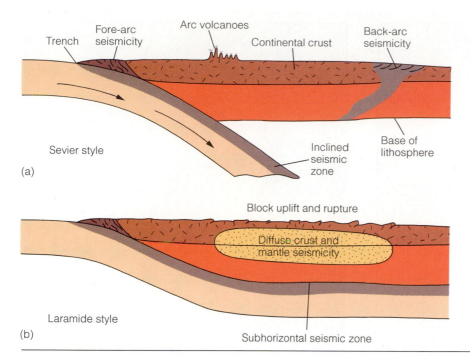

undergone a more complex tectonic evolution during the Cenozoic."[1]

The evolution of the North American Cordillera began during the Late Proterozoic when thick sequences of sediments were deposited along its west-facing continental margin (see Fig. 10.14). Sediment deposition continued during the Paleozoic, and during the Devonian, part of the region was deformed during the Antler orogeny. Beginning in the Late Jurassic and continuing into the Early Tertiary, deformation was more or less continuous as the Nevadan, Sevier, and Laramide orogenies progressively affected areas from west to east (see Figure 15.21). Both the Nevadan and Sevier orogenies were discussed in Chapter 15. The **Laramide orogeny** was a Late Cretaceous to Eocene event and will be considered in the following section.

Following the deformation of the Laramide orogeny, several other events occurred that contributed to the evolution of the Cordillera. For example, the structural style of the Basin and Range Province (Fig. 17.15) developed as a result of extensional tectonics after the compressional deformation of the Laramide orogeny. Vast outpourings of rhyolitic and especially basaltic volcanics occurred in the Pacific Northwest, and volcanism contin-

ues in the Cascades of California, Oregon, and Washington. The Cordillera continues to evolve, especially along its western margin, where the Juan de Fuca plate is being subducted and where the **San Andreas transform fault** cuts through coastal California (Fig. 17.15).

Laramide Orogeny

The **Laramide orogeny** differs from other orogenies in that deformation occurred much farther inland from the arc-trench system than is typical of these collisions. Also, deformation was not accompanied by significant batholithic intrusions. To account for these observations, geologists have modified the classic model for arc orogens. When oceanic lithosphere is subducted beneath continental lithosphere (◆ Fig. 17.16a) oceanic lithosphere descends at a steep angle, 30° or more, arc magmatism occurs inland from the trench, and the thick sediments deposited on the continental margin are deformed. In the Laramide style (◆ Fig. 17.16b), the subducted oceanic slab descends at a lower angle and moves nearly horizontally beneath the continental lithosphere, deforming cratonic crust far inland from the continental margin. Furthermore, arc magmatism seems to occur only when the descending plate penetrates as deep as the mantle, so in the Laramide style, magmatism would be suppressed.

During the Late Cretaceous, a subduction zone existed

[1] "Cenozoic Paleogeography of the Western United States," in *Cenozoic Paleogeography of the Western United States*, v. 3 (Pacific Section SEPM Paleogeography Symposium, 1979), p. 1.

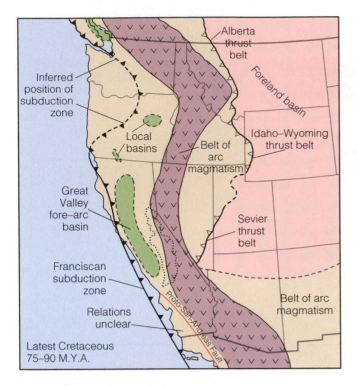

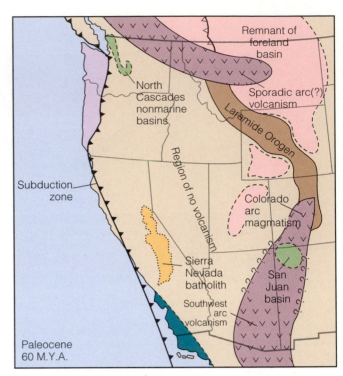

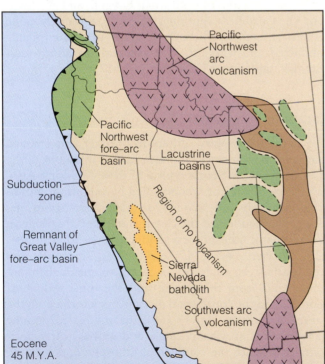

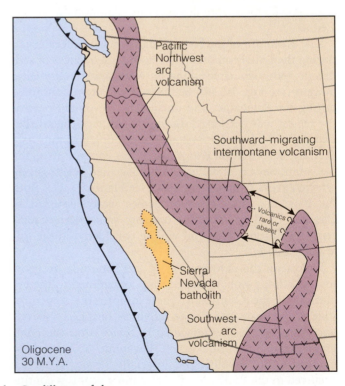

◆ **FIGURE 17.17** Paleotectonic and paleogeographic maps for the Cordilleran of the United States from latest Cretaceous to Oligocene time. The Laramide orogen is much farther inland than is typical of arc orogens. Also note that there was a region of little or no volcanism west of the Laramide orogen during the Paleocene and Eocene. Apparently, volcanism ceased in this area as the subducted Farallon plate moved nearly horizontally beneath the continent as shown in Figure 17.16.

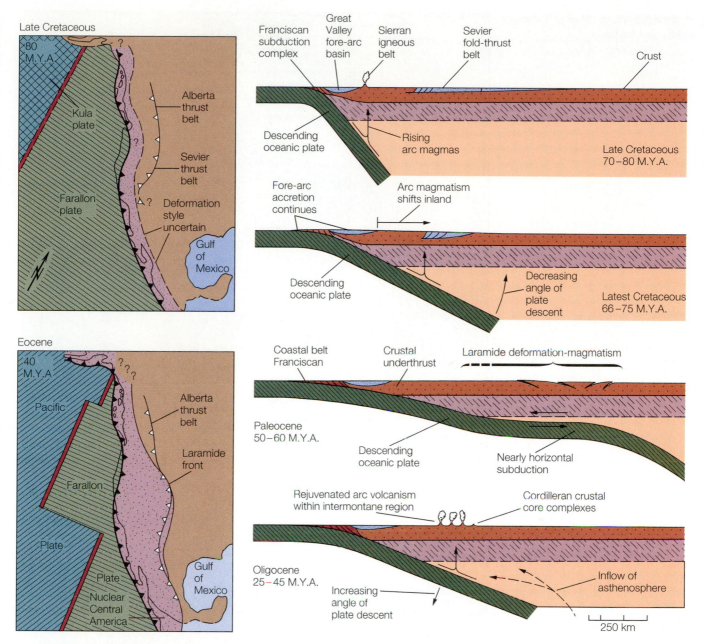

◆ **FIGURE 17.18** Schematic cross sections and maps showing the Laramide style of deformation. The Laramide orogeny was caused by subduction of the Farallon plate beneath North America. Notice especially that as the angle of subduction decreases, arc magmatism shifts inland and eventually ceases. Also, deformation occurs much farther inland when the subducted plate moves nearly horizontally beneath the continent.

along the entire west coast of North America, where the **Farallon plate** was consumed. The Farallon plate descended at about a 50° angle, and arc magmatism occurred 150 to 200 km inland from the trench (◆ Figs. 17.17, 17.18). By the Early Tertiary, the angle of sub-

duction apparently decreased, and the Farallon plate moved nearly horizontally beneath North America. As a result, arc magmatism shifted farther inland and eventually ceased, since the descending Farallon plate did not penetrate to the mantle. In fact, during Laramide defor-

◆ **FIGURE 17.19** Map of Laramide uplifts and basins.

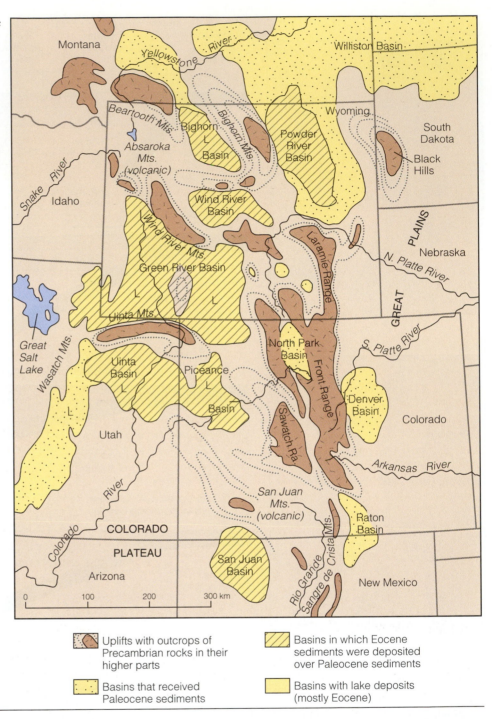

Uplifts with outcrops of Precambrian rocks in their higher parts

Basins that received Paleocene sediments

Basins in which Eocene sediments were deposited over Paleocene sediments

Basins with lake deposits (mostly Eocene)

mation, there was a region of little or no magmatism west of the main Laramide orogen (Fig. 17.17).

Another consequence of the decreasing angle of subduction was a change in the tectonic style of deformation. The fold-thrust tectonic style of the Sevier orogenic belt gave way to large-scale buckling and fracturing,

which produced fault-bounded uplifts (Fig. 17.18). Major mountain ranges with intervening intermontane basins formed along these faults. These intermontane basins were the sites of Early Tertiary deposition of sediments that were eroded from the uplifted blocks (◆ Fig. 17.19).

(a)

(b)

◆ **FIGURE 17.20** (a) The Lewis overthrust in Glacier National Park, Montana. Late Proterozoic strata of the Belt Supergroup rest upon Cretaceous strata. The trace of the fault is the nearly horizontal light line on the side of the mountain. (b) Chief Mountain, Montana, is an erosional remnant of the Lewis overthrust.

The Laramide orogen is centered mostly in the middle and southern Rockies in Wyoming and Colorado (Fig. 17.17). Nevertheless, Laramide compressional forces also deformed the Cordillera far to the north and south. In the northern Rockies of Montana and Alberta, Canada, large slabs of pre-Laramide strata were transported to the east along large overthrust faults. Along the Lewis overthrust of Montana, a large slab of Precambrian strata was moved at least 75 km eastward and now rests upon Cretaceous strata (◆ Fig. 17.20). In the Canadian Rockies of Alberta, thrust sheets piled one upon another in shinglelike fashion, resulting in complex structural relationships.

Deformation also occurred south of the main Laramide orogen. In the Sierra Madre Oriental of east-central Mexico, Late Jurassic to Late Cretaceous shelf carbonates and detrital rocks are now part of a major fold-thrust belt. Laramide structural trends can also be traced south into the area of Mexico City.

The Laramide style of deformation ended during the Eocene. Apparently, the cessation of Laramide deformation coincided with an increasing angle of descent of the Farallon plate (Fig. 17.18), and arc magmatism was re-established as the descending slab penetrated the mantle. The uplifted, fault-bounded blocks of the Laramide orogen continued to be eroded, and the debris filled the adjacent intermontane basins. By Late Tertiary time, the rugged, eroded mountains had been nearly buried in their own erosional debris, forming a vast plain across which streams flowed (◆ Fig. 17.21). During a renewed cycle of erosion, these streams stripped away much of the younger basin-fill sediment and eroded downward, incising their valleys into the uplifted blocks. The deep canyons cutting through the present ranges are the products of this cycle of erosion (◆ Fig. 17.22). The present-day elevations of these ranges are the result of Late Tertiary uplift; in some areas uplift continues to the present.

Cordilleran Volcanism

The vast batholiths of the Sierra Nevada in California, Idaho, and in the Coast Ranges of British Columbia were largely emplaced during the Mesozoic Era, but intrusive activity continued into the Tertiary (see Fig. 15.22). Numerous small plutons were emplaced, including copper- and molybdenum-bearing stocks in Utah, Nevada, Arizona, and New Mexico.

Tertiary extrusive volcanism occurred more or less continuously in the Cordillera, although it did vary in eruptive style and location (◆ Fig. 17.23). Eocene lava flows and sedimentary rocks composed of volcanic rock particles accumulated in the Yellowstone National Park region of Wyoming. Further south in Colorado, lava flows, tuffs, and large calderas characterize the Oligocene San Juan volcanic field. In Arizona, Pliocene to Quaternary volcanism built up the San Francisco Mountains (Fig. 17.23). Here, volcanism may have ceased as recently as 1,200 years ago.

In the Pacific Northwest, a large area is underlain by about 200,000 km^3 of Miocene basalt flows, the Columbia River basalts (◆ Fig. 17.24, p. 494). These flows is-

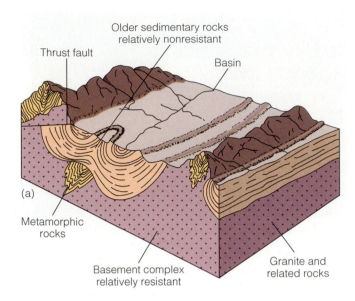

(a)

Thrust fault

Older sedimentary rocks
relatively nonresistant

Basin

Metamorphic
rocks

Basement complex
relatively resistant

Granite and
related rocks

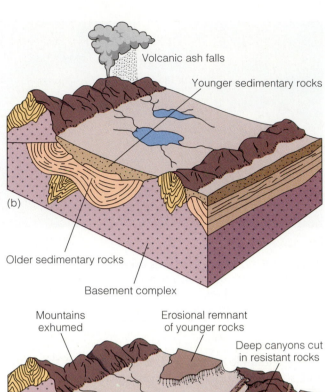

(b)

Volcanic ash falls

Younger sedimentary rocks

Older sedimentary rocks

Basement complex

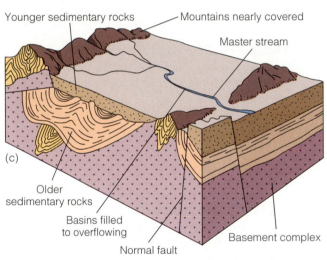

(c)

Younger sedimentary rocks

Mountains nearly covered

Master stream

Older
sedimentary rocks

Basins filled
to overflowing

Normal fault

Basement complex

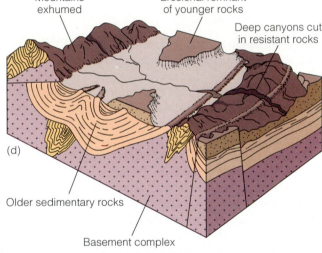

(d)

Mountains
exhumed

Erosional remnant
of younger rocks

Deep canyons cut
in resistant rocks

Older sedimentary rocks

Basement complex

◆ **FIGURE 17.21** (a) through (c) Sediments eroded from the uplifted blocks of the Laramide orogen filled the intermontane basins. (d) These basin sediments were partially excavated, and streams eroded deep canyons into the uplifted blocks.

◆ **FIGURE 17.22** View of the Wind River Canyon, Wyoming. The canyon was eroded into the Owl Creek Mountains as shown in Figure 17.21d.

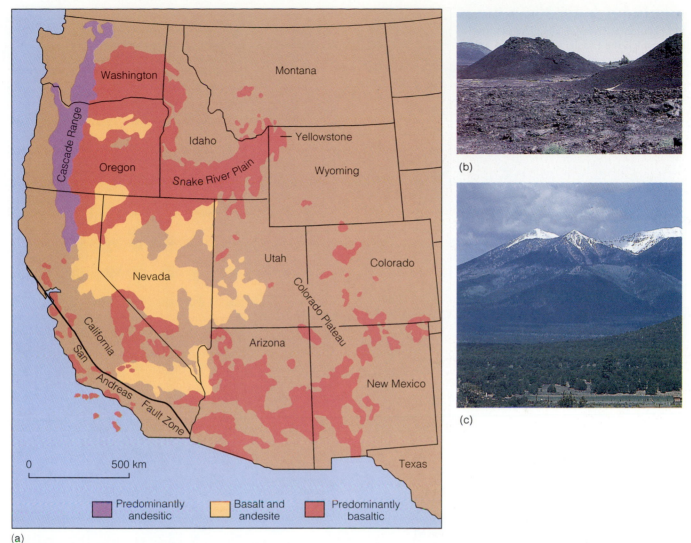

(a)

(b)

(c)

◆ **FIGURE 17.23** (a) The distribution of Cenozoic extrusive volcanics in the western United States. (b) Snake River Plain basalt flow with spatter cones on its surface in Craters of the Moon National Monument, Idaho. (Photo courtesy of Wayne Moore.) (c) Volcanic rocks in the San Francisco Mountains, Arizona.

sued from long fissures and resulted in overlapping flows that are now well exposed in the walls of the deep gorges cut by the Snake and Columbia rivers (Fig. 17.24). The magnitude of some of these eruptive events is difficult to imagine. For example, the single Roza flow covers 40,000 km² and has been traced more than 300 km west from its source. Peter Hooper of Washington State University says:

The enormity of this event is hard to visualize. A lava front about 30 m high, over 100 km wide, and at a temperature of 1100°C, advanced at an average rate of 5 km per hour.[2]

Most of the Columbia River basalts were erupted during a span of 3.5 million years and filled a basin with an aggregate thickness of 2,500 m at the center. Despite all the detailed studies, however, the relationship of this phenomenal eruptive event to plate tectonics remains uncertain.

The Snake River Plain (Fig. 17.15) is actually a depression in the Earth's crust that has been filled mostly by Pliocene and younger basalts (Fig. 17.23b). The volcanics are oldest in the southwestern part of the Snake River

---

[2] "The Columbia River Basalts," *Science* 215 (1982): 1466.

◆ **FIGURE 17.24** (a) The distribution of the Columbia River basalts. The fissures from which the basalts were erupted are shown by heavy lines. (b) The Columbia River basalts. (Photo courtesy of Ward's Natural Science Establishment, Inc.)

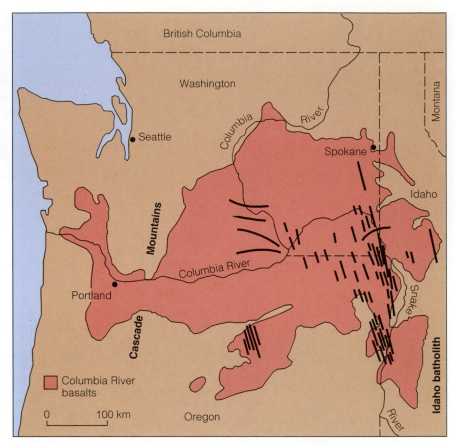

(a)

(b)

(a)

(b)

◆ **FIGURE 17.25** (a) Two basalt flows separated by the Tower Creek Conglomerate in Yellowstone National Park, Wyoming. (b) The Grand Canyon of the Yellowstone River is eroded into the hydrothermally altered Yellowstone Tuff.

Plain and become progressively younger toward the northeast, leading some geologists to propose that North America has migrated over a mantle plume. This plume, although its existence is debated, may now lie beneath Yellowstone National Park in Wyoming. Other geologists disagree with the plume hypothesis and propose that these volcanics were erupted along an intracontinental rift zone.

Bordering the Snake River Plain on the northeast is the Yellowstone Plateau (Fig. 17.23), an area of Late Pliocene and Quaternary rhyolitic and some basaltic volcanism (◆ Fig. 17.25). As just noted, a mantle plume may be located beneath this area. Some source of heat at depth is indicated by the current hydrothermal activity,

including one of the world's most famous geysers, Old Faithful (◆ Fig. 17.26). Geophysical evidence indicates, however, that the heat source may be only a body of intruded magma that has not yet completely cooled.

Some of the most majestic and highest mountains in the Cordillera are in the Cascade Range of northern California, Oregon, and Washington (Figs. 17.15, and ◆ 17.27). Twice during this century, Cascade volcanoes have erupted. Mount Lassen, in northern California, was active from 1914 to 1921, and more recently, Mount St. Helens, in Washington, devastated a large area when it erupted in May 1980. Cascade volcanism continues as a consequence of subduction of the Juan de Fuca plate (Fig. 17.15).

◆ **FIGURE 17.26** Old Faithful geyser, Yellowstone National Park, Wyoming.

(a)

(b)

(c)

◆ **FIGURE 17.27** Vocanism in the Cascade Range of northern California, Oregon, and Washington. (a) Mount Lassen in northern California, shown erupting in 1915, erupted from 1914 to 1921. (b) Crater Lake, Oregon. Crater Lake is a caldera that formed when the summit of Mount Mazama collapsed about 6,600 years ago. The small cone within the caldera is a cinder cone called Wizard Island. (c) The eruption of Mount St. Helens, Washington, on May 18, 1980. (Photo courtesy of the U.S. Geological Survey.)

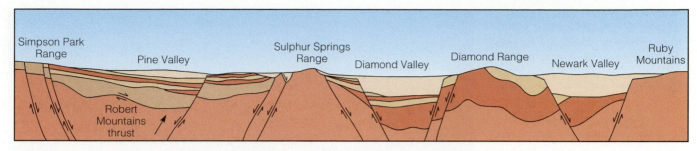

◆ **FIGURE 17.28** Cross section of part of the Basin and Range Province in Nevada. The ranges and valleys are bounded by normal faults.

The volcanoes of the Cascade Range were built by andesitic volcanism during the Pliocene, Pleistocene, and Recent. The internal structure of one of these Cascade volcanoes is exposed at Crater Lake, Oregon, where explosive eruptions about 6,600 years ago were followed by summit collapse and the formation of a large caldera (Fig. 17.27b).

Basin and Range Province

The continental crust of a large region centered in Nevada, but extending into adjacent states and northern Mexico, has been subjected to tensional forces and extended since the Late Miocene. Crustal extension in this region, the **Basin and Range Province** (Fig. 17.15), has yielded north-south oriented, normal faults. Differential movements on these faults have produced a kind of topography that geologists have come to call *basin-and-range structure,* consisting of mountain ranges separated by broad valleys (◆ Fig. 17.28).

Before block-faulting began, the Basin and Range Province was deformed in three successive orogenies: the Nevadan, the Sevier, and the Laramide. During the Early Tertiary, the entire area was an upland undergoing erosion; Early Miocene eruptions of rhyolitic lava flows and pyroclastic materials covered large parts of this region. Block-faulting and basaltic volcanism began during the Late Miocene. Large crustal blocks moved relatively upward along parallel normal faults and other crustal blocks moved downward along normal faults, forming the ranges and basins characteristic of this region (Fig. 17.28). As the ranges were uplifted, they were eroded, and the erosional debris was transported into the adjacent basins where it was deposited as thick alluvial fans and playa-lake deposits.

At its western margin, the Basin and Range Province is bounded by a large escarpment that forms the east face of the Sierra Nevada (◆ Fig. 17.29). The Sierra Nevada block was uplifted along normal faults during the Pliocene and Pleistocene and tilted to the west. Its crest now stands 3,000 m above the basins immediately to the east. Prior to uplift, the Basin and Range was in a subtropical climatic regime, but the rising Sierra Nevada created a rain shadow that caused the climate to become increasingly arid.

Several models have been proposed to account for basin-and-range structure. Among these are back-arc spreading, spreading at the East Pacific Rise (which is thought to now lie beneath North America), a mantle plume beneath the area, and deformation related to the San Andreas transform fault (◆ Fig. 17.30). There is little agreement on the mechanism, but we can be sure that whatever it is, it is still active.

Colorado Plateau

The Colorado Plateau (Fig. 17.15) is a vast area of deep canyons, broad mesas, volcanic mountains, and brilliantly colored rocks (◆ Fig. 17.31). Exposed strata range in age from Archean to Cenozoic. Remember from our discussions in Chapters 12 and 15 that the Colorado Plateau region was the site of deposition of extensive red beds during the Permian and Triassic. Many of these formations are now exposed in the uplifts and walls of canyons.

At the end of the Cretaceous, the Colorado Plateau was near sea level, as indicated by marine sedimentary rocks of that age. During the Early Tertiary, Laramide deformation produced broad anticlines and arches and basins. These basins became the depositional sites for sediments shed from adjacent highlands. A number of large normal

◆ **FIGURE 17.29** View of the western margin of the Basin and Range Province. The Sierra Nevada are bounded on the east by normal faults and rise 3,000 m above the basins to the east.

faults also cut the area, but overall deformation was far less intense than it was elsewhere in the Cordillera. During its Early Tertiary stage of development, the Colorado Plateau had no deep canyons, and the region was nearly encircled by mountains, making it a basin of internal drainage in which sediments accumulated.

Late Tertiary uplift elevated the region from near sea level to the 1,200 to 1,800 m elevations seen today. As uplift proceeded, deposition ceased, and erosion of the canyons began. There is considerable disagreement on the details of canyon cutting. For example, were the streams antecedent or superposed? If antecedent, the streams were already established in their present courses before the existing topography developed. Thus, they simply eroded downward as uplift proceeded. If superposed, the implication is that the entire area was buried by younger strata upon which streams flowed. With uplift, the streams stripped away these strata and eroded downward into the underlying strata.

Pacific Coast

The present plate tectonic elements of the Pacific Coast developed as a consequence of the westward drift of North America, the partial consumption of the Farallon plate, and the collision of North America with the Pacific-Farallon ridge. Prior to the Eocene, the entire Pacific Coast was bounded by a subduction zone that stretched from Mexico to Alaska (Fig. 17.17). Most of the Farallon plate was consumed at this subduction zone, and now only two small remnants exist—the Juan de Fuca and Cocos plates. As discussed earlier, continuing subduction of these small plates accounts for seismicity and volcanism in the Pacific Northwest and Central America, respectively.

Westward drift of North America also resulted in its collision with the Pacific-Farallon ridge and the origin of the Queen Charlotte and San Andreas transform faults (◆ Fig. 17.32). Since the Pacific-Farallon ridge was oriented at an angle to the margin of North America (Fig. 17.32), the continent-ridge collision occurred first in northern Canada during the Eocene and later during the Oligocene in southern California. In southern California, two triple junctions formed, one at the intersection of the North American, Juan de Fuca, and Pacific plates, and the other at the intersection of the North American, Cocos, and Pacific plates. Continued westward movement of North America over the Pacific plate caused the triple junctions to migrate, one to the north and the other to the south, giving rise to the **San Andreas transform fault** (Fig. 17.32). A similar occurrence along the coast of Canada resulted in the origin of the Queen Charlotte transform fault (Fig. 17.32).

Where the San Andreas transform fault cuts through coastal California, it forms a complex zone of shearing. Additionally, faults such as this are seldom straight; they curve and branch into smaller faults (◆ Fig. 17.33).

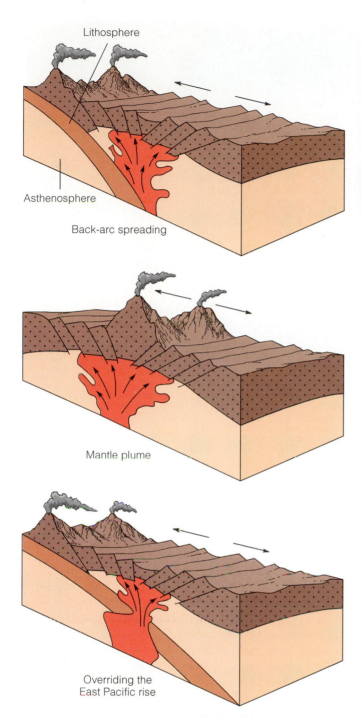

FIGURE 17.30 Geologists agree that tensional forces caused by crustal extension are responsible for basin-and-range structure. There is no agreement, however, on what caused the crustal extension. These illustrations show three models that have been proposed to account for crustal extension and the resulting basin-and-range structure.

◆ **FIGURE 17.31** Rocks of the Colorado Plateau. (a) The Grand Canyon, Arizona. (b) A volcanic neck in northern Arizona. (c) Arches National Park, Utah. Delicate Arch is shown.

Movements on complex fault systems of this type subject the crustal blocks within and adjacent to the shear zone to extensional and compressional stresses. Those areas subjected to extension form strike-slip basins, while the areas of compression are uplifted and supply sediments to the basins.

Cenozoic movements on the San Andreas transform fault account for the origin of numerous basins and uplifts in California, and in adjacent areas off the southern California coast (Fig. 17.33). Many of these fault-bounded basins subsided below sea level and soon filled with turbidites and other deposits. Many of these basins are areas of prolific oil and gas production.

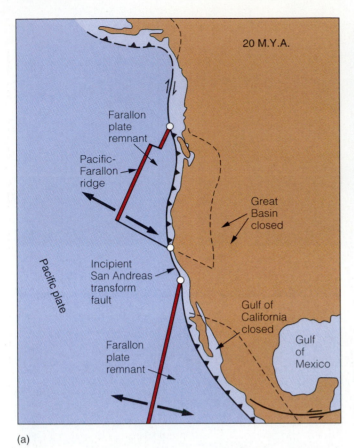

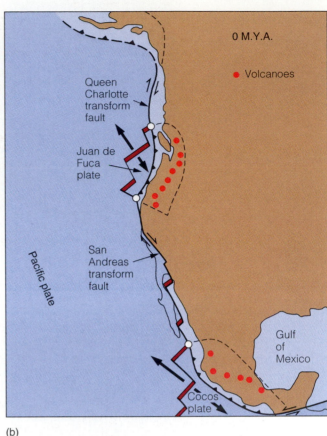

(a)

(b)

◆ **FIGURE 17.32** (a) and (b) The westward drift of North America caused it to collide with the Pacific-Farallon ridge. As North America overrode the ridge, the plate margin became bounded by transform faults rather than a subduction zone.

❖ THE INTERIOR LOWLANDS

The Interior Lowlands is a vast region bounded by the Cordillera, the Canadian Shield, the Appalachians, the Ozark and Interior Low Plateaus, the Ouachita Mountains, and the Gulf Coastal Plain (Fig. 17.15). During the Cretaceous, most of the western part of this region, an area called the Great Plains, was covered by the Zuni epeiric sea. By Early Tertiary time, the Zuni Sea had withdrawn from most of North America, but a sizable remnant arm of the sea was still present in North Dakota during the Paleocene Epoch. Sediments derived from Laramide highlands located to the west and southwest were transported to this remnant sea where they were deposited in marginal marine and marine environments.

Elsewhere within the western Interior Lowlands, sediment was also transported eastward from the Cor-

dillera. These sediments were deposited in terrestrial environments, mostly in fluvial systems, and form large, eastward-thinning wedges that now underlie the Great Plains (◆ Fig. 17.34). The Black Hills, the easternmost uplift associated with Laramide deformation, were a sediment source for fluvial and lacustrine deposits, which were later intricately eroded into the White River Badlands of South Dakota (Fig. 17.34a).

Igneous activity was not widespread in the western Interior Lowlands, but in some local areas it was significant. In northeastern New Mexico, for example, Late Tertiary extrusive volcanism produced volcanoes and numerous lava flows. As a matter of fact, explosive volcanism in this area may have produced the volcanic ash that fell far to the northeast in Nebraska (see Perspective 4.1). A number of small intrusive bodies were emplaced in Colorado, Montana, South Dakota, and Wyoming. One

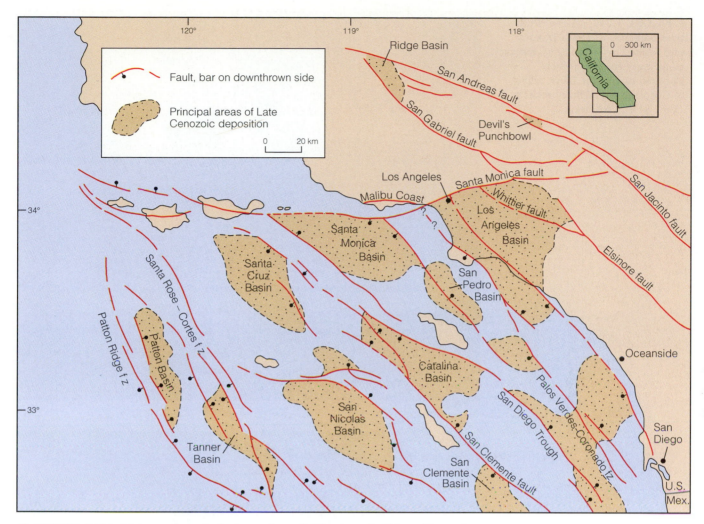

◆ **FIGURE 17.33** Major onshore faults in southern California and offshore Cenozoic basins. The principal areas of Late Cenozoic deposition are shown.

of several in the Black Hills of South Dakota-Wyoming is called Devil's Tower (see the Prologue).

Eastward, beyond the Great Plains, Cenozoic deposits, other than those of Pleistocene glaciers, are uncommon in the rest of the Interior Lowlands. Much of the Interior Lowlands was subjected to erosion during the Tertiary. Of course, the eroded material had to be deposited somewhere, and that was on the Gulf Coastal Plain (Fig. 17.15).

❖ THE GULF COASTAL PLAIN

Following the final withdrawal of the Cretaceous Zuni Sea, the Cenozoic **Tejas epeiric sea** (see Fig. 11.3) made a

brief appearance on the continent. But even at its maximum extent, this sea was largely restricted to the Atlantic and Gulf Coastal plains and parts of coastal California (◆ Fig. 17.35). In fact, its greatest incursion onto North America was in the area of the Mississippi Valley, where it extended as far north as southern Illinois. Epeiric seas were likewise restricted on the other continents except during the Early Tertiary of Europe.

Sedimentary facies development on the Gulf Coastal Plain was controlled largely by a regression of the Cenozoic Tejas epeiric sea. This sea extended far up the Mississippi River Valley during the Early Tertiary (Fig. 17.35), but then began its long withdrawal toward the Gulf of Mexico. Its regression, however, was periodically reversed by minor transgressions; eight transgressive-

◆ **FIGURE 17.34** (a) The White River Badlands of South Dakota were formed by erosion of Cenozoic sedimentary rocks that were deposited east of the Black Hills. (b) Cenozoic sedimentary rocks at Scott's Bluff, Nebraska.

(a)

(b)

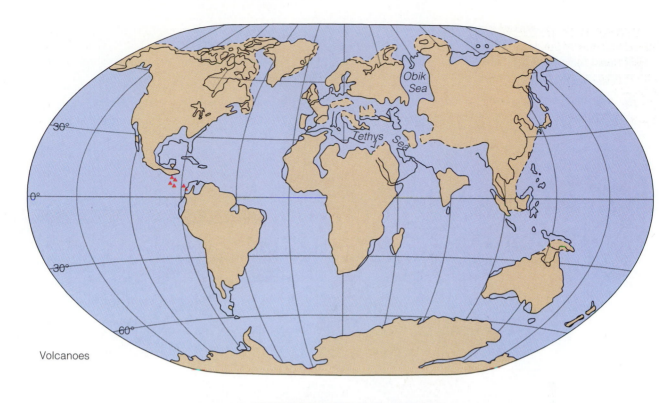

Volcanoes

(a)

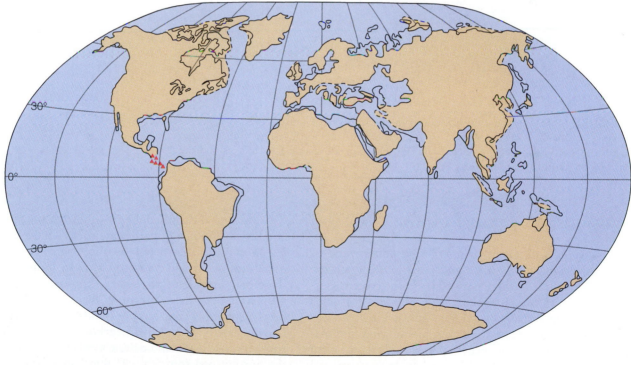

(b)

◆ **FIGURE 17.35** Cenozoic epeiric seas. (a) During the Early Cenozoic, the epeiric sea in North America covered the Atlantic and Gulf Coastal plains and extended far up the Mississippi River Valley. (b) During Late Cenozoic time, the epeiric sea was restricted to the Atlantic and Gulf Coastal plains. By this time the epeiric seas had largely withdrawn from the other continents as well.

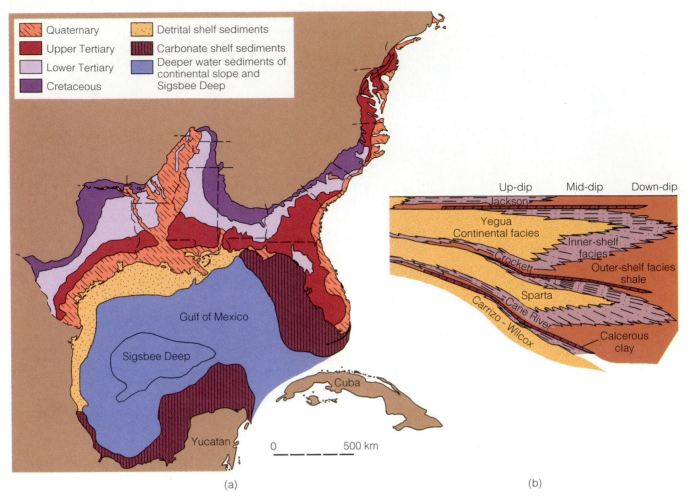

◆ **FIGURE 17.36** Cenozoic deposition on the Gulf Coastal Plain. (a) Depositional provinces in the Gulf of Mexico and surface geology of the northwestern Gulf Coastal Plain. (b) Cross section of the Eocene Claiborne Group showing facies changes and the seaward thickening of the deposits.

regressive episodes are recorded in Gulf Coastal Plain sedimentary rocks.

The general Gulf Coast sedimentation pattern was established during the Jurassic and persisted through the Cenozoic. Sediments were derived from the eastern Cordillera, western Appalachians, and Interior Lowlands and were transported toward the Gulf of Mexico. Deposition occurred in terrestrial, marginal marine, and marine environments. In general, the sediments form seaward-thickening wedges that grade from terrestrial facies in the north to progressively more offshore marine facies in the south (◆ Fig. 17.36). They also clearly record the shoreward and offshore migration of the shoreline (Fig. 17.36).

Gulf Coastal Plain sediments—mostly sand, silt, shale, and lesser amounts of carbonate rocks—have been studied in considerable detail because many contain large quantities of oil and natural gas. Even though many of these rock units have no surface exposures, thousands of wells have been drilled through them, so their compositions, geometries, and stratigraphic relationships are well known. Many of the hydrocarbon reservoirs are nearshore marine sands and deltaic sands that lie updip from the fine-grained source rocks that supply the hydrocarbons. Such oil and gas traps are called *stratigraphic traps* because they owe their existence to variations in the strata. The Eocene Wilcox Group is an excellent example. Detailed studies of subsurface rocks and outcrops in

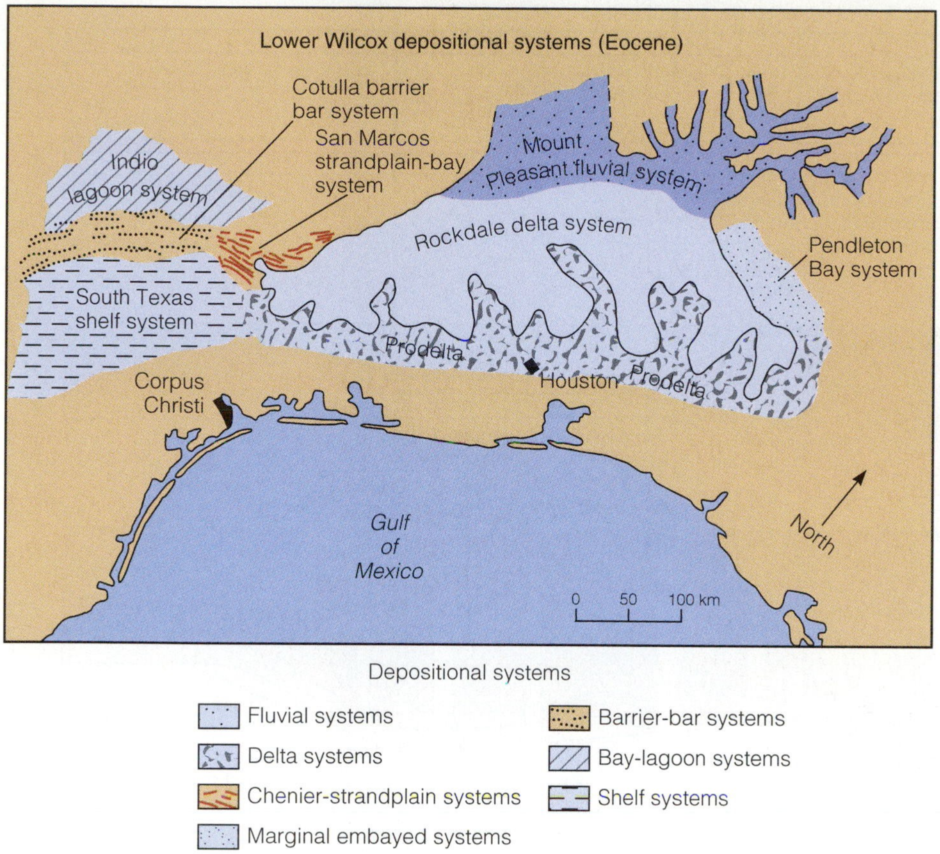

◆ **FIGURE 17.37** Facies of the Eocene Wilcox Group in the northwestern Gulf of Mexico.

Texas indicate deposition in fluvial, deltaic, barrier-island, and beach environments. In fact, a large area of what is now southern Texas was the site of a complex of depositional environments much like those of the present-day Mississippi River delta and adjacent areas (◆ Fig. 17.37).

Structural traps, in which hydrocarbons are trapped as a result of folding, faulting, or both are also common in the Gulf Coastal Plain. In the Gulf Coast, the Jurassic Louann Salt was deposited in the ancestral Gulf of Mexico when North America first separated from North Africa (see Fig. 15.5). Salt is a low-density sedimentary rock, and when buried beneath more dense sediment, it tends to rise toward the surface in pillars called **salt domes** (◆ Fig. 17.38). As the salt rises, it penetrates and deforms the overlying strata, creating structures that may trap oil or gas.

In addition to producing structures for the entrapment of oil and gas, the salt domes are natural resources in their own right. For example, the Avery Island salt dome in Louisiana has been mined for salt since the time of the Civil War. And a number of salt domes have a cap composed in part of sulfur, which is also mined. Salt domes have also been proposed as sites for the storage of nuclear waste products.

Much of the Gulf Coastal Plain was dominated by detrital sediment deposition during the Cenozoic. In the Florida section of the coastal plain and the Gulf Coast of Mexico, however, significant carbonate deposition occurred. A carbonate platform was established in Florida during the Cretaceous, and shallow-water carbonate deposition continued through the Early Tertiary. During the Miocene Epoch, parts of northern Florida formed islands that were populated by a variety of mammals including horses and camels. Although this is not the only Tertiary-aged mammalian fauna east of the Mississippi, it is the only one of Miocene age. Carbonate deposition continues in Florida at the present, but now it occurs only in Florida Bay and the Florida Keys.

Southeast of Florida, across the 80-km-wide Florida Strait, lies the Great Bahama Bank. This area has been a carbonate bank from the Cretaceous to the present.

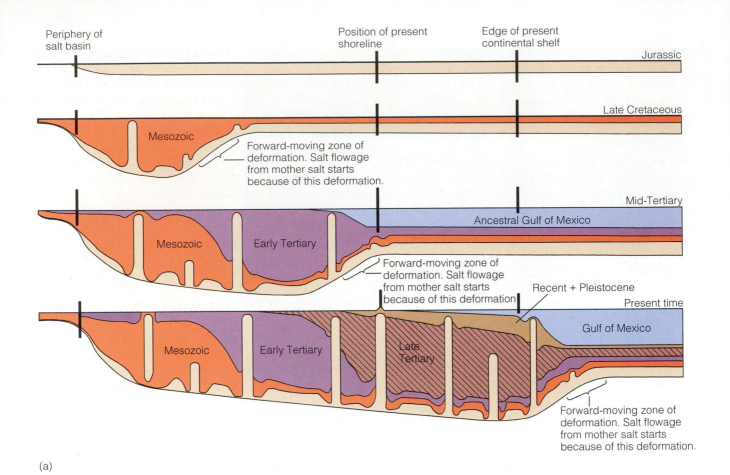

(a)

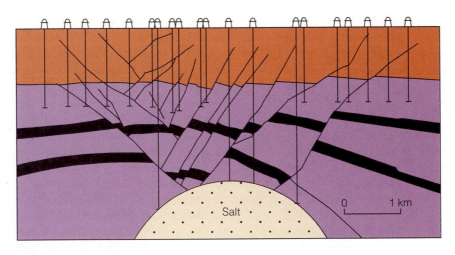

(b)

◆ **FIGURE 17.38** (a) Four stages in the rise of Gulf Coast salt domes. (b) Cross section of the top of the Heideberg salt dome, Mississippi, showing deformation of overlying strata.

Thick, shallow-water carbonates accumulated there, and the region has been used for a long time as a modern-day laboratory to help us understand the conditions under which limestone is deposited (◆ Fig. 17.39).

❖ EASTERN NORTH AMERICA

The eastern seaboard has been a passive continental margin since Late Triassic rifting separated North America from North Africa and Europe. Some seismic activity still occurs there; a major earthquake struck Charleston, South Carolina, on August 31, 1886, killing 60 people and causing $23 million in property damage. Overall, however, the region lacks the geologic activity characteristic of active, convergent margins.

Cenozoic Evolution of the Appalachians

The present distinctive topography of the Appalachian Mountains is the product of Cenozoic uplift and erosion.

◆ **FIGURE 17.39** Location of the Great Bahama Bank. This area is underlain by thick, shallow-water carbonate rocks. Carbonate deposition continues at the present.

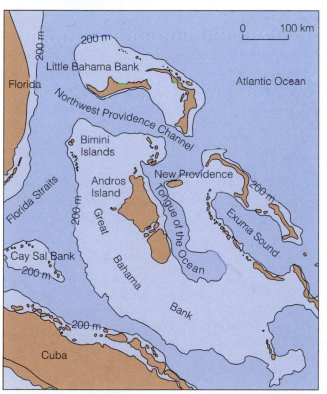

Recall that the Appalachians originally formed during the Late Paleozoic closure of the Iapetus Ocean, an event called the Allegheny orogeny. Block-faulting began during the Late Triassic in response to extensional forces related to the breakup of Pangaea (see Fig. 15.4). In short, in what is now the eastern part of North America, an episode of rifting began that was probably very similar to the rifting that is now occurring in the Red Sea and Gulf of Aden (Fig. 17.5). Since that time, the area has been a passive continental margin.

By the end of the Mesozoic, the Appalachian Mountains had been eroded to a plain (◆ Fig. 17.40). Cenozoic uplift rejuvenated the streams, which responded by renewed downcutting. As the streams eroded downward, they were superposed on resistant strata and cut large canyons across these strata. For example, the distinctive topography of the Valley and Ridge Province is the product of Cenozoic erosion and preexisting geologic structures. It consists of northeast-southwest trending ridges of resistant upturned strata and intervening valleys eroded into less resistant strata (Fig. 17.40). Large canyons were cut directly across the ridges, the result of superposition of streams.

The origin of erosion surfaces at different elevations in the Appalachians has been a subject of debate. Some geologists think these erosion surfaces are evidence of uplift followed by extensive erosion and renewed uplift. Others are convinced that they represent differential response to weathering and erosion. According to this view, a low-elevation erosion surface developed on softer strata and was eroded more or less uniformly. Higher surfaces represent the weathering and erosion response of more resistant strata.

The Atlantic Continental Margin

The Atlantic continental margin of North America includes the Atlantic Coastal Plain (Fig. 17.15) and extends seaward across the continental shelf, slope, and rise (◆ Fig. 17.41). It is a classic example of a passive continental margin. When Pangaea fragmented during the Early Mesozoic, North America separated from North Africa and Europe, continental crust was rifted, and a new ocean basin formed. The Atlantic continental margin possesses a number of Mesozoic and Cenozoic sedimentary basins that formed as a consequence of this rifting event (Fig. 17.41). Deposition in these basins began during the Jurassic, and even though sediments of this age are known only from a few deep wells, they are presumed to underlie the entire margin of the continent. The distribution of Cretaceous and Cenozoic

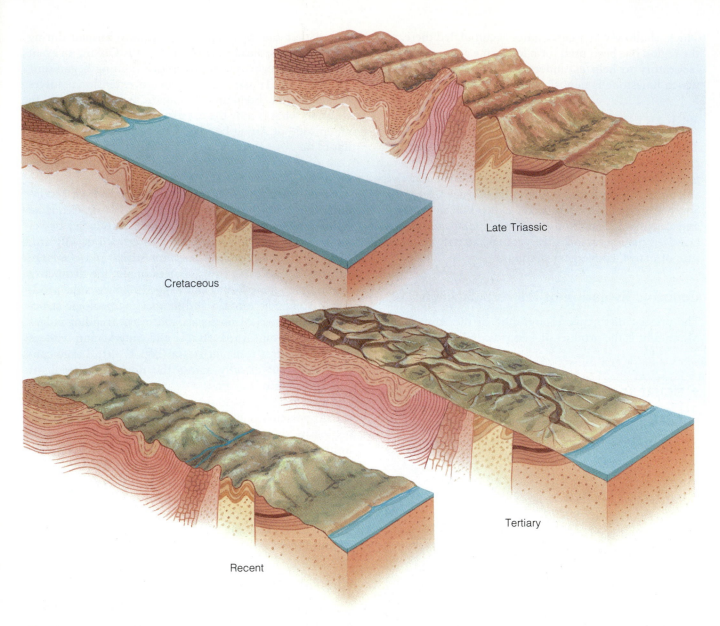

◆ **FIGURE 17.40** The origin of the present topography of the Appalachian Mountains. Erosion in response to Cenozoic uplift accounts for this topography.

sediments is better known because both crop out in the Atlantic Coastal Plain, and both have been penetrated by wells on the continental shelf.

The sediments of the Atlantic Coastal Plain were derived from the Appalachian Mountains, transported east, and deposited in fluvial, nearshore marine and shelf en-vironments. In Maryland, for example, the rocks composing the Calvert Cliffs on the western shore of Chesa-peake Bay represent marginal marine deposits (◆ Fig. 17.42). These deposits are interesting because they contain an assemblage of fossil vertebrates that includes sea cows, whales, sharks, bony fishes, birds, and reptiles. In

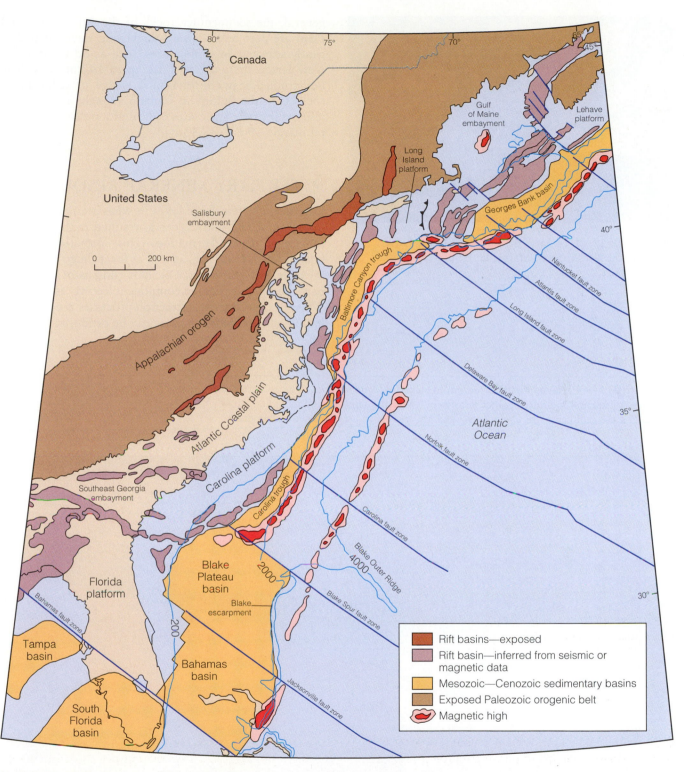

◆ **FIGURE 17.41** The major Mesozoic and Cenozoic basins of the North American continental margin.

◆ **FIGURE 17.42** Tertiary rocks exposed in the Calvert Cliffs on the Jett Property, Maryland. (Photo courtesy of Peter R. Vogt.)

general, the sedimentary rocks of the Atlantic Coastal Plain are part of a seaward-thickening wedge that dips gently seaward. In some places, such as off the coast of New Jersey, these deposits are up to 14 km thick. The best-studied seaward-thickening wedge of sedimentary rocks is in the Baltimore Canyon Trough, an area that also exhibits the structures typical of a passive continental margin (◆ Fig. 17.43).

❖ TERTIARY MINERAL RESOURCES

Deposition in two large lakes in what are now parts of Wyoming, Utah, and Colorado resulted in the origin of the Green River Formation. This formation is well known for its fossil fish, as well as for its huge reserves of oil shale and evaporites. Oil shale consists of small clay particles and an organic component known as *kerogen*. When the appropriate extraction processes are used, liquid oil and combustible gases can be produced from the

◆ **FIGURE 17.43** The continental margin in eastern North America. The coastal plain and the continental shelf are covered mostly by Cenozoic sandstones and shales. Beneath these sediments are Cretaceous-aged and probably Jurassic-aged sedimentary rocks.

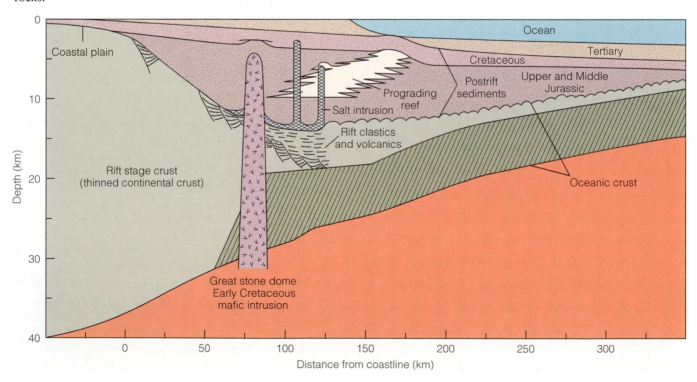

kerogen of oil shale. To be designated as true oil shale, however, the rock must yield at least 10 gallons of oil per ton of rock.

The Green River Formation oil shales yield 10 to 140 gallons of oil per ton of rock processed, and the total amount of oil recoverable with present processes is estimated at 80 billion barrels. Currently, however, no oil is produced from these oil shales because conventional drilling and pumping are less expensive than processing the shales. Oil shales occur on all continents, but the Green River Formation contains the world's most extensive deposits.

One of the Green River Formation evaporites, a sodium carbonate called *trona*, has been mined as a source of sodium compounds. The trona and associated evaporites such as rock salt are interpreted as deposits that accumulated in lakes during times of intense evaporation.

In 1989, more than half of Florida's total mineral value was obtained from mining phosphate-bearing rock in the Central Florida Phosphate District. Phosphorus from phosphate rock has a variety of uses in metallurgy, preserved foods, ceramics, and matches. The Florida-phosphate rock is mined from the Upper Miocene Bone Valley Member of the Peace River Formation (◆ Fig. 17.44); this rock unit also contains an interesting marine and nonmarine fossil fauna that includes such animals as stingrays and mastodons. Mining of this rock unit began during the late 1800s, and in 1989 total production was 38.2 million metric tons.

Diatomite is a sedimentary rock composed of the microscopic shells of diatoms, single-celled, marine and freshwater plants that secrete a skeleton composed of silica (SiO_2) (see Fig. 5.7). This rock, also called diatomaceous earth, is so porous and light that it will float on water when dry. It is used chiefly in gas purification and to filter a number of liquids such as molasses, fruit juices, water, and sewage. The United States is the leader in diatomite production, mostly from mines in California, Washington, and Oregon.

About 30% of the world's coal is in North America, and of that 97% is in the United States. Historically, most of the coal mined in this country has been Pennsylvanian-aged bituminous coal from the interior and Appalachian regions. However, huge deposits of lower-grade coals, mostly lignite and subbituminous, in the Northern Great Plains are becoming increasingly important resources. These coal deposits are Late Cretaceous to Early Tertiary and are most extensive in the Williston and Powder River basins of Montana, Wyoming, and North and South Dakota (see Fig. 12.22). In

◆ **FIGURE 17.44** Bone Valley Member of the Peace River Formation in the IMC Four Corners Mine, Polk County, Florida. (Photo courtesy of T. Scott, Florida Geological Survey.)

addition, to having a low sulfur content, some of these coals occur in beds 30 to 60 m thick!

Gold production from the Pacific Coast, particularly California, comes mostly from Tertiary and Quaternary gravels. The gold occurs as placer deposits, which formed as concentrations of minerals separated from weathered debris by fluvial processes. These placers were derived by erosion of Mesozoic-aged quartz veins in the Sierra Nevada batholith and adjacent country rocks, which were the host rocks for the gold (see Chapter 15 Prologue).

Reservoir rocks of Tertiary age contain a large proportion of the world's petroleum. Some of the petroleum production in the Persian Gulf region, where more than 50% of all petroleum reserves are located, comes from Tertiary reservoir rocks. In fact, oil production from Tertiary reservoir rocks occurs on all continents except Antarctica.

Barely 90 years after the first well was drilled in 1859, the United States became a net petroleum importer and now imports more than half of all the petroleum it consumes, much of it from the Persian Gulf region. Nevertheless, the United States is still the second leading producer, accounting for more than 18% of the total. Much of this production comes from Tertiary reservoirs of the Gulf Coast basin, which includes the Gulf Coastal Plain and the adjacent continental shelf. For example, the Eocene Wilcox Group of Texas (Fig. 17.37) contains considerable petroleum and natural gas. Much of the Gulf Coast petroleum is in structural traps related to salt domes (Fig. 17.38) and other structures.

Several Tertiary basins in southern California are also important areas of petroleum production. These fault-bounded basins formed as a consequence of shearing related to movements on the San Andreas fault (Fig. 17.33). Six oil accumulations in these basins exceed one billion barrels (1 barrel = 42 gallons).

A variety of other Tertiary mineral deposits are important. For example, the United States must import almost all manganese used in the manufacture of steel. The largest manganese deposits are in Lower Tertiary rocks of Russia. A number of small copper- and molybdenum-bearing stocks in Utah, Nevada, Arizona, and New Mexico were emplaced during the Tertiary. One Tertiary molybdenum deposit in Colorado accounts for much of the world production of this element. And Tertiary sand and gravel as well as evaporites, building stone, and clay deposits are quarried from many areas around the world.

Chapter Summary

1. The rifting of Pangaea that began during the Late Triassic continued through the Cenozoic and accounts for the present distribution of continents and oceans.

2. Cenozoic orogenic activity occurred mostly in two major belts—the Alpine-Himalayan orogenic belt and the circum-Pacific orogenic belt. Each belt is composed of smaller units called orogens.

3. The Alpine orogeny resulted from convergence of the African and Eurasian plates. Mountain building occurred in southern Europe, the Middle East, and northern Africa. Plate motions also caused the closure of the Mediterranean basin, which became a site of evaporite deposition.

4. India separated from Gondwana, moved north, and eventually collided with Asia, causing the uplift of the Himalayas.

5. Arc orogens characterize the western and northern Pacific Ocean basin. Back-arc spreading appears to be responsible for back-arc marginal basins such as the Sea of Japan.

6. Subduction of oceanic lithosphere occurred along the western margins of the Americas during much of the Cenozoic. This process continues beneath Central and South America, but the North American plate is now bounded mostly by transform faults.

7. The North American Cordillera is a complex mountainous region extending from Alaska into Mexico. Its Cenozoic evolution included deformation during the Laramide orogeny, extensional tectonics that formed the basin-and-range structures, intrusive and extrusive volcanism, and uplift and erosion.

8. Subduction of the Farallon plate beneath North America resulted in the vertical uplifts of the Laramide orogeny. The Laramide orogen is centered in the middle and southern Rockies, but Laramide deformation occurred from Alaska to Mexico.

9. Cordilleran volcanism was more or less continuous through the Cenozoic, but it varied in eruptive style and location. The Columbia River basalts represent one of the world's greatest eruptive events. Volcanism continues in the Cascades of the northwestern United States.

10. Tensional tectonics in the Basin and Range Province yielded north-south oriented, normal faults. Differential movement on these faults produced uplifted ranges separated by broad, sediment-filled basins.

11. The Colorado Plateau was deformed less than other areas in the Cordillera. Late Tertiary uplift and erosion were responsible for the present topography of the region.

12. The westward drift of North America resulted in its collision with the Pacific-Farallon ridge. Subduction ceased, and the continental margin became bounded by major transform faults, except where the Juan de Fuca plate continues to collide with North America.

13. Sediments eroded from Laramide uplifts were deposited in intermontane basins, on the Great Plains, and in a remnant of the Cretaceous epeiric sea in North Dakota.

14. Gulf Coastal Plain facies patterns were controlled by transgressions and regressions of the Cenozoic epeiric sea. A seaward-thickening wedge of sediments pierced by salt domes on the Gulf Coastal Plain contains large quantities of oil and natural gas.

15. Cenozoic uplift and erosion were responsible for the present topography of the Appalachian Mountains. Much of the sediment eroded from the Appalachians was deposited on the Atlantic Coastal Plain.

Important Terms

Alpine-Himalayan orogenic belt
Alpine orogeny
arc orogen
back-arc marginal basin
Basin and Range Province

circum-Pacific orogenic belt
collision orogen
Farallon plate
Laramide orogeny
North American Cordillera

Quaternary
salt dome
San Andreas transform fault
Tejas epeiric sea
Tertiary

Review Questions

1. An area of North America currently being deformed by tensional forces is the:
a. _____ Colorado Plateau; b. _B_ Basin and Range Province: c. _____ Columbia River basalts; d. _____ Interior Lowlands; e. _____ Atlantic Coastal Plain.

2. Numerous salt domes penetrate the Mesozoic and Cenozoic sedimentary rocks of the:
a. _____ northeastern United States; b. _____ Hudson Bay region of Canada; c. _____ Gulf Coastal Plain; d. _____ Badlands of South Dakota; e. _____ Tejas epeiric sea.

3. The Columbia Plateau is a vast area of Tertiary:
a. _____ limestone deposition; b. _____ oil shales; c. _____ basalts; d. _____ strike-slip faulting; e. _____ subduction.

4. Which of the following areas within the North American Cordillera was little deformed during the Laramide orogeny?
a. _____ Colorado Plateau; b. _____ Coast Ranges; c. _____ Laramide orogen; d. _____ Sierra Nevada; e. _____ San Andreas fault.

5. Other than Mount St. Helens, which Cascade Range volcano was active during this century?
a. _____ Mount Rainier; b. _____ Mount Hood; c. _____ Crater Lake; d. _____ Mount Garibaldi; e. _____ Mount Lassen.

6. Tertiary facies relationships of the Gulf Coastal Plain were controlled largely by:
a. _____ transgressions and regressions; b. _____ continental convergence; c. _____ seafloor spreading and subduction; d. _____ glaciation; e. _____ deformation, uplift, and erosion.

7. The Tertiary history of the Appalachians involved mostly:
a. _____ uplift and erosion; b. _____ subduction and island arc collision; c. _____ tension and block-faulting; d. _____ compression and folding; e. _____ carbonate deposition and origin of salt domes.

8. During the Tertiary Period, the _____ Sea was the last of the epeiric seas to invade North America.

a. _____ Sauk; b. _____ Alleghenian; c. _____ Tejas; d. _____ Antler; e. _____ Absaroka.

9. The San Andreas transform fault formed when the _____ plate was finally consumed by subduction beneath North America.
a. _____ Juan de Fuca; b. _____ Farallon; c. _____ Nazca; d. _____ Los Angeles; e. _____ Cocos.

10. Much of the present-day tectonic activity in the Middle East can be accounted for by the northward movement of:
a. _____ North America; b. _____ Eurasia; c. _____ the Arabian plate; d. _____ the Central Indian Ridge; e. _____ India.

11. Much of the sediment eroded from the Himalayas has been deposited in the Arabian Sea and Bay of Bengal as:
a. _____ submarine fans; b. _____ braided stream deposits; c. _____ deltas; d. _____ shallow marine carbonates; e. _____ back-arc marginal basin deposits.

12. Which of the following is an area of active volcanism?
a. _____ Rocky Mountains; b. _____ Great Plains; c. _____ Interior Lowlands; d. _____ Ozark Plateau; e. _____ Cascade Range.

13. Back-arc spreading, a rising mantle plume, and overriding the East Pacific Rise are explanations for the structure of the:
a. _____ Gulf Coastal Plain; b. _____ Basin and Range Province; c. _____ Grand Canyon; d. _____ Alpine-Himalayan orogenic belt; e. _____ Andes Mountains.

14. The Andes Mountains formed as the result of convergence between the South American and _____ plates.
a. _____ Farallon; b. _____ Kula; c. _____ Nazca; d. _____ Mid-Atlantic; e. _____ Caribbean.

15. Some geologists think that the hot spot now beneath Yellowstone National Park, Wyoming, was responsible for volcanism in the:
a. _____ Snake River Plain; b. _____ Colorado Plateau; c. _____ Laramide intermontane basins; d. _____ Juan de Fuca plate; e. _____ Central America orogen.

16. Which of the following statements is correct?
a. _____ the Laramide orogen is farther inland than is typical of most orogens; b. _____ volcanism occurs when a subducted plate moves beneath another at a low angle; c. _____ the Tejas epeiric sea covered most of North America during the Early Tertiary; d. _____ the Great Bahama Bank has been an area of Tertiary detrital sediment deposition; e. _____ the Gulf Coastal Plain Tertiary rocks are mostly carbonates.

17. Explain how rifting in East Africa, the Red Sea, and Gulf of Aden helps us better understand the origin of the Atlantic Ocean basin.

18. Briefly summarize the events responsible for the Alpine orogeny.

19. Explain why granitic intrusions are found in Tibet and why the highest parts of the Himalayas are composed of shallow-marine sedimentary rocks.

20. What is an arc orogen? How do back-arc marginal basins form? Give an example of a back-arc marginal basin.

21. What accounts for the fact that the North American Cordillera in the United States is more complex than other segments of the circum-Pacific orogenic belt?

22. How does the Laramide orogen differ from typical arc orogens?

23. Why was there no intrusion of batholiths and no extrusive volcanism west of the Laramide orogen during the Early Tertiary?

24. How can the Basin and Range Province be explained in the context of plate tectonics?

25. What sequence of events was responsible for the origin of the Queen Charlotte and San Andreas transform faults?

26. Laramide highlands shed large quantities of sediment. Where was this sediment deposited? Give specific examples of depositional sites.

27. Describe the sedimentary facies of the Gulf Coastal Plain. What event or process controlled facies patterns?

28. How do salt domes form, and why are oil and gas sometimes associated with salt domes?

29. Briefly outline the Cenozoic history of the Appalachian Mountains.

30. What type of continental margin is represented by the Atlantic continental margin of North America, and how did it form?

Additional Readings

Armentrout, J. M., M. R. Cole, and H. Terbest, Jr. 1979. Cenozoic paleogeography of the western United States. *Pacific Coast Paleogeography Symposium 3*. Los Angeles, Calif.: Society of Economic Paleontologists and Mineralogists.

Bonatti, E. 1987. The rifting of continents. *Scientific American* 256, no. 3: 96–103.

Dickinson, W. R. 1978. Plate tectonics of the Laramide orogeny. In *Laramide Folding Associated with Basement Block Faulting in the Western United States,* ed. V. Matthew. *Geological Society of America Memoir* 151: 355–65.

Flores, R. M., and S. S. Kaplan, eds. 1985. Cenozoic paleogeography of the west-central United States. *Rocky Mountain Paleogeography Symposium 3*. Denver, Colo.: Society of Economic Paleontologists and Mineralogists.

Hooper, P. R. 1982. The Columbia River basalts. *Science* 215: 1463–68.

Hsü, K. J. 1972. When the Mediterranean dried up. *Scientific American* 227, no. 6: 27–46.

McBirney, A. R. 1978. Volcanic evolution of the Cascade Range. *Annual Review of Earth and Planetary Sciences* 6: 437–56.

Molnar, P. 1986. The geologic history and structure of the Himalaya. *American Scientist* 74, no. 2: 144–54.

CHAPTER 18

Yosemite Falls, Yosemite National Park, California. (Photo courtesy of Richard L. Chambers.)

CENOZOIC GEOLOGIC HISTORY: QUATERNARY PERIOD

Prologue

Following the Great Ice Age, which ended about 10,000 years ago, a general warming trend occurred that was periodically interrupted by short relatively cool periods. One such cool period, from about 1500 A.D. to the mid- to late-1800s, was characterized by the expansion of small glaciers in mountain valleys and the persistence of sea ice at high latitudes for longer periods than had occurred previously. This interval of nearly four centuries is known as the **Little Ice Age.**

The climatic changes leading to the Little Ice Age actually began by about 1300 A.D. During the preceding centuries, Europe had experienced rather mild temperatures, and the North Atlantic Ocean was warmer and more storm-free than it is at the present. During this time, the Vikings discovered and settled Iceland, and by 1200 A.D., about 80,000 people resided there. They also discovered Greenland and North America and established two colonies on the former and one on the latter. As the climate deteriorated, however, the North Atlantic became stormier, and sea ice occurred further south and persisted longer each year. As a consequence of the poor sea conditions and political problems in Norway, all shipping across the North Atlantic ceased, and the colonies in Greenland and North America eventually disappeared.

During the Little Ice Age, many of the small glaciers in Europe and Iceland expanded and moved far down their valleys, reaching their greatest historic extent by the early 1800s. A small ice cap formed in Iceland where none had existed previously, and glaciers in Alaska and the mountains of the western United States and Canada also expanded to their greatest limits during historic time. Although glaciers caused some problems in Europe where they advanced across roadways and pastures, destroying some villages in Scandinavia and threatening villages elsewhere, their overall impact on humans was minimal. Far more important from the human perspective was that during much of the Little Ice Age the summers in northern latitudes were cooler and wetter.

Although worldwide temperatures were a little lower during this time, the change in summer conditions rather than cold winters, or glaciers caused most of the problems. Particularly hard hit were Iceland and the Scandinavian countries, but at times much of northern Europe was affected (◆ Fig. 18.1). Growing seasons were shorter during many years, resulting in food shortages and a

◆ FIGURE 18.1 During the Little Ice Age, many of the glaciers in Europe, such as this one in Switzerland, extended much farther down their valleys than they do at present. This view titled *The Unterer Grindelwald* was painted in 1826 by Samuel Birmann (1793–1847). (Photo courtesy of Offentliche Kunstsammlung, Kupferstichkabinett Basel.)

The geologic time scale chart at top of page showing:

Archean Eon	Proterozoic Eon	Phanerozoic Eon						
		Paleozoic Era						
Precambrian		Cambrian	Ordovician	Silurian	Devonian	Mississippian	Pennsylvanian	Permian
						Carboniferous		

2,500 M.Y.A.   570 M.Y.A.

number of famines. Iceland's population declined from its high of 80,000 in 1200 to about 40,000 by 1700. Between 1610 and 1870, sea ice was observed near Iceland for as much as three months a year, and each time the sea ice persisted for long periods, poor growing seasons and food shortages followed.

Exactly when the Little Ice Age ended is debatable. Some authorities put the end at 1880, whereas others think it ended as early as 1850. In any case, during the late 1800s, the sea ice was retreating northward, glaciers were retreating back up their valleys, and summer weather was becoming more stable.

❖ INTRODUCTION

The **Quaternary Period** is divided into two epochs, the **Pleistocene** and **Holocene** (or Recent). Together they represent the past 1.6 million years of Earth history. From the perspective of 4.6 billion years, these two epochs are but a mere moment in geologic time. If we were to scale down the geologic history of the Earth to a 24-hour day, this 1.6 million–year episode would represent only the last 38 seconds of the day! However, from our perspective, these last 38 seconds (1.6 million years) are extremely important. Not only did our own species, *Homo sapiens,* evolve but the Pleistocene was a time when large ice sheets advanced and retreated over the land surface, directly and indirectly forming much of our present-day topography. From evidence preserved in the rock record, times of glaciation in the geologic past appear to be relatively rare events; thus, the study of the Pleistocene may provide clues about the causes of previous glaciations.

It now seems hard to believe that so many competent naturalists of the last century were skeptical that widespread glaciation had occurred on the northern continents during the recent past. Many naturalists invoked the biblical flood to explain why large boulders occurred throughout Europe far from their source. Others believed the boulders were rafted to their present location by icebergs floating in floodwaters. It was not until 1837 that the Swiss naturalist Louis Agassiz argued that the large displaced boulders (called *erratics*), coarse-grained sedimentary deposits, polished and striated bedrock, and U-shaped valleys found throughout parts of Europe were the result of huge ice masses moving over the land. Based on Agassiz's observations and arguments, geologists soon accepted that an Ice Age had indeed occurred during the recent geologic past.

We know today that the last Ice Age began about 1.6 million years ago and consisted of several intervals of glacial expansion separated by warmer interglacial periods. Based on the best available evidence, it appears that the present interglacial period began about 10,000 years ago. Geologists do not know, however, whether we are still in an interglacial period or are entering another colder glacial interval.

The Onset of the Ice Age

The onset of glacial conditions really began about 40 million years ago when surface ocean waters at high southern latitudes rapidly cooled, and the water in the deep-ocean basins soon cooled to about 10°C colder than it had been previously. The gradual closing of the Tethys Sea during the Oligocene further limited the flow of warm water to the higher latitudes. By the Middle Miocene an Antarctic ice cap had formed, accelerating the formation of very cold waters. Following a brief warming trend during the Pliocene, ice sheets began forming in the Northern Hemisphere about 1.6 million years ago, and the Pleistocene Ice Age was underway (Table 18.1).

Phanerozoic Eon											
Mesozoic Era			Cenozoic Era								
Triassic	Jurassic	Cretaceous	Tertiary						Quaternary		
			Paleocene	Eocene	Oligocene	Miocene	Pliocene		Pleistocene	Holocene	

5
.A.

66
M.Y.A.

❖ PLEISTOCENE PALEOGEOGRAPHY AND CLIMATE

Paleogeography

While the Pleistocene Epoch is well known for glaciation, it was also a time of tectonic unrest. Folding, faulting, and uplifts were common occurrences, resulting from Pacific and North American plate interaction along the San Andreas fault system. Many of California's coastal oil- and gas-producing fold structures formed during the Pleistocene. Elevated marine terraces covered with Pleistocene sediments testify to repeated uplifts during this time (◆ Fig. 18.2).

Much of the Pleistocene volcanic activity in the western states of the United States and Mexico resulted from continued subduction along the western continental mar-

gin of North America. Crater Lake, Oregon (see Fig. 17.27b), which occupies the collapsed caldera of Mount Mazama; Yellowstone National Park, Wyoming; and Craters of the Moon National Monument, Idaho (◆ Fig. 18.3), are three well-known areas of North American Pleistocene volcanism. Elsewhere, volcanic activity was occurring around the rim of the Pacific Ocean basin—along South America's western margin and in Japan, the Philippines, and the East Indies—and in other locations such as Iceland, Spitzbergen, and the Himalayas as well.

Pleistocene tectonic activity took place in the Sierra Nevada in California, the Grand Tetons of Wyoming, and portions of the central and northern Rocky Mountains as well as in the Alpine-Himalayan orogenic belt (see Chapter 17). Many of these uplifts, particularly the Sierra Nevada, influenced weather conditions by creating rain shadows that led to the present-day deserts located on the landward sides of these mountains.

Climate

The climatic effects responsible for Pleistocene glaciation were, as one would expect, worldwide. Nevertheless, the world was not as frigid as it is commonly portrayed in cartoons and movies, nor was the onset of glacial conditions as rapid as many people believe. In fact, evidence from various lines of research indicates that the world's climate gradually cooled from the beginning of the Eocene through the Pleistocene (◆ Fig. 18.4, p. 522). Oxygen isotope data (the ratio of O^{18} to O^{16}) from deep-sea cores reveal that during the past 2 million years there were at least 20 major warm-cold cycles in which the world's mean temperature ranged from 6°C to 10°C. Furthermore, stratigraphic evidence indicates that there were at least four major episodes of Pleistocene glaciation in North America and six or seven major glacial advances and retreats in Europe.

◆ **FIGURE 18.2** Wavecut terraces on the west side of San Clemente Island, California. Each terrace represents a period when that area was at sea level. The highest terraces are now about 400 m above sea level. (Photo © John Shelton.)

TABLE 18.1   Summary of Events Leading to the Pleistocene Ice Age

Time (Millions of Years Ago)	Events
1.6	Ice Age begins. Ice sheets form in Northern Hemisphere.
5–3	Gulf Stream intensified. Isthmus of Panama closes.
6	Rapid cooling occurs.
15	Present abyssal oceanic circulation established. Iceland-Faroe Ridge sinks; North Atlantic water wells up around Antarctica; more moisture and copious snowpack. Antarctic ice cap develops.
30–25	Antarctica much colder but still no ice cap. Circum-Antarctic current established.
35–30	Partial circulation around Antarctica. Barriers in the western Pacific and closing of Tethys Sea in Middle East and Near East breaks up circum-equatorial circulation pattern.
38	First glaciers on Antarctica. Deep-water circulation speeds up. Rapid cooling of surface water in the south and of deep water everywhere.
48–45	Slight cooling takes place.
>50	No glaciers or ice cap on Antarctica. Deep water much warmer than at present; circulates slowly. Uniform climate and a warm ocean. Ocean waters flow freely around the world at the equator.

During times of glacier growth, those areas in the immediate vicinity of the glaciers experienced short summers and long, wet winters. Areas outside the glaciated regions experienced varied climates. During times of glacial growth, lower ocean temperatures reduced evaporation so that most of the world was drier than it is today. Some areas that are arid today, however, were much wetter during times of glacial growth. For example, the expansion of the cold belts at high latitudes caused the temperate, subtropical, and tropical zones to be compressed toward the equator, and the rain that now falls on the Mediterranean shifted so that it fell on the Sahara of North Africa enabling lush forests to grow in what is now desert. California and the arid Southwest were also

◆ **FIGURE 18.3** Volcanoes in Craters of the Moon National Monument, Idaho. (Photo courtesy of Wayne E. Moore.)

wetter because a high-pressure zone over the northern ice sheet deflected Pacific winter storms southward.

    <u>Pollen analysis</u> is an especially useful tool in determining past climates (◆ Fig. 18.5). Produced by the male reproductive bodies of seed plants, pollen grains have a resistant waxy coating that enables them to be easily preserved as microfossils. Because of its morphology, most pollen is transported by wind and eventually is deposited in streams, lakes, swamps, bogs, or the nearshore marine realm where it is incorporated into the accumulating sediments. Once the pollen is recovered from sediment samples, the family or type of plants or trees that produced it can usually be identified. By counting and statistically analyzing the types of pollen present, scientists can determine the general floral composition of the area and make some general climatic interpretations as well.

    Pollen diagrams (Fig. 18.5), tree-ring analysis (see Chapter 3), and measurements of the fluctuations of mountain glaciers have produced a wealth of data for determining the climatic history of the Northern Hemisphere since the last major ice sheets retreated 10,000 years ago. Pollen data indicate a continuous trend toward warmer, more temperate climatic conditions until around 6,000 years ago. In many areas the interval between 8,000 and 6,000 years ago was a time of very warm temperatures. After this warm period, conditions gradually became cooler and moister favoring the growth of mountain glaciers on the Northern Hemisphere continents. Careful studies of the deposits along the margins of present-day glaciers reveal that glaciers expanded several times during the last 6,000 years (a time called the

Neoglaciation). The last expansion, which occurred between 1500 and the mid- to late-1800s, was the *Little Ice Age* (see the Prologue).

    This brief overview of climatic conditions reveals just how complex climatic patterns were during the last 1.6 million years. Intermediate fluctuations were superimposed on large-scale warm-cold cycles, and those fluctuations in turn were composed of even shorter variations, such as the *Little Ice Age*, which had profound effects on the social and economic fabric of human society.

❖ PLEISTOCENE AND HOLOCENE GEOLOGIC HISTORY

Although there is still debate over which section should serve as the Pleistocene stratotype, the currently accepted date for the beginning of the Pleistocene Epoch is 1.6 million years ago. The Pleistocene-Holocene boundary, placed at 10,000 years ago, is based on a climatic change from cold to warmer conditions corresponding to the melting of the most recent ice sheets. This climatic change is well established by pollen and vegetation changes and variations in the ratio of O^{18} to O^{16} as preserved in the shells of various marine organisms.

Terrestrial Stratigraphy

Shortly after Louis Agassiz first proposed his theory for glaciation, research focused on reconstructing the history of the Ice Age. This work involved recognizing and mapping terrestrial glacial features and placing them in a stratigraphic sequence. From such glacial features as the distribution of moraines, erratic boulders, and drumlins (◆ Fig. 18.6, p. 524), geologists have determined that at their greatest extent, Pleistocene glaciers covered about three times as much of the Earth's surface as they do now and were up to 3 km thick (◆ Fig. 18.7, p. 525). Furthermore, detailed mapping of glacial features has revealed that several glacial advances and retreats occurred, not just a single advance and retreat.

    By mapping the distribution of terminal moraines and correlating the till deposits, geologists have determined that at least four major episodes of Pleistocene glaciation occurred in North America alone. Each of these glacial advances was followed by retreating glaciers and warmer climates. The four **glacial stages,** the *Wisconsin, Illinoian, Kansan,* and *Nebraskan,* are named for the states representing the farthest advance where deposits are well exposed. The three **interglacial stages,** the *Sangamon, Yarmouth,* and *Aftonian,* are named for localities of well-exposed interglacial soil and other deposits (◆ Fig. 18.8,

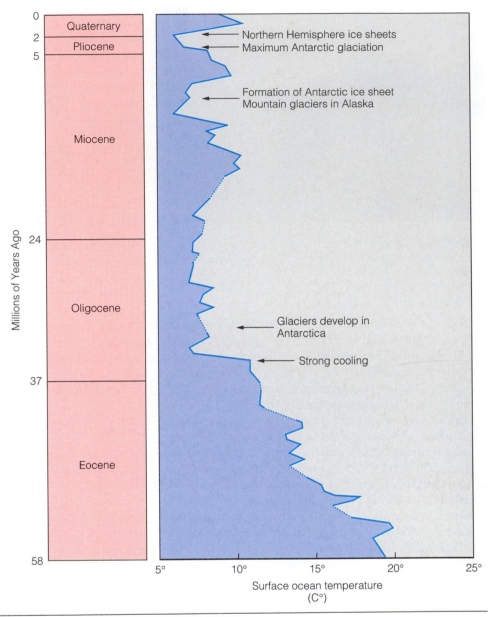

◆ **FIGURE 18.4** Fluctuations in the ratio of O^{18} to O^{16} isotopes from a sediment core in the western Pacific Ocean reveal changes in surface ocean temperatures during the past 58 million years. A change from warm surface waters to colder conditions occurred around 35 million years ago.

Millions of Years Ago

Quaternary
Pliocene
Miocene
Oligocene
Eocene

0
2
5

24

37

58

Northern Hemisphere ice sheets
Maximum Antarctic glaciation

Formation of Antarctic ice sheet
Mountain glaciers in Alaska

Glaciers develop in Antarctica

Strong cooling

5° 10° 15° 20° 25°
Surface ocean temperature (C°)

p. 525). Recent detailed studies of glacial deposits have indicated, however, that there were an as yet undetermined number of pre-Illinoian glacial events, and that the history of glacial advances and retreats in North America is more complex than previously believed. In view of these data, the traditional four-part subdivision of the Pleistocene of North America must be modified.

While at least four major Pleistocene episodes of glaciation are recognized in North America, six or seven major glacial advances and retreats are recognized in Europe, and at least 20 major warm-cold cycles can be detected in deep-sea cores. Why isn't there better correlation among the different areas if glaciation was a worldwide event? Part of the problem is that glacial deposits are typically chaotic mixtures of coarse materials that are difficult to correlate. Furthermore, advances and retreats of the ice sheets usually destroy the sediment left by the previous advance, thus obscuring earlier chronological evidence. Even within a single major glacial advance, several minor advances and retreats may have occurred. For example, careful study of deposits from the Wisconsin glacial stage reveals that at least four distinct

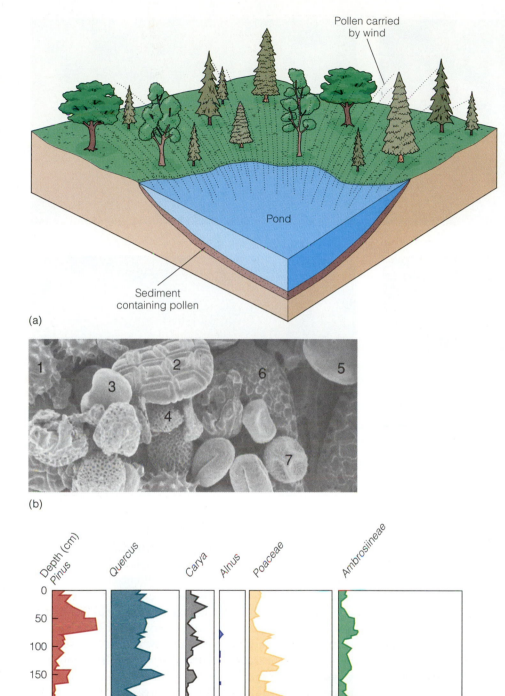

(a)

(b)

(c)

◆ **FIGURE 18.5** Fossil pollen can be used to determine the types of vegetation present and the general climatic conditions of an area during the past. (a) Pollen grains from a variety of nearby vegetation are carried by the wind and fall into a pond where they become part of the accumulating sediment. (b) Scanning electron micrograph of modern pollen grains. Included are (1) sunflower, (2) guajillo, (3) live oak, (4) white mustard, (5) little walnut, (6) lechuguilla, and (7) ash juniper. (Photo courtesy of Vaughn M. Bryant, Jr., Texas A and M University.) (c) Pollen diagrams showing pollen abundance for six different trees. The pollen was recovered from samples taken in the Ferndale Bog, Atoka County, Oklahoma. Changes in the pollen abundance reflect changes in the climate during the past 12,000 years at this locality. (Diagram courtesy of Richard G. Holloway.)

(a)

(c)

(b)

(d)

◆ **FIGURE 18.6** Features characteristic of glaciated areas.
(a) Glacial till is the unsorted sediment deposited by a glacier.
Till from the Mount Cook area, South Island, New Zealand.
(b) Glacial striations are scratches on a rock caused by the
abrasion of a moving glacier. Glacial striations and polish on
the surface of a basalt flow, Devil's Postpile, California.
(c) Drumlins are long, streamlined hills composed of till and
elongated parallel with the direction of glacial movement.
Drumlins in Antrim County, Michigan. (d) Erratics are
ice-transported boulders that have been carried far from their
original source by glaciers. Precambrian erratic on top of
Upper Cambrian sediments, South Hammond, New York.
(Photos a and d courtesy of R. V. Dietrich, photo c courtesy
of B. M. C. Pape.)

fluctuations of the ice margin occurred during the last
70,000 years in Wisconsin and Illinois (◆ Fig. 18.9).

Geologists using sophisticated radiometric dating techniques and detailed pollen studies can now construct an
accurate and detailed chronology for terrestrial Pleistocene
glacial and interglacial episodes. However, because of the
complexity of glacial deposits and the difficulty of separating similar-appearing terrestrial sediments, they cannot always recognize small-scale fluctuations in climate and correlate them with deposits in other areas.

Pleistocene Deep-Sea Stratigraphy

Until recently, the traditional view of Pleistocene chronology was based on terrestrial sequences of glacial sediments. During the early 1960s, however, new evidence
from ocean sediment samples indicated that numerous
climatic fluctuations had occurred during the Pleis-

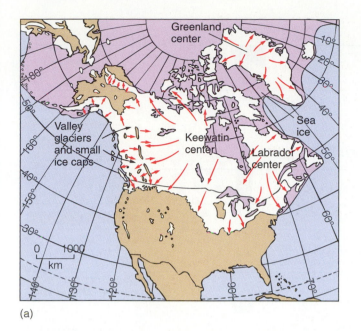

(a)

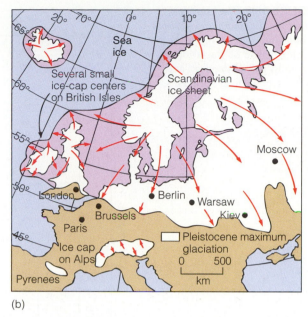

(b)

◆ **FIGURE 18.7** (a) Centers of ice accumulation and maximum extent of Pleistocene glaciation in North America. (b) Centers of ice accumulation and directions of ice movement in Europe during the maximum extent of Pleistocene glaciation.

tocene. These climatic fluctuations are determined from changes in surface ocean temperature recorded in the shells of planktonic foraminifera, which sink to the seafloor after they die and accumulate as sediment. These shells can be recovered for study by coring the ocean

sediments. After the sediment is washed through screens, the foraminiferal shells are concentrated, the different species are identified, and the coiling directions and O^{18} to O^{16} ratios of the shells are analyzed.

One way to determine past changes in ocean surface

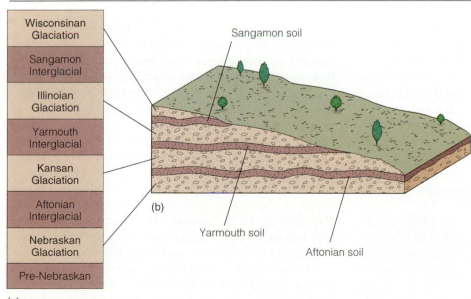

(a)

| Wisconsinan Glaciation |
| Sangamon Interglacial |
| Illinoian Glaciation |
| Yarmouth Interglacial |
| Kansan Glaciation |
| Aftonian Interglacial |
| Nebraskan Glaciation |
| Pre-Nebraskan |

(b)

Sangamon soil

Yarmouth soil

Aftonian soil

◆ **FIGURE 18.8** (a) Standard terminology for Pleistocene glacial and interglacial stages in North America. (b) A reconstruction showing an idealized succession of deposits and soils developed during the glacial and interglacial stages.

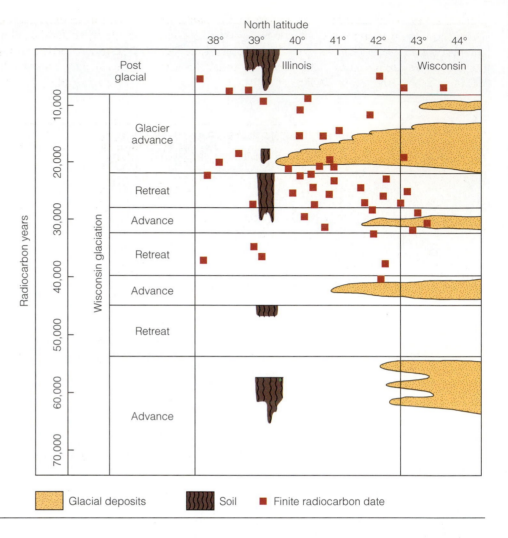

◆ **FIGURE 18.9** The complex fluctuations of climate and glaciers in Illinois and Wisconsin during the last major period of glaciation are preserved in glacial deposits and intervening nonglacial sediments.

temperatures is simply to determine whether the planktonic foraminifera were warm- or cold-water species. Many planktonic foraminifera are sensitive to variations in temperature and migrate to different latitudes when the surface water temperature changes. For example, the tropical species *Globorotalia menardii* is present or absent within Pleistocene cores depending on what the surface water temperature was at the time. During periods of cooler climate, it is found only near the equator, while during times of warming its range extends into the higher latitudes.

Some planktonic foraminifera species change the direction they coil during growth in response to temperature fluctuations. For example, the Pleistocene species *Globorotalia truncatulinoides* predominantly coils to the right in water temperatures above 10°C, but predominantly coils to the left in water below 8° to 10°C

(◆ Fig. 18.10a). On the basis of changing coiling ratios, detailed climatic curves have been constructed for the Pleistocene and earlier epochs (◆ Fig. 18.10b).

Climatic events can also be determined by analyzing changes in the O^{18} to O^{16} ratio preserved in the shells of planktonic foraminifera. The abundance of these two oxygen isotopes in the calcareous ($CaCO_3$) shells of foraminifera is related to the amount of dissolved oxygen in the seawater when the shell is secreted. The exact ratio of these two isotopes reflects the amount of ocean water stored in glacial ice. Seawater has a higher O^{18} to O^{16} ratio than glacial ice because water containing the lighter O^{16} isotope is more easily evaporated than water containing the O^{18} isotope. Therefore, Pleistocene glaciers are enriched in O^{16} relative to O^{18}, while the heavier O^{18} isotope is concentrated in seawater. The declining percentage of O^{16} and consequent rise of O^{18} in seawater during times of glaciation is preserved in the shells of planktonic foraminifera. Consequently, oxygen isotope fluctuations in planktonic foraminifera accurately reflect surface water temperature changes and thus climatic changes due to glaciation (◆ Fig. 18.10c).

Unfortunately, geologists have not yet been able to correlate these detailed climatic changes with corresponding changes recorded in the terrestrial sedimentary record. The time lag between the onset of cooling and the resulting glacial advance produces discrepancies between the

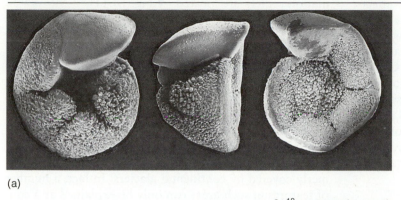
(a)

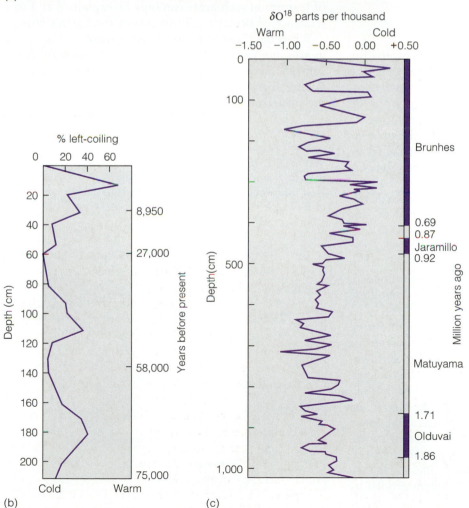

(b)

(c)

◆ **FIGURE 18.10** Two methods used to determine changes in ocean surface water temperatures. (a) Some planktonic foraminifera species coil in different directions depending on water temperatures. *Globorotalia truncatulinoides* coils to the left in water temperatures below 8° to 10°C. (b) Changes in abundance of coiling direction can be used to determine surface water temperatures and hence oceanic climatic conditions. These data, from a core in the Caribbean, show evidence of three intervals of mild climate during the Wisconsin glacial stage. (c) Changes in the O^{18} to O^{16} ratio as preserved in planktonic foraminifera shells also reflect surface water temperature fluctuations and thus climatic changes due to glaciation. (Photo courtesy of Dee Breger, Lamont-Doherty Geological Observatory.)

marine and terrestrial records. Thus, it is unlikely that all the minor climatic fluctuations recorded in deep-sea sediments will ever be correlated with the continental stratigraphic units.

❖ THE EFFECTS OF GLACIATION

Glaciation has had many direct and indirect effects. Movement of glaciers over the Earth's surface has altered the landscape. Sea level has risen and fallen with the melting and formation of glaciers, and these changes in turn have affected the margins of continents. Glaciers have also altered the world's climate, causing cooler and wetter conditions in some areas that are arid to semiarid today. In addition to the usual evidence of glacial activity (Fig. 18.6), one of the largest floods in history was caused by the breaking of a glacier-dammed lake, resulting in huge ripple marks in eastern Washington (see Perspective 18.1).

Changes in Sea Level

More than 70 million km³ of snow and ice covered the continents during the maximum glacial coverage of the Pleistocene. This had a major effect not only on the glaciated areas, but also on regions far removed from the glaciers. The storage of ocean waters in glaciers lowered sea level 130 m and exposed large areas of the present-day continental shelves, which were soon blanketed by vegetation. Indeed, a land bridge existed across the Bering Straits from Alaska to Siberia. Native Americans crossed the Bering land bridge, and various animals migrated between the continents; the American bison, for example, migrated from Asia. The British Isles were connected to Europe during the glacial intervals because the shallow floor of the North Sea was above sea level. When the glaciers disappeared, these areas were again flooded, drowning the plants and forcing the animals to migrate farther inland.

Lowering of sea level caused by glacier growth during the Pleistocene also affected the base level of most major streams. When sea level dropped, streams eroded downward as they sought to adjust to a new lower base level. Stream channels in coastal areas were extended and deepened along the emergent continental shelves. When sea level rose with the melting of the glaciers, the lower ends of stream valleys along the east coast of North America were flooded and are now important harbors, while just off the west coast, they form impressive submarine canyons. Great amounts of sediment eroded by

the glaciers were transported by streams to the sea and thus contributed to the growth of submarine fans along the base of the continental slope.

A tremendous quantity of water is still stored on land in present-day glaciers. If these glaciers should melt completely, sea level would rise by about 70 m, flooding all of the coastal areas where many of the world's large population centers are located (❖ Fig. 18.11).

Glaciers and Isostasy

When enough sediment or ice accumulates in a particular area, the Earth's crust responds by gradually subsiding under the weight of the load. Conversely, when the weight is removed by erosion of the sediment or melting of the ice, the crust slowly rises or rebounds. Such a phenomenon is known as **isostasy,** and there is no question that isostatic rebound has occurred in areas formerly covered by continental glaciers. In fact, a number of features in such areas can only be explained as a consequence of isostatic adjustments of the Earth's crust.

When the Pleistocene ice sheets formed and increased in size, the weight of the ice caused the crust to slowly subside deeper into the mantle. In some places, the Earth's surface was depressed as much as 300 m below the preglacial elevations. As the ice sheets retreated by

❖ **FIGURE 18.11** Large areas of North America—and all the other continents—would be flooded by a 70 m rise in sea level that would result if all the Earth's glacial ice were to melt.

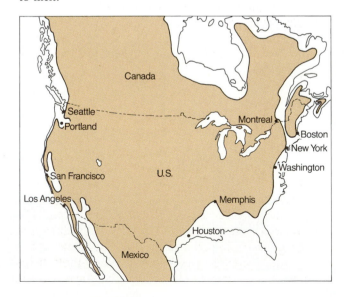

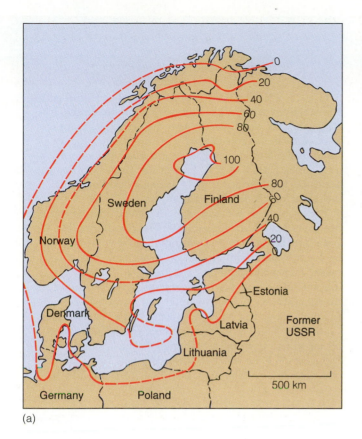

(a)

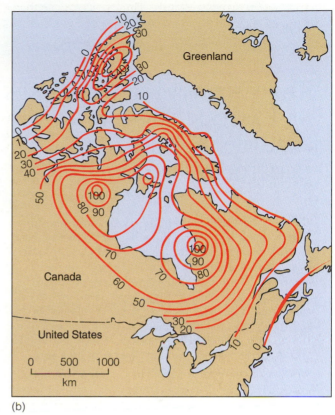

(b)

◆ **FIGURE 18.12** (a) Isostatic rebound in Scandinavia. The lines show rates of uplift in centimeters per century. (b) Isostatic rebound in eastern Canada in meters during the last 6,000 years.

melting, the downwarped areas gradually rebounded to their former positions. Evidence of isostatic rebound can be found in such formerly glaciated areas as parts of Scandinavia and the North American Great Lakes region (◆ Fig. 18.12).

Pluvial and Proglacial Lakes

The climate of the present-day southwestern United States is hot and dry, and the area has few major streams or lakes. During the Wisconsin glacial stage, however, there were many large lakes in what are now dry basins. These lakes formed as a result of greater precipitation and overall cooler temperatures (especially during the summer), which lowered the evaporation rate. At the same time, increased precipitation and runoff helped maintain high water levels. Lakes that formed during

those times are called **pluvial lakes,** and they correspond to the expansion of glaciers elsewhere (◆ Fig. 18.13a, p. 532). Ancient shoreline features such as wavecut cliffs, beaches, and deltas found today on the sides of the enclosing slopes are evidence of these former lakes (◆ Fig. 18.13b, p. 532). The largest of these lakes was Lake Bonneville, which attained a maximum size of 50,000 sq km² and a depth of at least 335 m. The vast salt deposits of the Bonneville Salt Flats west of Salt Lake City, Utah, formed as parts of this ancient lake dried up; Great Salt Lake is simply the remnant of this once great lake.

Another large pluvial lake (Lake Manly) existed in Death Valley, California, which is now the hottest, driest place in North America. During the Pleistocene, however, that area received enough rainfall to maintain a lake 145 km long and 178 m deep. When the lake evaporated, the dissolved salts were precipitated on the valley floor; some of these evaporite deposits, especially borax, which is used for ceramic glazes,

THE CHANNELLED SCABLANDS OF EASTERN WASHINGTON— A CATASTROPHIC PLEISTOCENE FLOOD

The term *scabland* is used in the Pacific Northwest to describe areas from which the surface deposits have been scoured, thus exposing the underlying rock. Such an area exists in a large part of eastern Washington where numerous deep and generally dry channels are present. Some of these channels, cut into basalt lava flows, are more than 70 m deep, and their floors are covered by gigantic "ripple marks" as much as 10 m high and 70 to 100 m apart. Additionally, a number of high hills in the area are arranged such that they appear to have been islands in a large braided stream.

In 1923, J Harlan Bretz proposed that the channelled scablands of eastern Washington were formed during a single, gigantic flood. Bretz's unorthodox explanation was rejected by most geologists who preferred a more traditional interpretation based on normal stream erosion over a long period of time. In contrast, Bretz held that the scablands were formed rapidly during a flood of glacial meltwater that lasted only a few days.

The problem with Bretz's hypothesis was that he could not identify an adequate source for his floodwater. He knew that the glaciers had advanced as far south as Spokane, Washington, but he could not explain how so much ice melted so rapidly. The answer to Bretz's dilemma came from western Montana where an enormous ice-dammed lake (Lake Missoula) had formed. Lake Missoula formed when an advancing glacier plugged the Clark Fork Valley at Ice Cork, Idaho, causing the water to fill the valleys of western Montana (◆ Fig. 1). At its highest level, Lake Missoula covered about 7,800 km² and contained an estimated 2,090 km³ of water (about

---

◆ **FIGURE 1** Location of glacial Lake Missoula and the channelled scablands of eastern Washington.

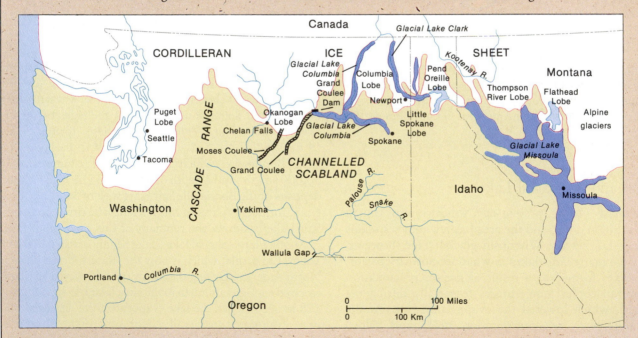

◆ **FIGURE 2** The horizontal lines on Sentinel Mountain at Missoula, Montana, are wavecut shorelines of glacial Lake Missoula. (Photo courtesy of P. Weis, USGS.)

◆ **FIGURE 3** These gravel ridges are the so-called giant ripple marks that formed when glacial Lake Missoula drained across this area near Camas Hot Springs, Montana. (Photo courtesy of P. Weis, USGS.)

42% of the volume of present-day Lake Michigan). The shorelines of Lake Missoula are still clearly visible on the mountainsides around Missoula, Montana (◆ Fig. 2).

When the ice dam impounding Lake Missoula failed, the water rushed out at tremendous velocity and drained south and southwest across Idaho and into Washington. The maximum rate of flow is estimated to have been nearly 11 million m^3/sec, about 55 times greater than the average discharge of the Amazon River. When these raging floodwaters reached eastern Washington, they stripped away the soil and most of the surface sediment, carving out huge valleys in solid bedrock. The currents were so powerful and turbulent they plucked out and moved pieces of basalt measuring 10 m across. Within the channels, sand and gravel were shaped into huge ridges, the so-called giant ripple marks (◆ Fig. 3).

Bretz originally believed that one massive flood formed the channeled scablands, but geologists now know that Lake Missoula formed, flooded, and re-formed at least four times and perhaps as many as seven times. The largest lake formed 18,000 to 20,000 years ago, and its draining produced the last great flood.

How long did the flood last and did humans witness it? It has been estimated that approximately one month passed from the time the ice dam first broke and water rushed out onto the scablands to the time the scabland streams returned to normal flow. No one knows for sure if anyone witnessed the flood. The oldest known evidence of humans in the region is from the Marmes Man site in southeastern Washington dated at 10,130 years ago, nearly 2,000 years after the last flood from Lake Missoula. However, it is now generally accepted that Native Americans were present in North America at least 15,000 years ago.

◆ **FIGURE 18.13** (a) Pleistocene pluvial lakes in the western United States. (b) Evidence of the extent of Lake Bonneville can be seen in the ancient shorelines cut into the slope of the Oquirrh Range, Utah.

(a)

fertilizers, glass, solder, and pharmaceuticals, are important mineral resources (◆ Fig. 18.14).

In contrast to pluvial lakes, which form far from glaciers, **proglacial lakes** are formed by the meltwater accumulating along the margins of glaciers. In fact, in many proglacial lakes, one shoreline is the ice front itself, while the other shorelines consist of moraines. Lake Agassiz, named in honor of the French naturalist Louis Agassiz, was a large proglacial lake covering about 250,000 km^2 of North Dakota and Manitoba, Saskatchewan, and Ontario, Canada. It persisted until the glacial ice along its northern margin melted, at which time the lake was able to drain northward into Hudson Bay.

Numerous proglacial lakes existed during the Pleistocene, but most of these eventually disappeared. Lake Agassiz and a number of smaller lakes drained when the glaciers disappeared, but others simply filled with sediment. Notable exceptions, however, are the Great Lakes, all of which first formed as proglacial lakes.

History of the Great Lakes

The Great Lakes of North America provide us with an excellent example of the direct effects of glaciation. Before the Pleistocene, no large lakes existed in the Great

(b)

◆ FIGURE 18.14 Twenty-mule teams carried borax out of Death Valley during the late 1800s. (Photo courtesy of United States Borax and Chemical Corporation.)

Lakes region, which was then an area of generally flat lowlands with broad stream valleys draining to the north (◆ Fig. 18.15). As the glaciers advanced southward, they eroded the stream valleys more deeply, forming what were to become the basins of the Great Lakes. During these glacial advances, the ice front moved forward as a series of lobes, some of which flowed into the preexisting lowlands where the ice became thicker and moved more rapidly. As a consequence, the lowlands were deeply eroded—four of the five Great Lakes basins were eroded below sea level.

At their greatest extent, the glaciers covered the entire Great Lakes region and extended far to the south (Fig. 18.7a). As the ice sheet retreated northward during the late Pleistocene, the ice front periodically stabilized, and numerous recessional moraines were deposited. By about 14,000 years ago, parts of the Lake Michigan and Lake Erie basins were ice-free, and glacial meltwater began forming proglacial lakes (◆ Fig. 18.16). As the retreat of the ice sheet continued—although periodically interrupted by minor readvances—the Great Lakes basins were uncovered, and the lakes expanded until they eventually reached their present size and configuration (Fig. 18.16). Currently, the Great Lakes contain nearly 23,000 km³ of water, about 18% of the water in all freshwater lakes.

Although the history of the Great Lakes just presented is generally correct, it is oversimplified. For instance, the areas and depths of the evolving Great Lakes fluctuated widely in response to minor readvances of the ice front. Furthermore, as the lakes filled, they spilled over the low-est parts of their margins, thus cutting outlets that partly drained them. And finally, as the glaciers retreated northward, isostatic rebound initially raised the southern parts of the Great Lakes region, greatly altering their drainage systems.

❖ CAUSES OF PLEISTOCENE GLACIATION

Thus far we have examined some of the effects of glaciation, but have not addressed the central questions of what causes large-scale glaciation and why so few episodes of widespread glaciation have occurred. For more than a century, scientists have been attempting to develop a comprehensive theory explaining all aspects of ice ages, but have not yet been completely successful. One reason for their lack of success is that the climatic changes responsible for glaciation, the cyclic occurrence of glacial-interglacial episodes, and short-term events such as the Little Ice Age operate on vastly different time scales.

Only a few periods of glaciation are recognized in the geologic record, each separated from the others by long intervals of mild climate. Such long-term climatic changes probably result from slow geographic changes related to plate tectonic activity. Moving plates can carry continents to high latitudes where glaciers can exist, provided that

◆ FIGURE 18.15 Theoretical preglacial drainage in the Great Lakes region. The divide separating the preglacial Mississippi and St. Lawrence drainage basins was probably near its present location. The future sites of the Great Lakes are outlined by dotted lines.

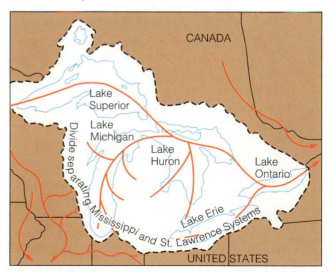

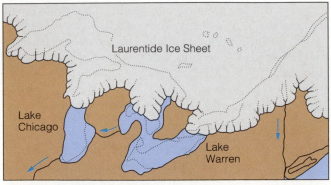

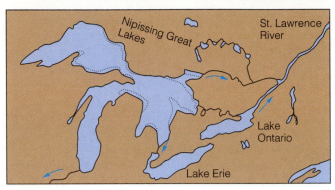

13,000 years ago

11,500 years ago

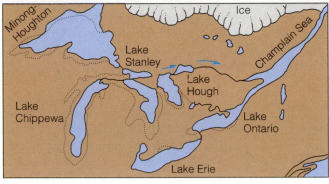

9,500 years ago

6,000 years ago

◆ **FIGURE 18.16** Four stages in the evolution of the Great Lakes. As the glacial ice retreated northward, the lake basins began filling with meltwater. The dotted lines indicate the present-day shorelines of the lakes.

they receive enough precipitation as snow. Plate collisions, the subsequent uplift of vast areas far above sea level, and the changing atmospheric and oceanic circulation patterns caused by the changing shapes and positions of plates also contribute to long-term climatic change.

Intermediate-term climatic events, such as the glacial-interglacial episodes of the Pleistocene, occur on time scales of tens to hundreds of thousands of years. The cyclic nature of this most recent episode of glaciation has long been a problem in formulating a comprehensive theory of climatic change.

The Milankovitch Theory

A particularly interesting hypothesis for intermediate-term climatic events was put forth by the Yugoslavian astronomer Milutin Milankovitch during the 1920s. He proposed that minor irregularities in the Earth's rotation and orbit are sufficient to alter the amount of solar radiation that the Earth receives at any given latitude and hence can affect climatic changes. Now called the **Milankovitch theory,** it was initially ignored, but has received renewed interest during the last 20 years.

Milankovitch attributed the onset of the Pleistocene Ice Age to variations in three parameters of the Earth's orbit (◆ Fig. 18.17). The first of these is orbital eccentricity, which is the degree to which the orbit departs from a perfect circle (Fig. 18.17a). Calculations indicate a roughly 100,000-year cycle between times of maximum eccentricity. This corresponds closely to 20 warm-cold climatic cycles that occurred during the Pleistocene. The second parameter is the angle between the Earth's axis and a line perpendicular to the plane of the ecliptic (Fig. 18.17b). This angle shifts about 1.5° from its current value of 23.5° during a 41,000-year cycle. The third

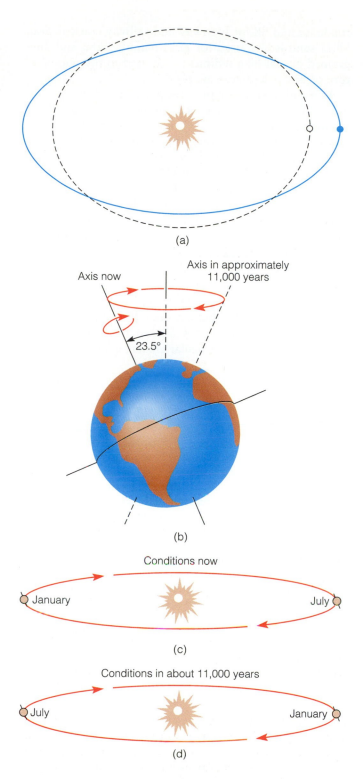

(a)

Axis now

Axis in approximately 11,000 years

23.5°

(b)

Conditions now

January

July

(c)

Conditions in about 11,000 years

July

January

(d)

◆ **FIGURE 18.17** (a) The Earth's orbit varies from nearly a circle (dashed line) to an ellipse (solid line) and back again in about 100,000 years. (b) The Earth moves around its orbit while spinning about its axis, which is tilted to the plane of the ecliptic at 23.5° and points toward the North Star. The Earth's axis of rotation slowly moves and traces out the path of a cone in space. (c) At present, the Earth is closest to the Sun in January when the Northern Hemisphere experiences winter. (d) In about 11,000 years, as a result of precession, the Earth will be closer to the Sun in July, when summer occurs in the Northern Hemisphere.

parameter is the precession of the equinoxes, which causes the position of the equinoxes and solstices to shift slowly around the Earth's elliptical orbit in a 23,000-year cycle (Fig. 18.17c and d).

Continuous changes in these three parameters cause the amount of solar heat received at any latitude to vary slightly over time. The total heat received by the planet, however, remains little changed. Milankovitch proposed, and now many scientists agree, that the interaction of these three parameters provides the triggering mechanism for the glacial-interglacial episodes of the Pleistocene.

Short-Term Climatic Events

Climatic events having durations of several centuries, such as the Little Ice Age, are too short to be accounted for by plate tectonics or Milankovitch cycles. Several hypotheses have been proposed, including variations in solar energy and volcanism.

Variations in solar energy could result from changes within the Sun itself or from anything that would reduce the amount of energy the Earth receives from the Sun. The latter could result from the solar system passing through clouds of interstellar dust and gas or from substances in the Earth's atmosphere reflecting solar radiation back into space. Records kept over the past 75 years, however, indicate that during this time the amount of solar radiation has varied only slightly. Thus, although variations in solar energy may influence short-term climatic events, such a correlation has not been demonstrated.

During large volcanic eruptions, tremendous amounts of ash and gases are spewed into the atmosphere where they reflect incoming solar radiation and thus reduce atmospheric temperatures. Small droplets of sulfur gases remain in the atmosphere for years and can have a significant effect on the climate. Several such large-scale volcanic events have been recorded, such as the 1815 eruption of Tambora, and are known to have had climatic effects. However, no relationship between periods of volcanic activity and periods of glaciation has yet been established.

❖ QUATERNARY MINERAL RESOURCES

Most people are probably not aware of the many mineral resources that formed as a result of Quaternary geologic events. Yet both directly and indirectly, glaciation played an important role in the formation and distribution of Quaternary mineral resources.

Sand and gravel deposits resulting from glacial activity are a valuable resource in many formerly glaciated areas of the world. In fact, such deposits are one of the most valuable nonmetallic mineral resources in many countries. Most Pleistocene sand and gravel deposits originated as postglacial floodplain or terrace gravels, outwash sediment, or esker deposits. The bulk of the sand and gravel in the United States and Canada is used as roadbase and fill for highway and railway construction. Silica sand is also used for glass manufacture, and fine-grained glacial lake sediments are used in the manufacture of bricks and ceramic products.

As mentioned in Chapter 16, placer deposits containing gold are another important mineral resource of the Quaternary. The California Gold Rush of the late 1840s was fueled by the discovery of gold in placer deposits of the American River. The discovery of placer gold deposits in the Yukon Territory of Canada was primarily responsible for the settlement of the area.

The periodic evaporation of pluvial lakes in the Death Valley region of California during the Pleistocene led to the concentration of many evaporite minerals such as borax. During the 1880s, borax was transported from Death Valley to Barstow, California, by the famous 20-mule team wagon trains (Fig. 18.14).

Another Quaternary mineral resource is peat, a vast potential energy resource that has been developed in Canada and Ireland. These peatlands formed from plant assemblages as the result of particular climate conditions.

Chapter Summary

1. The Quaternary Period consists of two unequal epochs: the Pleistocene Epoch, from 1.6 million years ago to 10,000 years ago, and the Holocene Epoch, from 10,000 years ago to the present.

2. In addition to glaciation, the Pleistocene was also a time of tectonic unrest during which folding, faulting, uplifts, and volcanism were common occurrences.

3. During the Pleistocene Epoch, glaciers covered about 30% of the land surface. About 20 warm-cold Pleistocene climatic cycles are recognized from paleontologic and oxygen isotope data derived from deep-sea cores.

4. Several intervals of widespread glaciation, separated by interglacial periods, occurred in North America. The other Northern Hemisphere continents were also affected by widespread Pleistocene glaciation.

5. Areas far beyond the ice were affected by Pleistocene glaciation: climate belts were compressed toward the equator, large pluvial lakes existed in what are now arid regions, and sea level was as much as 130 m lower than at present.

6. Loading of the Earth's crust by Pleistocene glaciers caused isostatic subsidence. When the glaciers disappeared, isostatic rebound began and still continues in some areas.

7. Major glacial intervals separated by tens or hundreds of millions of years probably occur as a consequence of the changing positions of tectonic plates, which in turn cause changes in oceanic and atmospheric circulation patterns.

8. Currently, the Milankovitch theory is widely accepted as the explanation for glacial-interglacial intervals.

9. The reasons for short-term climatic changes, such as the Little Ice Age, are not understood. Two proposed causes for such events are changes in the amount of solar energy received by the Earth and volcanism.

10. Mineral resources of the Quaternary are mainly sand and gravel along with some evaporite minerals such as borax.

Important Terms

glacial stage
Holocene Epoch
interglacial stage
isostasy

Little Ice Age
Milankovitch theory
Pleistocene Epoch
pluvial lake

pollen analysis
proglacial lake
Quaternary Period

Review Questions

1. The most recent ice age occurred during the:

   a. _____ Archean Eon; b. _____ Pleistocene Epoch;
   c. _____ Mesozoic Era; d. _____ Cambrian Period;
   e. _____ Tertiary Period.

2. How many years ago did the Pleistocene Epoch begin?
   a. _____ 6.6 million; b. _____ 2 million; c. _____ 1.6
   million; d. _____ 1 million; e. _____ 10,000.

3. In addition to glaciation, the Pleistocene is known for:
   a. _____ volcanism; b. _____ folding; c. _____
   faulting; d. _____ uplift; e. _____ all of these.

4. During the Pleistocene most of the world was
   _____ than it is today.
   a. _____ drier; b. _____ wetter; c. _____ warmer;
   d. _____ answers (a) and (b); e. _____ answers (b) and
   (c).

5. Pollen analysis can reveal past:
   a. _____ vegetation; b. _____ climate; c. _____
   methods of transportation of the enclosing sediment;
   d. _____ answers (a) and (b); e. _____ answers (b) and
   (c).

6. The cool period from about 1500 to the mid- to
   late-1800s is known as the:
   a. _____ Neoglaciation; b. _____ Little Ice Age;
   c. _____ Big Ice Age; d. _____ glacial maxima;
   e. _____ none of these.

7. Unsorted sediment directly deposited by a glacier is:
   a. _____ outwash; b. _____ loess; c. _____ till;
   d. _____ moraine; e. _____ drumlin.

8. Which of the following is not one of the North
   American glacial stages?
   a. _____ Wisconsin; b. _____ Nebraskan; c. _____
   Kansan; d. _____ Iowan; e. _____ Illinoian.

9. Which of the following is a North American interglacial
   stage?
   a. _____ Sangamon; b. _____ Würm; c. _____ Riss;
   d. _____ Mindel; e. _____ Gunz.

10. Which of the following can be used to determine ocean
    surface temperatures during the Pleistocene?
    a. _____ O^{18} to O^{16} isotope ratios; b. _____
    identification of planktonic foraminiferal species;
    c. _____ coiling ratios of planktonic foraminiferal
    species; d. _____ ocean chemistry; e. _____ all of
    these.

11. Lakes that formed far from glaciers during times of
    glaciation in what are now dry areas are known as:

    a. _____ proglacial; b. _____ pluvial; c. _____ playa;
    d. _____ peneplain; e. _____ none of these.

12. The phenomenon in which the crust sinks and then rises
    as a result of loading by glaciers followed by melting is:
    a. _____ isostasy; b. _____ progradation; c. _____
    proglaciation; d. _____ peneplanation; e. _____
    recumbency.

13. The formation of land bridges such as the one across the
    Bering Straits or between Europe and England during
    the Pleistocene resulted from:
    a. _____ a rise in sea level; b. _____ a drop in sea
    level; c. _____ isostatic adjustment of the crust;
    d. _____ tectonic activity; e. _____ none of these.

14. Which of the following is not one of the parameters
    Milankovitch used to explain the glacial-interglacial
    episodes of the Pleistocene?
    a. _____ volcanic eruptions; b. _____ orbital
    eccentricity; c. _____ the angle between the Earth's axis
    and a line perpendicular to the plane of the ecliptic;
    d. _____ precession of the equinoxes; e. _____ all of
    these.

15. Short-term climatic events can be explained by:
    a. _____ variations in solar energy; b. _____ volcanic
    eruptions; c. _____ plate collisions; d. _____ answers
    (a) and (b); e. _____ answers (b) and (c).

16. Which of the following is not an important Quaternary
    mineral resource?
    a. _____ gold placers; b. _____ sand; c. _____ gravel;
    d. _____ borax; e. _____ petroleum.

17. What direct evidence is there for Pleistocene glaciation?

18. How did glaciers indirectly affect areas far removed
    from the ice sheets?

19. Why don't the Pleistocene and Holocene chronologies
    based on terrestrial stratigraphy and deep-sea cores
    correlate in terms of the number of glacial-interglacial
    intervals?

20. How can planktonic foraminifera be used to determine
    past climatic conditions?

21. Why is pollen analysis such a good tool for
    reconstructing Pleistocene conditions?

22. Explain how the Milankovitch theory accounts for
    glacial-interglacial cyclicity.

23. Explain how pluvial and proglacial lakes differ.

24. Briefly discuss the geologic history of the Great Lakes.

25. What are the major Quaternary mineral resources and
    how did they form?

Additional Readings

Covey, C. 1984. The Earth's orbit and the ice ages. *Scientific American* 250, no. 2: 58–67.

Denton, G. H., and T. J. Hughes. 1981. *The last great ice sheets*. New York: John Wiley & Sons.

Flint, R. F. 1971. *Glacial and Quaternary geology*. New York: John Wiley & Sons.

Fulton, R. J., ed. 1989. *Quaternary geology of Canada and Greenland*. The geology of North America, vol. K-1. Ottawa, Canada: Geological Survey of Canada.

Morrison, R. B., ed. 1991. *Quaternary nonglacial geology: Conterminous U.S.* The geology of North America, vol. K-2. Boulder, Colo.: Geological Society of America.

Nilsson, T. 1983. *The Pleistocene: Geology and life in the Quaternary ice age*. Holland: D. Reidel Publishing Co.

Ruddiman, W. F., and H. E. Wright, Jr., eds. 1987. *North America and adjacent oceans during the last deglaciation*. The geology of North America, vol. K-3. Boulder, Colo.: Geological Society of America.

CHAPTER 19

Chapter Outline

Mural showing Pliocene mammals of the grasslands of western North America. The mammals shown include *Amebeledon,* a shovel-tusked mastodon (background); *Teleoceras,* a short-legged rhinoceros (left center); *Cranioceras,* a horned, hoofed mammal (left foreground); *Synthetoceras,* a hoofed mammal with a horn on its snout (right foreground); *Merycodus,* an extinct pronghorn (middle foreground); and three small mammals in the foreground. (Photo courtesy of the Smithsonian Institution. Reconstruction painting of *Amebeledon* and *Teleoceras* mammals by Jay H. Matternes © Copyright 1982.)

LIFE OF THE CENOZOIC ERA

Prologue

The evolutionary history of mammals is better known than the history of any other class of vertebrates. One of the reasons for this is that Cenozoic terrestrial deposits are more common than those of the Mesozoic and Paleozoic, and many of these deposits are accessible at the surface or in the shallow subsurface. And being younger, these deposits have had less time for contained fossils to be altered or destroyed. Accordingly, many mammals such as horses, camels, elephants, rhinoceroses, and whales have good fossil records.

Cenozoic terrestrial deposits are most widespread in western North America, thus most fossils come from this area. For example, Badlands National Park in South Dakota is noted for its numerous fossil mammals as well as its rugged, scenic topography (◆ Fig. 19.1). The deposits are part of a vast blanket of sediments of Late Eocene to latest Oligocene age that were deposited in meandering stream channels, on floodplains, and in small lakes.

Badlands develop in dry areas with sparse vegetation and nearly impermeable yet easily eroded rocks. Rain falling on such unprotected rocks rapidly runs off and intricately dissects the surface by forming numerous closely spaced, small gullies and deep ravines, thus yielding sharp angular slopes and steep pinnacles. The topography developed on the Brule Formation (Fig. 19.1) is an example of such intricate dissection. In marked contrast, the underlying Chadron Formation erodes to form smooth, rounded surfaces (Fig. 19.1).

The scenic badlands are reason enough to visit the park but it has more to offer; the rocks contain the most complete succession of mammal fossils known anywhere in the world. The Park Service has left a number of these fossil mammals exposed but protected for viewing by park visitors. Among these mammals are rodents, rabbits, doglike carnivores, saber-toothed cats, and various hoofed mammals including horses, camels, and the extinct oreodonts and titanotheres.

In the east, mammal fossil localities of Pleistocene age are fairly common, but those of Tertiary age are rare. There are, however, some notable exceptions. For example, the rocks composing the Calvert Cliffs in Maryland contain fossil sea cows and whales (see Fig. 17.42). Florida is notable in being the one area in the east with numerous terrestrial mammal fossil localities. Terrestrial mammal fossils are known from all Cenozoic epochs except the Paleocene and Eocene, although fossil whales are known from Eocene rocks.

◆ FIGURE 19.1 Outcrops of the Oligocene Brule Formation (on skyline) and Chadron Formation (foreground) in Badlands National Park, South Dakota. Notice that erosion of the Brule Formation yields sharp, angular slopes while smooth, rounded slopes develop on the Chadron Formation.

Archean Eon	Proterozoic Eon	Phanerozoic Eon						
		Paleozoic Era						
Precambrian		Cambrian	Ordovician	Silurian	Devonian	Mississippian	Pennsylvanian	Permian
						Carboniferous		

2,500
M.Y.A.

570
M.Y.A.

24
M.Y

One of the most notable Florida fossil mammal localities is the Thomas Farm quarry in Gilchrist County, from which fossil horses, camels, rhinoceroses, deerlike animals, and various carnivores have been recovered. During the Miocene, this area was apparently a sinkhole or cavern as indicated by numerous bat remains.

The Upper Miocene Bone Valley Member of the Peace River Formation in Polk County, Florida (see Fig. 17.44) has yielded giant sloths, sea cows, whales, and elephants in addition to the mammals from the Thomas Farm quarry. Pliocene mammal fossils are also well known, and during the Pleistocene, Florida had an extraordinary variety of mammals. In fact, the teeth and bones of Pleistocene elephants and various hoofed mammals are among the most common fossils found in the state (◆ Fig. 19.2).

❖ INTRODUCTION

The world's flora and fauna continued to change during the Cenozoic Era as more familiar types of plants and animals appeared. In this chapter we are concerned mainly with the adaptive radiation of mammals, especially some of the more familiar types such as carnivores, elephants, and hoofed mammals. Recall from Chapter 16 that mammals evolved from cynodonts during the Late Triassic but were small and not very diverse through the rest of the Mesozoic. Following the Mesozoic extinctions, however, mammals diversified and soon became the most abundant land-dwelling vertebrate animals.

Although we emphasize mammalian evolution in this chapter, one should be aware of other important events. The flowering plants continued to dominate land-plant communities, the present-day groups of birds appeared early in the Tertiary, and some marine invertebrates continued to diversify.

❖ MARINE INVERTEBRATES AND PHYTOPLANKTON

The Cenozoic marine ecosystem was populated mostly by those plants, animals, and single-celled organisms that survived the terminal Mesozoic extinction event. Gone were the ammonites, rudists, and most of the planktonic foraminifera. Cenozoic invertebrate groups that were especially prolific were the foraminifera, radiolarians, corals, bryozoans, mollusks, and echinoids. The marine invertebrate community in general became more provincial during the Cenozoic because of changing ocean currents and latitudinal temperature gradients. In addition, the Cenozoic marine invertebrate faunas became more familiar in appearance.

Entire families of phytoplankton became extinct at the end of the Cretaceous. Only a few species in each major

◆ **FIGURE 19.2** Fossilized bones and teeth of Pleistocene mammals from Florida.

Phanerozoic Eon										
Mesozoic Era			Cenozoic Era							
Triassic	Jurassic	Cretaceous	Tertiary						Quaternary	
			Paleocene	Eocene	Oligocene	Miocene	Pliocene		Pleistocene	Holocene

45
M.A.

66
M.Y.A.

group survived into the Tertiary. These species diversified and expanded during the Cenozoic, perhaps because of decreased competitive pressures. The coccolithophores, diatoms, and dinoflagellates all recovered from their Late Cretaceous reduction in numbers to flourish during the Cenozoic. The diatoms were particularly abundant during the Miocene, probably because of increased volcanism during this time. Volcanic ash provided increased dissolved silica in seawater and was used by the diatoms to construct their skeletons. Massive Miocene diatomites have been found in California (◆ Fig. 19.3).

The foraminifera comprised a major component of the Cenozoic marine invertebrate community. Though dominated by relatively small forms (◆ Fig. 19.4), it included some exceptionally large forms that lived in the warm waters of the Cenozoic Tethys Sea. Shells of these larger

forms accumulated to form thick limestones, some of which were used by the ancient Egyptians to construct the Sphinx and the Pyramids of Gizeh.

The corals were perhaps the main beneficiary of the terminal Cretaceous extinctions. Having relinquished their reef-building role to the rudists during the mid-Cretaceous, corals again became the dominant reef-builders during the Cenozoic. They formed extensive reefs in the warm waters of the Cenozoic oceans and were particularly prolific in the Caribbean and Indo-Pacific regions (◆ Fig. 19.5).

Other suspension feeders such as the bryozoans and crinoids were also abundant and successful during the Tertiary as well as the Quaternary. The bryozoans, in particular, were very abundant. Perhaps the least important of the Cenozoic marine invertebrates were the brachiopods, with fewer than 60 genera surviving today.

Just as during the Mesozoic, bivalves and gastropods were two of the major groups of marine invertebrates during the Tertiary, and they had a markedly modern appearance. Following the extinction of the ammonites and belemnites at the end of the Cretaceous, the Cenozoic cephalopod fauna consisted of nautiloids and shell-less cephalopods such as squids and octopuses.

The echinoids continued their expansion in the infaunal habitat and were particularly prolific during the Tertiary. New forms such as sand dollars evolved during this time from biscuit-shaped ancestors (◆ Fig. 19.6).

❖ CENOZOIC BIRDS

The first members of many living orders and families of birds such as owls, hawks, ducks, penguins, and vultures evolved during the Early Tertiary. Today, birds vary considerably in size and adaptations, but their basic skeletal structure has not changed significantly throughout the

◆ **FIGURE 19.3** Outcrop of diatomite from the Miocene Monterey Formation, Newport Lagoon, California.

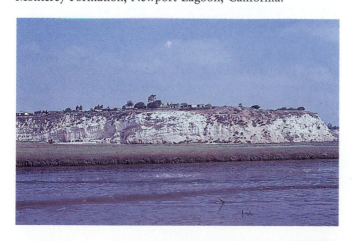

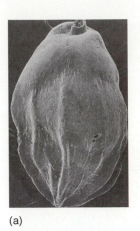
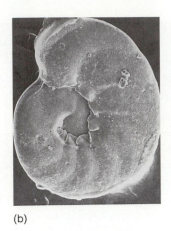
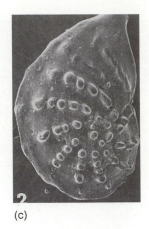
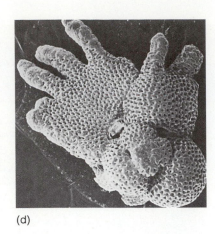

(a)   (b)   (c)   (d)

◆ **FIGURE 19.4** Foraminifera of the Cenozoic Era. (a) through (c) are benthonic forms. (a) *Uvigerina cubana,* Late Miocene, California. (b) *Cibicides americanus,* Early Miocene, California. (c) *Lenticulina mexicana,* Eocene, Louisiana. (d) A planktonic form, *Globigerinoides fistulosus,* Pleistocene, South Pacific Ocean. (Photos courtesy of B. A. Masters.)

Cenozoic. This uniformity is not surprising, because most birds are fliers, and adaptations for flying impose limitations on variations in structure.

Birds adapted to numerous habitats and increased in diversity through the Tertiary and into the Pleistocene. Since then, their diversity has decreased slightly. One of the more remarkable early adaptations was the development of large, flightless predatory birds. One of these flightless predators, *Diatryma,* was more than 2 m tall and had a huge head and beak, toes with large claws, and small vestigial wings (◆ Fig. 19.7). Its legs were massive and short, indicating that *Diatryma* did not move very fast, but the early mammals upon which it preyed were slow-moving as well. *Diatryma* and related genera were widespread during the Early Tertiary of North America and of Europe, but eventually became extinct; apparently, they were replaced by the developing mammalian carnivores.

Two remarkable flightless birds known only from Pleistocene-aged deposits are the moas of New Zealand and the elephant birds of Madagascar. The moas were up to 3 m tall, whereas the elephant birds were shorter but more heavily built, weighing as much as 500 kg. Both of these giant flightless birds became extinct soon after humans appeared in New Zealand and Madagascar.

Flightless birds notwithstanding, the true success story among birds belongs to the fliers. They did not undergo much structural change during the Cenozoic, but a bewildering array of adaptive types arose. In fact, birds have been as successful as mammals; they have exploited the aerial habitat as fully as the mammals have adapted to terrestrial habitats.

❖ THE AGE OF MAMMALS BEGINS

For more than 100 million years, mammals coexisted with dinosaurs; yet, their fossil record indicates that during this entire time they were neither diverse nor abundant (see Table 16–3). Even during the Late Cretaceous, very near the end of the Age of Reptiles, only eight families of marsupial and placental mammals existed. This situation was soon to change. Mesozoic extinctions eliminated the dinosaurs and many of their relatives, thereby creating numerous adaptive opportunities that were quickly exploited by mammals. The Age of Mammals had begun.

The Cenozoic evolutionary history of the mammals is better known than the history of any of the other classes of vertebrates. Two factors account for this. First, Cenozoic terrestrial deposits are more common than Mesozoic and Paleozoic deposits (see the Prologue). Secondly, mammal fossils are easier to identify. In Chapter 16 we noted that differentiation of teeth was a trend established in the therapsids. In true mammals the teeth are fully differentiated into distinctive types, and the chewing teeth, called **molars,** differ in each of the mammalian orders. In fact, a single mammal molar is commonly sufficient to identify the genus from which it came.

◆ **FIGURE 19.5** Corals such as this colonial scleractinian were the dominant reef-builders during the Cenozoic Era. (Photo courtesy of Sue Monroe.)

◆ **FIGURE 19.6** Echinoids were particularly abundant during the Tertiary, and new infaunal forms such as this sand dollar evolved from their Mesozoic biscuit-shaped ancestors. (Photo courtesy of Sue Monroe.)

◆ **FIGURE 19.7** *Diatryma*, which stood more than 2 m tall, was a large, flightless predatory bird of the Late Paleocene and Eocene of North America.

❖ PLACENTAL MAMMALS

Among living mammals only **monotremes** lay eggs, while marsupials and placentals give birth to live young. **Marsupials** are born in a very immature, almost embryonic condition, and then undergo further development in the mother's pouch. **Placentals,** on the other hand, have developed a different reproductive method. In these animals, the amnion of the amniote egg (see Fig. 14.16) has fused with the walls of the uterus, forming a *placenta*. Nutrients and oxygen are carried from mother to embryo through the placenta, permitting the young to develop much more fully before birth.

The phenomenal success of placental mammals is related in part to their reproductive method. A measure of this success is that more than 90% of all mammals, fossil and living, are placentals. Recall from Chapter 16 that placental mammals were present during the Late Cretaceous. The great adaptive radiation of placental mammals began with early shrewlike animals (◆ Fig. 19.8).

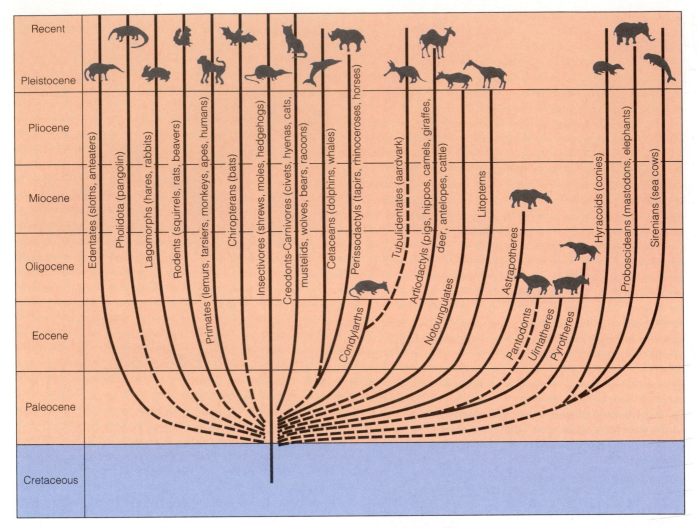

◆ **FIGURE 19.8** The adaptive radiation of placental mammals began with shrewlike ancestors. Diversification began during the Late Cretaceous and continued through the Cenozoic.

❖ DIVERSIFICATION OF PLACENTAL MAMMALS

A major adaptive radiation of mammals began during the Paleocene and continued through the Cenozoic Era. Paleocene mammalian faunas are considered archaic because they were composed of primitive mammals, some of which—including marsupials, insectivores, and the rodentlike multituberculates—were holdovers from the Mesozoic Era (◆ Fig. 19.9).

Thirteen new orders of mammals first appeared during the Paleocene (◆ Fig. 19.10). Among these were the first rodents, rabbits, primates, and carnivores, but many of the other new orders soon became extinct. Most of these Paleocene mammals, even those assigned to living orders, had not yet become clearly differentiated from their insectivore ancestors, and the differences between herbivores and carnivores were slight. Large mammals did not evolve until the Late Paleocene, and the first giant terrestrial mammals appeared during the Eocene (◆ Fig. 19.11).

Diversification continued during the Eocene, when nine more orders evolved; all but one of these orders still exist (Fig. 19.10). Most of the existing mammalian orders were present by Eocene time; yet if we could somehow go back and visit the Eocene, we would probably

♦ **FIGURE 19.9** The archaic mammalian fauna of the Paleocene Epoch included such animals as the multituburculate *Ptilodus* (right foreground), insectivores (right background), *Protictis,* an early carnivore, (left background), and the pantodont *Pantolambda* that stood about 1 m tall.

not recognize many of these animals. Some would be at least vaguely familiar to us, but the horses, camels, rhinoceroses, and elephants, for example, would bear little resemblance to their living descendants.

By Oligocene time, all of the living orders of mammals had evolved (Fig. 19.10), while a number of archaic mammals became extinct during the Late Eocene and Oligocene. Diversification continued during the Oligocene, but it was within the existing orders as more familiar families and genera appeared. Miocene and Pliocene mammals were mostly animals that we could easily identify. Perhaps a few would look rather odd to us, but we would nevertheless easily recognize them as horses, deer, cats, elephants, and so on. A few unfamiliar types of mammals still existed during the Pliocene, but

many of the genera then present were the direct ancestors of Pleistocene and Recent forms.

Rodents, Rabbits, and Bats

The adaptive success of rodents is manifested by the fact that the order Rodentia alone accounts for more than 40% of all living mammal species. Rodents are extremely diverse and have adapted to a wide range of environments. Following their appearance during the Paleocene, they rapidly diversified and soon occupied many of the microhabitats unavailable to larger animals.

Superficially, rabbits resemble rodents, but they differ in anatomical details and have an independent evolutionary history. Rabbits are known from the Paleocene but

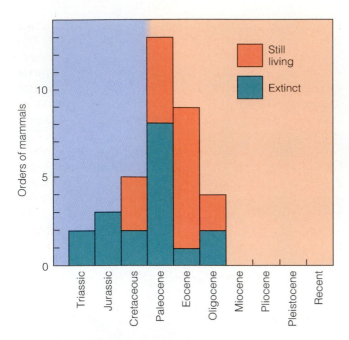

FIGURE 19.10 Times during which the mammalian orders first appeared in the fossil record. Most of the living orders appeared during the Paleocene and Eocene, but many Paleocene orders became extinct. All living orders of mammals had evolved by Oligocene time.

FIGURE 19.11 The uintatheres were Eocene rhinoceros-sized mammals with three pairs of bony protuberances on the skull and saberlike upper canine teeth. (Photo courtesy of the Field Museum of Natural History, Chicago. Neg. #CK46T.)

did not become abundant until the Oligocene. Long, powerful hind limbs for hopping and speed are the most obvious evolutionary trend in rabbits.

The oldest known fossil bat is a specimen from the Eocene Green River Formation of Wyoming (◆ Fig. 19.12). Apart from having forelimbs modified into wings, bats differ little from their insectivore ancestors. In fact, the bat's wing is simply the basic vertebrate forelimb modified to support a wing membrane. Unlike pterosaurs and birds, bats evolved to use the entire hand for wing support (Fig. 19.12; compare with Fig. 16.25).

Primates

The order **Primates** includes the tarsiers, lemurs, and lorises (the so-called lower primates) and living monkeys, apes, and humans (the higher primates). Much of the primate story is better told in Chapter 20 where we consider human evolution, so in this chapter we shall be brief. Primitive primates may have evolved by the Late Cretaceous, but they were undoubtedly present by the Early Paleocene. Most Paleocene primates were mouse-sized, long-snouted animals that looked much like the insectivores from which they evolved. By Eocene time,

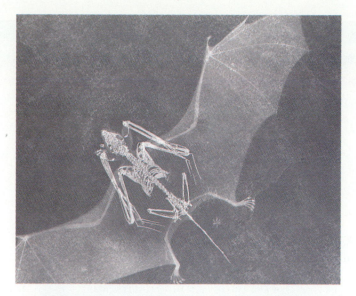

◆ **FIGURE 19.12** The oldest known fossil bat, from the Eocene Green River Formation, Wyoming. The bat's wings are shown as they would have appeared when the bat was alive. Notice that each wing is supported by four elongated fingers.

large primates had evolved, and fairly modern-looking lemurs and tarsiers are known from Asia and North America. By Oligocene time, primitive New and Old World monkeys had evolved in South America and Africa, respectively. The hominoids, the group containing apes and humans, evolved during the Miocene (see Chapter 20).

Carnivores

Many land-living carnivorous mammals depend on speed, agility, and intelligence to catch their prey and have teeth modified for a diet of meat. The canine teeth are well developed because they are used to make the kill, and all carnivores have specialized shearing teeth called **carnassials** (◆ Fig. 19.13). During the Paleocene, carnivorous mammals called *creodonts* and **miacids** made their appearance. Most were rather small animals with short, heavy limbs. They were not particularly fast runners, but, then, neither were their prey. Miacids were ancestral to all later members of the order Carnivora (◆ Fig. 19.14, p. 552). These weasel-like animals had well-developed carnassials, but they retained such primitive features as short limbs. One of the groups of carnivores that arose

from miacids includes the `cats, hyenas, and viverrids (mongooses, civet cats, and genet cats) (Fig. 19.14). Despite the obvious differences among these animals, studies of the fossil record and chromosomes of living species clearly indicate their close relationships. Cats have not changed much since they first appeared, although there have been some remarkable developments in the canine teeth (see Perspective 19.1).

The group of carnivores that includes dogs, bears, weasels, and pandas (Fig. 19.14) had better developed carnassials than miacids, and their limbs were longer. The braincase was larger, too, indicating a higher level of intelligence. The seals, sea lions, and walruses (Fig. 19.14) are adapted to an aquatic life, but their ancestry is not well documented by fossils. Aquatic adaptations include a streamlined body, a layer of blubber for insulation, and limbs modified into paddles.

Ungulates

Ungulate is an informal term referring to several types of mammals, including the orders Artiodactyla and Perissodactyla. **Artiodactyls,** the even-toed hoofed mammals, are by far the most diverse and abundant living ungulates; representative artiodactyls include cattle, sheep, goats, swine, antelope, deer, and camels. **Perissodactyls,** the odd-toed hoofed mammals, consist of only 16 living species of horses, rhinoceroses, and tapirs.

All ungulates are herbivores. The chewing teeth in mammals are the premolars and molars, but the ungulates show an evolutionary trend in the premolars called

◆ **FIGURE 19.13** Jaws and teeth of a cat. All members of the order Carnivora have large canine teeth and a specialized pair of shearing teeth called *carnassials* (blue).

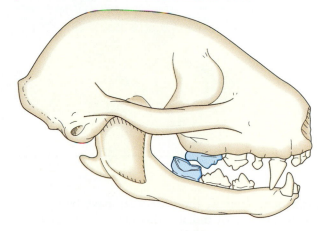

Saber-Toothed Carnivores

Saber-toothed cats of the Pleistocene, with canines 15 cm long, are the best-known carnivores with enlarged canine teeth (◆ Fig. 1). They were not, however, the only mammals with this specialization. Large, saberlike canines developed independently four times during the Cenozoic: once in an extinct group of mammals known as creodonts, once in a catlike marsupial in South America, and twice in the same family of cats. In all of these animals, skull modifications were necessary so that the mouth could be opened widely enough for the enlarged canines to be used effectively

◆ **FIGURE 1** *Smilodon*, a Pleistocene saber-toothed cat, is shown feeding on the carcass of a mammoth trapped in the La Brea Tar Pits of Southern California.

molarization. That is, the premolars have become enlarged to the size of molars; this gives the ungulates a continuous series of molarlike teeth for grinding vegeta-tion. Some ungulates are **grazers** and have a grass diet. However, grasses contain tiny particles of silicon dioxide and are very abrasive, so they wear down teeth. Accord-

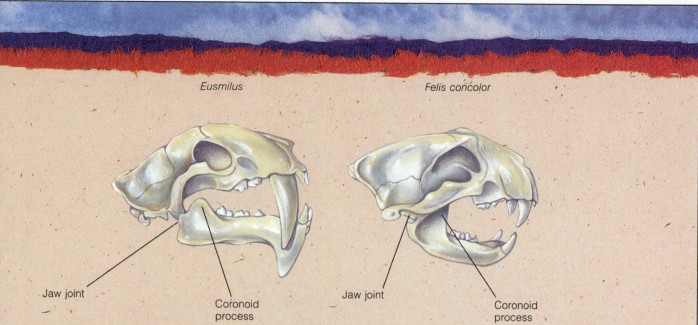

Eusmilus

Felis concolor

Jaw joint

Coronoid
process

Jaw joint

Coronoid
process

◆ **FIGURE 2** Comparison of the skulls of a present-day mountain lion, *Felis concolor* (right), and an Oligocene saber-toothed cat, *Eusmilus* (left). The skull of *Eusmilus* shows the basic modifications in saber-tooths that allowed the mouth to be opened widely enough for effective use of the enlarged canines. The jaw joint is lower and the coronoid process is shortened, allowing for longer muscle fibers. The carnassials are positioned closer to the jaw joint, and the face is rotated upward relative to the braincase.

(◆ Fig. 2). The skulls of all saber-tooths show a similar construction, which is not surprising because there are only a few ways to modify the basic mammalian skull to accommodate such large teeth.

Recall from Chapter 6 that different organisms develop similar features as a consequence of convergent evolution or parallel evolution. Convergent evolution accounts for enlarged canines in creodonts, marsupials, and cats because all are distantly related. However, the same feature developing independently twice in closely related members of a family of cats is an example of parallel evolution.

How saber-toothed carnivores used their canines is a debatable point. Pleistocene saber-toothed cats are commonly portrayed inflicting deep stab wounds on large prey animals such as elephants. This mode of predation seems unlikely because the curvature of the canines would have prevented stabbing thrusts even though the mouth could be opened very wide. It also seems unlikely that the canines were used to pierce the top of the victim's neck or skull; saberlike canines were too weak to penetrate or break bones.

Leonard Radinsky and Sharon Emerson have proposed that the saberlike canines were used to make a well-placed, shallow slash across the throat.[1] Present-day cats that prey on animals as large or larger than themselves bite the victim's throat and hold on until it suffocates. Perhaps saber-toothed carnivores also preyed on large animals, but simply inflicted a slash across the throat and then waited for the victim to die.

Perhaps the most curious aspect of saber-toothed carnivores is that none exist now. The development of saberlike canines must have been a successful adaptation, since it occurred four times, and saber-tooths of one kind or another existed from the Early Cenozoic until the end of the Pleistocene. The cause of their final extinction is unknown, but hypotheses for the extinction event at the end of the Pleistocene are discussed later in this chapter.

[1] See "The Late Great Sabertooths," *Natural History* 91, no. 4 (1982): 50–57.

ingly, the grazing ungulates have developed high-crowned teeth that are more resistant to abrasion (◆ Fig. 19.15). In contrast, those ungulates characterized as browsers, which eat the tender shoots, twigs, and leaves of trees and shrubs, did not develop high-crowned teeth.

Many ungulates live in open-grassland habitats and

◆ **FIGURE 19.14** Evolution of the carnivorous mammals. Miacid ancestors gave rise to all present-day placental carnivores. Some of the relationships shown here are well documented by fossils and by studies of present-day animals, but some relationships are yet to be firmly established.

depend on speed to escape predators. Adaptations for running include elongation of the bones of the palm and sole, which increases the length of the limbs (◆ Fig. 19.16). Also associated with running is the trend to reduce the number of bony elements in the limbs, espe-

◆ **FIGURE 19.15** Comparison of low-crowned and high-crowned teeth. Grazing ungulates developed high-crowned chewing teeth as an adaptation for eating abrasive grasses. The cusps of a high-crowned tooth are elevated into tall, slender pillars, and the entire tooth is covered by enamel and cement, both of which are hard substances.

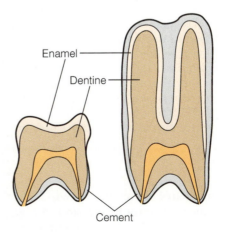

cially toes. The limbs of running ungulates are long and slender, ideally suited for speed. These animals walk and run on their toes, and each toe is covered by a hoof.

Not all ungulates are speedy runners. Some became very large, and their bulk alone was protection enough from predators. In this case massive limbs developed to support their great weight. Nevertheless, some large ungulates—rhinoceroses, for example—can run surprisingly fast, at least for short distances.

Artiodactyls—Even-Toed Hoofed Mammals

Rabbit-sized ancestral artiodactyls appeared during the Early Eocene, but at this early stage of development, they differed little from their ancestors. From these tiny ancestors, artiodactyls rapidly diversified into numerous families, many of which are now extinct (◆ Fig. 19.17). Among these extinct families were the piglike oreodonts (Fig. 19.17), which were common in North America until their extinction during the Pliocene. In addition, various other hoofed mammals lived in western North America (◆ Fig. 19.18).

Small, four-toed, ancestral camels appeared early in the diversification of artiodactyls (◆ Fig. 19.19). By the Oligocene, all camels were two toed, and during the Miocene, they diversified into several distinctive types. Among these types were giraffelike camels, gazellelike camels, and giant camels standing 3.5 m at the shoulder. Camels were abundant from Eocene to Pleistocene times in North America, and most of their evolution occurred on this continent. During the Pliocene Epoch, camels migrated to South America, and Asia, where their descendants survive. In North America, camels became extinct near the end of the Pleistocene.

The *bovids* (Fig. 19.17), the most diverse living artiodactyls, include cattle, bison, sheep, goats, and antelopes. Bovids evolved during the Miocene, but most of their diversification took place during the Pliocene. The most common Tertiary bovid of North America was the pronghorn, which roamed the western interior in vast herds (Fig. 19.18). Bovids evolved on the northern continents but have since migrated to southern Asia and Africa, where they are most common today.

Perissodactyls—Odd-Toed Hoofed Mammals

The living perissodactyls are the horses, rhinoceroses, and tapirs. These animals do not look much alike, but

◆ **FIGURE 19.16** Modifications in the limbs of ungulates. (a) *Oxydactylus,* a Miocene camel, shows the evolutionary trend of limb elongation. The bones between the wrist and toes and between the ankle and toes became longer, thereby increasing the length of ungulate limbs. (b) The ancestors of ungulates had five-toed feet, but as shown here, the trend in ungulates is toward a reduced number of toes; horses retain only one toe, pigs have four toes but with reduced side toes, rhinoceroses are three-toed, and camels retain two toes.

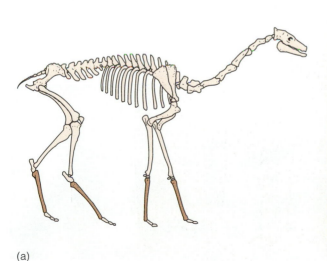

(a)

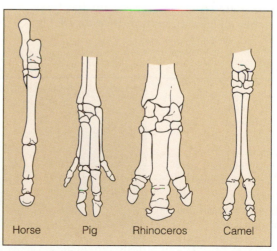

Horse   Pig   Rhinoceros   Camel

(b)

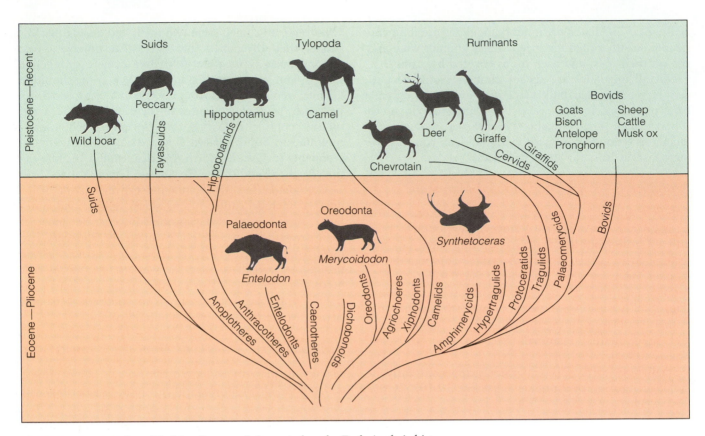

◆ **FIGURE 19.17** Simplified family tree of the artiodactyls. Early in their history, artiodactyls split into three major groups: the suids include the pigs, hippopotamuses, and extinct giant hogs; the tylopoda are represented by the camels; and the ruminants consist of the cud-chewing animals.

they and the extinct titanotheres and chalicotheres are united by several shared characteristics, and the fossil record indicates they all have a common ancestor (◆ Fig. 19.20).

Perissodactyls evolved during the Eocene and increased in diversity through the Oligocene, but they have declined markedly since then (◆ Fig. 19.21). Today, perissodactyls constitute a minor part of the fauna, and rhi-

◆ **FIGURE 19.18** This mural, painted by Charles Knight in 1930 for the American Museum of Natural History, shows various hoofed mammals living on the plains of western Nebraska during the Late Miocene. These animals include ancestors of the present-day pronghorn (left), giraffe camels (left background), the rhinoceros *Teleoceras,* and ancestors of the present-day horses (right foreground). (Photo © American Museum of Natural History, Trans. #4700(2).)

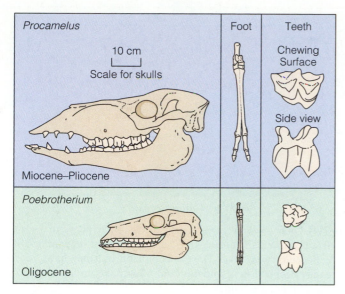

Procamelus

10 cm
Scale for skulls

Miocene–Pliocene

Poebrotherium

Oligocene

Foot

Teeth

Chewing
Surface

Side view

◆ FIGURE 19.19 These two genera of camels show the evolutionary trends of increase in size, elongation of the legs, reduction of lateral toes, and development of high-crowned teeth.

herbivores with complex three- or four-chambered stomachs. Ruminants use the same resources as perissodactyls, but use them more efficiently.

Evolution of Horses. The earliest member of the horse family was **Hyracotherium** (◆ Fig. 19.22) from the Early Eocene of Great Britain and western North America. About the size of a fox, it had four-toed front feet and three-toed hind feet. Each toe, however, was covered by a small hoof. *Hyracotherium* has few of the specializations we associate with horses (Table 19.1).

Since *Hyracotherium* was so unhorselike, how can we be sure it belongs to the horse family at all? Fortunately, the fossil record of horses is exceptionally good, and this record clearly shows that *Hyracotherium* is linked to the modern horse, *Equus,* by a series of intermediates (Fig. 19.22). However, the various trends in horse evolution (Table 19.1) did not all occur at the same rate. For example, molarization of the premolars was complete by the Oligocene, but the reduction of toes to one was not complete until the Pliocene.

We can recognize two distinct branches in the adaptive radiation of horses (Fig. 19.22a). Both have their ancestry in *Hyracotherium;* but one branch led to three-toed browsing horses, all of which are now extinct, and the other led first to three-toed grazing horses and finally to

noceroses appear to be on the verge of extinction. It has been suggested that competition with artiodactyls could explain the decline of perissodactyls. Most artiodactyls are called **ruminants,** meaning "cud-chewing." They are

◆ FIGURE 19.20 Evolution of perissodactyls. Adaptive radiation and divergence from a common ancestor explain why perissodactyls share several characteristics yet differ markedly in appearance.

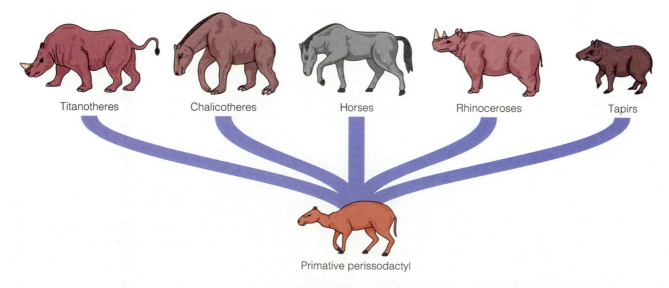

Titanotheres   Chalicotheres   Horses   Rhinoceroses   Tapirs

Primitive perissodactyl

one-toed grazers. Horse evolution occurred almost totally in North America, but a few genera migrated to the Old World, and during the Pliocene one genus even reached South America.

The appearance of grazing horses during the Miocene coincides with the appearance of the open-prairie, grassland habitat. Like all habitats this one had its special problems. For one thing, as mentioned earlier, grass contains silica and is very abrasive to the teeth, wearing them down rapidly. Horses and other ungulates adapted by developing high-crowned teeth that are more resistant to abrasion (Fig. 19.15). Also, speed was essential to escape predators in this habitat; thus, the limbs became longer, and the number of toes was reduced (Fig. 19.22b).

Merychippus is a good example of the early grazing horses (Fig. 19.22b). It was about the size of a present-day pony and had teeth well suited for grazing. Its limbs were long, and most of the weight was borne on the third toe, although side toes were still present. By Pliocene time, one-toed grazers had evolved (Fig. 19.22b). Compared with *Merychippus,* these horses were larger and had longer limbs and higher-crowned teeth. Modern *Equus* evolved during the Late Pliocene in North America and migrated to the Old World, where it still lives in the wild.

Other Perissodactyls. The other living perissodactyls— tapirs and rhinoceroses—both had Eocene ancestors that looked much like *Hyracotherium.* In fact, all of the earliest perissodactyls, including the extinct titanotheres and chalicotheres, are so similar that it is difficult to differentiate among them.

Both tapirs and rhinoceroses increased in size during the Cenozoic and became more abundant, diverse, and widespread than they are now. An Oligocene-Miocene hornless rhinoceros of Asia was the largest land mammal ever to exist (◆ Fig. 19.23). Most rhinoceros evolution occurred in the Old World, but North American rhinoceroses were quite common until they became extinct during the Late Pleistocene.

In addition to horses, rhinoceroses, and tapirs, small *Hyracotherium*-like ancestors also gave rise to titanotheres and chalicotheres (Fig. 19.20). Titanotheres existed only from the Early Eocene to Early Oligocene, but they evolved from small ancestors to giants about 2.5 m high at the shoulder (◆ Fig. 19.24). The fossil evidence indicates that chalicotheres were never very abundant, although the group existed from the Early Eocene to the Pleistocene. Miocene and later chalicotheres were large animals that superficially resembled horses, but their feet had claws rather than hoofs (◆ Fig. 19.25). The prevail-

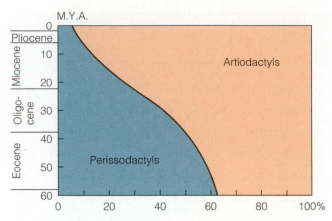

◆ **FIGURE 19.21** Perissodactyls and artiodactyls were relatively abundant during the Cenozoic. At the beginning of the Eocene, perissodactyls were dominant, but they have steadily declined in abundance and now make up less than 10% of ungulate faunas.

ing opinion favors the interpretation that these claws were used to hook and pull down branches.

Giant Mammals—Whales and Elephants

Living blue whales more than 30 m long and weighing more than 130 metric tons are very likely the largest animals ever to have existed. Some sauropod dinosaurs

TABLE 19.1 Trends Characteristic of the Evolution of Horses During the Cenozoic Era

1. Increase in size.
2. Lengthening of legs and feet.
3. Reduction of lateral toes, with emphasis on the middle toe.
4. Straightening and stiffening of the back.
5. Widening of the incisor teeth.
6. Molarization of the premolars.
7. Increase in height of the crowns of premolars and molars.
8. Increase in complexity of the crowns of premolars and molars.
9. Deepening of the front portion of the skull and the lower jaws to accommodate the high-crowned premolars and molars.
10. Lengthening of the face in front of the eye, also to accommodate the high-crowned premolars and molars.
11. Increase in size and complexity of the brain.

◆ **FIGURE 19.22** Evolution of horses. (a) Summary chart showing the recognized genera of horses and their evolutionary relationships. Note that during the Oligocene, two separate lines emerged, one leading to three-toed browsing horses and the other to one-toed grazers. (b) Simplified diagram showing some of the evolutionary trends from *Hyracotherium* to the modern horse, *Equus*. Important trends shown here include an increase in size, loss of toes, and development of high-crowned teeth with complex chewing surfaces.

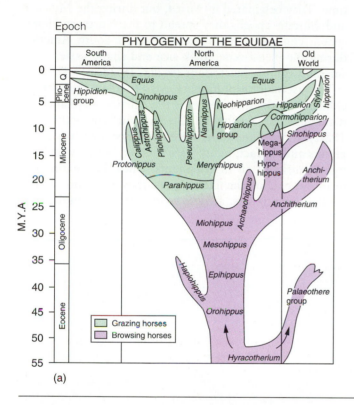

(a)

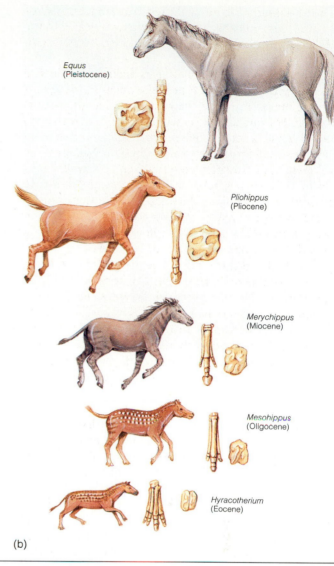

(b)

may have been of comparable size, but this is uncertain. The size of present-day elephants is impressive, too; they are fully as heavy as many large dinosaurs were. It is important to note, however, that at least half of the known dinosaur genera were larger than elephants.

The whales are the most fully aquatic mammals. Evolutionary trends that occurred during the transition to the aquatic habitat included modification of the front limbs into flippers, loss of the hind limbs, migration of the nostrils to the top of the head to form the blowhole, and development of a large horizontal tail fluke used for propulsion.

Whales evolved during the Early Eocene and by the Late Eocene had become diverse and widespread, and some had achieved large size (◆ Fig. 19.26, p. 560). Eocene whales still possessed vestigial rear limbs, had teeth resembling those of their land-dwelling ancestors, and were proportioned differently than living whales (Fig. 19.26). By the Oligocene Epoch, both groups of living whales, the toothed whales and baleen whales, had evolved.

Living elephants, represented by only two genera and two species, are the dwindling remnants of the once diverse **proboscideans** (order Proboscidea). During much of the Cenozoic, elephants were widespread, especially on the northern continents, but now they are restricted to Africa and southeast Asia.

One of the earliest elephants, *Moeritherium* from the Eocene, had small tusks but most closely resembled a

◆ **FIGURE 19.23** The rhinoceros *Paraceratherium* lived in Asia during the Oligocene and Miocene. This animal measured 5.4 m at the shoulder and may have weighed 30 metric tons.

hippopotamus and probably lived a semiaquatic life (◆ Fig. 19.27). By Oligocene time, elephants clearly showed the trend to large size and the development of a long proboscis and large tusks (Fig. 19.27). Mastodons with teeth adapted for browsing had evolved by Miocene time. They originated in Africa, but during the Miocene and Pliocene epochs they spread to the Northern Hemisphere continents. One genus reached South America during the Pleistocene. Tusk size and shape varied considerably among mastodons (◆ Fig. 19.28, p. 562).

The last major evolutionary event to affect the proboscidians was the divergence of the existing elephant and mammoth lines during the Pliocene and Pleistocene (Fig. 19.27). Most mammoths were no larger than living elephants and, in fact, many were smaller, but they had the largest tusks of all the elephants. Until their extinction near the end of the Pleistocene, they lived on all the northern continents, as well as in Africa and India.

◆ **FIGURE 19.24** Oligocene titanotheres, which were larger than rhinoceroses, evolved from sheep-sized Eocene ancestors similar to the one in the inset.

◆ FIGURE 19.25 Chaliocotheres evolved from small *Hyracotherium*-like ancestors (see inset) into large animals such as *Moropus* of the Miocene of North America and Asia. This restoration shows *Moropus* in a browsing position.

❖ MAMMALS OF THE PLEISTOCENE EPOCH

One of the most remarkable aspects of the Pleistocene mammalian fauna is that so many very large species existed. In North America, for example, there were mastodons and mammoths, giant bison, huge ground sloths, giant camels, and beavers nearly 2 m tall at the shoulder (◆ Fig. 19.29, p. 562). Kangaroos standing 3 m tall, wombats the size of rhinoceroses, leopard-sized marsupial lions, and large platypuses characterize the Pleistocene fauna of Australia. In Europe and parts of Asia lived cave bears, elephants, and the giant deer commonly called the Irish elk with an antler spread of 3.35 m (◆ Fig. 19.30, p. 563).

Many smaller mammals were also present, many of which still exist. The major evolutionary trend, however, was toward large body size. Perhaps this was an adaptation to the cooler temperatures of the Pleistocene. Large animals have proportionately less surface area compared to their volume and thus retain heat more effectively than do smaller animals.

The frozen remains of several Pleistocene mammals have been found in Alaska and Siberia. Probably the best-known frozen remains are the woolly mammoths of Siberia (see Fig. 4.23). More than three dozen frozen carcasses have been recovered, but only a few are fairly complete. Nevertheless, they provide a wealth of information unavailable from fossilized bones and teeth. Contrary to the popular myths that these animals were found in blocks of ice or even icebergs, all have been recovered from permafrost (permanently frozen soil).

In addition to mammals, some other Pleistocene vertebrate animals were of impressive proportions. We have already mentioned the giant moas of New Zealand and the elephant birds of Madagascar, but Australia also had giant birds standing 3 m tall and weighing nearly 500 kg and a lizard 6.4 m long and weighing 585 kg. The tar pits of Rancho La Brea in southern California contain the

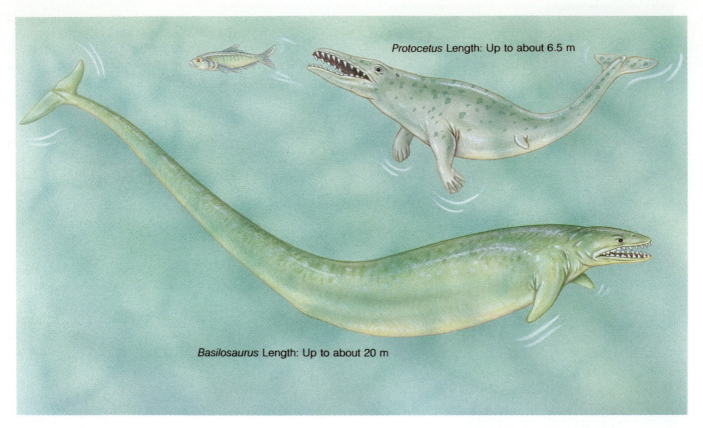

Protocetus Length: Up to about 6.5 m

Basilosaurus Length: Up to about 20 m

◆ **FIGURE 19.26** Restoration of two Eocene whales. *Protocetus* was about 6.5 m long and had teeth that closely resembled those of carnivorous condylarths. *Basilosaurus* was a giant whale 20 m long, with a long, slender body and a very small head.

remains of at least 200 kinds of animals. Many of these are fossils of dire wolves, saber-toothed cats, and other mammals, but there are also remains of many birds, especially birds of prey, and a giant vulture with a wingspan of 3.6 m (Fig. 19.29).

❖ PLEISTOCENE EXTINCTIONS

About 10,000 years ago almost all of the large terrestrial mammals of North America, South America, and Australia became extinct. Extinction events, both large and small, are part of the history of life and have been occurring since the Proterozoic Eon. This latest extinction event was modest by comparison to earlier ones, but it was unusual in that it affected, with relatively few exceptions, only large terrestrial mammals weighing more than 40 kg.

During the Late Pleistocene, North America lost approximately 33 out of 45 genera (73%) of its large terrestrial mammals, South America 46 out of 58 genera (80%), Australia 15 out of 16 genera (94%), Europe 7 out of 23 genera (30%), and Africa south of the Sahara only 2 out of 44 genera (5%). Clearly, Europe and Africa south of the Sahara were less affected by the extinction event.

What caused this latest extinction, and why did it seemingly affect only large mammals? The debate over this particular extinction event rages between those who believe that the large mammals became extinct because they could not adapt to the rapid climatic changes resulting from the termination of the last glaciation, and those who believe the mammals were killed off by human hunters, a hypothesis known as *prehistoric overkill*. The question is, did climate or humans cause the extinctions of the large mammalian fauna at the end of the Great Ice Age?

Those researchers who favor a climatic cause for the extinctions point to the rapid changes in climate and vegetation that occurred over much of the Earth's surface during the Late Pleistocene. As the various glaciers began retreating, the North American and northern Eurasian open-steppe tundras were replaced by conifer and broad-leaf forests as warmer and wetter conditions prevailed. The Arctic region flora changed from a productive herbaceous one that supported a variety of large mammals, including mammoths, to a relatively barren water-logged tundra that supported a far sparser fauna. The southwestern United States region also changed from a moist area with numerous lakes, where saber-tooths, giant ground sloths, and mammoths roamed, to a semiarid environment unable to support a diverse large mammalian fauna.

While rapid changes in climate with their accompanying effects on vegetation can certainly result in changes in animal populations, the hypothesis of climate as an agent of extinction presents several problems. First, why didn't the large mammals migrate to more suitable habitats as the climate and vegetation changed? After all, many other animal species did. For example, reindeer and the Arctic fox lived in southern France during the last glaciation and migrated to the Arctic when the climate became warmer. Why didn't the mammoths and other large animals that were adapted to cold climates simply migrate north?

The second argument against the climatic hypothesis is the apparent lack of correlation between extinctions and the earlier glacial advances and retreats throughout the Pleistocene Epoch. Previous changes in climate were not marked by episodes of mass extinctions.

Paul Martin of the University of Arizona at Tucson, the leading proponent of the prehistoric overkill hypothesis, argues that the mass extinctions in North and South America and Australia coincided closely with the arrival of humans in each area. According to Martin, hunters had a tremendous impact on the faunas of North and South America about 11,000 years ago because the animals had no previous experience with humans. The same thing happened much earlier in Australia soon after people arrived about 40,000 years ago. No large-scale extinctions occurred in Africa and most of Europe because animals in those regions had long been familiar with humans.

Several arguments can also be made against the prehistoric over-kill hypothesis. One problem is that archaeological evidence indicates that the early human inhabitants of North and South America, as well as Australia, probably lived in small, scattered communities, gathering food and hunting. It is hard to imagine how a few hunters could have decimated so many species of large mammals. On the other hand, humans have caused major extinctions on oceanic islands. For example, in a period of about 600 years after arriving in New Zealand, humans exterminated several species of large, flightless birds called moas.

A second problem is that present-day hunters concentrate on smaller, abundant, and less dangerous animals.

◆ FIGURE 19.27 Simplified family tree of the elephants. Increase in size, and development of large tusks and a long proboscis were some of the evolutionary trends in elephants.

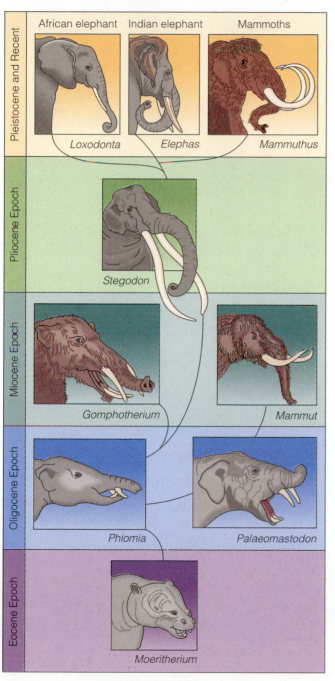

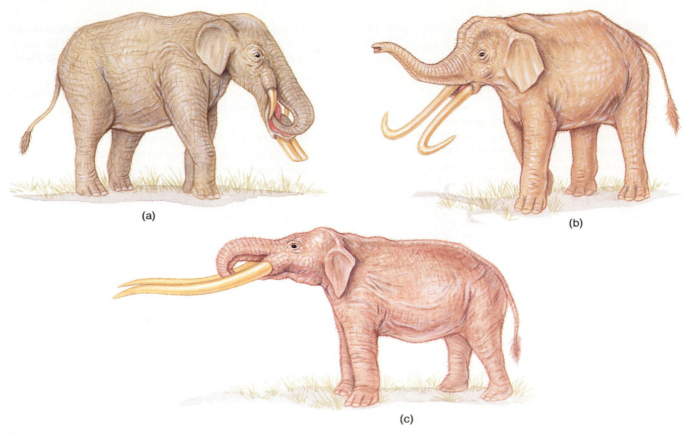

(a)

(b)

(c)

◆ **FIGURE 19.28** Restorations of three mastodons showing variation in tusk size and shape. (a) *Ambelodon* of the Late Miocene of North America had shovel-like lower tusks and small upper tusks. (b) *Stegodon* had tusks so closely spaced that there was no space between them for the trunk. This mastodon lived during the Pliocene and Pleistocene. (c) *Anancus* from France was a Pliocene-Pleistocene genus with straight tusks 3 m long.

◆ **FIGURE 19.29** A mural by Charles R. Knight showing Late Pleistocene mammals at Rancho La Brea, California. A giant ground sloth is trapped in a tar pit, while giant vultures, saber-toothed cats, and dire wolves wait nearby. A herd of mammoths is shown in the background. (Photo courtesy of the George C. Page Museum, artist Charles R. Knight.)

◆ **FIGURE 19.30** Restoration of the giant deer *Megalocereos giganteus,* commonly called the Irish elk. It lived in Europe and Asia during the Pleistocene. Large males had an antler spread of about 3.35 m. (Photo courtesy of the Field Museum of Natural History, Chicago, Neg. #CK30T.)

The remains of horses, reindeer and other small animals are found in many prehistoric sites in Europe, while large mammoth and woolly rhinoceros remains are scarce.

Finally, few human artifacts are found among the remains of extinct animals in North and South America, and there is usually little evidence that the animals were hunted. Countering this argument is the assertion that the impact on the previously unhunted fauna was so swift as to leave little evidence.

The reason for the extinctions of large mammals about 10,000 years ago is still unresolved and probably will be for some time. It may turn out that the extinction resulted from a combination of many different circumstances. Populations that were already under stress from climatic changes were perhaps more vulnerable to hunting when humans occupied new areas.

❖ CENOZOIC VEGETATION AND CLIMATE

Angiosperms, or flowering plants, evolved during the Cretaceous Period, and their diversification continued into the Tertiary. Some gymnosperms, especially conifers, remained abundant, and seedless vascular plants still occupied many habitats. Overall, the Cenozoic was a time when more and more familiar types of plants evolved.

Many of the Tertiary plants would be quite familiar to us, but their geographic distribution was markedly different than it is now. It has long been known that plant distribution is strongly controlled by climate, and for the Tertiary we see evidence of changing climatic patterns accompanied by shifting distributions of plants. Mean annual temperatures during the Paleocene and Eocene were high, but a marked temperature decrease occurred at the end of the Eocene (◆ Fig. 19.31).

Leaf structure is a good climatic indicator and is the basis for the inferences shown in Figure 19.31. Leaves with entire or complete margins characterize areas of high annual precipitation and high mean annual temperatures, whereas leaves with incised margins typify cooler, less humid regions. If a fossil flora has a high percentage of entire-margined leaves, the climate was probably wet and warm.

The types of plants found in fossil floras are also good climatic indicators. For example, Paleocene strata in the

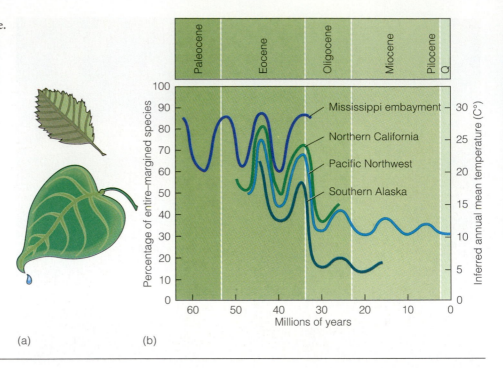

◆ **FIGURE 19.31** Cenozoic climate. (a) Leaves are good indicators of climate. Plants adapted to cool climates typically have small leaves with incised margins (top), while plants of humid, warm habitats have larger, entire-margined leaves, and many have pointed drip-tips (bottom). (b) Inferred climatic trends for four areas in North America based on the percentages of plant species with entire-margined leaves. Notice the sharp drop in mean annual temperature at the end of the Eocene.

(a) (b)

western interior of North America contain fossil ferns and palms, which are common to warmer, more humid climates. In fact, all of North America, including most of Alaska, had warm-temperate to subtropical climates throughout the Paleocene.

Subtropical conditions continued into the Eocene of North America, probably the warmest of all the Tertiary epochs. The fossil flora of the Eocene beds of the John Day Beds in Oregon includes tropical ferns, figs, and laurels (◆ Fig. 19.32), all of them plants found today only in the humid parts of Mexico and Central America. Yellowstone National Park, Wyoming, does not now have a climate in which avocado, magnolia, laurel, and fig trees could grow. Yet, during the Eocene Epoch, these kinds of trees grew there. Their presence indicates that the climate of Wyoming was considerably warmer during the Eocene than it is now.

A major climatic change occurred at the end of the Eocene Epoch (Fig. 19.31). Mean annual temperatures dropped as much as 7°C in about 3 million years. From the Early Oligocene on, mean annual temperatures have varied somewhat worldwide but overall have not changed much in the middle latitudes except during the Pleistocene.

A general decrease in precipitation over the last 25 million years took place throughout the midcontinent of North America. As the climate became drier, the vast

forests of the Oligocene gave way first to *savannah* conditions (grasslands with scattered trees) and finally to *steppe* environments (short-grass prairie of the desert margin). Herbivorous mammals quickly adapted to the savannah habitat by developing high-crowned teeth suitable for a diet of grass. Among the horses in particular, grazers became more common and browsers declined to extinction.

❖ INTERCONTINENTAL MIGRATIONS

The mammalian faunas of North America, Europe, and northern Asia exhibited many similarities throughout the Cenozoic. Even today Asia and North America are only narrowly separated at the Bering Strait, and several times during the Cenozoic the Bering Strait formed a land corridor across which mammals migrated. During the Early Cenozoic, a land connection existed between Europe and North America, allowing mammals to roam across all the northern continents. Many did; camels and horses are only two examples.

On the other hand, the southern continents were largely separate island continents during much of the Cenozoic. Africa remained fairly close to Eurasia, and, at times, faunal interchange between those two continents

◆ **FIGURE 19.32** Fossil alder leaf and maple fruit from the Oligocene John Day Beds, Oregon. (Photos courtesy of Steven R. Manchester.)

was possible. For example, elephants first evolved in Africa, but they migrated to all of the northern continents.

South America was an island continent from the Late Cretaceous until a land connection with North America was established about 5 million years ago. At that time, the South American mammalian fauna consisted of various marsupials and several orders of placental mammals that lived nowhere else in the world. These animals thrived in isolation and showed remarkable convergence with North American placentals.

When the Isthmus of Panama formed, however, migrants from North America soon replaced most of the indigenous South American mammals. Among the marsupials only opossums survived, and most of the placentals also died out. Even though the land connection allowed migrations in both directions, very few South American mammals successfully migrated northward (◆ Fig. 19.33).

Most of the living species of marsupials are restricted to the Australian region. Recall from Chapter 16 that marsupials occupied Australia before its complete separation from Gondwana, but apparently placentals, other than bats and a few rodents, never did.

Chapter Summary

1. Marine invertebrate groups that survived the Mesozoic extinctions continued to expand and diversify during the Tertiary.
2. Birds belonging to living orders and families evolved during the Early Tertiary Period.
3. The Paleocene mammalian fauna was composed of Mesozoic holdovers and a number of new orders. This was a time of diversification among mammals, and several orders soon became extinct. Most living mammalian orders were present by the Eocene.

4. Mammals have a good fossil record, and because of their varied teeth, mammal fossils are more easily identified than are fossils of other vertebrates. Consequently, the evolutionary history of mammals is known in considerable detail.
5. Placental mammals owe much of their success to their method of reproduction. Shrewlike placental mammals that appeared during the Cretaceous were the ancestral stock for the placental adaptive radiation of the Cenozoic.

◆ FIGURE 19.33 South America existed as an island continent during most of the Cenozoic Era. Its fauna evolved in isolation and consisted of numerous marsupial and placental mammals that were restricted to that continent. When the Isthmus of Panama formed during the Late Pliocene, migrations between North and South America occurred. Many types of placental mammals migrated south, and many South American mammals soon became extinct. Only a few mammals migrated north and successfully occupied North America.

6. Small mammals, especially rodents, occupy microhabitats unavailable to larger animals. Bats, the only flying mammals, have forelimbs modified for wing support, but otherwise they differ little from their insectivore ancestors.

7. Adaptations in carnivorous mammals include minor modifications of the limbs, well-developed canine teeth, and specialized shearing teeth called carnassials.

8. The perissodactyls and artiodactyls evolved during the Eocene. Ungulate adaptations include molarization of pre-molars for grinding up vegetation and limb modifications for speed.

9. Artiodactyls diversified throughout the Cenozoic and are now the most abundant ungulates. Perissodactyls were abundant and diverse during the earlier Tertiary but have declined to only 16 living species.

10. The evolutionary history of horses is particularly well documented by fossils. The earliest horse, *Hyracotherium*, and the present-day horse, *Equus*, differ considerably, but a

continuous series of intermediate fossils shows that they are related.

11. Horses, rhinoceroses, tapirs, titanotheres and chalicotheres diverged from a common ancestor during the Eocene.

12. Studies of fossils and living animals indicate that whales evolved from land-living mammals.

13. Elephants evolved from pig-sized ancestors, became diverse and abundant, especially on the northern continents, then dwindled to only two living species.

14. Flowering plants continued their dominance in land-plant communities. Subtropical to tropical climates prevailed through the Paleocene and Eocene in North America. Temperatures declined at the end of the Eocene, and during the Oligocene-Miocene epochs, the midcontinent region became increasingly arid.

15. Horses, camels, elephants, and other mammals spread across the northern continents because land connections between these continents existed at various times during the Cenozoic. The southern continents were mostly isolated, and their faunas evolving in isolation were unique.

16. The predominant evolutionary trend in Pleistocene mammals was toward giantism. Many of these large mammals became extinct at the end of the Pleistocene.

Important Terms

artiodactyl
browser
carnassials
grazer
Hyracotherium
marsupial

miacid
molar
molarization
monotreme
perissodactyl

placental
primate
proboscidian
ruminant
ungulate

Review Questions

1. An animal with teeth adapted for eating grass is a:
   a. _____ grazer; b. _____ primate; c. _____ proboscidean; d. _____ browser; e. _____ marsupial.

2. The specialized shearing teeth in carnivorous mammals are:
   a. _____ molars; b. _____ canines; c. _____ carnassials; d. _____ incisors; e. _____ premolars.

3. During the earliest Tertiary, the large, land-dwelling predators were probably:
   a. _____ miacids; b. _____ saber-toothed cats; c. _____ bryozoans; d. _____ tapirs; e. _____ flightless birds.

4. The most numerous and diverse living artiodactyls are:
   a. _____ camels; b. _____ rhinoceroses; c. _____ horses; d. _____ bovids; e. _____ hippopotamuses.

5. The oldest known horse is:
   a. _____ *Ceratotherium*; b. _____ *Pliohippus*; c. _____ *Paraceratherium*; d. _____ *Hyracotherium*; e. _____ *Moeritherium*.

6. The ancestors of all existing placental mammals were _____ animals:
   a. _____ shrewlike; b. _____ rhinoceros-sized; c. _____ flying; d. _____ burrowing; e. _____ aquatic.

7. Which of the following mammals migrated to North America from South America after the Isthmus of Panama formed?
   a. _____ tapirs; b. _____ camels; c. _____ opossums; d. _____ mastodons; e. _____ saber-toothed cats.

8. The ruminants are those artiodactyls that:
   a. _____ chew the cud; b. _____ are adapted for speed; c. _____ have carnassial teeth; d. _____ have a long, flexible nose; e. _____ are descended from *Hyracotherium*.

9. A marked temperature decrease occurred at the end of the:
   a. _____ Pleistocene; b. _____ Mesozoic; c. _____ Eocene; d. _____ Tertiary; e. _____ Miocene.

10. Which family of perissodactyls is extinct?
    a. _____ oreodonts; b. _____ tapirs; c. _____ titanotheres; d. _____ miacids; e. _____ proboscideans.

11. One feature of Eocene whales that indicates they had land-dwelling ancestors is:
    a. _____ large size; b. _____ elongate body; c. _____ vestigial rear limbs; d. _____ high-crowned chewing teeth; e. _____ enlarged incisors.

12. The large body size of Pleistocene mammals may have been an adaptation to:
    a. _____ increased predation; b. _____ more seasonal climates; c. _____ cooler temperatures; d. _____ higher elevations; e. _____ longer summers.

13. Increased volcanism and more dissolved silica in seawater probably account for the Miocene abundance of:

a. _____ foraminifera; b. _____ diatoms; c. _____ angiosperms; d. _____ scleractinian corals; e. _____ bryozoans.

14. All existing orders of mammals had appeared in the fossil record by the:
a. _____ Mesozoic; b. _____ Pliocene; c. _____ Oligocene; d. _____ Recent; e. _____ Paleocene.

15. Which of these mammals are most closely related?
a. _____ whales, elephants, and rhinoceroses; b. _____ primates, bats, and insectivores; c. _____ camels, horses, and pigs; d. _____ hyenas, cats, and mongooses; e. _____ birds, bats, and pterosaurs.

16. The largest land-dwelling mammal was a(n):
a. _____ hippopotamus; b. _____ deer; c. _____ mammoth; d. _____ wombat; e. _____ rhinoceros.

17. Which of the following is a hypothesis for Pleistocene extinctions?
a. _____ meteorite impact; b. _____ prehistoric overkill; c. _____ reduced area of continental shelves; d. _____ freezing; e. _____ extensive volcanism.

18. Although numerous adaptive types of birds exist, they vary little in basic structure. Why?

19. Briefly summarize the important Paleocene and Eocene evolutionary events among mammals.

20. Marsupials seem to have been successful only in Australia and South America. Explain. What happened to the South American marsupials near the end of the Cenozoic?

21. How do placental mammals differ from marsupials and monotremes?

22. Compare the wing structures of bats, birds, and pterosaurs.

23. How were the limbs and teeth of placental carnivores modified for a predatory life?

24. Why is the evolutionary history of mammals more fully known than the history of any other vertebrate class?

25. What are ungulates? How have their limbs been modified for speed?

26. What effect did the appearance of widespread grasslands during the Miocene have on the evolution of horses?

27. Explain how the fossil record demonstrates that animals as different as horses and rhinoceroses share a common ancestor.

28. Give a brief account of the evolutionary history of elephants.

29. How can the leaf structure of fossil plants be used to make inferences about ancient climates?

30. Discuss the evidence for the two hypotheses for Pleistocene extinctions.

Additional Readings

Adams, D. B. 1981. The nine lives of the sabercat. *Science 81* 2, no. 1: 42–47.

Carroll, R. L. 1988. *Vertebrate paleontology and evolution.* New York: W. H. Freeman and Co.

Colbert, E. H., and M. Morales. 1991. *Evolution of the vertebrates* (4th ed.) New York: John Wiley & Sons.

Kurten, B. 1988. *Before the Indians.* New York: Columbia University Press.

Marshall, L. G. 1988. Land mammals and the great American interchange. *American Scientist* 76: 380–88.

Park, A. 1990. Giants once ruled Australia, fossil discoveries reveal. *Smithsonian* 10, no. 10: 133–43.

Prothero, D. R. 1987. The rise and fall of the American rhino. *Natural History* 96, no. 8: 26–33.

Savage, R. J. G. 1986. *Mammal evolution: An illustrated guide.* New York: Facts on File Publications.

Simpson, G. G. 1951. *Horses.* Oxford: Oxford University Press.

Storch, G. 1992. The mammals of island Europe. *Scientific American* 266, no. 2: 64–69.

Sutcliffe, A. J. 1985. *On the trace of Ice Age mammals.* Cambridge, Mass.: Harvard University Press.

Trefil, J. 1991. Whale feet. *Discover* 2, no. 5: 44–48.

Wolfe, J. A. 1978. A paleobotanical interpretation of Tertiary climates in the Northern Hemisphere. *American Scientist* 66: 694–702.

CHAPTER 20

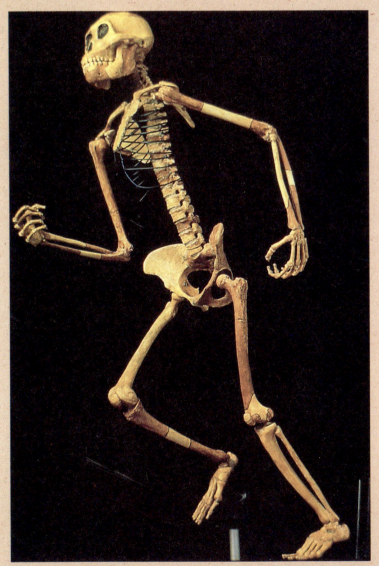

A reconstruction of Lucy's skeleton by Owen Lovejoy and his students at Kent State University, Ohio. Lucy, whose fossil remains were discovered by Donald Johanson, is an approximately 3.5-million-year-old *Australopithecus afarensis* individual. This reconstruction illustrates how adaptations in Lucy's hip, leg, and foot allowed a fully bipedal means of locomotion. (Photo © 1985 David L. Brill.)

EVOLUTION OF THE PRIMATES AND HUMANS

Prologue

A new hypothesis on the origin and evolution of modern humans has sparked an intense debate between traditional paleoanthropologists who work with fossil bones and molecular biologists who study the genetics of organisms. The molecular biologists claim that all humans evolved from a woman they call Eve, who lived in the hot savanna of Africa 200,000 years ago. Eve left as her legacy resilient genes that are carried by all living humans. This hypothesis has aroused considerable controversy because it challenges the traditional view of human evolution based on fossil evidence. According to the molecular biologists, modern humans did not evolve slowly in different parts of the world as many traditional paleoanthropologists believe, but instead evolved from Eve's family in Africa. Sometime between 180,000 and 90,000 years ago, a group of her offspring migrated from their homeland and replaced all other human groups that they encountered.

What makes the molecular biologists so sure that they are right? To answer that question, we need to look at an earlier battle between molecular biologists and paleoanthropologists over the time when ancestral humans and chimpanzees diverged. Prior to 1967, most paleoanthropologists believed ancestral humans and chimpanzees diverged at least 15 million years ago and possibly even earlier; this conclusion was based on bones of an apelike creature who seemed to be ancestral to humans but not apes. In 1967, geneticists began studying the molecular structure of a blood protein believed to change at a slow, steady rate as a species evolved. When they examined this blood protein in baboons, chimpanzees, and humans, the geneticists discovered major differences between the blood protein molecules of chimpanzees and baboons. This result was expected because according to the fossil evidence, the two species had been evolving separately for 30 million years. The difference between chimpanzees and humans, however, was very small, leading the geneticists to conclude that the two species diverged no more than a few million years ago, which was millions of years later than predicted by the fossil evidence. As more fossils were discovered, paleontologists realized that the divergence between chimpanzees and ancestral humans occurred more recently than they had believed.

The current Eve hypothesis arose from an examination of the DNA from the mitochondria of cells. This mitochondrial DNA is useful for tracing family relationships because, unlike nuclear DNA, mitochondrial DNA is inherited only from the mother and so preserves a family phylogeny (◆ Fig. 20.1). Mitochondrial DNA is altered only by mutations that are passed on to the next generation. Each random mutation therefore produces a new type of DNA that is as distinctive as a fingerprint.

Geneticists compared the mitochondrial DNA from many different infants and found surprisingly small but clear differences. There were two general categories of DNA: one type was found in some infants of recent African descent, and the second type was found in everyone else including other Africans. According to the geneticists, these findings mean that modern humans evolved in Africa; then at some point a group of Africans split off, forming a second branch of DNA that was carried and spread throughout the rest of the world.

That the mitochondrial DNA can be traced back to a single woman from whom we all descended is not surprising, considering the statistics of genetic inheritance. What surprised most paleoanthropologists about

		Phanerozoic Eon						
Archean Eon	Proterozoic Eon							
		Paleozoic Era						
Precambrian		Cambrian	Ordovician	Silurian	Devonian	Mississippian	Pennsylvanian	Permian
						Carboniferous		

2,500
M.Y.A.

570
M.Y.A.

245
M.Y.A.

the Eve hypothesis was how recently modern humans may have evolved. This age is calculated by counting the mutations that have occurred between the present and Eve's DNA. By looking at the DNA types that differ most from one another, and assuming mutations

occurred at a regular rate, geneticists have calculated how many times Eve's original DNA would have to have mutated to produce the different types; in other words, they have constructed a molecular clock. These calculations indicate Eve lived about 200,000 years ago.

Where does this leave the traditional paleoanthropologists who base their evolutionary schemes on fossil evidence? As would be expected, most are skeptical of the genetic data, particularly the molecular clock. By changing a few basic assumptions, one might use the geneticists' calculations to show that Eve evolved many hundreds of thousands of years ago, which would be more in line with the fossil data. In fact, it has been suggested that the computer program used to analyze the DNA data was misapplied, and that Eve could have originated in either Africa or Asia, or that there could even have been more than one Eve.

Paleoanthropologists point out that if modern humans did originate in Africa and spread throughout the world, then the earliest modern humans in other areas should have African features, a point that they believe is not supported by archaeological evidence from Asia or Australasia (Indonesia, New Guinea, and Australia). Furthermore, paleoanthropologists question how modern humans from Africa could have completely replaced all other human groups with whom they came in contact without leaving any evidence in the fossil record of the traits that made them successful or without any detectable genetic mixing between the two groups.

A recent discovery of several 350,000-year-old Asian fossil skulls indicates that a transition from the earliest humans to modern humans occurred in Asia at the same time as in Africa. This find lends strong support to the argument that the earliest Asian peoples were

◆ **FIGURE 20.1** DNA is the source of the genetic code contained in all humans. Nuclear DNA is a mixture of the genes from both parents. Mitochondrial DNA is inherited only from the mother and therefore is not a mixture of parental genes.

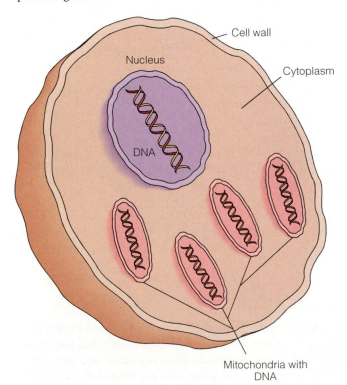

Cell wall

Nucleus

Cytoplasm

DNA

Mitochondria with DNA

Phanerozoic Eon									
Mesozoic Era			Cenozoic Era						
Triassic	Jurassic	Cretaceous	Tertiary					Quaternary	
			Paleocene	Eocene	Oligocene	Miocene	Pliocene	Pleistocene	Holocene

45
Y.A.

66
M.Y.A.

not replaced by more modern people from Africa as supporters of the Eve hypothesis believe. At this time, supporters of the Eve hypothesis are not yet ready to discard their hypothesis, but this new fossil evidence raises several important questions concerning the origin of modern humans.

❖ INTRODUCTION

In this final chapter we will examine the various primate groups, in particular, the origin and evolution of the *hominids,* which are humans and their extinct ancestors. The earliest fossil evidence of hominids is from rocks in eastern Africa and is dated at about 4 million years. These fossils reveal that our oldest known ancestors were individuals that walked fully upright and stood about 1 to 1.7 m tall and weighed between 25 and 60 kg. While these are the oldest hominid remains presently known, most paleoanthropologists believe that even older hominid fossils are still to be discovered and will push our ancestry farther into the past.

The study of the origins and evolutionary history of the human species has long fascinated people, but it was not until 1856 that workers discovered the first humanlike fossil bones in a quarry cave in the Neander Valley near Düsseldorf, Germany. Even in 1871, when Charles Darwin published *The Descent of Man,* the fossil evidence of humans and their relationship to the apes was still quite meager. Because the fossil record of hominids is sparse compared to that of other vertebrate groups, there has been, and continues to be, much controversy and debate over our history, even as new fossil discoveries are made and new techniques for scientific analysis are developed (see the Prologue). One of the reasons the study of hominids is so exciting is that new discoveries can completely change our view of hominid evolution.

We will discuss how these discoveries have changed ideas about our phylogeny (evolutionary history) and examine the current hypotheses of human origin and history.

❖ PRIMATES

Primates are difficult to characterize as an order because they lack the strong specializations found in most other mammalian orders. We can, however, point to several trends in their evolution that help to define primates and are related to their *arboreal* or tree-dwelling ancestry. These include a change in their overall skeletal structure and mode of locomotion, an increase in brain size, stereoscopic vision, evolution of a grasping hand with opposable thumb, and a shift toward smaller, fewer, and less specialized teeth. Not all of these trends occurred in every primate group, nor did they evolve at the same rate in each group. In fact, some primates have retained certain primitive features relating to their early evolutionary ancestry. For example, prosimians have a snout and five fingers and toes, whereas in other primate groups these features have been modified for a nonarboreal life-style.

The primate order can be divided into two suborders (❖ Fig. 20.2). The **prosimians** are primitive primates and include the tree shrews, lemurs, and tarsiers, while the **anthropoids,** which descended from the ancestors of prosimians, include monkeys, apes, and humans and their ancestors.

❖ PROSIMIANS

Prosimians range in size from that of a mouse up to a house cat. They are arboreal, have five digits on each hand and foot with either claws or nails, and are typically omnivorous. They have large, forwardly directed

◆ **FIGURE 20.2** Primates are divided into two suborders: (a) and (b) the prosimians and (c) through (f) the anthropoids. The anthropoids are further subdivided into three superfamilies: (c) New World monkeys, (d) Old World monkeys, and (e) and (f) great apes and humans. (Photo (a) © Ron Austing/Photo Researchers, Photo (b) © M. Loup/Jacama, Photo Researchers, Photo (c) © Renee Lynn/Photo Researchers, Photo (d) © Tim Davis/Photo Researchers, Photo (f) © Tom McHugh/Photo Researchers.)

(a)

(b)

ANTHROPOIDS

(c)

(d)

(e)

(f)

eyes specialized for night vision, and hence most are nocturnal (Figs. 20.2a and b). As their names implies (*pro* means before, and *simian* means ape), prosimians are the oldest primate lineage. They were abundant, diversified, and widespread in North America, Europe, and Asia during the Paleocene and Eocene (◆ Fig. 20.3). As the continents moved northward during the Cenozoic, and the climate changed from warm tropical to cooler mid-latitude conditions, the prosimian population decreased in both abundance and diversity. Tarsiers and lemurs, abundant in North America during the Eocene, were almost extinct by the Oligocene. The widespread Paleocene-Eocene prosimian populations of Europe and Asia began migrating southward during the Oligocene to the warmer latitudes of Africa, Asia, and the East Indies. Prosimians have largely been replaced by monkeys and apes and are currently found only in the tropical regions of Asia, India, Africa, and Madagascar.

❖ ANTHROPOIDS

The anthropoids evolved from one of the prosimian lineages sometime during the Late Eocene. By the beginning of the Oligocene (37 million years ago), anthropoids were already well established. As a matter of fact, we know little about their origin and early evolutionary history. Most of our knowledge about the early history of the group comes from fossils found in the Fayum Depression, a small desert area southwest of Cairo, Egypt. During the Oligocene, the Fayum Depression was a lush tropical rain forest that supported a diverse and abundant fauna and flora. Within this forest lived many different arboreal anthropoids, whose fossil remains provide us with a wealth of data concerning the early history of this group.

The anthropoids are divided into three superfamilies: the *New World monkeys,* the *Old World monkeys,* and the *hominoids.* The **New World monkeys** (monkeys of Central and South America) evolved independently from the Old World monkeys and hominoids, and their oldest fossil remains come from Oligocene beds from South America. New World monkeys are characterized by their prehensile tail, flattish faces, and widely separated nostrils and include the howler, spider, and squirrel monkeys (Fig. 20.2c).

The **Old World monkeys** are characterized by close-set downward-directed nostrils (like those of apes and humans), grasping hands, and a nonprehensile tail. They include the macaque, baboon, and proboscis monkey (Fig. 20.2d). The oldest fossils of this superfamily come from African Oligocene beds. Old World monkeys currently are widely distributed in the tropical regions of Africa and Asia.

The **hominoids** consist of three families: the great apes (pongids), which include the chimpanzees, orangutans, and gorillas (Fig. 20.2e and f); the lesser apes (hylobatids), which are the gibbons and siamangs; and the **hominids,** which are humans and their extinct ancestors. The oldest known hominoids were apelike creatures whose fossils, dated at about 25 million years old, mark the divergence of the hominoids from the Old World monkeys. During this time the movement of the continents resulted in pronounced climatic shifts. In Africa, Europe, Asia, and elsewhere, a major cooling trend began, and the tropical and subtropical rain forests slowly began to change to a variety of mixed forests and savannas and eventually to grasslands as temperatures and rainfall decreased.

The geologic and climatic changes occurring during this time profoundly affected the evolution of hominoids and other mammal groups. A newly formed land bridge between Eurasia and Africa allowed forest-living apes to migrate into Eurasia during the Miocene. As the climate continued to cool and become drier, the widespread forests were divided into smaller, isolated forests separated by zones of savannas and grasslands. Populations of apes became reproductively isolated from each other within the various forests leading to adaptive radiation and increased diversity among the hominoids. During this time, the lesser apes evolved from a lineage of early hominoids.

◆ **FIGURE 20.3** *Notharctus,* an Eocene lemur.

◆ **FIGURE 20.4** Probable life appearance of a dryopithecine. The fossil record of dryopithecines is poor, but fragmentary limb-bone fossils suggest they led a four-legged arboreal existence with limited activity on the ground.

At the present, fossil evidence indicates that the human family (Hominidae) did not evolve from the modern apes (Pongidae), but rather followed an independent line of evolution. Whether humans evolved from a single family of primitive apes is still not clear.

During the Miocene, two apelike groups evolved that gave rise to the modern great apes and perhaps even the earliest human ancestors. The first group, the **dryopithecines,** evolved in Africa during the Miocene about 20 million years ago and subsequently spread to Europe and Eurasia about 14 million years ago following the collision between Africa and Asia. The dryopithecines were a varied group in form, size, and life-style. Though no complete fossil skulls or skeletons have been found, partial skulls, jaws, and fragmentary limb-bone fossils indicate that they had a larger brain than their anthropoid ancestors, an apelike face, teeth, and jaw and ate more fruit and berries than leaves (◆ Fig. 20.4). Their limb structure also suggests a four-legged arboreal existence with limited activity on the ground. The dryopithecines were very abundant and diverse during the Miocene and Pliocene. Paleontologic evidence indicates that the dryopithecines were the evolutionary forerunners of the modern great apes and perhaps even the hominids.

The other group were the **ramapithecines,** which also lived during the Miocene. The ramapithecines were small creatures, a little over 1 m tall and between 20 and 70 kg in weight. Their faces were foreshortened and more hominid than apelike in appearance. They had rather small

canine teeth and molars with thick enamel and flat chewing surfaces (◆ Fig. 20.5). Based on skeletal features, it was widely believed in the 1960s that *Ramapithecus,* a genus of the ramapithecines, was the earliest hominid and thus the ancestor to humans. Reexamination of ramapithecine fossils has convinced most paleoanthropologists that *Ramapithecus* did not give rise to hominids, but rather was the ancestoral stock for the orangutans, a member of the great apes.

❖ HOMINIDS

The fossil record of the hominids extends back to only about 4 million years ago, while that of the dryopithecines and ramapithecines ends about 7 million years ago. Hominids are believed to have evolved from their hominoid ancestors during this approximately 3- to 4-million-year interval. Although hominids exhibit a variety of forms, several significant features distinguish them from other hominoids. The first is their manner of locomotion. Hominids are bipedal; that is, they have an upright posture that is reflected in the modification of their pelvis and limbs (◆ Fig. 20.6). The second feature that separates hominids from hominoids is a trend toward a large and internally reorganized brain (◆ Fig. 20.7). Other features include a reduced face and reduced canine teeth, omnivorous feeding behavior, and increased manual dexterity with its associated traits, the use and construction of sophisticated tools.

Many anthropologists believe that these common features of the hominids evolved as a response to major climatic changes that began during the Miocene and con-

◆ **FIGURE 20.5** Molars and upper jawbone of *Ramapithecus* showing the flat chewing surface of the molar teeth. (Photo courtesy of the Peabody Museum of Natural History.)

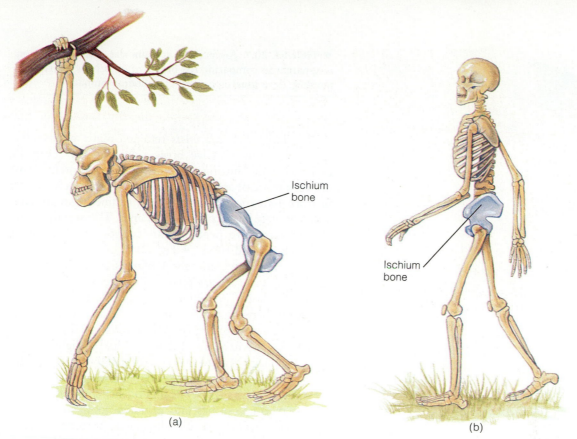

Ischium
bone

Ischium
bone

(a)

(b)

◆ **FIGURE 20.6** Comparison between quadrupedal and bipedal locomotion in gorillas and humans. (a) In gorillas, the ischium bone is long, and the entire pelvis is tilted toward the horizontal. (b) In humans, the ischium bone is much shorter, and the pelvis is vertical.

tinued into the Pliocene. During this time vast savannas replaced the African tropical rain forests where the prosimians and early anthropoids were so abundant. As the savannas and grasslands continued to expand, the hominids made the transition from true forest dwellers to life in an environment of mixed forests and grasslands.

Australopithecines

The oldest hominids belong to the genus *Australopithecus*. The first specimens of this genus were found in South Africa in the 1920s and 1930s by Raymond Dart and Robert Broom. Since then, other specimens have been found in Tanzania, Kenya, and Ethiopia by a number of different workers, the most famous probably being Louis and Mary Leakey and their son Richard, and Donald Johanson. Currently, four species of *Australopithecus* (*A. afarensis, A. africanus, A. robustus,* and *A. boisei*) are recognized (◆ Fig. 20.8).

Although there is still considerable debate, especially in light of a recent find of *Australopithecus boisei,* many paleontologists accept the evolutionary scheme shown in Figure 20.8 in which *Australopithecus afarensis* is considered the ancestor to both the later australopithecines and the genus *Homo.* We will consider this evolutionary scheme as well as an alternative one following a short discussion of the fossils themselves.

The earliest and most primitive of the **australopithecines** is *Australopithecus afarensis,* which lived from about 4 to 2.75 million years ago (◆ Fig. 20.9). Members of this species were fully bipedal (◆ Fig. 20.10) and exhibited great variability in size and weight, particularly between males and females. They ranged from just over 1 m to about 1.7 m tall and weighed between 25 and 60 kg. Their average brain capacity was larger than that of a chimpanzee—380 to 450 cubic centimeters (cc) versus 300 to 400 cc for a chimpanzee—but much smaller than that of a modern human (1,300 cc average).

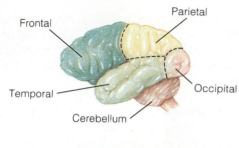

(a)

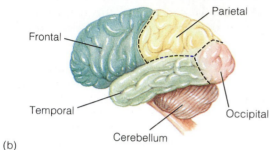

(b)

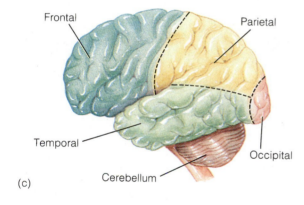

(c)

◆ **FIGURE 20.7** An increase in brain size and organization is apparent in comparing the brains of (a) a New World monkey, (b) a great ape, and (c) a modern human.

The skull of *A. afarensis* retained many apelike features, including massive brow ridges, a low forehead, and a forward-jutting jaw. Its teeth were intermediate between those of apes and humans with relatively smaller incisors than apes, relatively larger canines than humans, and relatively larger molars than apes. The heavily enameled molars were probably an adaptation to chewing fruits, seeds, and roots (◆ Fig. 20.11).

About 3 million years ago, *A. afarensis* was succeeded by another species, *Australopithecus africanus,* which lived from 3 to about 1.6 million years ago (◆ Fig. 20.12). The distinction between the two species is relatively minor and is mostly a matter of degree. *A. africanus* was slightly taller on average (1.4 m versus 1.2 m) and had a slightly larger cranial capacity (400 to 600 cc versus 380 to 450 cc). Furthermore, its face was slightly flatter, and its incisor teeth were relatively smaller. While no evidence shows that *A. africanus* modified stones into primitive tools, members of this species probably used sticks, leaves, or stones for food gathering or processing much as chimpanzees do today. The use of such tools would be impossible to decipher from the fossil record (◆ Fig. 20.13).

Both *A. afarensis* and *A. africanus* differ markedly from the two so-called robust species of *Australopithecus.* The first of these robust species, *A. robustus,* lived from about 2.3 to 1.3 million years ago, partially over-

◆ **FIGURE 20.8** The evolutionary scheme proposed by Donald Johanson and Tim White in which humans and the later australopithecines split from a common ancestor less than 3 million years ago.

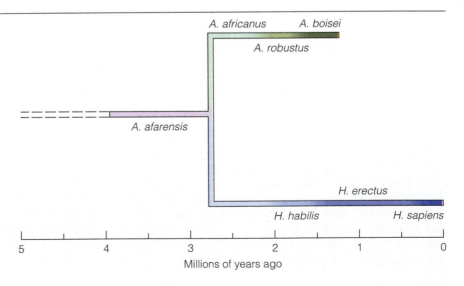

about 2.5 to 1.2 million years ago. *A. boisei* is very similar to *A. robustus* in possessing a ridge on the top of its skull, a flat face, and large, broad, flat molar teeth. These features, however, are much exaggerated in *A. boisei* (◆ Fig. 20.15, p. 582).

In the evolutionary scheme proposed by Donald Johanson and Tim White, *A. afarensis* is the single ancestral species that gave rise to all later hominids. This species split into two lineages about 2.5 million years ago (Fig. 20.8). One lineage contains all the australopithecines, arranged so they increase in robusticity through

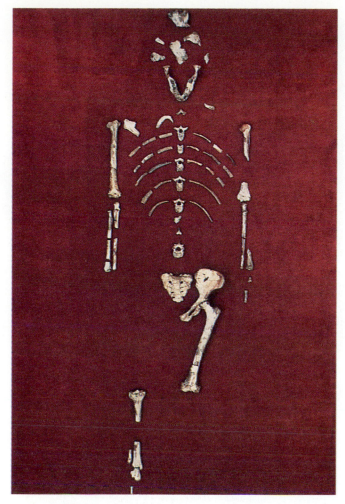

◆ **FIGURE 20.9** The skeletal remains of "Lucy," a nearly complete specimen of a 3.5-million-year-old *Australopithecus afarensis* female. (Photo courtesy of the Cleveland Museum of Natural History.)

◆ **FIGURE 20.10** Hominid footprints preserved in volcanic ash at the Laetoli site, Tanzania. Discovered in 1978 by Mary Leakey, these footprints proved that hominids were bipedal walkers at least 3.5 million years ago. The footprints of two adults and possibly those of a child are clearly visible in this photograph. (Photo © John Reader/Science Photo Library.)

lapping the geologic and geographic range of *A. africanus*. *A. robustus* was taller (1.5 m on average) and heavier (45 kg on average) than *A. africanus* and had a larger brain (500 to 600 cc). Members of *A. robustus* had massive skulls with flat faces and deep, strong jaws and very broad, flat molars, indicating they were primarily vegetarians. The crown of the skull had an elevated bony crest much like that of a gorilla, providing an additional area for the attachment of strong jaw muscles (◆ Fig. 20.14, p. 582).

The fourth species of *Australopithecus*, *A. boisei*, is also a robust form found in eastern Africa. It lived from

◆ **FIGURE 20.11** Recreation of a Pliocene landscape showing members of *Australopithecus afarensis* gathering and eating various fruits and seeds.

time, while the second lineage leads directly to humans.

The discovery of a 2.5-million-year-old skull of *A. boisei* casts doubt on such a simple evolutionary scheme. This skull displays many primitive features (found in the braincase and jaw joint) in common with *A. afarensis,* yet its derived features, mostly in the face and teeth, are shared with later specimens of *A. boisei,* the most specialized of the australopithecines. With these new facts,

Richard Leakey and Alan Walker have proposed an alternative evolutionary scheme in which there are two lineages of australopithecines, with *A. africanus* a contemporary of *A. robustus* and *A. boisei,* and not an ancestor (◆ Fig. 20.16). Furthermore, in this scheme the two australopithecine lineages and the human lineage evolve independently from ancestors whose fossils have not yet been discovered.

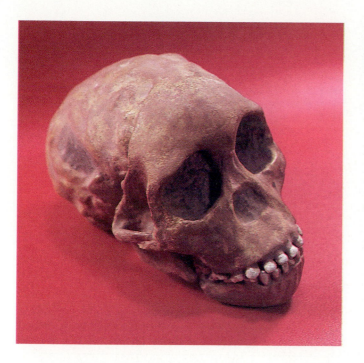

◆ **FIGURE 20.12** A reconstruction of the skull of *Australopithecus africanus*. This skull, known as the Taung child, was discovered by Raymond Dart in South Africa in 1924 and marks the beginning of modern paleoanthropology. (Replica courtesy of Carolina Biological Supply. Photo courtesy of Sue Monroe.)

The Human Lineage

Homo habilis

The earliest representative of our own genus *Homo* is a species named *Homo habilis*. Its remains were first found at Olduvai Gorge but are now known from South Africa, Kenya, and Ethiopia. Depending on which evolutionary scheme is accepted (Fig. 20.8 or 20.16), *H. habilis* evolved either from *A. afarensis* or from an as yet undiscovered early hominid about 2 million years ago and survived as a species until about 1.4 million years ago. A recent discovery of a 1.8-million-year-old skull and partial skeleton of *H. habilis* has cast new light on this earliest member of the human lineage (◆ Fig. 20.17). The

◆ **FIGURE 20.13** An *Australopithecus africanus* male digs into an insect mound with a stick, while a female harvests berries.

◆ **FIGURE 20.14** The skull of *Australopithecus robustus*. This species had a massive jaw, powerful chewing muscles, and large, broad, flat chewing teeth apparently used for grinding up coarse plant food.

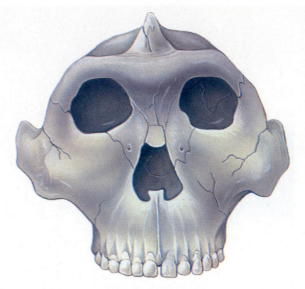

◆ **FIGURE 20.15** Diagrammatic reconstruction of the skull of *Australopithecus boisei*. Though similar to *A. robustus*, the features of the skull of *A. boisei* are more exaggerated.

head of *H. habilis* is rather like that of *Australopithecus* in that it has a flattish face and relatively large molars. However, its brain size is definitely much larger, averaging about 700 cc. What is most surprising about *Homo habilis* is how apelike in appearance it was; it had relatively long arms and a small body.

This recent find highlights the fact that the primitive body form characteristic of the australopithecines continued much longer in human evolution than was previously believed. The evolutionary transition between *H. habilis* and *H. erectus* appears to have occurred in a very short time, between 1.8 and 1.6 million years ago.

Homo erectus

In contrast to the australopithecines and *H. habilis*, which are unknown outside Africa, *Homo erectus* was a

◆ **FIGURE 20.16** In the evolutionary scheme proposed by Richard Leakey and his colleagues, humans and the two main australopithecine groups evolve from creatures that have not yet been discovered.

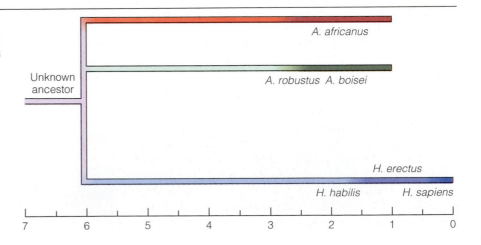

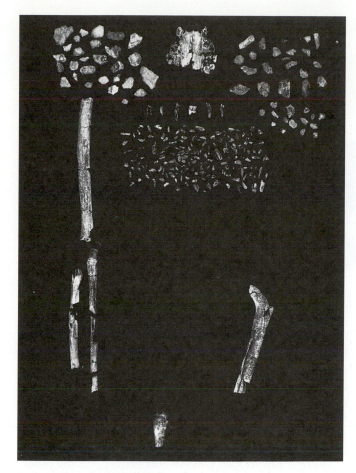

◆ **FIGURE 20.17** Fossil skull and skeleton fragments from a 1.8-million-year-old specimen of *Homo habilis*. (Photo courtesy of the Institute of Human Origins.)

forehead receded sharply behind strongly developed brow ridges (◆ Fig. 20.18). The teeth of *H. erectus* were smaller than those of australopithecines, but bigger than the teeth of modern humans. An African discovery of a 1.6-million-year-old boy indicates that *H. erectus* was comparable to modern humans in body size.

The archaeological record indicates that the members of *H. erectus* were tool makers. Their stone culture, known as the *Acheulean,* included the production of handaxes, flakes, and pointed scrapers. Furthermore, some sites show evidence that they used fire and lived in caves, an advantage for those living in more northerly climates (◆ Fig. 20.19).

Sometime between 300,000 and 200,000 years ago, *H. erectus* apparently evolved into *H. sapiens*. The transition between the two species seems to have been gradual, with a variety of types existing at the same time. These early or archaic members of *H. sapiens* had rounder and

◆ **FIGURE 20.18** A reconstruction of the skull of *Homo erectus,* a widely distributed species whose remains have been found in Africa, Europe, India, China, and Indonesia. (Replica courtesy of Carolina Biological Supply. Photo courtesy of Sue Monroe.)

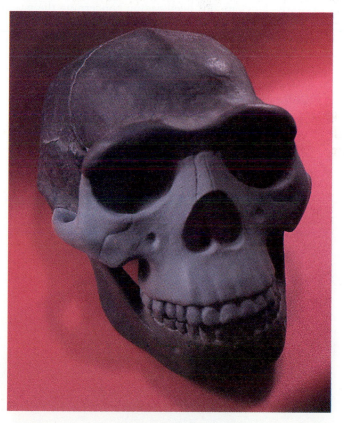

widely distributed species and the first to migrate from Africa during the interglacial periods of the Pleistocene. Specimens have been found not only in Africa, but in Europe, India, China ("Peking Man"), and Indonesia ("Java Man"). *H. erectus* evolved in Africa 1.8 million years ago and by 1 million years ago was present in southeastern and eastern Asia, where it survived until about 300,000 years ago.

Although *H. erectus* developed regional variations in form, the species differed from modern humans in several ways. The cranial capacity (800 to 1,300 cc), though much larger than that of *H. habilis,* was still less than in *Homo sapiens* (about 1,350 cc on the average). *H. erectus* possessed a long, low, thick-walled skull that was sharply angled at the back. Its face was massive, and the

◆ **FIGURE 20.19** Recreation of a Pleistocene setting in Europe in which members of *Homo erectus* are using fire and stone tools. (Photo courtesy of the Field Museum of Natural History, Chicago; Neg. A76851c.)

higher skulls, a more delicately structured face, and relatively small teeth and jaws, with some forms possessing a chin. The cranial capacity soon averaged about 1,350 cc (the same as modern humans), and these archaic forms evolved into modern humans.

Recently, a debate has been raging concerning the nature of *H. erectus* and whether it did, in fact, give rise to

◆ **FIGURE 20.20** Reconstructed Neanderthal skull. The Neanderthals were characterized by prominent heavy brow ridges and a weak chin. Their brain was also slightly larger on average than that of modern humans. (Photo courtesy of Department of Library Services, American Museum of Natural History, Neg. # 125964.)

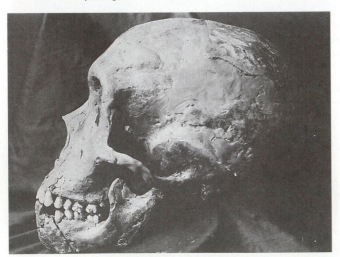

H. sapiens. There are now three competing hypotheses concerning the place *H. erectus* occupies in human evolution. One hypothesis advocates the traditional approach, with *H. erectus* evolving from *H. habilis* and then evolving into *H. sapiens.* A second hypothesis asserts that *H. erectus* split into two species, one of which became extinct and the other evolved into *H. sapiens.* This hypothesis uses cladistic analysis (see Perspective 6.1) to show that there were actually two species of *H. erectus,* one in Asia that became extinct and the other in Africa that evolved into *H. sapiens.* The third hypothesis seeks to eliminate *H. erectus* completely as a species and place its fossil remains within a diverse group of early *H. sapiens.* Which of these hypotheses eventually is accepted will be decided as new techniques are applied to the study of hominid fossils and new fossils are discovered.

Neanderthals

The most famous of all fossil humans are the **Neanderthals,** who inhabited Europe and the Near East between about 150,000 and 32,000 years ago (see Perspective 20.1). Some paleoanthropologists regard the Neanderthals as a variety or subspecies of our own species *(Homo sapiens neanderthalensis),* while others regard them as a separate species *(Homo neanderthalensis).* In any case, their name comes from the first specimens found in 1856 in the Neander Valley near Düsseldorf, Germany.

The Neanderthal's skeleton differs slightly from our own, with the most notable differences occurring in the features of the skull. The Neanderthal skull was long and

ANCIENT DNA

The newly emerging field of molecular archaeology and paleontology seeks to understand the molecular composition and genetics of ancient organisms by examining proteins and nucleic acids preserved in fossils. By comparing the molecular structures of blood proteins of different organisms, molecular biologists can determine how closely related various species are. One of the best examples of the usefulness of this technique is the research mentioned in the Prologue of this chapter where molecular biologists showed that differences between the blood protein molecules of chimpanzees and humans indicated they diverged from each other only several million years ago, not tens of millions of years ago as was originally believed based strictly on fossil evidence. Since then, molecular archaeology has expanded into many other areas, including the examination and cloning of DNA from preserved bodies.

The oldest known DNA comes from a 17-million-year-old magnolia tree, while the oldest human remains analyzed thus far are 8,000 years old. Until recently, fossil DNA from humans has come from mummies or frozen specimens. Surprisingly, tissues preserved in peat bogs are usually not suitable for study in spite of their excellent preservation because the tannic acid that cures the flesh also destroys the DNA. Scientists now believe that DNA can be extracted from the bones of ancient humans that have not yet become mineralized. Recently, biochemists at Los Alamos National Laboratory have been analyzing a fragment of a vertebra from a Neanderthal skeleton found at Shanidar, Iraq, and dated at about 50,000 years. Their work is still in the preliminary stage and no results have yet been published.

The study of DNA in ancient humans is turning up many surprises. For example, an analysis of DNA recovered from a 7,500-year-old man preserved in a peat pond in Windover, Florida, indicates that the genetics of the Windover population changed very little during the 1,000-year history of the burial grounds where 177 skeletons and 91 brains were recovered. The DNA could be analyzed because a limestone spring neutralized the tannic acid of the peat pond and thus allowed some DNA segments in the brains to be preserved. The homogeneity of DNA in these people may mean that the population was very inbred. Such isolation may have given rise to numerous languages, not unlike the situation found in the highland valleys of New Guinea where isolated populations have developed their own languages.

Other researchers are studying the population genetics of the peoples of North and South America as well as the effects of disease on ancient human populations. Furthermore, it is hoped that examination of fossilized DNA will confirm or reject the Eve hypothesis (see the Prologue). Molecular geneticists hope to settle this debate by analyzing the DNA preserved in fossil bones. Molecular archaeology promises to provide new insights into ancient humans, and it remains to be seen how far new techniques for extracting DNA can push back the molecular clock.

low with prominent heavy brow ridges, a projecting mouth, and a weak, receding chin (◆ Fig. 20.20). Its brain was large, actually slightly larger on average than our own, and somewhat differently shaped. The body of the Neanderthals was somewhat more massive and heavily muscled than ours, with rather short lower limbs, much like those of Eskimos and other cold-adapted people of today.

Of all the early hominids, none have been so maligned and misunderstood as the Neanderthals. The name alone conjures up images of brutishness and low intelligence. This unfortunate image has been fostered from reconstructions based on a nearly complete skeleton found in 1908 in the Carreze region of southwestern France. From this skeleton has come an image of a hulking, dim-witted brute who shuffled along with the bent-kneed walk of an ape. At the time of its discovery, however, people failed to realize that this was the skeleton of an individual who was stooped over because of severe hip arthritis, a broken rib, and diseased vertebrae. It was also missing most of its teeth due to the effects of gum disease.

Based on specimens from more than a hundred sites, we now know that the Neanderthals were not much different from us, only more robust. Europe's Neanderthals were the first humans to move into truly cold climates, enduring miserably long winters and short summers as they pushed north into tundra country (◆ Fig. 20.21).

The remains of Neanderthals are found chiefly in caves and hutlike rock shelters, which also contain a variety of specialized stone tools and weapons. Evidence such as

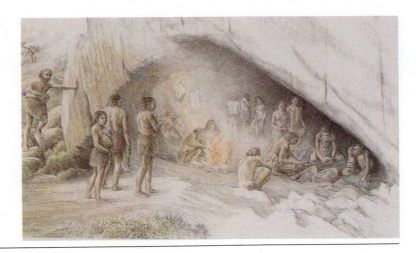

◆ **FIGURE 20.21** Archaeological evidence indicates that Neanderthals lived in caves and participated in ritual burials as depicted in this painting of a burial ceremony that occurred approximately 60,000 years ago at Shanidar Cave, Iraq. (Painting by Ronald Bowen, photo courtesy of Robert Harding Picture Library.)

the arthritic individual mentioned above indicates that Neanderthals commonly took care of their injured and buried their dead, frequently with such grave items as tools, food, and perhaps even flowers.

While the Neanderthals may have been socially and technologically less advanced than modern humans, they were certainly not dim-witted, brutish boors as they are frequently portrayed.

◆ **FIGURE 20.22** Recreation of a Cro-Magnon camp in Europe. Cro-Magnons were highly skilled hunters who formed living groups of various sizes.

◆ **FIGURE 20.23** Cro-Magnons were very skilled cave painters. Shown is a painting of a horse from the cave of Niaux, France. (Photo © Tom McHugh/Photo Researchers.)

Cro-Magnons

About 35,000 years ago, humans closely resembling modern Europeans moved into the region inhabited by the Neanderthals and completely replaced them. When and how this happened is still unknown. **Cro-Magnons,** the name given to the successors of the Neanderthals in France, lived from about 35,000 to 10,000 years ago; during this period the development of art and technology far exceeded anything the world had seen before.

Highly skilled nomadic hunters, Cro-Magnons followed the herds in their seasonal migrations. They used a variety of specialized tools in their hunts, including perhaps the bow and arrow (◆ Fig. 20.22). They sought refuge in caves and rock shelters and formed living groups of various sizes. Cro-Magnons were also cave painters.

Using paints made from manganese and iron oxides, Cro-Magnon people painted hundreds of scenes on the ceilings and walls of caves in France and Spain, where many of them are still preserved today (◆ Fig. 20.23).

With the appearance of Cro-Magnons, human evolution has become almost entirely cultural rather than biological. Humans have spread throughout the world by devising means to deal with a broad range of environmental conditions. Since the evolution of the Neanderthals about 150,000 years ago, humans have gone from a stone culture to a technology that has allowed us to visit other planets with space probes and land men on the Moon. It remains to be seen how we will use this technology in the future and whether we will continue as a species, evolve into another species, or become extinct as many groups have before us.

Chapter Summary

1. The primates probably evolved during the Late Cretaceous. Several trends help characterize primates and differentiate them from other mammalian orders. These include a change in overall skeletal structure and mode of locomotion, an increase in brain size, stereoscopic vision, and evolution of a grasping hand with opposable thumb.

2. The primates are divided into two suborders, the prosimians and anthropoids. The prosimians are the oldest primate lineage and include lemurs, tarsiers, and tree shrews.

The anthropoids include the New and Old World monkeys, apes, and hominids, which are humans and their ancestors.

3. The earliest known hominids are the australopithecines, a fully bipedal group that evolved in Africa nearly 4 million years ago. Currently, four australopithecine species are known: *Australopithecus afarensis, A. africanus, A. robustus,* and *A. boisei.*

4. There are two competing hypotheses concerning the relationship between the australopithecines and the lineage that gave rise to modern humans. One postulates that *A. afarensis* was the ancestor to all the other australopithecines and the human lineage. The other hypothesis holds that *A. afarensis* only gave rise to the other australopithecines and that the human lineage evolved independently from an as yet undiscovered ancestor.

5. The human lineage began about 2 million years ago in Africa with the evolution of *Homo habilis,* which survived as a species until about 1.4 million years ago. It evolved into *Homo erectus.*

6. *Homo erectus* evolved from *Homo habilis* about 1.8 million years ago and was the first hominid to migrate out of Africa. By 1 million years ago, *H. erectus* had spread to Europe, India, China, and Indonesia. *H. erectus* used fire, made tools, and lived in caves.

7. Sometime between 300,000 and 200,000 years ago, *Homo sapiens* evolved from *Homo erectus.* These early or archaic humans may be ancestors of Neanderthals.

8. The Neanderthals evolved about 150,000 years ago. Neanderthals were not much different from present-day humans, only more robust and with differently shaped skulls. They made specialized tools and weapons, apparently took care of their injured, and buried their dead.

9. The Cro-Magnons were the successors of the Neanderthals and lived from about 35,000 to 10,000 years ago. They were highly skilled nomadic hunters, formed living groups of various sizes, and were also skilled cave painters.

10. Modern humans succeeded the Cro-Magnons about 10,000 years ago and have spread throughout the world.

Important Terms

anthropoid
australopithecine
Cro-Magnon
dryopithecine

hominid
hominoid
Neanderthal
New World monkey

Old World monkey
primate
prosimian
ramapithecine

Review Questions

1. Which DNA is used as the basis for tracing family relationships and is central to the Eve hypothesis?
   a. _____ nuclear; b. _____ mitochondrial; c. _____ organelle; d. _____ plastid; e. _____ none of these.

2. Which of the following is not a hominoid?
   a. _____ chimpanzee; b. _____ orangutan; c. _____ gorilla; d. _____ gibbon; e. _____ prosimian.

3. What is the oldest primate lineage?
   a. _____ anthropoids; b. _____ hominoids; c. _____ hominids; d. _____ prosimians; e. _____ none of these.

4. Which of the following evolutionary trends characterize primates?
   a. _____ grasping hand with opposable thumb; b. _____ stereoscopic vision; c. _____ increase in brain size; d. _____ change in overall skeletal structure; e. _____ all of these.

5. The dryopithecines are currently believed to be possible ancestors to the:
   a. _____ great apes; b. _____ lesser apes; c. _____ hominids; d. _____ answers (a) and (c); e. _____ answers (b) and (c).

6. The oldest hominids belong to which genus?
   a. _____ *Australopithecus;* b. _____ *Homo;* c. _____

   Ramapithecus; d. _____ *Dryopithecus;* e. _____ *Pithacanthropus.*

7. The oldest hominid fossils are dated at about:
   a. _____ 7 million years; b. _____ 4 million years; c. _____ 1 million years; d. _____ 350,000 years; e. _____ 40,000 years.

8. The earliest known australopithecine is *Australopithecus* _____ .
   a. _____ *afarensis;* b. _____ *africanus;* c. _____ *boisei;* d. _____ *habilis;* e. _____ *robustus.*

9. Which genus do humans belong to:
   a. _____ *Australopithecus;* b. _____ *Dryopithecus;* c. _____ *Homo;* d. _____ *Pithacanthropus;* e. _____ *Ramapithecus.*

10. According to archaeological evidence, which were the first hominids to use fire?
    a. _____ *Australopithecus boisei;* b. _____ *Homo habilis;* c. _____ *Homo erectus;* d. _____ *Homo sapiens;* e. _____ *Homo neanderthalensis.*

11. Which were the first hominids to migrate out of Africa?
    a. _____ *Australopithecus robustus;* b. _____ *Australopithecus boisei;* c. _____ *Homo habilis;* d. _____ *Homo erectus;* e. _____ *Homo sapiens.*

12. Which extinct lineage of humans were skilled hunters and cave painters?
   a. _____ Neanderthals; b. _____ Cro-Magnons; c. _____ Archaic; d. _____ Acheuleans; e. _____ none of these.

13. To which hominid group do Java Man and Peking Man belong?
   a. _____ *Australopithecus robustus*; b. _____ *Australopithecus boisei*; c. _____ *Homo habilis*; d. _____ *Homo erectus*; e. _____ *Homo sapiens*.

14. According to the Eve hypothesis, all lineages of modern humans evolved about 200,000 years ago from a woman living in:
   a. _____ Africa; b. _____ Europe; c. _____ Asia; d. _____ Indonesia; e. _____ India.

15. Which of the following are not anthropoids?
   a. _____ Old World monkeys; b. _____ New World monkeys; c. _____ apes; d. _____ humans; e. _____ prosimians.

16. Discuss the differences between the prosimians and anthropoids.

17. How do hominoids differ from hominids?

18. Discuss the evolutionary history of the anthropoids.

19. Discuss the importance of geologic and climatic change to the evolution of primates during the Miocene.

20. What are the major evolutionary trends that characterize the primates and set them apart from the other orders of mammals?

21. Discuss the two hypotheses concerning the evolutionary history of the hominids.

22. Compare the various features of the four different species of australopithecines.

23. Discuss the evolutionary history of the genus *Homo*.

24. Discuss the current competing hypotheses concerning the validity of *Homo erectus* and whether it was ancestral to *Homo sapiens*.

25. Discuss the Eve hypothesis and explain why it is so controversial.

26. What are the main differences between the Neanderthals and Cro-Magnons?

Additional Readings

Blumenschine, R. J., and J. A. Cavallo. 1992. Scavenging and human evolution. *Scientific American* 267, no. 4: 90–97

Bower, B. 1992. Erectus unhinged. *Science News* 141, no. 25: 408–411.

Brace, C. L. 1991. *The stages of human evolution*. 4th ed. Englewood Cliffs, New Jersey. Prentice-Hall.

Brown, M. H. 1990. *The search for Eve*. New York: Harper and Row.

Johanson, D. C., and M. A. Edey. 1981. *Lucy, the beginnings of humankind*. New York: Simon and Schuster.

Johanson, D., and J. Shreeve. 1989. *Lucy's child: The discovery of a human ancestor*. New York: Avon Books.

Leakey, R. E. 1981. *The making of mankind*. New York: E. P. Dutton.

Lewin, R. 1987. *Bones of contention*. New York: Simon and Schuster.

Pfeiffer, J. 1985. *The emergence of humankind*. 4th ed. New York: Harper and Row.

Pilbeam, D. 1984. The descent of hominoids and hominids. *Scientific American* 250, no. 3: 84–96.

Ross, P. E. 1992. Eloquent remains. *Scientific American* 266, no. 5: 114–125.

Schopf, J. W., ed. 1992. *Major events in the history of life*. Boston, Mass.: Jones and Bartlett Publishers.

Shipman, P. 1986. Baffling limb on the family tree. *Discover* 7, no. 9: 86–93.

Theunissen, B. 1989. *Eugène Dubois and the ape-man from Java: The history of the first missing link and its discoverer*. Dordrecht, The Netherlands. Kluwere Academic Publishers.

Thomson, K. S. 1992. The challenge of human origins. *American Scientist* 80, no. 6: 519–22.

Thorne, A. G., and M. Wolpoff. 1992. The multiregional evolution of humans. *Scientific American* 266, no. 4: 76–83.

Weaver, K. F. 1985. The search for our ancestors. *National Geographic* 168, no. 5: 560–623.

Wilson, A. C., and R. L. Cann. 1992. The recent African genesis of humans. *Scientific American* 266, no. 4: 68–73.

Wolpoff, M., and A. Thorne. 1991. The case against Eve. *New Scientist* 130, no. 1774: 37–41.

ENGLISH-METRIC CONVERSION CHART

ENGLISH UNIT	CONVERSION FACTOR	METRIC UNIT	CONVERSION FACTOR	ENGLISH UNIT
Length				
Inches (in)	2.54	Centimeters (cm)	0.39	Inches (in)
Feet (ft)	0.305	Meters (m)	3.28	Feet (ft)
Miles (mi)	1.61	Kilometers (km)	0.62	Miles (mi)
Area				
Square inches (in^2)	6.45	Square centimeters (cm^2)	0.16	Square inches (in^2)
Square feet (ft^2)	0.093	Square meters (mi^2)	10.8	Square feet (ft^2)
Square miles (mi^2)	2.59	Square kilometers (km^2)	0.39	Square miles (mi^2)
Volume				
Cubic inches (in^3)	16.4	Cubic centimeters (cm^3)	0.061	Cubic inches (in^3)
Cubic feet (ft^3)	0.028	Cubic meters (m^3)	35.3	Cubic feet (ft^3)
Cubic miles (mi^3)	4.17	Cubic kilometers (km^3)	0.24	Cubic miles (mi^3)
Weight				
Ounces (oz)	28.3	Grams (g)	0.035	Ounces (oz)
Pounds (lb)	0.45	Kilograms (kg)	2.20	Pounds (lb)
Short tons (st)	0.91	Metric tons (t)	1.10	Short tons (st)
Temperature				
Degrees Fahrenheit (°F)	$-32° \times 0.56$	Degrees Celsius (Centigrade) (°C)	$\times 1.80 + 32°$	Degrees Fahrenheit (°F)

Examples: 10 inches = 25.4 centimeters; 10 centimeters = 3.9 inches
100 square feet = 9.3 square meters; 100 square meters = 1080 square feet
50°F = 10.08°C; 50°C = 122°F

ENGLISH-METRIC CONVERSION CHART

CLASSIFICATION OF ORGANISMS

Any classification is an attempt to make order out of disorder and to group similar items into the same categories. All classifications are schemes that attempt to relate items to each other based on current knowledge and therefore are progress reports on the current state of knowledge for the items classified. Because classifications are to some extent subjective, classification of organisms may vary among different texts.

The classification that follows is based on the five-kingdom system of classification of Margulis and Schwartz.* We have not attempted to include all known life forms, but rather major categories of both living and fossil groups.

KINGDOM MONERA

Prokaryotes
   Phylum Anaerobic Photosynthetic Bacteria—
      (Archean-Recent)
   Phylum Cyanobacteria—Blue-green algae or
      blue-green bacteria (Archean-Recent)

KINGDOM PROTOCTISTA

Solitary or colonial unicellular eukaryotes
   Phylum Acritarcha—Organic-walled unicellular
      algae of unknown affinity (Proterozoic-Recent)
   Phylum Bacillariophyta—Diatoms(Jurassic-Recent)
   Phylum Charophyta—Stoneworts (Silurian-Recent)
   Phylum Chlorophyta—Green algae
      (Proterozoic-Recent)

---

*Margulis, L., and K. V. S. Schwartz, 1982. *Five Kingdoms*. New York: W. H. Freeman and Co.

Phylum Chrysophyta—Golden-brown algae, silicoflagellates and coccolithophorids (Jurassic-Recent)
Phylum Euglenophyta—Euglenids (Cretaceous-Recent)
Phylum Myxomycophyta—Slime molds (Proterozoic-Recent)
Phylum Phaeophyta—Brown algae, multicellular, kelp, seaweed (Proterozoic-Recent)
Phylum Protozoa—Unicellular heterotrophs (Cambrian-Recent)
   Class Sarcodina—Forms with pseudopodia for locomotion (Cambrian-Recent)
      Order Foraminifera—Benthonic and planktonic sarcodinids most commonly with calcareous tests (Cambrian-Recent)
      Order Radiolaria—Planktonic sarcodinids with siliceous tests (Cambrian-Recent)
Phylum Pyrrophyta—Dinoflagellates (Silurian?, Permian-Recent)
Phylum Rhodophyta—Red algae (Proterozoic-Recent)
Phylum Xanthophyta—Yellow-green algae (Miocene-Recent)

KINGDOM FUNGI

Phylum Zygomycota—Fungi that lack cross walls (Proterozoic-Recent)
Phylum Basidiomycota—Mushrooms (Pennsylvanian-Recent)
Phylum Ascomycota—Yeasts, bread molds, morels (Mississippian-Recent)

KINGDOM PLANTAE

Photosynthetic eukaryotes

Division* Bryophyta—Liverworts, mosses, hornworts (Devonian-Recent)

Division Psilophyta—Small, primitive vascular plants with no true roots or leaves (Silurian-Recent)

Division Lycopodophyta—Club mosses, simple vascular systems, true roots and small leaves, including scale trees of Paleozoic Era (lycopsids) (Devonian-Recent)

Division Sphenophyta—Horsetails, scouring rushes, and sphenopsids such as the Carboniferous *Calamites* (Devonian-Recent)

Division Pteridophyta—Ferns (Devonian-Recent)

Division Pteridospermophyta—Seed ferns (Devonian-Jurassic)

Division Coniferophyta—Conifers or cone-bearing gymnosperms (Carbonaceous-Recent)

Division Cycadophyta—Cycads (Triassic-Recent)

Division Ginkgophyta—Maidenhair tree (Triassic-Recent)

Division Angiospermophyta—Flowering plants and trees (Cretaceous-Recent)

KINGDOM ANIMALIA

Nonphotosynthethic multicellular eukaryotes (Proterozoic-Recent)

Phylum Porifera—Sponges (Cambrian-Recent)

Phylum Archaeocyatha—Extinct spongelike organisms (Cambrian)

Phylum Cnidaria—Hydrozoans, jellyfish, sea anemones, corals (Cambrian-Recent)

Class Hydrozoa—Hydrozoans (Cambrian-Recent)

Order Stromatoporoida—Extinct group of reef-building organisms (Cambrian-Cretaceous)

Class Scyphozoa—Jellyfish (Proterozoic-Recent)

Class Anthozoa—Sea anemones and corals (Cambrian-Recent)

Order Tabulata—Exclusively colonial corals with reduced to nonexistent septa (Ordovician-Permian)

Order Rugosa—Solitary and colonial corals with fourfold symmetry (Ordovician-Permian)

Order Scleractinia—Solitary and colonial corals with sixfold symmetry. Most colonial forms have symbionic dinoflagellates in their tissue. Important reef builders today (Triassic-Recent)

Phylum Bryozoa—Exclusively colonial suspension feeding marine animals that are useful for correlation and ecological interpretations (Ordovician-Recent)

Phylum Brachiopoda—Marine suspension feeding animals with two unequal sized valves. Each valve is bilaterally symmetrical (Cambrian-Recent)

Class Inarticulata—Primitive chitino-phosphatic or calcareous brachiopods that lack a hinging structure. They open and close their valves by means of complex muscles (Cambrian-Recent)

Class Articulata—Advanced brachiopods with calcareous valves that are hinged (Cambrian-Recent)

Phylum Mollusca—A highly diverse group of invertebrates (Cambrian-Recent)

Class Monoplacophora—Segmented, bilaterally symmetrical crawling animals with cap-shaped shells (Cambrian-Recent)

Class Amphineura—Chitons. Marine crawling forms, typically with 8 separate calcareous plates (Cambrian-Recent)

Class Scaphopoda—Curved, tusk-shaped shells that are open at both ends (Ordovician-Recent)

Class Gastropoda—Single shelled generally coiled crawling forms. Found in marine, brackish, and fresh water as well as terrestrial environments (Cambrian-Recent)

Class Bivalvia—Mollusks with two valves that are mirror images of each other. Typically known as clams or oysters (Cambrian-Recent)

Class Cephalopoda—Highly evolved swimming animals. Includes shelled sutured forms as well as non-shelled types such as octopus and squid (Cambrian-Recent)

Order Nautiloidea—Forms in which the chamber partitions are connected to the wall along simple, slightly curved lines (Cambrian-Recent)

Order Ammonoidea—Forms in which the chamber partitions are connected to the wall along wavy lines (Devonian-Cretaceous)

---

*In botany, division is the equivalent to phylum.

Order Coleoidea—forms in which the shell is reduced or lacking. Includes octopus, squid, and the extinct belemnoids (Mississippian-Recent)

Phylum Annelida—Segmented worms. Responsible for many of the Phanerozoic burrow and trail trace fossils (Proterozoic-Recent)

Phylum Arthropoda—The largest invertebrate group comprising about 80% of all known animals. Characterized by a segmented body and jointed appendages (Cambrian-Recent)

Class Trilobita—Earliest appearing arthropod class. Trilobites had a head, body, and tail and were bilaterally symmetrical (Cambrian-Permian)

Class Crustacea—Diverse class characterized by a fused head and body and an abdomen. Included are barnacles, copepods, crabs, ostracodes, and shrimp (Cambrian-Recent)

Class Insecta—Most diverse and common of all living invertebrates, but rare as fossils (Silurian-Recent)

Class Merostomata—Characterized by four pairs of appendages and a more flexible exoskeleton than crustaceans. Includes the extinct eurypterids, horseshoe crabs, scorpions, and spiders (Cambrian-Recent)

Phylum Echinodermata—Exclusively marine animals with fivefold radial symmetry and a unique water vascular system (Cambrian-Recent)

Subphylum Crinozoa—Forms attached by a calcareous jointed stem (Cambrian-Recent)

Class Crinoidea—Most important class of Paleozoic echinoderms. Suspension feeding forms that are either free-living or attach to sea floor by a stem (Cambrian-Recent)

Class Blastoidea—Small class of Paleozoic suspension feeding sessile forms with short stems (Ordovician-Permian)

Class Cystoidea—Globular to pear-shaped suspension feeding benthonic sessile forms with quite short stems (Ordovician-Devonian)

Subphylum Homalozoa—A small group with flattened, asymmetrical bodies with no stems. Also called carpoids (Cambrian-Devonian)

Subphylum Echinozoa—Globose, predominantly benthonic mobile echinoderms (Ordovician-Recent)

Class Helioplacophora—Benthonic, mobile forms, shaped like a top with plates arranged in a helical spiral (Early Cambrian)

Class Edrioasteroidea—Benthonic, sessile or mobile, discoidal, globular or cylindrical shaped forms with five straight or curved feeding areas shaped like a starfish (Cambrian-Pennsylvanian)

Class Holothuroidea—Sea cucumbers. Sediment feeders having calcareous spicules embedded in a tough skin (Ordovician-Recent)

Class Echinoidea—Largest group of echinoderms. Globe or disk shaped with movable spines. Predominantly grazers or sediment feeders. Epifaunal and infaunal (Ordovician-Recent)

Subphylum Asterozoa—Stemless, benthonic mobile forms (Ordovician-Recent)

Class Asteroidea—Starfish. Arms merge into body (Ordovician-Recent)

Class Ophiuroidea—Brittle star. Distinct central body (Ordovician-Recent)

Phylum Hemichordata—Characterized by a notochord sometime during the life history. Modern acorn worms and extinct graptolites (Cambrian-Recent)

Class Graptolithina—Colonial marine hemichordates having a chitinous exoskeleton. Predominantly planktonic (Cambrian-Mississippian)

Phylum Chordata—Animals with notochord, hollow dorsal nerve cord, and gill slits during at least part of their lifecycle (Cambrian-Recent)

Subphylum Urochordata—Sea squirts, tunicates. Larval forms have notochord in tail region.

Subphylum Cephalochordata—Small marine animals with notochords and small fish-like bodies (Cambrian-Recent)

Subphylum Vertebrata—Animals with a backbone of vertebrae (Cambrian-Recent)

Class Agnatha—Jawless fish. Includes the living lampreys and hagfish as well as the extinct armored ostracoderms (Cambrian-Recent)

Class Acanthodii—Primitive jawed fish with numerous spiny fins (Silurian-Permian)

Class Placodermii—Primitive armored jawed fish (Silurian-Permian)

Class Chondrichthyes—Cartilaginous fish such as sharks and rays (Devonian-Recent)

Class Osteichthyes—Bony fish (Devonian-Recent)

Subclass Actinopterygii—Ray-finned fish (Devonian-Recent)

Subclass Sarcopterygii—Lobe-finned, air-breathing fish (Devonian-Recent)

Order Crossoptergii—Lobe-finned fish
that were ancestral to amphibians
(Devonian-Recent)
Order Dipnoi—Lungfish
(Devonian-Recent)
Class Amphibia—Amphibians. The first terrestrial
vertebrates (Devonian-Recent)
Subclass Labyrinthodontia—Earliest
amphibians. Solid skulls and complex tooth
pattern (Devonian-Triassic)
Subclass Salientia—Frogs, toads, and their
relatives (Triassic-Recent)
Subclass Condata—Salamanders and their
relatives (Triassic-Recent)
Class Reptilia—Reptiles. A large and varied
vertebrate group characterized by having scales
and laying an amniote egg (Pennsylvanian-
Recent)
Subclass Anapsida—Reptiles whose skull has a
solid roof with no openings
(Pennsylvanian-Recent)
    Order Cotylosauria—Captorhinomorphs
    or earliest reptiles (Pennsylvanian-
    Triassic)
    Order Chelonia—Turtles
    (Triassic-Recent)
Subclass Euryapsida—Reptiles with one opening
high on the side of the skull behind the eye.
Mostly marine (Permian-Cretaceous)
    Order Protorosauria—Land living
    ancestral euryapsids (Permian-
    Cretaceous)
    Order Placodontia—Placodonts. Bulky,
    paddle-limbed marine reptiles with
    rounded teeth for crushing mollusks
    (Triassic)
    Order Ichthyosauria—Ichthyosaurs.
    Dolphin-shaped swimming reptiles
    (Triassic-Cretaceous)
Subclass Diapsida—Most diverse reptile class,
characterized by two openings in the skull
behind the eye. Includes lizards, snakes,
crocodiles, thecodonts, dinosaurs, and
pterosaurs (Permian-Recent)
    Infraclass Lepidosauria—Primitive diapsids
    including snakes, lizards, and the large
    Cretaceous marine reptile group called
    mosasaurs (Permian-Recent)
    Order Mosasauria—Mosasaurs
    (Cretaceous)

Order Plesiosauria—Plesiosaurs
(Jurassic-Cretaceous)
Order Squamata—Lizards and snakes
(Triassic-Recent)
Order Rhynchocephalia—The living
tuatara *Sphenodon* and its extinct
relatives (Jurassic-Recent)
Infraclass Archosauria—Advanced diapsids
(Triassic-Recent)
Order Thecodontia—Thecodontians were
a diverse group that was ancestral to
the crocodilians, pterosaurs, and
dinosaurs (Permian-Triassic)
Order Crocodilia—Crocodiles, alligators,
and gavials (Triassic-Recent)
Order Pterosauria—Flying and gliding
reptiles called pterosaurs
(Triassic-Cretaceous)
Infraclass Dinosauria—Dinosaurs
(Triassic-Cretaceous)
Order Saurischia—Lizard-hipped
dinosaurs (Triassic-Cretaceous)
    Suborder Theropoda—Bipedal
    carnivores (Triassic-Cretaceous)
    Suborder Sauropoda—Quadrupedal
    herbivores, including the largest
    known land animals
    (Jurassic-Cretaceous)
Order Ornithischia—Bird-hipped
dinosaurs (Triassic-Cretaceous)
    Suborder Ornithopoda—Bipedal
    herbivores, including the duck-billed
    dinosaurs (Triassic-Cretaceous)
    Suborder Stegosauria—Quadrupedal
    herbivores with bony spikes on
    their tails and bony plates on their
    backs (Jurassic-Cretaceous)
    Suborder Pachycephalosauria—Bipedal
    herbivores with thickened bones of
    the skull roof (Cretaceous)
    Suborder Ceratopsia—Quadrupedal
    herbivores typically with horns or a
    bony frill over the top of the neck
    (Cretaceous)
    Suborder Ankylosauria—Heavily
    armored quadrupedal herbivores
    (Cretaceous)
Subclass Synapsida—Mammal-like reptiles
with one opening low on the side of the
skull behind the eye (Pennsylvanian-Triassic)

Order Pelycosauria—Early mammal-like reptiles including those forms in which the vertebral spines were extended to support a "sail" (Pennsylvanian-Permian)

Order Therapsida—Advanced mammal-like reptiles with legs positioned beneath the body and the lower jaw formed largely of a single bone. Many therapsids may have been endothermic (Permian-Triassic)

Class Aves—Birds. Endothermic and feathered (Jurassic-Recent)

Class Mammalia—Mammals. Endothermic animals with hair (Triassic-Recent)

Subclass Prototheria—Egg-laying mammals (Triassic-Recent)

Order Docodonta—Small, primitive mammals (Triassic)

Order Triconodonta—Small, primitive mammals with specialized teeth (Triassic-Cretaceous)

Order Monotremata—Duck-billed platypus, spiny anteater (Cretaceous-Recent)

Subclass Allotheria—Small extinct early mammals with complex teeth for grinding food (Jurassic-Eocene)

Order Multituberculata—The first mammalian herbivores and the most diverse of Mesozoic mammals (Jurassic-Eocene)

Subclass Theria—Mammals that give birth to live young (Jurassic-Recent)

Order Symmetrodonta—Small, primitive Mesozoic therian mammals (Jurassic-Cretaceous)

Order Upantotheria—Trituberculates (Jurassic-Cretaceous)

Order Creodonta—Extinct ancient carnivores (Cretaceous-Paleocene)

Order Condylartha—Extinct ancestral hoofed placentals (ungulates) (Cretaceous-Oligocene)

Order Marsupialia—Pouched mammals. Opossum, kangaroo, koala (Cretaceous-Recent)

Order Insectivora—Primitive insect-eating mammals. Shrew, mole, hedgehog (Cretaceous-Recent)

Order Xenungulata—Large South American mammals that broadly resemble pantodonts and uintatheres (Paleocene)

Order Taeniodonta—Includes some of the most highly specialized terrestrial placentals of the Late Paleocene and Early Eocene (Paleocene-Eocene)

Order Tillodontia—Large, massive placentals with clawed, five-toed feet (Paleocene-Eocene)

Order Dinocerata—Uintatheres. Large herbivores with bony protuberances on the skull and greatly elongated canine teeth (Paleocene-Eocene)

Order Pantodonta—North American forms are large sheep to rhinoceros-sized. Asian forms are as small as a rat (Paleocene-Eocene)

Order Astropotheria—Large placental mammals with slender rear legs and stout forelimbs and elongate canine teeth (Paleocene-Miocene)

Order Notoungulata—Largest assemblage of South American ungulates with a wide range of body forms (Paleocene-Pleistocene)

Order Liptoterna—Extinct South American hoofed-mammals (Paleocene-Pleistocene)

Order Rodentia—Rodents. Squirrel, mouse, rat, beaver, porcupine, gopher (Paleocene-Recent)

Order Lagomorpha—Hare, rabbit, pika (Paleocene-Recent)

Order Primates—Lemur, tarsier, loris, monkey, human (Paleocene-Recent)

Order Edentata—Anteater, sloth, armadillo, glyptodont (Paleocene-Recent)

Order Carnivora—Modern carnivorous placentals. Dog, cat, bear, skunk, seal, weasel, hyena, raccoon, panda, sea lion, walrus (Paleocene-Recent)

Order Pyrotheria—Large mammals with long bodies and short columnar limbs (Eocene-Oligocene)

Order Chiroptera—Bats (Eocene-Recent)

Order Dermoptera—Flying lemur (Eocene-Recent)

Order Cetacea—Whale, dolphin,
porpoise (Eocene-Recent)
Order Tubulidentata—Aardvark
(Eocene-Recent)
Order Perissodactyla—Odd-toed
ungulates (hoofed placentals). Horse,
rhinoceros, tapir, titanothere,
chalicothere (Eocene-Recent)
Order Artiodactyla—Even-toed
ungulates. Pig, hippo, camel, deer, elk,
bison, cattle, sheep, antelope,
entelodont, oredont (Eocene-Recent)
Order Proboscidea—Elephant,
mammoth, mastodon (Eocene-Recent)
Order Sirenia—Sea cow, manatee,
dugong (Eocene-Recent)

Order Embrithopoda—Known primarily
from a single locality in Egypt. Large
mammals with two gigantic bony
processes arising from the nose area
(Oligocene)
Order Desmostyla—Amphibious or
seal-like in habit. Front and hind
limbs well developed, but hands and
feet somewhat specialized as paddles
(Oligocene-Miocene)
Order Hyracoidea—Hyrax
(Oligocene-Recent)
Order Pholidota—Scaly anteater
(Oligocene-Recent)

GLOSSARY

A

Absaroka sequence A widespread sequence of Pennsylvanian and Permian sedimentary rocks bounded above and below by unconformities; deposited during a transgressive-regressive cycle of the Absaroka Sea.

absolute dating The process of assigning an actual age to geologic events. Various radioactive decay dating techniques yield absolute ages. (*See relative dating*)

Acadian orogeny A Devonian orogeny in the northern Appalachian mobile belt resulting from a collision of Baltica with Laurentia.

acanthodian Any of the fish first having a jaw or jawlike mechanism; a class of fishes (class Aconthodii) appearing during the Early Silurian and becoming extinct during the Permian.

acondrite A type of stony meteorite lacking condrules; composition similar to that of terrestrial basalt.

acritarchs Organic walled microfossils that probably represent the cysts of planktonic algae; appeared in the fossil record about 1.4 billion years ago and became abundant during the Late Proterozoic through Devonian.

adaptive radiation The adaptation of species of related ancestry to various aspects of the environment; the branching out of organisms of related ancestry into new habitats.

Alleghenian orogeny Pennsylvanian to Permian orogenic event during which the Appalachian mobile belt from New York to Alabama was deformed; occurred in the area of the present-day Appalachian Mountains.

allele Alternative form of a gene controlling the same trait.

allopatric speciation Model for the origin of a new species from a small population that has become geographically isolated from its parent population.

alluvial fan A cone-shaped deposit of alluvium; generally deposited where a stream flows from mountains onto an adjacent lowland.

alpha decay A type of radioactive decay involving the emission of a particle consisting of two protons and two neutrons from the nucleus of an atom; emission of an alpha particle decreases the atomic number by two and atomic mass number by four.

Alpine-Himalayan belt A major linear belt of deformation extending from the Atlantic Ocean eastward across southern Europe and northern Africa, through the Middle East, and into southeast Asia; one of two major Mesozoic-Cenozoic orogenic belts. (*See circum-Pacific belt*)

Alpine orogeny A Late Mesozoic-Cenozoic episode of mountain building affecting southern Europe and northern Africa.

altered remains Fossil remains that have been changed from their original composition or structure or both. (*See unaltered remains*)

amniote egg An egg in which the embryo develops in a liquid-filled cavity called the amnion. The embryo is also supplied with a yolk sac and a waste sac. The amniote egg is shelled in reptiles, birds, and egg-laying mammals and is retained but modified in all other mammals.

anaerobic A term referring to organisms that are not dependent on oxygen for respiration.

analogous organs Body parts, such as wings of insects and birds, that serve the same function, but differ in structure and development. (*See homologous organs*)

Ancestral Rockies Late Paleozoic uplift in the southwestern part of the North American craton.

angiosperm Vascular plants having flowers and seeds; the flowering plants.

angular unconformity An unconformity below which older strata dip at a different angle (usually steeper) than the overlying younger strata. (*See disconformity and nonconformity*)

anthropoid Any member of the primate suborder Anthropoidea; includes New World and Old World monkeys, apes, and humans.

Antler orogeny A Late Devonian to Mississippian orogeny that affected the Cordilleran mobile belt; deformation extended from Nevada to Alberta, Canada.

Appalachian mobile belt A mobile belt located along the eastern margin of the North American craton; extends from Newfoundland to Georgia; probably continuous to the southwest with the Ouachita mobile belt.

arc orogen An area of deformation, such as an island arc, that results from subduction of an oceanic plate; characterized by deformation and igneous activity.

archaeocyathid A benthonic sessile suspension feeder that lived during the Cambrian and constructed reeflike structures.

Archean Eon A part of Precambrian time beginning 3.8 billion years ago, corresponding to the age of the oldest known rocks on Earth, and ending 2.5 billion years ago. (*See Proterozoic Eon*)

artificial selection The practice of selective breeding of plants and animals for desirable traits.

artiodactyl Any member of the order Artiodactyla, the even-toed hoofed mammals. Living artiodactyls include swine, sheep, goats, camels, deer, bison, and musk oxen.

asthenosphere Part of the upper mantle lying below the lithosphere; behaves plastically and flows.

atom The smallest unit of matter that retains the characteristics of an element.

atomic mass number The total number of protons and neutrons in the nucleus of an atom.

atomic number The number of protons in the nucleus of an atom.

aulacogen A sediment-filled, inactive rift of a triple junction that formed above a rising mantle plume.

australopithecine A term referring to several extinct species of the genus *Australopithecus* that existed in South and East Africa during the Pliocene and Pleistocene epochs.

autotrophic Describes organisms that synthesize their organic nutrients from inorganic raw materials; photosynthesizing bacteria and plants are autotrophs. (*See heterotrophic*)

B

back-arc marginal basin A basin formed on the continentward side of a volcanic island arc; thought to form by back-arc spreading; the site of a marginal sea, e.g., the Sea of Japan.

Baltica One of six major Paleozoic continents; composed of Russia west of the Ural Mountains, Scandinavia, Poland, and northern Germany.

banded iron formation (BIF) Sedimentary rocks consisting of alternating thin layers of silica (chert) and iron minerals (mostly the iron oxides hematite and magnetite).

barrier island An elongate sand body oriented parallel to a shoreline, but separated from the shoreline by a lagoon.

Basin and Range Province An area centered on Nevada but extending into adjacent states and northern Mexico; characterized by Cenozoic block-faulting.

bedding (stratification) The layering in sedimentary rocks. Layers less than 1 cm thick are laminae, whereas beds are thicker.

bedding plane The bounding surface that separates one layer of strata from another.

benthos Any organism that lives on the bottom of seas or lakes; may live upon the bottom or within bottom sediments.

beta decay A type of radioactive decay during which a fast-moving electron is emitted from a neutron and thus is converted to a proton; results in an increase of one atomic number, but no change in atomic mass number.

Big Bang A model for the evolution of the universe from a dense, hot state followed by expansion, cooling, and a less dense state.

biogenic sedimentary structure Any structure in sedimentary rocks produced by the activities of organisms, e.g., tracks, trails, burrows. (*See trace fossil*)

biostratigraphic unit A unit of sedimentary rock defined by its fossil content.

bioturbation The process of churning or stirring of sediments by organisms.

biozone The fundamental biostratigraphic unit (e.g., range zone and concurrent range zone).

bipedal Walking on two legs as a means of locomotion.

body fossil The actual remains of any prehistoric organism; includes shells, teeth, bones, and, rarely, the soft parts of organisms. (*See trace fossil*)

bonding The process whereby atoms are joined to other atoms.

bony fish A class of fishes (class Osteichthyes) that evolved during the Devonian; the most common fishes; characterized by an internal skeleton of bone; divided into two subgroups, the ray-finned fishes and lobe-finned fishes.

brachiopod Any member of a group of bivalved, suspension-feeding, marine, invertebrate animals.

browser An animal that eats tender shoots, twigs, and leaves. Compare with *grazer*.

C

Caledonian orogeny A Silurian-Devonian orogeny that occurred along the northwestern margin of Baltica resulting from the collision of Baltica with Laurentia.

Canadian shield The Precambrian shield of North America; exposed mostly in Canada, but outcrops occur in Minnesota, Wisconsin, Michigan, and New York.

captorhinomorph The oldest known reptiles; evolved during the Early Pennsylvanian; ancestors of all other reptiles, thus commonly called the stem reptiles.

carbon 14 dating An absolute dating method that relies upon determining the ratio of C^{14} to C^{12} in a sample; useful back to about 70,000 years ago; can be applied only to organic substances.

carbonaceous chondrite A type of stony meteorite; same as ordinary chondrites except they contain about 5% organic compounds including inorganically produced amino acids.

carbonate mineral A mineral that contains the negatively charged carbonate ion $(CO_3)^{-2}$ (e.g., calcite $[CaCO_3]$ and dolomite $[CaMg(CO_3)_2]$).

carbonate rock A rock containing predominately carbonate minerals (e.g., limestone and dolostone).

carnassials A pair of specialized upper and lower shearing teeth in members of the order Carnivora.

carnivore-scavenger Any animal that depends on other animals, living or dead, as a source of nutrients.

cartilaginous fish Fishes such as living sharks, rays, and skates, and their extinct relatives that have a skeleton composed of cartilage.

cast A replica of an object such as a shell or bone formed when a mold of that object is filled by sediment or mineral matter. (*See mold*)

catastrophism The concept proposed by Baron Georges Cuvier that explained the physical and biological history of the Earth by a series of sudden, widespread catastrophes.

Catskill Delta The Devonian clastic wedge that was deposited adjacent to the highlands that formed during the Acadian orogeny.

chemical sediment/sedimentary rock Sediment/rock formed by precipitation of minerals derived from the ions taken into solution during weathering.

China One of six major Paleozoic continents; composed of all of southeast Asia, including China, Indochina, part of Thailand, and the Malay Peninsula.

chondrite A stony meteorite that contains chondrules.

chondrule A small, round mineral body formed by rapid cooling; found in chondritic meteorites.

chordate All members of the phylum Chordata; characterized by a notochord, a dorsal, hollow nerve cord, and gill slits at some time during the animal's life cycle.

chromosome Complex, double-stranded, helical molecule of deoxyribonucleic acid (DNA); specific segments of chromosomes are genes.

circum-Pacific belt One of two major Mesozoic-Cenozoic orogenic belts; located around the margins of the Pacific Ocean basin; includes the orogens of South and Central America, the Cordillera of western North America, and the Aleutian, Japan, and Philippine arcs. (*See Alpine-Himalayan belt*)

clastic texture Sedimentary rocks consisting of the broken particles of preexisting rocks or organic structures such as shells are said to have a clastic texture.

clastic wedge An extensive accumulation of mostly clastic sediments eroded from and deposited adjacent to an uplifted area; clastic wedges are coarse-grained and thick near the uplift and become finer-grained and thinner away from the uplift, e.g., the Queenston Delta.

collision orogen An orogen produced as a result of a collision of two continents, e.g., the collision of India with Asia that resulted in the formation of the Himalayas.

compound A substance resulting from the bonding of atoms of two or more different elements.

concurrent range zone A type of biozone established by plotting the overlapping ranges of fossils that have different geologic ranges; the first and last occurrences of fossils are used to establish concurrent range zone boundaries.

conformable The relationship between beds in a sedimentary sequence containing no discontinuities. (*See unconformity*)

conodont A small, toothlike fossil composed of calcium phosphate. Conodont elements were located behind the mouth of an elongate animal and probably functioned in feeding. Geologic range is Cambrian through Triassic.

contact metamorphism Metamorphism in which a magma body alters the adjacent country rock.

continental-continental plate boundary A type of convergent plate boundary along which two continental lithospheric plates collide, e.g., the collision of India with Asia.

continental drift The theory that the continents were once joined into a single landmass that broke apart with the various fragments (continents) moving with respect to one another; the theory of continental movement proposed by Alfred Wegener in 1912.

covalent bond A bond formed by the sharing of electrons between atoms.

convergent evolution The development of similarities in two or more distantly related organisms as a consequence of adapting to a similar lifestyle, e.g., ichthyosaurs and porpoises. (*See parallel evolution*)

convergent plate boundary The boundary between two plates that are moving toward one another; three types of convergent plate boundaries are recognized. (*See continental-continental plate boundary, oceanic-continental plate boundary, and oceanic-oceanic plate boundary*)

coprolite Fossilized feces.

Cordilleran orogeny A protracted episode of deformation affecting the western margin of North America from Jurassic to Early Cenozoic time; typically divided into three separate phases called the Nevadan, Sevier, and Laramide orogenies.

Cordilleran mobile belt A mobile belt in western North America bounded on the west by the Pacific Ocean and on the east by the Great Plains; extends north-south from Alaska into central Mexico. (*See North American Cordillera*)

core The interior part of the Earth beginning at a depth of about 2,900 km; probably composed mostly of iron and nickel; divided into an outer liquid core and an inner solid core.

correlation Demonstration of the physical continuity of rock units or biostratigraphic units in different areas, or demonstration of time equivalence as in time-stratigraphic correlation.

craton A name applied to the relatively stable part of a continent; consists of a Precambrian shield and a platform, a buried extension of a shield; the ancient nucleus of a continent.

cratonic sequence A widespread sequence of sedimentary rocks bounded above and below by unconformities; deposited during a transgressive-regressive cycle of an epeiric sea, e.g., the Sauk sequence.

Cretaceous Interior Seaway An interior seaway that existed during the Late Cretaceous; formed as northward-transgressing waters from the Gulf of Mexico joined with southward-transgressing water from the Arctic; effectively divided North America into two large landmasses.

Cro-Magnon A race of *Homo sapiens* that lived mostly in Europe from 35,000 to 10,000 years ago.

cross-bedding Strata containing beds that were deposited at an angle to the surface upon which they were accumulating, as in desert dunes, are said to be cross-bedded.

crossopterygian A specific type of lobe-finned fish; possessed lungs; ancestral to amphibians.

crust The outer part of the Earth; the upper part of the lithosphere; separated from the mantle by the Moho; divided into continental and oceanic crust.

crystalline solid A solid in which the constituent atoms are arranged in a regular, three-dimensional framework.

crystalline texture A texture of rocks consisting of an interlocking mosaic of mineral crystals.

Curie point The temperature at which iron-bearing minerals in a cooling magma attain their magnetism.

cyclothem A vertical sequence of cyclically repeated sedimentary rocks resulting from alternating periods of marine and nonmarine deposition; commonly contain a coal bed.

cynodont A type of therapsid (advanced mammal-like reptile); the ancestors of mammals.

D

daughter element An element formed by the radioactive decay of another element, e.g., argon 40 is the daughter element of potassium 40.

delta An alluvial deposit at the mouth of a river.

depositional environment Any area in which sediment is deposited; a depositional site that differs in physical aspects, chemistry, and biology from adjacent environments.

desiccation crack A crack formed in clay-rich sediments in response to drying and shrinkage.

detrital sediment/sedimentary rock Sediment/rock consisting of detritus, the solid particles of preexisting rocks.

disconformity An unconformity above and below which the strata are parallel. (*See angular unconformity and nonconformity*)

divergent evolution The diversification of a species into two or more descendant species. (*See adaptive radiation*)

divergent plate boundary The boundary between two plates that are moving apart; new oceanic lithosphere forms at the boundary; characterized by volcanism and seismicity.

DNA (deoxyribonucleic acid) The substance of which chromosomes are composed; the genetic material of all organisms except bacteria.

Doppler effect The apparent change in frequency of a wave resulting from motion of the source of the wave, the receiver, or both.

drift A collective term for all sediment deposited by glacial activity, including material deposited directly by glacial ice (till) and material deposited in streams derived from the melting of ice (outwash).

dryopithecine Any of the members of a Miocene family of apelike primates; possible ancestors of apes and humans. (*See ramapithicine*)

E

ectotherm An animal having a variably body temperature. Ectotherms acquire heat from their environment. (*See endotherm*)

Ediacaran faunas A collective name for all Late Proterozoic faunas containing animal fossils similar to those of the Ediacara fauna of Australia.

electromagnetic force A combination of electricity and magnetism into one force; binds atoms into molecules.

electron capture decay A type of radioactive decay involving the capture of an electron by a proton and its conversion to a neutron; results in a loss of one atomic number, but no change in atomic mass number.

element A substance composed of all the same atoms; it cannot be changed into another element by ordinary chemical means.

Ellesmere orogeny A Mississippian orogeny affecting the northern margin of Laurentia.

encounter theory A theory for the evolution of the solar system; involves a star passing close to the Sun and thus pulling away gaseous filaments that later accreted into solid bodies.

endotherm An animal whose body temperature is maintained within narrow limits by its metabolic processes. (*See ectotherm*)

epeiric sea A broad shallow sea that covers part of a continent; six epeiric seas covered parts of North America during the Phanerozoic Eon, e.g., the Sauk Sea.

eukaryotic cell A type of cell with a membrane-bounded nucleus containing chromosomes; also contains such organelles as plastids and mitochondria that are absent in prokaryotic cells. (*See prokaryotic cell*)

eupantothere Any member of a group of mammals that included the ancestors of both marsupial and placental mammals.

evaporite A sedimentary rock that forms by inorganic chemical precipitation of minerals from solution.

F

Farallon plate A Late Mesozoic-Cenozoic oceanic plate that was largely subducted beneath North America; remnants of the Farallon plate are the Juan de Fuca and Cocos plates.

felsic magma A type of magma containing more than 65% silica and considerable sodium, potassium, and aluminum, but little calcium, iron, and magnesium. (*See mafic and intermediate magma*)

fission track dating The process of dating samples by counting the number of small linear tracks (fission tracks) that result from damage to a mineral crystal by rapidly moving alpha particles generated by radioactive decay of uranium.

fluvial A term referring to streams, stream action, and the deposits of streams.

foliated texture A texture of metamorphic rocks in which platy and elongate minerals are arranged in a parallel fashion.

formation The basic lithostratigraphic unit; a unit of strata that is mappable and that has distinctive upper and lower boundaries.

fossil The remains or traces of prehistoric life preserved in rocks of the Earth's crust. (*See body fossil and trace fossil*)

Franklin mobile belt The most northerly mobile belt in North America; extends from northwestern Greenland westward across the Canadian Arctic islands.

G

gene The basic unit of inheritance; a specific segment of a chromosome. (*See allele*)

gene pool The sum total of all alleles of all genes available to an interbreeding population.

geologic record The record of past events preserved in rocks.

geologic time scale A geologic chart arranged such that the designation for the earliest part of geologic time appears at the bottom, and progressively younger designations appear in their proper chronologic sequence.

geology The science concerned with the study of the Earth; includes studies of Earth materials (minerals and rocks), surface and internal processes, and Earth history.

glacial stage A time of extensive glaciation. At least four glacial stages are recognized in North America, and six or seven are recognized in Europe.

***Glossopteris* flora** A Late Paleozoic flora found only on the Southern Hemisphere continents and India; named after its best known genus, *Glossopteris*.

Gondwana One of six major Paleozoic continents; composed of the present-day continents of South America, Africa, Antarctica, Australia, and India, and parts of other continents such as southern Europe, Arabia, and Florida; began fragmenting during the Triassic Period.

graded bedding A type of sedimentary bedding in which an individual bed is characterized by a decrease in grain size from bottom to top.

grain size A term relating to the size of the particles making up sediment or sedimentary rock.

granite-gneiss complex One of the two main types of rock bodies characteristic of areas underlain by Archean rocks.

graptolite A small, planktonic animal belonging to the phylum Hemichordata. Graptolites make good guide fossils. Geologic range is from Cambrian through Mississippian.

gravity The attractive force that acts between all objects, e.g., between the Earth and the Moon.

grazer Any animal that crops low-growing vegetation, especially grasses.

greenstone belt A linear or podlike association of volcanic and sedimentary rocks particularly common in Archean terranes; typically synclinal and consists of lower and middle volcanic units and an upper sedimentary rock unit.

Grenville orogeny An area in the eastern United States and Canada that was accreted to Laurentia during the Late Proterozoic.

guide fossil Any easily identifiable fossil that has a wide geographic distribution and short geologic range; used to determine the geologic ages of strata and to correlate strata of the same age.

gymnosperm The flowerless, seed-bearing land plants.

H

half-life The time required for one-half of the original number of atoms of a radioactive element to decay to a stable daughter product, e.g., the half-life of potassium 40 is 1.3 billion years.

herbivore An animal that is dependent on plants as a source of nutrients.

Hercynian orogeny Pennsylvanian to Permian orogeny in the Hercynian mobile belt of southern Europe.

heterotrophic Describes organisms such as animals that depend on preformed organic molecules from the environment as a source of nutrients. (*See autotrophic*)

hiatus The interval of geologic time not represented by strata in a sequence of strata containing an unconformity.

Holocene Epoch The last of two epochs comprising the Quaternary Period. Began 10,000 years ago.

hominid Abbreviated form of Hominidae; the family of hominoids to which humans belong. Such bipedal primates as *Australopithecus* and *Homo* are hominids. (*See hominoid*)

hominoid Abbreviated form of Hominoidea, the superfamily of primates to which apes and humans belong. (*See hominid*)

homogeneous accretion A model for the differentiation of the Earth into a core, mantle, and crust; holds that the Earth was originally compositionally homogeneous, but heated up, allowing heavier elements to sink to the core.

homologous organs Body parts in different organisms that have a similar structure, similar relationships to other organs, and similar development, but do not necessarily serve the same function, e.g., the wing of a bird and the forelimbs of whales and dogs. (*See analogous organs*)

hot spot Localized zone of melting below the lithosphere; detected by volcanism at the surface.

hypothesis A provisional explanation for observations. Subject to continual testing and modification. If well supported by evidence, hypotheses are then generally called theories.

Hyracotherium A small Early Eocene mammal; the oldest genus assigned to the horse family (family Equidae).

I

Iapetus Ocean A Paleozoic ocean basin that separated North America from Europe; the Iapetus Ocean began closing when North America and Europe began moving toward one another, and it was eliminated when the continents collided during the Late Paleozoic.

ichthyosaur Any of the porpoise-like, Mesozoic marine reptiles.

igneous rock Any rock formed by cooling and crystallization of magma, or formed by the consolidation of volcanic ejecta such as ash.

inheritance of acquired characteristics A mechanism proposed by Jean Baptiste de Lamarck to account for evolution; states that characteristics acquired during an organism's lifetime could be passed on to its offspring.

inhomogeneous accretion A model for the differentiation of the Earth into a core, mantle, and crust by the sequential condensation of core, mantle, and crust from hot nebular gases.

interglacial stage A time between glacial stages when glaciers cover much less area and global temperatures are warmer than during a glacial stage.

intermediate magma A magma having a silica content between 53% and 65% and an overall composition intermediate between felsic and mafic magmas.

intracontinental rift A linear zone of deformation within a continent produced by tensional forces; characterized by normal faults and volcanism, e.g., the East African Rift System.

ionic bond A bond that results from the attraction of positively and negatively charged ions.

iron meteorite A type of meteorite consisting of iron and nickel.

isostasy Theoretical concept of the Earth's crust "floating" on a dense underlying layer; areas of less dense continental crust rise topographically above more dense oceanic crust.

isotope All atoms of a chemical element have the same number of protons in the nucleus, but may have variable numbers of neutrons; those with different numbers of neutrons are isotopes of that element, e.g., carbon 12 and carbon 14.

J

Jovian planet Any of the four planets (Jupiter, Saturn, Uranus, and Neptune) that resemble Jupiter. They are all large and have low mean densities, indicating that they are composed mostly of lightweight gases, such as hydrogen and helium, and frozen compounds, such as ammonia and methane.

K

Kaskaskia sequence A widespread sequence of Devonian and Mississippian sedimentary rocks bounded above and below by unconformities; deposited during a transgressive-regressive cycle of the Kaskaskia Sea.

Kazakhstania One of six major Paleozoic continents; a triangular-shaped continent centered on Kazakhstan (part of Asia).

L

labyrinthodont Any of the amphibians, from the Devonian to the Triassic, characterized by the labyrinthine wrinkling and folding of their teeth.

Laramide orogeny The Late Cretaceous to Early Cenozoic phase of the Cordilleran orogeny; responsible for many of the structural features of the present-day Rocky Mountains. In contrast to the preceding Nevadan and Sevier orogenies, the Laramide orogeny deformed the margin of the craton.

Laurasia A Late Paleozoic, Northern Hemisphere continent composed of the present-day continents of North America, Greenland, Europe, and Asia.

Laurentia The name given to a Proterozoic continent that was composed mostly of North America and Greenland, parts of northwestern Scotland, and perhaps parts of the Baltic shield of Scandinavia.

lava Magma at the Earth's surface.

lithification The process by which sediment is transformed into sedimentary rock.

lithology The physical characteristics of rocks.

lithosphere The outer, rigid part of the Earth consisting of the upper mantle, oceanic crust, and continental crust; lies above the asthenosphere.

lithostratigraphic unit A unit of sedimentary rock, such as a formation, defined by its lithologic characteristics rather than its biologic content or time of origin.

Little Ice Age An interval of nearly four centuries (1300 A.D. to the mid- to late 1800s) during which glaciers expanded to their greatest historic extents.

living fossil A living organism that has descended from ancient ancestors with little apparent change.

lobe-finned fish A type of fish in which the fin contains a series of articulating bones and is attached to the body by a fleshy shaft; one of the two major subgroups of bony fishes.

M

mafic magma A type of magma containing between 45 and 52% silica and proportionately more calcium, iron, and magnesium than a felsic magma.

magnetic anomaly Any change, such as a change in average strength, of the Earth's magnetic field.

magnetic reversal The phenomenon of the complete reversal of the north and south magnetic poles.

mantle The zone surrounding the core and comprising about 83% of the Earth's volume; it is less dense than the core and is thought to be composed largely of peridotite.

marine regression The withdrawal of the sea from a continent or coastal area resulting in the emergence of land as sea level falls or the land rises with respect to sea level.

marine transgression The invasion of coastal areas or much of a continent by the sea resulting from a rise in sea level or subsidence of the land.

marsupial Any of the pouched mammals such as opossums, kangaroos, and wombats. At present, marsupials are common only in Australia.

mass extinction A time during which extinction rates are greatly accelerated thus resulting in a marked decrease in the diversity of organisms, e.g., the terminal Cretaceous extinction event.

meiosis A type of cell division during which the number of chromosomes is reduced by one-half. The cell division process that yields sex cells, sperm and eggs in animals, and pollen and ovules in plants (*See mitosis*)

metamorphic rock Any rock type altered by high temperature and pressure and the chemical activities of fluids is said to have been metamorphosed, e.g., slate, gneiss, marble.

meteorite A mass of matter of extraterrestrial origin that has fallen to the Earth's surface.

miacid Small, short-limbed, Early Tertiary carnivorous mammals; ancestors of all later members of the order Carnivora.

micrite Microcrystalline calcium carbonate mud; a limestone composed of micrite.

microplate Small lithospheric plate.

midcontinent rift A Late Proterozoic intracontinental rift within Laurentia; contains thick accumulations of extrusive igneous rocks and detrital sedimentary rocks.

Milankovitch theory A theory that explains cyclic variations in climate as a consequence of irregularities in the Earth's rotation and orbit.

mineral A naturally occurring, inorganic, crystalline solid having characteristic physical properties and a narrowly defined chemical composition.

mitosis A type of cell division during which the two cells resulting from division receive the same number of chromosomes as the parent cell possessed. (*See meiosis*)

mobile belt Elongated area of deformation as indicated by folds and faults; generally located adjacent to a craton, e.g., the Appalachian mobile belt.

modern synthesis A synthesis of the ideas of geneticists, paleontologists, population biologists, and others to yield a neo-Darwinian view of evolution; includes chromosome theory of inheritance, mutation as a source of variation, and gradualism.

molarization An evolutionary trend in the hoofed mammals (ungulates) during which the premolars became more molarlike.

molars The teeth of mammals used for grinding and chewing.

mold A cavity or impression of an organism or part thereof in sediment or sedimentary rock, e.g., a mold of a clam shell. (*See cast*)

monomer A comparatively simple organic molecule, such as an amino acid, that is capable of linking with other monomers to form polymers. (*See polymer*)

monotreme The egg-laying mammals; only two types of monotremes now exist, the platypus and spiny anteater of the Australian region.

morphology The form or structure of an organism or any of its parts.

mosaic evolution The concept that all features of an organism do not evolve at the same rate; all organisms are mosaics of characteristics some of which are retained from the ancestral condition, and some of which are more recently evolved.

multicellular organism Any organism consisting of many cells as opposed to a single cell; possesses cells specialized to perform specific functions such as reproduction and respiration.

mutation Any change in the genetic determinants or genes of organisms; some of the inheritable variation in populations upon which natural selection acts arises from mutations in sex cells.

N

natural selection A mechanism proposed by Charles Darwin and Alfred Russell Wallace to account for evolution; natural selection results in the survival to reproductive age of those organisms best adapted to their environment.

Neanderthal A type of human that inhabited Europe and the Near East from 150,000 to 32,000 years ago; considered by some to be a variety or subspecies (*Homo sapiens neanderthalensis*) of *Homo sapiens* and by some as a separate species (*Homo neanderthalensis*).

nekton Actively swimming organisms, e.g., fishes, whales, and squids. (*See plankton*)

Neptunism The concept held by Abraham Gottlob Werner and others that all rocks were precipitated from a worldwide ocean.

Nevadan orogeny Late Jurassic to Cretaceous phase of the Cordilleran orogeny; most strongly affected the western part of the Cordilleran mobile belt.

New World monkey Any of the monkeys native to Central and South America. A superclass of the anthropoids.

nonconformity An unconformity in which stratified sedimentary rocks above an erosion surface overlie igneous or metamorphic rocks. (*See angular unconformity and disconformity*)

nonfoliated texture A metamorphic texture in which there is no discernible preferred orientation of mineral grains.

nonvascular Plants lacking specialized tissues for conducting fluids.

North American Cordillera A complex mountainous region in western North America extending from Alaska into central Mexico. (*See Cordilleran mobile belt*)

nucleus The central part of an atom consisting of one or more protons and neutrons.

O

oceanic-continental plate boundary A type of convergent plate boundary along which oceanic lithosphere and continental lithosphere collide; characterized by subduction of the oceanic plate beneath the continental plate and by volcanism and seismicity.

oceanic-oceanic plate boundary A type of convergent plate boundary along which two oceanic lithospheric plates collide.

Old World monkey Any of the monkeys native to Africa, Asia, and southern Europe. A superfamily of the anthropoids.

ophiolite A sequence of igneous rocks thought to represent a fragment of oceanic lithosphere; composed of peridotite overlain successively by gabbro, sheeted basalt, and pillow basalts.

ordinary chondrite The most abundant type of stony meteorite; composed of high-temperature ferromagnesian minerals such as olivine and some pyroxenes.

organic evolution *See theory of evolution.*

organic reef A wave-resistant limestone structure with a structural framework of animal skeletons, e.g., stromatoporoid reef or coral reef.

ornithischian Any dinosaur belonging to the order Onithischia; characterized by a birdlike pelvis. (*See saurischian*)

orogen A linear part of the Earth's crust that was deformed during an orogeny. (*See orogeny*)

orogeny The process of forming mountains, especially by folding and thrust faulting; an episode of mountain building, e.g., the Acadian orogeny.

ostracoderm The "bony-skinned" fish; first appeared during the Late Cambrian and thus are the oldest known vertebrates; characterized by a lack of jaws and teeth and presence of bony armor.

Ouachita mobile belt A mobile belt located along the southern margin of the North American craton; probably continuous with the Appalachian mobile belt.

Ouachita orogeny An orogeny that deformed the Ouachita mobile belt during the Pennsylvanian Period.

outgassing The process whereby gases derived from the Earth's interior are released into the atmosphere by volcanic activity.

outwash A term for the sediment deposited in streams that issue from glaciers. (*See drift*)

P

paleoclimate A climate that existed during the past.

paleocurrent A term referring to an ancient current direction; determined by measuring the orientations of various sedimentary structures such as crossbedding.

paleogeography The study of the Earth's geography throughout geologic time; paleogeography may be determined for the entire Earth, such as the position of continents through time, or for a local area.

paleomagnetism The remanent magnetism in ancient rocks that records the direction and strength of the Earth's magnetic field at the time of their formation.

Pangaea The name proposed by Alfred Wegener for a supercontinent that existed at the end of the Paleozoic Era and consisted of all landmasses of the Earth.

Panthalassa Ocean The Late Paleozoic worldwide ocean that surrounded the supercontinent Pangaea.

parallel evolution The development of similarities in two or more closely related but separate lines of descent as a consequence of similar adaptations. (*See convergent evolution*)

parent element An unstable element that by radioactive decay is changed into a stable daughter element (*See daughter element*)

pelycosaur Pennsylvanian to Permian "finback reptiles"; possessed some mammalian characteristics.

period The fundamental unit in the hierarchy of time units; a part of geologic time during which a particular sequence of rocks designated as a system was deposited.

perissodactyl Any member of the order Perissodactyla, the odd-toed hoofed mammals; living perissodactyls include horses, rhinoceroses, and tapirs. (*See artiodactyl*)

photochemical dissociation A process whereby water molecules in the upper atmosphere are disrupted by ultraviolet radiation, yielding oxygen (O_2) and hydrogen (H).

photosynthesis The metabolic process of synthesizing organic molecules from water and carbon dioxide, using the radiant energy of sunlight captured by chlorophyll-containing cells.

phyletic gradualism An evolutionary concept holding that a species evolves gradually and continuously through time to give rise to new species. (*See punctuated equilibrium*)

placental Any of the mammals that have a placenta to nourish the embryo; fusion of the amnion of the amniote egg with the walls of the uterus forms the placenta; most mammals, living and fossil, are placentals.

placoderm The "plate-skinned" fishes; Late Silurian through Permian; characterized by jaws and bony armor especially in the head-shoulder region.

planetesimal An asteroid-sized body that along with other planetesimals aggregated to form protoplanets.

plankton Animals and plants that float passively, e.g., phytoplankton and zooplankton. (*See nekton*)

plate tectonics (plate tectonic theory) The theory that large segments of the outer part of the Earth (lithospheric plates) move relative to one another; lithospheric plates are rigid and move over the asthenosphere, which behaves much like a very viscous fluid.

playa lake Broad, shallow, temporary lake that forms in an arid region and quickly evaporates to dryness.

Pleistocene Epoch The first of two epochs comprising the Quaternary Period. Commonly called the Ice Age. Occurred between 1.6 million and 10,000 years ago.

plesiosaurs A group of Mesozoic, marine reptiles.

pluton An intrusive igneous body that forms when magma cools and crystallizes within the Earth's crust.

plutonic (intrusive igneous) rock Igneous rock that crystallizes from magma intruded into or formed in place within the Earth's crust.

pluvial lake Any lake that formed beyond the areas directly affected by glaciation during the Pleistocene as a result of increased precipitation and lower evaporation rates, e.g., Lake Bonneville.

point bar The sediment body deposited on the gently sloping side of a meander loop.

pollen analysis Identification and statistical analysis of pollen from sedimentary rocks; such analyses provide information about ancient floras and climates.

polymer A comparatively complex organic molecule, such as nucleic acids and proteins, formed by monomers linking together. (*See monomer*)

Precambrian A widely used informal term referring to all rocks stratigraphically beneath strata of the Cambrian System and to all geologic time preceding the Cambrian Period.

Precambrian shield A vast area of exposed Precambrian rocks on a continent; areas of relative stability for long periods of time, e.g., the Canadian shield.

primary producer Those organisms in a food chain, such as green plants and bacteria, upon which all other members of the food chain depend directly or indirectly; those organisms not dependent on an external source of nutrients. (*See autotrophic*)

Primates The order of mammals that includes prosimians (lemurs and tarsiers), monkeys, apes, and humans; characteristics include large brain, stereoscopic vision, and grasping hand.

principle of cross-cutting relationships A principle used to determine the relative ages of events; holds that an igneous intrusion or fault must be younger than the rocks that it intrudes or cuts.

principle of fossil succession The principle based on the work of William Smith that holds that fossils, and especially assemblages of fossils, succeed one another through time in a regular and determinable order.

principle of inclusions A principle that holds that inclusions, or fragments, in a rock unit are older than the rock unit itself, e.g., granite fragments in a sandstone are older than the sandstone rock unit.

principle of lateral continuity A principle that holds that sediment layers extend outward in all directions until they terminate.

principle of superposition A principle that holds that younger strata are deposited on top of older layers.

principle of uniformitarianism A principle that holds that we can interpret past events by understanding present-day processes; based on the assumption that natural laws have not changed through time.

proboscidian The elephants and their extinct relatives (order Proboscidea); characterized by an elongate or snout-like feeding organ (a trunk).

proglacial lake A lake formed by meltwater accumulating along the margins of a glacier.

progradation The seaward (or lakeward) advance of the shoreline as a result of nearshore sedimentation.

prokaryotic cell A type of cell having no nucleus and lacking such organelles as plastids and mitochondria; cells of bacteria and cyanobacteria (blue-green algae) are prokaryotic. (*See eukaryotic cell*)

prosimian Any member of the primate suborder Prosimii; includes tree shrews, lemurs, tarsiers, and lorises; commonly called lower primates.

Proterozoic Eon That part of Precambrian time beginning 2.5 billion years ago and ending 570 million years ago. (*See Archean Eon*)

pterosaur Any of the Mesozoic flying reptiles.

punctuated equilibrium An evolutionary concept that holds that a new species evolves rapidly, perhaps in a few thousands of years, then remains much the same during its several millions of years of existence.

pyroclastic material Fragmental material that has been explosively ejected from a volcano.

Q

quadrupedal Referring to locomotion on all four legs; as opposed to bipedal.

quartzite-carbonate-shale assemblage A suite of sedimentary rocks characteristic of passive continental margins, but also known from intracratonic basins and back-arc basins.

Quaternary A term for a geologic period or system comprising all geologic time or rocks from the end of the Tertiary to the present; consists of two epochs or series, the Pleistocene and the Recent (Holocene).

Queenston Delta The clastic wedge resulting from the erosion of the highlands formed during the Taconic orogeny; deposited on the west side of the Taconic Highlands.

R

radioactive decay The spontaneous decay of an atom by emission of a particle from its nucleus (alpha and beta decay) or by electron capture, thus changing the atom to an atom of a different element.

ramapithecine A family of Miocene apelike primates; possible ancestors of orangutans. (*See dryopithecine*)

range zone Biostratigraphic unit defined by the occurrence of a single type of organism such as a species or genus.

ray-finned fish A subclass (Actinopterygii) of the bony fish (class Osteichthyes). A term describing the way the fins are supported by thin bones that project from the body.

red beds Sedimentary rocks, mostly sandstone and shale, with red coloration due to the presence of ferric oxides.

reef *See organic reef.*

regional metamorphism Metamorphism that occurs over a large area and is usually the result of tremendous temperatures, pressures, and deformation within the deeper parts of the Earth's crust.

relative dating The process of determining the age of an event relative to other events; involves placing geologic events in their correct chronologic order, but involves no consideration of when the events occurred in terms of numbers of years ago. (*See absolute dating*)

ripple mark Wavelike (undulating) structure produced in granular sediment such as sand; formed by wind, unidirectional water currents, or wave currents.

rock A consolidated aggregate of minerals or particles of other rocks; although an exception to this definition, coal, natural glass, and aggregates of shells are also considered rocks.

rock cycle A sequence of processes through which Earth materials may pass as they are transformed from one rock type to another.

rock-forming mineral A common mineral that makes up a significant portion of a rock.

rounding The process by which the sharp corners and edges of sedimentary particles are abraded during transport.

ruminant Any of the cud-chewing placental mammals such as deer, cattle, camels, bison, sheep, and goats.

S

sabkha The term applied to coastal supratidal flats, especially those in arid regions.

salt dome A structure resulting from the upward movement of a mass of salt through overlying layers of sedimentary rocks. Oil and gas fields are commonly associated with salt domes.

San Andreas transform fault A major transform fault extending through part of California; connects with spreading centers in the Gulf of Mexico and in the Pacific Ocean off the northwest coast of the United States.

Sauk sequence A widespread sequence of sedimentary rocks bounded above and below by unconformities; deposited during a latest Proterozoic to Early Ordovician transgressive-regressive cycle of the Sauk Sea.

saurischian Any dinosaur belonging to the order Saurischia; characterized by a lizardlike pelvis. (*See ornithischian*)

scientific method A logical, orderly approach that involves data gathering, formulating and testing of hypotheses, and proposing theories.

seafloor spreading The theory that the seafloor moves away from spreading ridges and is eventually subducted and consumed at convergent plate margins.

sedimentary facies Any aspect of a sedimentary rock unit that makes it recognizably different from adjacent sedimentary rocks of the same, or approximately same, age, e.g., a sandstone facies.

sedimentary rock Any rock composed of sediment. The sediment may be particles of various sizes such as gravel or sand, the remains of animals or plants as in coal and some limestones, or chemical compounds extracted from solution by organic or inorganic processes.

sedimentary structure Any structure in sedimentary rock such as cross-bedding, desiccation cracks, and animal burrows.

sediment-deposit feeder Any animal that ingests sediment and extracts the nutrients from it.

seedless vascular plant A type of land plant with vascular tissues for transport of fluids and nutrients throughout the plant; reproduces by spores rather than seeds, e.g., ferns and horsetail rushes.

sequence stratigraphy The study of rock relationships within a time-stratigraphic framework of related facies bounded by widespread unconformities.

Sevier orogeny The Cretaceous phase of the Cordilleran orogeny that affected the continental shelf and slope areas of the Cordilleran mobile belt.

Siberia One of six major Paleozoic continents; composed of Russia east of the Ural Mountains, and Asia north of Kazakhstan and south of Mongolia.

silica tetrahedron The basic building block of all silicate minerals. It consists of one silicon atom and four oxygen atoms.

silicate A mineral containing silica that consists of silicon and oxygen atoms.

solar nebular theory A theory for the evolution of the solar system from a rotating cloud of gas.

Sonoma orogeny A Permian-Triassic orogeny caused by the collision of an island arc with the southwestern margin of North America.

sorting In sediment or sedimentary rock a term referring to the degree to which all particles are about the same size; e.g., well-sorted, poorly sorted.

species A population of similar individuals that in nature can reproduce and produce fertile offspring.

stony meteorite (**stone**) Meteorite composed of silicate minerals.

stony-iron meterorite (**stony-iron**) Meteorite composed of iron-nickel and silicate minerals.

stratigraphy A branch of geology; concerned with the composition, origin, and areal and age relationships of stratified rocks.

stromatolite A structure in sedimentary rocks, especially limestones, produced by entrapment of sediment grains on sticky mats of photosynthesizing bacteria; a biogenic sedimentary structure.

strong nuclear force The force that holds protons together within the nucleus of an atom.

subduction zone An elongate, narrow zone at a convergent plate boundary where an oceanic plate descends relative to another plate, e.g., the subduction of the Nazca plate beneath the South American plate.

submarine fan A cone-shaped sedimentary deposit that accumulates on the continental slope and rise.

Sundance Sea A wide seaway that existed in western North America during the Middle Jurassic Period.

suspension feeder An animal that consumes microscopic plants, animals, or dissolved nutrients from water.

system The fundamental unit in the time-stratigraphic hierarchy of units; the Devonian System refers to rocks deposited during a specific interval of geologic time, the Devonian Period.

T

Taconic orogeny An Ordovician orogeny that resulted in deformation of the Appalachian mobile belt.

Tejas epeiric sea A Cenozoic epeiric sea that was largely restricted to the Atlantic and Gulf Coastal plains and parts of coastal California, but did extend into the continental interior in the Mississippi Valley.

terrestrial planet Any of the four innermost planets (Mercury, Venus, Earth, and Mars). They are all small and have high mean densities, indicating that they are composed of rock and metallic elements.

Tertiary A term for a geologic period or system comprising all geologic time or rocks from the end of the Cretaceous to the beginning of the Quaternary. The Tertiary consists of five epochs or series: Paleocene, Eocene, Oligocene, Miocene, and Pliocene.

thecodontian An informal term for a variety of Permian and Triassic reptiles that had teeth set in individual sockets. Small, bipedal thecodontians are the probable ancestors of dinosaurs.

theory An explanation for some natural phenomenon that has a large body of supporting evidence; to be considered scientific, a theory must be testable, e.g., plate tectonic theory.

theory of evolution The theory that all living things are related and that they descended with modification from organisms that lived during the past.

therapsid Permian to Triassic reptiles that possessed mammalian characteristics and thus are called mammal-like reptiles; one group of therapsids, the cynodonts, gave rise to mammals.

thermal convection cell A type of circulation of material in the asthenosphere during which hot material rises, moves laterally, cools and sinks, and is reheated and reenters the cycle.

till All sediment deposited directly by glacial ice.

time-stratigraphic unit A unit of strata that was deposited during a specific interval of geologic time, e.g., the Devonian System, a time-stratigraphic unit, was deposited during that part of geologic time designated as the Devonian Period.

time transgressive Refers to a rock unit that was deposited in an environment that shifted with time, as during a marine transgression. Thus, the age of the rock unit varies over its geographic extent.

time unit Any of the units such as eon, era, period, epoch, and age used to refer to specific intervals of geologic time.

Tippecanoe sequence A widespread sequence of sedimentary rocks bounded above and below by unconformities; deposited during an Ordovician to Early Devonian transgressive-regressive cycle of the Tippecanoe Sea.

trace fossil Any indication of prehistoric organic activity such as tracks, trails, burrows, borings, or nests. (*See body fossil*)

Transcontinental Arch An area consisting of several large islands extending from New Mexico to Minnesota that was above sea level during the Cambrian transgression of the Sauk Sea.

transform plate boundary Plate boundary along which plates slide past one another, and crust is neither produced nor destroyed; a type of strike-slip fault along which oceanic ridges are offset.

tree-ring dating The process of determining the age of a tree or wood in structures by counting the number of annual growth rings.

trilobite A group of benthonic, detritus-feeding, extinct marine invertebrate animals (phylum Arthropoda), having skeletons of an organic compound called chitin.

triple junction A point where three plates meet, e.g., the triple junction formed at the intersection of the East African Rift System and the rifts in the Red Sea and Gulf of Aden.

trophic level The complex interrelationships among producers, consumers, and decomposers in a community of organisms. There are several trophic levels of production and consumption through which energy flows in a feeding hierarchy.

U

unaltered remains Fossil remains that retain their original composition and structure. (*See altered remains*)

unconformity An erosion surface that separates younger strata from older rocks. (*See angular unconformity, disconformity, and nonconformity*)

ungulate An informal term referring to the hoofed mammals, especially the orders Artiodactyla and Perissodactyla.

V

varve A couplet of sedimentary laminae representing deposition that occurred in one year, e.g., the dark (winter) and light (summer) layers in varved sediments deposited in glacial lakes.

vascular A term referring so some land plants possessing specialized tissues for transporting fluids.

vertebrate Any animal having a segmented vertebral column; members of the subphylum Vertebrata; includes fishes, amphibians, reptiles, mammals, and birds.

vestigial structure A body part that serves little or no function, e.g., the dewclaws of dogs; a vestige of a structure or organ that was well developed and functional in some ancestor.

volcanic island arc A curved chain of volcanic islands parallel to a deep-sea trench where oceanic lithosphere is subducted causing volcanism and the origin of volcanic islands.

volcanic (extrusive igneous) rock An igneous rock that forms when magma is extruded onto the Earth's surface and cools and crystallizes or when pyroclastic materials become consolidated.

W

Walther's law The observations by Johannes Walther serve as the basis for a principle that holds that facies deposited in adjacent environments will be superposed one upon the other when the environments migrate laterally; in a conformable vertical sequence of facies, the same facies that occur vertically will replace one another laterally.

weak nuclear force A force responsible for the breakdown of atomic nuclei, thus producing radioactive decay.

Wilson cycle The relationship between mountain building (orogeny) and the opening and closing of ocean basins.

Z

Zuni sequence An Early Jurassic to Early Paleocene sequence of sedimentary rocks bounded above and below by unconformities; deposited during a transgressive-regressive cycle of the Zuni Sea.

ANSWERS TO MULTIPLE-CHOICE REVIEW QUESTIONS

CHAPTER **1**

1. c; 2. d; 3. a; 4. d; 5. b; 6. d; 7. e; 8. a; 9. e; 10. d; 11. c; 12. c; 13. b; 14. d; 15. a.

CHAPTER **2**

1. b; 2. e; 3. c; 4. c; 5. b; 6. b; 7. d; 8. a; 9. d; 10. d; 11. c; 12. d; 13. b; 14. c; 15. c; 16. c; 17. a; 18. e; 19. d.

CHAPTER **3**

1. d; 2. b; 3. c; 4. a; 5. b; 6. a; 7. b; 8. c; 9. e; 10. b; 11. e; 12. d; 13. c; 14. a.

CHAPTER **4**

1. b; 2. d; 3. b; 4. c; 5. a; 6. d; 7. b; 8. a; 9. c; 10. e; 11. b; 12. e; 13. a; 14. a.

CHAPTER **5**

1. c; 2. e; 3. a; 4. d; 5. b; 6. a; 7. e; 8. c; 9. b; 10. a; 11. a; 12. d; 13. d; 14. c.

CHAPTER **6**

1. b; 2. d; 3. b; 4. c; 5. d; 6. b; 7. b; 8. c; 9. a; 10. e; 11. e; 12. a; 13. b; 14. d.

CHAPTER **7**

1. d; 2. e; 3. e; 4. c; 5. b; 6. a; 7. c; 8. b; 9. d; 10. a; 11. d; 12. d; 13. e; 14. b; 15. e; 16. b; 17. c.

CHAPTER **8**

1. d; 2. c; 3. d; 4. b; 5. e; 6. b; 7. e; 8. a; 9. c; 10. b; 11. e; 12. a; 13. d; 14. b; 15. e; 16. a; 17. e; 18. a.

CHAPTER **9**

1. b; 2. e; 3. e; 4. c; 5. b; 6. a; 7. a; 8. d; 9. a; 10. c; 11. c; 12. a; 13. b.

CHAPTER **10**

1. b; 2. d; 3. c; 4. b; 5. c; 6. b; 7. a; 8. e; 9. c; 10. c; 11. b; 12. d; 13. d.

CHAPTER **11**

1. e; 2. d; 3. a; 4. b; 5. c; 6. e; 7. c; 8. b; 9. a; 10. b; 11. d; 12. a; 13. e; 14. a; 15. e.

CHAPTER **12**

1. c; 2. c; 3. e; 4. b; 5. e; 6. a; 7. b; 8. a; 9. c; 10. a; 11. d; 12. d; 13. c; 14. b; 15. e.

CHAPTER **13**

1. b; 2. a; 3. d; 4. a; 5. e; 6. c; 7. d; 8. e; 9. c; 10. e; 11. a; 12. c; 13. b; 14. d; 15. e; 16. a; 17. e.

CHAPTER **14**

1. c; 2. d; 3. a; 4. a; 5. d; 6. e; 7. b; 8. d; 9. b; 10. c; 11. d; 12. c; 13. a; 14. b; 15. c; 16. e; 17. b; 18. c; 19. a; 20. d; 21. a; 22. b.

CHAPTER **15**

1. d; 2. e; 3. c; 4. c; 5. a; 6. e; 7. c; 8. d; 9. b; 10. d; 11. e; 12. a; 13. e; 14. b; 15. a; 16. c; 17. b; 18. e.

CHAPTER **16**

1. a; 2. c; 3. d; 4. c; 5. b; 6. d; 7. a; 8. b; 9. b; 10. a; 11. c; 12. b; 13. d; 14. c.

Chapter 17

1. b; 2. c; 3. c; 4. a; 5. e; 6. a; 7. a; 8. c; 9. b; 10. c; 11. a;
12. e; 13. b; 14. c; 15. a; 16. a.

Chapter 18

1. b; 2. c; 3. e; 4. a; 5. d; 6. b; 7. c; 8. d; 9. a; 10. e; 11. b;
12. a; 13. b; 14. a; 15. d; 16. e.

Chapter 19

1. a; 2. c; 3. e; 4. d; 5. d; 6. a; 7. c; 8. a; 9. c; 10. c; 11. c;
12. c; 13. b; 14. c; 15. d; 16. e; 17. b.

Chapter 20

1. b; 2. e; 3. d; 4. e; 5. d; 6. a; 7. b; 8. a; 9. c; 10. c; 11. d;
12. b; 13. d; 14. a; 15. e.

INDEX

continents, 404–5, 464
tectonics/sedimentation in western
North America, 416–23

K

Kaapvaal craton, 261
Kaibab-Defiance-Zuni uplift, 329, 332
Kaibab Limestone, 291
Kangaroo rat, parallel evolution of,
155–56, 157
Kangaroos, 559
Kansan glacial stage, 521, 525
Kant, Immanuel, 210, 211
Karst topography, 332
Kaskaskia Sea, 309, 324, 327
Kaskaskia sequence, 324–27
Kayenta Formation, 415, 416
Kazakhstania, 294–95, 296–97
Kelvin, Lord, 60–61, 62
Kerogen, 510–11
Key beds, 99
Keystone thrust fault, 421
Kimberlite pipes, 426–27
Kingdom (classification), 162, 164
Kiwi, 167, 196
Klamath island arc, 335
Klamath Mountains, 427
Komodo dragon, 458
Kona Dolomite, 273
Kontinen, Asko, 259
Kuiper, Gerard P., 212
Kupferschiefer, 341

L

La Brea Tar Pits, 86, 87, 550, 559–60
Labyrinthodonts, 382–83, 385, 416,
438
Laccoliths, 31–32, 34
Lagoons, 118, 119, 127, 128, 129,
130–31, 132, 361–62
barrier reefs and, 304, 333
Lake Bonneville, 529, 532
Lake Manly, 529
Lake Missoula, 530–31
Lamarck, Jean Baptiste de, 144–45, 146
Lamar River Formation, 92
Laminae, 114
Lampreys, 319, 377
Land plants. *See* Plants
Laplace, Pierre Simon de, 210, 211
Lapworth, Charles, 92, 95
Lapworthella, 351
Laramide orogeny, 417, 418–19, 422,
487–91

Late Paleozoic, 319–42, 344–45
Absaroka sequence, 327–33
Kaskaskia sequence, 324–27
microplates and formation of
Pangaea, 340–41
mineral resources, 341–42
mobile belts, 333–40
North America, history of, 321
paleogeography, 320–21, 322, 323
Lateral continuity, principle of, 56, 57,
79
Lateral gradation, 79, 80
Lateral stratigraphic relationships, 75
deltas and, 123
depositional environments and, 132,
133
Latimeria, 161, 377
Laurasia, 177
Mesozoic, 403, 405, 464
Proterozoic, 267, 269, 270
Laurasia
collision with Gondwana, 322, 323,
330, 341
formation of, 320, 322, 323, 339,
340
Laurentia
Early Paleozoic, 294–95, 296–97,
302, 309
Early Proterozoic, 261–65
Late Paleozoic, 320, 322, 339, 340
Middle Proterozoic, 265–67
Lava, 6, 31
pillow, 183
Lava flows, 31, 32, 33, 267, 268, 403,
411
dating sedimentary rocks from, 102
in greenstone belts, 239–40
magnetism in, 178–79, 180
on the Moon and other planets, 173,
219, 220, 221
at plate boundaries, 185, 186
in sequence of strata, 74, 75–76
Lead, 197, 311, 342
Leaves
as climate indicators, 563, 564
evolution of, 388–89
Leakey, Louis, 577
Leakey, Mary, 577, 579
Leakey, Richard, 577, 580, 582
Lemurs, 548, 549, 573, 575
Leonardo da Vinci, 73, 83, 166
Leptotriticites tumidus, 365
Lesquereux, L., 319
Lesser apes, 575

Lewis overthrust, 491
Lichins, 280
Life/life-forms. *See also* specific
life-forms
Archean, 248–53
Cenozoic, 541–65
extinctions. *See* Extinctions
Mesozoic, 433–67, 468–69
origin of, 248–51
Paleozoic invertebrates, 349–66,
368–69
Paleozoic plants, 386–95, 396–97
Paleozoic vertebrates, 373–86,
396–97
Proterozoic, 276–85
submarine hydrothermal vents and,
250–51
Lignin, 388
Limestone, 8, 30, 40, 47, 111, 301,
302, 326, 331, 342
in carbonate depositional
environments, 129–30
fossils in, 112, 113
Limonite, 39, 426
Lingula, 101, 102, 160, 435
Linnaean system of classification,
161–63
Linnaeus, Carolus, 162, 163
Lipalian interval, 283
Liquid, 23
Lithification, 6, 39, 110
Lithium, 254, 287
Lithology, 99, 111–13
Lithosphere, 10, 11–12, 183, 187
Lithostatic pressure, 42
Lithostratigraphic units, 94, 95, 96
and biostratigraphic units, 97
Little Belt Mountains, 282
Little Bighorn, Battle of, 238
Little Dal Group, 282
Little Ice Age, 517–18, 521, 535
Living fossils, 160, 161
Lizard-hipped dinosaurs. *See*
Saurischian dinosaurs
Lizards, 454
Lobe-finned fish, 377, 379–80, 382
Lodgepole Formation, 96
Lorises, 548
Louann Salt, 132, 426, 505
Lovejoy, Owen, 570
Lower (time-stratigraphic term), 97
"Lucy" (*A. afarensis* skeleton), 570,
579
Lung fish, 377, 379–80, 438

CREDITS

TABLE OF CONTENTS

Chapter 2: Photo courtesy of Richard L. Chambers. Chapter 6: Granger Collection, New York. Chapter 7: Photo courtesy of NASA. Chapter 8: Photo courtesy of the Palomar Observatory, California Institute of Technology. Chapter 12: Photo courtesy of Geoffrey Playford, University of Queensland, Brisbane, Australia. Chapter 13: Photo courtesy of the Smithsonian Institution, Trans. 86–1 3471A. Chapter 14: Photo courtesy of NAGT. Chapter 16: Carlyn Iverson. Chapter 19: Carlyn Iverson. Chapter 20: Carlyn Iverson.

CHAPTER 1

1.2 and 1.15: Precision Graphics. 1.8: Victor Royer. 1.9, 1.11, 1.12, and 1.14: Carlyn Iverson. 1.10: Rolin Graphics. 1.13: Carto-Graphics.

CHAPTER 2

2.1a: Publication Services. Reprinted with permission from B. J. Skinner, "Mineral Resources of North America," *Geology of North America,* v.A, 1989, p. 577 (Fig. 2). 2.2 and 2.3: Layout by Georg Klatt, final inking by Elizabeth Morales-Denney. 2.4, 2.5, 2.6, 2.8, 2.9, 2.12, 2.14, 2.22, 2.26, and 2.27a: Precision Graphics. 2.7: Precision Graphics. From Dietrich and Skinner, *Gems, granites, and gravels: Knowing and using rocks and minerals,* 1990, p. 39 (Fig. 3.4), reprinted with permission of Cambridge University Press. 2.17: Precision Graphics. Modified from R. V. Dietrich, *Geology and Michigan: Forty-nine Questions and Answers:* 1979.

CHAPTER 3

3.1, 3.8, 3.9, 3.10b, 3.11, 3.13: Precision Graphics. 3.5: Publication Services. 3.10a: Precision Graphics. From Don L. Eicher, *Geologic Time,* 2e, ©1976, p. 120. Reprinted by permission of Prentice Hall, Englewood Cliffs, New Jersey. 3.14: Precision Graphics. Reprinted with permission from Stokes and Smiley, *An Introduction to Tree Ring Dating,* ©The University of Chicago Press. Perspective 3.1, Figure 1: Publication Services. Reprinted with permission from J. W. Wells, "Coral Growth and Geochronometry," *Nature,* v. 197, no. 4871, 1963, p. 949 (Fig.1).

CHAPTER 4

4.3, 4.4b, 4.4c, 4.7, 4.9, 4.10, 4.12, 4.13, 4.15, 4.16, 4.30, 4.32d, 4.33, 4.38, 4.39, 4.40, 4.41, and 4.43: Publication Services. 4.5, 4.6, 4.8a, 4.8c, 4.8e, 4.11, 4.14, 4.28, 4.32a, and 4.34: Precision Graphics. 4.19c: Carlyn Iverson. 4.29: Carto-Graphics. Reprinted with permission from L. W. Mintz, *Historical Geology,* 3rd ed., ©1981, Charles E. Merrill Publishing Company. 4.35: Publication Services. From *History of the Earth: An Introduction to Historical Geology,* Second Edition, by Bernhard Kummer. ©1970 by W. H. Freeman and Company, reprinted by permission. 4.36: Precision Graphics. 4.36a Reprinted with permission from B. Rascoe, Jr., "Regional stratigraphic analysis of Pennsylvanian and Permian rocks in western mid-continent," *AAPG,* v. 46, no. 8, 1962, p. 1356 (Fig. 7), American Association of Petroleum Geologists, and 4.36b reprinted with permission from M. H. Rider, *The Geological Interpretation of Well Logs,* Revised Edition 1991, p. 2 (Fig. 1.2). 4.37a: Precision Graphics. From *Elements of Petroleum Geology,* by R. C. Selley, ©1985 by W. H. Freeman and Company, reprinted by permission. 4.42: Publication Services. Reprinted with permission from Hsu, Kelts, and Valentine, "Resedimented facies in Ventura Bay, CA, and model of longitudinal transport of turbidity currents," *AAPG,* v. 64, no. 7, 1980, pp. 1038–1039 (Fig. 3), American Association of Petroleum Geologists.

CHAPTER 5

5.2: Precision Graphics. Reprinted with permission from M. B. Cita, "Mediterranean evaporite: Paleontological arguments for a deep-basin dessication model," in C. W. Drooger (ed.), *Messinian Events in the Mediterranean,* Geodynamics Scientific Report No. 7, 1973, p. 212 (Fig. 3), Elsevier Science Publishers. 5.3 and 5.28: Publication Services. 5.8, 5.9, 5.10, 5.11, 5.12a, 5.12c, 5.14, 5.17b, 5.18b, 5.19a, 5.19b, 5.20a, 5.21, 5.22, 5.23, 5.24, and 5.27a: Precision Graphics. 5.15: Carlyn Iverson. 5.16b and 5.16d: Precision Graphics. Reprinted with permission from J. R. L. Allen, "Review of the Origins and Characteristics of Recent Alluvial Sediments," *Sedimentology,* 5, 1965, Blackwell Scientific Publications, Ltd. 5.17c: Precision Graphics. From Walker and Cant, "Sandy Fluvial Systems," in *Facies Models,* Geoscience Canada, Reprint Series 1, 1984, p. 73 (Fig. 2); Reproduced with permission of the Geological Association of Canada. 5.25: Carto-Graphics. From Davies and Gorsline, "Oceanic Sediments and Sedimentary Processes," in Riley and Chester (eds.), *Chemical Ocean-*

ography, v. 5, 2nd ed., reprinted with permission of Academic Press, Inc., London. 5.27b: Precision Graphics. From Blatt, Middleton, and Murray, *Origin of Sedimentary Rock,* 2nd ed., ©1980, p. 498, reprinted by permission of Prentice Hall, Englewood Cliffs, New Jersey. 5.29a and 5.29b: Publication Services. Reprinted with permission from Young, Fiddler, and Jones, "Carbonate Facies in Ordovician of Northern Arkansas," *AAPG,* v. 56, no. 1, 1972, pp. 72, 78, American Association of Petroleum Geologists. 5.30: Carto-Graphics. Reprinted with permission from *Geologic Atlas of the Rocky Mountain Region,* 1972, p. 252, the Rocky Mountain Association of Geologists. 5.31: Carto-Graphics. Reprinted with permission from R. J. Weimer, "Upper Cretaceous Stratigraphy, Rocky Mountain Area," *AAPG,* v. 44, no. 1, 1960, pp. 2, 5, 9, American Association of Petroleum Geologists. 5.32: Carto-Graphics. Reprinted with permission from *Geologic Atlas of the Rocky Mountain Region,* 1972, pp. 216, 221, the Rocky Mountain Association of Geologists. Perspective 5.1, Figure 1a: Precision Graphics. Reprinted with permission from J. M. Coleman, *Deltas: Processes of Deposition and Models for Exploration,* ©1976, p. 37, Continuing Education Publication Company, Inc. Perspective 5.1, Figure 1b: Carto-Graphics. Reprinted with permission from Fisk, McFarlan, Kolb, and Wilbert, "Sedimentary Framework of the Modern Mississippi Delta," *Journal of Sedimentary Petrology,* v. 24, 1954, p. 77, Society for Sedimentary Geology. Perspective 5.1, Figure 2: Carto-Graphics. Reprinted with permission from Kolb and Van Lopick, "Depositional environments of the Mississippi River deltaic plain, southeastern Louisiana," in *Deltas in Their Geologic Framework,* M. L. Shirley (ed.), 1966, p. 22 (Fig. 2), ©The Houston Geological Society. Perspective 5.1, Figure 3a and 3b: Carto-Graphics. Reprinted with permission from Fisher, Brown, Scott, and McGowen, *Delta Systems in the Exploration for Oil and Gas—A Research Colloquium,* 1969 (Fig. 47). Table 5.1: Reprinted with permission from R. C. Selley, "Concepts and methods of sub-surface facies analysis," *AAPG* continuing education course note series #9, 1978, American Association of Petroleum Geologists.

Chapter 6

Table 6.1, 6.9, 6.14, 6.22, 6.24, 6.30, 6.31, and Perspective 6.1, Figure 2: Publication Services. 6.2, 6.7, 6.8a, 6.11, 6.12, 6.15, 6.16, 6.23a, and 6.28: Precision Graphics. 6.3, 6.6, 6.17, 6.18, 6.19, 6.25, 6.26, and Perspective 6.1, Figure 3b: Carlyn Iverson. 6.10: Precision Graphics. Reproduced by permission of the Society for the Study of Evolution, from *Evolution,* v. 3, pp. 238–265. 6.13: Publication Services. From *Principles of Paleontology,* 2nd ed., by D. M. Raup and S. M. Stanley. Copyright ©1978 by W. H. Freeman and Company, reprinted by permission. 6.20: Precision Graphics. Reprinted with permission from L. S. Dillon, *Evolution: Concepts and Consequences,* 2nd. ed., 1978, p. 250 (Fig. 12–6). 6.21: Precision Graphics. Reproduced by permission of the Society for the Study of Evolution from *Evolution,* v. 28, p. 448. 6.27: John and Judy Waller. Perspective 6.1, Figure 1: Publication Services. Reprinted with permission from Starr and Taggart, *Biology: The Unity and Diversity of Life,* 1989, p. 556 (Fig. 37.2a), ©Wadsworth Publishing Company. Perspective 6.1, Figure 3a: Publication Services. Reprinted with permission from Weishampel, Dodson, et. al., *Dinosauria: Paleobiology of the Terrible Lizards,* ©1990 The Regents of the University of California. Perspective 6.2, Figure 2: Carlyn Iverson. Reprinted with permission from O. C. Marsh, "Recent Polydactyle Horses," *American Journal of Science,* v. XLIII, no. 256, 1982, pp. 339–355.

Chapter 7

7.4: Carto-Graphics. Reprinted with permission of MacMillan Publishing Company from R. J. Foster, *General Geology,* 4th ed., p. 351 (Fig. 20–2), Copyright ©1983 by Merrill Publishing Company. 7.5, 7.6a, 7.6b, 7.8a, 7.15, 7.20, 7.24, and 7.26: Precision Graphics. 7.7: Carto-Graphics. Reprinted with permission from Cox and Doell, "Review of Paleomagnetism," *GSA,* v. 71, 1960, p. 758 (Fig. 33), Geological Society of America. 7.8b: Publication Services. Reprinted with permission from A. Cox, "Geomagnetic Reversals," *Science,* v. 163, p. 240 (Fig. 4), Copyright ©1992 by the AAAS. 7.11: Rolin Graphics. From *The Bedrock Geology of the World* by Larson and Pitmann III, Copyright ©1985 R. Larson and W. C. Pitmann III, reprinted by permission of W. H. Freeman and Company. 7.12, 7.22 (right), 7.23, and 7.27: Rolin Graphics. 7.14, 7.17, 7.18, 7.19, 7.22 (left), and Perspective 7.1, Figure 1: Carlyn Iverson. 7.16a and 7.21: Carto-Graphics. 7.16b and 7.25a: Publication Services. 7.25b: Precision Graphics. Reprinted with permission from A. Hallam, *Scientific American,* v. 227, no. 5, 1972, p. 62 and with permission from Lee Boltin.

Chapter 8

8.1: Publication Services. Reprinted with permission from Henbest and Couper, *The Restless Universe,* 1982, Published by George Philip and Son, Ltd., ©Nigel Henbest and Heather Couper. 8.2, 8.3, 8.8, 8.11, and 8.16: Publication Services. 8.4: Publication Services. Reprinted with permission from Snow, *The Dynamic Universe,* 1985, ©West Publishing Company. 8.6a: John and Judy Waller. 8.6b, 8.12a, 8.13, and 8.14: Precision Graphics. 8.7: Publication Services. Reprinted with permission from Snow, *The Dynamic Universe,* 1985, ©West Publishing Company. 8.9: Publication Services. Reprinted with permission from Snow, *The Dynamic Universe,* 1985, ©West Publishing Company. 8.10: Precision Graphics. From Eicher and McAlester, *History of the Earth,* ©1980, p. 11. Reprinted by permission of Prentice Hall, Englewood Cliffs, New Jersey. Perspective 8.1, Figure 2: Publication Services. Reprinted with permission from Snow, *The Dynamic Universe,* 1985, ©West Publishing Company.

Chapter 9

Table 9.1: Publication Services. Reprinted with permission from H. L. James, "Stratigraphic Commission: Note 40—Subdivision of Precambrian: An Interim Scheme to be Used by USGS," *AAPG,* v. 56, no. 6, 1972, p. 1130 (Table 1), and from Harrison and Peterman, "North American Commission on Stratigraphic Nomenclature: Report 9—Adoption of Geochronometric Units for Divisions of Precambrian Time," *AAPG,* v. 66, no. 6, 1982, p. 802 (Fig. 1), American Association of Petroleum Geologists. 9.2, Table 9.2, 9.16, 9.19, and Timeline: Publication Services. 9.3a, 9.3b, 9.9a, 9.11, 9.15, and Perspective 9.2, Figure 1: Precision Graphics. 9.4: Carto-Graphics. Reprinted with permission from A. M. Goodwin, "The Most Ancient Continental Margins," in *The Geology of Continental Margins,* Burk and Drake (eds.), 1974, p. 768 (Fig. 1), ©Springer-Verlag. 9.6: Carto-Graphics. Reprinted with permission from K. C. Condie, *Archean Greenstone Belts,* ©1981, p. 2, (Fig. 1), Elsevier Science Publishers. 9.7: Carto-Graphics. Reprinted with permission of the Geological Association of Canada. 9.8: Publication Services. Reprinted from K. C. Condie, *Plate Tectonics and Crustal Evolution,* 2nd ed., ©1982, p. 218, with permission from Pergamon Press, Ltd., Headington Hill Hall, Oxford OX3 OBW, UK. 9.10: Carto-Graphics. Reprinted with permission from Condie and Hunter, "Trace Element Geochemistry of Archean Granitic Rocks from the Barberton Region, South Africa," *Earth and Planetary Science Letters,* v. 29, ©1976, p. 398 (Fig. 7), Elsevier Science Publishers, and with permission from Eriksson, "Tidal Deposits from the Archean Moodies Group, Barberton Mountain Land, South Africa," *Sedimentary Geology,* v. 18, ©1977, p. 278 (Fig. 23), Elsevier Science Publishers. 9.12: Precision Graphics. Reprinted with permission from Condie and Hunter, "Trace Element Geochemistry of Archean Granitic Rocks from the Barberton Region, South Africa," *Earth and Planetary Science Letters,* v. 29, ©1976, p. 398 (Fig. 7), Elsevier Science Publishers, and with permission from Eriksson, "Tidal Deposits from the Archean Moodies Group, Barberton Mountain Land, South Africa," *Sedimentary Geology,* v. 18, ©1977, p. 278 (Fig. 23), Elsevier Science Publishers. 9.13: Publication Services. Reprinted from K. C. Condie, *Plate Tectonics and Crustal Evolution,* 2nd ed., ©1982, p. 87, with permission from Pergamon Press, Ltd., Headington Hill Hall, Oxford OX3 OBW, UK. 9.14: Publication Services. Reprinted with permission from Dickinson and Luth, "A Model for Plate Tectonic Evolution of Mantle Layers," *Science,* v. 174, no. 4007, 1971, p. 402 (Fig. 1), ©1992 by the AAAS. 9.17: Publication Services. Reprinted with permission from S. L. Miller, "The Formation of Organic Compounds on the Primitive Earth," in *Modern Ideas of Spontaneous Generation,* Nigrelli (ed.), Annuals of the New York Academy of Sciences, v. 69, Art. 2, Aug. 30, 1957, p. 261 (Fig. 1). Perspective 9.1, Figure 1: Carto-Graphics. Reprinted with permission from A. N. Strahler, *Physical Geography,* 1975, p. 486, ©1975 by Arthur N. Strahler. Perspective 9.2, Figure 2: Publication Services. Reprinted with permission from Corliss, Baross, and Hoffman, "A hypothesis concerning the relationship between submarine hot springs and the origin of life on Earth, *Oceanologica Acta,* v. 4, 1981, suppl. C4, p. 62.

CHAPTER 10

Table 10.1, Table 10.3, 10.26, and Timeline: Publication Services. Reprinted with permission from Kontinen, *Precambrian Research,* ©1987, Elsevier Science Publishers. 10.2: Carto-Graphics. Reproduced with permission from P. F. Hoffman, "United Plates of America, The Birth of a Craton: Early Proterozoic Assembly and Growth of Laurentia," *Annual Review of Earth and Planetary Sciences,* v. 16, p. 544, ©1988 by Annual Reviews, Inc. 10.3, 10.28f, and Perspective 10.1, Figure 2: Precision Graphics. 10.4: Publication Services. Reprinted with permission from Hoffman and Bowring, "Short-Lived 1.9 Fa Continental Margin and its Destruction, Wopmay Orogen, Northwest Canada," *Geology,* v. 12, 1984, p. 70 (Fig. 2), Geological Society of America. 10.5: Carto-Graphics. Reprinted with permission from Medaris, Byers, Mikelson, and Shanks (eds.), "Proterozoic anorogenic granite plutonism of North America," *Proterozoic Geology: Selected Papers from an International Symposium,* p. 135 (Fig. 1). 10.7: Publication Services. Reprinted with permission from J. C. Green, *Tectonophysics,* v. 94, p. 414 (Fig. 1), ©1983, Elsevier Science Publishers. 10.8a: Publication Services. Reprinted with permission from Daniels, *Geology,* GSA Memoir 156, 1982, the Geological Society of America. 10.9: Publication Services. Reprinted with permission from R. P. E. Poorter, "Precambrian Paleomagnetism of Europe and the Position of the Balto-Russian Plate Relative to Laurentia," in *Precambrian Plate Tectonics,* A. Kroner (ed.), p. 603 (Fig. 24−2), ©1981, Elsevier Science Publishers. 10.10: Publication Services. Re-

printed from K. C. Condie, *Plate Tectonics and Crustal Evolution,* 2nd ed., ©1982, p. 228, with permission from Pergamon Press Ltd., Headington Hill Hall, Oxford OX3 OBW, UK. 10.11: Carto-Graphics. Reprinted with permission from Ian W. D. Dalziel, "Pacific margins of Laurentia and East-Antarctica-Australia as a conjugate rift pair: Evidence and implications for an Eocambrian supercontinent," *Geology,* v. 19, June 1991, p. 600 (Fig. 3), Geological Society of America. 10.12: Carto-Graphics. Reprinted with permission from G. M. Young, "Tectono-Sedimentary History of Early Proterozoic Rocks of the Northern Great Lakes Region," in *Early Proterozoic Geology of the Great Lakes Region,* Medaris (ed.), GSA Memoir 160, 1983, p. 16 (Fig. 1), Geological Society of America. 10.14 and 10−29a: Carto-Graphics. 10.17a: Publication Services. Reprinted with permission from D. A. Lindsey, "Glacial Marine Sediment in the Precambrian Gowganda Formation at Whitefish Falls, Ontario," *Paleogeography, Paleoclimatology, Paleoecology,* v. 9, 1971, p. 9 (Fig. 2). 10.17b: Publication Services. Reprinted with permission from Young and Nesbitt, "The Gowganda Formation in the Southern Part of the Huronian Outcrop Belt, Ontario, Canada," *Precambrian Research,* v. 29, p. 296 (Fig. 18), ©1985, Elsevier Science Publishers. 10.18 and 10.19: Publication Services. Reprinted with permission from L. A. Frankes, *Climates Throughout Geologic Time,* pp. 39 and 88, ©1979, Elsevier Science Publishers. 10.21: Precision Graphics. Reprinted with permission from A. M. Goodwin, "Distribution and Origin of Precambrian Banded Iron Formation," *Revista Brasileira de Geosciencias,* v. 12 (1−3), 1982, p. 462 (Fig. 4). 10.23: Precision Graphics. From Lynn Margulis, "Symbiosis and Evolution," copyright ©1971 by Scientific American, Inc. All rights reserved. 10.31: Publication Services. Reprinted with permission from J. Eichler, "Origin of Precambrian Banded Iron Formations," in *Handbook of Strata-Bound and Strataform Ore Deposits,* v. 7, p. 162 (Fig. 1), ©1976, Elsevier Science Publishers. Perspective 10.1, Figure 1: Precision Graphics. From *Symbiosis in Cell Evolution: Life and its Environments on the Early Earth,* by Lynn Margulis. Copyright ©1981 by W. H. Freeman and Company, reprinted by permission.

CHAPTER 11

11.1, 11.4, 11.7, and 11.10: Carto-Graphics. 11.2: Carto-Graphics. Topography reprinted with permission from Bambach, Scotese, and Ziegler, "Before Pangaea: The Geographies of the Paleozoic World," *American Scientist,* v. 68, no. 1, 1980. 11.3: Publication Services. Reprinted with permission from L. L. Sloss, "Sequences in the cratonic interior of North America," *GSA Bulletin,* v. 74, 1963, p. 110 (Fig. 6), Geological Society of America. 11.5, 11.6a, 11.8a, 11.9b, 11.12, 11.13, and 11.14: Precision Graphics. 11.11a: Precision Graphics. Reprinted with permission from Mesolella, Robinson, McCormick, and Ormiston, "Cyclical Deposition of Silurian Carbonates and Evaporites in Michigan Basin," *AAPG Bulletin,* v. 58, no. 1, 1974, p. 40, American Association of Petroleum Geologists. Perspective 11.2, Figure 1a and Timeline: Publication Services.

CHAPTER 12

12.1a, 12.1b, 12.2a, and 12.2b: Carto-Graphics. Topography reprinted with permission from Bambach, Scotese, and Ziegler, "Before Pangaea: The Geographies of the Paleozoic World," *American Scientist,* v. 68, no. 1, 1980. 12.3, 12.6a, and Timeline: Publication Services. 12.4,

12.5, 12.8, 12.10, 12.13a, 12.15, and 12.16: Carto-Graphics. 12.7, 12.11a, 12.11c, 12.13b, 12.20, and 12.21: Precision Graphics. 12.19: Precision Graphics. From Briggs and Roeder, *A Guidebook to the Sedimentology of Paleozoic Flysch and Associated Deposits, Ouchita Mountains and Arkoma Basin, Oklahoma, c.1975*, (Fig. 6), Reprinted by permission of the Dallas Geological Society. 12.22: Rolin Graphics. From the U.S. Geological Survey.

CHAPTER 13

13.1, 13.5a, and 13.9: Precision Graphics. 13.4, 13.13, Table 13.2, and Timeline: Publication Services. 13.6: Carlyn Iverson. 13.7: Publication Services. Reprinted with permission from H. Tappan, "Microplankton, Ecological Succession and Evolution, Part H," in *Proceedings of the North American Paleontological Convention*, E. Yochelson (ed.), 1970–71, p. 1064 (Fig. 2). 13.22: Publication Services. Reprinted with permission from Raup and Sekoski, "Mass Extinctions in the Marine Fossil Record," *Science*, v. 215, 1982, p. 1502, ©1982 by the AAAS.

CHAPTER 14

Table 14.1: Publication Services. Reprinted with permission from Gensel and Andrew, "The evolution of early land plants," *American Scientist*, v. 75, no. 5, 1987, p. 480 (Fig. 2). 14.1b, 14.8, 14.10, 14.12, 14.14, 14.15, 14.19, 14.20, 14.23, 14.25b, 14.26, and 14.28: Carlyn Iverson. Table 14.2, 14.4, 14.7, and Timeline: Publication Services. 14.3: Precision Graphics. Adapted with permission from drawing by Ralph Buchsbaum in V. Pearse, J. Pearse, M. Buchsbaum, and R. Buchsbaum, *Living Invertebrates*, 1987, Boxwood Press, reprinted with permission. 14.5: Precision Graphics. Reprinted with permission from Romer, *The Vertebrate Story*, ©The University of Chicago Press. 14.9, 14.11, 14.13, 14.16, 14.18, and 14.24: Precision Graphics. 14.17: Precision Graphics. Reprinted with permission from Stearn, Carroll, and Clark, *Geological Evolution of North America*, 3rd ed., 1979, p. 271 (Fig. 13–18), Copyright ©1979, John Wiley and Sons. 14.21: Precision Graphics. Reprinted with permission from L. E. Graham, "The origin of the life cycle of land plants," *American Scientist*, v. 73, no. 2, 1985, p. 181 (Fig. 2).

CHAPTER 15

15.2, 15.17, 15.23, and 15.24a: Precision Graphics. 15.3: Carto-Graphics. Reprinted with permission from Dott and Shaver, "Modern and Ancient Geosynclinal Sedimentation," *S.E.P.M. Special Publication*, No. 19, 1974, p. 50, Society for Sedimentary Geology. 15.4: Carto-Graphics. Reprinted with permission from Dietz and Holden, "Reconstruction of Pangaea: Breakup and Dispersion of Continents, Permian to Present," *Journal of Geophysical Research*, 1970, v. 75, no. 26, pp. 4939–4956, ©1970 by the American Geophysical Union. 15.5: Carto-Graphics. Reprinted with permission from K. Burke, "Atlantic Evaporites formed by evaporation of water spilled from Pacific, Tethyan, and southern oceans," *Geology*, v. 3, 1975, p. 614 (Fig. 1), The Geological Society of America. 15.6: Carto-Graphics. Reprinted with permission from Dietz and Holden, "Reconstruction of Pangaea: Breakup and Dispersion of Continents, Permian to Present," *Journal of Geophysical Research*, 1970, v. 75, no. 26, pp. 4939–4956, ©1970 by

the American Geophysical Union. 15.7: Carto-Graphics. Reprinted with permission from P. L. Robinson, "A problem of faunal replacement of Permo-Triassic Continents," *Paleontology*, v. 14, Part 1, 1971, p. 153. 15.8, 15.9, 15.10, 15.11, and 15.26a: Carto-Graphics. 15.12, 15.18a, 15.21, 15.22, Table 15.1, and Timeline: Publication Services. 15.16: Precision Graphics. Reprinted with permission from D. G. Bebout and R. G. Loucks, "Lower Cretaceous Reefs, South Texas," *AAPG Memoir* 33, 1983, p. 441, American Association of Petroleum Geologists. 15.19: Publication Services. Reprinted with permission from Schweickert and Cowan, "Early Mesozoic Tectonic Evolution of the Western Sierra Nevada, California," *GSA Bulletin*, v. 86, 1975, p. 1334 (Fig. 3), ©1975, The Geological Society of America. 15.20a: Rolin Graphics (Top) and Carlyn Iverson (Bottom), Reprinted with permission from B. M. Page, "Effects of Late Jurassic-Early Tertiary Subduction in California," in *Late Mesozoic and Cenozoic Sedimentation and Tectonics in California*, p. 66 (Fig. 5–9), San Joaquin Geological Society (Short Course 1977). 15.27: Publication Services. Reprinted with permission from R. L. Armstrong, "Sevier Orogenic Belt in Nevada and Utah," *GSA Bulletin*, v. 79, 1968, p. 446 (Fig. 5), ©1968 The Geological Society of America. 15.28: Publication Services. Reprinted with permission from Zvi Ben-Avraham, "The Movement of Continents," *American Scientist*, v. 69, no. 3, 1981, p. 298 (Fig. 9). 15.29: Precision Graphics. Data from Saudi Aramco.

CHAPTER 16

16.5, 16.13, 16.14, 16.17, 16.25c, 16.30, and Perspective 16.1, Figure 2: Precision Graphics. 16.11: Carlyn Iverson. Reprinted with permission for E. H. Colbert, *Evolution of the Vertebrates*, 3rd ed., p. 54, (Fig. 19), ©1980 John Wiley and Sons, Inc. 16.15, 16.16, 16.18, 16.19, 16.21, 16.22, 16.23, 16.25a, 16.25b, 16.26, 16.27a, 16.27b, 16.37, Perspective 16.1, Figure 1, and Perspective 16.2, Figure 1: Carlyn Iverson. 16.24, Table 16.4, and Timeline: Publication Services. 16.25d: Precision Graphics. Reprinted with permission from W. Langston, "Pterosaurs," *Scientific American*, copyright ©1981 by Scientific American, Inc. All rights reserved. 16.28: Precision Graphics. Reprinted with permission from R. Monastersky, "Chinese Bird Fossil: Mix of Old and New," *Science News*. v. 138, 1990, p. 247. 16.31: Precision Graphics. Reprinted with permission from J. A. Hopson, "The Mammal-Like Reptiles: A Study of Transitional Fossils," *The American Biology Teacher*, v. 49, no. 1, 1987, pp. 18, 22, (National Association of Biology Teachers: Reston, VA). 16.32: Precision Graphics. From *Synapsida: A New Look into the Origin of Mammals*, by John C. McLoughlin. Copyright ©1980 by John C. McLoughlin. Used by permission of Viking Penguin, a division of Penguin Books USA, Inc. 16.33: Precision Graphics. Reprinted with permission from J. A. Hopson, "The Mammal-Like Reptiles: A Study of Transitional Fossils." *The American Biology Teacher*, v. 49, no. 1, 1987, pp. 18, 22, (National Association of Biology Teachers: Reston, VA). 16.34: Publication Services. Reprinted with permission from J. Lillegraven, et. al, *Mesozoic Mammals: The First Two-Thirds of Mammalian History*, copyright ©1979, The Regents of the University of California. 16.35: Carlyn Iverson. Adapted with permission from J. Benes, *Prehistoric Animals and Plants*, 1979, pp. 174–175, illustration by Zdenek Burian, published by Hippocrene Books, Inc. 16.36: Precision Graphics. Reprinted with permission from Kermack and Kielan-Jaworowska, "Therian and Non-Therian Mammals," in *Early Mammals*, D. M. Kermack and K. A. Kermack (eds.), ©1971 Academic Press, Inc. 16.38: Publication Services. Reprinted with permission from Montanari, et. al., "Spheroids at the Cretaceous Boundary are Altered Impact Droplets of Basaltic Composition," *Geology,* v. 11, 1983, p. 669 (Fig. 2), Geological Society of America. 16.40: Publication Services. Reprinted with permission from Raup and Sepkoski, "Testing for Periodicity of Extinction," *Science,* v. 241, 1988, p. 95 (Fig. 2), Copyright ©1992 by the AAAS.

CHAPTER 17

17.3, 17.12, and Timeline: Publication Services. 17.4: Publication Services. Reprinted with permission from Gordon and Jurdy, *Cenozoic Global Plate Motions,* v. 91, no. B12, pp. 12, 394–5, 1986, ©by the American Geophysical Union. 17.5a and 17.5b: Precision Graphics. Reprinted with permission from Guennoc and Thisse, "Origins of the Red Sea Rift, the Axial Troughs, and their Mineral Deposits," v. 51, 1982, Editions BRGM BP 6009 45060, Orleans Cedex 2, and with permission from Baker, Mohr, and Williams. *Geology of the Eastern Rift System of Africa.* Special Paper 136, 1972, p. 2 (Fig. 1), Geological Society of America. 17.6: Carto-Graphics. Reprinted with permission from A. M. Spencer (ed.), *Mesozoic-Cenozoic Orogenic Belts,* 1974, pp. xii-xiii, The Geological Society Publishing House, Bath, England. 17.7a: Carto-Graphics. Reprinted with permission from Smith and Woodcock, "Tectonic Synthesis of the Alpine-Mediterranean Region: A Review in Alpine-Mediterranean Geodynamics," *Geodynamic Series 7,* 1982, p. 16, ©by the American Geophysical Union. 17.8: Precision Graphics. Modified from A. Heim, 1919, *Geologie der Schweiz,* C. H. Tauchnitz: Leipsig. 17.9: Publication Services. Reprinted with permission from Dewey and Bird, "Mountain Belts and the New Global Tectonics," *Journal of Geophysical Research,* v. 75, no. 14, 1970, p. 2627, ©by the American Geophysical Union. 17.10: Carlyn Iverson. From P. Molnar, "The Geologic History and Structure of the Himalayas," *American Scientist,* v. 74, no. 2, 1986, pp. 148–149, Reprinted by permission of *American Scientist,* journal of Sigma Xi, The Scientific Research Society. 17.11: Carto-Graphics. Reprinted with permission from A. D. Miall, *Principles of Sedimentary Basin Analysis,* ©1984, Springer-Verlag, Seacaucus, New Jersey. 17.13: Rolin Graphics. 17.14: Carto-Graphics. Reprinted with permission from Malfait and Dinkelman, "Circum-Caribbean Tectonic and Igneous Activity and the Evolution of the Caribbean Plate," *GSA Bulletin,* v. 83, 1972, pp. 254–257, 1972, The Geological Society of America. 17.15: Carto-Graphics. Compiled from several sources. 17.16: Publication Services. Reprinted with permission from Dickinson and Snyder, "Plate Tectonics of the Laramide Orogeny," in *Laramide Folding Associated with Block Faulting in the Western United States, GSA Memoir 151,* 1978, p. 359, The Geological Society of America. 17.17 and 17.18: Carto-Graphics. Reprinted with permission from W. R. Dickinson, "Cenozoic Plate Tectonic Setting of the Cordilleran Region in the U.S.," in *Cenozoic Paleogeography of the Western U.S., Pacific Coast,* Symposium 3, 1979, pp. 10–11 and pp. 2, 4. 17.19: Carto-Graphics. From P. B. King. *The Evolution of North America,* 1977, p. 116 (Fig. 74), ©1977 by Princeton University Press, Reprinted with permission of Princeton University Press. 17.21: Precision Graphics. Reprinted with permission from D. L. Blackstone, Jr., *Traveler's Guide to the Geology of Wyoming:* Geological Survey of Wyoming Bulletin 67, 1988, pp. 43–44. 17.23a: Publication Services. Reprinted with permission from Christiansen and Lipman, "Cenozoic Volcanism and Plate-Tectonic Evolution of the Western United States," *Philosophical Transaction,* The Royal Society, London, v. A271, 1972, p. 259. 17.24a: Publication Services. Reprinted with permission from

P. R. Hooper, "The Columbian River Basalts," *Science,* v. 215, 1982, p. 1465 (Fig. 3), ©1992 by the AAAS. 17.28: Precision Graphics. 17.30: Precision Graphics. Reprinted with permission from J. H. Stewart, *GSA Memoir 152,* 1978, p. 24 (Fig. 1–16), The Geological Society of America. 17.32: Publication Services. Reprinted with permission from W. R. Dickinson, "Cenozoic Plate Tectonic Setting of the Cordilleran Region in the U.S.," in *Cenozoic Paleogeography of the Western United States,* Pacific Coast, Symposium 3, 1979, p. 2 (Fig. 1). 17.33 and Perspective 17.1, Figure 1: Carto-Graphics. 17.35: Carto-Graphics. From C. G. Adams, "An Outline of Tertiary Paleography," in *The Evolving Earth,* L. R. M. Cocks (ed.), 1981, pp. 224, 228, reprinted with permission of Cambridge University Press. 17.36a: Publication Services. Reprinted with permission from Walker and Coleman, "Atlantic and Gulf Coastal Province," in *DNAG,* Centennial Special, Vol. 2, *Geomorphic Systems of North America.* 1987, p. 52 (Fig. 1). 17.36b: Publication Services. Reprinted with permission from S. W. Lowman, "Sedimentary Facies in the Gulf Coast," *AAPG,* v. 33, no. 12, 1949, p. 1972 (Fig. 23), American Association of Petroleum Geologists. 17.37: Carto-Graphics. Reprinted with permission from Fisher and McGowen. "Depositional systems in Wilcox Group (Eocene) of Texas and their relation to occurrence of oil and gas," *AAPG,* v. 53, no. 1, 1969, p. 42 (Fig. 7), American Association of Petroleum Geologists. 17.38a: Publication Services. Reprinted with permission from M. A. Hanna, "Salt Domes: Favorite Home for Oil," *Oil and Gas Journal,* v. 57, no. 6, 1959, p. 140. 17.38b: Publication Services. Reprinted with permission from D. J. Hughes, "Faulting Associated with Deep-Seated Salt Domes in the Northeast Portion of the Mississippi Salt Basin," Gulf Coast Association of Geological Societies, v. 10, 1960, p. 161 (Fig. 5). 17.39: Publication Services. Reprinted with permission from Friedman and Sanders, *Principles of Sedimentology,* 1978, p. 371, Copyright ©1978 John Wiley and Sons, Inc. 17.40: Carlyn Iverson. From D. Johnson, *Stream Sculpture on the Atlantic Slope,* 1931, Columbia University Press. 17.41: Carto-Graphics. Reprinted with permission from Grow and Sheridan, "U.S. Atlantic Continental Margin: A Typical Atlantic Type or Passive Continental Margin," *The Geology of North America,* v. 1–2, pp. 2.4: USGS. 17.43: Precision Graphics. Reprinted with permission from Grow and Sheridan, "U.S. Atlantic Continental Margin: A Typical Atlantic Type or Passive Continental Margin," *The Geology of North America,* v. 1–2, pp. 2.4: USGS. Perspective 17.1, Figure 1 insets: Carlyn Iverson.

CHAPTER 18

Table 18.1: Publication Services. Reprinted with permission of Cambridge University Press. 18.4: Publication Services. Reprinted with permission from Skinner and Porter, *Physical Geology,* 1987, p. 577, Copyright ©1987, John Wiley and Sons, Inc. 18.5a, 18.8, and 18.17: Precision Graphics. 18.5c: Publication Services. Reprinted with permission of Dr. Richard G. Holloway. 18.7a, 18.7b, and 18.13b: Rolin Graphics. 18.9: Publication Services. Adapted with permission from Willman and Frye, 1970, "Pleistocene Stratigraphy of Illinois," *Illinois Geological Survey Bulletin 94.* 18.10b: Publication Services. Reprinted with permission from Wollin, Ericson, and Ewing, "Late Pleistocene climates recorded in Atlantic and Pacific deep-sea sediments," in Turkekian (ed.), *The Late Cenozoic Glacial Ages,* 1971, p. 208, ©Yale University Press. 18.10c: Publication Services. Reprinted with permission from Cline and Hays (eds.), and J. van Donk, "0[18] record of the Atlantic Ocean for the entire Pleistocene Epoch,"

GSA Memoir 145, 1976, p. 150 (Fig. 1), The Geological Society of America.   18.11: Michael Thomas Associates.   18.12a: Publication Services. Reprinted with permission from B. Gutenberg, *Physics of the Earth's Interior,* p. 194, ©1959 The Academic Press: Orlando, FL.   18.12b: Publication Services. From J. T. Andrews, *Canadian Geological Survey Paper 68–53,* 1969, p. 53, Geological Survey of Canada, Department of Energy, Mines, and Resources. Reproduced with the permission of the Minister of Supply and Services, Canada, 1992.   18.13a: Publication Services. From Wright, Jr. and Frey (eds.), *Quaternary of the United States,* p. 266 (Fig. 1), Copyright ©1965 by Princeton University Press, reprinted with permission of Princeton University Press.   18.15 and Timeline: Publication Services.   18.16: Publication Services. From V. K. Prest, *Economic Geology Report 1,* 5th ed., 1970, pp. 90–91 (Fig. 7–6), Geological Survey of Canada: Department of Energy, Mines, and Resources, reproduced with permission of the Minister of Supply and Services, Canada, 1992.   Perspective 18.1, Figure 1: Carto-Graphics.

Chapter 19

19.7, 19.9, 19.23, 19.24, 19.25, 19.26, 19.28, and Perspective 19.1, Figure 1: Carlyn Iverson.   19.8: Publication Services. Reprinted with permission from E. H. Colbert, *An Outline of Vertebrate Evolution;* Carolina Biology Reader Series, No. 131, Copyright ©1983 Carolina Biological Supply Co., Burlington, North Carolina, U.S.A.   19.10, 19.17, 19.33, and Timeline: Publication Services.   19.13, 19.14, and 19.20: Precision Graphics.   19.15: Precision Graphics. From Alfred S.

Romer, *The Vertebrate Body,* 3rd ed., Fig. 224, copyright ©1962 by Saunders College Publishing, a division of Holt, Rinehart, and Winston, Inc., reprinted by permission of the publisher.   19.16: Precision Graphics. Reprinted with permission from O. A. Peterson, "Osteology of *Oxydactulas,*" *Annals of Carnegie Museum,* v. 2, no. 3, 1904, ©Carnegie Museum of Natural History, Pittsburgh, PA.   19.19: Precision Graphics, Reprinted with permission of The Natural History Museum, London.   19.21: Publication Services. Modified with permission from R. J. G. Savage, *Evolution: An Illustrated Guide,* 1986, p. 188, Facts of File Publications, and by courtesy of the Natural History Museum, London.   19.22a: Publication Services. Reprinted with permission from B. J. MacFadden, "Patterns of Phylogeny and Rates of Evolution in Fossil Horses," *Paleobiology,* v. 11, no. 3, 1985, p. 247 (Fig. 1), ©The Paleontological Society.   19.27: Precision Graphics. Reprinted with permission from L. S. Dillon, *Concepts and Consequences,* 1978, p. 256 (Fig. 12–13).   19.31: Precision Graphics. From J. A. Wolfe, "A Paleobotanical Interpretation of Tertiary Climates in the Northern Hemisphere," *American Scientist,* v. 66, no. 6, 1978, p. 695. Reprinted by permission of *American Scientist,* journal of Sigma Xi, The Scientific Research Society.   Perspective 19.1, Figure 2: Carlyn Iverson. From drawing by Douglas L. Cramer, courtesy of *Natural History Magazine.*

Chapter 20

20.1: Precision Graphics.   20.3, 20.4, 20.6, 20.7, 20.13, 20.15, and 20.22: Carlyn Iverson.   20.8, 20.16, and Timeline: Publication Services.   20.11: Darwen and Vally Hennings.